Prehistoric landscape development and human impact in the upper Allen valley, Cranborne Chase, Dorset

For John 'Snails' Evans

(John was the key mentor in persuading both Charles French and Mike Allen to do environmental archaeology and especially snail analysis many years ago! His career is chronicled in Allen (2006))

An imaginary view southwestwards along the line of the Dorset Cursus from Bottlebush to Gussage Cow Downs (Dave Bennett).

McDONALD INSTITUTE MONOGRAPHS

Prehistoric landscape development and human impact in the upper Allen valley, Cranborne Chase, Dorset

By Charles French, Helen Lewis, Michael J. Allen, Martin Green, Rob Scaife & Julie Gardiner

with contributions by Richard Bradley, Rosamund Cleal, Natasha Dodwell, Chris Gaffney and GSB Prospection, Kasia Gdaniec, Mike Ladle, Mark Maltby (with Victoria Ford & Kate Mason), Lesley McFadyen, Jacqueline I. McKinley, Vanessa Muros, Rog Palmer, Abigail Rothwell, Edgar Samarasundera, Cathy Stoertz & Lidia Żuk

illustrations by Matthew Brudenell, Crane Begg, Nick Griffiths, Jo Richards, Dave Bennett, Keith Faxon, Charles French, Helen Lewis, Martin Green, Rob Goller, GSB Prospection, Lesley McFadyen, Rog Palmer, David Redhouse, Dave Robinson, Rob Scaife, Cathy Stoertz, Fraser Sturt & Paul White

photography by Charles French, Martin Green, Gwil Owen, David Cousins, Paul Kitching, David Parry & David Webb

Published by:

McDonald Institute for Archaeological Research
University of Cambridge
Downing Street
Cambridge, UK
CB2 3ER
(0)(1223) 339336
(0)(1223) 333538 (General Office)
(0)(1223) 333536 (FAX)
dak12@cam.ac.uk
www.mcdonald.cam.ac.uk

Distributed by Oxbow Books
United Kingdom: Oxbow Books, 10 Hythe Bridge St, Oxford, OX1 2EW, UK.
Tel: (0)(1865) 241249; Fax: (0)(1865) 794449; www.oxbowbooks.com
USA: The David Brown Book Company, P.O. Box 511, Oakville, CT 06779, USA.
Tel: 860-945-9329; Fax: 860-945-9468

ISBN: 978-1-902937-47-2
ISSN: 1363-1349 (McDonald Institute)

Edited for the Institute by Graeme Barker (*Series Editor*) and Dora A. Kemp (*Production Editor*).

Cover illustration: *Oblique aerial view westwards from Wyke Down across the Allen valley to Gussage Down (top left). Fir Tree Field lies near top left with barrows 35 and 36 in the left foreground and the Wyke Down henges slightly above centre.*

Printed and bound by Short Run Press, Bittern Rd, Sowton Industrial Estate, Exeter, EX2 7LW, UK.

Contents

Contributors

Michael J. Allen
Allen Environmental Archaeology, Redroof, Green Rd, Codford, Wilts., BA12 0NW.
Email: aea@themolluscs.com

Crane Begg
Cam ARC, 15 Trafalgar Way, Bar Hill, Cambridge, Cambs., CB3 8SQ.
Email: crane.begg@cambridgeshire.gov.uk

Richard Bradley
School of Human & Environmental Sciences, University of Reading, Whiteknights, PO Box 227, Reading, Berks., RG8 6AB.
Email: R.I.Bradley@reading.ac.uk

Matthew Brudenell
Cambridge Archaeological Unit, Department of Archaeology, University of Cambridge, Downing Street, Cambridge, Cambs., CB2 3DZ.
Email: mjb73@cam.ac.uk

Rosamund Cleal
Alexander Keiller Museum, High St, Avebury, Wilts., SN8 1R.
Email: Rosamund.Cleal@nationaltrust.org.uk

David Cousins
18 Queen Mary Road, Salisbury, Wilts., SP2 9LD.
Email: cuz@waitrose.com

Natasha Dodwell
Cambridge Archaeological Unit, Department of Archaeology, University of Cambridge, Downing Street, Cambridge, Cambs., CB2 3DZ.
Email: nd207@cam.ac.uk

Charles French
Department of Archaeology, University of Cambridge, Downing St, Cambridge, Cambs., CB2 3DZ.
Email: caif2@cam.ac.uk

Chris Gaffney
Sunny Leigh, Cliffe Lane, Thornton, Bradford, West Yorks., BD13 3DX.
Email: c.gaffney@yahoo.co.uk

Julie Gardiner
c/o Allen Environmental Archaeology, Redroof, Green Rd, Codford, Wilts., BA12 0NW.
Email: aea@themolluscs.com

Kasia Gdaniec
Cambridgeshire City Council, Shire Hall, Castle Hill, Cambridge, Cambs., CB3 0AP.
Email: kasia.gdaniec@cambridgeshire.gov.uk

Rob Goller
Wessex Archaeology, Portway House, Old Sarum Park, Salisbury, Wilts., SP4 6EB.
Email: r.goller@wessexarch.co.uk

Martin Green
Down Farm, Sixpenny Handley, Salisbury, Wilts., SP5 5RY.

Nick Griffiths
9 Riverside, Wilton, Salisbury, Wilts., SP2 0HW.

GSB Prospection
Cowburn Farm, Market St, Thornton, Bradford, West Yorks., BD13 3HW.
Email: gsb@gsbprospection.com

Mike Ladle
23 Shaw Drive, Wareham, Dorset, BH20 7BS.
Email: mikeladle@hotmail.com

Helen Lewis
School of Archaeology, University College Dublin, Belfield, Dublin 4, Ireland.
Email: helen.lewis@ucd.ie

Mark Maltby
School of Conservation Sciences, Christchurch House, Talbot Campus, Poole, Dorset, BH12 5BB.
Email: mmaltby@bournemouth.ac.uk

Lesley McFadyen
School of Archaeology and Ancient History, University of Leicester, University Rd, Leicester, Leics., LE1 7RH.
Email: lm127@le.ac.uk

Jacqueline I. McKinley
Wessex Archaeology, Portway House, Old Sarum Park, Salisbury, Wilts., SP4 6EB.
Email: jackie@jimckinley.demon.co.uk

Vanessa Muros
Institute of Archaeology, University College London, 31–34 Gordon Square, London, WC1H 0PY.

Gwil Owen
Department of Archaeology, University of Cambridge, Downing St, Cambridge, Cambs., CB2 3DZ.
Email: gjo1@cam.ac.uk

Rog Palmer
20 Gunhild Way, Cambridge, Cambs., CB1 4QZ.
Email: rog.palmer@ntlworld.com

David Parry
11 Bryanston Street, Blandford Forum, Dorset, DT11 7AZ.
Email: bryanston11@btopenworld.com

David Redhouse
Department of Archaeology. University of Cambridge, Downing St, Cambridge, Cambs., CB2 3DZ.
Email: dir21@cam.ac.uk

Jo Richards
44 West End, Brampton, Huntingdon, Cambs., PE28 4SD.
Email: Arriscott@ntlworld.com

Abigail Rothwell
c/o School of Conservation Sciences, Christchurch House, Talbot Campus, Poole, Dorset, BH12 5BB.

Edgar Samarasundera
Institute of Archaeology, University College London, 31–34 Gordon Square, London, WC1H 0PY.
Email: epsamarsundera@yahoo.co.uk

Rob Scaife
School of Geography, University of Southampton, 2035 Shackleton Building, Highfield, Southampton, Hants., SO17 1BJ.
Email: R.Scaife@soton.ac.uk

Cathy Stoertz
Aerial Study, RCHME-NMRC, Kemble Dr, Swindon, Wilts., SN2 2G7.
Email: swincsd@rchme.gov.uk

Fraser Sturt
Centre for Maritime Archaeology, University of Southampton, Avenue Campus, Highfield Southampton, Hants., SO17 1BF.
Email: f.sturt@soton.ac.uk

Paul White
Wessex Archaeology, Portway House, Old Sarum Park, Salisbury, Wilts., SP4 6EB.
Email: pwhite@wessexarch.co.uk

Lidia Żuk
Institute of Prehistory, Adam Mickiewicz University, ul. św. Marcin 78, 61-809 Poznań, Poland.
Email: lidkazuk@amu.edu.pl

Figures

Tables

Acknowledgements

The assistance and access to land permitted by Mr Philip Rymer and the Shaftesbury Estate, Robert Antell and Bob Rowe is gratefully acknowledged. Manuel Arroyo-Kalin, Dave Robinson, Matt Brudenell, Fraser Sturt, Gianna Ayala, Brian Pitman, Lesley McFadyen, Julie Boreham, Jake Keen, Gillian Wallace, John Tate, David Parry, Jim Russell, Keith Faxon, Tom Blacklock, Paul Kitching, Karen Milek, Kasia Gdaniec, Susanne Makenbeck, Chris Stevens and Dave Bennett are all thanked for their stalwart help in the field. A number of others have provided valuable specialist assistance, namely Alan Clapham (University of Cambridge; plant botany), Rog Palmer and Lidia Żuk (Air Photo Services; aerial mapping), Julie Gardiner (lithics), Mike Ladle (shells), Rosamund Cleal and Kasia Gdaniec (pottery), all at GSB Prospection of Bradford (geophysical survey) but especially Chris Gaffney, Louis Harvey, Christine Martinez, Dan Shield and David Weston, Christian Pierce and Colin Shell (radar), Paul White (GIS), David Redhouse (topographical mapping), Matthew Brudenell (illustrations) and Crane Begg (illustrations and terrain mapping). Julie Boreham of the McBurney Geoarchaeology Laboratory, Department of Archaeology, University of Cambridge, is also thanked for producing the thin-section slides. Paul Gosling, Duncan Coe, Amanda Chadburn and David Field of English Heritage, Laurence Keen and Claire Pinder of Dorset County Council, and Peter Woodward of Dorchester Museum have been very supportive throughout. David Cousins, Paul Kitching, David Webb and David Parry are thanked for additional photography. Finally Professors Richard Bradley, John Barrett, Martin Bell and the late John Evans are greatly thanked for being long-standing supporters of the project.

The principal team members were Charles French (University of Cambridge), Helen Lewis (University College, Dublin), Michael J. Allen (AEA: Allen Environmental Archaeology), Rob Scaife (University of Southampton), Martin Green (Down Farm) and Julie Gardiner (Wessex Archaeology). The research was funded by the Arts and Humanities Research Board, the British Academy, the McDonald Institute for Archaeological Research and a private donor.

Much of the original artwork was re-drawn and scanned by Matthew Brudenell and Crane Begg, and the photographs developed and scanned by Gwil Owen, without whom the volume would not have been possible, so a great many thanks to each of you!

English Heritage, the Marc Fitch Fund, the Council for British Archaeology, the McBurney Geoarchaeology Laboratory, University of Cambridge, and the McDonald Institute for Archaeological Research, University of Cambridge, all supported the publication of this monograph, and their generosity is very much appreciated.

Also, a host of people helped the project in all manner of ways: Anita Becker, Mike Becker, Helen Brickell, Gill Broadbent, Corrine Bungay, Steve Bungay, Derek Bunting, Pam Bunting, Laura Burnett, Hayley Clark, Penny Copeland, the Fields at Pennywort Cottage (Bowerchalke), Rob Goller, Karin Green, Susanne Hakenbeck, Diana Hall, Alan Hawkins, Anita Hawkins, Jen Heathcote, Paul Heron, Liz James, Mark Knight, Gary Martin, Bob Middleton, Karen Milek, Judith Neath, Len Norris, Frank Peach, Brian Popple, The Roebuck Inn (Sixpenny Handley); Jim Russell, Anne Shields, Steve Smith, Chris Stevens, Gill Swanton, Mike Tizzard, Ben Urmston, Gerry Wareham, Janice Wareham and John Winterbottom.

Acknowledgements by individual authors

Dr Michael J. Allen

Sampling for snails was undertaken in the field by the writer assisted by the field team. Samples were processed at Wessex Archaeology by Sarah Wyles and Hayley Clark under the supervision of Sarah Wyles. The data was entered by Sarah Wyles into the Wessex mollusc data base, who prepared the histogram outlines. The histograms were produced at Wessex Archaeology by Rob Goller.

Edgar Samarasundera

With great thanks to NERC whose funding of my PhD made the GIS modelling work possible.

Dr Helen Lewis

I would especially like to thank Vlad Stukalov, the University of Oxford Research Laboratory for Archaeology and the History of Art School of Archaeology, and Department for Continuing Education, Chris Doherty, Mark Pollard, Quinton Carroll, the East Dorset Antiquarian Society and The President and Fellows of Newnham College.

Abstract

On- and off-site palaeoenvironmental and archaeological investigations of the upper Allen valley of Dorset conducted in 1998–2003 revealed a model of landscape development which contrasts with those previously put forward for the region. A programme of off-site geoarchaeological survey, aerial photographic survey and palynological analyses of two relict palaeochannel systems, along with sample investigations of two Neolithic long barrows, a Neolithic enclosure, the Dorset Cursus, four Bronze Age round barrows, several prehistoric lithic/settlement scatters, and an Iron Age bank and ditch system have been combined with inter-regional summaries of the archaeological excavation results of Martin Green and others as well as the extensive molluscan records compiled by Michael Allen to re-examine the prehistoric landscape dynamics in the study area. New data, especially palynological, molluscan and soil micromorphological, suggest that woodland development in the earlier Holocene of this landscape was more patchy and variable over space and time than the presumed model of full climax deciduous woodland. Finally, an attempt has been made to interrogate and visualize some of this new data using GIS-based simulation modelling techniques, which has potentially provided strong independent corroboration.

On the downland and in the valley bottom occupied by the Dorset Cursus in the upper part of the Allen valley, there appear to have been substantial open areas already present in the Mesolithic. In this valley head area, soil change had already occurred by the fourth millennium BC *from more organic- and clay-enriched brown earths to thin rendzinas associated with established grassland, thus slowing and altering the soil development trajectory of this part of the downlands. Consequently, many areas perhaps never developed thick, well-structured, clay-enriched soils (or argillic brown earths) as is suggested for the downs under forest, but rather thin brown earths. This landscape development route as well as human management led to the concomitant establishment of a predominantly grassland landscape by the Neolithic period.*

In contrast, further downstream in the Knowlton to Wimborne Minster part of the study area, a different trajectory appears to have been followed. Here, mixed deciduous woodland did develop throughout the Mesolithic and earlier Neolithic, and witnessed its first clearance inroads in the mid- to late fourth millennium associated with a coincident elm decline. This southern part of the study area continued to become much more open over the next three millennia and was dominated by grassland. Evidence of increasing arable land use does not seem to have occurred significantly until later prehistoric and earlier historic times and again in recent late twentieth-/twenty-first-century times. Indeed, geomorphological survey and ditch-fill studies throughout the whole study area suggest that neither significant nor extensive colluviation took place in the valley until historic times, with aeolian evidence for bare or tilled earth appearing only occasionally in later prehistoric times.

The combination of the presence of thinner and less well-developed soils over some large areas of downland by the Neolithic and long-established grassland environments removes the necessity for envisaging extensive soil erosion and thick aggraded deposits in the valley bottom occurring during the Neolithic and Bronze Age, as has been proposed elsewhere for the Wessex chalk downlands. The investigations suggest that if there were major changes in vegetation and soil complexes related to changing land-use practices in the Wyke Down area, these had already occurred by the Neolithic rather than in the Bronze Age as suggested by previous researchers, and the area has remained relatively stable since. In other parts of the study area to the south, the more 'conventional route' of deciduous woodland development into the Neolithic and subsequent gradual clearance, predominantly for grassland, did take place.

Chapter 1

Exploring the Prehistoric Landscape of Wessex: Old Issues and New Approaches

Charles French, Helen Lewis & Martin Green

Palaeoenvironmental and archaeological fieldwork focusing on the later Neolithic and earlier Bronze Age periods in the upper Allen valley area of Cranborne Chase in Dorset (Fig. 1.1) has produced new land-use data with which to re-examine landscape models put forward by Barrett and Bradley who researched the same area in the 1980s (Barrett 1994; Barrett *et al.* 1991a,b). The new research project focused on a *c.* 8 by 2 km swathe of the upper Allen valley in the northern-most corner of Dorset between Sixpenny Handley and Knowlton, and across the valley from Wyke and Bottlebush Downs in the north, to Gussage Cow and Harley Downs to the south (Fig. 1.1). This landscape is a 'classic' example of the chalk downlands in southern England, as well as being a typical 'Wessex' prehistoric landscape. It contains numerous prehistoric monuments of all kinds present in close proximity and in different parts of the valley system, as well as several new and unique sites excavated by Green (2000), the results of which are also contained in this volume.

The volume is organized into four chapters. The first chapter sets out the rationale and methodologies behind the project before examining the current view of the Holocene archaeology and landscape history of the area in Chapter 2. Chapter 3 presents the main evidence gathered in this project, and is related to Martin Green's excavations data sets. These are further described in Appendix 4. Chapter 4 discusses pre-historic landscape development and change for this part of Cranborne Chase, and then relates this story to the palaeoenvironmental record for the wider chalk downland region and beyond in England.

Project rationale

Charles French & Helen Lewis

This project was designed to investigate recurrent signatures of land-management practices in the geo-archaeological and ecological records in the chalk downland region of southern England, using the upper Allen River valley of Cranborne Chase as a pilot-testing area. The new data collected have furthered our understanding of the interactions between prehistoric settlement and land use, the monumental landscape and landscape/environmental change.

A principal emphasis of the project was the prospection for, and analysis of buried soils. Buried soils constitute the only readily available and wide-spread reservoirs of evidence for prehistoric land use in the otherwise severely plough-denuded Wessex region. They also contain time-depth information in terms of landscape and land-use change, but this record has rarely been systematically exploited, especially in regard to site-specific, pre-monument land use. Buried soils under monuments and associated colluvial and alluvial deposits were investigated using a combined approach of soil stratigraphic and micromorphological study with pollen and molluscan analyses, in order to document landscape changes in the Neolithic and earlier Bronze Age. All of these methods have been used in previous research in the area (Allen *et al.* 1990; Barrett *et al.* 1991a,b; Cleal *et al.* 1994), but only very rarely have they been combined (Macphail 1993; Cleal *et al.* 2004). Problems of interpretation, re-working and preservation, especially in the unique environments underneath monuments such as barrows in the chalk downlands (calcareous parent material with overlying weakly acidic barrow deposits), ensure that molluscan and soil pollen methods on their own can only partially address the history of landscape changes in the region. Soil micromorphological studies elsewhere (e.g. Gebhardt 1992; Courty *et al.* 1989; French 2003) suggest that it is possible through a pedogenic approach and by analogy with experimental results to identify the nature of prehistoric landscapes and land-use practices, as these are reflected in relict pedofeatures found in modern soil horizons. Furthermore, it is possible to

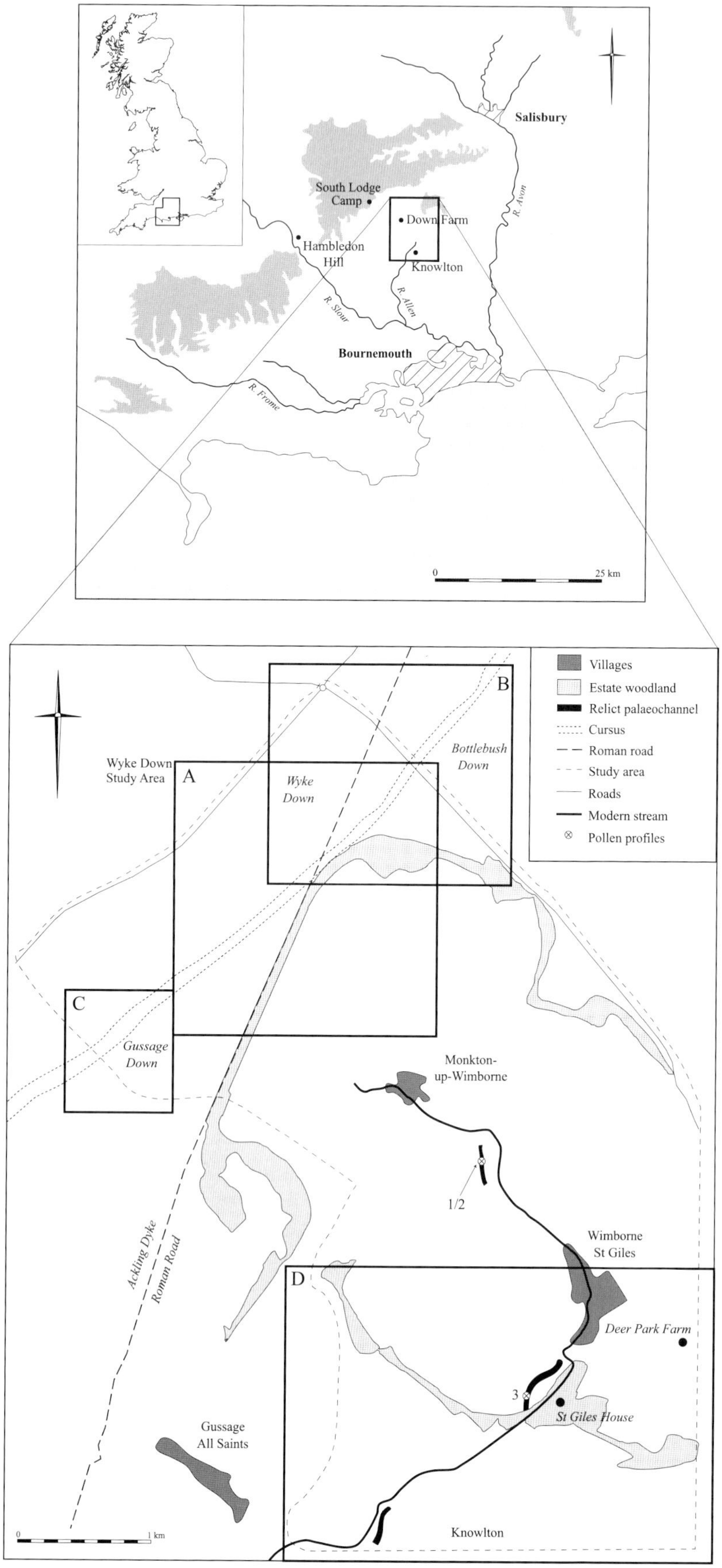

examine these practices on both a site-specific and a wider regional scale over time. We believe that only through the systematic comparison of the results of all of these methods can a well-defined and detailed history of the Late Neolithic and earlier Bronze Age landscape in this region be developed.

The importance of systematically examining buried soils in the chalk downland region for indicators of landscape and land-use changes lies in the fact that many past and current models of social practices in prehistoric southern England as a whole are based primarily on the archaeological remains of this area (Barrett 1994; Barrett *et al.* 1991a,b; Bradley 1978), and that land use has been related to social organization at both small and large scales, to settlement patterns, and to cosmology in interpretations of prehistoric communities in this landscape, as well as in societies around the world. Barrett (1994) suggests that the fourth millennium BC was marked by long fallow systems in a partially cleared landscape, with extensive mixed agriculture practised by a mobile community. Long-term land-use and landscape-perception changes, relating the agricultural system and the funerary monuments of the region to the development of land tenure, are said to have culminated in the evolution in the second millennium BC of a short fallow and more agrarian system, with traction based tillage, 'permanent' settlement and land division. Despite the long-standing popularity of models such as this, which invoke land-use and subsistence changes within explanations of the archaeological record in the region, very little systematic effort has been expended to document the proposed changes themselves. This project set out to examine a number of broadly contemporaneous buried soils

Figure 1.1. *Location map of the study area, with the detailed areas for Wyke Down (A), Bottlebush Down (B), Gussage Down (C) and the Knowlton environs (D) shown (C. Begg).*

and sedimentary deposits in one area of the region, namely in the upper Allen river valley, and to provide an opportunity to assess, test and/or refine such models on both site-specific and sub-regional scales.

The project is also an attempt to address key interpretative problems associated with each technique when used in chalk downland contexts. For example, to observe accurately the pedological effects of agricultural land use in thin section through soil micromorphology is not always easy or straightforward (Gebhardt 1992; Macphail *et al.* 1990; Carter & Davidson 1998; Macphail 1998; Lewis 1998b; Usai 2001). The recognition and interpretation of true woodland molluscan faunas is also fraught with interpretative difficulties, especially when dealing with landscapes which may have been in a state of transition from woodland to open ground and/or which may have been subject to periods of decalcification (Bell & Johnson 1996; Cameron 1978; Evans 1991). The interpretation of pollen assemblages from buried land surfaces and soils, especially from calcareous soils and sediments, may be affected by a variety of taphonomic problems (Dimbleby & Evans 1974; Dimbleby 1985; Crabtree 1996). Moreover, there is a dearth of sub-regional pollen sequences for the chalklands of southern England, as few suitable deposits with good preservation have been located. Only by repeatedly using these approaches in combination, from similar contexts with fine sample resolution, will these specialisms provide reliable results applicable to archaeological questions about this landscape. Thus the results of this approach will be of wide use to the whole archaeological community involved with the interpretation of past landscapes. Finally, the land-use model(s) developed through this research project may have a predictive value for other chalk downland areas of the Wessex region, for example the Stonehenge environs (Richards 1990; Cleal *et al.* 1995), and will undoubtedly raise questions about the nature of landscape development in these areas.

This project was also mindful of new advances in the use of geographical information systems and applications of erosion-modelling concepts to both past (e.g. Ayala 2003; Ayala & French 2003) and present-day landscapes (Selby 1993, 356–76). One of the big unanswered questions in this area of chalk downland, as with most of Wessex, is where have all the thick forest soils gone to? Whilst this project did not recover the necessary data in sufficient intensity over the study area to permit applications of erosion formulae in a GIS framework, it did generate sufficient information on soil and colluvial soil depths in different locations in this landscape to postulate the extent and thickness of soil cover that could have existed on the chalk downland slopes in the study area. The project also produced sufficiently intensive and expansive molluscan and palynological data to create new vegetational models using experimental GIS techniques. In turn, this led to some new thinking on the nature of soil development in the upper Allen River valley, questioning the trajectories of long-perceived models of climax soil development on chalk substrates, and the effects of human land use on the soil and vegetation cover observed through other palaeoenvironmental techniques.

The archaeological background of the Wyke Down study area

Martin Green & Helen Lewis

A major component of our fieldwork has been the continued study of four square kilometres centred on Wyke Down (Figs. 1.1A & 1.2). This area, crossed by the Dorset Cursus, contains the densest concentration of pre-Iron Age earthworks of comparable size in the region, over eighty, and has seen a recent history of intensive research (Barrett *et al.* 1991a,b; Green 2000; Allen 2002a). The main thrust of our current work has been to build upon and expand our knowledge of the landscape history and prehistoric land use of this area primarily by auger survey and selective excavation. An important component of this work has been the completion of the analysis of the crucial sequence provided by the Fir Tree Field shaft excavation (Green & Allen 1997; Allen & Green 1998).

This dramatic clustering of monuments, we would argue, relates to a large extent to the remarkable geomorphological phenomena present in this very specific part of the upper Allen valley. First, the river itself at this point flows below ground, except during seasonal unpredictable flooding events when significant areas of the valley floor can be under water. The west side of the valley is floored, for a distance of about a kilometre, by a series of major mounds and depressions known as naleds. These were formed under periglacial conditions when melt-water run-off moved vast amounts of chalky sludge downstream. Some of this material built up around ice bodies created by springs then emanating from the valley floor. Subsequent melting left a series of depressions, partly enclosed by mounds of chalky gravels, which had built up around the former obstructions. A trench dug through one of the smaller mounds in 1976 revealed it was composed of over 4 metres of bedded chalky gravels known as coombe rock (Catt *et al.* 1980). Some of the larger mounds are likely to be composed of over twice that amount. Despite the ravages of centuries of arable agriculture these features are still striking

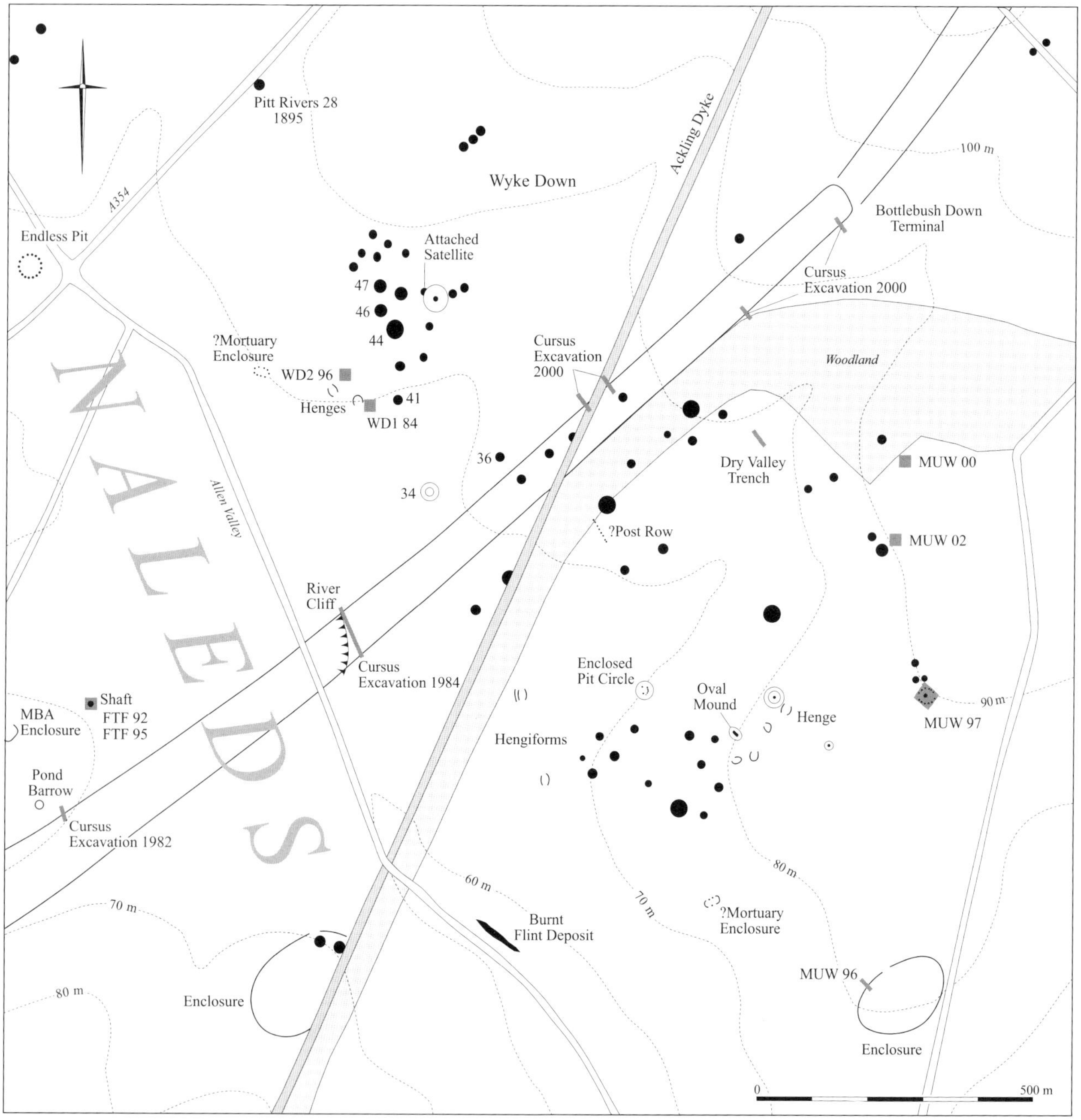

Figure 1.2. *The archaeological sites in the upper Allen valley at Down Farm, Wyke and Bottlebush Downs and Monkton-up-Wimborne Down showing the location of the sites investigated by excavation in this project and through previous work (C. Begg after M. Green).*

today and resemble prehistoric barrows and embanked enclosures. The fact that these structures have only previously been recorded in East Anglia attests to the special nature of this area.

Further unusual phenomena created during this period of intense periglacial activity include a steep river cliff forming part of the northeastern valley side for a distance of about 100 metres situated just downslope from the 1984 Dorset Cursus excavations of Bradley (Barrett *et al.* 1991a) (Fig. 1.2). During flooding events a lake forms immediately below the cliff, and the latter when viewed from a distance bears a striking resemblance to a long barrow. It should perhaps not surprise us that immediately above this cliff a major concentration of Neolithic activity has been located (Gardiner 1985; Bradley

1991a, 35–58; 1991b). Other features likely to have been at least partly formed at this time include the natural shafts or sink-holes in Fir Tree Field and Endless Pit. Although from excavation we know the former eventually collapsed about 7000 years ago, it is likely to have already substantially formed underground at a much earlier date by the release of meltwater. The much larger feature, Endless Pit, provides a natural terminal point to the northern extent of the naleds (Fig. 1.2) and gives its name to the cross roads at this location. A trench cut into the feature in 2002 (Green forthcoming) was abandoned at a depth of 3 m when soil horizons still producing modern material were reached. It is likely that this still visible feature contains a much longer sequence of infilling than the shaft in Fir Tree Field, which is known to have been completely infilled by around 2000 BC (Green & Allen 1997). Environmental data recovered from the earliest levels of the latter (see Allen below) have revealed evidence for the opening up of the landscape during the second half of the fifth millennium BC.

The Mesolithic background

Previous fieldwork in the Wyke Down study area revealed significant Mesolithic activity concentrated in the valley floor (Arnold *et al.* 1988; Green 2000). One site, St Giles Field, was located on a naled mound with a stratified sequence — to date a unique survival for a surface flint scatter on Cranborne Chase (Catt *et al.* 1980). The flintwork recovered from the limited investigation was comparable with the nearby site of Downton (Higgs 1959) and a date in the sixth millennium BC is most likely. Indeed, comparable although less-concentrated Mesolithic activity above the river cliff in Chalk Pit Field is associated with two radiocarbon determinations in the sixth millennium BC. The remaining Mesolithic activity all appears to relate to the latter half of the fifth millennium BC including the majority of the natural infilling recorded in the Fir Tree Field shaft. Here a particularly valuable late fifth millennium BC sequence, encompassing the Mesolithic/Neolithic transition, was revealed by nine radiocarbon dates (Allen & Green 1998). These were bracketed by dates of 4470–4240 cal. BC, obtained from a roe deer skeleton at a depth of 7 m in chalk rubble, to 4340–3990 cal. BC from a cow/auroch scapula at a depth of 2 m on the base of grey silty layer 7 (Table 1.1). Contained within the fills of this sequence were small but significant amounts of animal bone, worked flints (including an associated group of seven microliths, five of rod form) and comminuted charcoal. Of the three remaining Mesolithic sites in the Wyke Down study area, Yard Field has produced a rod microlith and numbers of small cores indicative of a similar late date with the other two sites producing no significant diagnostic forms to date.

It seems clear that towards the end of the fifth millennium BC the area had seen significant human activity, enough to have produced a substantially open landscape, attractive to the first farming groups arriving in the area at about the same time.

Neolithic and earlier Bronze Age monument complexes and landscapes

The area has a large number of well known Neolithic and earlier Bronze Age sites (Figs. 1.3–1.5), many of which form complexes including phases from both, and monument biographies that carry on well into historic periods. The main types of sites known in the Neolithic include long barrows, and mortuary structures, the Dorset Cursus, henges and hengiforms, round barrows, pit groups and artefact scatters. Within the earlier Bronze Age, mirroring the national picture, a more limited range of sites includes round barrows of various types, pit groups and artefact scatters.

The Fir Tree Field shaft provides crucial information, this time concerning the earliest Neolithic occupation. At the interface between the last Mesolithic layer 7 and the succeeding layer 6b (Figs. 1.2 & 3.14), some evidence for human disturbance was revealed, including a hearth. This yielded a radiocarbon date of 4050–3790 cal. BC and from the same layer nearby, further charcoal produced a second date of 3980–3710 cal. BC (Table 1.1). Slightly above this level in layer 6a a sherd of plain bowl pottery and the butt end of a polished flint axe were recovered. A radiocarbon date derived from a pig femur from this layer produced a date of 4240–3980 cal. BC. As this date is similar to those obtained from the underlying Mesolithic layer 7, it is probable that this bone was moved during the disturbance noted above.

However, the two dates mark the earliest evidence for Neolithic occupation in the study area and tie in well with the date obtained from antler found on the old land surface beneath the Thickthorn long barrow of 4040–3810 cal. BC (Drew & Piggott 1936; Bradley & Entwhistle 1985). This old land surface also yielded worked flints including a late microlith. It is striking that the earliest dates for the Neolithic in the region correspond to areas where significant Late Mesolithic activity had already taken place.

Following initial colonization in the Neolithic, evidence for stabilization is manifest towards the middle of the fourth millennium BC with the beginning of monument building. In Cranborne Chase a number of long barrows and the great complex of

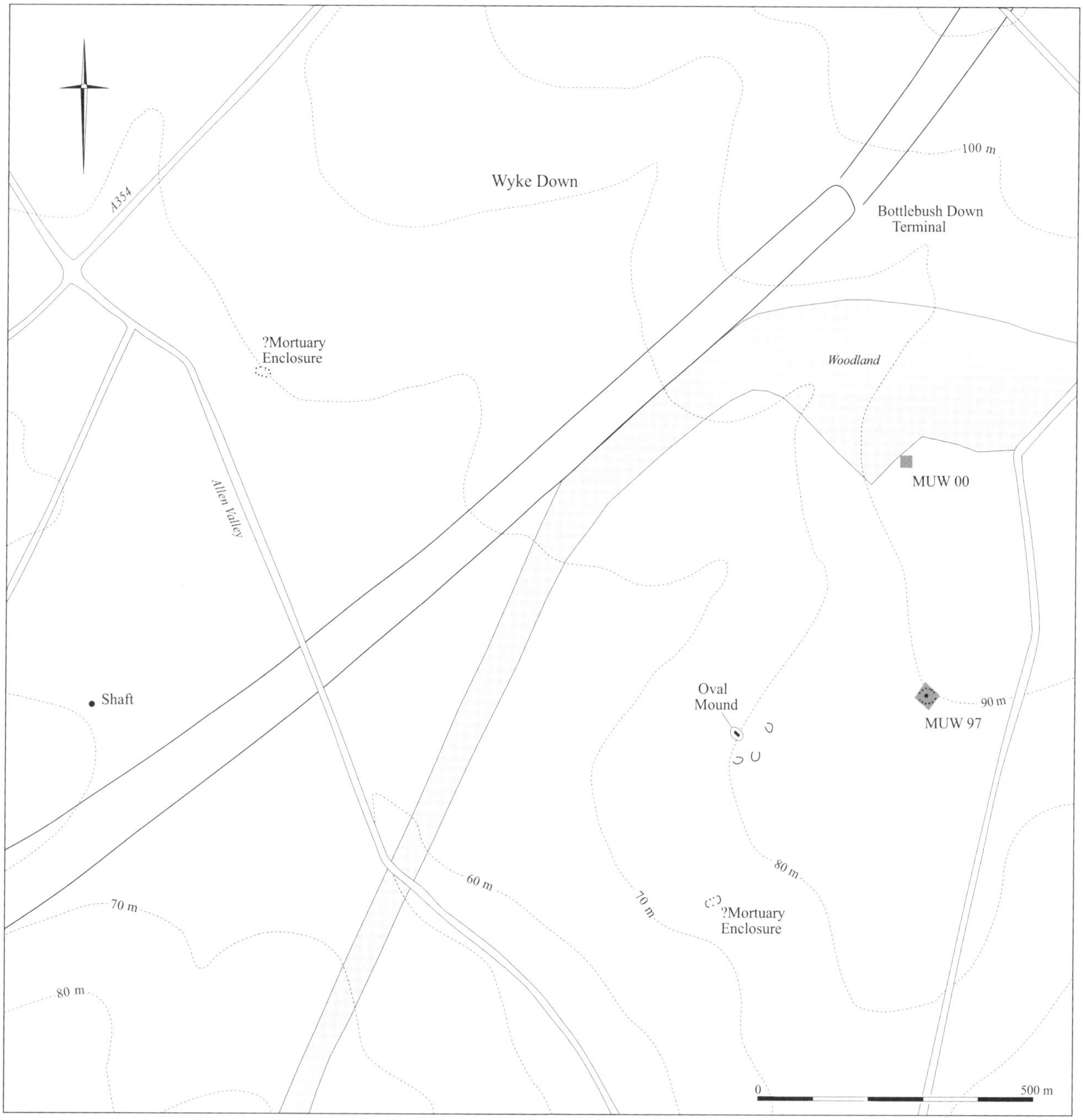

Figure 1.3. *The main sites of the Middle Neolithic period in the upper Allen valley at Down Farm, Wyke and Bottlebush Downs and Monkton-up-Wimborne (C. Begg after M. Green).*

earthworks on Hambledon Hill were under construction if not already completed. Many of the long barrows in the study area are associated with the Dorset Cursus. Several of these are clearly earlier than this vast earthwork, including two long barrows which it incorporates, whilst others appear to be later embellishments along its course (Bradley 1986; Barrett *et al.* 1991a). A number of the long barrows have distinctive U-shaped or almost completely enclosing ditches, giving rise to the form being termed 'Cranborne Chase type' (Ashbee 1970). Included in this category are two excavated examples — Wor Barrow (Pitt Rivers 1898) and Thickthorn Down 163a (Drew & Piggott 1936). At Wor Barrow this later type of long barrow produced a small number of exclusively adult male burials. This find contributes to the increasing body of evidence recovered nationally which suggests that social stratification begins towards the middle of the Neolithic. Thickthorn Down 163a provides comparative environmental evidence, with the molluscan assemblage

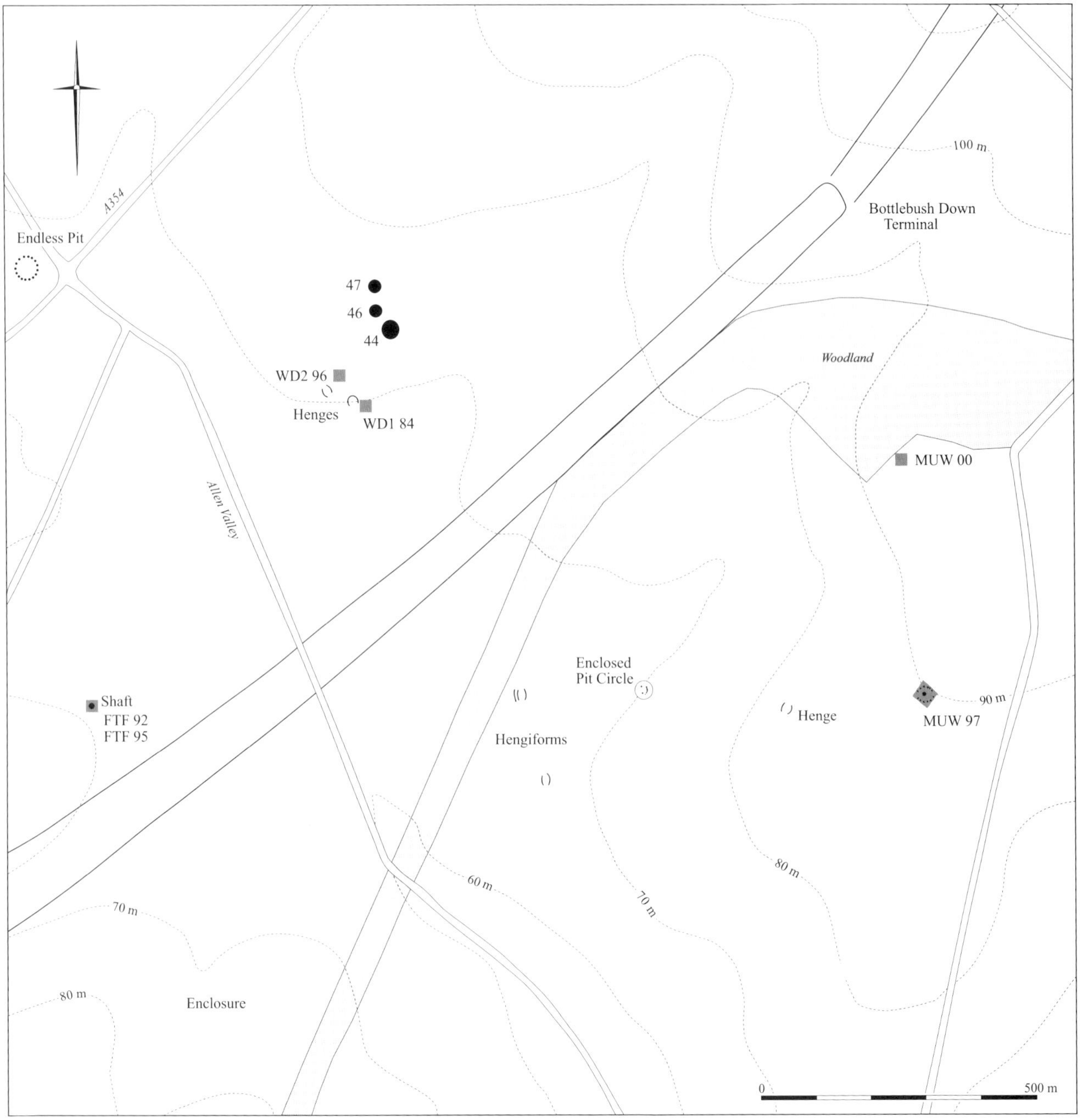

Figure 1.4. *The main sites of the later Neolithic period in the upper Allen valley at Down Farm, Wyke and Bottlebush Downs and Monkton-up-Wimborne (C. Begg after M. Green).*

suggesting open grassland (Bradley & Entwistle 1985). A number of broadly contemporary rectilinear 'mortuary' structures are known in the area and one at Monkton-up-Wimborne was sampled during this project (see below).

As the Dorset Cursus crosses the Allen valley it seems to deliberately incorporate a number of the natural phenomena (Figs. 1.2 & 1.7–1.10). Firstly it passes within 130 m of the Fir Tree Field shaft which at this stage was starting to develop a stabilization horizon at the layer 6/5 interface (Fig. 3.14). It then bisects the valley where it includes a significant tranche of the naleds. The northern Cursus ditch does cut through some of the larger naleds and may have needed more detailed planning to keep it on line, particularly with regard to the careful incorporation of the river cliff (Fig. 1.8). Cropmark evidence clearly shows the lines of the Cursus ditches unbroken as they cross the course of the Allen River proving that even in ancient times, the river above ground was only a

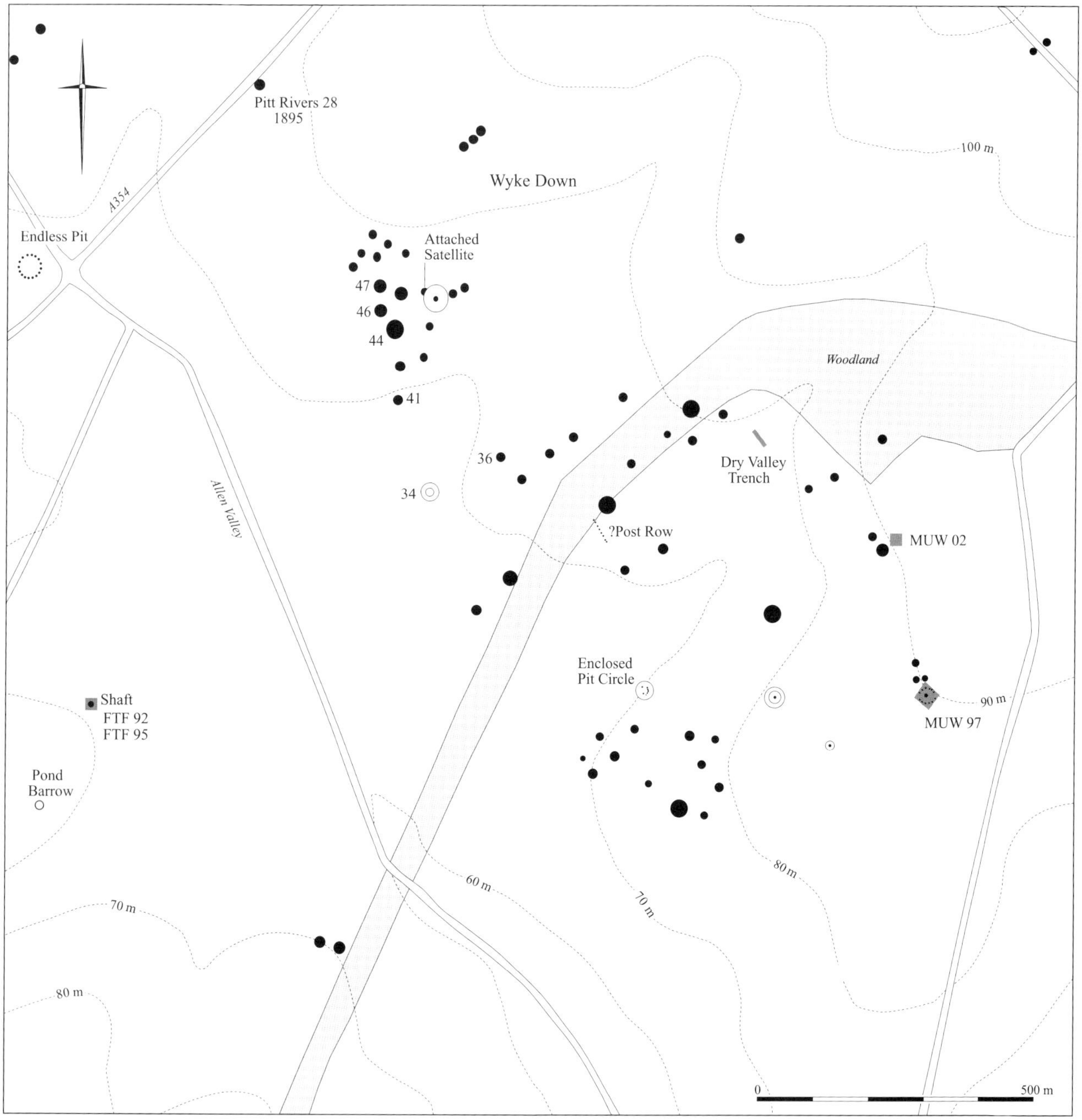

Figure 1.5. *The main sites of the Early Bronze Age period in the upper Allen valley at Down Farm, Wyke and Bottlebush Downs and Monkton-up-Wimborne (C. Begg after M. Green).*

seasonal phenomenon at this point. Once across the river course the ditches enclose the river cliff which as we have already noted bears a striking similarity to a long barrow when approached from the southwest. From the top of the cliff looking southwest is an uninterrupted view to the long barrow cresting Gussage Down, behind which, on a clear day, the mid-winter sun can be observed to set. At certain times the seasonal lake below the cliff would appear. Also contained within the Cursus earthwork, it may have appeared to observers that the workings of nature were somehow being controlled by the monument itself.

The Dorset Cursus was built in two separate sections of approximately equal length. The earlier 'Gussage' part of the Cursus incorporates the northern long barrow, Gussage 14, on Gussage Cow Down (Figs. 1.1C & 3.1) before crossing the Allen valley and terminating on Bottlebush Down where the Gussage and Pentridge lengths of ditch and bank meet (Figs. 1.1B & 1.2). Previous to our work only three small lengths of ditch had been excavated on this first section, from which Plain Bowl pottery, human and animal bones and good molluscan data were recovered (Barrett *et al.* 1991a,b). The later unexcavated 'Pentridge' Cursus incorporates long barrow Pentridge 19 within its northern bank before terminating close to Martin Down. Practically the whole length of the Dorset Cursus is ploughed except the southwestern terminal on Thickthorn Down. An apparent length of preserved 'bank' on Bottlebush Down has been shown through our excavations to be a 'ghost' (see below). Here, the natural chalk had survived to a higher level due to having been formerly protected by the now obliterated bank.

An extensive flint scatter spreads from the cliff and northern Cursus ditch over 200 m to the east, well beyond the limits of the southern ditch (Bradley 1991a, 59–69). Significant amounts of this material were found in the secondary silts of the northern ditch, where it was associated with Peterborough Ware. A radiocarbon sample obtained from animal bone in the primary silt produced the date 3370–2920 cal. BC providing a *terminus ante quem* for this material. This scatter, the most extensive by far of any located within or close to the Cursus, was subjected to detailed analysis by Gardiner (1985). The location of this large site, when linked to the discovery of a large number of 'higher-status' artefacts, suggests it primarily developed to play a significant role in the functioning of the Cursus shortly after it was completed.

Heading northeast, the first phase Cursus monument is today overlain by the Ackling Dyke before its terminal point is reached on Bottlebush Down (Fig. 1.2). In this area we excavated four slots, two in each Cursus ditch, to add to those cut in the 1980s (Figs. 1.2 & 3.33). These will be described in detail later but for the moment one observation is clear — with one exception all the excavated ditch profiles are remarkably similar with no evidence of recutting. The one exception is the southern ditch dug in 1984 (Barrett *et al.* 1991a). Here the section revealed a comparatively wide and shallow ditch with evidence of at least two phases of recutting. However, the surficial geology at this point consists of relict clay-with-flints overlying coombe rock. At the junction between the two a severely eroded and disintegrated surface was uncovered, full of solution pipes, making the recognition of the ditch sides and base particularly difficult. It seems likely that this most atypical profile is the result of under-digging and that the true ditch parameters were not reached.

The Gussage part of the Cursus terminates some 400 m below the crest of the next ridge. Extending from this point on a different alignment, presumably to avoid the mass of Penbury Knoll, the secondary Pentridge Cursus crosses the River Crane and ends close to Bokerley Dyke. As Bradley (1986) has pointed out the apparent premature termination of the Gussage part of the Cursus is to retain the skyline position of the northern Gussage Down long barrow. Thus from this point observers would have an uninterrupted view along half of the monument's course, and weather permitting, witness the setting of the midwinter sun. The completion of this great earthwork at this most particular location seems to have imbued the immediate surrounding landscape with sacred status. To the south and west of this point large numbers of burial and ritual monuments were created, eighty known in our study area alone, which would have all been visible to those observers at the terminal.

Although developing over some considerable time, perhaps as much as two millennia, this concentration of monuments still represents a huge investment of planned effort. Even seemingly simple monuments such as round barrows, upon excavation, can reveal long and complex histories involving major phases of re-modelling (e.g. see Wyke Down barrow 34 below). From our investigations we have some idea of the basic chronology of these developments. Most of the monuments in the study area are known only from aerial survey, the majority consisting of ring ditches which without excavation cannot be dated more accurately than to the second or third millennium BC.

However, a number of sites revealed from the air are sufficiently distinctive in form to give us a clue as to their relative chronology, and when added to the evidence gleaned from fieldwalking, geophysical survey and excavation a broad four phase development of this sacred landscape can be outlined:

I *Middle Neolithic:* sites likely to be broadly contemporary with the Dorset Cursus (Fig. 1.3);
II *Later Neolithic:* sites likely to be contemporary with the use of Grooved Ware (Fig. 1.4);
III *Earlier Bronze Age:* sites likely to be contemporary with the use of Beakers, Food Vessels and Collared Urns (Fig. 1.5);
IV *Middle Bronze Age:* sites likely to be contemporary with Deverel-Rimbury ceramics (Fig. 1.6).

Suggested Phase I developments include three possible rectilinear enclosures of the type often referred to as 'mortuary enclosures'. One of these, Monkton-up-Wimborne (MUW 00) (Figs. 1.3 & 3.58) was examined during our survey (see below) and a *terminus post quem* date was provided by Beaker material found in the final ditch silts. About 400 m due south of this site lies the remarkable Monkton-up-Wimborne pit circle/shaft monument (MUW 97) (Figs. 1.3 & 3.61). Fully excavated and described below (see Green this volume), the radiocarbon dating revealed broad contemporaneity with the Cursus. The monument consists of 14 separate oval pits, a large central pit with multiple burials and a deep shaft (Green 2000, fig. 52, pl. 8; & this volume). Within the filling of the shaft was a series of special deposits including a decorated chalk object found near its base (Fig. A4.29). The bone-isotope measurements carried out on the burials suggested that the people had moved into the area from a different part of the country, possibly the Mendips. The location of these four sites within the overall monument distribution pattern is quite striking as they occupy peripheral spatial positions suggesting the parameters of the sacred landscape were defined early in the sequence.

Just to the west of Monkton-up-Wimborne pit circle/shaft monument there is a cluster of features which belong to Phase I including a double-ditched oval mound (Fig. 1.3). This monument bears a striking resemblance to the excavated double-ditched oval mound at Barrow Hills, Radley (Bradley 1992; Barclay & Halpin 1999). Similarly a component of a sacred complex which included other rare types such as pit circle hengi-forms and pond barrows, the extensively excavated Barrow Hills landscape provides a remarkable parallel to our region.

Additional features in our cluster, possibly of Phase I date, include a 'U'-shaped enclosure remarkably similar to Phase 2 of the Radley monument. Two further oval or 'U'-shaped enclosures just to the north and west of that previously described are both unfortunately partly obscured by a hedge.

Within the filling sequence of the Fir Tree Field shaft, Phase I is represented by a long stabilization horizon (layer 5) associated with Peterborough Ware and radiocarbon dates derived from charcoal, bracketing the period 3350–2910 cal. BC (Table 1.1). A similar date from a sheep bone was obtained from the lower part of the succeeding layer 4 with further sherds of Peterborough Ware. This layer contained much weathered chalk in contrast to the rich organic component of layer 5, suggesting the breaking up of pasture for arable cropping occurred at this stage.

Henge (Phase II) monuments cluster in two main groups — around the Dorset Cursus and at Knowlton (Figs. 1.1C & D, 1.4, 1.10 & 3.79). The latter area includes three definite henges, one of truly monumental proportions, two probable and several new possibilities that have been mapped and described by K. Stoertz (see below). Several have seen some geophysical survey but the only excavation information comes from a trench through the ditch and bank of the southern circle (Burrow & Gale 1995; Burrow 1996) and from a pipe trench cut through the outer ditch of the Great Barrow, which may be a henge itself (Field *et al.* 1964; Green 2000). The Dorset Cursus cluster includes numbers of henges/hengi-forms of smaller proportions. These include two of Class I and II types respectively, excavated on Wyke Down in 1983/4 and 1996 (Barrett *et al.* 1991a,b; Green 2000 & this volume). The henge ditches were constructed by digging a series of closely spaced oval pits and both were associated with Grooved Ware. In addition, the basal fill of the Wyke Down henge 2 ditch gave a radiocarbon date of *c.* 2930–2750 cal. BC (Table 1.1). Both sites also shared a series of 'special deposits' within their ditch fillings, suggesting a similar, ritual, use. Round barrows of certain Neolithic date are known from excavation (Barrett *et al.* 1991a) and other probable examples have been recorded through aerial photography and our own geophysical surveys of a number of the barrows on Wyke Down (see below; Figs. 3.34, 3.35 & A2.12). Further double- and triple-ditched barrows, including barrow 34 (French *et al.* 2000; this volume) suggest successive phases of enlargement perhaps starting in the Neolithic. Other Neolithic sites include two circular post-built timber structures, probably houses, adjacent to Wyke Down henge 2 (see below), pit groups and major flint scatters particularly on the clay-with-flints to the north of the main study area (Barrett *et al.* 1991a; Green 2000). Environmental evidence from all these sites is available for comparative purposes and is further discussed by Allen (this volume).

There are dozens of Bronze Age barrows (Phase III) in the study area (Figs. 1.5, 1.10 & 3.79). These are of all types including pond barrows. Although it appears from our survey fieldwork that many of these barrows have seen some excavation in the past (evidence of trenching), there is no published or known record for several of the main groups. This includes the Wyke Down group investigated during this project, and from Knowlton, where two main groups (northern and southern) and an avenue of barrows exist together with more dispersed mounds amongst the main monumental complex. There are of course, numerous comparable barrow cemeteries elsewhere in the Wessex chalklands and the local Oakley Down barrow cemetery (Colt Hoare 1812) includes several standard variations of mounds. The Oakley Down cemetery was particularly rich in artefacts of local and foreign make, including linen cloth, bronze daggers, amber, faience and shale ornaments. These were associated with primary and secondary inhumations and cremations which were frequently discovered placed in pits below the mounds. The prehistoric burials date from the Beaker period through to the Middle Bronze Age.

Although our trenches in the barrow mounds and across ditches have been small, with the exception of the total excavation of the Monkton-up-Wimborne pond barrow (MUW 02), there is a similar variation in monument type, with one disc barrow (Wyke Down 41), a bowl barrow with a single ditch (Wyke Down barrow 36), one double ditched, multi-phased barrow with a stake revetment circle (Wyke Down barrow 34), and one successively enlarged barrow within a single ring-ditch (Wyke Down barrow 44). The fairly good preservation of turves seen in those with larger mounds (Wyke Down barrows 34 & 36) suggest levels of artefact preservation similar to that found at Oakley Down could still be present. However, many of the barrows have suffered severely from root and rabbit damage as well as from recent and historic cultivation. Indeed four of the five barrows investigated in this project (Wyke Down barrows 34, 41 & 44 and the Monkton-up-Wimborne pond barrow) have shown substantial alteration in form and preservation owing to these factors and many in the immediate area have been ploughed flat and remain only as ring-ditches (Fig. 3.34).

Other earlier Bronze Age types of sites include some 'flat graves' that may represent destroyed barrows, pit groups, surface artefact scatters and two possible megalithic structures. These latter structures are represented by a find recorded by Colt Hoare (1812) of a sarsen setting around long barrow 23 at Pentridge, and a new find of a possible burial cist at Knowlton (Lewis *et al.* 2000). This new find consists of two large heathstone slabs, one decorated with four concentric pecked circles (Fig. A4.33), which were ploughed up in the vicinity of three ring ditches just to the northeast of the Knowlton southern henge.

The study area also includes two excavated pond barrows (Fig. 1.5), a rare form easily levelled through arable farming. These were only revealed due to the continuing long-term detailed study of this landscape and could easily have been missed by more rapid survey. Other activities recorded during Phase III include pit groups, probably the deeper surviving elements of settlements, and evidence of re-use of earlier Phase I and II monuments including a burial added to the central pit at the Monkton-up-Wimborne pit circle/shaft monument which has a radiocarbon date of 1890–1610 cal. BC (Table 1.1). The final sequence within the Fir Tree Field shaft consisted of deposits which were also of Phase III date. A nearby round house yielded a Beaker sherd from one of the postholes and may also belong to this phase.

Within the wider area during Phase IV a restructuring of the landscape is evident (Barrett *et al.* 1991a). Aspects of this are present in our much smaller detailed study area and comprise the construction of the settlement buildings at Down Farm and the later digging of the surrounding enclosure earthwork (Barrett *et al.* 1991a, 185). The enclosure lies close behind a major lynchet (Bowen 1990, fig. 1) which follows the soils interface between thin rendzinas and deeper valley deposits for a distance of almost a kilometre. This major boundary may even have originated in the later Bronze Age with traces of fields close to the Down Farm settlement both to the north and south of this boundary. These fields are misaligned to the main

Table 1.1. *Available (after Allen, in Green 2000, 154–5) and new (this project) radiocarbon dates for a range of archaeological sites in the upper Allen valley.*

Site/Context	Laboratory reference	Radiocarbon date (years BP)	cal. BC (2 sigma)
Down Farm enclosure:			
Upper midden, enclosure ditch	BM-2577	2980±50	1400–1040
Upper midden, enclosure ditch	BM-1853N	2980±50	1400–1040
Upper midden, enclosure ditch	BM-1854R	3030±110	1520–940
Upper midden, enclosure ditch	BM-1852N	3160±40	1520–1320
Middle midden, enclosure ditch	BM-1850R	2900±160	1510–800
Lower midden, enclosure ditch	BM-1851R	2950±110	1440–900
Wyke Down:			
Henge 2, basal ditch fill	Wk-18750	4262±32	2930–2860, 2810–2750
House 1, post-hole	Wk-18751	4203±33	2900–2830, 2820–2670
House 2, post-hole	Wk-18752	4117±40	2880–2570
Monkton-up-Wimborne pit/shaft complex:			
Burial group	OxA-8035	4585±50	3500–3100
'Adam' burial	OxA-8034	3180±40	1530–1400
base of chalk platform	Wk-18753	4427±42	3330-3210, 3190–3150, 3130–2910
shaft, layer 10	Wk-18754	4674±37	3630–3590, 3530–3360
Monkton-up-Wimborne pond barrow:			
central cremation	Wk-18749	3423±50	1890–1610
Cursus (Chalkpit Field):			
Middle secondary fill	OxA-624	4570±120	3640–2920
Lower secondary fill	OxA-625/6	4575±77	3520–3040
Surface of primary fill	BM-2438	4490±60	3370–2920
Surface of primary fill	OxA-627	6900±100	5890–5560
Primary fill	OxA-628	5160±140	5640–5140
Fir Tree Field shaft:			
Layer 2a, weathering cone	OxA-7985	3775±45	2460–2040
Layer 5, weathering cone	OxA-8007	4405±45	3310–2910
Main shaft at –3.5 m	OxA-8012	5315±45	4330–4000
Knowlton South Circle:			
Ditch	Beta-141096	3890±60	2570–2190

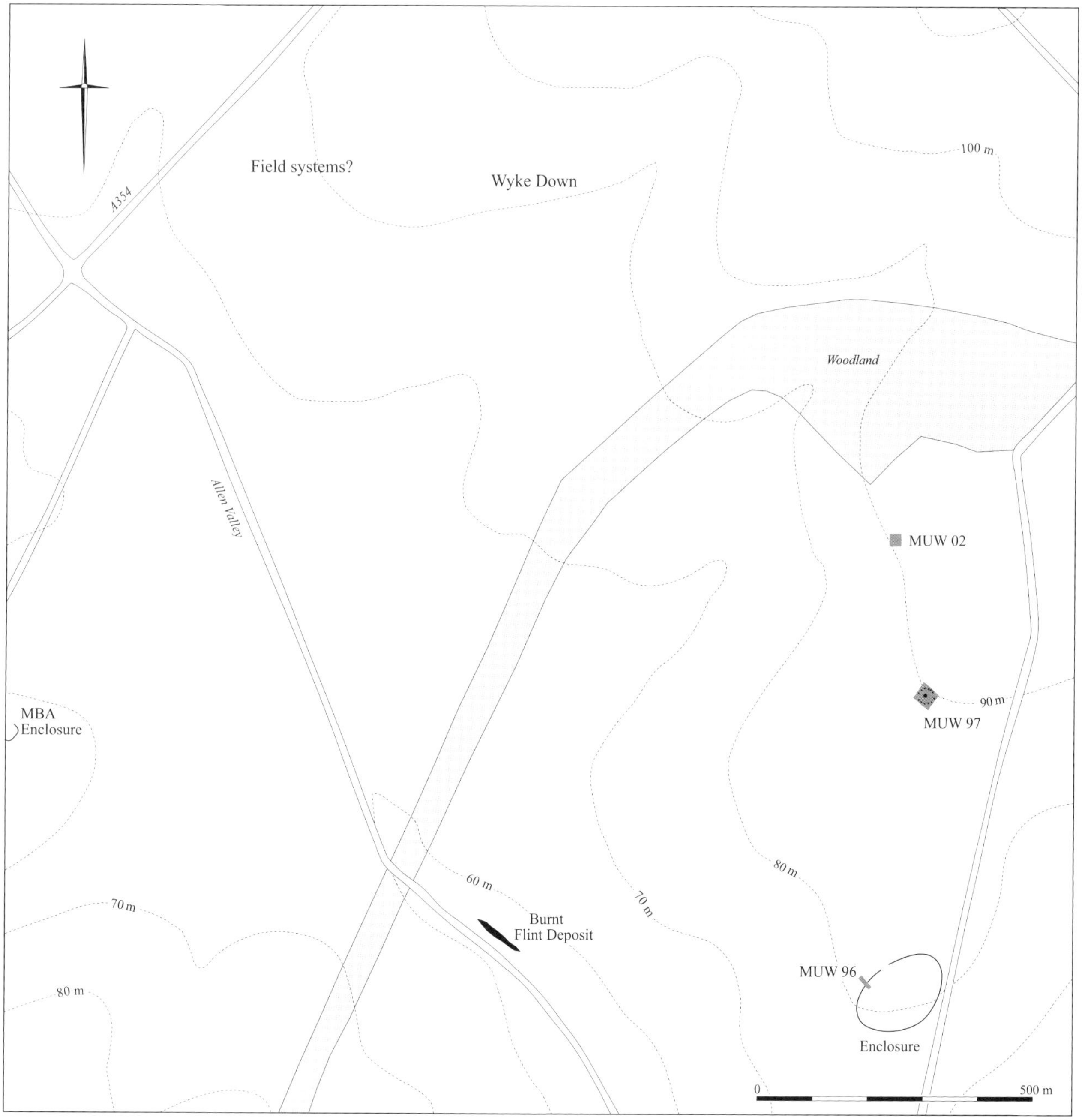

Figure 1.6. *The main sites of the Middle Bronze Age period in the upper Allen valley at Down Farm, Wyke and Bottlebush Downs and Monkton-up-Wimborne (C. Begg after M. Green).*

Gussage Down system and are likely to be earlier and broadly contemporary with the settlement. Further field systems lie to the northwest of the main barrow cluster on Wyke Down but remain undated.

Burial and funerary evidence during Phase IV consists to date, with one exception, of secondary activity within earlier monuments. The ring ditch associated with the Down Farm settlement, which lies just outside our study area, had been re-modelled at least twice before it was used as a cemetery in the Middle Bronze Age (Barrett *et al.* 1991a). Traces of plough-damaged secondary cremations have been recorded from barrows 57 and 58 on Wyke Down (Grinsell 1982) and final phases of activity took place at the Monkton-up-Wimborne pond barrow and pit circle/shaft complex. The adjacent ring ditch examined during the latter excavation (see below) currently remains the only ceremonial monument known to have been constructed during Phase IV in the detailed study area.

Figure 1.7. *Oblique aerial view westwards from Wyke Down across the Allen valley to Gussage Down (top left). Fir Tree Field lies near top left with barrows 35 and 36 in the left foreground and the Wyke Down henges slightly above centre (M. Green).*

Later Bronze Age and Iron Age monuments and landscape

In addition to the many burials dating to the Middle and later Bronze Age found in local barrow cemeteries (with some, such as Barrow Pleck, appearing to date only from this later period), a number of Bronze Age enclosures and field systems have been excavated (Fig. 1.6). In the immediate area, the Middle Bronze Age enclosure at Down Farm (in Fir Tree Field) encompassed the remains of at least six round houses and a long house, and produced evidence for a mixed farming economy with arable crops and animal husbandry (Barrett *et al.* 1991a,b). The Middle Bronze Age enclosure at Martin Down and the Middle/Late Bronze Age Angle Ditch enclosure (Pitt Rivers 1898) are both nearby, and these may relate to larger, linear land divisions (the former, for instance, is associated with Grim's Ditch). The famous South Lodge Camp site (Pitt Rivers 1898; Barrett *et al.* 1991a), an enclosure associated with Barrow Pleck cemetery, provides a nearby comparative complex. Most of these sites have produced some useful palaeoenvironmental data including new samples from Down Farm for our present project (Barrett *et al.* 1991b; Allen this volume).

A large deposit of burnt flint in the valley floor, over 100 m in extent, was probably produced through long episodes of 'trough cooking', although industrial activity cannot be ruled out. Associated with sherds of later Bronze Age date it is likely that a contemporary settlement lies nearby. Indeed, during the examination of the earlier Iron Age settlement enclosure Monkton-up-Wimborne 1996 (Fig. 1.2), a shallow ditch partly cut by the enclosure was traced running parallel to it for a short distance. A sherd recovered from this ditch suggests an earlier phase of later Bronze Age activity prior to the construction of the much larger Iron Age enclosure.

Developments in the Iron Age include the building of the large oval settlement enclosure (Monkton-up-Wimborne 1996; see Appendix 4.12) and a site which is its mirror image 850 m to the west on the other side of the valley (Figs. 1.2 & 3.77). Aerial photographic evidence reveals the former enclosure to be

Figure 1.8. *The Dorset Cursus looking northeast, crossing the Allen Valley and incorporating the river cliff (centre top) in Chalk Pit Field, Down Farm. Note the sinuous course of the northern, probable two-phase, ditch where it crosses the naleds.*

cut by a double-ditched boundary earthwork running northwards for a distance of over 800 m roughly parallel to an existing track. This links to an area of settlement 100–200 m due east of the Monkton-up-Wimborne pond barrow (MUW 02), which has produced surface finds of primarily later Iron Age.

During the Iron Age the area saw a large number of settlement enclosures created within an increasingly ordered landscape of field systems and managed woodland (Bowen 1990; Green 2000). These include a number of settlements, field systems and banjo enclosures (Pitt Rivers 1888; Wainwright 1968; 1979; Barrett *et al.* 1991a). The area is probably a sub-Durotrigan political entity involved in the continental trade centred at Hengistbury Head to the south (Corney 1991). There are no developed hillforts in the immediate vicinity, although several are located nearby, some of which (e.g. Hambledon Hill, Hod Hill and Winkelbury) have seen detailed study (Richmond 1968; Pitt Rivers 1888). A few settlement enclosures have been excavated in recent times (e.g. Wainwright 1968; 1979; Green 1985), and the picture emerges from this work of a well-populated landscape. Our present data when combined with this earlier work have provided much palaeoenvironmental information with which to model, in some detail, changing land use and landscape during this period.

The local hillforts normally see prehistoric activity back to the Neolithic or Bronze Age; some, such as Whitsbury, appear to be part of the larger Bronze Age enclosures mentioned above. In general, the area appears to have been quite densely occupied with farming communities in later prehistoric times (Green 2000). Many of the Iron Age sites in the region have seen relatively recent excavation, and thus provide a good deal of information regarding the landscape during this period, including land-use and environmental data.

Romano-British and later historic periods

Many of the local native Iron Age settlements continue in use throughout the Romano-British period and are best represented by the sites excavated by Pitt Rivers (1887; 1888) at Woodcuts and Rotherley, which were

Figure 1.9. *View northeastwards along the line of the Dorset Cursus from Gussage Cow Down, across the corner of Fir Tree Field and across Wyke and Bottlebush Downs with tree-capped Penbury Knoll in the distance (C. French).*

illuminatingly reinterpreted by Hawkes & Piggott (1947). A few such sites were slighted during the construction of the Ackling Dyke Roman road (Fig. 1.2) which runs from Badbury Rings in the south to Old Sarum in the north and was sectioned during our project on Wyke Down (see below). This road is almost certainly a military construction, as is Bokerley Dyke, a massive defensive earthwork, possibly constructed over a much earlier 'border', which still marks the county boundary between Dorset and Hampshire (Pitt Rivers 1898; Bowen 1990).

Several villas have been found in the area, for example at Minchington (Sparey-Green 1998), Rockbourne and Tarrant Hinton with a probable Romano-Celtic temple complex known on Gussage Cow Down (Green 2000, 135). Extensive 'ribbon' development, partly investigated by Pitt Rivers, is known to exist along the Ackling Dyke at Woodyates (Pitt Rivers 1887; 1888; Bowen 1990). During his excavations there, Pitt Rivers (1887) uncovered a number of burials, several of which were associated with large rectilinear ditched funerary enclosures.

The presence of the springs and a seasonal lake at the Wyke Down end of the valley makes this a likely place for significant Roman remains although little evidence has been found to date. Roman temples, shrines and other ritual sites are well known in limestone regions of Wessex and elsewhere, frequently placed close to springs. These important locations often reveal a long history from Iron Age ritual use to interesting early medieval activities (e.g. Marcham/Frilford in Berkshire) (Gosden & Lock 2003; Hingley 1985; Buxton 1921; Yeates 2005, 109–85; Lewis 2003). Although the Roman remains at Minchington are currently described as a possible villa, they have not been extensively investigated. The additional presence of late prehistoric boundaries/borders and possible routeways in the area, plus the important Ackling Dyke Roman road, only adds to the impression that there should be a Roman ritual site nearby. This aspect of the landscape is ripe for future study.

Although there are relatively few Saxon remains known (mostly burials), several of the local villages betray Saxon origins. Medieval remains include the Norman church, graveyard and settlement at Knowlton, parts of the Sixpenny Handley church and some houses and earthworks around that village, and the nearby settlement at Brockington. Fieldwork in both

Figure 1.10. *Oblique aerial view of the Wyke Down barrow group, Dorset cursus, henge 1 (under excavation) and the Ackling Dyke taken from the north in 1984, looking south (M. Green).*

the upper Allen and Gussage valleys has revealed extensive dispersed settlement associated with the remains of strip field systems (Green 2000, fig. 104). Earthwork traces in other nearby valleys suggest a similar pattern would emerge if these areas were studied to the same degree.

The Shaftesbury Estate which encompasses most of our survey area appears to have been established in the early seventeenth century. Although dating is still unclear, some of the field layouts and drainage systems encountered in our valley survey around St Giles House and Knowlton are clearly associated with the Shaftesbury Estate, while others may have earlier origins.

The palaeoenvironmental background

Charles French & Helen Lewis

Most of the existing detailed palaeoenvironmental work is based on the analysis of molluscan assemblages and, more rarely, on palynological data from ditch fills and buried soils from a number of monuments (e.g. Cranborne Chase, Hambledon Hill, Wessex Linear Ditch project, King Barrow Ridge) (Barrett *et al.* 1991a,b; Entwistle & Bowden 1991; Cleal & Allen 1994; Allen 1994; 1995a–d; 1997a–c; 1998). This evidence has been coupled with more generalized colluvial sequences (e.g. Strawberry Hill; Heytesbury) and with the regional pollen record in order to reconstruct sequences of early prehistoric land use (Allen 1992; 1995b). When specific monuments have been investigated (Cleal & Allen 1994; Allen *et al.* 1995), only rarely has micromorphological analysis of the buried palaeosols accompanied archaeological, molluscan and occasionally palynological studies (Fisher 1982; 1991; Macphail 1993; Scaife 1994a,b). The free-standing sequences at each monument investigated require corroboration and integration with the wider archaeological and palaeoenvironmental record of the region, along with testing of previous palaeoenvironmental results through the application of finer sampling intervals and in light of recent disciplinary/technological developments and ecological and experimental studies (Allen *et al.* 1993).

Prior to this research, the state of knowledge concerning the environmental evidence for land use during the Neolithic to the Middle Bronze Age in Wessex may be summarized as follows. The regional pollen record shows a decline in forest cover over time, and a change from small, short-term clearings to larger and longer-term clearings by the second millennium BC (Scaife 1982; 1987b; Thorley 1981; Kerney *et al.* 1964). The implied connection (inferred from a small amount of archaeological evidence, supplemented by the use of ethnographic analogy) is increasing human clearance of woodland and scrub vegetation, and greater maintenance of open areas for settlement and pastoral and arable use (Barrett 1994; Bradley 1978). Associated with this record is some evidence in certain parts of the chalk downland indicative of a period of accelerated erosion and colluviation processes during the second millennium BC (Allen 1992). This is often cited also as evidence for the intensification of agricultural land use, but a variety of factors may have been responsible for this erosion such as soil destabilization owing to forest clearance, intensification of arable practices in general, 'evolutionary' technological changes in tillage implements, or even extensive abandonment or lack of management of much of this landscape, none of which has been previously addressed in detail (see below). The molluscan record is the most comprehensive (Allen 1992; 1995a–d; 1997a–c; 1998; Bell 1983; Evans 1972; Green & Allen 1997; Kerney *et al.* 1964), and suggests an increasing opening-up of woodland over these periods, with more frequent occurrence of grassland areas and disturbed, possibly tilled, land.

Erosion models

Many authors writing on the prehistoric landscape of Wessex suggest that soil erosion occurred on a grand scale in later prehistoric times, and especially from later Bronze Age times (Barrett 1994; Bell 1992; Boardman 1992; Catt 1978). Deforestation throughout the Neolithic and Bronze Age, associated with the uptake of land for arable and pastoral purposes, and the intensification of settlement, is thought to have led to both destabilization and degradation of early Holocene forest brown earth soils, and the accumulation of hillwash deposits in dry valleys and as alluvium in the main river floodplains. This process was believed to be probably most marked in later prehistoric and medieval times. For example, some authors have suggested colluvial/hillwash erosion on an extensive scale, with Catt (1978; Catt *et al.* 1980) suggesting at least 1–2 m and Boardman (1992) suggesting about a metre of soil loss since the early Holocene. The latter has been observed in specific parts of Wiltshire (Allen 1992) and Sussex (Bell 1992), but this hypothesis has not been evaluated systematically in much of the region. Indeed, many of the areas that have been assessed, such as around Maiden Castle (Evans & Rouse 1991) and in the Kennet valley (Evans *et al.* 1993), show erosion occurring much later, in either later prehistoric or historic times, or reveal none at all, such as in Stonehenge Bottom (Allen 1997a), or show localized erosion in unexpected areas such as in the River Avon valley just east of Durrington Walls (M. Allen, in Cleal *et al.* 2004).

In most cases, soil erosion and degradation is believed to be clearly associated with the growing and harvesting of cereal crops (cf. Barrett 1994). With clearance, former thick woodland soils lose their binding root systems, suffer losses of organic and nutrient status, lie with exposed surfaces on slopes for several months of the year, and are physically disturbed by the action of rainsplash, frost, clearance and cultivation activities. The coincident effects of erosion and degradation take hold quickly and may be irreversible through normal land-management practices. This is perceived as the general model of soil formation for southern England. Finding reliable proof of this soil history is, however, much aggravated by modern farming methods, which truncate and transform the soils in question, with the only intact buried soils being found beneath upstanding (usually scheduled) monuments. Nevertheless, models of soil and sedimentary history developed in specific landscapes are still commonly applied using a broad brush to regional archaeological landscape interpretations. These assumed sequences of soil and vegetational development and change need to be questioned and examined from a soil process point of view, and on a case-by-case basis, before models of prehistoric landscape development and the role of land-use practices like clearance and agriculture in this are proposed. This is where the combination of off-site geoarchaeological survey, on-site soil micromorphological studies and re-examining soil development models is crucial to suggesting new trajectories of land use, soil development and landscape change in the study area that may be of wider applicability.

Methodological approaches

Charles French & Helen Lewis

Aims

The overall aims of the project were three-fold and interlinked. First, a selection of upstanding prehistoric monuments, consisting of four barrows of the Wyke Down barrow group, the Dorset Cursus and the Gussage Cow Down later prehistoric field systems and Neolithic long barrows (Figs. 1.2, 3.1 & 3.34) were investigated by targeted excavation to allow environmental sampling for pollen, molluscan and soil micromorphological analyses. This aimed to establish chronologically controlled palaeoenvironmental sequences for the upper Allen valley. Second, a geomorphological survey of the upper Allen valley was carried out through aerial photographic and systematic augering surveys to allow geographical information system mapping of erosion versus aggradation zones in the landscape. A selection of sediment burial zones were sampled by trenching to extract pollen, soil and molluscan samples from zones of colluvial and alluvial aggradation, old land surfaces, and relict channel systems in the valley bottom. The aim of this work was to assess processes of change and degradation in the landscape; this also allowed prospection for off-site buried landscapes containing new palaeoenvironmental data that would contribute to the creation of new sub-regional palaeoenvironmental sequences. Third, the results were combined to produce new models of land-use change and landscape development from the earlier Holocene and later in the study area, to re-examine existing land-use models at a variety of different scales — for England as a whole, the chalklands of southern England, Cranborne Chase and the upper Allen valley study area itself.

Within these aims were also some more specific targeted questions. These included:

- Could waterlogged repositories of Holocene vegetation evidence be found in this otherwise well drained and calcareous landscape?
- What was the common modal soil type of the region that developed in the earlier Holocene? And

in particular could brown forest soils be located, preserved or identified?

- Could specific evidence be found that related to the impacts of deforestation and arable and pastoral agriculture?
- Could deposits of colluvium and alluvium be related to past erosion events or particular periods of greater/lesser intensity of disturbance to the landscape?
- What would be the best suite of methodological approaches to use to assess late glacial and Holocene landscape change in the chalk downland area of Wessex?

Field methodology

Our approach in the field used a combination of conventional archaeological and palaeoenvironmental methodologies but at two levels of resolution — specific site-based sampling investigations for the recovery of archaeological and palaeoenvironmental data, and more extensive off-site geoarchaeological survey for the recovery of soil, soil erosion and vegetation data.

As a rule, the actual methodologies employed in a field project are defined in the specific design brief and are rarely available for others to see. In this instance, it is worthwhile setting out the methodological approach adopted for this project, as follows:

For the investigation of the monuments, the following techniques were used:

1. selection of prehistoric monuments of major periods in different parts of the valley (i.e. on brow, upper slopes, lower slopes and valley bottom) for survey and appropriate sampling by all members of the project team;
2. for the most part the monuments are scheduled and therefore require scheduled monument consent before any work begins; prospective sites must be identified, and the project design and access to them discussed, first with the relevant English Heritage inspector and ancient monuments warden, the land owner, and then an application must be made to the Secretary of State for consent;
3. once consent was obtained, an initial hand-auger survey was used to construct monument profiles, to confirm the presence/absence and thickness of buried soil survival beneath the mound or bank and on the inner berm of ring ditches, and to ascertain the depth of infill deposits in ditches;
4. topographical survey and plan of each monument at a scale of 1:20;
5. magnetometer and resistivity survey of each monument to reveal structural components; this was occasionally augmented by ground-penetrating radar survey (see Pierce & Shell, in French *et al.* 2000, 63);
6. a full photographic record of each monument being examined as it survives today;
7. selection of areas for sample excavation on the basis of 1–6 above and in discussion with the English Heritage inspector and warden, and the land owner; sample excavations targeted ditches, banks and mounds, and were normally *c.* 1 × 2 m, 2 × 2 m and 2 × 3 m in size;
8. description of all excavated sections, recorded at 1:10 and photographed;
9. collection of series of samples for molluscan, soil micromorphological, plant macro-fossil and palynological analyses from the buried soil, ditch and bank/mound material, and if possible, samples for radiocarbon assay and optically stimulated luminescence (OSL);
10. re-instatement of all sample excavation trenches, with every effort made to follow the pre-existing contours;
11. assessment of the present condition of the monument and the sequence of past land-use history of the site with respect to its current scheduled status; this new information could aid in future management decisions taken by both the responsible national and county curators;
12. integration of the data from the new sample investigations of monuments with the results of the concurrent and extensive excavation programme of prehistoric sites in the study area by Martin Green.

For off-site investigations, the following procedures were employed:

1. seek the land owners' permission for access;
2. full aerial photographic mapping of the study area, delimiting zones of thin and thick soil cover with respect to the known archaeological record;
3. hand-augered borehole transects across the downland and hand-/power-augered transects across the valley floor, with *c.* 100 m between auger holes and transects set at *c.* 500 m interval widths; where palaeochannel systems were identified in the valley floor, the auger holes were placed at 10 m intervals in transects set *c.* 25 m apart (Fig. 1.11);
4. create an erosion assessment map for the region using data derived from 2 and 3;
5. target relict meander loops in the river valley bottom for augering with a Russian corer to recover cores for palynological analysis and radiometric dating;
6. on the basis of 1 to 3, identify zones of colluvial aggradation or lynchet systems for sample excavation using small trial trenches (1 × 2 m), again with land owners' permission.

From each site investigated, the buried soil was always sampled for soil micromorphology in at least two locations beneath the surviving mounds and/or upcast banks. In addition, where possible turf lines or soil-like material had either been dumped into or accumulated in the associated ditches, these horizons were also sampled for micromorphological analysis.

The methods used for the production of soil thin sections were those of Murphy (1986), and the descriptions were based on the internationally agreed terminology as set out in Bullock *et al.* (1985) and Stoops (2003). Thin sections were produced by Julie Boreham at the McBurney Geoarchaeology Laboratory, Cambridge, and analysed at that laboratory and at the Oxford Research Laboratory for Archaeology and the History of Art, using Laborlux 12 Pol and Optiphot microscopes and a Wild/Leica macroscope. Micromorphology results are discussed with the site descriptions to which they refer; detailed descriptions by site are presented in Appendix 4.

How the project proceeded

After preliminary reconnaissance of the upper Allen valley in the summer of 1997, the initial key into the problem and the landscape was felt to be the investigation of the Wyke Down barrow group near the head of the valley, and the prospection for deeper, better-preserved soils and sediments via an extensive augering programme. As most upstanding prehistoric monuments in the region are protected by scheduling, it was felt that the ear-marking of several previously damaged barrows would provide an acceptable access into these protected sites and their underlying old land surfaces. Accordingly, the first small grant application to the British Academy was designed to test the approach outlined above on two antiquarian and plough-damaged barrows to enable the retrieval and analysis of palaeoenvironmental samples with minimal disturbance to the barrow mounds and ring ditches. Two barrows of the Wyke Down group (34 and 41) (Figs. 3.34, 3.39–3.42 & 3.46–3.48) were the first to be sample excavated after a full programme of topographical, geophysical and augering survey. The stratigraphy of each small trench was recorded, and samples taken as appropriate for molluscs, soils, pollen, macro-fossils and radiocarbon assay. Working in teams of two people, this never took less than five days per barrow.

Once this relatively quick and inexpensive method had been tested and proved viable, a larger grant application was prepared for the Arts and Humanities Research Board. This extended the same approach to the investigation of another two barrows of the Wyke Down group (36 & 44), plus associated extensive geophysical and augering surveys, the Dorset Cursus where it crosses Wyke and Bottlebush Downs and is crossed by the Roman Ackling Dyke, and two Neolithic long barrows on Gussage Cow Down towards the southern end of the Dorset Cursus on the southern side of the upper Allen valley. As it turned out, additional sites became available and were added to this original list; these included a Neolithic subrectangular enclosure on the eastern edge of Bottlebush Down (excavated jointly with Martin Green), situated near to the Neolithic pit circle and shaft complex of Monkton-up-Wimborne (Green 2000), a set of banks, ditches and lynchets associated with the large Iron Age site on Gussage Cow Down, and the University of Bournemouth's trench through the South Circle at Knowlton. These sites were investigated in the late summers of 2000, 2001 and 2002, along with associated topographical, geophysical and augering surveys.

At the same time, three other lines of research were being developed. First, there were the on-going, separately funded series of large-scale excavations of a series of prehistoric sites and monuments in the study area by Martin Green. In particular, Green had investigated a late Upper Palaeolithic lithic scatter site at Deer Park Farm, a Late Neolithic pit complex at Monkton-up-Wimborne, a Late Neolithic henge and associated settlement on Wyke Down, a Mesolithic to Beaker shaft in Fir Tree Field and a number of smaller prehistoric settlement/field enclosure types of site in the Down Farm/Wyke Down area (see Green below). All of these sites provided pertinent data, especially molluscan and faunal remains evidence related to vegetation sequences and land use.

Second, separate funding was obtained from the McDonald Institute for Archaeological Research to support a valley-wide geoarchaeological survey (see Chapter 2). In particular, this project aimed to find areas of well-preserved palaeosols, erosion deposits and waterlogged infills of relict palaeochannels. This involved aerial photographic mapping of the erosion and aggradation zones in the upper Allen valley between Sixpenny Handley and Knowlton (Figs. 2.1, 2.2 & 2.9), and the hand-augering transect survey of the same area. Some 880 boreholes were made and recorded, with selected test pits dug in several of the dry valleys, and core sampling for pollen from two major relict stream channels that were discovered through the valley-wide augering survey (Fig. 1.11).

Third, towards the end of the period of excavations and valley geoarchaeological survey, further funding was obtained from the McDonald Institute to support fieldwalking survey. This was aimed to especially target the southern end of the valley study area around Knowlton where much less was known about the contemporary settlement record than the

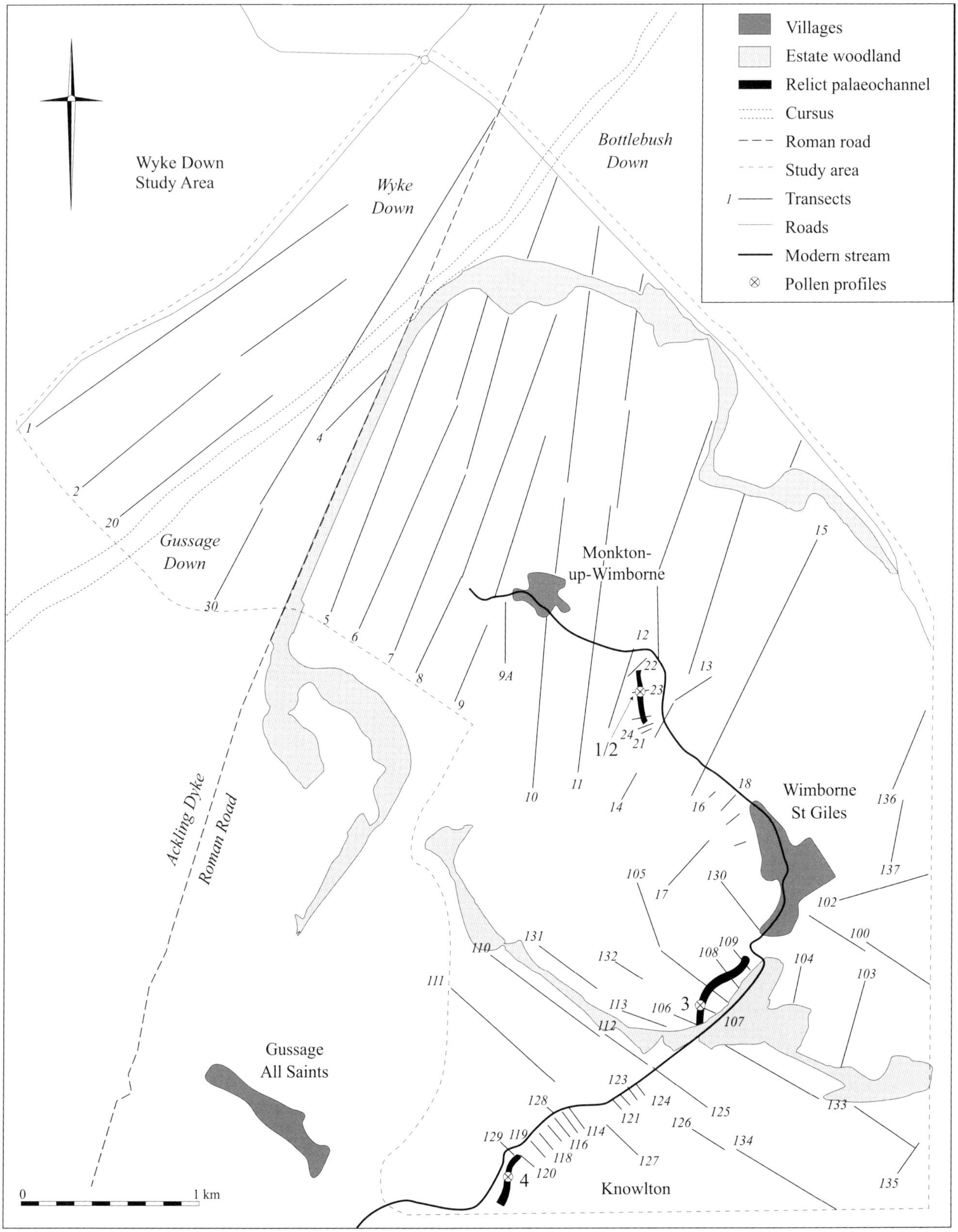

Figure 1.11. *The position of the auger transects, relict palaeochannel systems (in solid black), the location of the pollen profiles, and the main site areas investigated (C. Begg after H. Lewis).*

northern Down Farm end (Green 2000) of the study area. Eight possible locations of potential settlement were identified, and these were sample investigated in 2003 (see Chapter 3).

Specific issues

The project set out to critically examine some widely held views about the nature of Neolithic and Bronze Age chalk downland landscape development and land use, with the upper Allen valley on the southern edge of Cranborne Chase as a case-study area. These can be summarized briefly in the following model: extensive Mesolithic woodland with associated thick argillic brown earth soil development became extensively and increasingly exploited for monument building and agricultural activities leading to soil degradation and erosion on a large scale over the later Neolithic and Bronze Age periods (Bradley 1978; Catt 1978; Barrett *et al.* 1991a; Fisher 1991; Allen 1992; Bell 1992; Barrett 1994). In particular, the Middle Bronze Age transition has been envisaged as a major period of intensification of agricultural land use with the concurrent development of the first enclosed farmstead sites with attendant field systems, such as at Down Farm and South Lodge (Barrett *et al.* 1991a, 183–211) and Angle Ditch and Martin Down (Pitt Rivers 1898), associated with both widespread and increasing impacts of soil erosion and colluviation (Allen 1992; Barrett 1994; Bradley 1978).

Assessing land-use and agricultural practices within this landscape through approaches focused on settlement and burial data make it difficult to interpret the relationship between changes such as sedentism in the Middle Bronze Age and the proposed agricultural changes mentioned previously (Lewis 1998b). It was felt that any relationship with land-use change would be best explored through a study focused on environmental approaches to landscape, particularly since so few pre-Middle Bronze Age settlement contexts are known. Of the various reference and regional palaeoenvironmental data available, the molluscan record is by far the most comprehensive. In contrast, other categories of data such as soil and palynological data are fewer and far between for the chalk downlands as a whole, and this data is considered in greater detail in Chapter 4. There are a few snapshots of soil data available for the study area (Fisher 1991) and the wider region (Fisher 1982), regarding the degradation of brown forest earths on chalk, as well as Macphail's investigations of buried brown earths beneath specific Neolithic monuments such as Hazelton North (1990) and Easton Down (1993), all of which is further discussed in Chapter 4. The molluscan data suggest a gradual opening-up of woodland over the Neolithic and Bronze Age periods with the landscape becoming predominantly open grassland with small areas of disturbed, possibly ploughed land by the Middle–Late Bronze Age (Allen 1992; 1995b; 1997a; 1998; Bell 1983; Evans 1972; Entwistle & Bowden 1991; Evans & Jones 1979; Kerney *et al.* 1964). There is also the possibility of a slight re-growth in woodland conditions occurring in the later Neolithic (Allen 1997b), and the deposition of windblown deposits in the Middle–Later Bronze Age. This data has recently been summarized by Michael J. Allen (2000a, 36–49; 2002a), and is re-assessed in this volume (in Chapters 2 & 4). Each of these trends requires interrogation and corroboration from other strands of evidence, and that is what this volume sets out to do.

Chapter 2

The Holocene History of the Upper Allen River Valley: Variation at the Sub-regional Scale

Charles French, Helen Lewis, Michael J. Allen, Rob Scaife, Rog Palmer, Cathy Stoertz & Lidia Żuk

A series of studies was carried out with the aim of reconstructing Holocene valley history, including sedimentation and soil surveys, and a series of pollen analyses. Sedimentation and soil studies were carried out through an extensive augering programme recording soil profiles, parent material and valley deposits, plus an aerial photographic study of alluvial and colluvial deposition along the entire Allen Valley from Sixpenny Handley to just south of Knowlton. Pollen studies were carried out on palaeochannels identified through the surveys, as well as on comparative developer funded sites in Wimborne Minster.

Terrain analysis

Charles French & Helen Lewis

The Allen River is a bificated tributary stream of the River Stour. The northern branch of the river emerges about halfway down the upper Allen valley study area just to the southeast of Wyke Farm as a series of springs which turn into a shallow and narrow meandering stream channel. Here the Devensian gravels are just beneath the modern turf/topsoil. Most of the route of the channel to the south through the study area has been straightened and canalized with the establishment of the Shaftesbury estate in the mid-seventeenth century. The southern stream rises at Manor Farm, Gussage St Michael, and joins the northern bification of the River Allen just to the south of Gussage All Saints where they merge into one channel, just outside and to the southwest of the study area.

The off-site augering survey of the valley comprised a systematic series of some 880 boreholes located every 100 m on transects set approximately 500 m apart (Fig. 1.11). This survey has indicated a generally low level of colluvial deposition and/or survival within both the tributary dry valleys and the main valley floor, and no alluvial deposition present in the valley bottom to the west of Wimborne St Giles.

The survey found ubiquitous modern rendzina soils ploughed (Ap/C) and under pasture (Ah/C) over the chalk parent material (Figs. 2.3 & 2.5). These range in thickness from 12 to 25 cm. Argillic brown earth soils (Ah/Bt/C) survive occasionally on the upper to mid-slopes of downs currently under pasture, where there are coincident outcrops of either clay-with-flints or the gravelly Reading Bed geology (Figs. 2.1, 2.2 & 2.5). These occur in the eastern and southeastern parts of the study area. Argillic brown earths were also found over chalk in forested zones at the edge of the Shaftesbury estate woodland ride. In addition, two areas with deep peat deposits over chalk were found and mapped; these represent two palaeochannels to the west and east of Wimborne St Giles, which were sampled for pollen analysis at three locations (see below) (Figs. 1.11, 2.4 & 2.18).

The northern part of the valley, near Down Farm and Sixpenny Handley (Figs. 1.2, 1.7 & 2.1) was generally characterized by very localized clay surficial deposits on the tops of some downs, under argillic brown earths, with chalk being exposed on the crests and down the slopes, under rendzinas, into the shallow valley bottom, where gravel fans were encountered under stony rendzina soils. In this upper part of the valley, most soils were covered by arable fields, with some reclaimed pasture and zones of woodland. As the valley profile changed to the south (Figs. 2.1 & 2.11), chalk parent material with rendzinas dominated at all points of the topography, under arable fields except in the woodland ride of the Shaftesbury estate, with soils becoming thicker rendzinas in the wet pasturelands in the valley bottoms. This middle part of the survey area is complicated by canalization of the river; this is also the location of one peat-filled palaeochannel located by aerial photography and augering. Around the

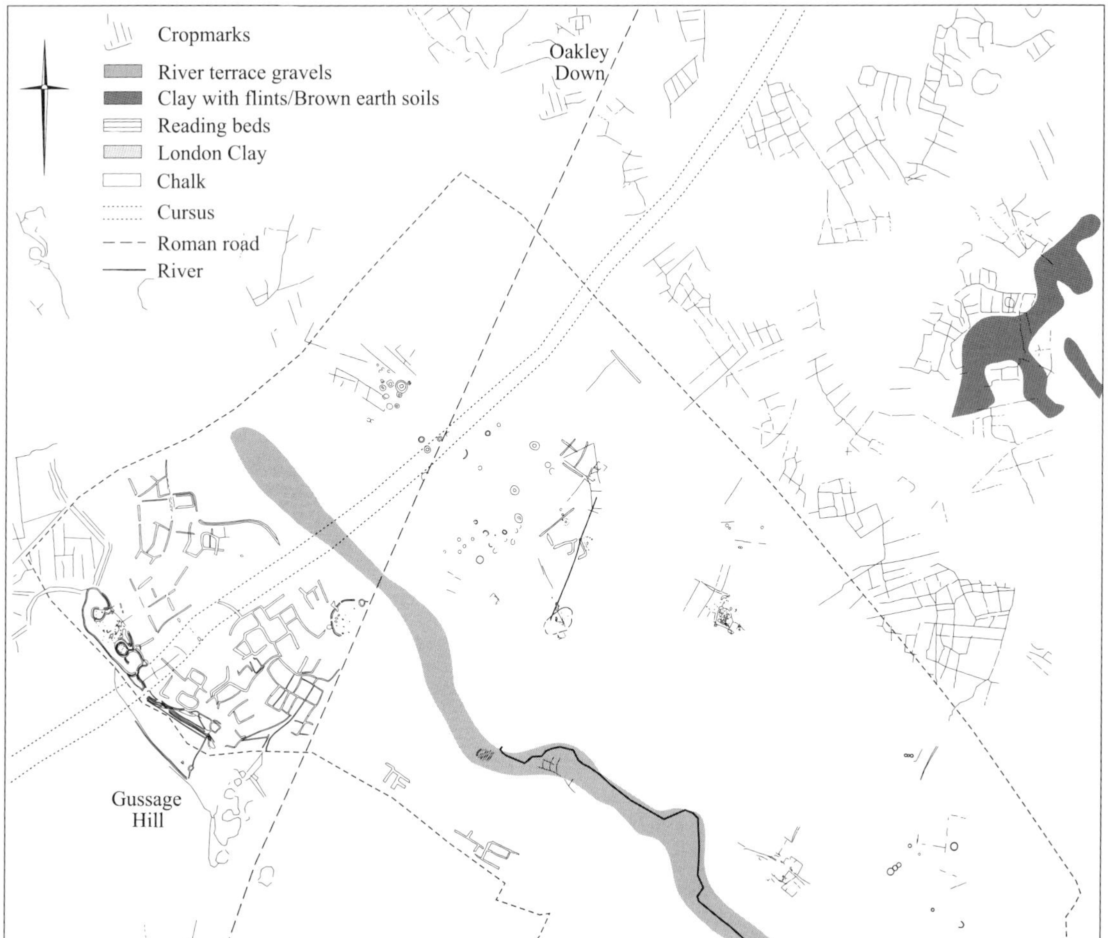

Figure 2.1. *The cropmark data set against the geology for the study area (after Bowen 1990 and M. Green) (R. Palmer).*

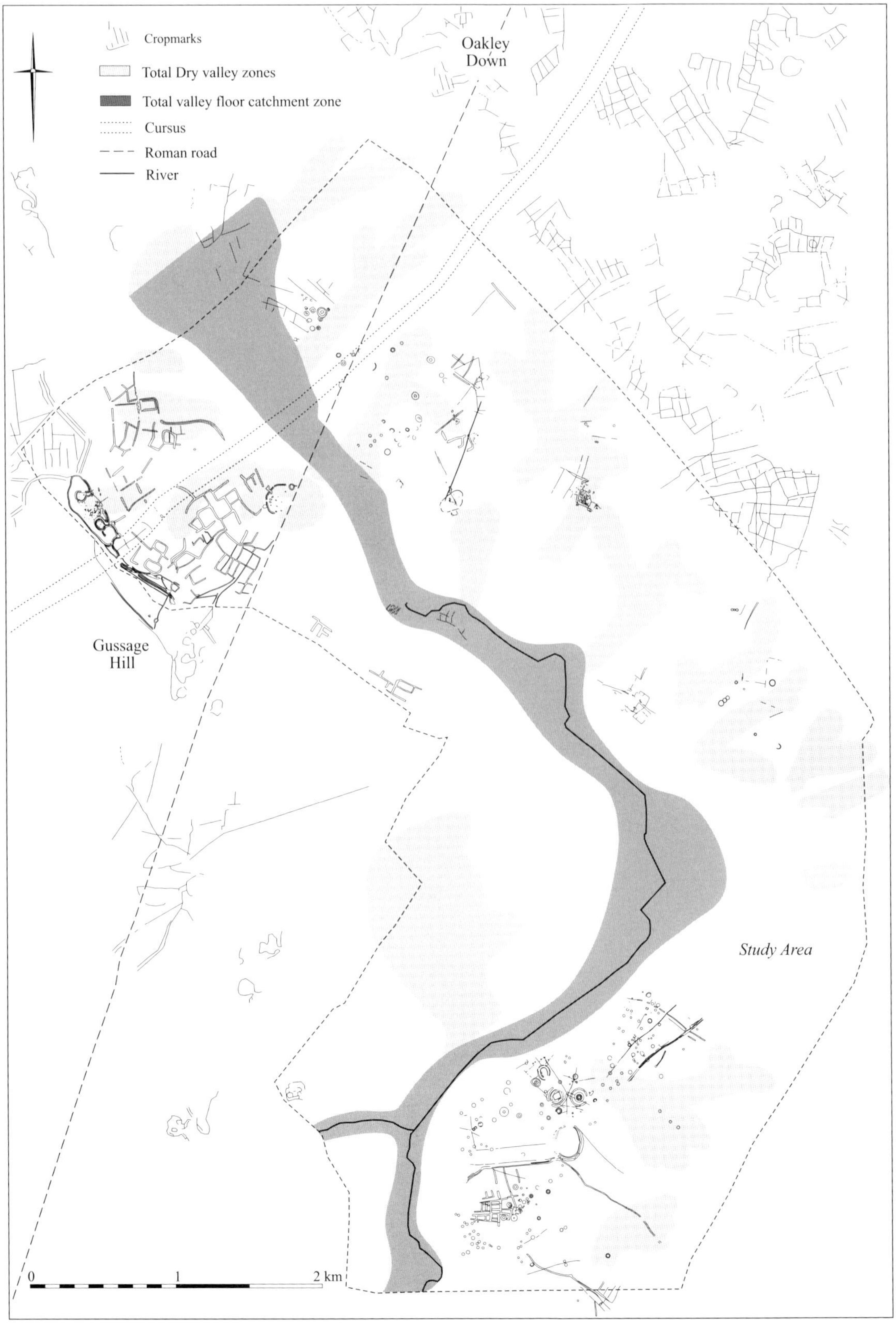

Figure 2.2. *The colluvial (pale shaded) and alluvial cover (dark shaded) set against the crop/soil mark evidence as derived from the augering and aerial photographic survey (C. Begg after R. Palmer & Bowen 1990, fig. 1; with permission of English Heritage).*

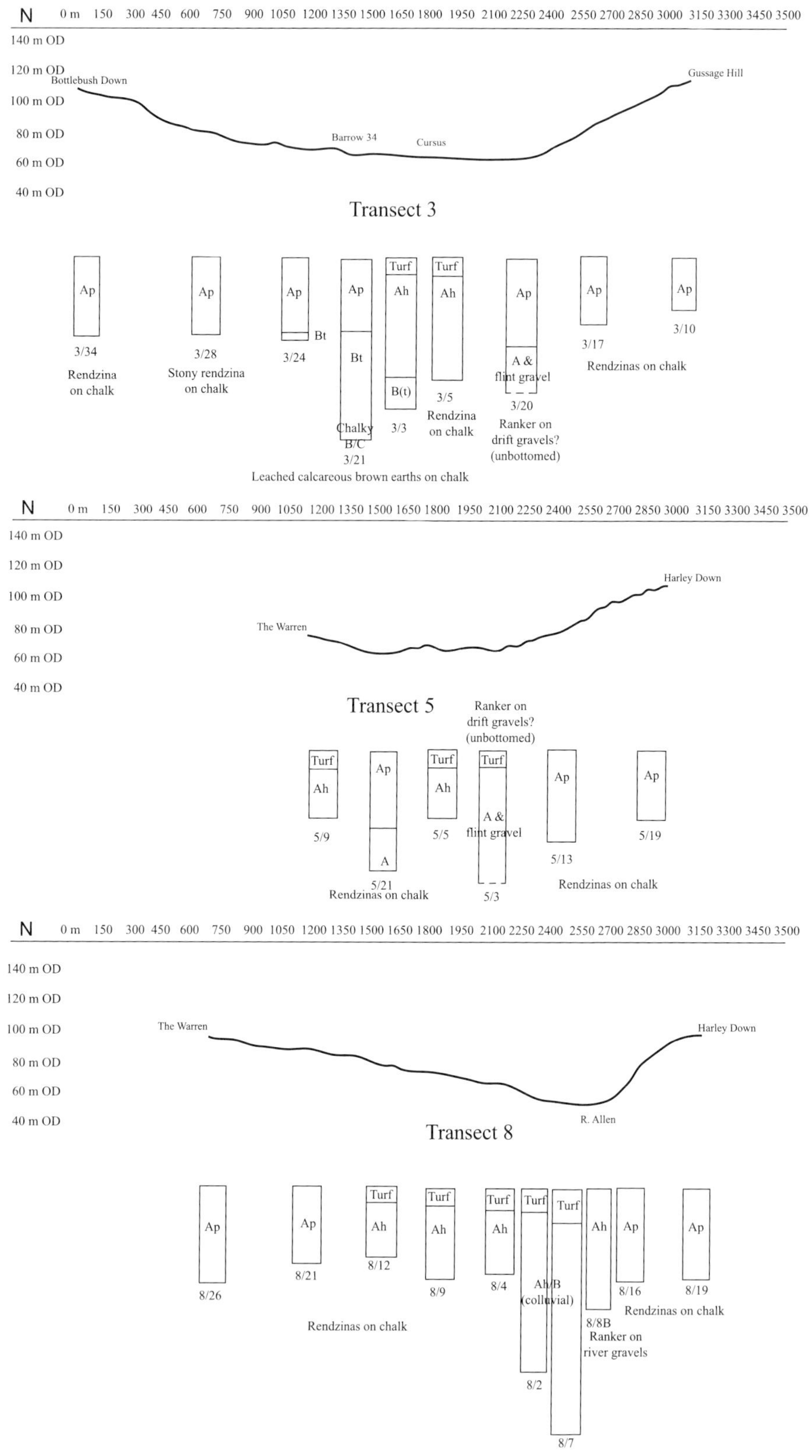

Figure 2.3. *Example valley cross-sections from augering; transects 3, 5 and 8, the latter profile with colluvium (H. Lewis).*

Knowlton area the river system becomes more significant, with the valley narrowing before it opens out again where the upper Allen joins the southern/Gussage valley tributary. Most soils on the slopes are rendzinas on chalk, with arable fields on the slopes and pasture fields lower in the valley with argillic brown earths on the Reading Beds to the southeast.

The auger survey was augmented by aerial photographic mapping, targeting areas of soil erosion and aggradation, as well as zones of maximum extent of potential soil erosion (Figs. 2.1–2.5 & 2.9). This work suggests that soil erosion was primarily confined to the now dry, tributary valleys that dissect the downland slopes of the main valley (Żuk & Palmer 2001a). There is rarely more than 50–70 cm of accumulation of colluvial calcareous silt loam to clay loam material, suggesting that soil erosion is confined to certain parts of the landscape and may be the result of low intensity and infrequent events in an otherwise stable landscape. It is hard to envisage the catastrophic erosion that other authors (Bell 1983; Catt 1978) have postulated ever having happened in this part of Wessex. These occasional loamy colluvia overlie buried brown earth soil profiles of 30–35 cm in thickness (Figs. 2.7 & 2.8). Two of these were excavated for sampling and to obtain dating information.

One dry valley section investigated was situated about 500 m to the west of the Monkton-up-Wimborne Neolithic enclosure (see below) (Figs. 1.2 & 2.7). There were about 30 cm of colluvial silt loam and chalk rubble found beneath the modern ploughsoil, and this had accumulated on a *c.* 35 cm thick brown earth profile. This buried soil exhibited some evidence of brown earth horizonation (Ah, A1 & B horizons). Dry sieving of this soil revealed a wide variety of artefacts ranging in date from Neolithic blade flint to nineteenth-century glazed pottery, with soil faunal mixing processes and colluviation seriously affecting the buried profile. It also contained an impoverished land molluscan fauna typical of hillwash deposits (see Allen below).

Micromorphological analysis of samples from this profile revealed quite a well-developed brown earth comprising an upper organic A horizon and a weathered or cambic B horizon. There was a calcareous silt component in the fine groundmass and in the pore space which suggests long-term but small contributions of fine material into this soil associated with a minor influence of hillwash and bioturbation. Besides the instance of modern argillic brown earth mentioned above in the Shaftesbury woodland ride, this soil was the best-developed soil profile found in the project on the ubiquitous chalk substrate; indeed every soil found beneath prehistoric monuments was thinner and less well-developed. Although this is only one 'snapshot', there are some hints of earlier better-developed brown earths evident in several profiles such as beneath the bank of the Knowlton South Circle, Wyke Down round barrows 41 (on coombe rock over chalk) and 36, and beneath Gussage Cow Down long barrow 78 (see below). These certainly hint at brown

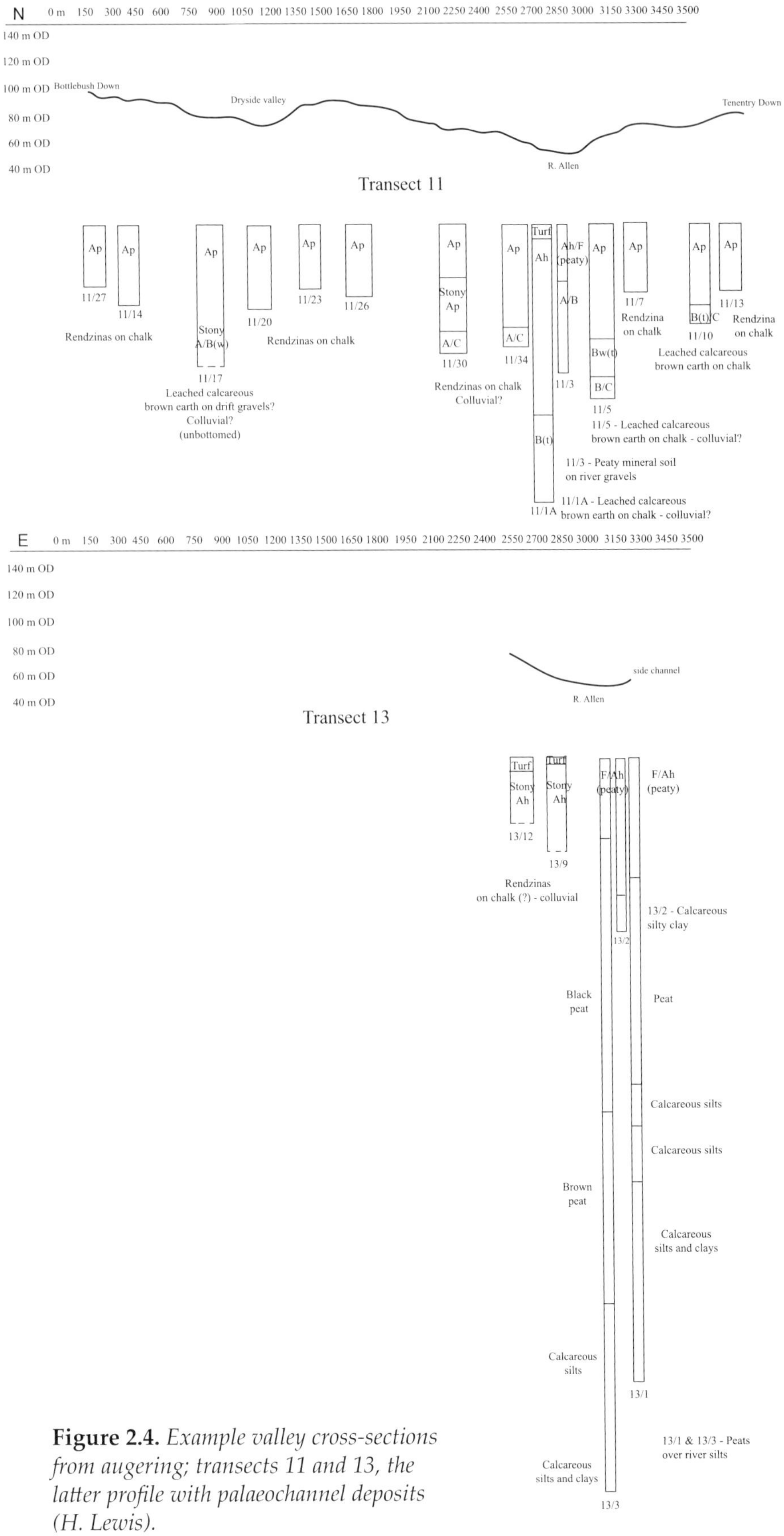

Figure 2.4. *Example valley cross-sections from augering; transects 11 and 13, the latter profile with palaeochannel deposits (H. Lewis).*

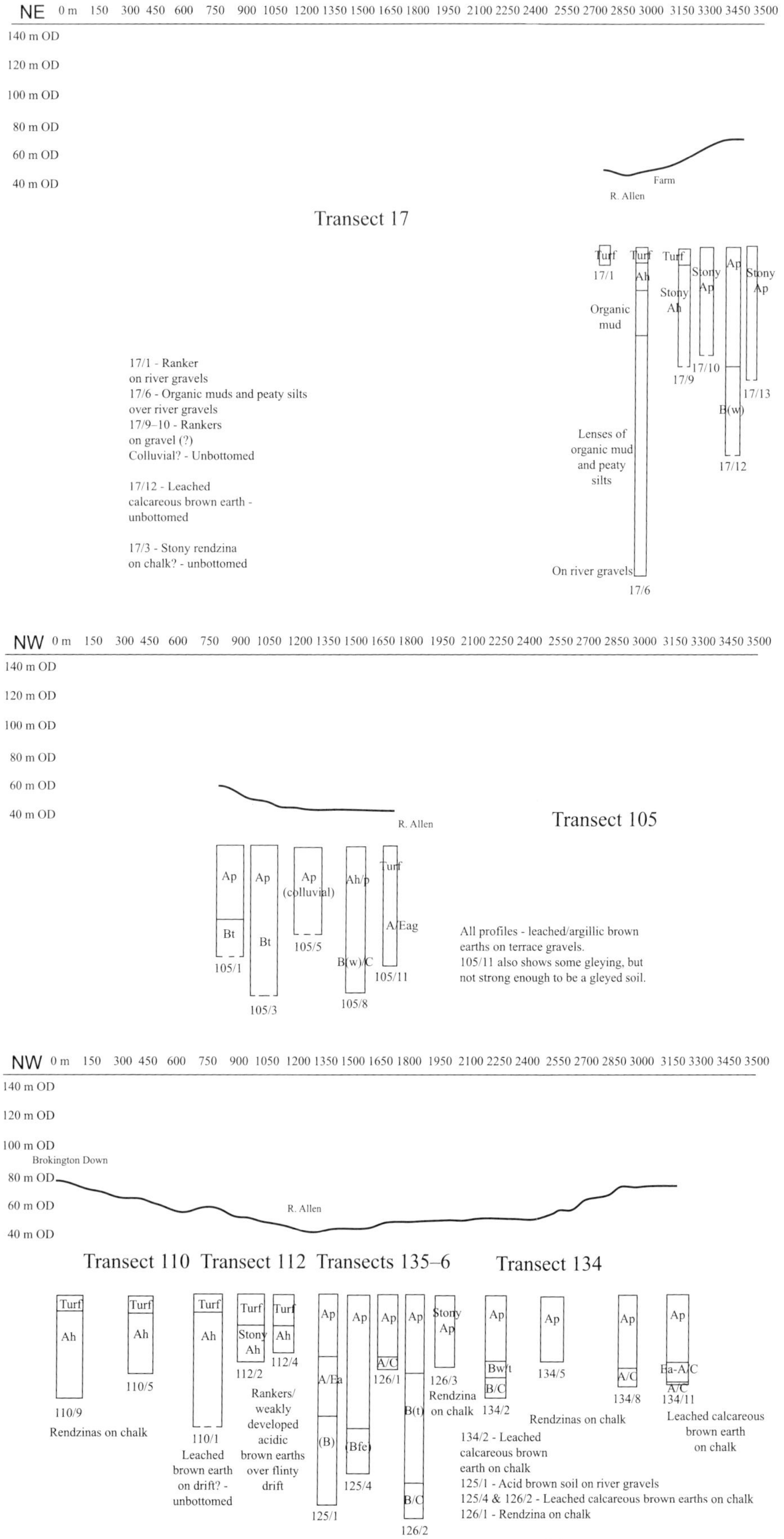

earths being once more common on the chalk downland in the Neolithic of this area than it seems at first glance, with the otherwise ubiquitous rendzinas perhaps best viewed as 'degraded' brown earths that have undergone longer-term human manipulation and management in the past.

The off-site geomorphological survey and the sample excavation of eleven archaeological sites have revealed only a very few examples of relict brown forest earth with well developed argillic (Bt) horizons (argillic brown earths) and brown earth soils, but much more commonly rendzina soil profiles. The few buried argillic profiles range from *c.* 20–50 cm in thickness (notably at Deer Park Farm; see below) (Fig. 2.6). The more common buried rendzina profiles are generally less than 15 cm in thickness and range in date from pre-early Neolithic to the Roman period (e.g. at the Wyke Down round barrows, beneath the banks of the Dorset Cursus and sealed by Gussage Cow Down long barrow 78 and Iron Age bank systems as well as the Roman Ackling Dyke) (Figs. 3.9, 3.10, 3.52 & 3.55). The ubiquitous modern ploughsoil rendzina profiles range in thickness from about 12 to 25 cm. Occasionally, there are *c.* 50–75 cm accumulations of calcareous silt loam colluvial material overlying buried brown earth soil profiles of about 30–35 cm in thickness (Figs. 2.7 & 2.8). These occur in the now dry valleys which

Figure 2.5. *Example valley cross-sections from augering; transects 17, 110, 112, 125–6 with brown earths, 105 and 134 (H. Lewis).*

Figure 2.6. *The buried soil profile at Deer Park Farm with* in situ *flints (M. Green).*

Figure 2.7. *Dry valley section just west of Monkton-up-Wimborne 2000 site showing the palaeosol beneath thin colluvial deposits (C. French).*

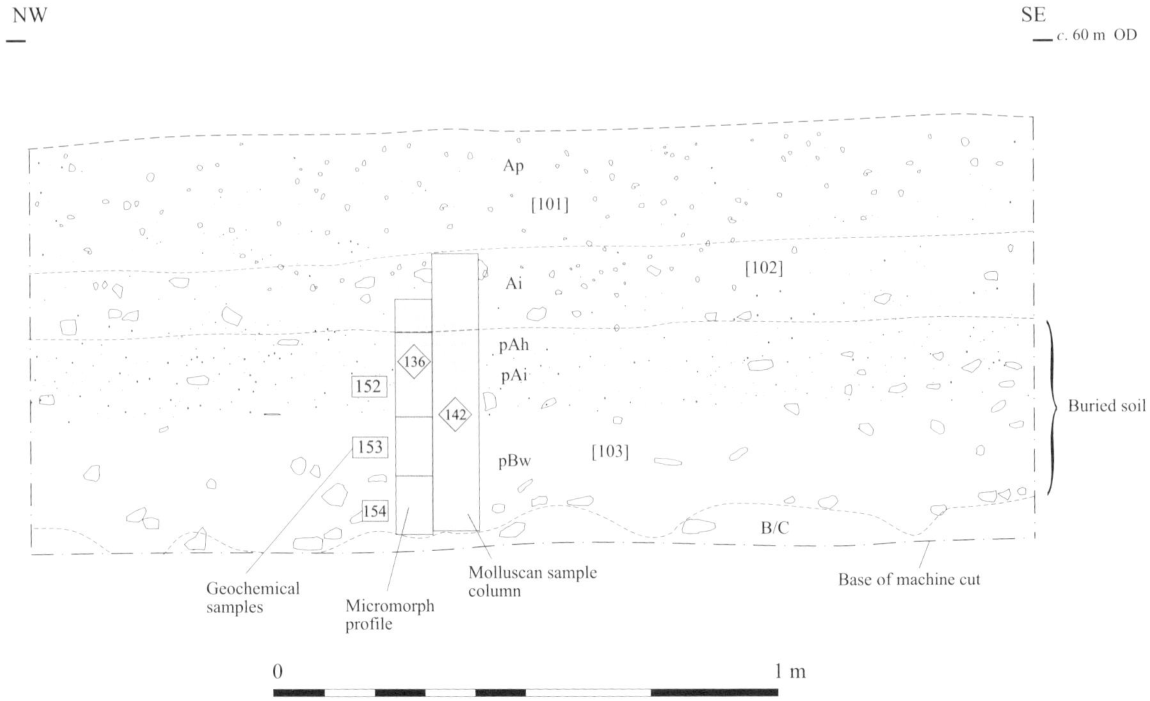

Figure 2.8. *Dry valley section with sample locations just west of Monkton-up-Wimborne 2000 site showing the palaeosol beneath thin colluvial deposits (M. Brudenell after C. French).*

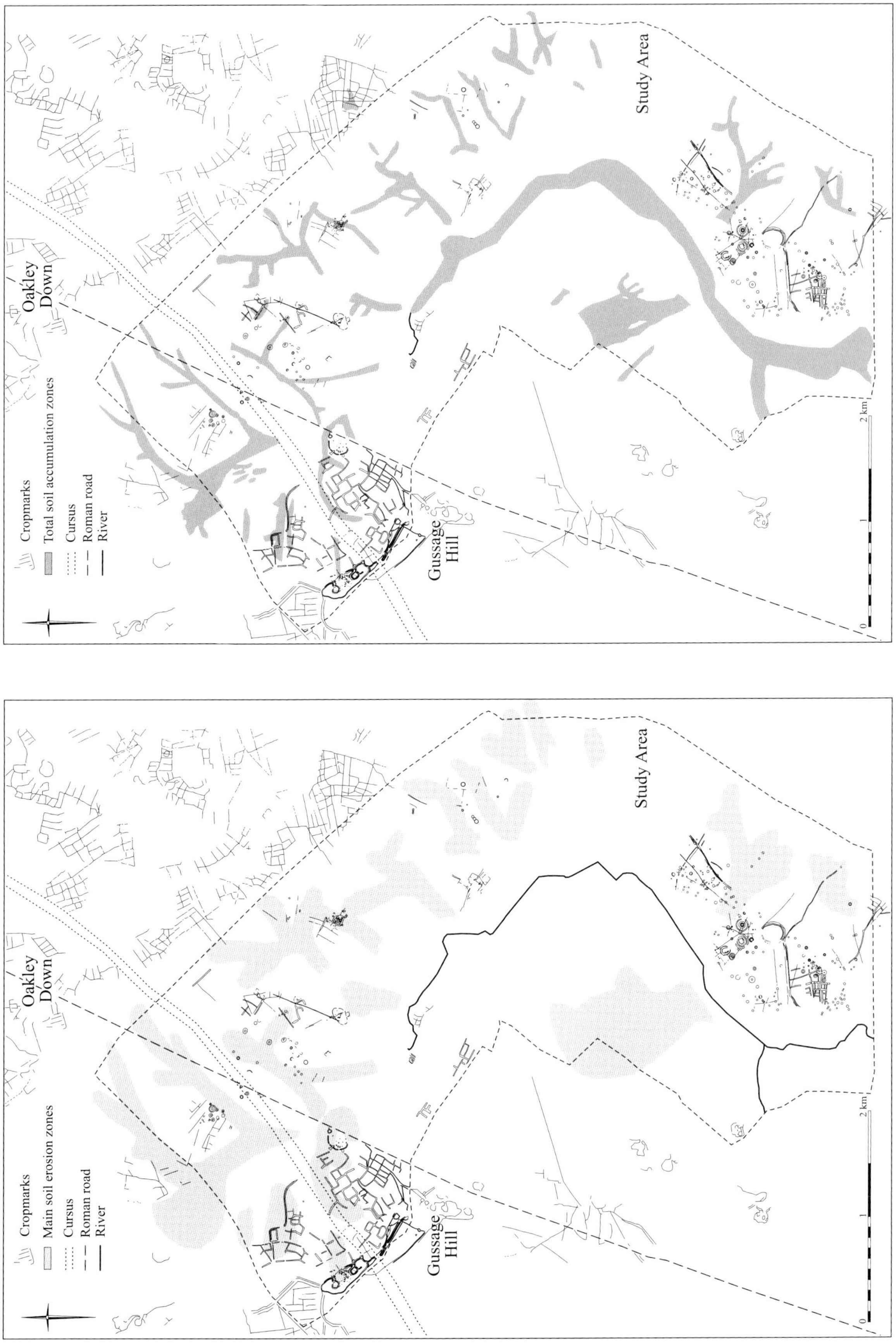

Figure 2.9. *The zones of soil accumulation (on the right) versus the potential maximum zones of slope erosion (on the left) (C. Begg after R. Palmer).*

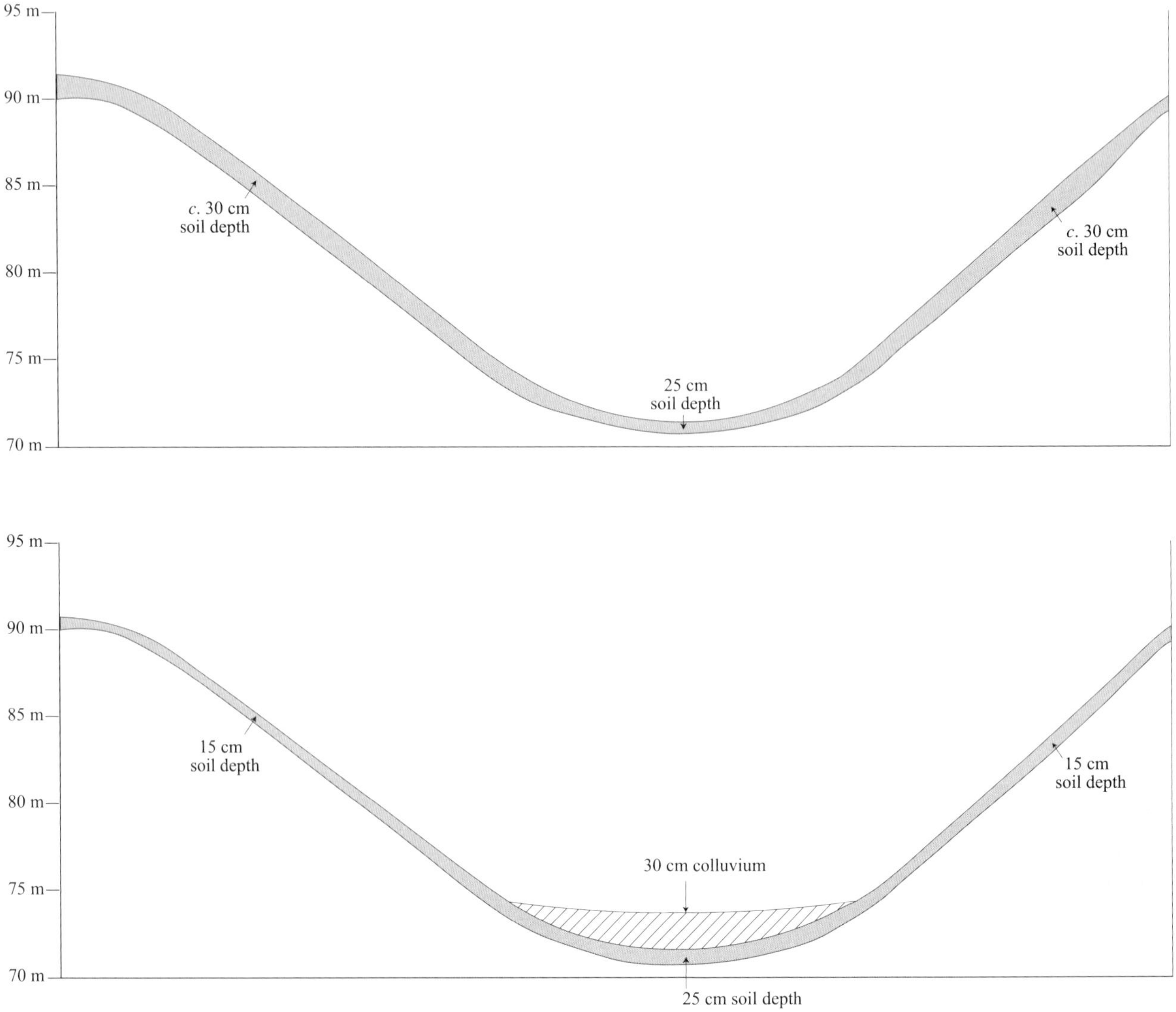

Figure 2.10. *Schematic section of the early versus late Holocene soil and colluvial cover in the upper Allen valley study area as determined by auger survey (C. Begg after C. French).*

interrupt the downland slopes on either side of the upper Allen valley. Indeed artefacts found within the colluvium in the dry valleys suggest an historic period date for this hillwash phase, but this requires extensive further investigation from archival records and further dating.

The combination of aerial photographic mapping, systematic augering survey across the valley and trial trench testing of the dry valleys has demonstrated that hillwash/colluvial deposition is seen only occasionally and is very localised on the lower chalk downland slopes and in the river valley floor. The information from buried soils under monuments and their ditch fills of varying ages agrees with this scenario. Moreover, where the dry valley sections exhibit a thin colluvial silt loam and chalk rubble accumulation on a buried brown earth soil profile (Fig. 2.10), there is nothing to suggest that this hillwash deposition is anything but relatively recent, that is of historic to recent date. A similar situation is also seen in the valley bottom and flood-plain edge profiles where prehistoric peat deposits are buried by thin horizons of overbank alluvial material which contain thin laminae of calcareous silt loam. This intermittent deposition of colluvial material is presumably derived from disturbance related to human activities and fine run-off upstream and uphill. The mechanisms, extent, intensity and dating of the colluvial material in some of the dry valleys and the flood-plain edge, and its relationship to the creation of a later peat-filled palaeochannel require more investigation and accurate dating.

Rather unexpectedly, there do not appear to be any great accumulations of eroded soil/sediment material anywhere in the upper Allen valley system in the study area (Figs. 2.1, 2.2 & 2.9), nor any of demonstrably great antiquity. This situation is markedly different when compared to other parts of the southern English chalklands, for example in the Sussex Downs (Bell 1992) and elsewhere in the Wiltshire/Salisbury Plain area (Allen 1992; 1994) where extensive and thick deposits of colluvial material are dated to Beaker and later prehistoric periods.

As a spin-off from the augering survey, several new artefact scatter sites were noted in the valley, mainly to the northwest and north of the Knowlton henge complex. The distribution of the find spots is striking (Fig. 3.80) as, with the exception of those found at Knowlton itself, all of the sites are on the opposite side of the river to the monument groups. In addition, they are located in places where no sites have been previously recorded on the Dorset County Council Sites and Monuments Record, and where no major cropmarks are seen on aerial photographs. These potential settlement sites are on the upper and southern edge of the flood plain. Reconnaissance and systematic artefact collection has been carried out on some of these sites and is reported on below (see Chapter 3).

Aerial photographic survey and erosion mapping

Lidia Żuk & Rog Palmer
with a contribution by Cathy Stoertz

The aerial photographic survey was a three-part project: to identify valley sediments, to map archaeological features at 1:10,000 and so update, as necessary, the earlier work of Colin Bowen (1990) and to examine photographs covering artefact scatters that had been located by field walking. The archaeological evidence has been enhanced by the inclusion of 1:2500-level mapping undertaken by Cathy Stoertz (English Heritage) for a Bournemouth University project (Stoertz 2005; Gale in prep.).

Archaeological and natural features from aerial photographs

In suitable cultivated soils, sub-surface features — including archaeological ditches, banks, pits, walls or foundations — may be recorded from the air in different ways in different seasons. In spring and summer these may show through their effect on crops growing above them. Such indications tend to be at their most visible in ripening cereal crops, in June or July in this part of Britain, although their appearance cannot accurately be predicted and their absence cannot be taken to imply evidence of archaeological absence. In winter months, when the soil is bare or crop cover is thin (when viewed from above), features may show by virtue of their different soils. On the Wessex chalk, ancient field systems in modern arable land may be at their most visible in such conditions. Upstanding remains, which may survive in unploughed grassland, are also best recorded in winter months when vegetation is sparse and a low angle of the sun helps pick out slight differences of height and slope.

Natural deposits can cause similar differences in crop growth and may also appear as colour differences in bare winter soils. Dry valley deposits in the Wyke Down area were mapped from air photographs as the first stage of this current project (Palmer 2000; French *et al.* 2003). The visible edges and extents of deep soil areas tend to vary from year to year with the amount of ground moisture content.

Land use

Most modern fields in the area have been in arable use since the early 1940s and archaeological features therein are mostly leveled by ploughing. Some of the larger sites — henges and burial mounds — remain as slight earthworks in pasture or small islands of protected ground. Within the area there is a little woodland, a small area of parkland and pasture fields that fringe the river Allen. These form zones in which aerial survey is unlikely to detect archaeological information.

Types of aerial photographs

The most immediately informative aerial photographs of archaeological subjects tend to be those resulting from observer-directed flights. This activity is usually undertaken by an experienced archaeological observer who will fly at seasons and times of day when optimum results are expected. Oblique photographs, taken using a hand-held camera, are the usual products of such investigation. Although oblique photographs are able to provide a very detailed view, they are biased in providing a record that is mainly of features noticed by the observer, understood, and thought to be of archaeological relevance. To be able to map accurately from these photographs it is necessary that they have been taken from a sufficient height to include surrounding control information.

Vertical photographs cover the whole of Britain and can provide scenes on a series of dates between (usually) 1946–7 and the present. Many of these vertical surveys were not flown at times of year that are best to record crop and soil differences that may indicate the presence of sub-surface features. Some, however, are very informative. Vertical photographs are taken by a camera fixed inside an aircraft with its exposures timed to take a series of overlapping views that can be examined stereoscopically. They are often of relatively small scale and their interpretation requires higher perceptive powers and a more cautious approach than that necessary for examination of obliques. Use of these small-scale images can also lead to errors of location and size when they are rectified or re-scaled to match a larger-scale map.

Photographs came from the Cambridge University Collection of Aerial Photographs (CUCAP) and the National Monuments Record: Air Photographs, Swindon (NMRC). They included those resulting from observer-directed flights and routine vertical surveys. Photographs examined for valley sediments and archaeological survey were held in those collections in June 2000. Work on the scatter areas examined additional photographs accessioned by June 2003. A full list of photographs examined is held as part of the project archive.

Interpretation and mapping

All photographs were examined by eye and under slight (2×) magnification, viewing them as stereoscopic pairs when possible. Interpretations, made at 1:10,000, were marked on overlays to individual prints. These were then scanned and transformed to match a 1:10,000 base map using the specialist program AirPhoto (Scollar 2002). Transformed files were set as a background layers in AutoCAD Map, where features were overdrawn using standard colour conventions. Lidia Żuk carried out most of the archaeological photo interpretation and mapping, supervised by Rog Palmer, during a placement at Air Photo Services Cambridge. Mapping by Cathy Stoertz was undertaken at 1:2500 using interpretative overlays of key prints and transforming those using the specialist program AERIAL 5 (Haigh 1989). Editing and drawing was completed in AutoCAD Map 2.

Figure 2.11. *Oblique aerial view southeastwards down the Upper Allen valley towards Wimborne St Giles showing the relict palaeochannel and terrace deposits to the right of the modern road in the centre ground. Wyke Farm lies centre bottom (M. Green).*

Valley sediments

Procedure

Vertical photographs were inspected to find those sorties that best recorded soil differences and those were then examined stereoscopically where possible. This method of viewing creates a three-dimensional view and so helps a photo interpreter to read the topography. Deeper soil shows as a darker-toned band in winter photographs and can also affect cereal growth in summer. Such features were mapped as 'valley sediments'.

The scale of the working map (1:25,000) did not require or allow great accuracy in transferring the information from photographs to map and all information was sketched using modern fields and boundaries to determine size and location. Information from the best photographs was first added to the map, and other photographs then examined to confirm (if possible) the extent of sediments mapped or to add supplementary information.

Results

A disconnected series of valley sediments was mapped which show what may be a number of former tributaries of the River Allen (Figs. 2.2 & 2.11). The fragmentary result is due to the fact that land use on the dates of photography did not provide a uniformly clear view of the ground. Fields of pasture, for example, will not usually show differences in soil depths although, when the photographs are viewed stereoscopically, the shape of a dry valley may be identified and continuity of sediment from adjacent arable land may be assumed.

Work elsewhere in England, in association with developer-funded evaluations, has shown the shallowness of some features mapped as 'deeper soil'. Depth of sediment is dependent on a number of variables

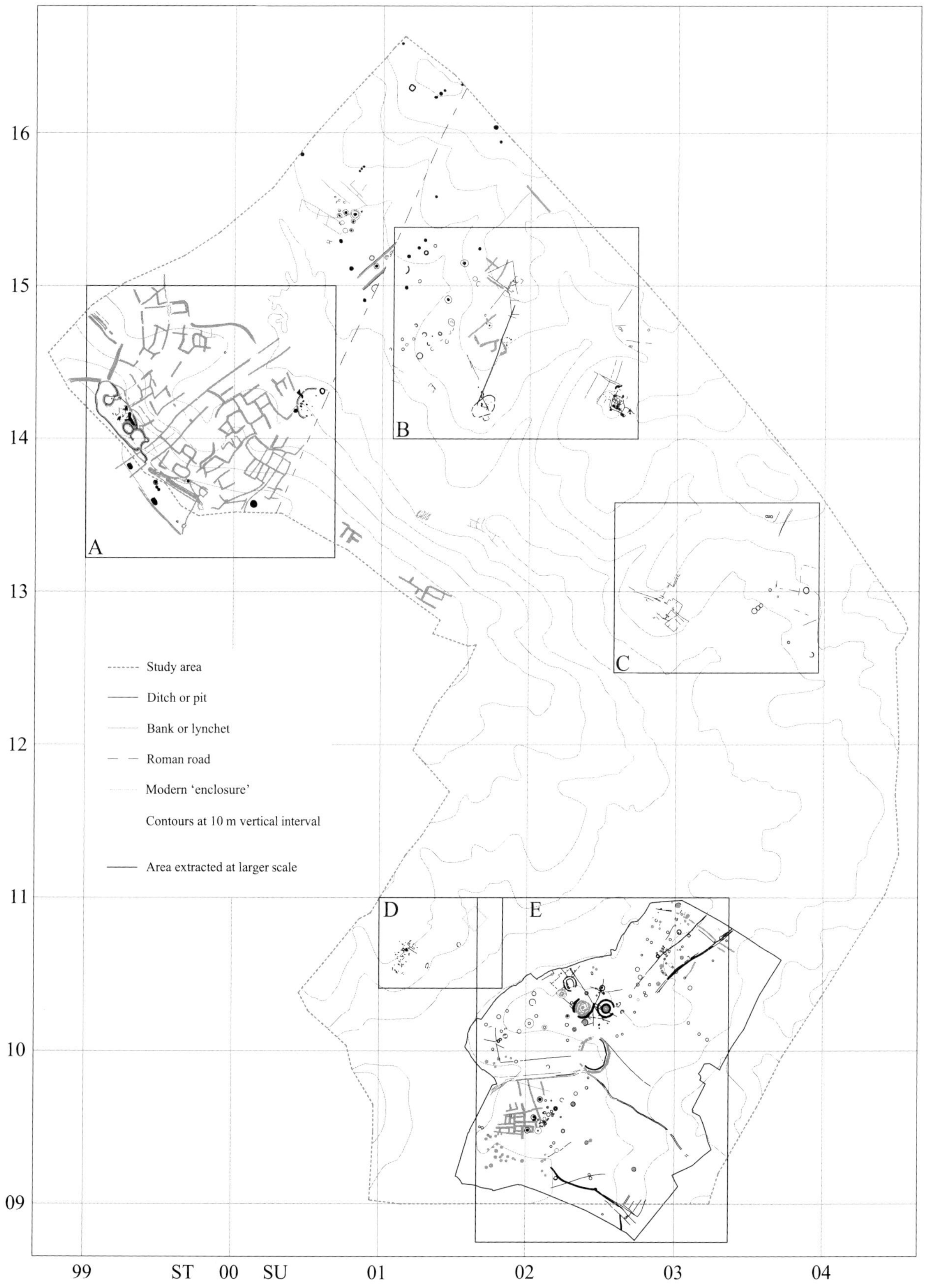

Figure 2.12. *The aerial photographic record set against the topography of the upper Allen valley study area (R. Palmer).*

including local relief and past natural and anthropogenic activities. Actual depth of deposits is impossible to determine from air-photographic evidence alone.

Archaeology

General description of the study area

Wyke Down falls within the archaeological landscape that was surveyed and analysed by Bowen (1990) as the first part of the proposed Inventory of Hampshire by RCHME. Bowen's work, dealing with earthwork remains and aerial photographic evidence tends to focus on later prehistory. Features from earlier periods — Neolithic to Middle Bronze Age — are discussed by Barrett *et al.* (1991a) following their programme of excavations in Cranborne Chase to which Martin Green (2000) has continued to contribute.

The Dorset Cursus and several long barrows are the earliest features of which some evidence still remains in earthwork form. Later Neolithic activity has been found in excavations of hengiform sites and their adjacent occupation features (Green 2000) and the apparent ritual focus in the area may be seen to continue in the high number of Bronze Age burials. From the later Bronze Age onwards much of Cranborne Chase appears to have become a farmed landscape on which well-developed lynchets indicate arable land. These are also associated with Iron Age enclosure (Barrett *et al.* 1991a, 228–36) and were likely to have continued in use through Romano-British times although the Roman road, Ackling Dyke, cuts through parts of the field system at Oakley Down to the north of Wyke Down (Bowen 1990, fig. 1).

Comments on specific areas

Figure 2.12 shows the project area and all features in its northern part that were identified on aerial photographs. The southern part of the study area includes the henge monuments and adjacent features at Knowlton. These were mapped by Cathy Stoetz (English Heritage) and have been added to the other mapping. Modern land use is not indicated in Figure 2.12 but is likely to account for some of the apparently blank areas that are within woods, parkland or otherwise managed in ways that do not allow detection of buried features from the air. Five smaller areas have been printed at about 1:10,000 to aid brief discussion of their archaeological features.

Extract A (Fig. 2.13)

A. This area includes the pair of banjo enclosures on Gussage Hill (ST 992142) that had been photographed in 1924 as earthworks (Crawford & Keiller 1928, 112–14). These had been levelled by 1943 when they were recorded on RAF verticals. The banjo enclosures overlook ground to the northeast where, down-slope from them, traces of Celtic fields have been mapped, parts of which took the earlier Dorset Cursus as an axis. The banjo enclosures have an associated compound within which a cluster of pits may indicate an area of settlement of unknown date. At its north and south ends the encircling compound is linked with multiple linear earthworks that appear to have been a major land boundary that is also marked by differences in land use: to the north are Celtic fields, to the south no Celtic fields have been identified within 2 km (Bowen 1990, area plan 2).

B. The multiple banks extend about 400 m to the southeast where there are other embanked enclosures on the north side. On the south side two single banks form a large triangular enclosure, within which are long and round barrows. One of the banks continues out of the study area and links with three other banjo enclosures, two of which appear to have an associated compound similar to that on Gussage Hill (Bowen 1990). These sites overlook ground to the east. North of Gussage Hill the multiple linear ditches extend in two directions and are associated with smaller linear ditches and field traces at ST990145 (Bowen 1990) that were not identified on the photographs examined for this project.

   Together these enclosures and linear ditches may suggest dividing up the land or management of slopes by a local community.

C. At SU005142 is a large enclosing feature, possibly once complete but later cut by the Roman road (Ackling Dyke). Internal and external pits suggest this to have been a settlement site at some period. A small penannular ditched enclosure lies mostly within the larger feature but is partly superimposed on or by its bank. Artefacts of Early Iron Age to Romano-British in date were recovered from this site by field walking (M. Green pers. comm.).

Extract B (Fig. 2.14)

D. The large enclosure at SU017142 with internal and external features was sectioned by Martin Green and found to be of Iron Age date (M. Green, Monkton-up-Wimborne 1996, this volume). A straight or almost straight double-ditched track is superimposed on, or by the enclosure. This track has been recorded for some 700 m and can be traced from the enclosure to another area of ditched features (E) to the north.

E. The track from D joins with a group of rectangular narrow-ditched enclosures, which pre-date the modern road running northwest from Monkton-

Figure 2.13. *The aerial photographic record for the Gussage Cow Down area, extract A (L. Żuk & R. Palmer).*

up-Wimborne. It is unclear from the aerial photographs whether the track was contemporary with this group of features or those at its apparent southern end (D). Their plan-forms are sufficiently dissimilar to suggest they are not of the same period or were designed for different purposes.

F. The barrow group at SU013146, on the west side of the track, includes one (SU01221474) with a causewayed inner ditch. A similar example is at SU03751267 (Extract C: K).

G. The small pit circle with a large internal pit at SU01721474 was excavated by Martin Green, shown to be a form of henge monument (Green 2000, 77–8, fig. 52 & this volume).

H. The group of pits as SU01841463 may include an arc of a circle. Further aerial photographs or field

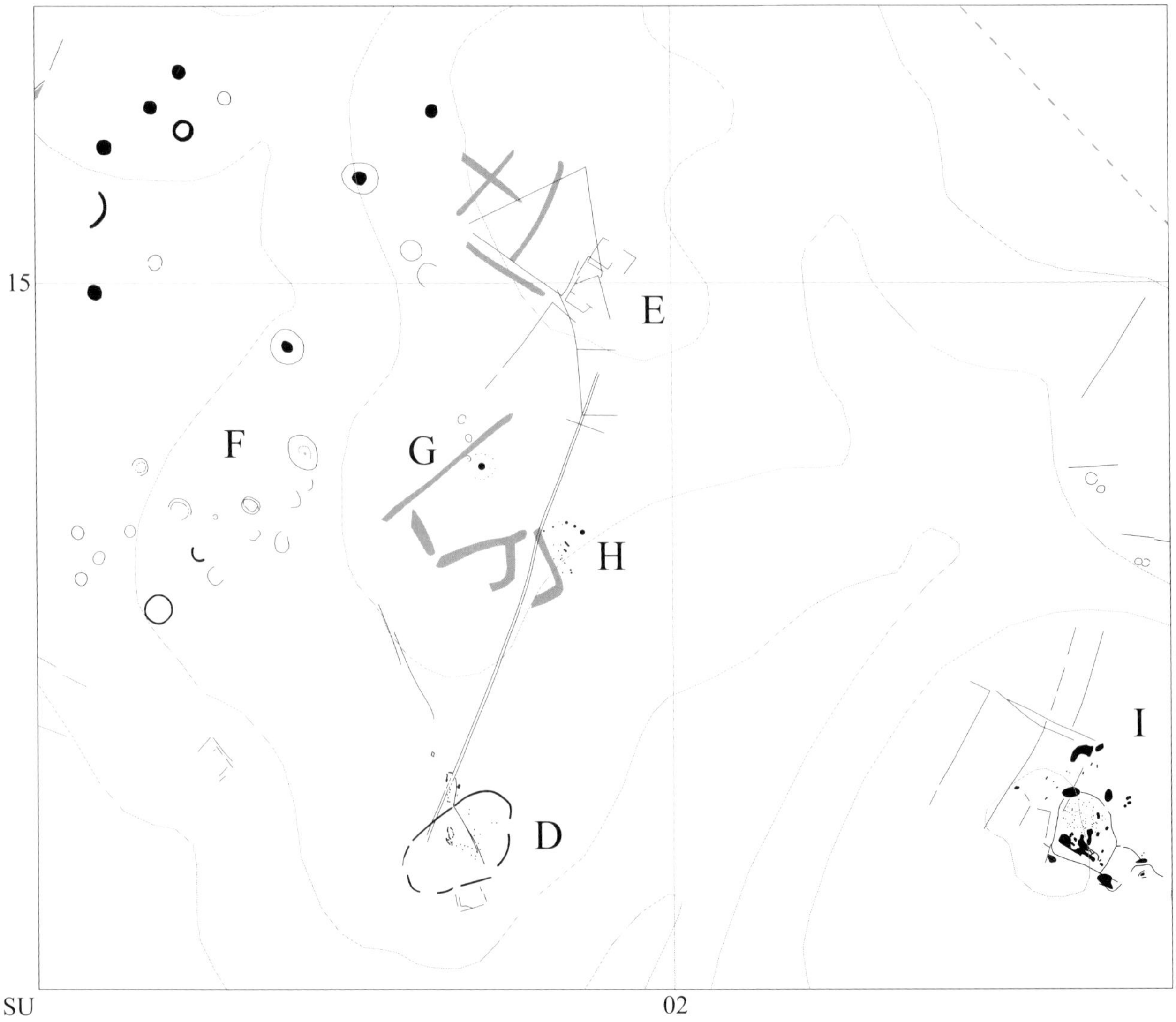

Figure 2.14. *The aerial photographic record for the Bottlebush Down and Monkton-up-Wimborne area, extract B (L. Żuk & R. Palmer).*

investigation may be necessary to clarify this.

I. A D-shaped enclosure with internal and external pits and hollows is at SU026142. The pits and hollows are likely to indicate settlement remains which may or may not be contemporary with the enclosure. Much Late Iron Age/Romano-British material was recovered from this site by field walking (M. Green pers. comm.). To the north and west are ditched boundaries that may remain from recent field divisions. These show another phase of activity in this area as one is superimposed on or by the D-shaped enclosure.

Extract C (Fig. 2.15)

J. The small group of rectilinear enclosures at SU029128 have what may be a double-ditched track entering them from the southeast. The slightly funnel-like plan of this track may indicate there to have been land for stock keeping south of the enclosures on the east bank of the River Allen where no ditches or Celtic fields have been identified.

K. The scatter of ring ditches include two groups of three that are conjoined (SU03551289 and SU03621349). Another ring ditch (SU03751267) has an inner causewayed ditch and is similar to that mapped at SU01221474 (Extract B: F).

Extract D (Fig. 2.16)

L. In two adjacent modern fields centred on SU011106 and SU015107 a scatter of probable pits, ditches and several small ring ditches or arcs of circles

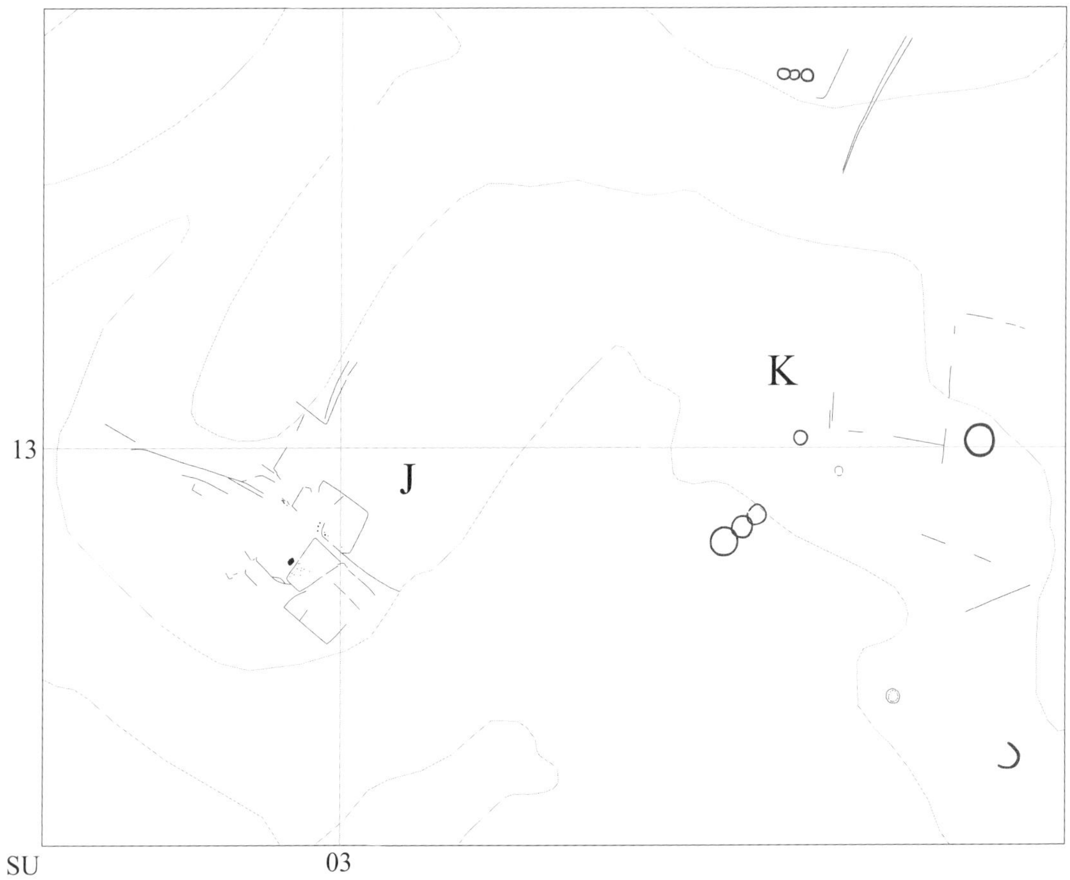

Figure 2.15. *The aerial photographic record for the area northeast of Wimborne St Giles, extract C (L. Żuk & R. Palmer).*

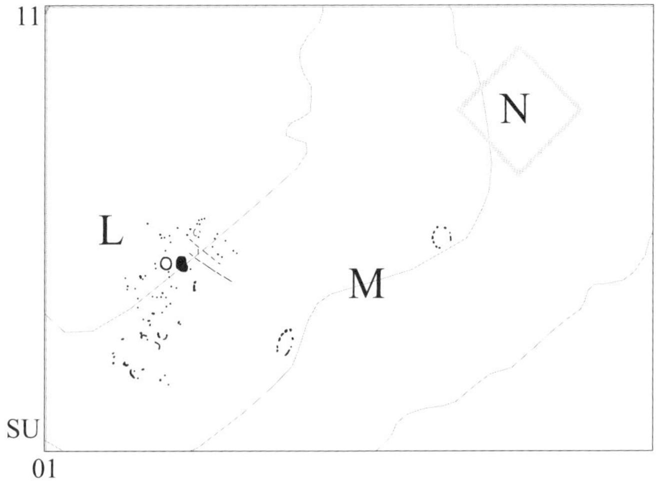

Figure 2.16. *The aerial photographic record for the area northwest of Knowlton, extract D (R. Palmer).*

may indicate a former settlement. This may be a promising area for future field walking.

M. Two pit-defined oval features that seem likely to be of Neolithic or Bronze Age date were photographed on one date only. On the map one of these appears very elongated and is at approximately SU01311055. The shape, and probably the exact position, is likely to be inaccurate as this feature lay outside the control points used to transform the photograph. Its shape was expected to be more circular but without better-controlled photographs the present plan is the best that can be offered. The other, at SU01541069, has been mapped from a different photograph and lies within the control points. Its location and shape should be as accurate as use of a 1:10000 map base allows (OS 2006).

N. This feature is discussed below as part of scatter site SU017108.

Figure 2.17. *The aerial photographic record for the Knowlton area, extract E (R. Palmer after C. Stoertz).*

Extract E (Fig. 2.17)
The Knowlton area has been mapped by Cathy Stoertz as part of a separate project. As that work was undertaken at 1:2500, more detail was identified and mapping is at a higher level and thus merits its inclusion as a separate contribution (below).

Scatter sites
A specific and separate examination of aerial photographs was made at the sites of five artefact scatters recorded during field survey (Palmer 2003). The aerial photographs examined were from CUCAP, NMRC and included RAF verticals purchased by the project. The NMRC collection included some informative photographs taken, or added to the library, after the first survey was completed (Żuk & Palmer 2001a). Sites are discussed in national grid order.

- SU004140. The scatter is immediately west of a now-levelled enclosure. The most informative APs showed the enclosure with internal pits, but no features were identified outside the enclosure from which artefacts may have been derived.
- SU005154. The field containing the scatter was pasture in the 1940s but was converted to arable by 1954. Photographs taken in early spring when local soils and crops are likely to be relatively sensitive to buried features show no recognizable archaeological information other than traces of lynchets that had previously been mapped.
- SU017108 (Extract D: N). The field appears to have been pasture in 1953 although it showed marks of cutting or harvesting in 1947. No earthworks were apparent on early photographs. However, in 2001 English Heritage photographed a large broad-ditched enclosure between two scatter areas. The site appeared to be of genuine archaeological origin but was slightly suspect as one side lay parallel to a modern road and the enclosure was central to the field's width. The non-archaeological date was confirmed by checking the Millennium photography (www.multimap.com) which showed the square to have been the site of a recent event — either a country fair or an unusual form of crop trials.
- SU027105. The 1:2500 survey by Stoertz (Extract E) covers this scatter area and no additional information was seen during further examination of photographs.
- SU024113. No observer-directed photographs had been targeted on this site but verticals taken in April 1947 show it to be in pasture. The time of year and crop make it most unlikely that any evidence of sub-surface features would be visible.

It is not surprising to find an absence of crop- or soil-marked evidence at lithic/pottery scatter sites in this part of England. One aspect of the Hambledon Hill Survey, undertaken in the late 1970s, included examination of aerial photographs of scatter sites. Some of the photographs were specially taken at opportune dates in the hope of identifying features (pits at the very least) that may complement the information from field walking (Palmer & Oswald forthcoming). Results were entirely negative as was also the case in Julian Richards's Stonehenge environs survey (J. Richards pers. comm.) other than the assumption that the majority of features recorded from the air (Celtic fields, linear ditches and enclosures) were of, or later than, the later Bronze Age (Richards 1990, 277).

The reason for this absence of evidence is likely to relate to local soils. In the Hambledon area all flint scatters were located on clay-with-flints and crops on this heavy soil are unlikely to indicate variations in sub-surface depth (e.g. pits and ditches) other than in conditions of extreme drought.

Aerial photographic survey of Knowlton Circles
Cathy Stoertz

Aerial photographic interpretation and mapping within the southern part of the Allen Valley study area was carried out by English Heritage in 2001 (Figs. 1.2 & 2.17). That survey, summarized here, was commissioned by Bournemouth University to accompany a monograph on field investigations around the Knowlton henges during 1993–97, whose publication is expected (Gale in prep.). The survey covers just over 3 km^2, centred on the Knowlton Circles henge complex and extending roughly 1 km to the northeast and 1 km to the southwest to encompass an extensive spread of round barrows.

The henges and round barrows at Knowlton were described by RCHME in the fifth volume of their Inventory of Historical Monuments in the County of Dorset, published in 1975. Much of the information, particularly the inventory of round barrows, was derived from aerial photographs, as was the rather schematic plan that accompanied the text (RCHME 1975, 114). The compilers of the Inventory had access only to photographs up to about 1974: the current transcription has added a considerable amount of detail to the Inventory results, primarily because more aerial photographs have been taken during the intervening years.

Archaeology
The archaeological features around Knowlton span several millennia, from the Neolithic to the medieval periods. The earliest and latest, the henge complex and the ruins of a medieval church, represent ritual and funerary monuments, while Iron Age or Romano-British

boundaries and field systems indicate an agricultural landscape.

Detailed archaeological investigations have so far focused on the henges and the larger round barrows. The interpretation and chronology suggested for those features recorded solely from aerial photographs is based on widely accepted morphological criteria, but has yet to be confirmed by material evidence. Further research and field analysis is required to test and refine the hypotheses presented below.

The earliest monuments at Knowlton are Neolithic ceremonial and funerary structures. Four henge monuments — the Church Circle, the South Circle, the North Circle and the Old Churchyard — provide a focal point for this part of the ancient landscape. The Church Circle (SU023102), at the centre of the complex, is the best preserved of the group, with a classic henge bank surviving to a height of about 2 m. The North Circle (SU022104) appears as a double-ditched elongated oval, open at its south end, whose similarity to the nearby Wor Barrow suggests a link with earlier burial mounds (Gale in prep.). The Old Churchyard (SU022103) comprises a square bank with rounded corners enclosed by a double circular outer ditch — although this enclosure is generally included in discussions of the henge group, its interpretation and date require further investigation. The South Circle (SU024099) is the largest monument in the group, but only its northwest quadrant has not been levelled by ploughing.

The fifth monument in the Knowlton group is the Great Barrow (SU025102), the largest round barrow in Dorset, which lies just to the east of the Church Circle. Aerial photographic evidence shows that the mound is enclosed by two ditches. The outer ditch, 120 m in diameter, is segmented or causewayed in a manner strongly reminiscent of a henge. The significance of the interrupted ditch, the sequence of construction and the relationship between the barrow, the outer ditch and the henges all pose unresolved questions: it is possible that at least one phase of this monument is Neolithic.

Within 700 m of the henge group are five small features whose morphology suggests that they may belong to the Neolithic ritual landscape, but for which no material dating evidence exists.

A. Two, within 250 m of the henge group (SU024105, SU026103), are oval enclosures which resemble mortuary enclosures identified elsewhere (Loveday & Petchey 1982; Malim 1999). A third (SU029105) is a sub-rectangular enclosure. It was not possible to tell, from the aerial photographic evidence, whether any of these enclosures had ever been covered by a mound or associated with an external or internal bank.
B. West of the henges is a possible long barrow comprising a pair of short parallel ditches enclosed by a circular ditch. This feature is very similar to a site at Huggate, on the Yorkshire Wolds, which the Ordnance Survey depicted as a round barrow but which aerial photographs showed to include a pair of parallel ditches when it had been levelled, suggesting that a long barrow lay beneath, or had been converted into, the round barrow (Stoertz 1997, 23 & fig. 8).
C. About 710 m to the east of the henges are two parallel, narrow banks now levelled by ploughing. They do not seem to be part of the later prehistoric pattern of linear boundaries (see below) but instead show similarities to Neolithic monuments such as cursuses or mortuary enclosures. Parallels have been suggested with long monuments in the Paris basin of northern France, which date to the Middle Neolithic (Stoertz & Gale in prep.; Kinnes 1999).

At Knowlton 178 round barrows were recorded by the aerial photographic survey and it is likely that more existed in antiquity. Although sometimes referred to as the Northern and Southern cemeteries, the most recent aerial photographic evidence suggests that the barrows comprise a single large group distributed along the northwest facing slopes overlooking the River Allen, between 60 m and 50 m OD. Fewer than ten round barrows exhibit any surface relief — the vast majority has been identified on aerial photographs by the characteristic ring ditches.

D. Included among the many round barrows at Knowlton are 46 small circular features, more than half of which appear towards the south edge of the survey area. These have been interpreted as Bronze Age round barrows and are visible as mound traces rather than the more usual ring ditches. There remains the possibility that at least some represent chalkland geological features, such as sink holes, but the presence of mound material within ring ditches is known on aerial photographs elsewhere in this area.
E. Seventy-six round barrows, over one-third of the total, have diameters of 15 m or less. The very small barrows appear in two main clusters towards the northern and southern extents of the survey area. They are arranged in tightly-spaced groups and, particularly in the south, appear to have been inserted into spaces between larger round barrows, suggesting a later date than the larger barrows. A smaller more dispersed group lies close to the main henge complex.

The date of the small barrows requires further investigation, because very small round barrows have been recorded elsewhere in England in Bronze Age, Iron Age, Roman and early medieval contexts. A post-Bronze Age date for the small round barrows at Knowlton may be suggested by the presence, in the same area, of possible Iron Age square barrows (see below) — one of the square barrow cemeteries at Garton Slack, on the Yorkshire Wolds, includes very small round barrows that represent a very late phase of the Iron Age burial rite (Stead 1991; Stoertz 1997).

The monumental landscape of the Neolithic and Bronze Age appears to have given way to one of primarily agricultural and domestic use in the Iron Age and Romano-British periods, but the presence of a handful of possible square barrows and related structures suggests a continuation of ceremonial and funerary elements in these later phases.

F. The survey recorded six small rectilinear enclosures which are similar in size and shape to Iron Age square barrows: the largest is about 15 x 15 m, and the smallest about 5 x 5 m. Two are located at the northwest edge of the area, north of the henge group, about 175 m from the River Allen at 50 m OD (SU022105; SU023105). Two are on higher ground (about 55 m OD) near the northern extent of the survey area, adjacent to a probable late prehistoric linear boundary (SU030106, SU030107). Nearby, a pair of parallel ditches (SU032107, near H) may represent the remnants of an additional square barrow. At the southern extent of the area, two possible square barrows appear to be associated with larger enclosures and a trackway.

G. Two further small rectangular enclosures have been recorded in the area between the South Circle, Church Circle and the Great Barrow. Although similar in size to square barrows, their interpretation is both unclear and intriguing. Each enclosure has internal pits (three in one case, four in another) and one appears to lie within a fragmentary outer enclosure. Their location within a multi-period ceremonial and funerary landscape suggests ritual associations, possibly of Iron Age or Romano-British date, but this interpretation is tentative at best and remains to be explored.

Several long linear features, mainly defined by ditches, form a rough grid pattern which extends over the whole survey area and suggests a series of land divisions or boundaries. Some appear to be aligned on the henges and, at several points, the course of a linear feature crosses over a round barrow, suggesting that construction of the boundaries post-dates that of the barrows and larger monuments.

H. In the northwest, a linear earthwork 605 m long runs across the slope perpendicular to the River Allen. Perpendicular to it, and parallel to the river, are two long ditches, each just under 700 m long. The latter appears to be a double-ditched trackway with a central hollow. Both run along the land contours, parallel to the River Allen and seemingly aligned on the South Circle.

The grid pattern of linear ditches or boundaries continues around, and to the south of, the South Circle while smaller, more fragmentary ditches can be seen in the area surrounding the henge group.

I. Immediately to the south of the South Circle is a multiple-ditched linear feature, part of the general grid pattern, with a total length of 1580 m. Its approximate mid-point abuts the South Circle. This feature fits within the suggested pattern of Iron Age or Romano-British boundaries, although it was assigned a medieval date by the RCHME Dorset Inventory (1975, 116):

> Two almost parallel ditches ... follow a sinuous course across a gentle W slope of the Chalk. Their line is continued W ... by a broad hollow-way probably of medieval or later date. Whether there is any connection between it and the ditches has not been established.

Aerial photographic evidence from 1976, 1989 and 1995 clearly shows that a connection does exist between the two parts, adjacent to the South Circle, and that the 'broad hollow way' and the 'two almost parallel' eastern ditches are elements of the same boundary or trackway. The western part is defined by at least two and in some places three or four ditches. A bank on the north side and a hollow along the centre were still visible in low relief on the 1989 photos, but are not depicted here for reasons of clarity.

The medieval date suggested by RCHME appears to be based on the identification of this feature as part of the bounds of Cranborne Chase, described in an Assize roll of 1279 (Dayrell-Reed 1931). However, its use as a landmark in defining the Chase could easily indicate an earlier date as, in other parts of England (e.g. the Yorkshire Wolds), medieval boundaries were sometimes laid out along prehistoric earthworks (Stoertz 1997, 40, 65). The presence of a field system just to the south, and additional linear boundaries and trackways on a similar alignment to the north and south, suggests that a late prehistoric date for this linear feature should be considered.

J. At the southern extent of the survey area is a trackway whose course is roughly parallel to the eastern part of feature I. It is 695 m long, defined by ditches and a central hollow, and appears to belong to the general grid pattern seen to the north. This trackway's eastern end appears to be associated with ditched rectilinear enclosures or fields.

K. A compact group of embanked or lyncheted rectangular fields, covering about 10 hectares, lies immediately to the south of, and is apparently bounded by, the western part of the linear feature which abuts the South Circle. The fields here are slightly smaller than those to the north and east of the Gussage St Michael banjos (ST995145 & SU000140; Fig. 2.13, extract A), and have never been photographed as earthworks. The field system appears to fit broadly within the grid pattern of the presumed Iron Age boundaries, and to respect the positions of several large round barrows.

L. About 950 m to the southeast is a group of fragmentary ditched rectilinear enclosures or fields, associated with the southern trackway (J) and two possible square barrows (F) (see above).

The ruins of a twelfth-century church stand at the centre of the Church Circle, within a subrectangular enclosure defined by a slight bank which fits closely within the inner ditch of the henge. This enclosure appears to be a churchyard boundary, and is probably medieval in date, but could be later.

Palynological analyses

Rob Scaife

The augering survey of the upper Allen river valley made two major discoveries in the form of the location of two relict palaeochannel systems (Fig. 1.11; Table 2.1). In combination with previous investigations, it appears that there are at least four major palaeochannel systems in the valley between Down Farm and Wimborne Minster that existed prior to the present river course, over a distance of some 15 km, and a date range from the late Upper Palaeolithic to medieval times.

The first system is located just to the west of Wimborne St Giles on the southern side of the modern flood plain of the River Allen (Figs. 1.11:1,2 & 2.18). The relict channel survives as two small oxbows up to 3 m deep and about 30 m wide, with a completely waterlogged profile composed of a basal calcareous silt 'mud' with abundant included fine chalk fragments overlain by a sequence of well-preserved detrital and wood peats. Palynological analysis of the basal peat and contact zone with the underlying chalky silt includes evidence for an open and herbaceous plant dominated landscape which rapidly becomes wooded,

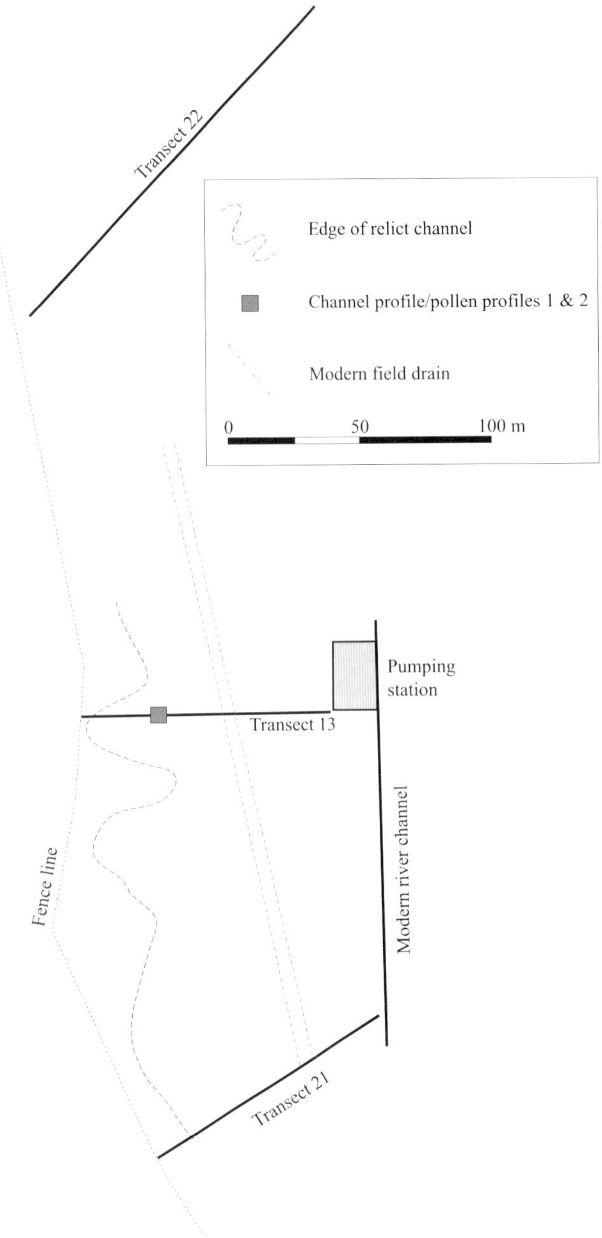

Figure 2.18. *Location of the Wimborne St Giles pollen profiles 1 and 2 in the palaeochannel system observed in transect 13.*

and demonstrates that the fills of this channel are of late Devensian and early Holocene age, that is relating to the upper Palaeolithic and Early Mesolithic periods. This has been confirmed by radiocarbon assay which gives a date range for the lower peat sequence of this palaeochannel of between 11,820–10,710 cal. BC (Beta-168611; 11,120±170 BP) and 6580–6400 cal. BC (Beta-168610; 7630±60 BP) (Table 2.1) (Stuiver *et al.* 1998; Stuiver & van der Plicht 1998; Pretoria Calibration Procedure, Talma & Vogel 1993; INTCAL98 data base).

Table 2.1. *Radiocarbon dating of the peats in the palaeochannels in the upper Allen valley. (Note that calibrations are derived from Stuiver* et al. *1998, Stuiver & van der Plicht 1998, Pretoria Calibration Procedure, Talma & Vogel 1993, and the INTCAL98 data base.)*

Location	Profile	Laboratory reference	Conventional radiocarbon age	cal. BC (2 sigma)
Wimborne St Giles	1, base	Beta-168611	11,120±170 BP	11,820–11,740, 11,530–10,870 & 10,780–10,710 cal. BC
Wimborne St Giles	2, middle	Beta-168610	7630±60 BP	6580–6400 cal. BC
St Giles House	Tr. 106, base	Beta-189162	2800±40 BP	1030–840 cal. BC
Knowlton	Tr. 151, top	Beta-195862	2570±40 BP	810–760 & 640–560 cal. BC
Knowlton	Tr. 151, base	Beta-189164	3490±40 BP	1910–1700 cal. BC
Wimborne Minster	Allenbourne, middle	Beta-195863	4780±40 BP	3650–3510 & 3420–3390 cal. BC
Wimborne Minster	Allenbourne, base	Beta-189166	7000±40 BP	5980–5780 cal. BC

The second palaeochannel system (Fig. 1.11; Table 2.1) is located just downstream of Wimborne St Giles where there is a wide (*c.* 50–75 m) and shallow (<1.5 m), gently meandering system in a relatively wide flood plain. This was sampled at two locations, in front of St Giles House (Fig. 1.11:3) and to the north of the Knowlton henge complex (Fig. 1.11:4), and these appear to date to Bronze and Iron Age times (second and first millennia BC). In addition, another profile through an earlier Holocene palaeochannel system that was discovered in Wimborne Minster in 1994 (Allenbourne car park) (Scaife 1994a), was re-analysed. The Allenbourne channel system dated from the latest Mesolithic/earliest Neolithic to the elm decline later in the Neolithic (*c.* 6000 to 3500 cal. BC), and effectively continued to infill after the earliest channel system went out of use in the upper part of the Allen valley around Wimborne St Giles.

Finally, another much later palaeochannel system, Pippins Stream, which had also been discovered previously in Wimborne Minster (Scaife 1994a), was re-analysed. Although no material suitable for radiocarbon assay was salvageable from this core, the pollen assemblage is suggestive of a medieval to more recent date (see below).

The palaeoecology of the southern English chalkland

Because of the paucity of peat accumulations in the present-day downlands of southern England and the poor or absent pollen preservation where they do exist, the past vegetation and development of the chalklands has remained enigmatic. Most palaeoenvironmental data has come from examination of molluscan assemblages which, although providing evidence of environmental change, cannot provide specific information on vegetation communities. In some circumstances, pollen has been preserved in sealed palaeosols underlying colluvial deposits as at Brook, Kent where periglacial solifluction deposits have preserved an Allerød palaeosol (Kerney *et al.* 1964), and under prehistoric field monuments which provide a limited insight into the on-site habitat immediately prior to burial. Dimbleby & Evans (1974) and Dimbleby (1985) in their classic papers demonstrated the potential and value of integrating both lines of evidence, and that pollen preservation can occur in calcareous soils in certain circumstances. Subsequent studies have demonstrated this, including Silbury Hill (Dimbleby 1999 unpublished), Stonehenge (Scaife 1995), the Jurassic limestone at Churn Valley, Cirencester, Ashton Keynes (Scaife 1999), Hazelton, Gloucestershire (Scaife 1990), Apes Down and Gallibury Down, Isle of Wight (Scaife 1980; 1984; 1994a,b; 2003b).

Past researchers have sought to examine the developing habitats of the Jurassic limestone and Cretaceous chalk lithologies with varying degrees of success. Godwin's (1962) pioneer work at Wingham and Frogholt, Kent, discussed what he felt to be deforestation through human impact on the nearby Kentish downland. However, better understanding of pollen taphonomy, and especially pollen production, transport and representation in pollen spectra suggest that his results may portray the vegetation much more local to the sample site. Thorley's work in Sussex at Amberley Wild Brooks and Lewes (Thorley 1971a,b; 1981), whilst the sites are closely adjacent to the Sussex Downs, may similarly over represent the autochthonous vegetation, especially during phases of alder carr woodland at the expense of the palaeovegetation of the chalk interfluves. Scaife (1980) and Waton (1982a,b) have similarly carried out detailed surveys of peat and pollen sites which are also closely adjacent to the downlands of the Isle of Wight, Dorset and Hampshire. These were undertaken with the realization that small valley peat, and in the case of Rimsmoor, Dorset (Waton & Barber 1987), a doline, can have local or restricted pollen catchments.

Data from these studies have increased our knowledge of the long-term downland history to the extent that it is accepted that these areas were once occupied by dominant woodland communities whose development followed the typical Holocene dynamic/seral changes after the close of the Devensian cold stage. This contrasts with much earlier ideas (of the late nineteenth and twentieth centuries) that the downlands were a 'natural' phenomenon utilized by prehistoric communities because of their openness and ease of access. This latter concept has, however, been revived with a degree of controversy. Bush (1988; 1989) and Bush & Flenley (1986), from data obtained in a small mire in the Yorkshire Wolds, have suggested that there was a continuity of open herb (grass) communities from the close of the Devensian but within a developing wooded environment. Although Thomas (1989) has contested this, further pollen data from southern England has suggested that areas of open grass-dominated herb communities may have remained well into the Holocene, rather than the normal scenario of woodland seral changes to the climax of the middle Holocene (Flandrian Chronozone II). Waller & Hamilton (1998) have examined the archaeology and palaeoenvironment of the Sussex, Mount Caburn area of the chalk downland and also concluded continuance of herb communities.

Polliniferous sites continue to be found on calcareous lithologies which increasingly demonstrate the complexity of the origins and development of the downlands as we see them today, especially in relation to the progressive anthropogenic impact on these areas. Of note relating to the late Devensian and early Holocene periods are profiles examined from Holywell Combe, Folkestone, Kent (Kerney *et al.* 1980; Preece & Bridgland 1999). Durrington Walls, known for its Neolithic occupation, has also provided pollen data from a late prehistoric palaeosol underlying hillwash/colluvium and from peat deposits in the Avon flood plain (Scaife 1994a,b; 2004). These demonstrate early Holocene woodland development of pine and hazel dominance and subsequently woodland clearance and open habitats during the Neolithic and Bronze Age. Similarly, rare flood-plain peat and mineral sediments adjacent to Clatterford Roman Villa, Carisbrooke, Isle of Wight, in the core of the island's chalkland, have provided information on the downland habitat specific to the Romano-British period (Scaife 2001).

During survey of the River Allen valley from Cranborne Chase to Wimborne Minster, a range of features including flood-plain peat and palaeosols has been examined for pollen preservation with the aim of studying the vegetation development of this important downland/chalk region. Although sub-fossil pollen is absent in many of these contexts, useful sites have been found which go some way to achieving the goals of this study. Of local and regional importance also are the pollen sequences from the New Forest (Barber & Clarke 1987), Testwood-Rownhams, Hampshire (R. Scaife pers. comm.), and the Isle of Wight (Scaife 1980; 1982; 1987a). Earlier analyses at Wareham and Nursling (Seagrief 1959), Cranes Moor (Seagrief 1960) and Elstead, Surrey (Seagrief & Godwin 1960), although dated, must be regarded as pioneer studies in the southern region. All of these have contributed to our wider understanding of the vegetation history of central southern England.

Pollen method

Cores from the River Allen flood-plain sediments described were obtained using both a Russian/Jowsey corer and samples from old land surfaces underlying the prehistoric monuments, obtained directly from excavated sections or using steel monolith profiles. Sediment descriptions and sub-sampling for pollen analysis and radiocarbon dating were carried out in the laboratory. Standard techniques were used for the extraction of the sub-fossil pollen and spores (Moore & Webb 1978; Moore *et al.* 1991). Samples of 1–2 ml were used in this analysis. Identification and counting was carried out using an Olympus biological research microscope fitted with phase contrast and Leitz optics. A pollen sum of up to 600 grains per level (where preservation made this possible) of dry land taxa plus all extant spores and pollen of marsh taxa (largely Cyperaceae) and fern spores, along with miscellaneous pre-Quaternary palynomorphs, were counted. Pollen diagrams have been plotted using *Tilia* and *Tilia* Graph (Figs. 2.19–2.24). Percentages have been calculated as follows:

Sum =	% total dry land pollen (tdlp)
Marsh/aquatic =	% tdlp+sum of marsh/aquatics
Spores =	% tdlp+sum of spores
Miscellaneous =	% tdlp+sum of miscellaneous taxa

Because *Alnus* (alder) is a strong local element influencing the overall pollen spectra it has been excluded from the main pollen sum in the case of one site (Allenbourne Park) and calculated as a percentage of the sum plus *Alnus* (Janssen 1969). Taxonomy follows that of Stace (1991) and Moore *et al.* (1991), modified according to Bennett *et al.* (1994) for pollen types and Stace (1991) for plant descriptions. These procedures were carried out in the Palaeoecology Laboratory of the School of Geography, University of Southampton.

The River Allen Valley from Cranborne Chase to Wimborne Minster: the site locations and results

A stratigraphical survey of the sediment architecture of the River Allen valley (French *et al.* 2000; 2003) located a number of palaeochannels, possibly meander cut-offs, filled with peat and organic muds. Pollen results from five peat profiles and one soil sequence are discussed here. These sites fall into four broad groups. First, Wimborne St Giles sections come from a possible sub-surface hollow in a palaeomeander on the flood plain of the River Allen (Wimborne St Giles 1) and a second deeper profile through the fills of the palaeochannel (Wimborne St Giles 2; see below). Pollen analysis and radiocarbon dating demonstrate that the fills of this channel are of late Devensian and early Holocene age, that is, relating to the Upper Palaeolithic and Early Mesolithic periods. Second, are the more recent (late prehistoric and historic age) organics sequences at St Giles House and Knowlton a small distance down-stream of Wimborne St Giles (see below). Third are studies undertaken due to recent developments in the centre of Wimborne Town (flood relief scheme) which facilitated sampling and analysis of peat and sediments at two localities adjacent to the River Allen. One of these has provided palaeovegetation data which extends back to the middle Holocene (the Allenbourne car park); the second profile is thought to be of medieval date (Pippins Bridge, Wimborne; see below). Finally, attempts at gaining pollen data from the well-preserved buried soils examined during the Cranborne Chase programme have largely proved unsuccessful. However, some pollen was found preserved in the buried soils underlying two of the Bronze Age mounds (Wyke Down barrows 34 & 41) (see Scaife in Chapter 3).

Group 1: The Wimborne St Giles palaeochannel/meander system
The two profiles taken from the same palaeochannel about 5 m apart (Figs. 1.11, 2.19 & 2.20) represent the oldest sediments located which yielded pollen.

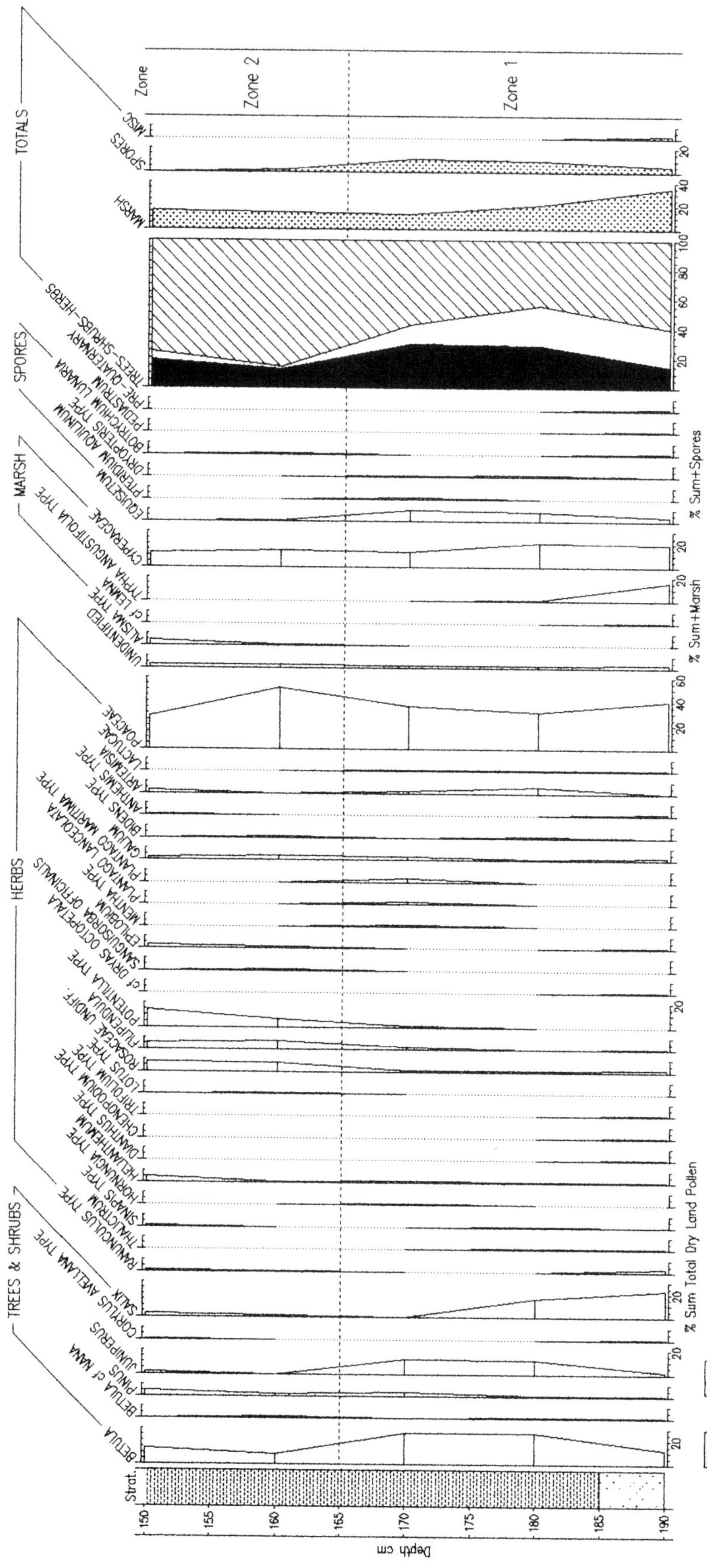

Figure 2.19. *Pollen diagram of the late glacial/early Holocene palaeochannel fill at Wimborne St Giles profile 1 (R. Scaife).*

These span the late upper Palaeolithic, late Devensian cold stage (probably Windermere interstadial and Loch Lomond Re-advance Zones II and III) and subsequent early Holocene (Flandrian Chronozone Ia–c) Pre-Boreal and Boreal periods.

Wimborne St Giles 1

Two pollen assemblage zones can be recognized from this analysis (Fig. 2.19). These are characterized from the base upwards as follows:

Zone 1: 190 cm to 165 cm (4 levels). Betula-Juniperus-Salix-Poaceae. *Betula* is the dominant tree (to 28%) with only sporadic occurrences of *Pinus*. Shrubs comprise *Juniperus* (to 18%) and *Salix* (23% at base and decreasing). *Betula* cf. *nana* is present. Herbs are dominant (70%) with notably Poaceae (35-40%), marsh taxa — Cyperaceae (17%) and *Typha angustifolia* type (16% in the basal level). Of note are *Thalictrum, Helianthemum*, cf. *Dryas octopetala, Plantago maritima* type and *Artemisia* (these are suggestive of late-glacial open herbaceous floral communities) and the fern spores *Botrychium lunaria* and *Equisetum*.

Zone 2: 165 cm to 150 cm (2 levels). Poaceae-Cyperaceae. This zone is characterized by a reduction in tree and shrub pollen (notably *Betula, Juniperus* and *Salix*). *Betula nana* and *Pinus* remain sporadically. Herb totals increase (75–80%) with increases of Poaceae (to 60%), Rosaceae undiff. (70%), *Filipendula* and *Potentilla* type (14%). Marsh taxa: Cyperaceae remain dominant with some increase of *Alisma, Plantago aquatica* type.

Wimborne St Giles 2

This palaeochannel appears as two small oxbow meanders some 30 m wide which contain up to 3 m of sediment fill. These remain completely waterlogged and comprise basal calcareous silt containing abundant fine chalk fragments and occasional flints. This is overlain by detrital fen with occasional wood remains and layers of organic mud and fine-grained calcareous material (possibly tufa). Radiocarbon dating and pollen analysis show that these channel fills are of late Devensian and early Holocene age, relating to the upper Palaeolithic and Early Mesolithic archaeological periods.

Radiocarbon dating: The basal/lowest peat has provided a date of a calibrated range of dates between 11,820–10,710 cal. BC (Beta-168611; 11,120±70 BP) and at the mid-point of peat formation a date of 6580–6400 cal. BC (Beta-168610; 7630±60 BP) (Table 2.1) (Talma & Vogel 1993: INTCAL98 data base).

Pollen data: Six local pollen assemblage zones have been recognized from this meander cut-off/palaeochannel (Fig. 2.20). These are characterized from the base of the profile (194 cm) upwards.

Zone ALL: 1: 185 cm to 195 cm (2 levels). Thermophiles-Poaceae-Cyperaceae. These two basal levels of the profile are at the transition of the peat into the underlying gravel valley fills. *Corylus avellana* type is present in both samples being especially important at 192 cm (to 48%) with *Ulmus* (3%), *Quercus* (6%), *Tilia* (2%) and *Alnus* (12%). Poaceae (60%), *Plantago lanceolata* (11%) and Cyperaceae (33%) are the dominant herbs. There are a greater number of *Pteridium aquilinum* and monolete *Dryopteris* spores.

Zone ALL: 2: 185–126 cm. A gap in the pollen diagram (148–180 cm) is due to the 'sloppiness' of the sediments; peat did not remain in the peat corer. Palynologically the zone is dominated by *Betula* (to 80%). There is a single large peak of *Salix* and *Juniperus* (to 20%) at 180 cm prior to the expansion of *Betula*. Herbs remain dominated by Poaceae (30%) and Cyperaceae (25–30%). Of particular note also are peaks of *Helianthemum* and *Filipendula* (11%). There are few spores. Thermophiles noted in zone 1 are absent.

Zone ALL: 3: Poaceae-Cyperaceae-*Pediastrum. 126 cm to 96 cm:* This zone corresponds with a stratigraphical change to organic muds from underlying peat. *Betula* values of the preceding zone are reduced (to 20%). There are small increases of *Corylus avellana* type (6%) and *Salix* (7%). Herb assemblages are diverse with higher values of *Potentilla* type (4%), *Artemisia* (peaks to 7%) and dominant Poaceae (to 55%). Marsh taxa are dominated by Cyperaceae (peak to 40%), *Alisma* type (4%). Spores include *Pteridium* aquilinum and *Dryopteris* type (expanding from this zone). Of note is the importance of algal *Pediastrum* (to 70%).

Zone ALL: 4: Juniperus-Filipendula-Poaceae-Cyperaceae. *96 cm to 66 cm: Betula* remains as zone 3. *Pinus*, however, expands (av. 10%). The zone is defined by the expansion of *Juniperus* (to 38%). Herbs remain dominated by Poaceae (*c.* 50%) and *Filipendula* which expands to 35% at the base of the zone. Cyperaceae (*c.* 50–60%) remains important with some increase of *Typha angustifolia* type (5%). Values of *Pediastrum* are reduced to absence. This corresponds with a return to peat after the organic muds of zone 3.

Zone ALL: 5: 66 cm to 44 cm. Pinus is declining to low values with *Corylus avellana* type increasing to its highest values (60%). There are also reductions in *Salix* and *Juniperus* and in the general numbers and importance of herbs. *Dryopteris* (monolete) type reaches its highest values (to 35–40%). *Polypodium vulgare* and *Pteridium* become more consistent (<5%).

Zone ALL: 6: 44 cm to 32 cm. Ulmus and *Quercus* peak (18% and 19%) with *Tilia* (40%) and *Alnus* (15%). *Betula, Pinus* (<5%) and *Corylus avellana* type (to 7%) decline and *Salix* (<5%) decline. Within the herbs there are increases of *Plantago lanceolata* (10%), Lactucoideae (17%), Poaceae (40%). Cereal type is present.

Group 2: St Giles House and Knowlton

These two sites near Knowlton and in front of St Giles House (Fig. 1.11) date to the later prehistoric periods.

St Giles House, Wimborne St Giles

Site location: A second palaeochannel was located downstream of Wimborne St Giles, on the flood plain of the River Allen near St Giles House (Fig. 1.11). This is a wide channel (*c.* 50–70 m) which meanders within the relatively broad flood plain at this point in the valley. The stratigraphy comprises shallow peats resting on chalk bedrock and overlain by alluvial sediments.

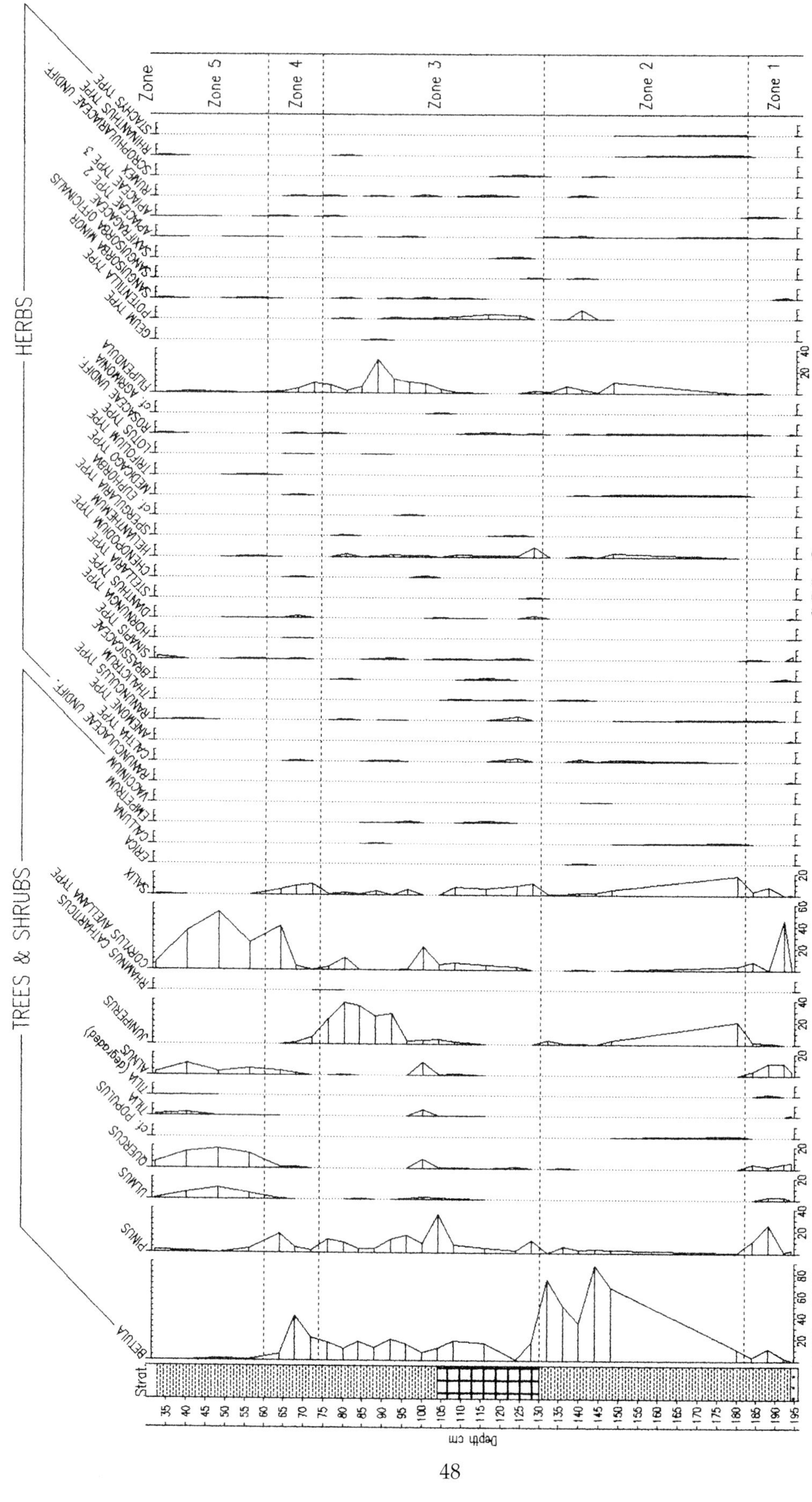

Figure 2.20. *Pollen diagram of the early Holocene palaeochannel fill at Wimborne St Giles profile 2 (R. Scaife).*

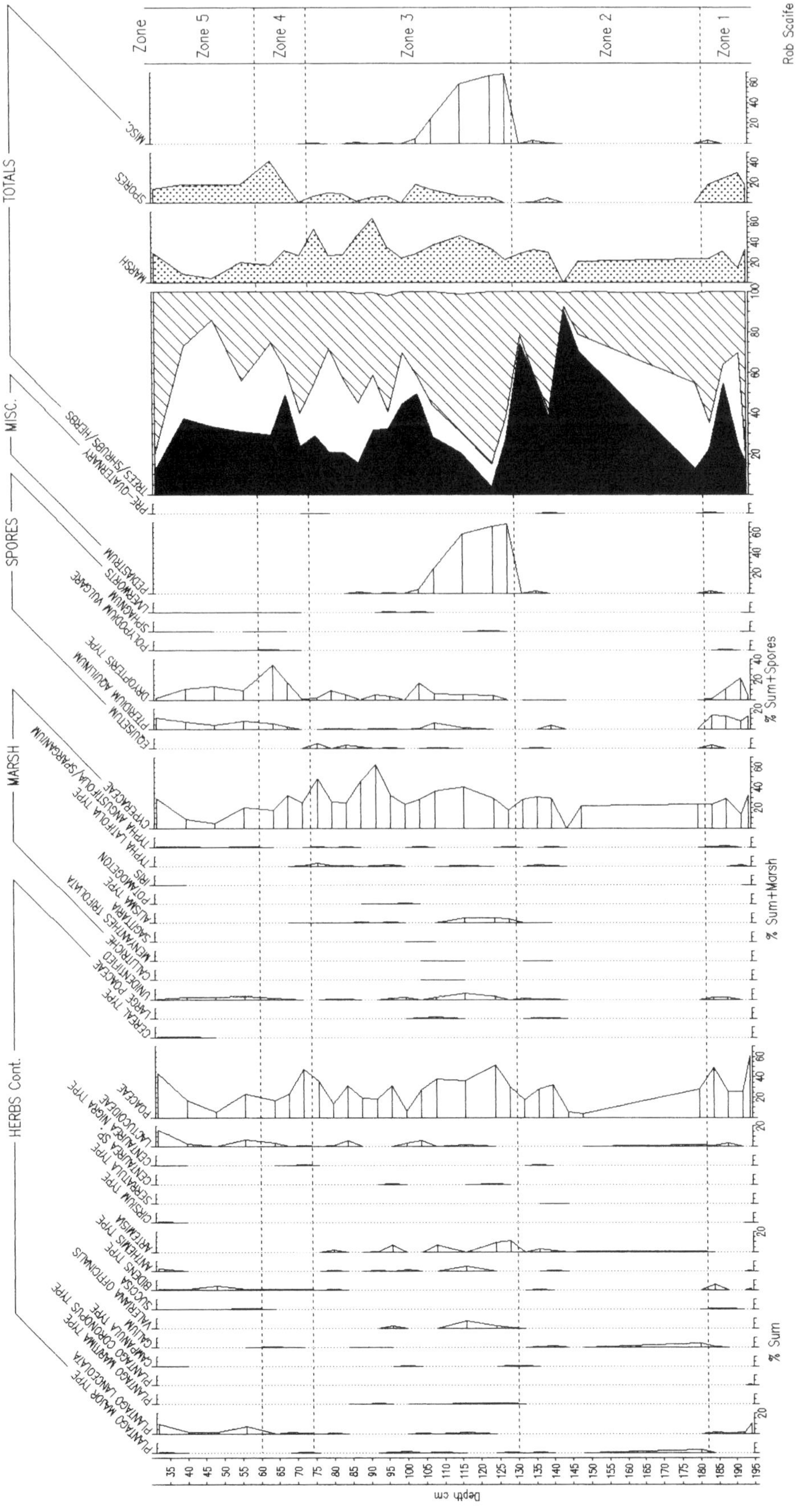

Figure 2.20. *(cont.)*

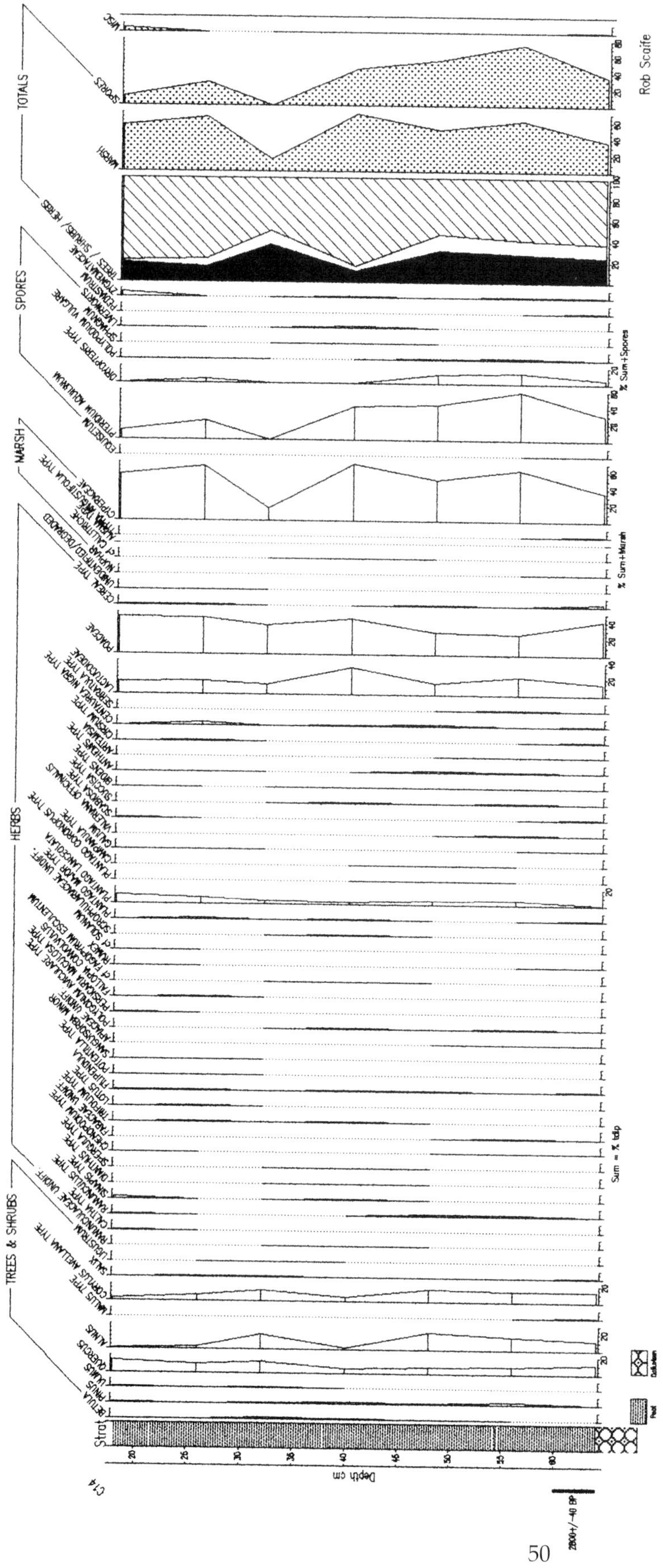

Figure 2.21. *Pollen diagram of the palaeochannel fill at St Giles House (R. Scaife).*

Radiocarbon dating: A radiocarbon date of 1030–840 cal. BC (2800±40 BP; Beta-189162) was obtained for the base of the peat at 60–66 cm (Table 2.1). This sequence thus appears to overlap in time with the upper part of the Knowlton sequence as discussed below, and pertains to the later Bronze Age and earlier Iron Age periods.

Pollen data: Pollen zonation has not been carried out on this shallow profile because, although there are some changes in the pollen spectra, this profile is relatively homogeneous and does not warrant zonation (Fig. 2.21). Where these changes occur, they relate to possible reduction in *Ulmus* and *Corylus* and increases of *Ulmus* in the upper levels. These are, however, tentative. Consequently, the palynological characteristics of the profile as a whole are as follows.

Trees/shrubs: These are subordinate to herbs (20–30% and 10%, respectively). *Quercus* (10–12%), *Alnus* (to 20%) and *Corylus avellana* type (15% and declining upwards). There are sporadic occurrences of *Betula, Pinus, Ulmus* (in the upper levels from 35 cm) and *Salix, Ligustrum*.

Herbs: These are dominant with Poaceae (to 45%), Lactucoideae (peak to 35%) and *Plantago lanceolata* (increasing to 10%). In addition there is a relatively diverse assemblage of herbs which include cultigens — cereal type and *Fagopyrum esculentum* (26 cm). There are also a range of ruderals and segetals present which include cf. *Spergula, Polygonum aviculare* type, *Fallopia convolvulus* and other taxa which are not differentiable to species or community level. There are pastoral taxa including Poaceae, *Plantago lanceolata* and *Sanguisorba minor*.

Cultigens: In addition to small percentages/values of cereal type (3% at base) is *Fagopyrum esculentum* (buckwheat).

Marsh and aquatic: Cyperaceae are dominant (to 65% sum + marsh) with only sporadic occurrences of aquatic taxa (*Nuphar, Callitriche* and *Pediastrum*) and reed-swamp taxa (*Typha/Sparganium* type).

Spores: Apart from a small number of Liverwort and *Sphagnum* spores, ferns are important. *Pteridium aquilinum* is important in the lower levels (to 60% at 56 cm). There are lesser numbers of monolete *Dryopteris* type spores.

Knowlton

This site is located on the River Allen flood plain about 1.5 km downstream of the St Giles House profile and immediately to the northwest of the Late Neolithic henge complex at Knowlton (Fig. 1.11).

Radiocarbon dating: Two local pollen assemblage zones have been recognized in this flood-plain peat sequence. Two radiocarbon dates were obtained from this sequence: 1910–1700 cal. BC (3439±40 BP; Beta-189164) at the base of the peat profile at 100–108 cm, and 810–760 and 640–560 cal. BC (2570±40 BP; Beta-195862) from towards the top of the peat profile at 48–55 cm (Table 2.1). This therefore places the initiation of peat formation during the Early Bronze Age and continuing into the Early Iron Age. Thus, this profile provides a valuable data set pertaining to the vegetation and environment during the later Neolithic, the whole of the Bronze Age period and just into the earlier Iron Age in the upper Allen valley. The upper part of this sequence overlaps with the Knowlton sequence as discussed above. Overall, the pollen spectra (14 levels) show presence of trees and shrubs but with a substantial importance of herbs (Fig. 2.22).

L.P.A.Z. KN: 1 128 cm to 94 cm. *Quercus* (18%), *Alnus* (to 58%) and *Corylus avellana* type (to 20%). There are only sporadic occurrences of other tree and shrub taxa. Herbs are dominated by Poaceae (to 43%) with some peaks of *Plantago lanceolata* and Lactucae at the base. Marsh taxa consist of Cyperaceae (30%) with occasional *Typha angustifolia* type. Spores of ferns include monolete form (*Dryopteris* type and *Polypodium vulgare*) and *Pteridium aquilinum* (expanding to 30%). There are small numbers of algal *Pediastrum*. A radiocarbon date of 1910–1700 cal. BC (3490±40 BP; Beta-189164) at 108–100 cm suggests that this zone spans the very Late Neolithic and Early Bronze Age periods.

L.P.A.Z. KN: 2 94 cm to 30 cm. This zone is delimited by an expansion of Lactucoideae (to 45% at the top of the profile). This may in part be due to the effects of differential pollen preservation at this site. Trees and shrubs remain similar to the preceding zone with *Quercus* and *Corylus avellana* the principal taxa of drier ground, and *Alnus* which shows slightly lower percentages corresponding with expansion of Cyperaceae. The herb flora remains diverse and important, dominated by Poaceae (to 45%). As with the preceding zone, there are only sporadic/single occurrences of Cereal pollen. Spores of *Pteridium aquilinum* are important (47%) with *Dryopteris* type and fewer *Polypodium vulgare*. There is a substantial peak of *Pediastrum* in sediments at 44 cm. The radiocarbon date of 810–760 and 640–560 cal. BC (2570±40 BP; Beta-195862) (Table 2.1) from towards the top of the peat profile at 48–55 cm suggests that this zone covers the middle–later Bronze Age and earlier Iron Age periods.

Group 3: Wimborne Minster town centre flood relief scheme
Site location: Pollen analysis was carried out as part of Wimborne Minster town centre flood relief scheme from two sections which are some 300 metres apart in the town centre, that is, from the north end of the Allenbourne car park (Fig. 2.23) and from Pippins Stream (Fig. 2.24).

The stratigraphy of these sites shows a complex pattern of changing fluviatile conditions with periods of relative stability when fen peat formed on the fringes of the flood plain to freshwater alluvial sediments containing mollusca. From inspection of the two pollen sequences, it is clear that the ages of the two sequences of peats and sediments analysed and the environments represented by the pollen spectra are markedly different. A radiocarbon date of 5980–5780 cal. BC (7010±40 BP; Beta-189166) was obtained for the lower peats of the Allenbourne car park (Table 2.1). This, therefore, demonstrates a middle Holocene (or Atlantic) Age for the earliest organic peat/sediments. It was not possible, unfortunately, to obtain enough organic material from the cores of site 2 (Pippins Stream) for radiocarbon dating. These sequences, however, differ markedly in their floristic content, their dating and the environment which they illustrate. The principal characteristics of these pollen sequences are as follows:

The Allenbourne car park

The stratigraphy of this sequence comprises *c.* 130 cm of organic peat, fluviatile silts and sands containing freshwater molluscs. The top of the pollen column (Fig. 2.23) was 67 cm below the present land surface. The site is immediately adjacent to the present-day channel of the River Allen and underlying the modern wall of the flood relief scheme. The stratigraphy of the pollen sequence is described as follows:

0–12 cm	Grey alluvial clay/silt; monocotyledon root penetration from above
12–52 cm	Fibrous monocotyledonous sedge fen peat
52–74 cm	Grey organic silts containing freshwater molluscs
74–82 cm	Monocotyledonous/detritus fen peats
82–100 cm	Sands containing flints to 5 mm and some organic inclusions and freshwater molluscs
100–126 cm	Fibrous monocotyledonous peat; dark, detrital and well humified; resting directly on gravel base
126 cm	Fluviatile gravels resting on bedrock

Radiocarbon dating: Wood from near the base of the lower peat profile at 110–114 produced a date of 5980–5780 cal. BC (7000±40 BP; Beta-189166), and a date of 3650–3510 and 3420–3390 cal. BC (4780±40 BP; Beta-195863) was obtained from mid-profile at 78–80 cm (Table 2.1). Thus this profile is interpreted to span the later Mesolithic and the whole Neolithic period.

Pollen data: Four pollen assemblage zones and a number of sub-zones within them have been recog-

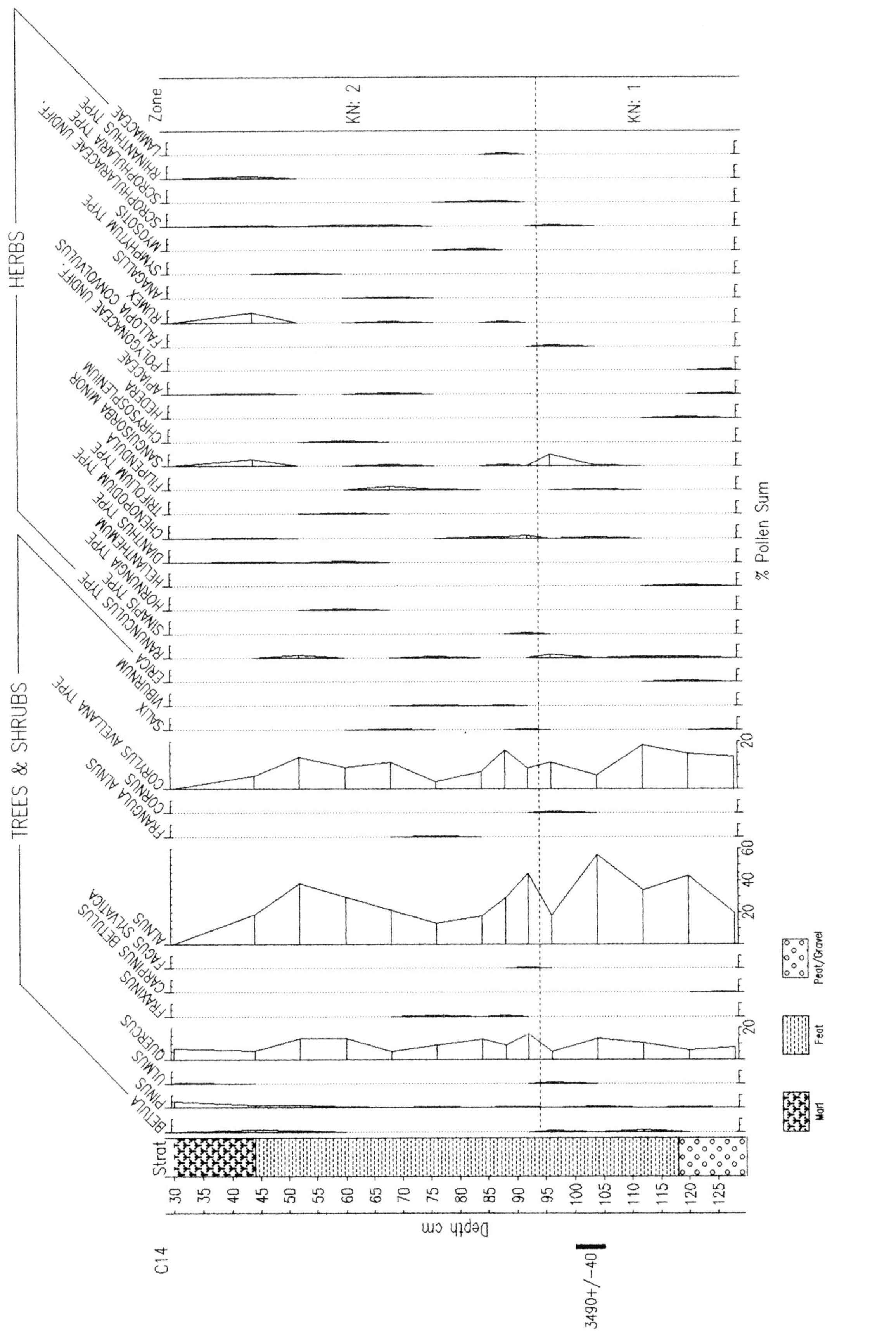

Figure 2.22. *Pollen diagram of the palaeochannel fill at Knowlton (R. Scaife).*

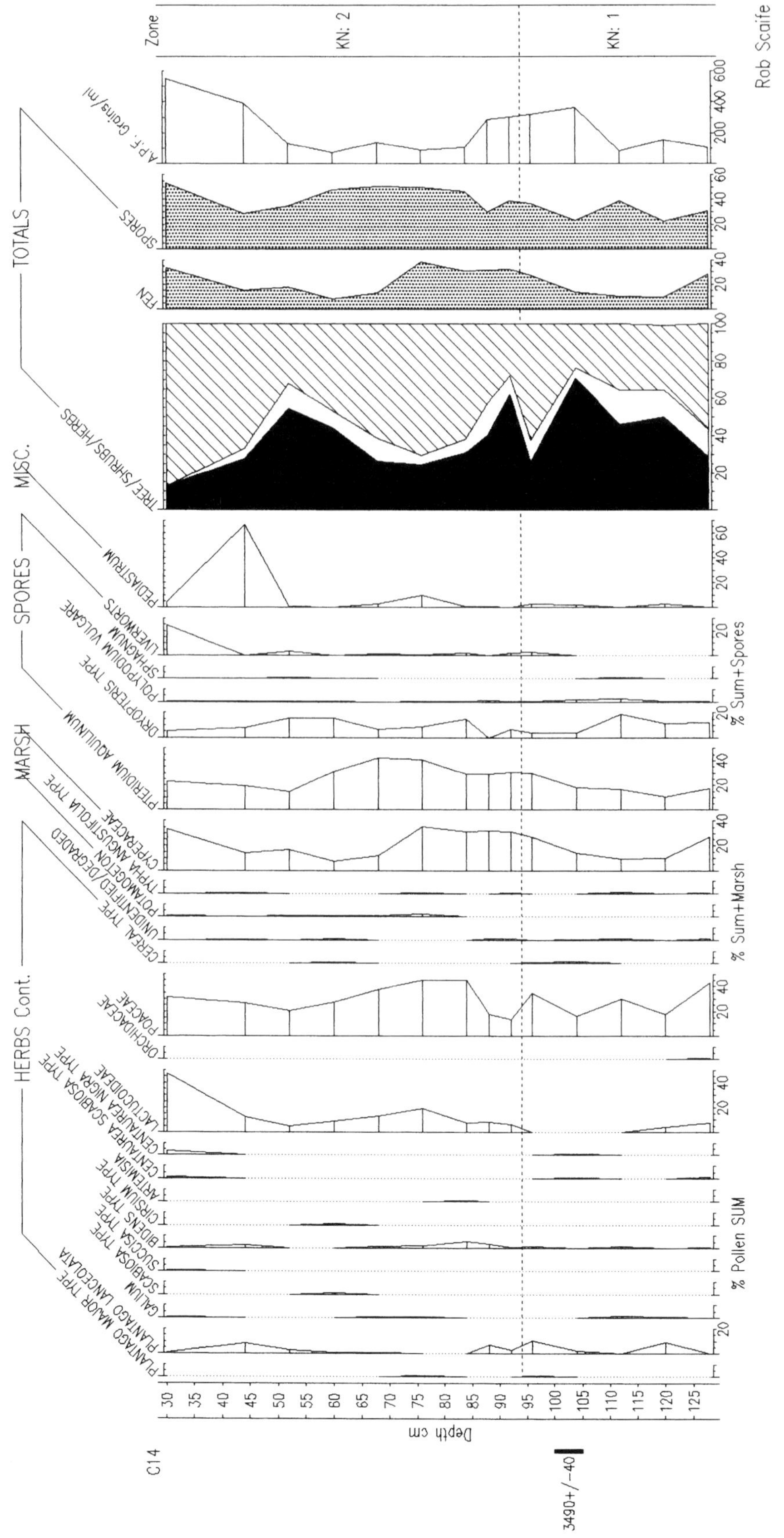

Figure 2.22. (cont.)

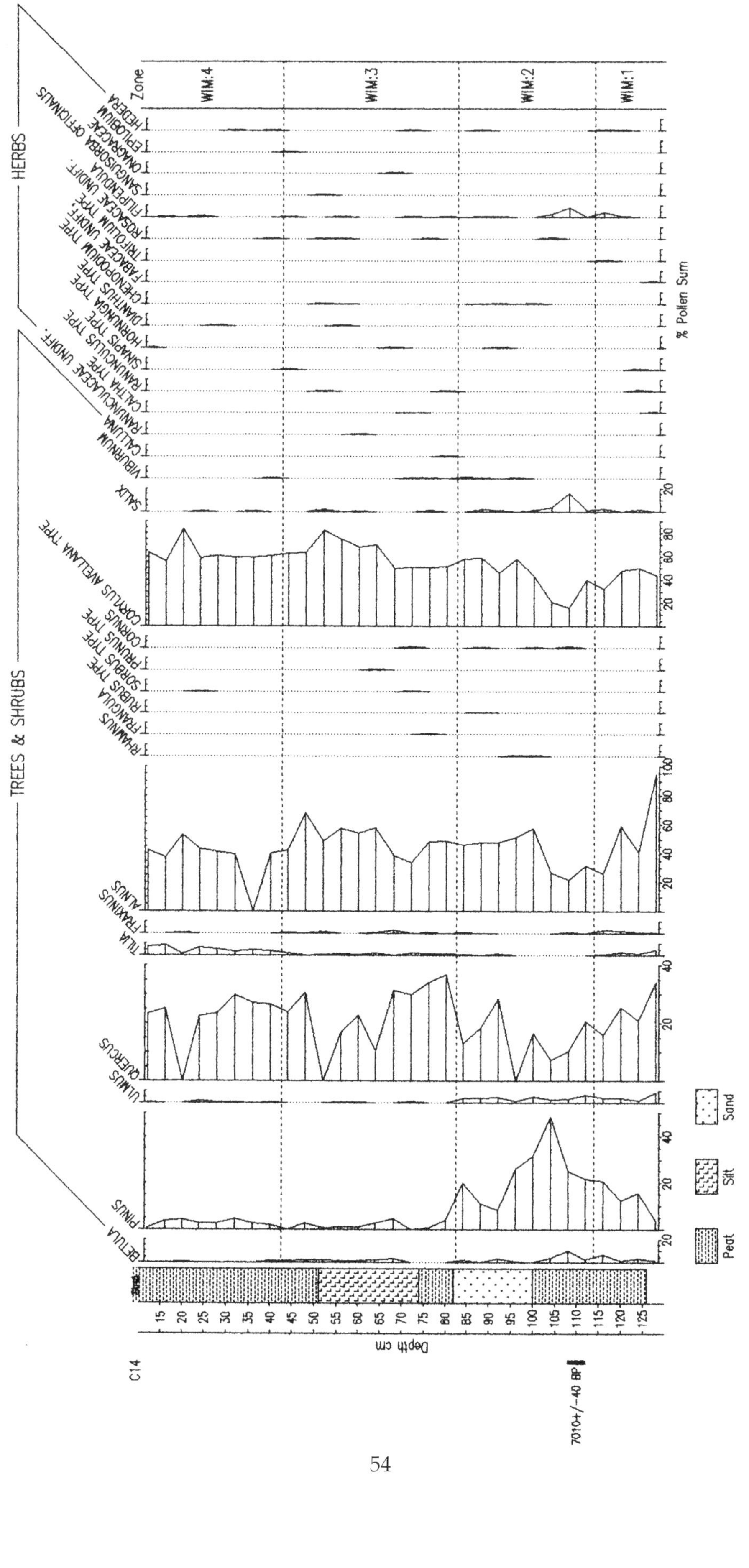

Figure 2.23. *Pollen diagram of the palaeochannel fill at Allenbourne car park, Wimborne Minster (R. Scaife).*

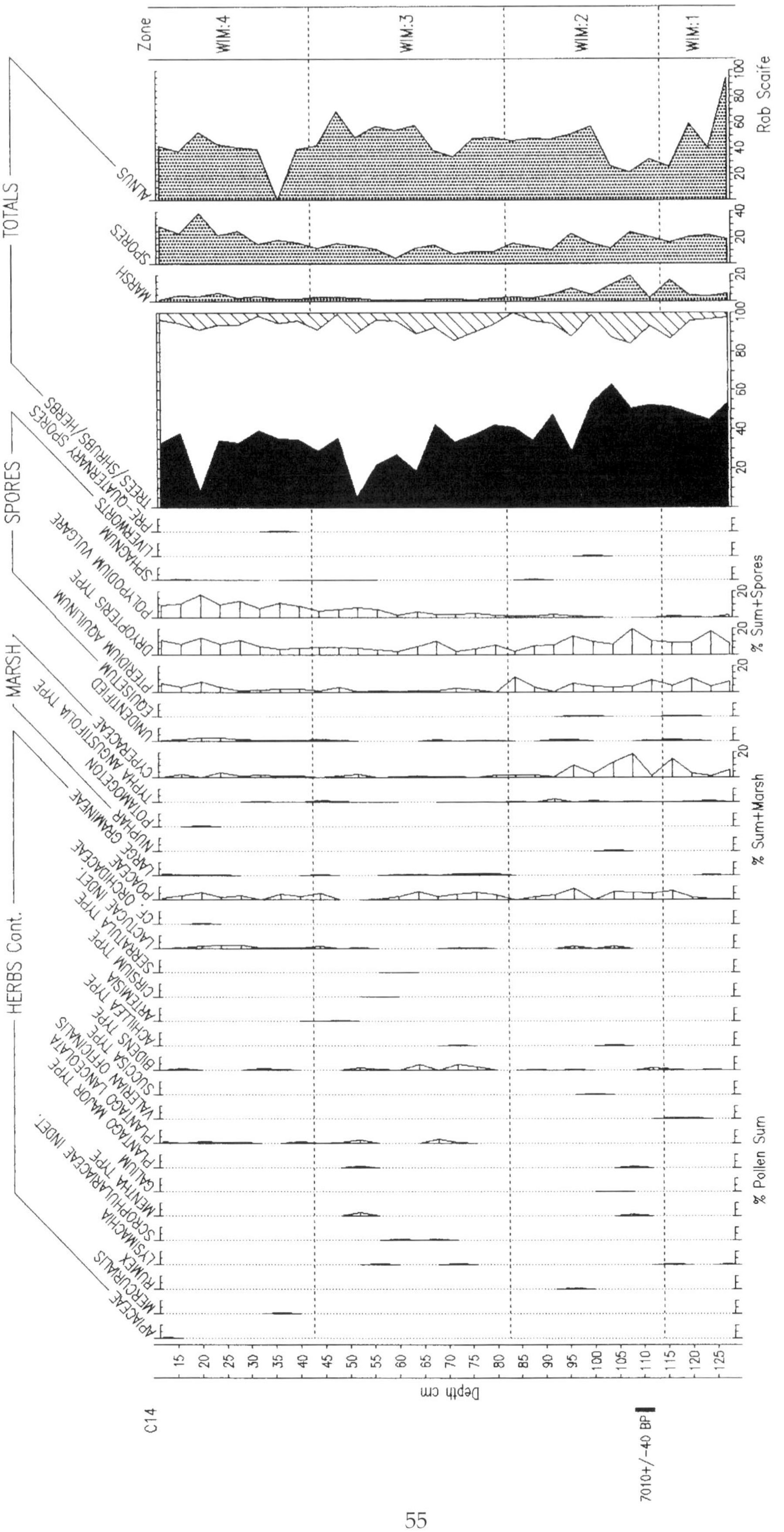

Figure 2.23. (cont.)

nized (Fig. 2.23; WIM: 1–4). These are characterized as follows from the base of the section upwards.

WIM: 1: 128 cm to 114 m. This basal zone is characterized by high total arboreal percentages. The principal taxa include *Pinus* (to 20%), *Quercus* (to 35%), *Alnus* with highest values at the base of the section (95% sum + *Alnus*) and *Corylus avellana* type (to 50%). Other tree taxa are also of note and include *Betula* (to 12%), *Ulmus* (to 10%), *Tilia* (5%) and *Fraxinus* (to 5%). Herbs taxa are fewer than in subsequent pollen zones. *Filipendula ulmaria* (meadow sweet), Poaceae (grasses) and Cyperaceae (sedges) are dominant and increasing in value towards the top of the zone. Pteridophytes are represented by *Equisetum* (horsetails), *Pteridium aquilinum* (bracken), *Polypodium vulgare* (common polypody) and monolete *Dryopteris* type. The radiocarbon date would suggest that this part of the sequence spans the later Mesolithic period.

WIM: 2: 114 cm to 82 cm. Tree and shrub pollen remain dominant (*Pinus, Quercus, Alnus*) but with some increases in herb diversity and percentages. *Tilia* and *Fraxinus* decline to absence in the middle of the zone and re- occur at the top. *Betula* declines progressively. In contrast *Pinus* values increase sharply to a peak of 50% at 104 cm. Shrubs are dominated by *Corylus avellana* type (to 60%) with a marked peak of *Salix* (17% at 108 cm). *Rhamnus, Rubus* type, *Cornus* and *Viburnum* are noted. Herbs are dominated by Poaceae (8%) and a peak of *Filipendula* 98%) at 108 cm. Aquatic and marginal aquatic taxa include Cyperaceae (to 20%), *Nuphar* (yellow water lily) and *Typha angustifolia* type. Spores of ferns continue as in the preceding zone.

WIM: 3: 82 cm to 42 cm. This pollen zone is delimited by a marked decrease in pollen of *Pinus* (to <10%), *Ulmus* (to <1%) and the occurrence of large Poaceae (cereal type), *Plantago lanceolata* (ribwort plantain) and a number of other 'weed' taxa. In addition to the above, *Tilia, Fraxinus* and *Corylus* become progressively more important whereas *Betula* shows some decline initially. Sporadic occurrences of the shrubs *Frangula, Sorbus* type, *Prunus* type, *Cornus* and *Salix* are noted. Spores remain dominated by *Pteridium aquilinum, Dryopteris* type and *Polypodium vulgare*. The former shows some decline. Immediately above the main zone boundary by the rise in importance of cereal type/large Poaceae and *Plantago lanceolata* (p.a.s.z.). From the radiocarbon date taken at 78–80 cm, the beginning of this pollen zone is well within the middle Neolithic period.

WIM: 4: 42 cm to 12 cm. *Quercus* (to 30%), *Alnus* (to 50%) and *Corylus avellana* type (to 87%) are dominant. However, *Tilia* increases in importance from zone 2 (to 10% in the upper-most level). *Pinus* similarly shows a minor increase. Herbs are generally fewer in this zone but *Plantago lanceolata*, cereal type remains in small numbers. An expansion of Lactucoideae (Liguliflorae) occurs. Spores of *Polypodium vulgare, Pteridium aquilinum* and *Dryopteris* type increase. *Sphagnum* although sporadic are more frequent than in preceding zones.

Pippins Stream

Site location: Pollen sampling was carried out on the edges of the mill-race channel where earlier assessment work had shown that organic sediments existed. A number of boreholes were made and it became clear that stratigraphy varied markedly over horizontal distances of some few metres. The most representative section was examined and sampled for pollen analysis. Unfortunately, no organic material for radiocarbon dating was saved.

The stratigraphy recorded was as follows:

0–24 cm	Pale organic mud
24–32 cm	Fine/medium sand containing freshwater molluscs
32–64 cm	Grey organic silts
64–76 cm	Black, well-humified sedge fen peat
76–84 cm	Grey organic silts and freshwater muds containing molluscs
84–86 cm	Freshwater molluscs in grit resting on bedrock

Pollen data: This pollen sequence (Fig. 2.24) differs markedly from the Allenbourne car park profile (see above). Here, tree and shrub pollen are generally less important and in contrast, herb taxa are dominant both in diversity and numbers. Three pollen assemblage zones (PS: 1–3) have tentatively been recognized within this sequence. These have been delimited from the base at 84 cm as follows:

P.A.Z. PS: 1: 84 cm to 62 cm. This basal assemblage zone is characterized by the highest values of tree pollen in this stratigraphical sequence. *Quercus* (to 20%), *Alnus* (to 30%) and *Corylus avellana* type (to 32%) are dominant. The latter has marked peaks in percentages at the top of the zone. Small numbers of *Betula, Pinus, Fraxinus, Fagus* and a minor peak of *Tilia* are noted. Herbs are dominated by Poaceae (highest values at the base to 50%), cereal type (to 12%), Lactucoideae (15%). Remaining herbs are diverse with *Ranunculus* type, *Sinapis* type, *Filipendula ulmaria, Rumex* sp., *Plantago lanceolata, Bidens* type being most important. Marginal aquatic and aquatic plants are represented by Cyperaceae, *Typha angustifolia* type and *Potamogeton* type. Spores of ferns include *Pteridium aquilinum, Polypodium vulgare* and monolete *Dryopteris* type. Higher values of *Polypodium vulgare* are noted in this zone.

PS: 2: 62 cm to 28 cm. This pollen assemblage zone is delimited by a decrease in the arboreal taxa noted in P.A.Z: 1. This is especially marked with *Alnus* and *Corylus avellana* type which decrease to <10%. A single peak of *Salix* (willow) is noted at 32 cm. In contrast there is a substantial increase in herb pollen. Poaceae is dominant (to 56%) with cereal type (15%), Lactucae (23%), *Plantago lanceolata* (6%), *Bidens* type (6%), *Ranunculus* type, *Chenopodium* type, *Sinapis* type and an overall increase in herbaceous diversity. Of note is *Centaurea cyanus* (blue cornflower) throughout. Marginal aquatics remain dominated by Cyperaceae (25%) with *Typha angustifolia/Sparganium* type present.

PS: 3: 28 cm to 1 cm. This pollen assemblage zone is tentatively delimited by some increase in arboreal pollen percentages and Ericaceae (*Erica* and *Calluna*). Thus, *Quercus* and *Alnus* appear to expand to *c.* 10%. Herbs remain little changed from P.A.Z: 2 displaying a diverse range of herbs of ruderal and segetal habitats. These are dominated by Poaceae, cereal type, *Plantago lanceolata, Bidens* type and Lactucoideae. A small number of algal *Pediastrum* spores are noted throughout the spectra indicating a freshwater habitat.

The palaeovegetation of the River Allen valley

The systematic examination of the Allen Valley long profile stratigraphy and sedimentary archive, has provided material from the late Devensian interstadial and Devensian/Holocene transition until the historic period. Although the sites analysed are geographically separated, they do provide information on the changing vegetation of the region. It must be noted, however, that such pollen data will relate to a much more local zone, with the majority of pollen coming from within a few hundreds of metres around the sample sites (e.g. Tauber 1965; 1967; Edwards 1982). In spite of this a changing picture of the changing habitats has been built up which provides useful insight into the changing habitats of the prehistoric and historic archaeological periods.

The late Devensian (or late glacial period)

Sediments of this age recovered from the Allen Valley at Wimborne St Giles (above) provide data on the habitat at the close of late Devensian at *c.* 11,000 to 10,000 BP (Upper Dryas: Loch Lomond stadial). The vegetation is characterized by open herbaceous communities with areas of Juniper (*Juniperus communis*) scrub (possibly the prostrate form) with possible scatters of birch scrub woodland. There are typical late glacial indicator plants with dwarf birch (*Betula nana*), possible *Dryas octopetala* (mountain avens), rock rose (*Helianthemum*), *Sanguisorba officinalis* (greater burnett), *Artemisia* (mugworts) and the halophyte *Plantago maritima.* The high values of *Salix* (willow) at 180–190 cm may derive from dwarf *Salix* (*S. herbacea)* or shrub types. Overall, the vegetation communities suggested by the pollen include dwarf shrubs, possible birch scrub, disturbed ground, aquatic and marsh, tall pasture and halophytes, as follows:

Dwarf shrubs:	*Betula nana; Juniperus communis,* ?*Salix*
Scrub/park woodland:	*Betula;* ?*Salix*
Long-distance transport:	*Pinus*
Disturbed ground/soliflucted?:	Chenopodiaceae, *Artemisia*
Short turf grassland:	*Helianthemum, Botrychium lunaria,* cf. *Dryas octopetala,* Poaceae, *Plantago lanceolata, Potentilla* type
Marsh:	Cyperaceae; Poaceae *Alisma* type, *Typha* spp.
Aquatics:	*Lemna, Pediastrum.*
Tall herbs:	*Filipendula, Thalictrum, Plantago lanceolata,* Poaceae, *Sanguisorba officinalis.*

Unfortunately, pollen does not often allow identification to species level and the above are suggested growth habitats. It is understood that some of these taxa may relate to one or more vegetation communities and habitats. Taxa such as Poaceae, Cyperaceae, *Epilobium* type may thus relate to a diverse range for growth situations and cannot be characterized.

The dating of the Wimborne St Giles Profile 1 is enigmatic. Two possibilities exist, both relating to environmental change during the late Devensian. First, it is possible that the profile relates to the end of Zone II, the Allerød/Windermere interstadial at *c.* 11,000 BP, that is when birch scrub with juniper was being replaced by open tundra-type vegetation. Alternatively, the change may represent the transition at the Devensian/Holocene boundary at *c.* 10,000 BP. The former is perhaps more likely since higher values of *Pinus* and lesser dominance and diversity of herbs than might be expected, and the base of the profile yielded a radiocarbon date of 11,280±170 BP (Beta-168611), whose calibrations centre around 11,000 cal. BC (Table 2.1).

The early Holocene (Flandrian Chronozone Ia–c): The Pre-Boreal and Boreal

After temperature amelioration began from about 10,000 years ago, there is evidence for rapidly changing (seral) vegetation changes of the early Holocene (Flandrian Chronozones Ia–Ic). At Wimborne St Giles Profile 2, the high values of *Betula* in pollen zone 2 and subsequent expansion of *Pinus, Quercus, Ulmus* and *Corylus avellana* type suggest that the profile is typical of the early Holocene (Flandrian Chronozone Ia), or the pre-Boreal and Boreal periods.

Pollen zone ALL: 1 is, however, similarly anomalous. In the basal peat levels at the interface with underlying marl there are peaks of thermophiles — *Ulmus, Quercus, Alnus* and *Corylus.* These two samples were taken from the core and prepared at different times in order to see if contamination had taken place. This, however, appears not to be the case and the pollen is present in the basal peat. It seems unlikely that later pollen has been transported at or along the peat/sediment interface and derived pollen from earlier interglacial deposits seems a more likely explanation. This will be a focus for further investigation.

There is, however, a distinct and typical peak of *Juniperus* which marks the onset of temperature amelioration at *c.* 10,000 BP. The accuracy of this date is hindered by the radiocarbon plateau spanning this important climatic transition. The expansion of juniper was followed by rapidly colonizing birch on the interfluves and willow in the wetter valley bottoms. A gap in the pollen profile between 180 and 150 cm was due to loss of material from the peat corer due to sloppiness and waterlogging of the peat matrix. At the top of zone 2/base of zone 3, pine values expand along with the first (rational limits) of oak and slightly after, elm and hazel. This suggests the progressive expansion/migration and colonization of early Holocene vegetation, that is, an environment of dynamic biogeographical changes (Birks 1989).

These changes equate with Flandrian Chronozone Ia for the birch peak (zone ALL: 2) and Flandrian Ib for pine with oak and elm (the early Boreal). A number of interesting aspects of the vegetational

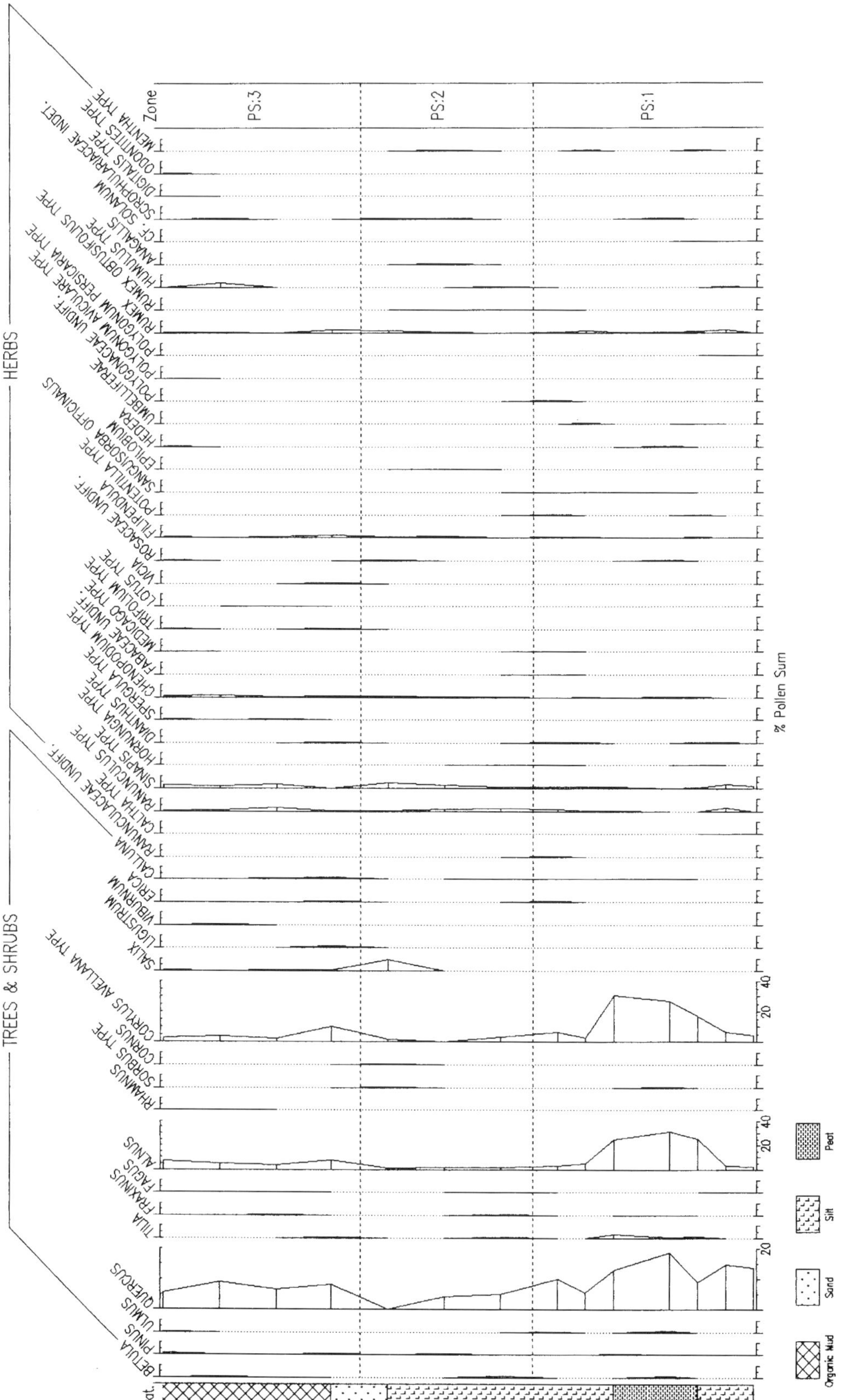

Figure 2.24. *Pollen diagram of palaeochannel fill at Pippin's Stream, Wimborne Minster (R. Scaife).*

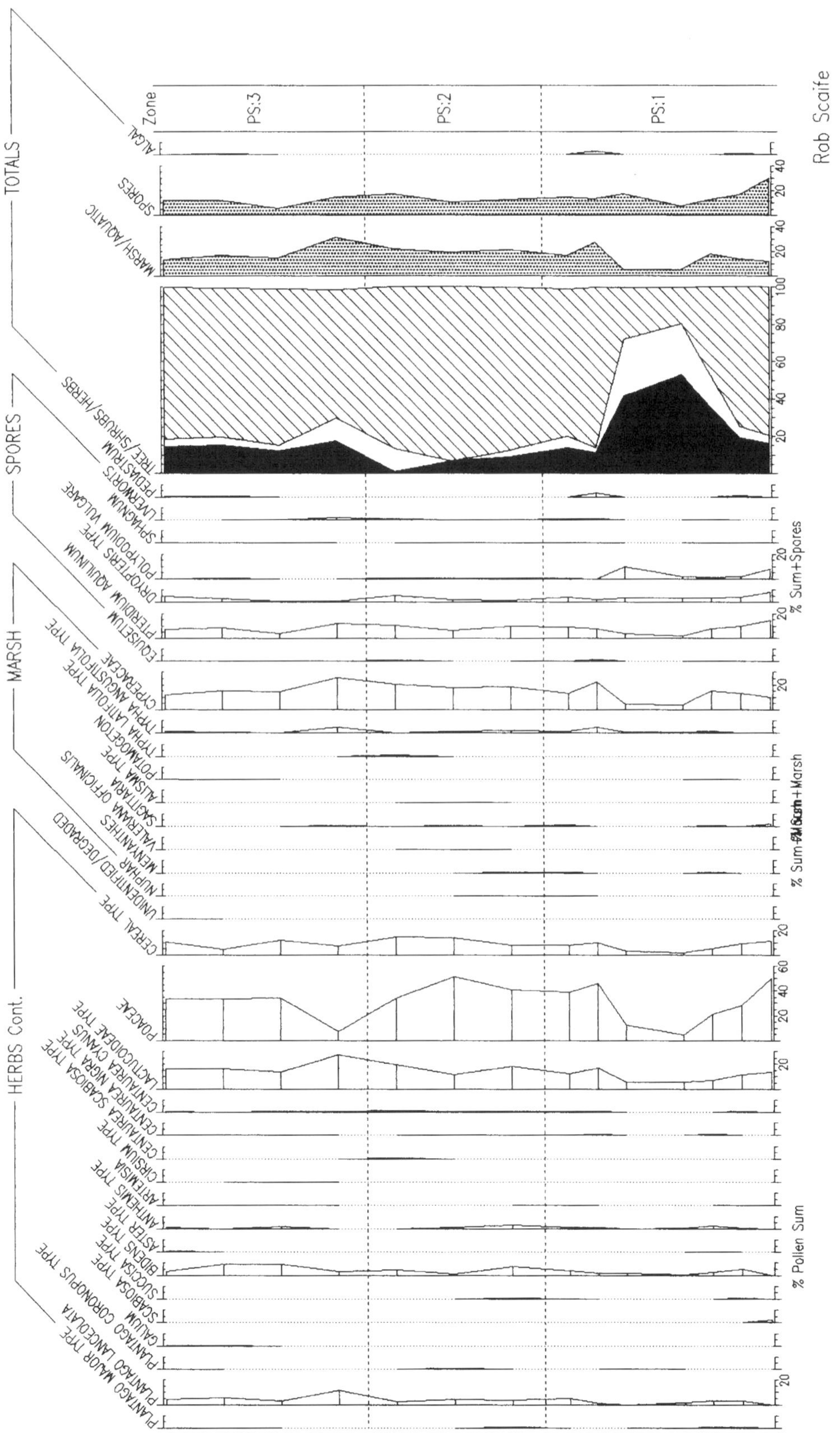

Figure 2.24. *(cont.)*

history are apparent. Firstly, the values of pine are somewhat lower compared with other pollen data for this period (e.g. locally Scaife 1980; 1982; 1987b; Barber & Clarke 1987; Haskins 1978). In contrast, herb pollen and diversity are much greater than normally encountered. The latter shows the contribution of herbaceous communities for the late Devensian and although dominated by Poaceae there are elements of communities described for profile 1. There are also plants characteristic of calcareous areas or grassland including the continuance of *Helianthemum* and *Sanguisorba minor*. Such a diversity of herbs contrasts with the usual upland floras of Boreal and Atlantic periods. This suggests that perhaps the local chalkland remained as refugia for herbaceous plants, not being completely dominated by forest. This has also been controversially discussed for the Yorkshire Wolds (Bush 1988; Bush & Flenley 1987; Thomas 1989). Of note also is a second high peak of *Juniperus*, again suggesting this. Between 125–110 cm, organic muds (at the base of zone 3) contain high values of *Pediastrum* and *Alisma* type, *Callitriche* and *Menyanthes trifoliata* suggesting a phase of wetter marsh perhaps with standing water. *Salix* (a largely under-represented taxon) remains relatively important, and along with sedges, suggests a willow fringed fen or sedge swamp.

In zone 4/5 there is a further expansion of elm, oak and hazel but with declining pine. There are also increases in *Plantago lanceolata* (ribwort plantain) and the presence of cereal type pollen. There appears to be a major hiatus in the peat accumulation, the lower zones 1, 2 and 3 fill the palaeochannel whereas zone 5 is a spread of later, Flandrian Chronozone 3 (sub-Boreal and sub-Atlantic) peat which covers the flood plain. This is further evidenced in the peat profile and pollen sequence from St Giles House.

The Middle Holocene (Flandrian Chronozone II): The Atlantic

Fortuitously, the sequence of peat and alluvial sediments obtained from the Allenbourne car park, Wimborne town centre, has provided the only data pertaining directly to the Holocene period after that discussed above. A radiocarbon date of 5980–5780 cal. BC (7010±40 BP; Beta-189166) comes from the lower peats of pollen zone WIM: 1 (130–100 cm) (Table 2.1), that is during a period of peat accumulating under typical alder-willow carr woodland. The latter, being entomophilous, is under-represented in the pollen spectra whereas *Alnus*, being anemophilous, is largely over-represented. Other constituents of this local fen carr community comprised marginal/aquatic taxa such as *Caltha* (marsh marigold), *Typha angustifolia* type (reed mace and burr reed), *Filipendula* (meadowsweet), Poaceae (grasses) and Cyperaceae (sedges). Woodland growing on the interfluves is of particular interest with *Quercus* (oak), *Ulmus* (elm), *Tilia* (lime/lindens), *Fraxinus* (ash) and *Corylus avellana* (hazel) present. *Tilia* is diagnostic for the middle Holocene (as seen here) and frequently attains relatively high values which indicate dominance or co-dominance with *Quercus* over large areas (Moore 1977; Greig 1982; Scaife 1987b; 2000), a feature which lasted until well into the later prehistoric period when widespread deforestation took place (Turner 1962; Baker *et al.* 1978).

Of significance are the relatively high values of *Pinus* (pine) which increase to a peak at 104 cm. Although pine is anemophilous, is dispersed over long distances and is frequently over-represented in pollen spectra, the values found here are enough to suggest that at least some pine was growing in the local area. Although pine was a dominant element in the early Holocene (Flandrian chronozone Ib & Ic) in southern England (i.e. *c.* 9000 to 7500 BP), its presence in later periods has remained enigmatic. Here, the peats of zone WIM: 1 and WIM: 2 are of middle Holocene, Atlantic age (Flandrian Chronozone II). Godwin (1975) has demonstrated that pine remained growing on the sandy soils of the Brecklands during the mid- and later Holocene. Suggestions have also been made that pine remained in isolated areas of central southern England on similarly acid sandy substrates (Haskins 1978; Scaife 1980; 1987; Cameron & Scaife 1988; Scaife 1991). Plant macrofossil (especially charcoal) evidence has also been noted from a number of archaeological sites in the region (Allen 1994; Allen & Scaife in press). It therefore seems probable that local stands of pine woodland remained on edaphically suited areas of Dorset and Hampshire well into the middle and later Holocene. As with the Poole Harbour area (Cameron & Scaife 1998; Scaife 1991) relatively high pine pollen percentages are found here which provide further evidence of this.

The Neolithic and Early Bronze Age periods

Data come from three sites which span the Neolithic period (Allenbourne car park) and Bronze Age (Allenbourne car park; Knowlton and St Giles House).

The Neolithic: In the Allenbourne profile at 82 cm (zone WIM: 2/3 boundary) there is a significant change in the pollen spectra with the decline of *Ulmus* and *Pinus* pollen (pollen zone 2/3). This change is interpreted as the 'primary *Ulmus* decline'. There has been much discussion of this significant event in northwest European pollen sequences and it is clear from numerous radiocarbon dates that it occurred more or less synchronously during the period 5500–5000 BP (Smith &

Pilcher 1973). Reasons for this reduction in elm pollen are varied, ranging from climatic change, clearance for agriculture, use of leaves for feeding stall-kept animals (Troels-Smith 1960) and disease spread by insect vectors (see Smith 1970; 1981; Scaife 1987a for discussion). The latter now gains most favour in view of recent (1970s) destruction caused by the Dutch elm bark beetle (*Scolytus scolytus*) carrying fungus. Girling (1988) showed elegantly that *Scolytus* sp. was present in the mid-Holocene woodland and that the onset of the Neolithic and subsequent opening of the woodland canopy allowed the rapid spread of the disease. It has been shown from other sites in southern England (Peglar 1993; 2006, 26–9; Waller 1994, 91–4) and especially the Isle of Wight that the primary elm decline is associated in pollen diagrams with the first evidence of cereal pollen (Scaife 1980; 1982; 1987a). At Allenbourne, this similarly appears to be the case with the elm pollen decline followed immediately by the first continuous records of cereal pollen (zone WIM: 3 sub-zone 1). The latter are identified by large diameter, typically >45 μm grains, with large pores/annulus, thick pollen wall and distinct columellae. The small peak of cereal pollen is also typically associated with weed taxa such as *Plantago lanceolata* (ribwort plantain) and *Bidens* type (which includes many taxa of the Asteraceae). The values for tree pollen remain largely unchanged except for *Ulmus* and *Pinus*, and it is suggested that arable cultivation was on a small scale. Such localized agriculture has often been described in terms of shifting cultivation (*landnam* clearances of Iversen 1949). Here, however, the cereal cultivation is not associated with increases in *Betula* scrub and wild grass pollen as noted for example at Gatcombe Withy Bed, Isle of Wight (Scaife 1980; 1987a). It is possible that at Wimborne, cultivation was on a small scale, perhaps in close proximity to settlement. It must, however, be noted that there is a possibility that the arable and associated pollen may have been transported from some distance where more extensive woodland disturbance may have taken place. This is not considered likely given the dominance of local woodland which would have made the effective pollen catchment small. It is concluded that we are here seeing the first incidence of Neolithic cultivation on a small scale in a local landscape which remained predominantly wooded. After an initial peak in cereals, pollen becomes sporadic but consistent throughout the profile indicating the continuity of local cultivation of crops.

The decline in *Pinus* pollen at this time (WIM: 3/4) is interesting. High *Pinus* values are typical of the flora in the early Holocene (Boreal) of this region and, as noted above, there is evidence for its continuance into the middle Holocene (Flandrian chronozone II: the Atlantic period). The decline of pine at *c.* 5000 BP with the elm decline is also seen at Gatcombe Withy Bed, Isle of Wight where the decline is dated at 4850 ±50 BP (SRR-1338). This is incongruous since with opening of woodland, *Pinus* pollen values might be expected to rise because of increased long distance component (i.e. pine being anemophilous and transported over long distances). This is not the case and other causes must be considered for its decline and the possibility of selective felling of pine for construction purposes, by virtue of its long and straight trunks, may be considered (Scaife 1980; 1987b; Allen & Scaife 2007). After this reduction in pine its values are commensurate with pollen derived from long distance sources.

Although there is evidence for the first cereal cultivation, mixed deciduous woodland also remained after *c.* 5500–5000 BP. This was dominated by *Quercus* (oak) with *Corylus avellana* (hazel). Here, values are relatively low compared to many sites spanning this period. However, being entomophilous, its pollen is poorly disseminated (Andersen 1970; 1973) and the dense character of the alder carr would probably be the result of screening of the mire from pollen rain (Tauber 1965). This similarly applies to *Fraxinus* (ash) and even though pollen values are relatively small for these taxa, their importance at this time should not be underestimated.

Subsequent to this period of Neolithic activity, there is some evidence of woodland regeneration in the Allenbourne profile with *Fraxinus* again becoming more important (zone 3) and *Tilia* in zone 4. As noted earlier, however, the complex river flood-plain environment with changes from peats to alluvial sediments may in part be responsible for changing sources of pollen input. Whilst the increase of secondary ash (and elm) woodland is frequently seen in the immediate post elm decline period, the expansion of *Tilia* may be due to changing taphonomic factors. Although it is not clear at what date the organic peat finished accumulating in the Allenbourne palaeochannel, on the basis of the radiocarbon dating available, it seems likely that this was during the Late Neolithic or Early Bronze Age. This is based on the fact that woodland remained consistent and dominant and that *Tilia* was increasing in importance. The latter frequently shows a marked decline in the later Bronze Age period in southern England, and this is not evident in this sequence.

A more detailed representation comes from Knowlton, where peat started forming during the Late Neolithic before *c.* 3500 BP. Perhaps unsurprisingly given the important archaeology of this date in this region, the pollen data suggest that the environment was a composite of woodland but probably with substantial areas of open land. Whereas lime woodland may be expected in many areas of southern Britain at this time, this had been apparently been removed during the Neolithic. This is likely because of the preference of *Tilia* for well-drained soils which would have been easier to clear than heavier lower valley side soils and also more suited to agriculture. Such removal is not, however, unknown for the Neolithic (Scaife 2000). Oak and hazel woodland remained on drier interfluve soils and possibly drier areas of the flood plain. The valley bottom at Knowlton was grass-sedge fen-type habitat with areas of fringing alder in drier areas which were not subject to more than two to three months of waterlogging during the winter months. There is little evidence for arable cultivation during this period with only occasional cereal pollen recovered. It is most probable that pasture was prevalent in the cleared areas and whilst many of the grassland/pasture type herbs may also come from the on-site flood-plain community, the presence of ribwort plantain (*Plantago lanceolata*) is perhaps evidence of the status of the drier ground.

The cause of peat accumulation at this time may in itself be a response to woodland removal. Reduction of woodland will have reduced evapo-transpiration causing higher groundwater tables locally and also increased surface run-off, and have produced anaerobic conditions suited to peat/organic accumulation (Moore & Wilmott 1973; Scaife 1980).

Late Bronze Age to the historic period

Extensive cross valley profiling of the Allen flood plain revealed peat, although shallow, in many areas of the flood plain of the River Allen such as St Giles House and the Pippins Bridge profile of Wimborne Minster town centre. A radiocarbon date of 1030–840 cal. BC (2800 BP; Beta-189162) (Table 2.1) from the St Giles House sequence places the start of peat accumulation in the Late Bronze Age, thus placing the overlying peats within the very late prehistoric and probably early historic periods. Pollen data show typical features of the Iron Age, Romano-British and later periods with relatively small numbers of trees and shrubs of which oak and hazel were most common in drier habitats and alder remaining in isolated valley fen and fringing the banks of the river. Although there is a small amount of cereal pollen, the majority of herb taxa indicate a largely grassland/pasture habitat, at least in proximity to the St Giles House site. This appears to be a continuation of the activity noted throughout the Bronze Age at Knowlton.

The absence in any quantity of trees in general and particularly lime (*Tilia*), where radiocarbon measurements have not been obtained (Pippins Bridge), clearly indicates a post-lime decline age, that is post-Neolithic to Middle–Late Bronze Age. Furthermore, absence of pine and elm in quantity in the upper levels of the St Giles House, Knowlton and Pippins Bridge profiles suggest a date prior to the plantations of exotic pine at *c.* AD 1750–1800 (Haskins 1978; Long *et al.* 1999; Scaife 1980; 2000) and also the maturation of hedgerow elms resulting from the enclosure movement.

Although no radiocarbon dating could be carried out on the Pippins Bridge sequence, the assemblages appear to be typical of the medieval period and prior to *c.* AD 1700–1750 (see above). The pollen spectra in general show a small arboreal component with, in contrast, herb taxa dominant. The latter show a marked diversity of plants typical of waste ground (ruderals) and arable cultivation (segetals) typical of this period. As with the older sediments of the Allenbourne profile (above), the stratigraphy at Pippins Bridge varies between organic peat accumulated under sedge fen and fen carr conditions, organic muds laid down in still water and sands deposited under higher energy fluviatile conditions. Thus, the taphonomy of the pollen recovered may vary between strong local elements found in the peats of pollen assemblage zone 1 and the possibility of longer distance transport by water.

The basal sands and silts at Pippins Bridge contain freshwater molluscs before merging into organic fen type peats above. This stratigraphical change is mirrored in the pollen sequence with a change from marginal aquatic vegetation (*Potamogeton*, Cyperaceae, *Sagittaria* and *Typha angustifolia*/*Sparganium* type) to alder carr. However, percentages of *Alnus* are lower than might be expected if extensive carr woodland was growing on the site (see for example pollen zone WIM: 1). It is likely, therefore, that small numbers of alder were growing along the edges of the River Allen channel and on drier areas of the flood plain. Dryland vegetation during this phase shows the highest percentages of tree and shrub pollen in this diagram with *Quercus* and *Corylus avellana* type dominant. This woodland was, however, possibly of only limited extent in the local area since herb pollen is dominant throughout. Small numbers of poor pollen producers/dispersers including *Tilia, Fraxinus* and *Fagus* are perhaps of some significance indicating occasional local growth of these trees in the nearby region. Subsequently, in pollen zones PIP: 2 and PIP: 3 all tree and shrub taxa decline to relatively small values indicating a general absence of woodland in the near region. Small increases in percentages in pollen zone PIP: 3, may result from taphonomic changes with sedimentary changes. It is unfortunate that sufficient material for radiocarbon dating was not available from the lowest levels of the Pippins Bridge profile.

Herb pollen is dominant throughout all pollen zones at Pippins Bridge and St Giles House. The local depositional environment of the former is evidenced by aquatic and marginal aquatic plants which include *Potamogeton*, *Nuphar*, *Menyanthes* (bog bean), *Sagittaria sagittifolia* (arrow read), *Alisma* cf. *Plantago lanceolata* (water plantain), *Typha latifolia* (great reed-mace), *Typha angustifolia/Sparganium* type (bur reed and lesser reed-mace) and Cyperaceae (sedges). These plants and the character of the organic muds and silts of pollen zones PS: 2 and PS: 3 indicate deposition in freshwater of slow flowing character and possibly in a mill pond within medieval Wimborne.

Dryland herbs in all pollen zones show a strong presence of cereal pollen of possibly associated segetals and ruderals. Because of the similarity of pollen morphology in many pollen taxa, it is not possible to establish with certainty many of the species (or even genera) which may have been associated with arable cultivation. However, taxa such as *Centaurea cyanus* (blue cornflower), Brassicaceae (*Sinapis* type; charlock), Chenopodiaceae (goose-foot), *Spergula* type (spurrey), *Polygonum* spp. (knotweeds) and Lactucoideae (possibly hawkbit, hawk's-beard and sow thistles) are typical of plants which, in this situation, may be attributed to arable environments. Additionally there are weeds such as *Plantago media/major* type, typical of waste/disturbed ground, which may also be associated with arable disturbed ground. The diversity of these herb taxa and presence of types such as *Centaurea* spp. with pollen which is generally poorly dispersed, suggests that extensive arable cultivation was taking place during the period of deposition of these sediments. Two factors must, however, be considered. First, that the pollen may have been transported fluvially from some distance upstream of the site examined and second, that these pollen may have been derived from crop-processing activities from mills along the stream. Substantial percentages of Poaceae and sporadic occurrences of herbs not attributable to cereal and waste-ground habitats may derive from grassland habitats, possibly flood-plain grassland or pasture and it is possible that these pollen spectra provide evidence for both arable and pastoral agriculture. *Humulus/Cannabis* type (hop and hemp, respectively) occurs in all three pollen zones. Unfortunately, pollen preservation did not allow identification to generic level and it remains unclear whether this type reflects hemp cultivation or growth of wild hops along the river valley. However, it is highly probable, given the agricultural environment shown, that the former is represented since its cultivation was widespread in medieval and post-medieval periods.

The environment was thus clearly agricultural with arable cultivation important with pasture. Oak and hazel woodland remained and was most probably managed coppice with standards. Alder, where it occurs, comes from trees fringing the valley or from extensive carr woodland at a distance, whereas lime, ash and beech all found in these later sediments may represent local/ephemeral growth of these trees. Overall, in spite of some evidence of cereal cultivation and associated arable weeds, pasture types dominate. Grass with Lactucoideae (dandelion types) is dominant, with ribwort plantain and other typical pasture types present (*Ranunculus* type, Fabaceae and Asteraceae type).

Conclusions

The pollen data analysed at present are unusual in that they come from a calcareous, chalkland habitat, an environment which is not usually conducive to pollen preservation. Extensive survey by project members has located isolated peat filled palaeochannels and areas of flood-plain peat on the upper River Allen flood plain. The organic fills of the palaeochannels appear to be of late Devensian and early Holocene age whilst broader spreads of organics found in some localities are more recent (late prehistoric and historic). Earlier pollen work (Scaife 1994a), necessitated by new flood defences along the River Allen within the middle of Wimborne town centre, also revealed peat and alluvial sediments which fortuitously span the middle (Atlantic; Late Mesolithic), later Holocene (late prehistoric periods; Flandrian Chronozone III) and historic periods.

Of particular significance is the evidence for early Holocene woodland development from juniper, birch, pine, hazel and oak/elm. However, it is also suggested here that the chalkland habitat remained a substantially open environment allowing continuity of many herbs well into the middle part of the early Holocene (Boreal period; Early Mesolithic) and later phases of juniper expansion. This contrasts with the usual absence of herb pollen found in Boreal pollen spectra.

Middle Holocene (Late Mesolithic) woodland saw the continuity of pine after *c.* 7000 BP along with typical Atlantic dominance/importance of lime/linden, oak, elm, hazel and alder along the flood plain. The former is similarly enigmatic and provides further evidence for pine perhaps remaining on the poorer sandy soils of the Tertiary and river terraces in the region. This contrast with most regions of southern England where pine was largely ousted by deciduous woodland by *c.* 7000 BP.

The later Holocene, i.e. Flandrian Chronozone III (the Neolithic and Bronze Age), is heralded by a significant decline in *Ulmus* pollen seen in the Allenbourne profile and is attributed to the frequently described 'Primary Elm Decline' at *c.* 5500–5000 BP. Typically, this is also associated with the first indications of cereal pollen and associated weeds and thus, Neolithic arable activity in the region. The presence of pine during the middle Holocene has been noted as important and furthermore, its decline at the time of the elm decline is similarly enigmatic and is tentatively attributed to selective felling for construction of buildings by the Neolithic. Neolithic activity at least in this area seems, however, to have been localized and the overall effect on woodland dominance was negligible. However, there is some evidence for Late Neolithic woodland regeneration.

Data pertaining to the Bronze Age period at both Knowlton and St Giles House demonstrates that substantial woodland clearance had taken place, especially of lime/linden woodland which remained in many areas of southern England. Remaining woodland consisted of alder in and along the fringes of the valley with oak and hazel on drier areas of the flood plain and valley interfluves. Open zones were of grassland (possibly pasture) with little evidence for cereal cultivation.

Pippins Stream in Wimborne town centre has provided sediments which are though to be of medieval age and as may be expected for this period there is a paucity of trees but a rich herb flora.

From these sites, a picture of the changing vegetation of this valley within one of the most important areas of calcareous lithology has been built up. Whilst the pattern of changes for many of the periods is typical of the country and non-calcareous regions of southern Britain as a whole, there are some interesting contrasts and aspects of vegetation history which are viewed as important as well as controversial.

The development of the upper Allen valley: variation at the sub-regional scale

Charles French, Rob Scaife & Helen Lewis

When exploring a landscape for ancient land-use information, it is important to develop a picture of what the landscape looked like over time in the past, and to utilize the data 'sinks' available in the area. The Wyke Down project carried out a borehole survey along with aerial photographic terrain analysis mapping of the upper Allen Valley, to locate areas of colluvial and alluvial deposition and to map the various soil types found today. This was done both to put the valley into the context of models of the erosional history of the southern English chalklands and the relationship of this with ancient land use, and to identify the location of colluvial and alluvial deposits for environmental archaeological sampling. In the process of doing the survey, a number of new artefact scatter locations were also recorded (see below).

In general, this survey and associated palynological study are significant in revealing a general lack of eroded colluvial material dating to prehistoric times in the upper Allen valley, and an absence of well developed brown forest earth soils except in rare instances. Brown earths and rendzinas appear to be ubiquitous on the chalk by the later Neolithic period (see below), with only pockets of brown forest earths present on the clay-with-flints on the high downs to the northeast of the study area or on the Reading Beds in the southeast of the study area (Fig. 2.1). One argillic brown earth was found through augering in the modern forest around the Shaftesbury Estate on chalk. The existence of this profile, especially in light of the rest of the valley soil results, suggests that this woodland is very long-lived.

The pollen data from the uppermost part of the Allen valley west of Wimborne St Giles indicates the presence of a substantially open environment, with continuity of many herbs well into the middle part of the early Holocene (Boreal or Mesolithic period), and the profiles do not show any periods of significant soil erosion. The tree pollen results suggest the progressive expansion, migration and colonization of early Holocene vegetation. The lower values of pine and greater herb diversity than normally encountered in this part of the study area region for this time contrast with the usual upland floras of the Boreal and Atlantic periods, and suggests that parts of the local chalkland remained as refugia for herbaceous plants throughout the Mesolithic period. Moreover, these indications of greater and longer-lived openness were not evident down valley to the east of Wimborne St Giles at Knowlton and Wimborne Minster where there was progressive and full mixed deciduous woodland development in the earlier Holocene.

Later in the Mesolithic period there was a further expansion of elm, oak and hazel set against declining pine, along with an increase in ribwort plantain and the presence of cereal type pollen. There then appears to have been a major hiatus in peat accumulation before renewed peat formation across the whole flood plain, which began towards the end of the Neolithic and beginning of the Bronze Age, probably just before 4000 BP. The later prehistoric period palaeochannel systems at Knowlton (Late Neolithic, Bronze Age and Early Iron Age) and St Giles House (later Bronze and Early Iron Age) indicate that land use was clearly agricultural with pasture and some oak and hazel woodland, possibly managed coppice with standards. In spite of some evidence of cereal cultivation and associated arable weeds, pasture pollen types dominate. All of these channel peats appear to have accumulated with relatively small amounts of seasonal soil movement and deposition occurring in the active flood plain of the day.

For the historic period as represented by the Pippins Bridge sequence, the landscape is seen to be more or less completely open with a small arboreal component and dominant herb taxa indicative of waste ground and arable land. In contrast to the prehistoric period, there is much variation in the sediment complex for the first time seen in the peat associated with sedge fen and fen carr conditions along the river's edge, organic muds laid down in still, standing water, and sands deposited under higher energy fluviatile conditions. The erosion modeling and test excavations suggest that the main period of slope wash in the valley system was the medieval period and later, although this requires confirmation through a future dating programme.

This picture of substantial zones of long-term grassland with woodland in some areas and more extensive woodland in others, and relatively little soil disturbance throughout the prehistoric periods is clearly supported by the evidence from the excavations, the subject of the next chapter. The implications of all these results are further developed in Chapter 4.

Chapter 3

Studying Land Use from Monuments: Archaeology and Environment on Bottlebush, Wyke and Gussage Cow Downs

Charles French, Helen Lewis, Martin Green, Michael J. Allen, Rob Scaife, Julie Gardiner & Edgar Samarsundera

In terms of visible monuments and settlement/field system areas in this landscape, there are marked concentrations to either end of the study area, on Wyke and Gussage Cow Downs to the northwest and around the Knowlton henge complex to the south (Figs. 1.2, 2.17 & 3.1). This is undoubtedly a skewed picture as Martin Green's recent discovery and excavation of several sites in the intervening area has demonstrated (Green 2000, and see below). Overall, the project's sample investigations of nine prehistoric sites are complimented by a record of over 25 excavations already undertaken in the same study area within the past two decades (Green 2000) (Fig. 1.2), of which several are reported on below. The most important excavation data is briefly summarized below, essentially organized from south to north across the western end of the study area, with a new survey and test excavations in the Knowlton area described in the following chapter.

Gussage Cow Down

Trenches were excavated into two Early Neolithic long barrows with the aim of investigating pre-Neolithic buried soil and ditch fill evidence. In addition, the Gussage Cow Down Iron Age bank and ditch system was investigated to gain information on the Iron Age and later landscape.

Gussage Cow Down long barrows 294 and 78

Charles French & Helen Lewis

The mound of long barrow 294 on Gussage Cow Down is trapezoidal in shape, with a wider eastern and narrower western end, and has two straight-sided ditches to either side, approximately 6 m in width and some 2.15–2.5 m deep, a typical Wessex bayed long barrow (Figs. 3.1–3.4). The wider, eastern end of the mound has been disturbed by a modern fence, hedge line and farm track. Forty-three years previous to our excavations, Grinsell (1959, 79) recorded that the mound stood some three feet (*c*. 90 cm) high, 210 feet long (*c*. 63 m) and 85 feet wide (*c*. 25.5 m). Today, the mound survives to a similar length and width, but there is no surviving mound left, only a slight (*c*. 25–35 cm high) contour rise of the higher chalk subsoil surface beneath the modern ploughsoil where the mound would have been.

Both ditches, each some 65 m in length, revealed similar stratigraphic sequences in the 2 m slot excavated in each of the extant ditches (Fig. 3.4). A thick deposit (*c*. 40–80 cm) of weathered chalk rubble from the upper edges of each ditch had accumulated directly on the flat base of each ditch. There was no primary soil or organic accumulation on the bases of the ditches, suggesting that the ditches began to infill almost as soon as they were dug. Given the extremely small percentage of the ditches that was actually excavated in these trenches, it is significant that four 'caches' of flint were found in primary levels (Fig. 3.5). These discrete flint assemblages comprised *c*. 200–700 freshly struck flints each, with large blades, bladelets, waste flakes and fine spalls, all suggestive of primary production flakes. Similar 'nests' of flint were found in the base of the ditches at every 6 feet at the Thickthorn Down long barrow at the southern end of the Dorset Cursus (Drew & Piggott 1936). Above this primary activity was a standstill horizon composed of a thin (<15 cm) layer of organic silt loam with fine chalk rubble. This initial infilling and stabilization sequence follows the observed scenario in the Overton Down experimental earthwork, which by implication suggests that this stage may have

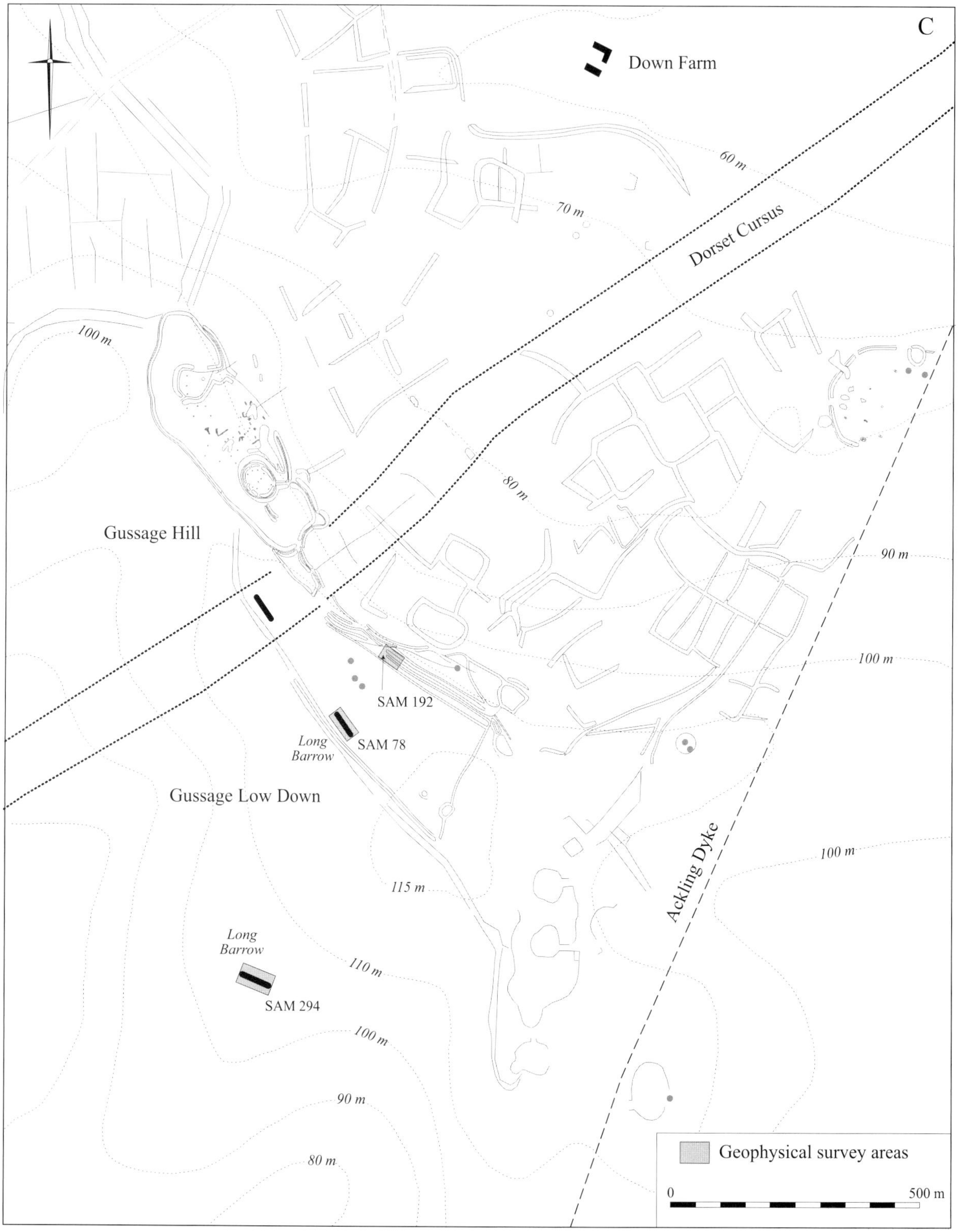

Figure 3.1. *Location map of the sites investigated on Gussage Cow Down (SAM sites 78, 192 and 294) set against the earthworks of the Dorset Cursus, the Gussage Down Iron Age complex and the Roman Ackling Dyke (Fig. 1.1:C) (C. Begg, after Bowen 1990).*

taken only six years, and certainly no more than 30 years (Bell *et al.* 1996, figs. 7.5 & 14.2). The subsequent secondary fill is composed of a dark brown silt with frequent angular flint and chalk rubble, the latter becoming more abundant towards the top of the secondary fill. This material probably represents mound material that was physically dragged off the monument through some disturbance activity, apparently in prehistory. The tertiary fills comprise a greyish-brown silt loam with abundant fine chalk fragments with a few Romano-British and medieval pottery sherds. This represents the base of a rendzina soil developed under pasture in the historic period. Finally, the modern ploughsoil with common angular flint and chalk cobbles completes the fill sequence; this represents further destruction of the remaining mound since this monument was taken into cultivation from World War II onwards.

In thin section, the organic fills contain frequent rock fragments, and show evidence of earthworm activity, mixing and rooting. Rare very tiny clean clay fragments attest to the presence somewhere nearby of clay-rich horizons (soil horizons formed under forest, or clay geology); the clay appears most similar to that seen in soil Bt horizons. The ditch fills show some earthworm sorting (rock fragments sorted by size), although this is imperfect, and frequent chalk rock fragment inclusions. Rare micro-charcoal is seen, although there are rare chalky soil aggregates with more frequent charcoal inclusions; these charcoal fragments are very fine and linear, similar to charcoal seen in trampled surfaces on other sites. This is seen at the primary–secondary fill transition, and suggests some of the ditch fill originates from an activity area nearby.

Figure 3.2. *View of the Gussage Cow Down long barrow 294 under excavation from the south (C. French).*

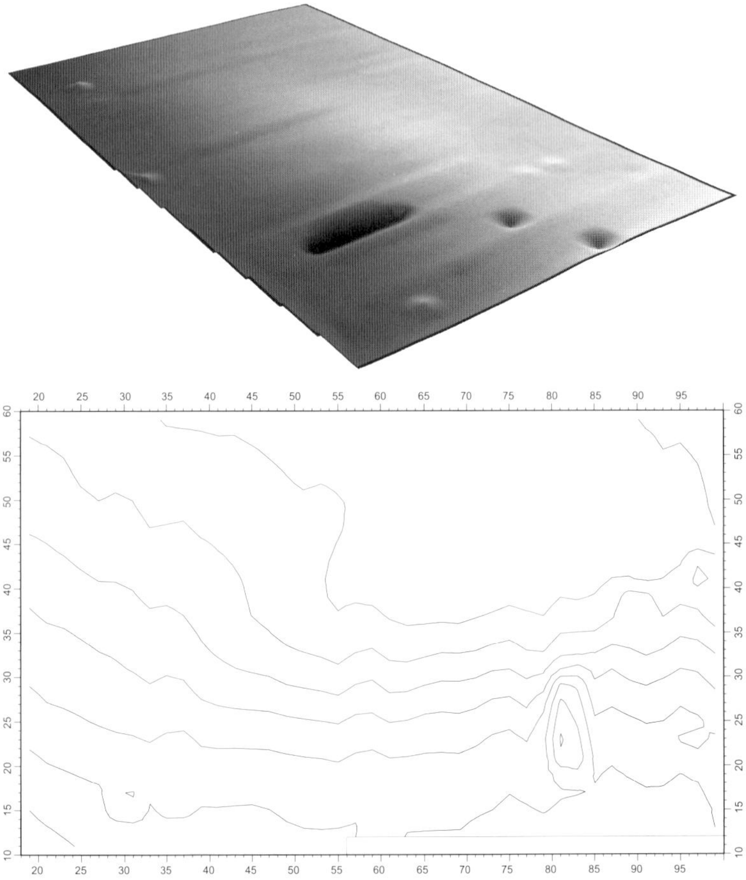

Figure 3.3. *Three-dimensional model and topographical plot of the Gussage Cow Down long barrow 294 (D. Redhouse).*

Gussage Cow Down long barrow 78 (Figs. 3.1, 3.6–3.9) is much better preserved. It is about 50 m in length, *c.* 22 m in width and survives to a height of about 4 m above the modern ground surface. It is surrounded by a partially infilled, horseshoe-shaped ditch, and is considered to be a bayed long barrow of the Thickthorn Down type. Augering and topographical survey suggest that this ditch is *c.* 6–8 m in width, over 4m in depth, and at least 1.7–2 m of ditch deposit still survive *in situ.* The northeastern end of the mound was severely affected by 1930s quarrying for chalk. Nonetheless, the geophysical survey (magnetometer and resistivity)

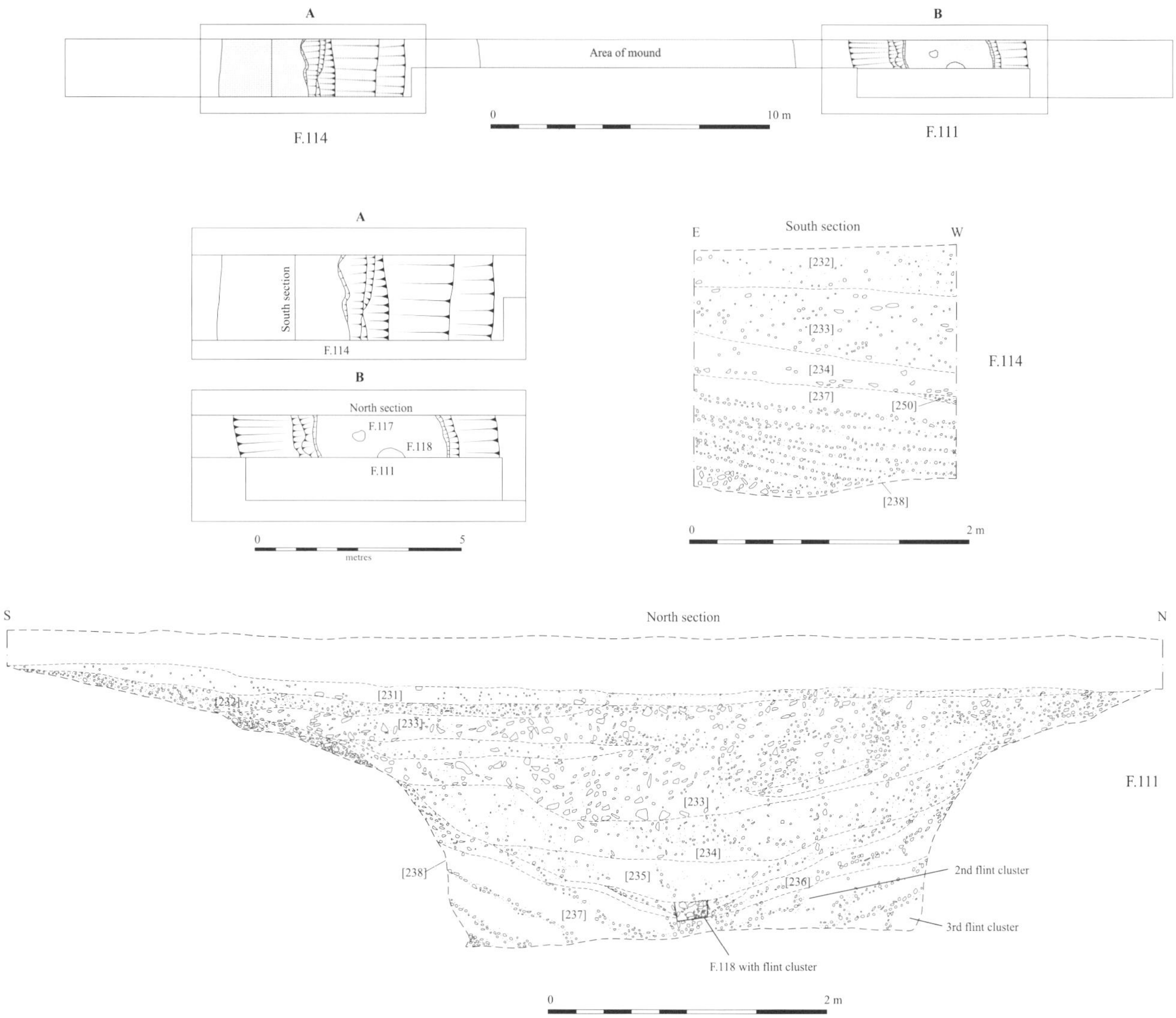

Figure 3.4. *Plan and sections of the test excavations of the Gussage Cow Down long barrow 294 (M. Brudenell after H. Lewis & C. French).*

suggests that linear anomalies along the long axes of the mound could represent constructional features.

A sample trench through the damaged eastern area of the mound revealed a full stratigraphic sequence through two superimposed chalk rubble mounds each with chalk block revetments or footings (Figs. 3.8 & 3.9). These were located directly on a thin (<18 cm thick) but well-preserved rendzina soil profile composed of a very dark brown to black, stone and chalk-free, silty clay loam turf line developed on dark greyish-brown silt loam with fine chalk fragments (lower A

Figure 3.6. *The Gussage Cow Down long barrow 78 from the north (C. French).*

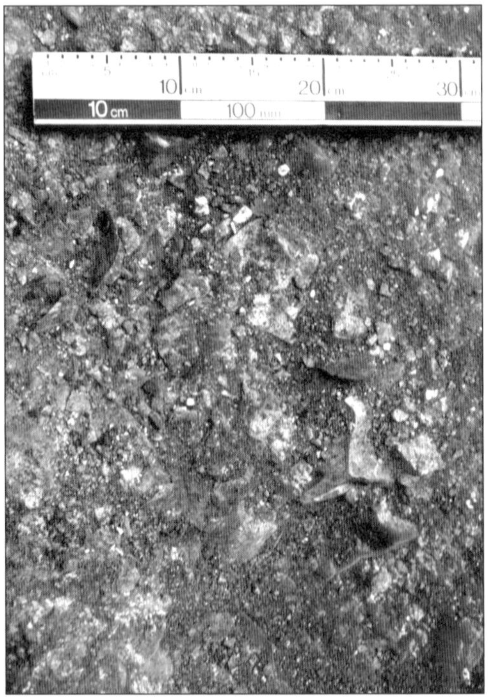

Figure 3.5. *Flint cache in the primary fill of the Gussage Cow Down long barrow 294 (C. French).*

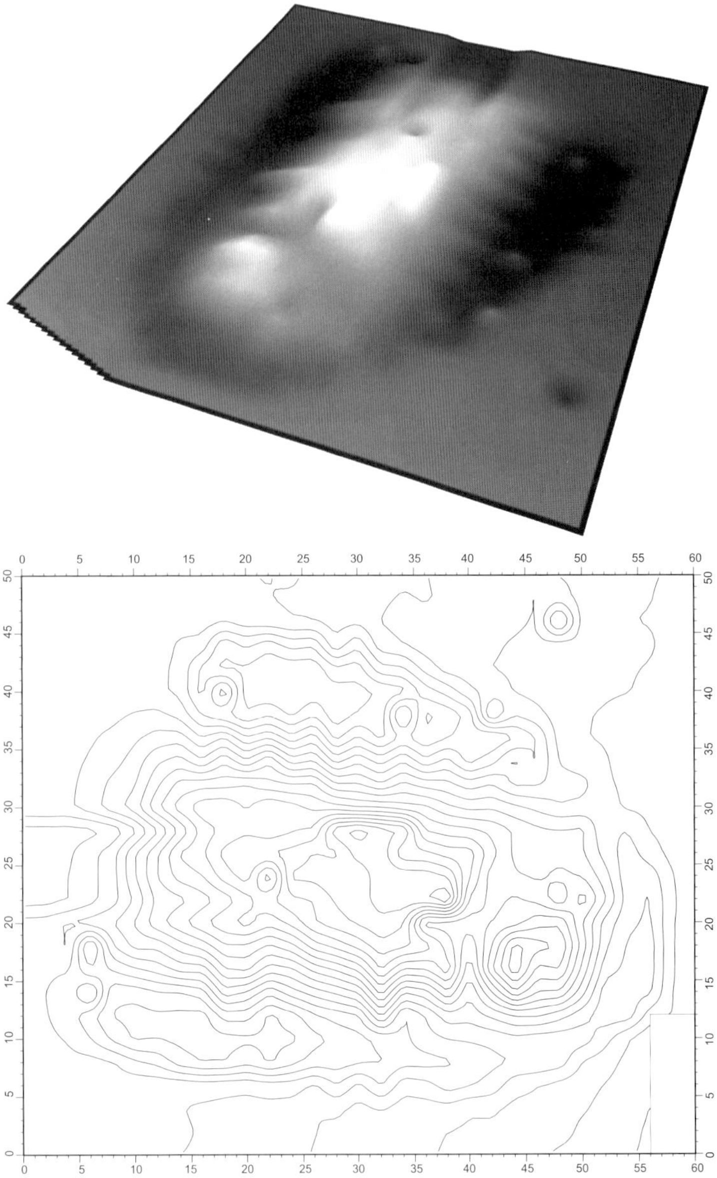

Figure 3.7. *Three-dimensional model and topographical plot of the Gussage Cow Down long barrow 78 (D. Redhouse).*

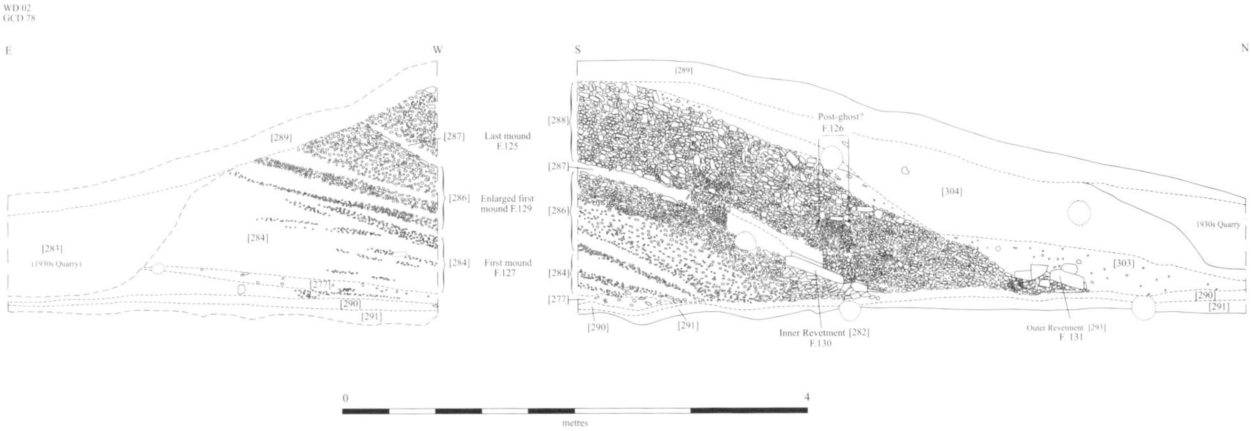

Figure 3.8. *The mound profiles of the Gussage Cow Down long barrow 78 showing the two phases of mound construction and the early twentieth-century quarry disturbance of the eastern end of the mound (M. Brudenell after L. McFadyen & C. French).*

Figure 3.9. *The main mound profile of the Gussage Cow Down long barrow 78 as excavated with the thin turf line and rendzina defined at the base of the chalk rubble mound (C. French).*

and weathered B/C horizon) (Fig. 3.9). Very fine iron mottling seen in the buried soil thin sections suggests fine rooting; similar types of mottles are often found in the upper layers of grass-covered soils and in turves used in barrows, and suggest that at some stage this was a grassland soil. A turf sampled from the mound had a similar nature, but was slightly more sandy. Only rare micrite-rich clay coatings were seen microscopically, with no further clay pedofeatures and a crystallitic groundmass; this grassland soil appears to have been a stable, shallow calcareous rendzina for a very long time before the long barrow was built. This shows that this locality had been unforested for quite some time before the Early Neolithic construction of the long barrow.

A comparative structural discussion of the Gussage Cow Down long barrows

Lesley McFayden

Double construction

Before setting out a comparative structural discussion of the Gussage Cow Down monuments, it is important to note the association of many long barrows in pairs. The Gussage Cow Down long barrows 77 and 78 were constructed in close proximity to each other and similarly the Thickthorn Down long barrow, mentioned later in the text, existed as one of a pair.

Mound architecture

The Gussage long barrows 77 and 78 were of 'Cranborne Chase type' (after Ashbee 1970, 15; Barrett *et al.* 1991a, 36) and demonstrate similar structural complexity to the Thickthorn Down long barrow at Gussage St Michael (excavated by Drew & Piggott in 1933: see Drew & Piggott 1936) and Wor Barrow on Handley Down (excavated by Pitt Rivers in 1895; see Pitt Rivers 1898).

The mound of Gussage long barrow 78 exhibited two phases of construction. The first mound was made up of thin layers of fine, compressed chalk rubble and redeposited turf. Its outer edge was retained by a large chalk block and flint nodule-lined footing. Two supporting posts were recorded in the make-up of the mound on the inside of the footing, and a cluster of worked flints was found within the upper surface of the buried soil just to the outside of this revetment. A larger and higher second mound was built of large block chalk material, and this incorporated the earlier mound into its construction. It also had a revetment at its outer basal edge composed of chalk blocks.

In their re-assessment of Thickthorn Down long barrow, Bradley & Entwistle (1985, 174) identified a series of vertical breaks in the stratigraphy of the longitudinal section of the long barrow from both drawings and photographs. They argued that some kind of shuttering, panelling or post-revetment would have had to be in place in order for materials to have kept their distinct vertical form. There were nine breaks on the long axis of the barrow and these were recognized in lower and higher parts of the mound (i.e. evidence for an axial divide). No vertical distinctions were identified in the transverse section. In contrast, vertical distinctions were recorded in the transverse section of Gussage long barrow 78 (Fig. 3.8), and in two cases there was the ghost or void of a supporting post. These distinctions were evident in low and high parts of the mound construction. There is also evidence from the longitudinal section that materials were built inwards, that is the tip lines in the mound construction went against the grain of the natural contour. This is significant as it indicates that the upcast mounds of Thickthorn Down and Gussage Cow Down long barrow 78 were detailed and complex constructions and did not show straightforward dumping from the central core of the barrow outwards.

This complexity of the build is very similar to the turf, chalk and wood structures of the Beckhampton Road and South Street long barrows in northern

Wiltshire (Ashbee *et al*. 1979). The use of fresh, large chalk block material in the later phase of the architecture at Gussage long barrow 78 is also comparable to the upper and outer areas of the Easton Down long barrow in Wiltshire (Whittle *et al*. 1993). The use of the fresher, blockier chalk in these parts of the build cannot simply be explained in terms of this material having been extracted from the deeper (and so it is argued later) parts of the surrounding ditch. Firstly, long barrow ditches were cut as a series of inter-connecting pits and so block chalk would have been encountered at many times during the construction process. Secondly, Barrett has argued that care was taken over the kinds of material that were used in creating outer appearances to round barrows (Barrett 1994), and I would argue that there is a similar dynamic at work during particular points in the history of long barrow construction.

It is interesting to note that there were further similarities between internal features and the kinds of material used in their construction, there was an earlier palisade trench filled with flint nodules at Wor Barrow (Pitt Rivers 1898, 65) and a flint nodule-lined footing at Gussage long barrow 78.

Pit-cut ditches

In the photographs of Thickthorn Down and Wor Barrow (Drew & Piggott 1936, pl. xx; Pitt Rivers 1898, pl. 253), the ditches are very sinuous and organic in their shape. The butt ends were markedly circular and distinct as if these areas were constructed through the cutting of discrete pits. It is possible that the ditches were in fact strings of pits, as with the Windmill Hill causewayed enclosure (Whittle *et al*. 1999) and the Beckhampton enclosure (Gillings *et al*. 2002). Indeed, Drew and Piggott wrote that: 'The whole appearance was closely comparable to the ditches of the Neolithic causewayed camps' (Drew & Piggott 1936, 81).

Gussage long barrow 294 was 63 m in length and 17–25 m in width and the long axis was orientated in an east–west direction. It was flanked by two ditches on either side. On excavation (French & Lewis, above and appendix) it was found that the visible 'mound' corresponded to a slight contour rise of the higher chalk subsoil surface beneath the modern ploughsoil, the mound material having been ploughed out. There was, however, evidence that both ditches had been cut as a series of inter-cutting pits as within the base of the southern ditch three interconnecting pits could be identified. In the longitudinal section of the southern ditch (Fig. 3.4), there were contrasting tip lines (west to east and east to west) within the primary fill and these connected to the circular cuts that were remarked upon in the base. The latitudinal section across the northern ditch indicated a pit (F.118) cut into the primary fills and containing a cluster of worked flint. There was a second cluster of flints located in contexts 236/7 approximately 1m to the east of F.118 which appeared to be also cut through the primary fills. A third cluster of worked flint was located against the northern edge, at the base, where there looked to be a circular cut to a pit within the base of the ditch at this point (Fig. 3.5).

Material culture

Four clusters of worked flint were found in primary levels of the ditches of barrow 294 (French & Lewis, above and appendix). Two clusters were located directly on the base of the ditch, one on the northern side of the northern ditch and the other in the centre of the southern ditch. The other two clusters of worked flint were recorded from pits which cut through the primary chalk rubble fill in the central zone of the northern ditch. In this volume, Julie Gardiner has reported (below) that in terms of raw material the flint was very fresh and of good quality. Over 43 per cent of the total assemblage was recovered from the primary silts and consisted entirely of waste material from knapping, including many small chips and spalls. Gardiner has written that 'the material appears to be a series of deliberate dumps of knapping debris made either during or very shortly after construction of the monument' (see Gardiner this volume). One of the pits (F.118) which cut into the primary silts produced 508 struck flints including five cores, and again Gardiner considers this material 'to be a deliberate dump of knapping debris'. Drew and Piggott stated that six 'nests' of worked flint were recovered from the bottom of the ditch at Thickthorn Down, and that there was a further 'nest' in the primary silt (with an average of 82 pieces of worked flint in each 'nest') (Drew & Piggott 1936, 82). They recorded that: '... it seems fairly certain from the number of cores found, and the "nests" of flakes on the bottom of the ditch, that the knapping of the bulk of the flints found was done on or near the site' (Drew & Piggott 1936, 91).

Pollard has argued that these clusters of worked flint were not evidence of single working areas but that these materials created a network of 'groups of debitage' within long barrow architecture (Pollard 1993). In particular, he noted the regular spacing of the deposits within the primary ditch fills at Thickthorn Down (Pollard 1993), and commented that these areas of flint-working were in association with deposits of other kinds (in particular deposits produced from the butchery of animal bones and working of antler). For example, groups of debitage on the base of the ditch at

Figure 3.10. *The Iron Age triple ditch and bank system of Gussage Cow Down site 192 showing the outer bank, middle ditch and middle bank (C. French).*

Thickthorn Down occurred alongside deposits of antler and worked chalk (Pollard 1993; Drew & Piggott 1936).

As with worked flint, Pollard has demonstrated a formal quality to the deposition of faunal assemblages within ditch contexts. This is important work, for as Pollard argues, with the exception of the work carried out by Thomas (1991), these contexts have received little attention since the intentional nature of their deposition is rarely appreciated. Pollard has also drawn attention to the small bone groups from ditch ends at the Horslip and South Street long barrows in north Wiltshire (Pollard 1993).

Pollard (1993) and Thomas (1991) have demonstrated that the working of flint and the butchery of specific animals within these areas of construction should not be ascribed a secondary status (*contra* Piggott 1954), or, as Pollard has noted, an assumption that the inclusion of material culture within these areas was some kind of equivalent practice to the deposition of human remains (*contra* Bradley 1984; Barrett *et al.* 1991a; and discussed in Pollard 1993).

Whilst it is important to give dimension to this material culture, and to presence these materials in dynamic ways, I am concerned that in specifying exclusively a practice of deposition, these accounts ultimately remain complicit with the dominant accounts they are attempting to subvert. For encounters are expressed here only in terms of flint and animal bone (material culture) being placed in long barrows (architecture). My problem is that by dividing deposition from construction we seem to have lost sight of the specific material and historical conditions that were an intimate part of the context of this past activity.

Material culture as architecture

Why is the working of flint and the butchery of particular animals separated from the constructional process? Both Thomas (1991) and Pollard (1993) have spent a long time arguing for a more inclusive understanding of prehistoric productive technology, as a media that incorporated all aspects of social life. Why should this technology, or ways of working, be separated from the constructional process and only be reintegrated through the medium of depositional activity? By figuring freshly worked flints as an event in construction, these complex assemblages invite us to take flint working seriously as a productive activity that was a part of construction.

If the Thickthorn Down and Gussage Cow Down long barrow ditches were created through a series of inter-cutting pits, then it is more likely that conditions were messy and complex. For example, at these sites, as one pit was cut, the activity involved in constructing it created spoil which partly entered into the previously excavated pit. The action of pit cutting was a part of architectural practice. If we consider the conditions that would have been negotiated in creating successive pits, no one example would have been sculpted to squeaky clean chalk and have maintained that pristine state for material culture to then have been deposited in to it. Rather, within these areas of construction, where pits were being cut, spoil was being thrown backwards into previous cuttings and slump was occurring, flint was being knapped and the material incorporated as part of an ongoing assemblage of activities.

Gussage Cow Down Iron Age bank and ditch system

Charles French

The east–west oriented triple bank and ditch system sampled is part of the extensive Gussage Cow Down Iron Age settlement complex (SAM 192), which is located just below the top of the down overlooking the upper Allen valley (Fig. 3.1). It is associated with an extensive lynchet and field system on its northern,

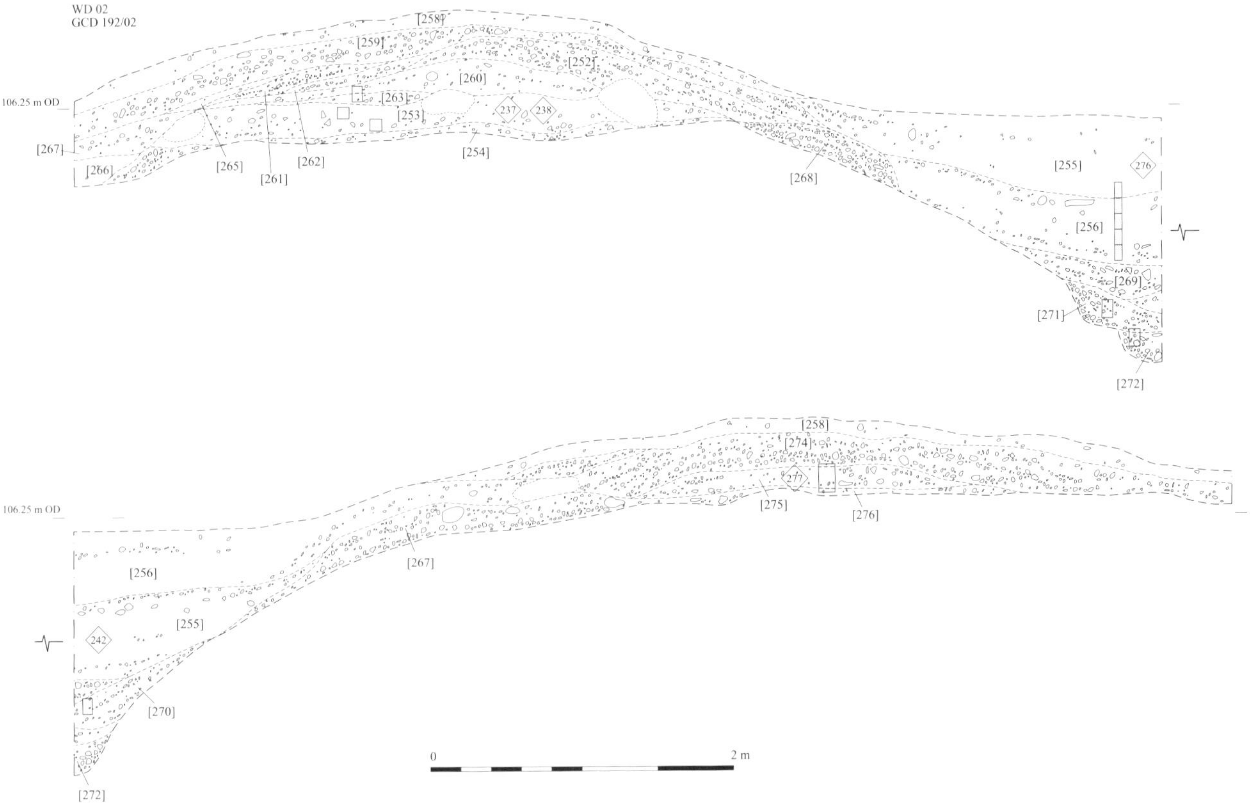

Figure 3.11. *Section through the northern Iron Age bank and ditch system of Gussage Cow Down site 192 (M. Brudenell).*

downslope side, which is all oriented at right angles to the much earlier Dorset Cursus (Fig. 3.1).

Trenching revealed the following sequence: a shallow ditch, a well-preserved bank with upcast from both sides constructed on a complete *in situ* rendzina soil profile, a deep V-shaped ditch with a thick, homogeneous, pale brown calcareous silt secondary fill, slight upcast bank preserving a thin rendzina soil profile, and a *c.* 6 m wide, flat, linear zone which was possibly a green lane, and a very slight upcast bank (Figs. 3.10 & 3.11). The silt-dominated fill of the northern, V-shaped ditch was very well sorted and homogeneous, suggesting a windblown origin from bare topsoils from adjacent chalk downland.

The important points are the presence of the rendzina beneath the upcast banks and the probable windblown material in the largest ditch. In contrast to the rendzina beneath the nearby long barrow 78, the rendzina beneath the Iron Age bank is much more depleted of organic matter and less structured, but is of a similar thickness (*c.* 15–20 cm). Thin-section analysis (sample 273) shows a subangular to angular blocky clay loam, earthworm and mite turbated but not well sorted, with inclusions of chalk, shells and flint. Although organic inclusions are low, there is occasional micro-charcoal; indeed, one ped which appears to come from elsewhere has 30 per cent shredded tiny linear charcoal inclusions. These are often found in contexts such as trampled floors, and must be introduced in this location, possibly from an activity area nearby. One thin section included a quartz sand-tempered pot fragment which appears to have microinclusions of 'grog' of a different fabric. The V-shaped ditch was almost entirely filled with dense, homogeneous, stone-free, brown silt loam. This may represent a wider phenomenon, but is more likely to indicate local aggradation of windblown material from nearby arable fields later in the Iron Age. Similar, but thinner, deposits of windblown material were found in the tertiary fill of the Neolithic enclosure ditch on Bottlebush Down (MUW 00) and in the lower tertiary fill of Bronze Age barrow 36 in the Wyke Down group (investigated in 2000 by this project) (French *et al.* 2000).

The later prehistoric or Iron Age sites tend to be just off the uppermost part of the down, especially on Gussage Cow Down to the south. This could have something to do with land organization at the time, these perhaps being the last areas to be fully cleared, or could have a defensive aspect, or could suggest that these were already the most marginal areas in terms of micro-climate and agricultural use and therefore used for settlements so as not to occupy relatively better land elsewhere on the downs, and/or some combination of these and other unknown factors.

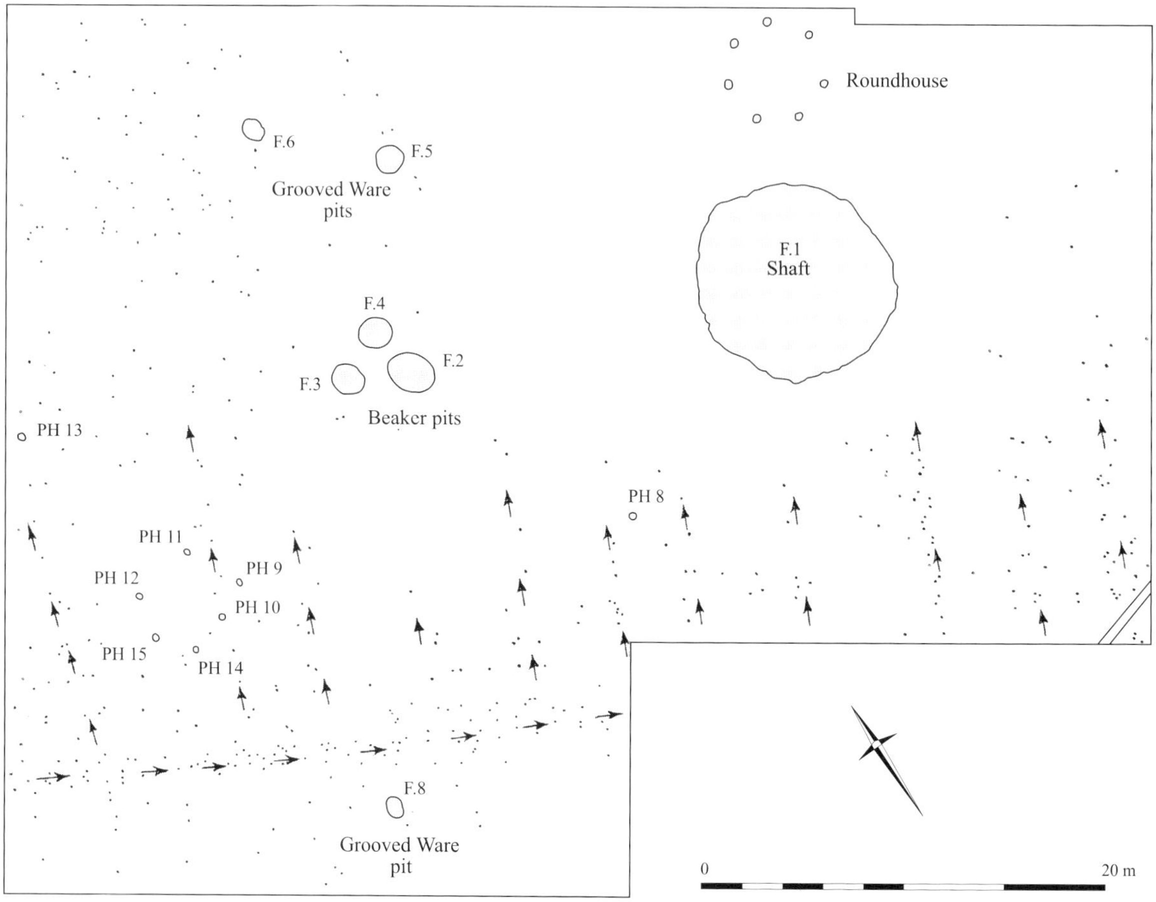

Figure 3.12. *Fir Tree Field shaft: location plan of the shaft and the associated pits, postholes and round house (N. Griffiths after M. Green).*

Down Farm
A number of small assessment/test excavations were undertaken of sites around Down Farm to augment the auger survey and obtain further palaeo-environmental data, including targeted fieldwork in Fir Tree Field, Gussage Down and Endless Pit (Fig. 1.2). The ditch of the Middle Bronze Age enclosed farmstead (or Down Farm) (Green 2000) was sectioned by hand for molluscan sampling and analysis, a selection of three 'Celtic field' lynchets on Gussage Down were trial trenched by machine to check for the presence of buried soils, and a 35 m trial trench was placed across Endless Pit at the head of the upper Allen valley by Martin Green.

Fir Tree Field shaft (see Appendix 4.2)
MARTIN GREEN
Summary
This remarkable structure was excavated to a depth of 13.2 m between 1992–94 with augering suggesting the base lay some 12 m beyond, below the current water table. During 1995–96 some 2000 square metres were stripped around the periphery of the shaft revealing further features (Figs. 3.12–3.13).

Two interim reports were subsequently published (Green & Allen 1997; Allen & Green 1998). The shaft was created by the collapse of an underground cavern probably originally created by melt-water released at the end of the last Ice Age. From radiocarbon determinations we know the collapse did not take place until much later, in the fifth millennium BC following thousands of years of mainly, subterranean, erosion. It subsequently gradually infilled itself almost entirely by natural processes over the following 2500 years incorporating a range of artefacts and ecofacts from the later Mesolithic to the earliest Bronze Age. This provided the opportunity to sample an unprecedented environmental sequence which not only encompassed the Mesolithic/Neolithic transition but the Neolithic/Bronze Age one as well (Allen this volume). Excavation around the periphery of the shaft produced structures of Neolithic and Beaker date clearly indicating the importance with which the feature was regarded.

The shaft sequence
The suite of seventeen radiocarbon determinations (Allen & Green 1998; Green 2000, 152–5) included six within the main coarse rubble fill of the shaft with the remainder coming from the weathering cone which consisted of a series of calcareous silty clay loams. The six dates obtained from the shaft infill covered about 4 m of the infill from a depth of 7 m to 3.05 m spanning fewer than 500 radiocarbon years, with the probability distributions indicating a likely span of 200–300 years (Allen & Green 1998). Further they are very late, or transitional Mesolithic in date (4350–4100 cal. BC). The rate of infill indicates very rapid and constant accumulation. Eleven results were obtained from the weathering cone and this series produced one obvious outlier (the determination from the dog/wolf bone in layer 5, OxA-8008), otherwise a coherent set of results relating to their stratigraphic relationships was obtained (Fig. 3.14). The dates from the basal 2 m of the weathering cone (layers 8 & 7) all cluster within a maximum period of less than 350 years (4340–3990 cal. BC) at the Late Mesolithic/Neolithic transition. Worked flint from layer 7 was clearly of Mesolithic type and included a tight group of seven microliths (Fig. A4.3). These were found so close together that it is likely they entered the shaft in a hafted state as part of an offering or hunting loss. Significantly five of the microliths were of rod form, characteristic of the latest Mesolithic industries in Britain. This group is the latest dated example of such an industry yet recorded in this country (Barton & Roberts 2004).

The clustering of the first twelve radiocarbon samples (Allen & Green 1998) — that is late Mesolithic to Early Neolithic — is surprisingly tight and indicates relatively rapid sedimentation of both the shaft and lower weathering cone. The filling of the latter seems to occur as rapidly as the chalk rubble infill of the shaft. This might to an extent be exacerbated by both human activity, as evidenced by the increase of finds, and the possibility of gradual sediment subsidence as accumulation and infill took place. The rate of infill only decreases where *in situ* human activity is attested; i.e. the Early Neolithic hearth in layer 6b (average OxA-8009 and OxA-8010 as part of the same population equals 3990–3780 cal. BC) and the Peterborough Ware associated buried soil layer 5. The radiocarbon date OxA-7981 from layer 6a, taken from a pig bone, would appear to be too early. Certainly the layer produced a sherd of plain bowl pottery and a butt of a polished flint axe but also three microliths. The bone may possibly be derived from disturbance of the underlying layer 7. The radiocarbon sample from layer 6 did not date but the layer was clearly associated with several Peterborough Ware sherds. The overlying

Figure 3.13. *View of the Fir Tree Field shaft as excavated with the roundhouse in the background (M. Green).*

buried soil layer 5, represents a major slow-down in the rate of infill and produced three radiocarbon dates. With the exception of the obvious outlier OxA-8008, the two remaining results from this buried soil (OxA-7986, 3350–3030 cal. BC and OxA-8007, 3310–2910 cal. BC) are statistically indistinguishable at the 95 per cent confidence level. This is consistent with both the interpretation of this as a buried soil, and the occurrence of a high level of artefacts including fifty sherds of Peterborough Ware, one of the highest concentrations in the area.

The determination from a sheep/goat bone in layer 4 above the buried soil is also statistically indistinguishable at the 95 per cent confidence level from a sample (OxA-8007) in the latter; we cannot however, be confident that the sheep/goat bone dates layer 4 as it might be residual from the activity associated with the Peterborough Ware associated soil. A few sherds of Beaker found towards the top of layer 4 would suggest a later date for the accumulation. The

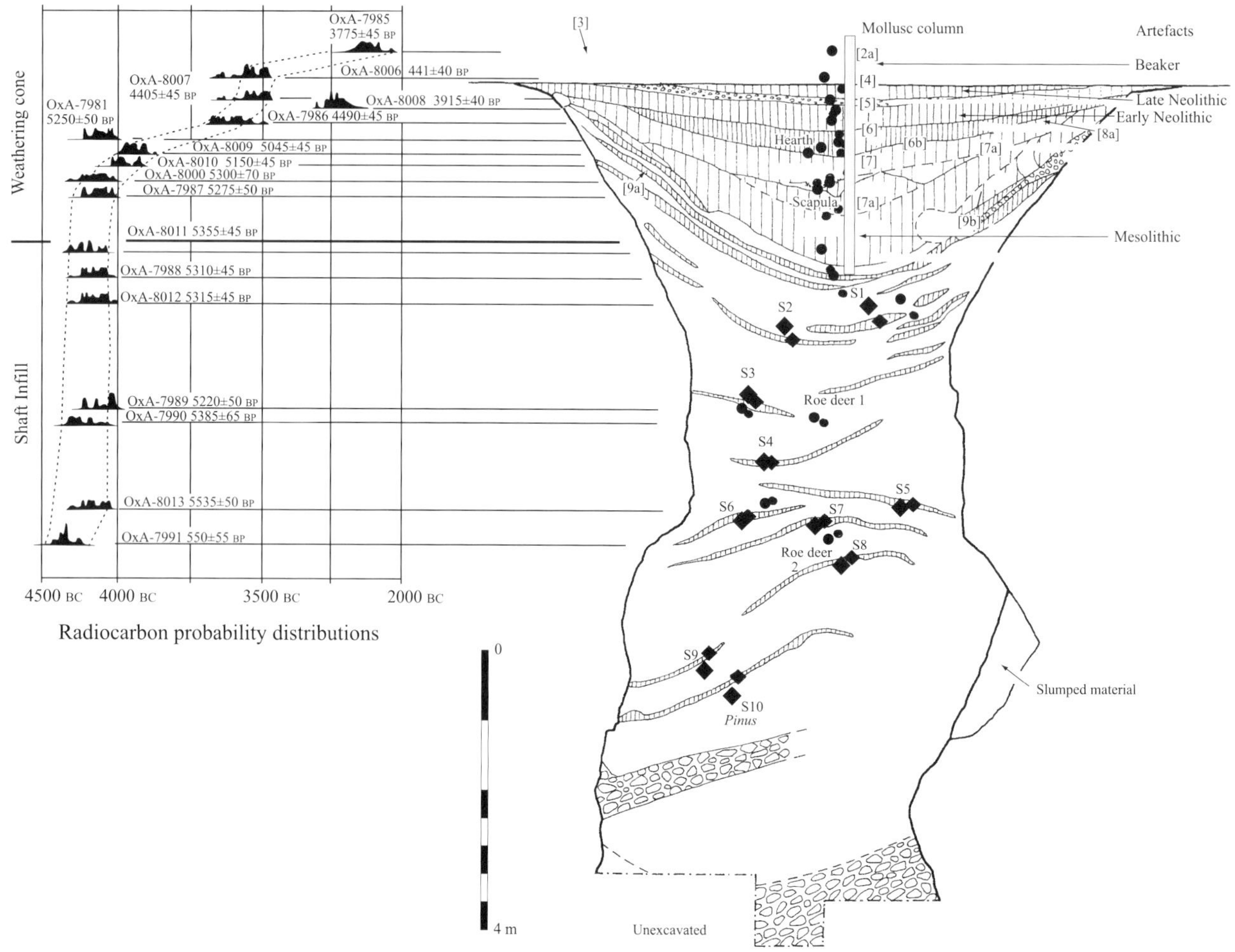

Figure 3.14. *Fir Tree Field shaft: section through the shaft and radiocarbon data (N. Griffiths after Allen & Green).*

succeeding layer 3 consisted of a spread of deliberately introduced chalk rubble, tightly rammed, and perhaps represents a symbolic capping or sealing of the shaft. Bracketed by Beaker sherds both above and below, this layer was clearly deposited in the Beaker phase and was most likely derived from three large pits found close by and to be discussed later. The final layer, 2, accumulates in the hollow left after the gradual settling of the shaft fills and contained 141 Beaker sherds associated with a radiocarbon date of 2460–2040 cal. BC (Cleal this volume).

The shaft environs

After cessation of work on the shaft, following exposure of the current summer water table, work for the next two seasons in 1995/6 concentrated on revealing over 2000 square metres around the periphery of the shaft (Figs. 3.12 & 3.13). This resulted in the discovery of a number of Neolithic and Beaker period features including a roundhouse, scattered postholes, stake-hole fence lines and pits (Figs. 3.12, 3.15–3.18). These will be discussed in turn.

Lying just 3 m to the northeast of the shaft a small roundhouse just under five metres in diameter and consisting of seven postholes was revealed. The wider gap between postholes 6 and 7 would suggest this was the entrance, opening to the southeast. Very few finds came from the postholes but included a probable sherd of Beaker (see Cleal p. 291). Although this sherd is insufficient by itself to convincingly date the roundhouse, it is significant that no pottery later than Beaker was found during the excavations, despite sieving 25 per cent of the plough soil overlying the roundhouse. This seems particularly noteworthy when a Middle Bronze Age settlement is known to lay only 140 m away (Barrett *et al.* 1991a). Allowing for an outer wall line perhaps 1.5 m outside the surviving inner post ring, the roundhouse would have been less than 2 m from the shaft. However, if we assume a Beaker date, the shaft would have been entirely filled by this stage and capped by the packed rubble. It seems likely that the rich assemblage of Beaker pottery (141 sherds) (see Cleal, App. 4.2) and almost a 1000 worked flints (see Green, App. 4.2) in the final fill of the shaft, were derived from domestic activity associated with this building.

Less than 14 m to the northwest of the shaft three large pits (Figs. 3.12 & 3.16), in a 'clover leaf' arrangement were uncovered. The largest pit in the group, F.2, was oval and approximately 2.25 × 1.75 × 1.30 m deep. The primary fill consisted of a layer of dark ashy soil rich in debris. Overlying this was a deposit of coarse chalk rubble subsequently overlain by a series of silty calcareous lenses. The pit was rich in artefacts, producing 210 potsherds, a bone pin (Fig. A4.1) and almost 500 worked flints.

The basal deposit was particularly rich and contained 59 per cent of the pottery and 70 per cent of the flint. The chalk rubble produced few finds with the secondary silts producing the bulk of the remainder. Amongst the pottery were sherds of a very fine Wessex/Middle Rhine Beaker found both in the secondary silts and the primary deposit. A substantial part of the same vessel was found in the nearby pit F.3 which was also oval and approximately 1.75 × 1.40 × 1.15 m deep. Cut in to the floor was a much smaller pit or posthole about 0.5 m wide × 0.4 m deep. Within the very loose chalky silt filling the feature were many amphibian bones and the very large sherd of the Wessex/Middle Rhine Beaker previously mentioned, placed vertically rim down with the other end projecting above the floor of the larger pit. The middle silting in this pit comprised bands of chalk rubble interspersed with dark silty lenses. This pit was also rich in artefacts producing 124 sherds, a bone pin (Fig. A4.1) and 253 worked flints. The basal fill, although of a different nature to that in F.2, also contained the largest quantity of the finds with approximately half of the pottery and flint artefacts.

The other pit in the cluster, F.4, was smaller and circular, 1.5 m in diameter by about 0.9 m deep. It had a basal deposit similar to that of F.2 but lacking the ash. Also rich in artefacts it produced 67 sherds, the tip of a bone pin (Fig. A4.1) and 520 flint artefacts including 51 scrapers (Fig. A4.5). Once again the majority of these were found in the basal deposit which produced 54 per cent of the pottery and 89 per cent of the flintwork.

The large size of these pits, their dating and proximity to the shaft, strongly suggests they provided the chalk rubble (layer 3) which capped the shaft. Their subsequent infilling with soils rich in artefacts and the structured deposit cut into the floor of pit F.3 would suggest they were consequently treated in a most careful and particular way.

Approximately 10 m to the north of this pit group two further pits (F.5 & F.6) were revealed (Figs. 3.12 & 3.15–3.16). Pit F.5, was circular, 1.4 m in diameter by about 0.8 m deep. It contained a similar basal fill to F.4 with a series of alternate rubbly and dark silty lenses above. Artefacts were much less abundant than in the pit group just described and yielded 18 sherds and 150 worked flints. Surprisingly the basal deposit produced only 17 per cent of the pottery and 7 per cent of the flint. Amongst the pottery were sherds of Peterborough Ware, Beaker and with the remainder undiagnostic. A few sherds of Peterborough Ware were found in each of the three pits in the cluster but were clearly residual to the Beaker-dominated assemblages. The sherds of Beaker found in this pit combined with the similarity of profile and filling, clearly dates this pit to the Beaker phase as well.

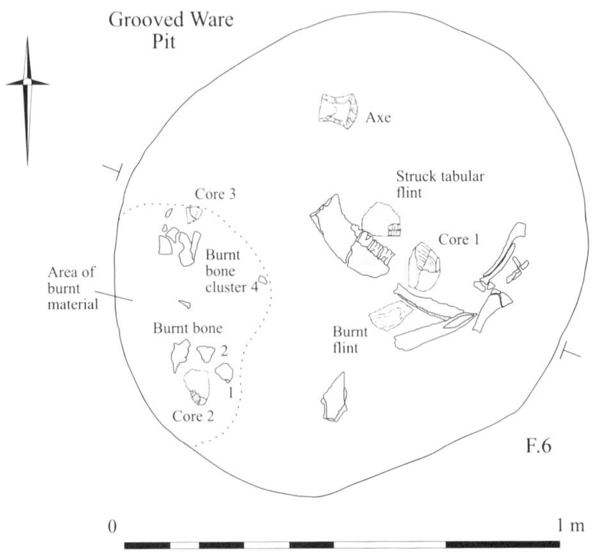

Figure 3.15. *Fir Tree Field Shaft: the Grooved Ware pit F.6 outside the shaft, showing the locations of the artefact clusters (N. Griffiths after M. Green).*

Pit F.6 which lay nearby was clearly of different character and date. Oval in plan 1.1 × 0.9 m and only 0.35 m in depth with a shallow bowl profile. The fill was primarily a homogeneous dark loamy textured soil, slightly ashy towards the base with little chalk rubble. Near the base a clear structured deposit of animal bones and worked flint had been deposited (Green 2000, fig. 44c). Amongst the former was a cow jaw and from the latter an extraordinary broken double splayed flint axe (Fig. A4.5, no. 1) of unique form. At a higher level in the pit a very crude flint axe (Fig. A4.5, no. 2) and three sherds of Grooved Ware were found. Some 30 m to the south of this feature a pit of very similar character was uncovered, F.8. This oval pit measured 0.95 × 0.75 × 0.42 m deep. It was principally filled with a dark homogeneous ashy textured soil with much small

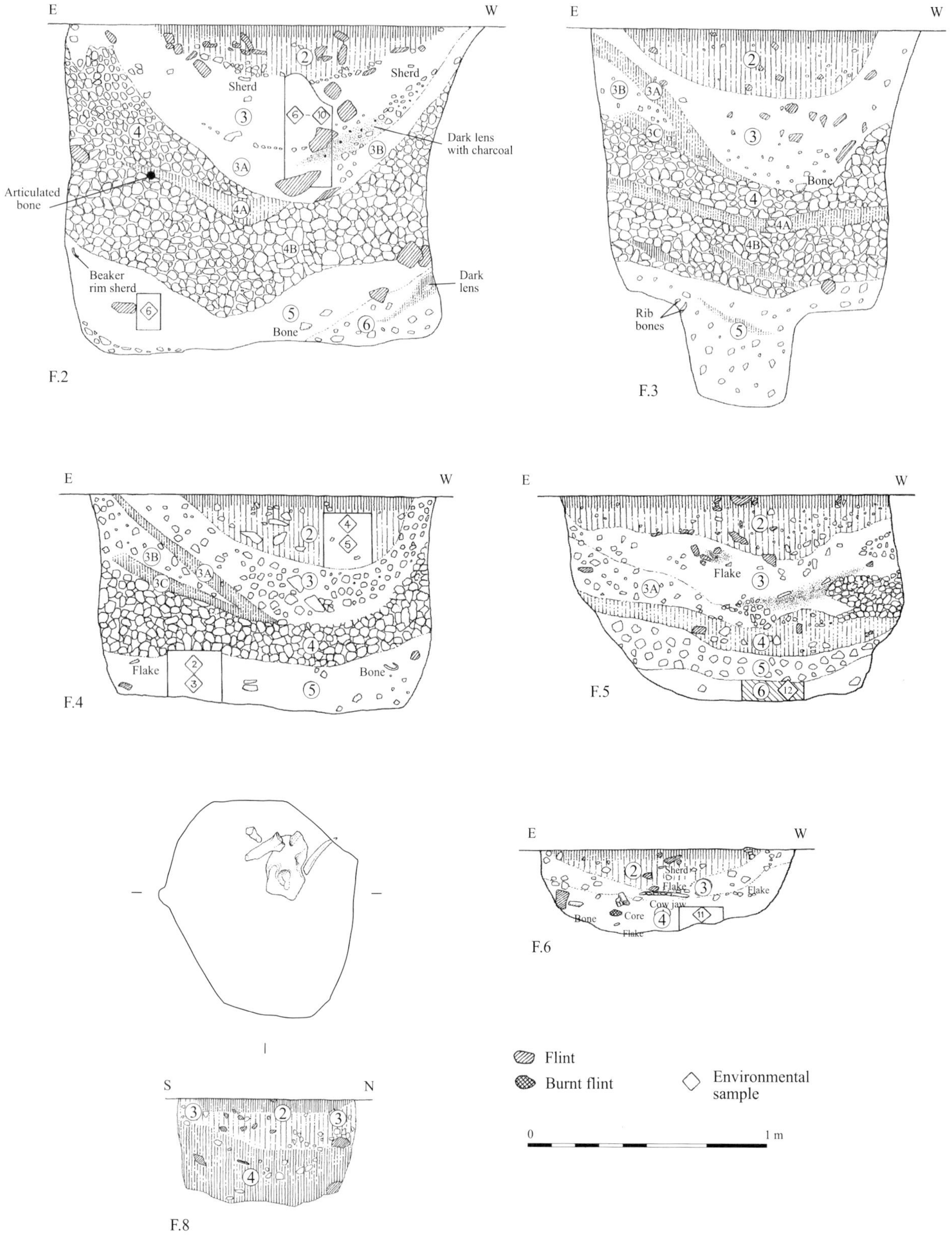

Figure 3.16. *Fir Tree Field shaft: Beaker pits F.2–F.5 and Grooved Ware pits F.6 and F.8 uncovered outside the shaft (N. Griffiths after M. Green).*

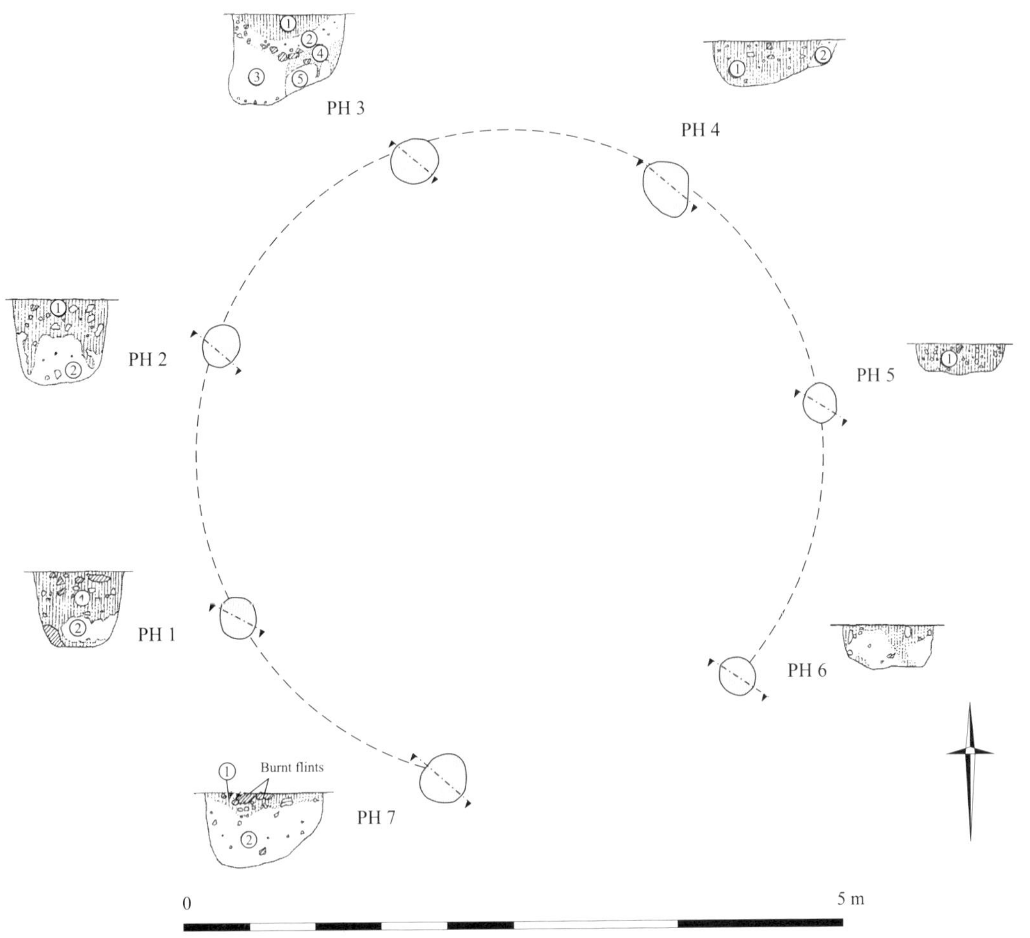

Figure 3.17. *Fir Tree Field shaft: the round house just to the northeast of the shaft and the posthole sections (N. Griffiths after M. Green).*

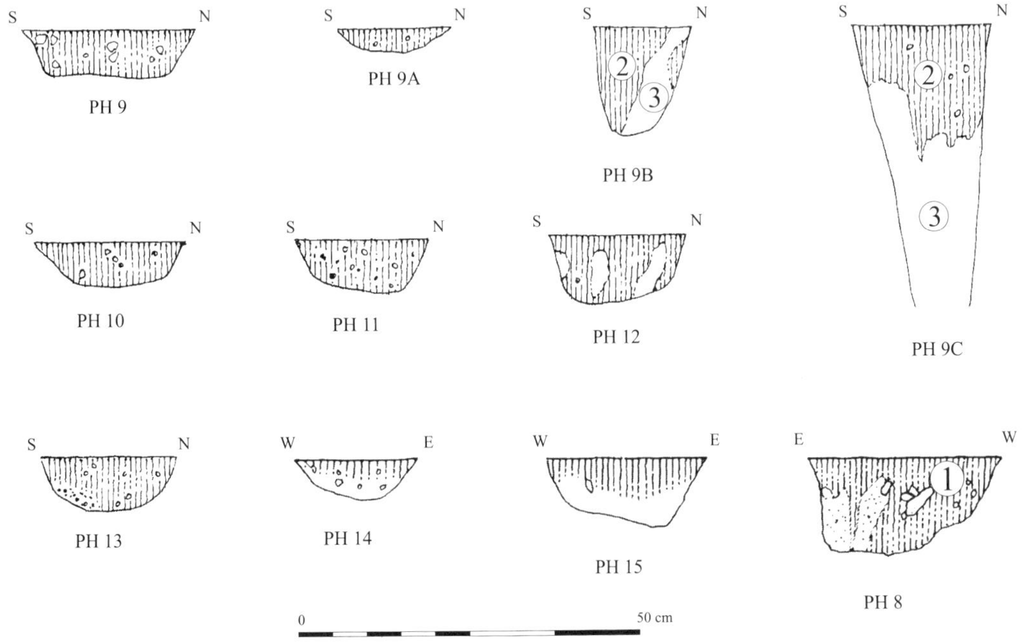

Figure 3.18. *Fir Tree Field shaft: additional posthole sections (N. Griffiths after M. Green).*

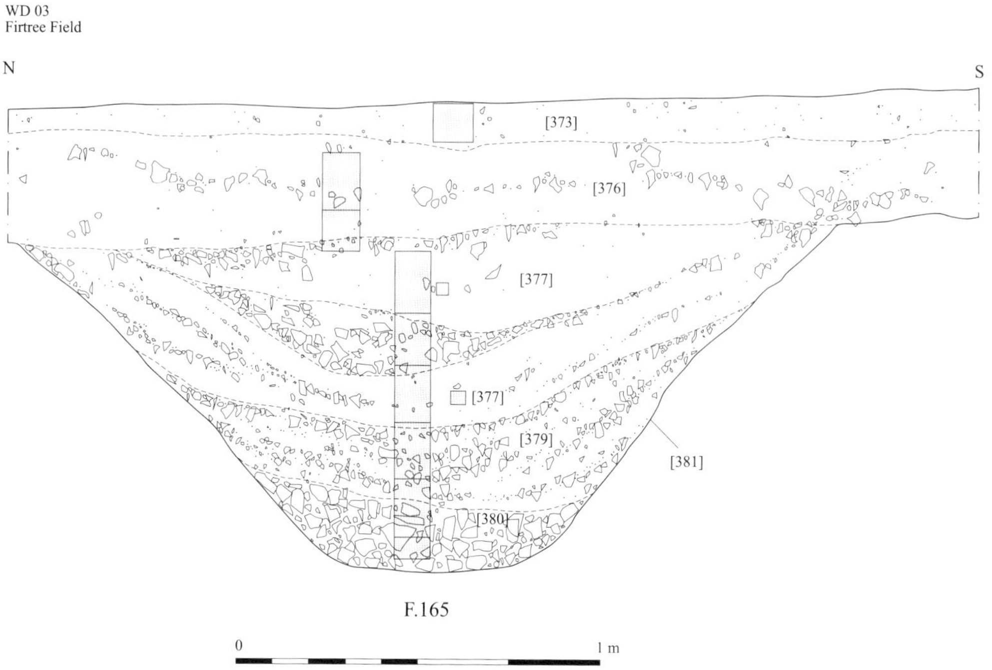

Figure 3.19. *Section through the Middle Bronze Age enclosure ditch in Fir Tree Field (M. Brudenell & F. Sturt).*

burnt flint and charcoal flecks. Carefully placed on the pit floor towards its western edge was a remarkable structured deposit (Green 2000, fig. 44b) consisting of a large flint nodule, a red deer antler tine and two cow bones. The nodule was clearly carefully selected from the nearby valley gravel deposits and was perforated by a number of natural openings. The juxtapositioning of the objects, particularly the nodule and antler tine, gave the appearance of a tusked animal skull, with the holes in the nodule simulating the eye sockets, nasal and spinal column openings. Further finds within this layer were sherds of Grooved Ware, a bone awl (Fig. A4.1) and rib knife and 571 worked flints including four serrated flakes. About 10 m to the north of F.8 was an oval setting of six degraded postholes (postholes 9–12, 13 & 14) some 4 × 5 m in diameter (Fig. 3.18). Producing no diagnostic finds it is undatable but its proximity to the pits makes it tempting to suggest it may be the remains of a Neolithic structure.

Pits F.6 and F.8 share much in common with the pits found underlying the Middle Bronze age settlement excavated only 140 m away to the southwest (Barrett *et al.* 1991a). Here pits of similar size, profile, and fillings contained a number of remarkable structured deposits. These included settings with antlers, boars tusks, stone axes, Grooved Ware, remarkable natural flints and a cow skull and antler (Green 2000, fig. 43 & 44a), reminiscent of the setting in F.8 just described.

Large numbers of stakeholes were recorded during the work on the shaft environs (Fig. 3.12), particularly to the south and many appeared to be part of much repaired fences. Just north of pit F.8, part of a spinal northwest/southwest fence was uncovered crossing the stripped area for a distance of 30 m. Set out perpendicular from this were a series of 11 parallel fences revealed in the stripped area, ranging in distances apart from 4–7 m. Remarkably, no stakeholes were found close to or crossing the shaft, particularly to the north around the roundhouse. This wide margin would suggest the fence builders were aware of the shaft, even if perhaps it was substantially or entirely infilled by this stage. It seems highly likely therefore that the fences were divisions constructed in the later Neolithic/Beaker phase.

Clearly this further evidence provided by the pits and fences, emphasizes the special way with which this sacred land next to the shaft and Cursus was treated in the third millennium BC.

Fir Tree Field Middle Bronze Age enclosure

(see Appendix 4.3)

Charles French & Julie Gardiner

Green excavated the Down Farm Middle Bronze Age settlement enclosure nearly thirty years ago but fortunately left a length of the enclosure ditch unexcavated for posterity (Green 2000, 106–8, fig. 73). This has enabled us to open up another section of the ditch to re-examine the profile (Fig. 3.19), obtain additional dating evidence and collect contiguous samples for molluscan analysis. In particular, the section exhibits a substantial standstill horizon at about the halfway mark which may equate to a period of soil accumulation and turf growth over the ditch and associated, former chalk rubble bank. The new molluscan data will be used to complement and corroborate the known land use history of this part of the upper Allen valley and to provide comparative data to all of the other near contemporary sites investigated in the near vicinity (Allen 2002a).

An assemblage of 37 worked flints was recovered from this enclosure ditch section. These consist of 34 flakes, 1 piece of irregular waste and 2 end scrapers. Most of the flakes are heavy and squat with numerous hinge fractures. The raw material is local, probably originating from the ditch itself. Most has been subject to plough damage though the six flakes, one irregular piece and one end scraper from the lower ditch fill above the basal chalk rubble are fresh and sharp. There are blade-like flakes among this material but it is all crudely struck using a hard hammer and is probably of Bronze Age date. The Ah horizon and upper ditch fills produced over 2 kg of burnt flint with only 50 g in the lower fill.

Gussage Down lynchets

CHARLES FRENCH

An extensive system of later prehistoric field boundaries remain as ploughed lynchets on the southern slope of the upper Allen valley in the vicinity of Down Farm on Gussage Down (Figs. 2.12 & 2.13). Although extensively plough damaged, three accessible lynchets were trial trenched by machine in the field to the south of Fir Tree Field on Gussage Down in order to assess their preservation and investigate any old land surfaces that might survive. Unfortunately, but as expected, these lynchets only survive as dispersed linear bands of slightly thicker and darker soil (+5 cm over a *c.* 5–10 m width). There was no buried soil or old land surface surviving.

Endless Pit, Thorney Down Farm

MARTIN GREEN, MICHAEL J. ALLEN & CHARLES FRENCH

Given the extraordinary palaeoenvironmental and archaeological information retrieved from the shaft in the northeast corner of Fir Tree Field (Green 2000; Green & Allen 1997; Allen & Green 1998), an examination of Endless Pit (Fig. 1.2), more or less at the head of the Allen valley, was long overdue. Accordingly, a multi-technique geophysical survey was undertaken by Paul Cheetham, School of Conservation Sciences, University of Bournemouth, which suggested that this was an embanked archaeological feature, possibly even a large pond barrow. This was then followed up by a trial trench excavation across its northwestern half and power augering to establish the infilling sequence.

This investigation suggested that it was not an archaeological construction as such but a substantial natural feature (cf. Green & Allen 1997) which has been open and infilling since at least late glacial times. The encircling 'bank/ditch' is a natural feature in the valley coombe rock, and is probably also sculpted by ploughing. Nonetheless, there do appear to be some windblown ash and charcoal deposits contained in the natural shaft itself, possibly associated with adjacent prehistoric land use, and a sequence of pollen samples was taken which it is hoped will provided further sub-regional vegetational history data to augment that from Monkton-up-Wimborne, some 4 km downstream (French *et al.* 2000; 2003; French & Lewis 2005).

Wyke Down

Wyke Down was the core area of the site investigations, with four round barrows and nearby settlement remains being explored and sampled.

Wyke Down henge 2 and the associated Late Neolithic settlement (WD2 96) (see Appendix 4.4)

MARTIN GREEN

This, the second henge to be revealed on Wyke Down was spotted as an indistinct crop mark from the air by Green in 1995 (Fig. 3.20). It lies on a south-facing slope at 70 m OD some 200 m above the floor of the Allen valley. Subsequently confirmed by geophysical survey undertaken by Steve Burrow (1996), this survey additionally revealed a causeway to the south. This faced towards the previously excavated henge Wyke Down 1 (WD 1) which lay only 40 m away. Excavation in 1996 confirmed the monument was a henge

Figure 3.20. *Oblique aerial view of Wyke Down henge 2 with henge 1 visible at the top right (P. Kitching).*

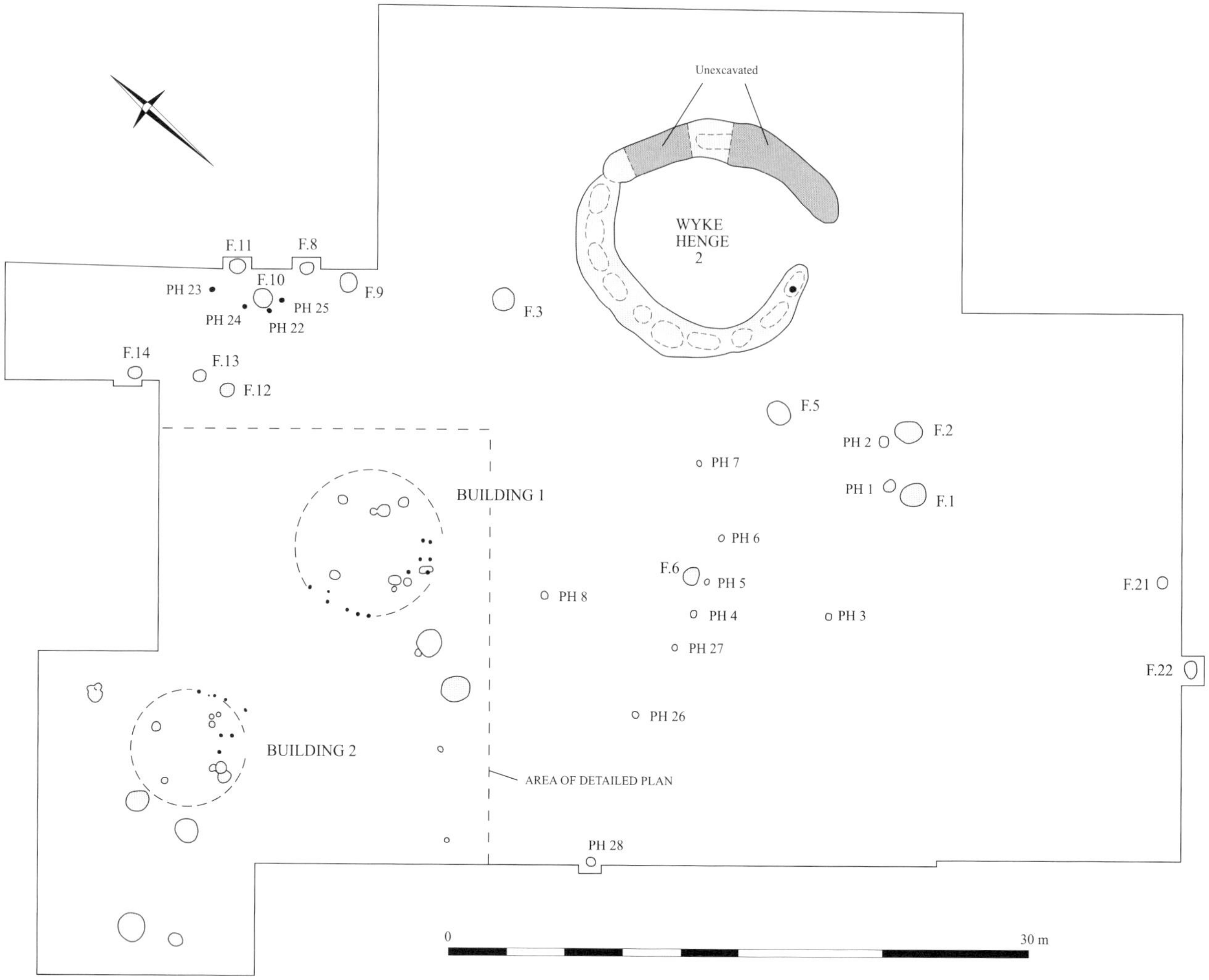

Figure 3.21. *Wyke Down henge 2 and associated structures and features (WD 2 1996): overall plan (N. Griffiths after M. Green).*

which had been constructed in two major segments — a semi-circular western side and a banana-shaped eastern side (Fig. 3.21 & A4.10). Where the two met at the northern end, a second narrow causeway was revealed just below the level of the natural chalk which gave a NNW/SSE axis to the monument. Both ditch segments had been constructed by digging a series of oval pits, the causeways for which survived to a maximum height of 0.56 m above the ditch floor, which itself was dug to a maximum depth of 1.75 m. The entirely excavated west side revealed the outlines of ten pits while the east side was left substantially unexcavated with only one trench bottomed to reveal a single causeway. Its maximum diameter from the outer ditch edges was 12 m.

Outside the monument to the west concentrations of pits and postholes led to an expansion of the excavation (Figs. 3.21 & 3.22). These features proved to be exclusively Neolithic in date, with many yielding sherds of Grooved Ware (Figs. A4.12–A4.19). Prominent within this area were two Neolithic buildings both associated with a pairs of pits. Charcoal from a post burnt *in situ* in a posthole associated with each structure allowed two radiocarbon determinations: House 1: 2900–2830 (28.5 per cent probability) and 2820–2670 (66.9 per cent probability) cal. BC (4203±33 BP; Wk-18751), and House 2: 2880–2570 cal. BC (4117±40 BP; Wk-18752). These two determinations would appear to overlap with each other, although house 2 maybe slightly later than house 1, and with the henge ditch itself, although it may be slightly earlier in terms of construction. Note that the limits to the excavation were arbitrary and in no way defined the full extent of this area of settlement activity.

The western section of the henge was completely excavated in a series of seven trenches (Figs. 3.21, 3.25, 3.26 & A4.10). The eastern side was left substantially unexcavated with only one of the three trenches,

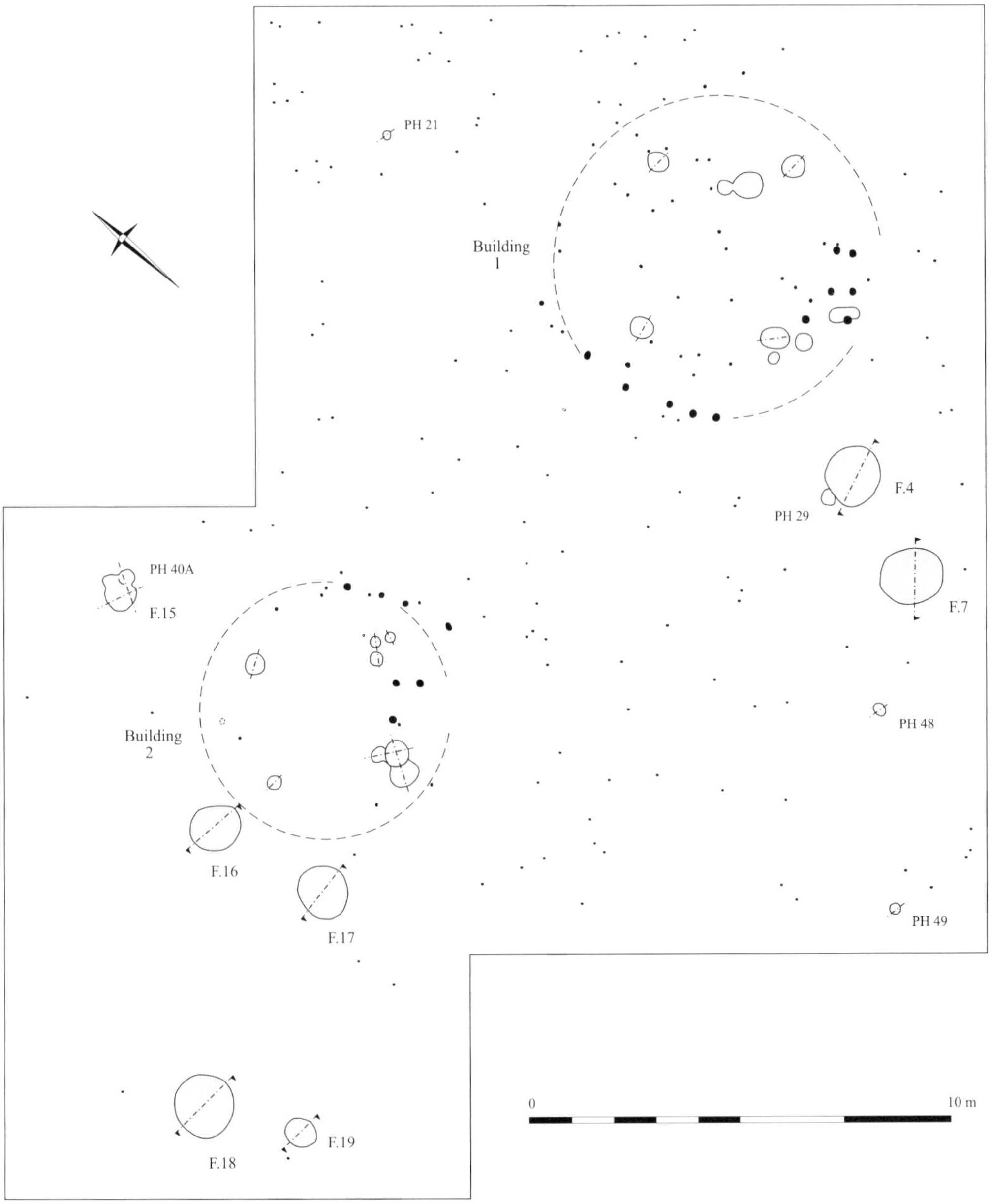

Figure 3.22. *Wyke Down 2 (WD 2 1996): plan of structures 1 and 2 (M. Brudenell after M. Green).*

Trench B being bottomed (Fig. 3.26). Trench E (north-east terminal) was excavated to a depth of 50 cm and Trench I only removed layer 2 (Fig. A4.10).

The complete excavation of the western henge ditch revealed it comprised 10 pits with a maximum depth of 1.75 m, separated by remnants of chalk baulks surviving to a maximum height of 0.56 m above the pit bases. The single trench bottomed on the east side also revealed a baulk showing the same construction technique was employed. The lower primary fills consisted of coarse chalk rubble inter-spersed with occasional organic-rich lenses (layer 5) except in the southwestern terminal pit exposed in trench G. Here, a deliberate dump of organic-rich soil had covered the floor to a depth of 0.2 m and a remarkable, almost complete Grooved Ware vessel, was incorporated (Fig. A4.14, P41). A single radio-carbon date has been obtained from charcoal in this context of 2930–2860 (86.6 per cent probability) and 2810–2750 (8.8 per cent probability) cal. BC (4262±32 BP; Wk-18750). Other deposits placed on the pit floors comprised occasional animal bones, a chalk ball (Fig. A4.20, c1) and scraps of Peterborough Ware at the northwest terminal. Upper primary fills consisted of finer chalk rubble with thicker organic-rich lenses (layer 4) with lower secondary fills composed of grey silts (layer 3). Cut through the top of this layer at the southwest terminal to a depth of *c.* 50 m, and sealed by layer 2, was a pit or posthole (F.20). Containing a homogeneous fill of greyish-brown silt, it stopped short of the special organic-rich deposit on the floor of the ditch previously described. Containing sherds of Grooved Ware it was clearly a re-marking of this significant terminal. Excavation of the opposing southeastern terminal pit was restricted to the removal of layer 2, sufficient to see that there was no corresponding feature across the entrance. At layer 3/2 interface within trench H significant amounts of pottery, fired clay, poorly preserved animal bone and worked flint including two fine oblique arrowheads (Fig. A4.11, nos. 10 & 11) were uncovered. This midden deposit corresponded to two of the underlying causewayed pits which formed part of the ditch, and lies close to the settlement area described below.

The upper secondary fill was very dark brown decalcified silt with much natural flint, and contained both Grooved Ware and Beaker pottery. The interior of the henge was largely featureless apart from a scatter of fifteen stakeholes which did not produce a coherent pattern.

Owing to the discovery of features outside the henge to the west, the excavation was considerably expanded in this direction. This resulted in the discovery of 22 pits and 49 postholes, most containing diagnostic later Neolithic material (Fig. 3.21). Two groups of postholes were clearly the remains of two buildings. The limits of this settlement zone were not reached and remain to be defined by future excavation.

Two pits F.21 and F.22 were the first features encountered. F.21 produced no diagnostic material but F.22 contained a very rich Grooved Ware assemblage in a predominantly dark organic fill. Some 15 m away two postholes (postholes 1 & 2) and two shallow pits (F.1 & F.2) formed a clear setting. F.1 produced two oblique arrowheads and the setting may represent the surviving elements of an otherwise destroyed structure (Fig. 3.31). Just to the north of this was another pit, F.5, close to the southeastern henge entrance which produced the skeleton of a neo-natal pig. To the west of this a line of five postholes aligned on the southern henge entrance survives for a distance of ten metres. Just to the north of this line a further pit, F.6, contained a basal deposit of a part polished flint 'ball' and a dog skull (Green 2000, fig. 44d). Not far from the northwest entrance to the henge a pit, F.3, with a similar spatial relationship to F.5 produced a deposit of animal bone, primarily consisting of skull fragments of cow, red deer and pig. Some seven metres to the northwest of this pit was a cluster of features, only partly revealed, which included 8 pits. One of these, F.8, contained a placed deposit of artefacts including two arrowheads and other objects (Figs. 3.30 & A4.11).

To the southeast of this cluster and some 12 m to the west of the henge the first building (1) was uncovered (Fig. 3.23), consisting of an internal rectangular setting of 4 posts (postholes 12–15), covering an area of some 3.25 × 4 m. To the southeast of this setting was a small rectangular structure about 1 × 0.75 m which, together with an elongated double posthole to the southwest, forms a clear entrance to the building. To the west lay an arc of small postholes (36–40). A circle drawn from the centre point of the large rectangular element on a 4 m radius incorporates the surviving post arc on the western side, strongly suggesting this is the surviving remains of the outer wall line of the building. The overall diameter suggested (8 m) would also incorporate the entrance structure which bears a similar alignment to the nearby henge. A shallow conjoined internal feature (posthole 16/17) contained much charcoal and scorched chalk suggesting it may have been a hearth or oven base. Within the surviving post pipes of the building much charred material was evident suggesting it may have burnt down.

Significant finds from the postholes included Grooved Ware pottery, amongst which was a very large body sherd from posthole 15 (Fig. A4.15, no. 44) and a body sherd with concentric circles from posthole 12 (Fig. A4.17, no. 74). A piece of a scallop shell also came from posthole 15 and remarkably, fragments of the actual walls themselves were retrieved from a number of the postholes. This material consisted mainly of a chalk-based 'render', about 1 cm thick, with a smooth outer surface bearing traces of a lime-wash finish and with impressions of round wood/reed wattle on the inside (Figs. A4.20 & A4.21). A few fragments had been impressed or incised on the outer surface suggesting parts of the exterior walls were decorated. Further evidence of the walls consisted of a number of fragments of a much coarser clay-based daub including a fragment from the outer wall line. As the finer chalk based render was retrieved entirely from the internal postholes of the building it seems likely the four main uprights formed the corners of an inner room with finely finished and perhaps decorated interior surface with an encircling wall of coarser construction.

Only 7 m to the west a second smaller building (2) of very similar ground plan to building 1 was uncovered. Consisting of an internal square four post setting three metres in diameter with three postholes of an entrance structure (postholes 41–43), similar to building 1, surviving to the southeast. A circle drawn from the centre of this building on a 3 m radius not only incorporates this entrance structure but also an arc of small postholes (postholes 45–47) which

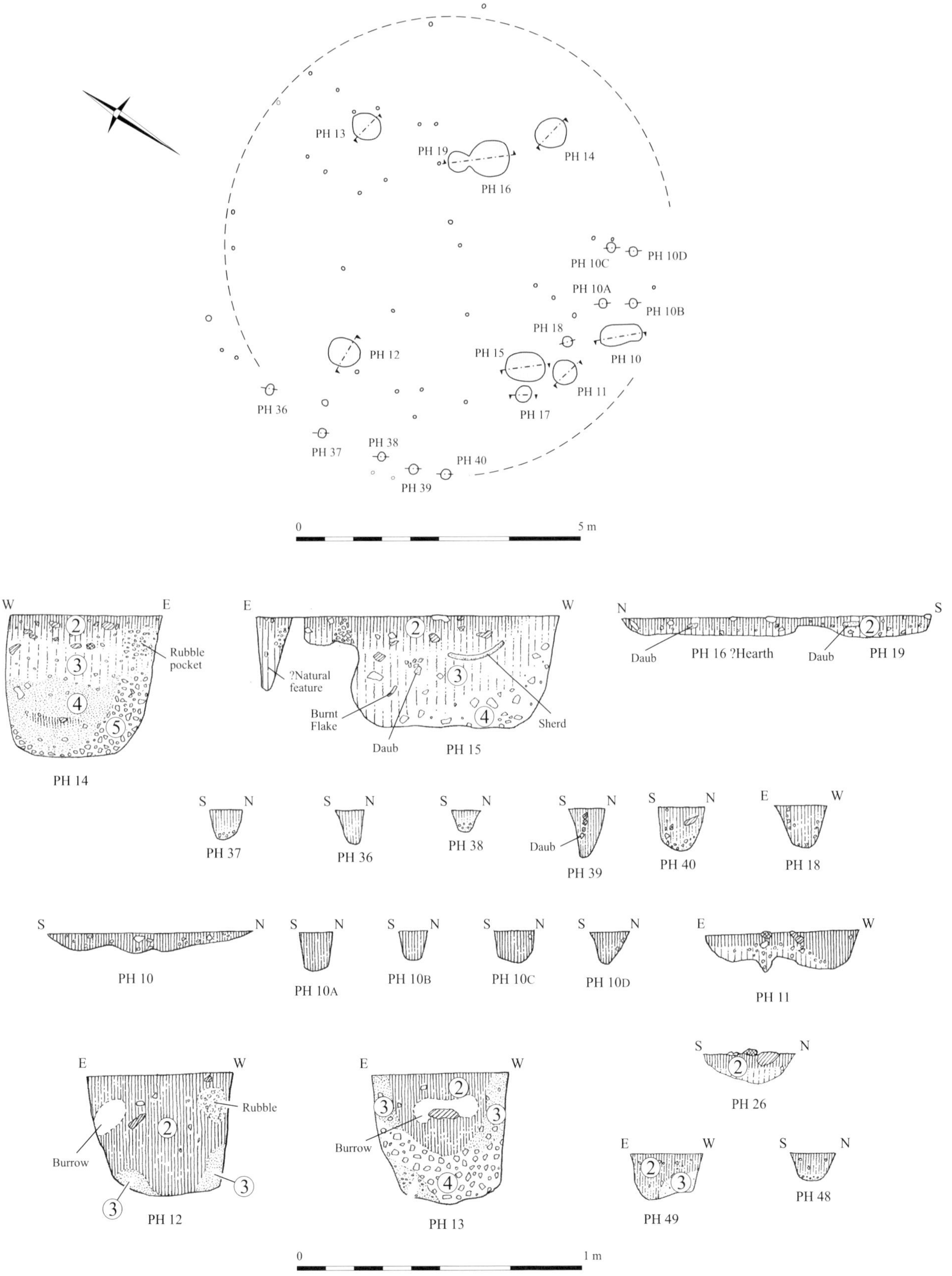

Figure 3.23. *Wyke Down 2 (WD 2 1996): plan and sections of house 1 (M. Brudenell after M. Green).*

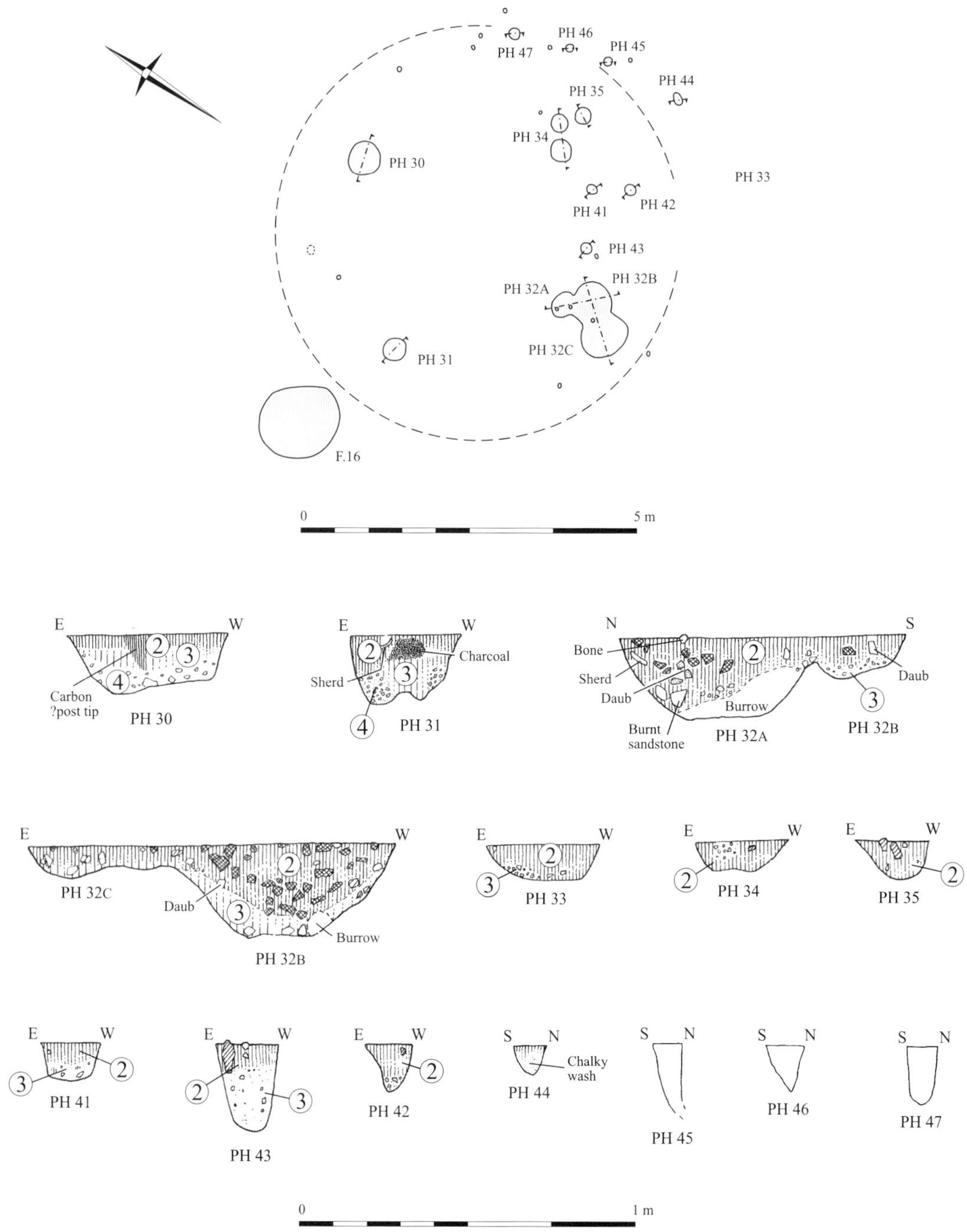

Figure 3.24. *Wyke Down 2 (WD 2 1996): plan and sections of house 2 (M. Brudenell after M. Green).*

represent the only surviving part of the outer wall of the building. Similar finds were made to building 1 and included striking spatial associations of marine shells in posthole 32 and concentrically decorated Grooved Ware in posthole 31 (Fig. A4.15, P42). A pair of pits lay close to the west side of each building and produced further sherds of Grooved Ware, daub, animal bone and worked flint. A single cow horn was recovered from each of the three pits (F.15–F.17) closest to building 2. Further pits lay near to building 2 with F.18 producing a carved chalk disc (Fig. A4.20, C2). Although specific finds such as the aforementioned can be regarded as deposits of ritual significance the bulk of the material from the pits suggests more mundane deposition through domestic activity.

Parallels for these buildings are scarce but include two structures found only 32 km to the north at Durrington Walls. One was found within the great

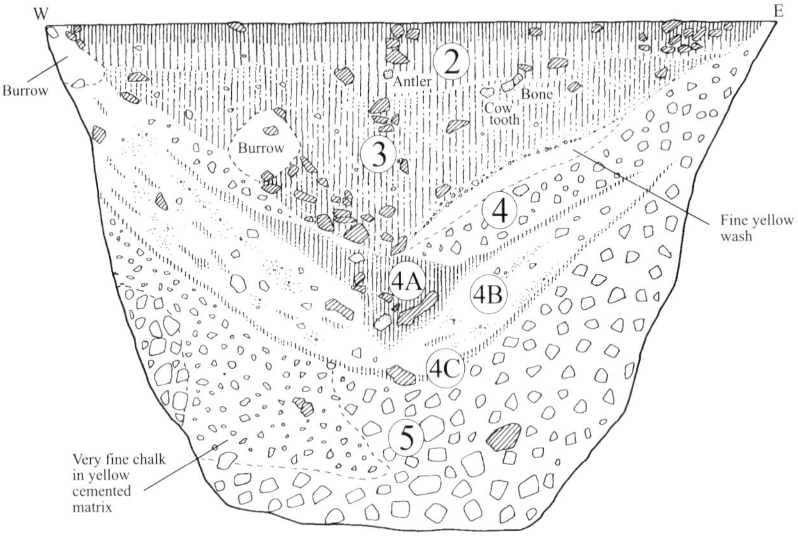

Section C

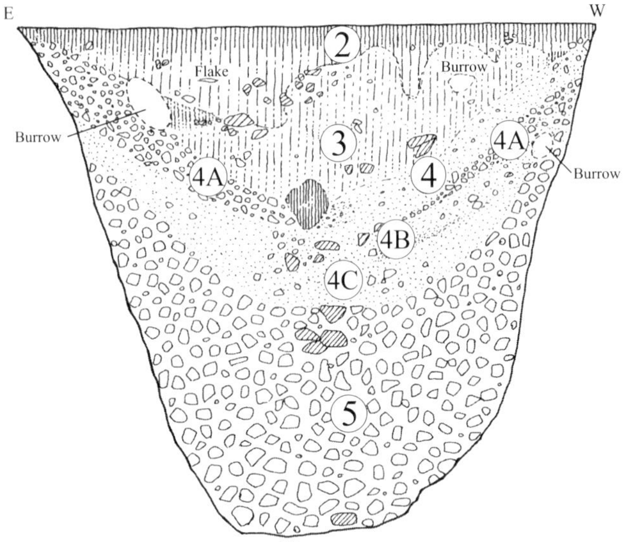

Trench F

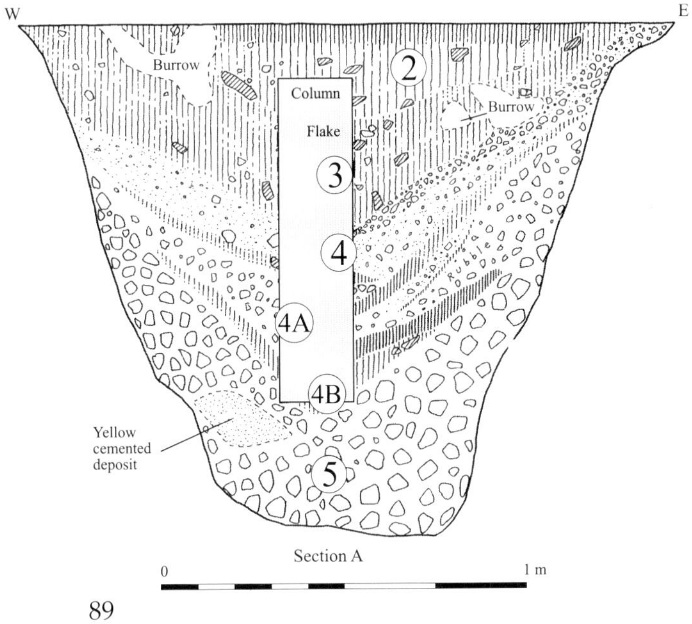

Section A

Figure 3.25. *Wyke Down henge 2: ditch sections A, C and F, west side (N. Griffiths after M. Green).*

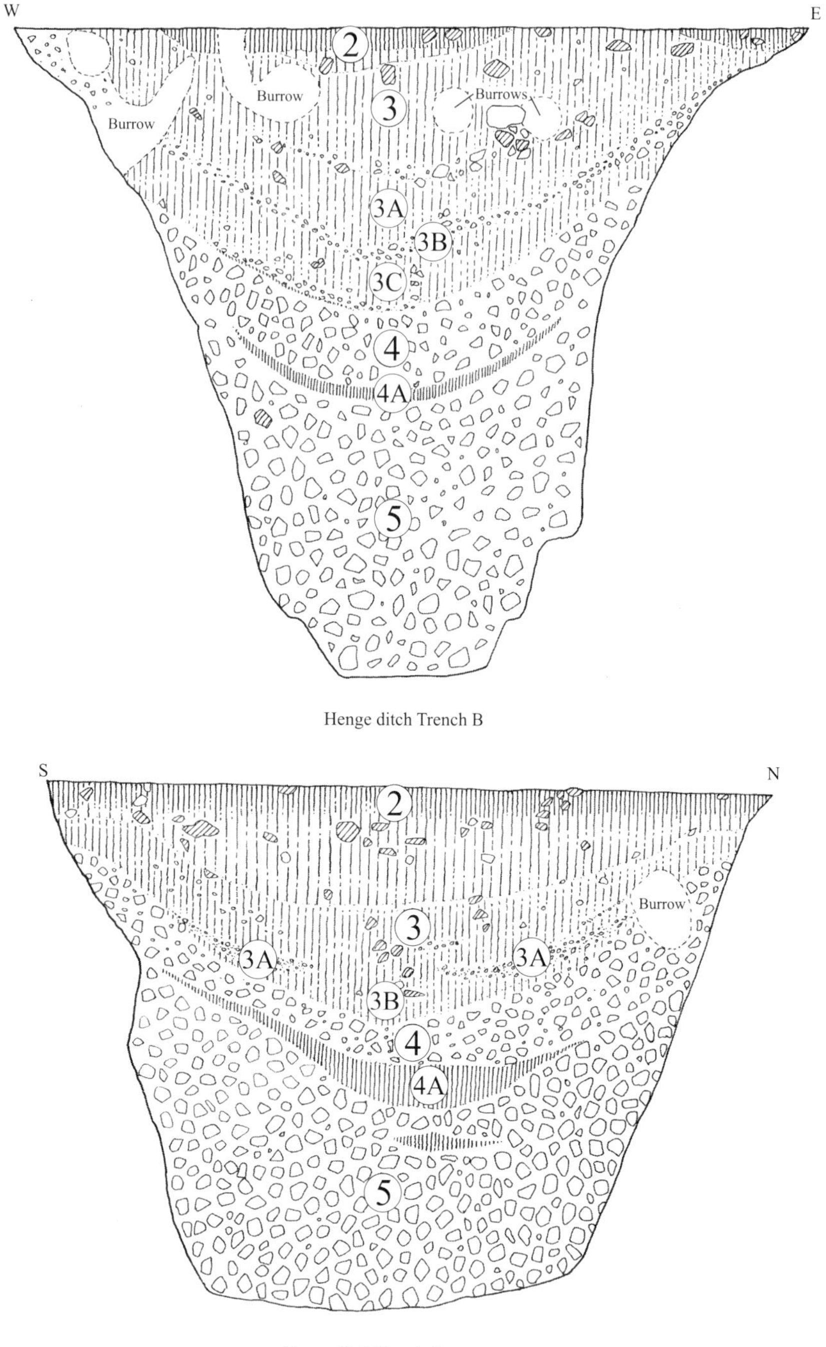

Figure 3.26. *Wyke Down henge 2: ditch sections B east side and E the northwest terminal (N. Griffiths after M. Green).*

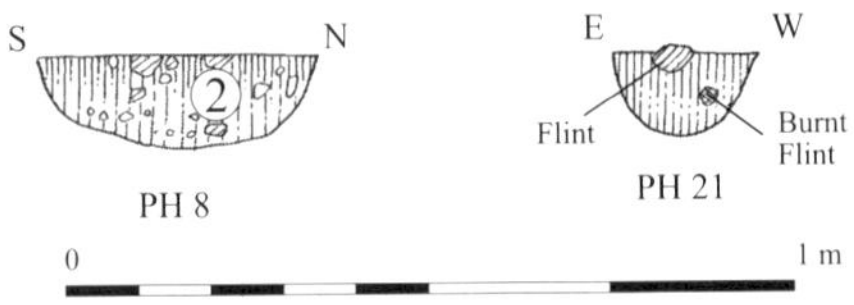

Figure 3.27. *Wyke Down henge 2: sections of isolated postholes (N. Griffiths after M. Green).*

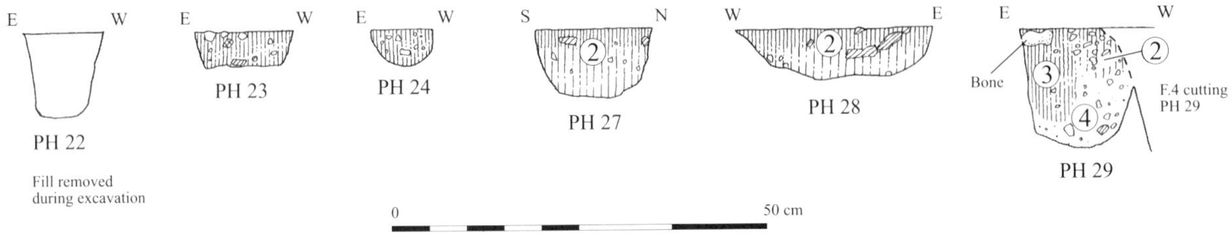

Figure 3.28. *Wyke Down henge 2: sections of postholes 22–4, 27–9 of the Grooved Ware settlement (N. Griffiths after M. Green).*

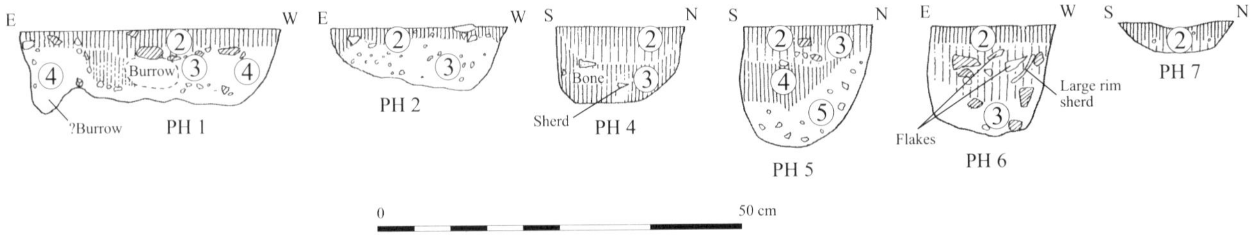

Figure 3.29. *Wyke Down henge 2: sections of postholes 1–2, 4–7 of the Grooved Ware settlement (N. Griffiths after M. Green).*

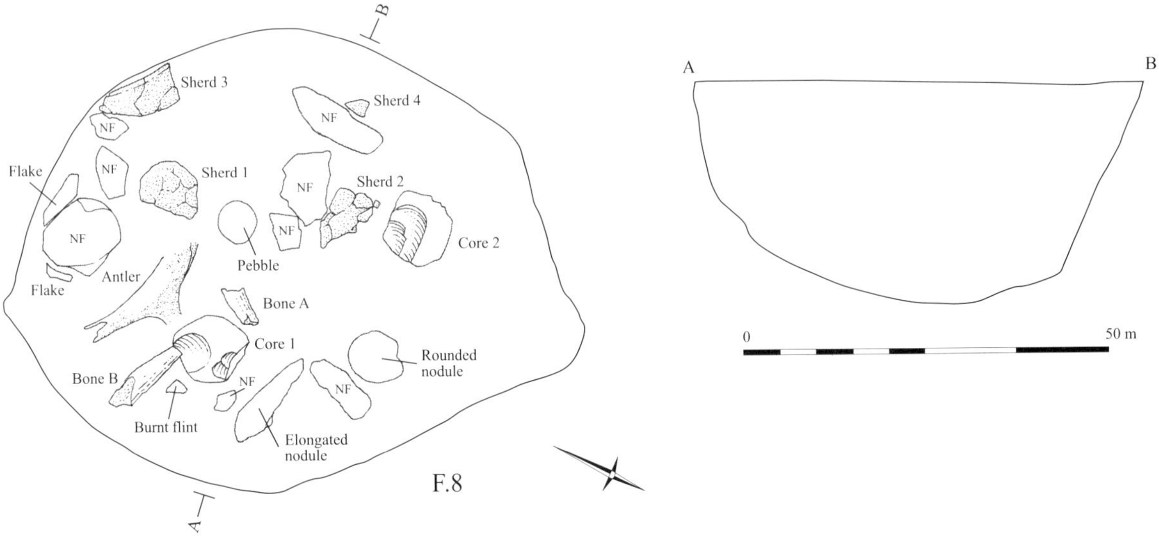

Figure 3.30. *Wyke Down henge 2: plan of the artefact distribution in the Grooved Ware pit F8 and its profile. NF = Natural flint (N. Griffiths after M. Green).*

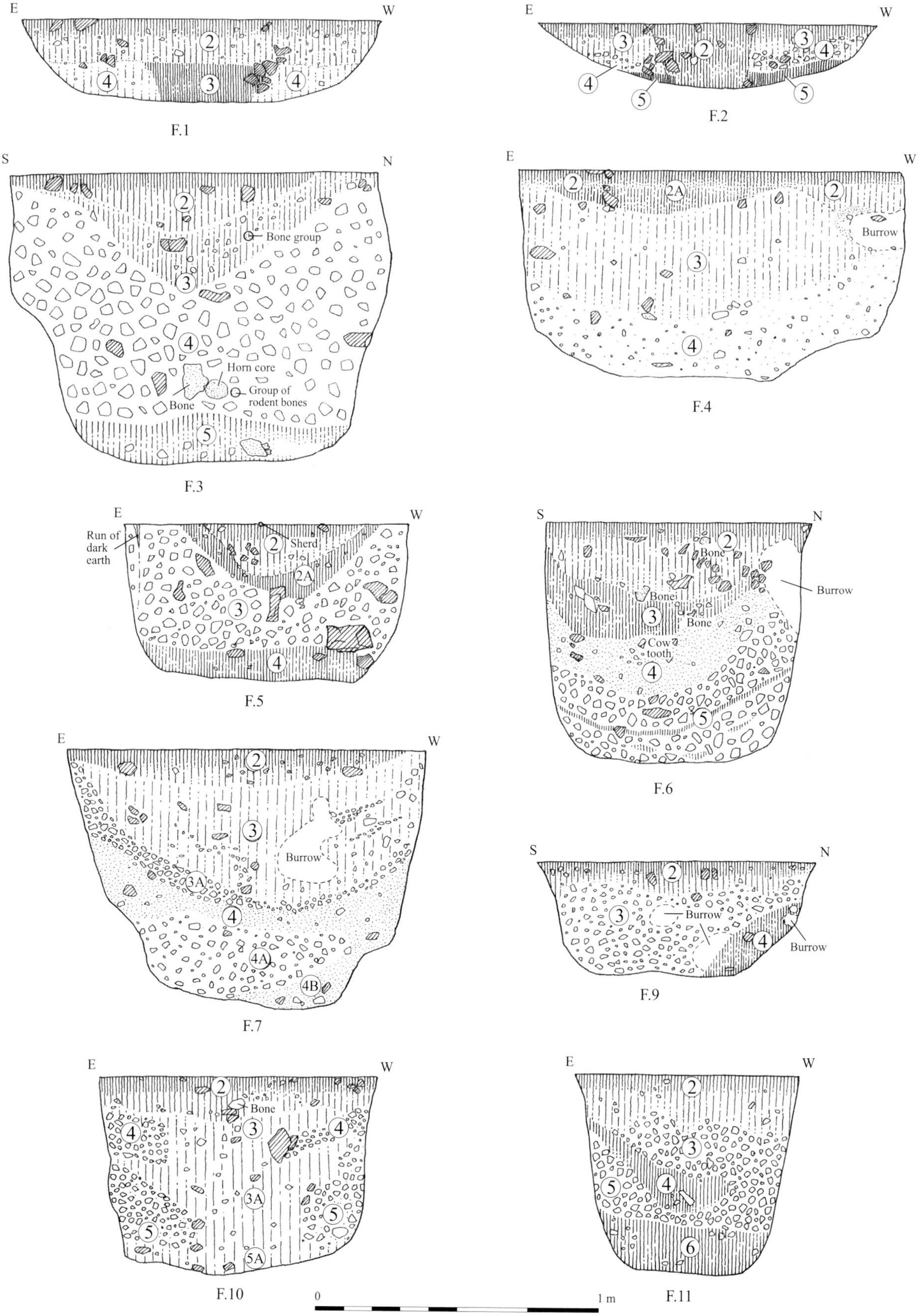

Figure 3.31. *Wyke Down henge 2: sections of pits F.1–F.7 and F.9–F.11 of the Grooved Ware settlement (N. Griffiths after M. Green).*

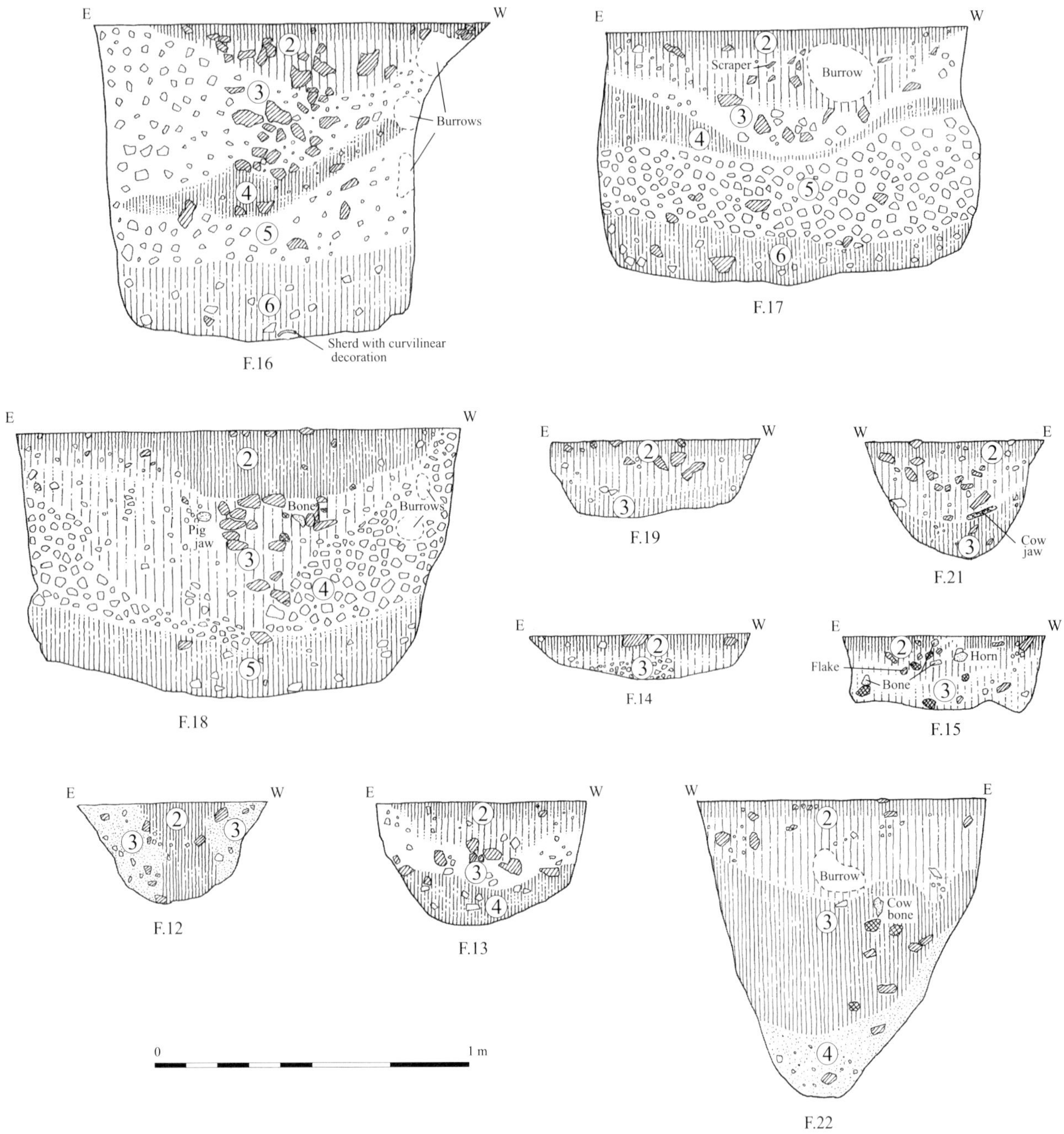

Figure 3.32. *Wyke Down henge 2: sections of pits F.12–F.19 and F.21–F.22 (N. Griffiths after M. Green).*

henge monument at Durrington Walls and is known as the north circle (Wainwright & Longworth 1971) and the other was found just to the south underlying a round barrow (Cunnington 1929) and was recently re-interpreted by Josh Pollard (1995). Although both are nearly twice the size of those on Wyke Down they are strikingly similar in ground plan, sharing substantial internal square settings with slighter built outer walls. Other comparisons include broadly contemporary stone built houses found on Orkney such as those found at Barnhouse (Richards 2005). Here internal rectangular settings were enclosed by outer circular walls. Even some stone circles (e.g. The Stones of Stenness) enclose rectangular or square settings, suggesting a formal or ritualized combination of elements. It appears we have at Wyke Down part of an unenclosed later Neolithic settlement, of unknown extent, which included its own ceremonial structures.

The Wyke Down round barrow group

Charles French, Helen Lewis & Michael Allen

In terms of visualizing the monuments on Wyke Down itself, the Bronze Age round barrows (Figs. 1.2 & 3.33) were compared to the underlying geology (Fig. 3.34), their slope and aspect visualized with respect to the dry valley bisecting the down (Fig. 3.35), as well as the upstanding mounds versus flat ring-ditches. All of the round barrow sites on Wyke Down are best seen and most prominent from the lower slopes of the valley, not from higher up on the tops of the downs, and occupy slight ridges or prominences to either side of the dry valley. Across the study area, all prehistoric monuments except for the Dorset Cursus (which crosses the whole valley/down-scape) and long barrows tend to be located mid-slope (Fig. 1.2). In addition, they are positioned on the margins and upslope of any zones of soil accumulation (Fig. 2.9). Their slope

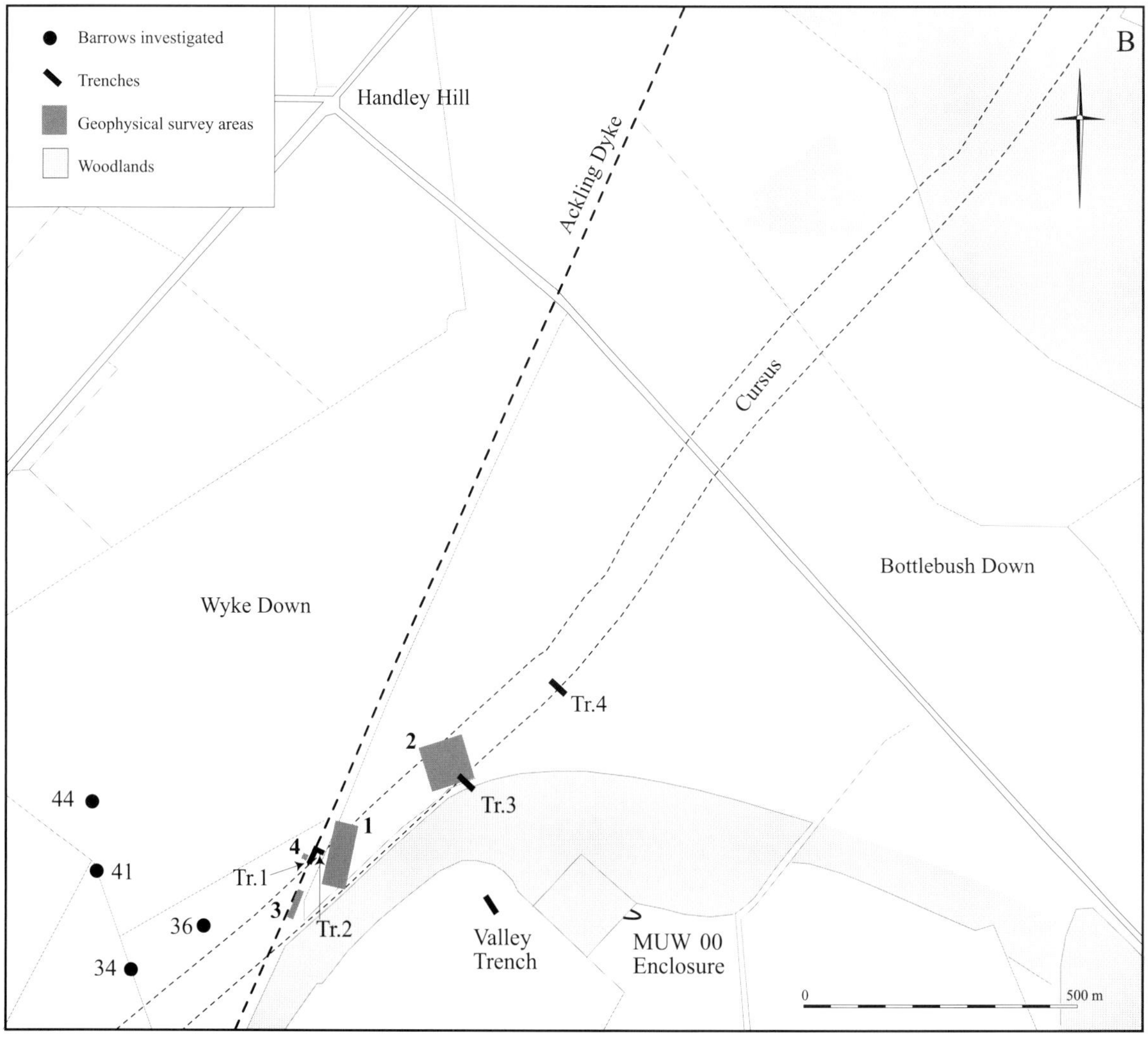

Figure 3.33. *Location map of the trenches cut through the Wyke/Bottlebush Downs part of the Dorset Cursus, as well as the location of the Monkton-up-Wimborne Neolithic enclosure (MUW 00) and the soil test trench through the adjacent dry valley (C. Begg).*

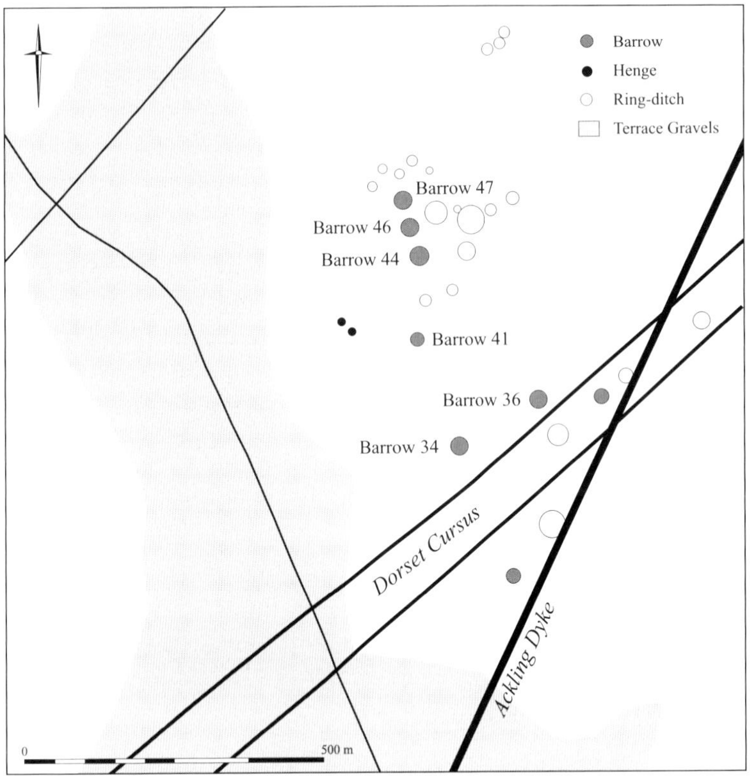

Figure 3.34. *The location of the four upstanding barrows (34, 36, 41 & 44) of the Wyke Down group and the two henge monuments (WD 1 & 2) that were excavated set against the background geology with the Devensian terrace gravels speckle shaded and all other areas chalk except the dry valley in the upper right which has an overlay of coombe rock (C. Begg after P. White).*

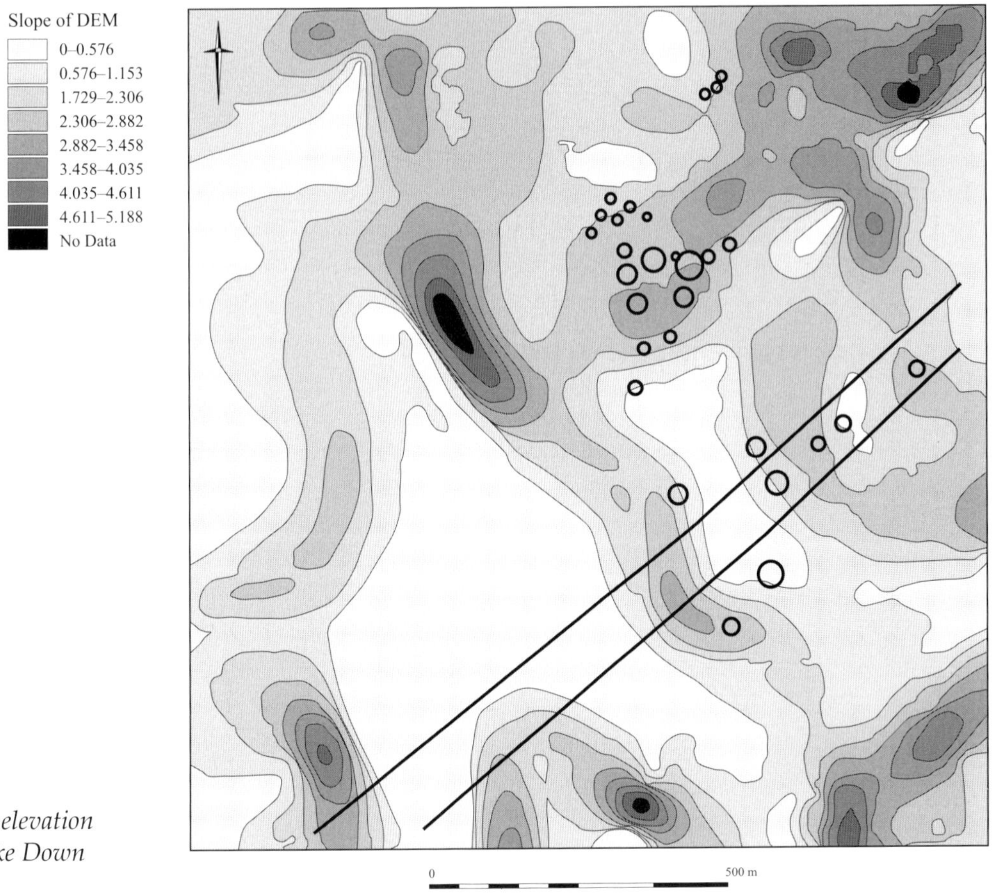

Figure 3.35. *The digital elevation model (DEM) of the Wyke Down barrow group (P. White).*

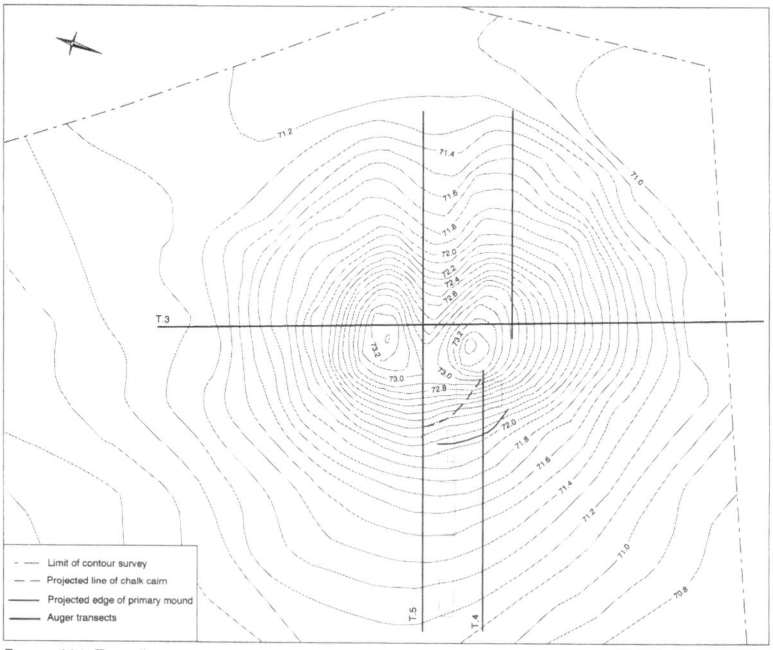

Barrow 034- Topsoil contours (height in m. O.D.)

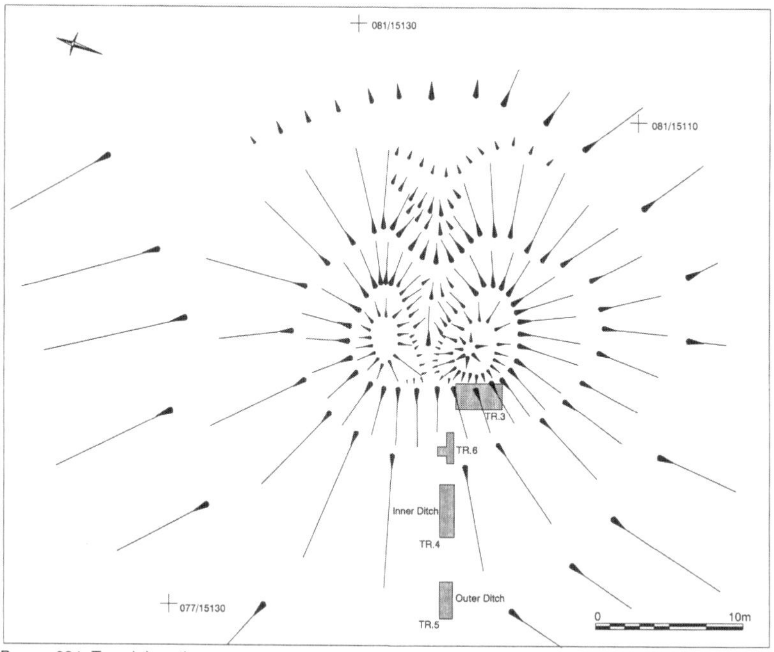

Barrow 034- Trench Location

Figure 3.36. *Barrow 34 topsoil contour plot and trench location plan (P. White).*

position puts them in an 'erosion' zone topographically (regarding slopewash), and this is supported by the discovery of thin rendzina soils found beneath the Bronze Age barrows investigated.

The Wyke Down round barrow group consists of seven upstanding bowl barrows and at least another 22 ring-ditches (Fig. 3.34). Four upstanding barrows were sample excavated (Figs. 3.36–3.50), and these and most of the remainder of the group were surveyed using magnetometry (French. *et al.* 2000, fig. 4) (Figs. A2.8–A2.14). Sample excavation of the ring-ditches and barrow mounds was used to obtain new molluscan and soil data in particular. The barrow ring-ditches are unremarkable in their fills, except for zones of very fine calcareous silt loam in the upper secondary fill, which is probably representative of windblown material derived from open arable soil in the immediate vicinity at some point in the later Bronze Age. This fine calcareous silt of probable windblown origin was also observed to compose the secondary fill of an Iron Age ditch on Gussage Cow Down (see SAM 192, above). The ditches also show a pattern of immediate back-filling with relatively clean chalk rubble, suggesting that this practice, often seen in Neolithic monuments, was also carried out in the Early Bronze Age.

Unfortunately, pollen survival was very poor (see Scaife below), but the molluscan assemblages from these barrows were predominately of open country species with no clear well-defined zones of major vegetation regeneration detected in any ditch sequence (Allen 1998; 2000a; 2003). These analyses demonstrated that prior to the construction of the Bronze Age barrows, a largely open, clear-felled landscape existed in the upper Allen valley. The local land-use history through the Bronze Age as defined by the assemblages from the ditch fills, indicates maintenance of open environments, and that the only scrub vegetation regeneration that occurred, if any, was probably local to the barrow ditches and was not part of a wider landscape phenomenon.

The mound sequences and buried soils beneath them produced interesting and complimentary data. Each mound was composed of a central chalk or turf cairn that was then capped by chalk rubble and/or redeposited turves. In one case, a large chalk cairn had evidently been exposed at the surface for some time (barrow 36). The buried soils are composed of a turf and weathered subsoil horizon indicative of a thin (<15 cm), rendzina soil profile (Figs. 3.41, 3.44, 3.47 & 3.49). The implication is that the soil was already severely denuded and altered by the Early Bronze Age and was open grassland. But the turves in the mounds suggest that this former topsoil was more organic, clay-rich and better structured than the buried, and arguably contemporary, rendzina beneath the barrows. These features hint at the presence of former brown earths in the area previously. In addition, the strong presence of oriented clay within the turves of the mounds may suggest a minor phase of immediately pre-barrow ground disturbance and limited wind/water erosion of fine soil material downslope that became incorporated in the turves used to construct the barrow mound. This could have been associated with nearby turf stripping on a large scale for barrow construction purposes as much as ploughing.

The conventional magnetometer and resistivity surveys of barrows 34 and 41 served to pin-point the position of the ring-ditches, the extent of the surviving mounds, and areas of disturbance with great accuracy (Figs. A2.8–A2.11) (GSB 1998; 1999). The magnetometer survey of the intervening area of land revealed no definite archaeological features, but a very shallow dry valley is present (Figs. A2.8 & A2.9). The assessment use of ground penetrating radar survey on barrow 34 (by Christian Pierce and Colin Shell) also located the outer and inner ring-ditches with great accuracy, but appeared to over-emphasize the depth of the ditches substantially (French *et al.* 2000, fig. 8). Moreover, the in-field radar print-out suggested the shape of the ditch cuts, and gave hints at multi-horizonation within the ditch fills that were not completely verified in the excavated sections. It is suggested that the radar was confused by bedding disconformities within the chalk subsoil itself (C. Pierce pers. comm.). The radar survey also recognized changes in the composition and form of the mound which were proved by excavation to be entirely accurate. Nonetheless, the identification of the composition of the mound material and the nature

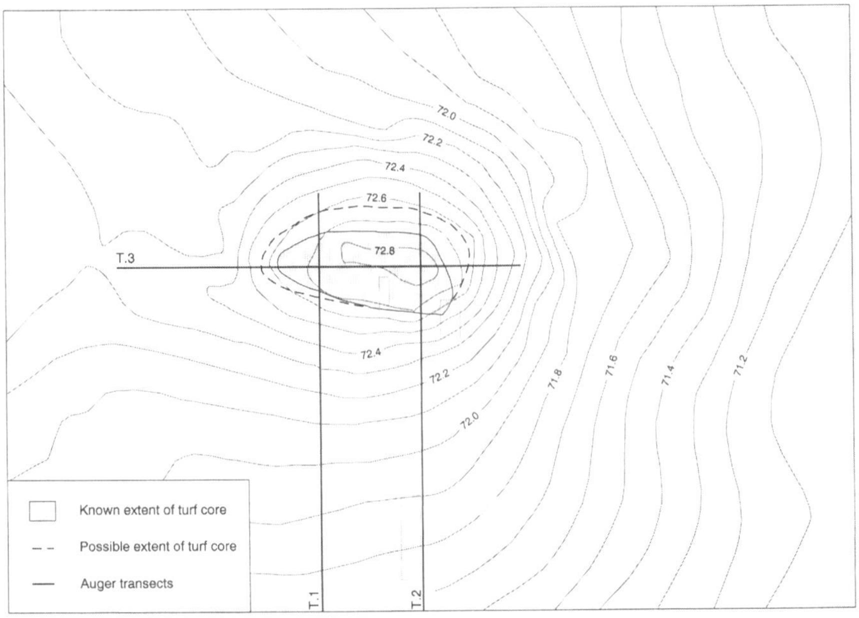

Barrow 041- Topsoil contours (height in m. O.D.)

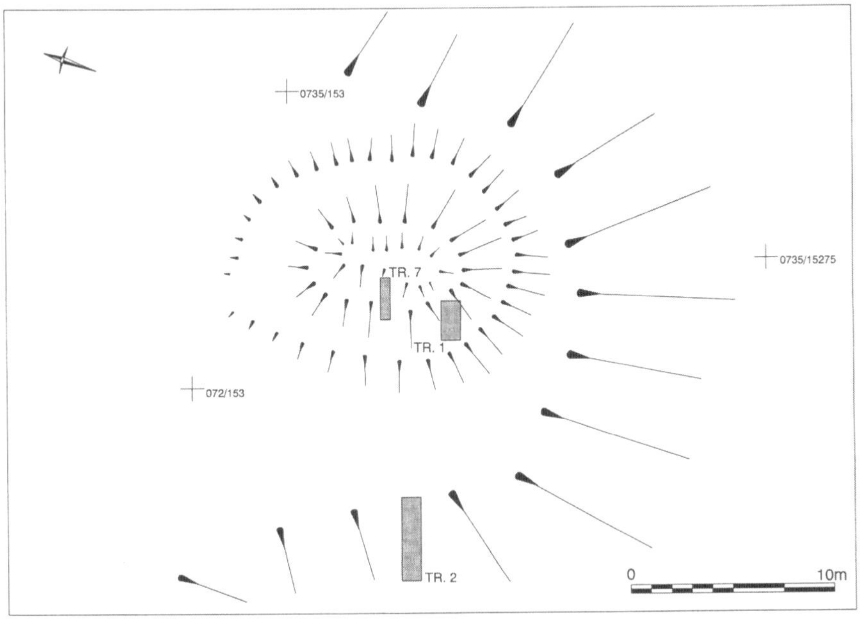

Barrow 041- Trench Location

Figure 3.37. *Barrow 41 topsoil contour plot and trench location plan (P. White).*

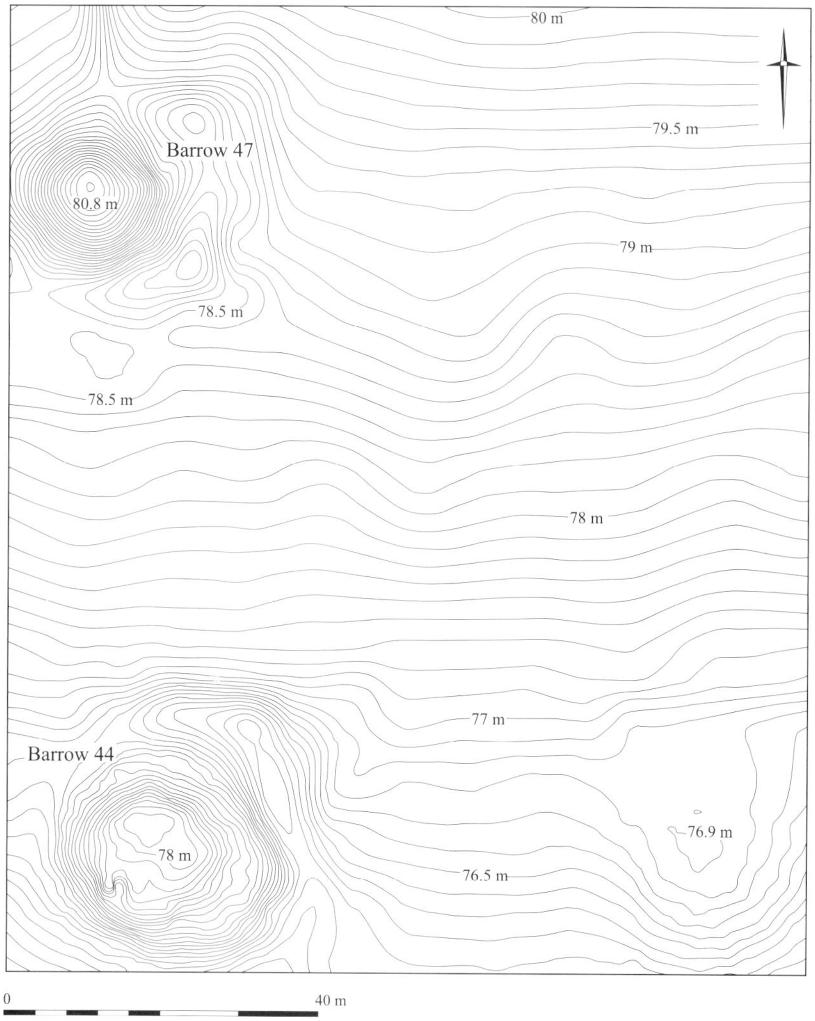

Figure 3.38. *Contour plot of the main Wyke Down barrow group including barrows 44 and 47 (M. Brudenell after P. White).*

of the change could only be seen with the hindsight of excavation.

The low barrow (41) (Figs. 3.37, 3.46–3.48 & A2.10) turned out to be much destroyed/denuded, and to a large extent this had already occurred in the past, rather than more recently as had been the common perception. Today there is a small, low (*c.* 1 m high, but this may be accentuated as the mound is situated on the top centre of a low knoll), turf mound surviving in the centre of the area enclosed by the ring-ditch. The ring-ditch is about 3.5 m wide, 1.6 m deep, with an internal diameter of about 19 m and an external diameter of about 26 m. Within the fill of the ring-ditch itself, the inner third was composed of densely-packed chalk rubble. This would appear to be a deliberate backfill deposit into the ring-ditch. One could speculate that the chalk acted as either a cap to the turf core, or more probably as a circular bank on the inner edge of the ring-ditch, which had been deliberately put back into the ditch after a very short period (as this chalk rubble is very clean and soil-free). The remaining ring-ditch, unaffected by this deliberate backfilling, was narrower (*c.* 2.5 m) but of the same depth as the first ditch, and remained open to infill naturally. This evidence suggests that barrow 41 has probably been two slightly modified versions of a disc barrow.

On the other hand, barrow 34 (Figs. 3.34, 3.36 & 3.39–3.43) exhibited a more complicated and diverse sequence. Despite the presence of a large antiquarian trench (which may be early twentieth century in date, and is regretably unrecorded), the mound of this barrow appears to be substantially intact and standing to a

Figure 3.39. *Barrow mound 34 under excavation with the sample trenches through the inner ditch and mound, and the antiquarian trench visible in the centre of the extant mound (C. French).*

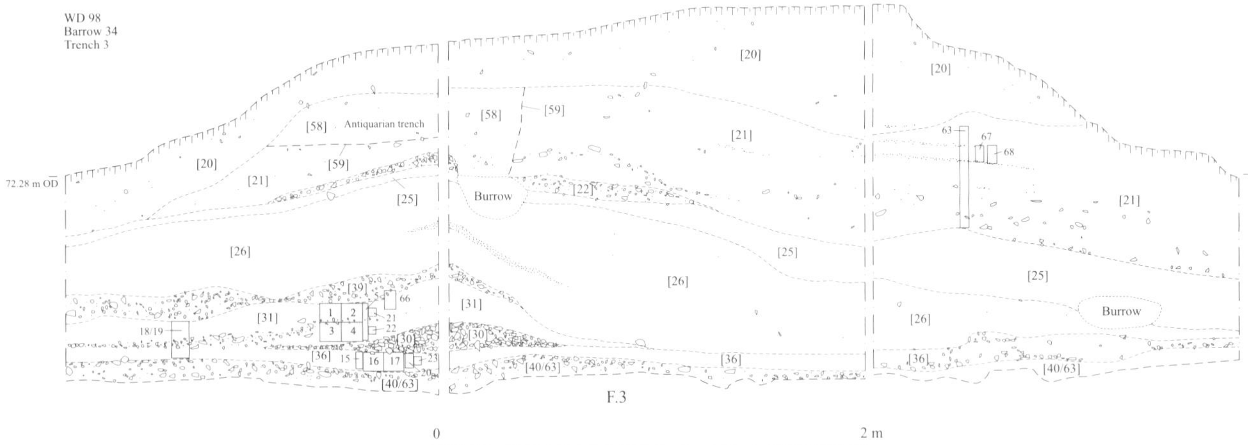

Figure 3.40. *Section through the mound in barrow 34, Trench 3 (M. Brudenell after H. Lewis & C. French).*

Figure 3.41. *Section through the chalk and turf core of the mound in barrow 34 and the thin rendzina palaeosol at the base of the mound (C. French).*

height of about 2 m above the present-day ground surface. One trench was excavated through the southeastern quadrant of the mound and revealed the following sequence. First, a small (*c.* 2 × 3 m), sub-rectangular chalk cairn was erected on a thin (<10 cm), truncated old land surface. This cairn would appear to mark the centre of the subsequent mounds. It was covered by one or two layers of turf. It was decided not to disturb the cairn as it is probably preserving a central and primary interment. The inner ring-ditch (of about 18 m internal diameter) was then dug (or deepened?), and the chalk subsoil upcast from this to make a primary chalk mound of about 16 m diameter placed over the small chalk and turf cairn. The surface of this chalk mound was initially left bare and underwent a period of weathering sufficient to discolour the upper 5–8 cm of the chalk, and saw the accumulation of large quantities of worked flint, probably worked *in situ*. Then, a more substantial turf mound was created, at least doubling the height of the mound (to about 2 m), with its outer edge vertical sided and retained by a slight wooden revetment of stakes (slender fence post size, <10 cm), with each stake placed about 40 cm apart, presumably retaining a woven wattle-work fence, located about 1 m in from the inner edge of the ring-ditch (in Trench 6). The inner ring-ditch (in Trench 4) had become completely infilled by the time the outer, slighter, ring-ditch was dug (with a *c.* 23 m internal diameter, *c.* 24.5 m external diameter). The mixed upcast of turves, soil and chalk rubble produced by the digging of this second ditch was thrown-up against the outside of the wattle fence around the first barrow, which was now leaning outwards, thus increasing the diameter of the barrow mound by about 3 m and creating a gently sloping outer edge to the barrow, but still leaving a wide berm (2.5–3 m).

In terms of the flint assemblages from the mounds and ring-ditches of the four barrows investigated, which are further discussed by Gardiner in Appendix 4.6, the assemblages are predominantly waste material, with minor amounts of retouched implements and cores (Tables A4.25 & A4.26). In contrast to the abundant flint present, just six sherds of pottery were recovered, one from barrow 41 and five from barrow 34, all from secondary or tertiary contexts (K. Gdaniec pers. comm.). Only one sherd of possible Beaker pottery was found in the upper secondary fill

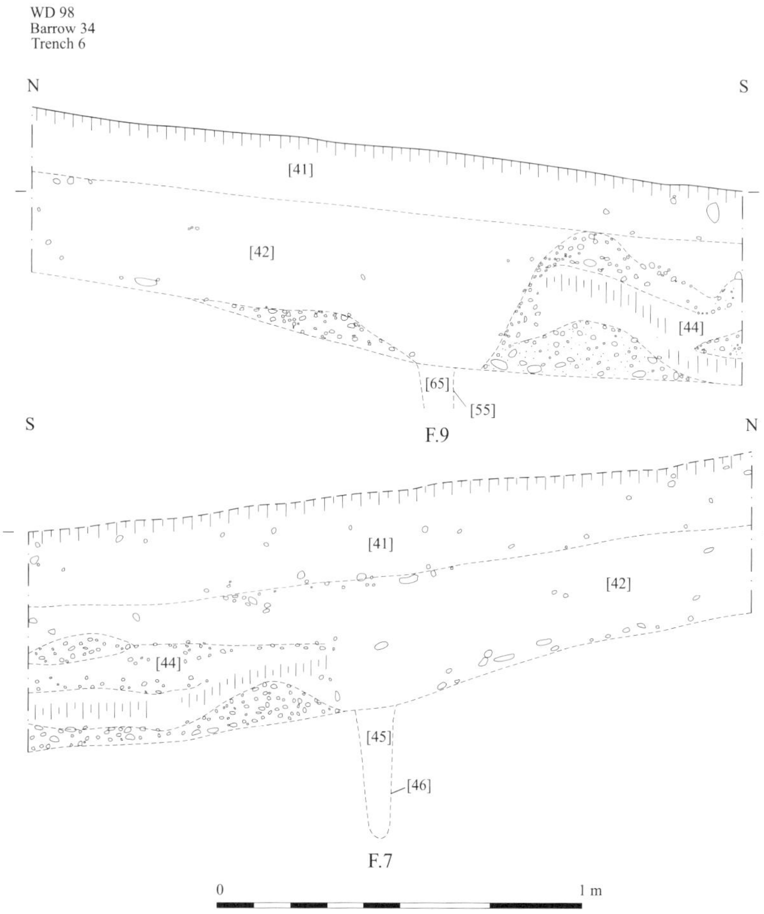

Figure 3.42. *Section through the outer edge of the surviving mound of barrow 34 showing the postholes of the wooden revetment of the primary enlarged chalk and turf mound in Trench 6 (M. Brudenell after C. French).*

Table 3.1. *Micromorphology sample descriptions for Wyke Down round barrows.*

Sample	Context	Depth (cm)	Field description
WD 18 WD 19	Barrow 34, Trench 3, [31]	2–16	Turves/buried soil (upper). Very dark brown silty clay loam, fine columnar/blocky structure, rare chalk fragments (10 YR 2/2).
WD 20	Barrow 34, Trench 4, [36]	24–28	Lower buried soil. Very dark brown (10 YR 2/2) silty (clay) loam. Occasional chalk fragments (<2 cm), increasing in abundance as [36] merges with [63]. A–A/C of rendzina.
WD 66	Barrow 34, [31]		Turf over chalk core [30]. Very dark brown silty clay loam, fine columnar/blocky structure, rare chalk fragments (10 YR 2/2).
WD 67 WD 68	Barrow 34, [21]		Mound/turves. Very dark greyish-brown (10YR 3/2) silty clay loam, occasional to common gravel/flint pebbles (<3 cm). In several places turves divided by thin (<2 cm) horizontal iron pans (orange). Each sample is an entire turf in the turf stack, with mottling on either side of it.
WD 54 WD 55 WD 56	Barrow 41, Trench 1, [13]		Reddish-brown (5YR 4/4)–dark brown (10YR 3/3) silty clay-silty clay loam, frequent chalk rubble (*c.* 50%), occasional flint cobbles (<10%). Old A–B horizons.
WD 69 WD 70	Barrow 41, Trench 1, [6]		Mottled chalky turf core. Pale grey-brown (10YR 7/2–10YR 5/3), orange (Fe) mottling, silt loam.

of the ring-ditch of barrow 41. At barrow 34, there was one sherd of either Beaker or Collared Urn with deeply incised, filled lozenges, and three small sherds of Beaker pottery (one with rows of possibly impressed bird bone decoration, one with comb-impressed decoration, and one rusticated sherd) in the upper secondary and tertiary fills of the inner ring-ditch of barrow 34, and one very abraded piece of possible Beaker from the secondary fill and one simple, flat-topped rim sherd of possibly Neolithic, sand- and quartz-tempered, pottery from the tertiary fill of the outer ring-ditch.

There is little specific dating evidence for either mound. The general characteristics of the flint assemblage suggest a mature, probably Middle Bronze Age date for the material manufactured from the freshly quarried flint. The small, but mainly Beaker, pottery assemblage is all in secondary contexts, that is it was already present when the barrows were constructed, and is therefore suggestive of a post-Early Bronze Age date. Nonetheless, there is sufficient charcoal recovered from the old land surface, primary turves and surface of the primary chalk mound of barrow 41 that radiocarbon assay will be possible, and provide *terminus post quem* dating evidence.

Soil micromorphology of the buried soils under the round barrows

Helen Lewis & Charles French

Samples for soil micromorphological analyses were taken from the buried soils and turves of all four round barrows trenched. The following summary description in Table 3.1 is supplemented by photomicrographs in Figure 3.101 (d–f) and by detailed descriptions in Appendix 4.6.

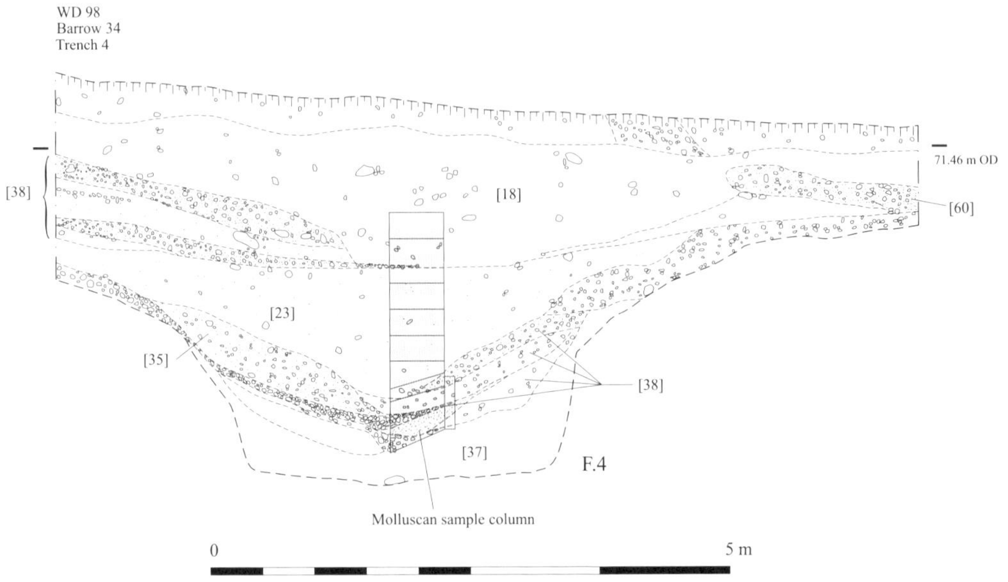

Figure 3.43. *Section through the inner ring-ditch of barrow 34 (M. Brudenell after H. Lewis).*

Barrow 34

The sample excavation through both phases of barrow 34 revealed a buried turf and weathered subsoil horizon comprising a thin (<15 cm) rendzina type of soil profile (Ah, B/C, C) (Figs. 3.40, 3.41 & 3.101d & e). It contains no relict features of any B horizon material and is strongly earthworm reworked, but is not perfectly sorted with respect to its chalk fragment component. These features suggest that this was a grassland soil with either some early disturbance which has left no trace, or which saw some continued input of fine chalk through physical disturbance. The presence of this thin rendzina suggests that this location had somewhat degraded grassland soils by the Bronze Age. The turves that comprise the bulk of the mound contain a significant clay content, possibly suggesting an origin from an argillic brown earth type of soil, as known from elsewhere in the valley (see above). In general, despite some possible disturbance, all of the profiles show the development of long-term grassland.

The early ditch fills have a fine silt (aeolian) component, which might suggest some bare earth around, and may relate to the disturbance seen in the buried soil. Both molluscan and pollen evidence (below) supports the interpretation of grassland, with a sequence of established grassland with some loose soil (but no clear arable indicators) giving way to locally shadier conditions (possibly simply longer grass around the ditches/monuments). The general sequence seen suggests that both barrows were built during a time of established open conditions, with some possible bare earth or arable farming around.

Barrow 36

The mound and buried soil sequence was superficially similar to that observed for barrow 34 (Figs. 3.44 & 3.45). Indeed, the mound was composed of turves which showed some evidence of compaction, and the base of which was marked by much oxidation mottling. Unusually though, there were coatings, infills and hypo-coatings of clay and dusty clay with low birefringence in the void space within the turves of the mound (Fig. 3.101f). This could suggest that the turves, prior to being cut and placed in the barrow mound, were subject to the downhill input of fine eroded material from a more clay-rich soil further upslope on the down. This might be indirect evidence of the upper part of the down being ploughed and/or with exposed and bare soil surfaces open to rainsplash and sheet erosion in the earlier Bronze Age.

The buried soil profile was characterized by a thin, compacted turf with an evident horizontal line of manganese mottling which appears to be a rooting line at the base of the turf topsoil, much like that visible beneath most of the barrows investigated on Wyke Down. This was developed on a sandy clay loam with evident oxidation mottling. Unusually though, this lower horizon of the buried soil contained a relatively abundant clay content (*c.* 40 per cent) with abundant pure and occasional dusty clay coatings throughout. In addition, it exhibited a few distinct patches of calcareous sandy clay loam soil of up to *c.* 2.5 mm in size, and iron and manganese mottling. The much higher clay component might well suggest that this soil had once been a clay enriched or argillic brown earth type

Figure 3.44. *Section through the mound of barrow 36 with the thin rendzina defining below as the dark Organic (turf) band (C. French).*

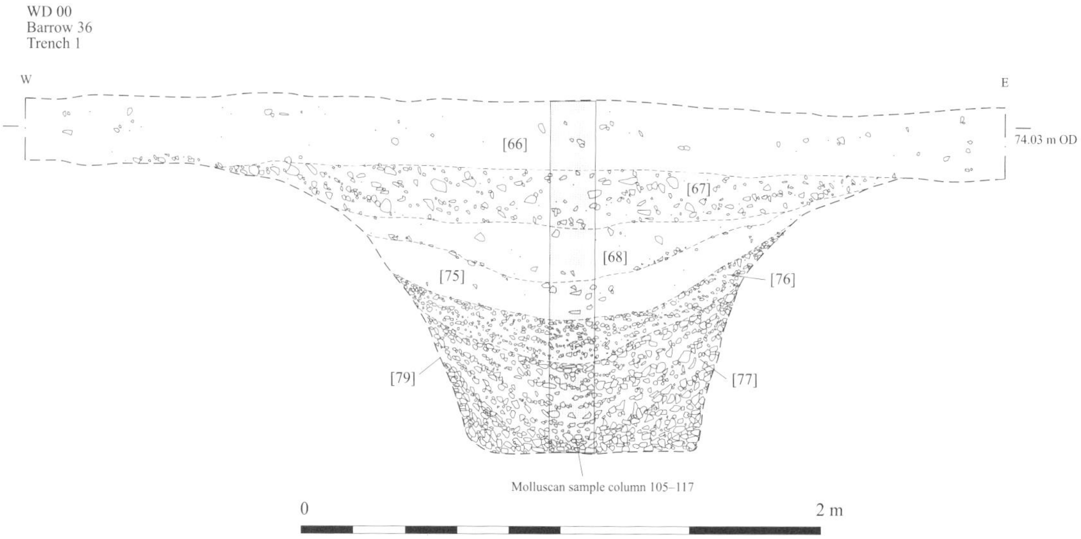

Figure 3.45. *Section through the inner ring-ditch of barrow 36 (M. Brudenell after C. French).*

of soil which had subsequently seen dramatic pedogenetic alteration due to being maintained as long-term grassland prior to burial by the turf barrow mound.

Barrow 41

The section beneath barrow 41 reveals a different story (Figs. 3.46 & 3.47). Beneath the turf mound is an irregular horizon of large, compacted flint cobbles with numerous flint flakes, probably some kind of late glacial drift or solifluction deposit. The flakes presumably come from the mound (moved down through earthworm sorting of the soil) or possibly from some activity before the mound was constructed. Almost no soil cover appears to have survived on these cobbles beneath the turf core of the barrow mound except for a thin turf line, implying both poor soil development and severe truncation in pre-barrow times. Underlying this cobble horizon is a series of nearly contiguous solution hollows with palaeosol material preserved within them exhibiting two horizons. The upper horizon is a reddish-brown silty clay loam with small, sub-angular blocky ped structure and the lower horizon is a dark brown silt clay loam. Most of the clay in these horizons is lo-

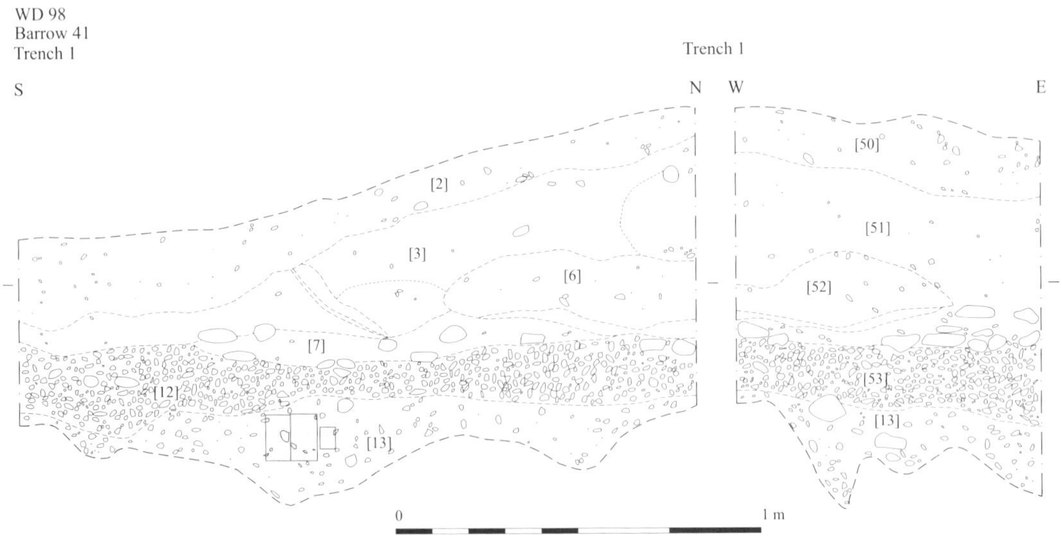

Figure 3.46. *Section through the surviving mound of barrow 41 showing the coombe rock/solifluction deposit and interglacial buried soil beneath (M. Brudenell after H. Lewis).*

cated in the pore spaces, and not incorporated within the fine groundmass, suggesting clay translocation associated with the presence and disturbance of a much thicker original soil (argillic brown earth development), since lost, and/or a great deal of later disturbance of the barrow mound itself occurred. The latter has certainly clearly happened, resulting in a small degraded turf mound.

The translocation of clay and the inclusion of silt and organic matter particles creating a 'dusty' aspect, as seen in the buried soil horizons under barrow 41, are generally related to soil disturbance (Macphail *et al.* 1987), and this could be related to a number of events in this case. These include drift deposition, pre-barrow land use and/or erosion, barrow construction or even modern plough destruction of the mound itself. In summary, this buried profile appears to represent soil formation in an interstadial phase within the last glacial period with strong evidence of much more recent disturbance.

The observable horizon sequence found within the turves of the barrow 41 mound suggests that they originated from an argillic brown earth type of profile (Lewis 1998a). The turves in the barrow were also very clayey, again with frequent clay coatings suggestive of disturbance. In consultation with Richard Macphail, it appears that the turves also show a sequence which suggests previous erosion and/or turf stripping before the later stripping involved in the inclusion of the turves in the barrow. The turves possess a thin (<3 cm) turf horizon (Ah) developed on a horizon depleted of clay and organic matter (Eb). The presence of this leached horizon suggests these sods developed as part of an argillic brown earth (overlying a clay-rich

Figure 3.47. *Section through the surviving mound of barrow 41 showing the coombe rock/solifluction deposit and interglacial buried soil beneath (C. French).*

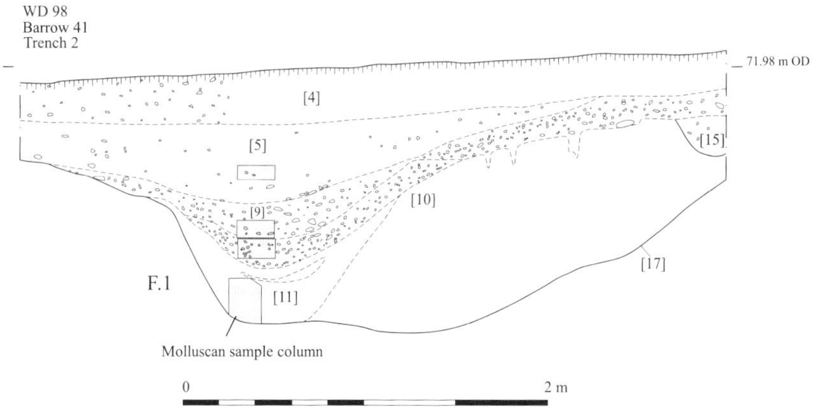

Figure 3.48. *Section through the largely chalk back-filled ring-ditch of barrow 41 (M. Brudenell after C. French).*

Bt horizon), and suggests that at one time some depth of topsoil, now replaced by a thin turf, overlay the original profile. This depth was apparently removed, through stripping or tilling, or was eroded away, with the later thin turf horizon developing some time not long before final stripping for barrow construction. The identification of an argillic brown earth within the turves suggests that they originally overlay a clayey parent material, and it is tempting to suggest that they may have originated overtop of a material such as the flinty drift or solifluction deposit found underneath the barrow – the lack of a topsoil overlying the cobble layer makes it even more tempting to suggest that the turves were dug up from the very spot on which the barrow mound was built.

The profile truncation seen in the mound material itself could also be associated with turf and soil stripping for earlier monument construction in the vicinity (Wyke Down henges 1 and 2, or other barrows in the Wyke Down group), or the stripping and stacking of the turves from the very spot chosen for the building of barrow 41 at a slightly later date.

Barrow 44

Similarly to barrow 36, the lower mound material and thin buried soil of barrow 44 (Fig. 3.49) was also a clay loam. As in the turf mound of barrow 36, there was also evidence of clay intercalation within the turves of the mound, and the strong clay element in this material and the underlying buried soil suggests that this was a clay enriched brown earth type of soil previously to becoming a grassland rendzina type.

These repeated signatures of a pre-barrow, thin turf and rendzina profile actually being indicative of a clay-enriched brown earth soil when examined in thin section is the best evidence yet that there has been major pedogenesis on Wyke Down prior to the Early Bronze Age, or the end of the third/beginning of the second millennia BC. Although one cannot transpose these results to the whole of Cranborne Chase, there are certainly similar hints of this occurring elsewhere at several locations in the study area. For example, there are surviving brown earths (Figs. 2.7 & 2.8) and rare thick argillic brown earths found in the vicinity of Wimborne St Giles (in the auger survey; see above) (Fig. 2.5) and of the very early Holocene at Deer Park Farm in the southern part of the study area (Fig. 2.6).

Pollen analyses from the Wyke Down barrows

Rob Scaife

Sampling for pollen analysis was carried out on barrows 34 and 41, ostensibly as a test of pollen preservation. Monoliths and sequential samples taken from open/excavated sections (Figs. 3.41 & 3.46) were obtained from the old land surface/buried soil, the mound structure including possible turves, and from the surrounding ditches. It was anticipated that any pollen found would provide information relating to the Bronze Age habitat and environment prior to construction of the burial mounds. Although pollen preservation is poor or marginal in calcareous soils, it has been shown that this is variable and that useful information can in some cases be obtained despite calcareous soil and subsoil conditions (Dimbleby & Evans 1977; Dimbleby 1985).

The following samples were selected for assessment of pollen presence or absence and content:

Barrow 34 Trench 3 (31): 2–4 cm
Barrow 34 Trench 3 (31): 12–14 cm
Barrow 34 Trench 3 (31): 14–16 cm
Barrow 34 Trench 3 (31): 8–10 cm
Barrow 34 Trench 3 (36): 24–27 cm
Barrow 34 Trench 3 (37): 1.14–1.16 m
Barrow 34 Trench 3 (38): 1.50–1.55 m
Barrow 41 Trench 1 (13): sample 57
Barrow 41 Trench 1 (13): sample 58

Figure 3.49. *Section through the mound and buried rendzina soil of barrow 44 (C. French).*

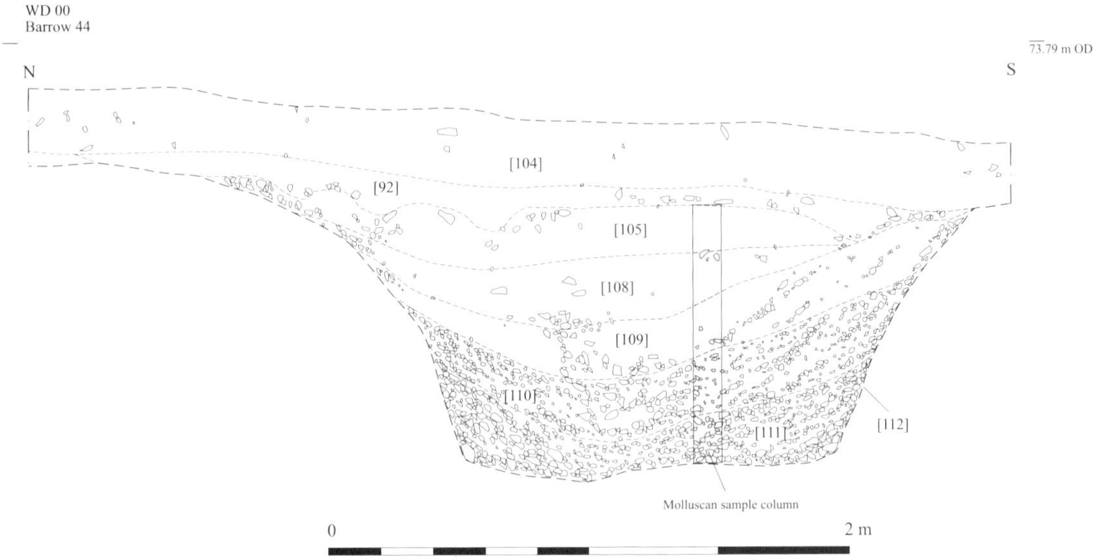

Figure 3.50. *Section through ring-ditch of barrow 44 (M. Brudenell after H. Lewis).*

Results

Overall, pollen preservation was extremely poor with the majority of samples not containing any preserved pollen and spores. Where present, there is clear evidence of differential preservation of those taxa with robust exines and typically spores of ferns which also remain for long periods in oxidized soils.

The buried soil underlying barrow mound 34 contained spores of *Pteridium aquilinum* (bracken), *Dryopteris* type (monolete) spores (typical ferns, e.g. male fern) and *Polypodium vulgare* (common polypody fern). These were the most abundant plant microfossils and are clearly present due to differential preservation of these robust types. It can be noted, however, that *Polypodium vulgare* is perhaps an indicator of earlier woodland on whose trees it was growing (epiphytically). It was particularly present at 12–14 cm. Pollen taxa, where present, are largely those types with thick exines, typically over represented in poor pollen preserving situations. These include Lactucae (dandelion types) and *Bidens* type (e.g. daisy but includes many taxa). Apart from these, the only other taxa recorded include sporadic/occasional grains of *Corylus avellana* (1.14–1.16 m) and Poaceae (grasses) at 12–14 cm.

The buried soil underlying barrow mound 41 was less well-preserved and there was the possibility of contamination from the present surface through root and worm penetration. However, the well sealed, apparently late Devensian deposit found underlying a flint layer did contain greater numbers (but still sparse by normal standards) of pollen grains. This was particularly the case for sample 57 context 13 which produced dominant Poaceae (15 grains), with sporadic occurrences of *Sinapis* type (charlocks), *Anthemis* type (daisy family) a single grain of *Quercus* (oak) and the spores of *Dryopteris* type, *Polypodium vulgare* and *Pteridium aquilinum*.

The pollen content was disappointing from this site given the success of earlier analysis of such soils preserved below and in earthen structures on the chalk (Dimbleby & Evans 1974; Scaife 1984; 1990). This is in spite of the substantial thickness of the buried rendzina soils which in places were decalcified. It is perhaps this maturity of the soil which was detrimental to pollen preservation by virtue of worm sorting/soil turn over and oxidation. A similar situation was found in the analysis of very well-preserved Bronze Age soils at Kingbarrow Ridge near Stonehenge (Scaife & Allen 1994).

From the results obtained, very little interpretation of the vegetation can be made. The presence of grasses and herbs in the late Devensian sediments is commensurate with the cold and open herbaceous environment during Zone III (Loch Lomond stadial). A single grain of *Quercus* (oak) is, however enigmatic for this period and may be either from long-distance transport or from contamination via more recent root channels. Pollen from the Bronze Age soil, whilst degraded, contains Lactucae (i.e. dandelion, sow thistles, hawkbit). This might imply a herbaceous habitat and particularly grassland (although some Lactucae as sow-thistles favour disturbed ground). As noted, *Polypodium vulgare* is typically an epiphyte in mature woodland and its presence here may be seen as an indicator of such although the only tree/shrub pollen found is a single grain of hazel.

Molluscan analyses from the round barrows

Michael J. Allen

A series of 29 samples for land snail analysis were taken from the ditch, palaeosol and mound contexts of barrows 34, 36, 41 and 44 (Figs. 3.96 & 3.97; Tables 3.14–3.16). There was variable shell survival, consequently 2 kg size samples were processed. The molluscan record from these barrows has provided a consistent picture of the contemporary environment during and after barrow construction. The palaeosol beneath barrow 41 exhibited the poorest molluscan preservation of any context examined, exhibiting an impoverished open-country fauna, possibly indicative of cold climate, periglacial-type conditions existing in late glacial times. There were also relatively very low numbers of open-country snails in the fill of the ring-ditch, dominated by *Vallonia costata*. It is likely that there has been some decalcification of the ditch deposits subsequent to their infilling.

On the other hand, all the contexts in barrow 34 exhibited good preservation of molluscs. Each major context was consistently dominated by the open-country species *Vallonia costata, V. excentrica, Helicella itala* and *Vertigo pygmaea*. This is strongly indicative of well-drained, close-cropped, grass sward being established by the Early Bronze Age in this area. There is little evidence of woodland or shaded contexts existing in the near vicinity, but the secondary fill of the inner ditch did exhibit a more restricted range of open-country snails, with the lower secondary fill containing the only shade-loving species present (mainly *Carychium tridentatum* and *Punctum pygmaeum*). However, the presence of these species along with the catholic species *Trichia hispida* and *Cochlicopa* sp. proably reflect the establishment of longer (?ungrazed) grass on and around the barrow, or possibly in the ditch itself. These assemblages are mirrored to an extent by those in the outer ditch, although the diversity of shade-loving species of Clausiliidae may reflect changes in the wider landscape rather than the barrow itself.

Bottlebush Down

Four sample sections were excavated through the northern and southern Dorset Cursus ditch and bank on the *c.* 1 km stretch between the Bottlebush Down terminal and the Roman Ackling Dyke (Figs. 3.33, 3.51–3.57). These were situated to the north of the sections through the Cursus ditches excavated by Bradley in 1982 and 1984 (in Barrett *et al.* 1991a) (Fig. 1.2).

The Dorset Cursus

Charles French

Unlike those sections previously excavated by Richard Bradley (Barrett *et al.* 1991a, figs. 2.12 & 2.13), the four sample sections revealed primary and secondary fills that were dominated by clean chalk rubble (Figs. 3.33, 3.51 & 3.55–3.57). The surviving bank was much denuded in height (<30 cm) and ard marks surviving in the upper surface of the subsoil in one location suggest that it had been ploughed in prehistoric or Early Roman times, possibly even related to the construction of the Ackling Dyke. The chalk back-fills suggest that the Bottlebush Down length of cursus bank was deliberately slighted back into the ditch shortly after the monument was constructed. This is in contrast to the other fill sequences observed further to the south in the same

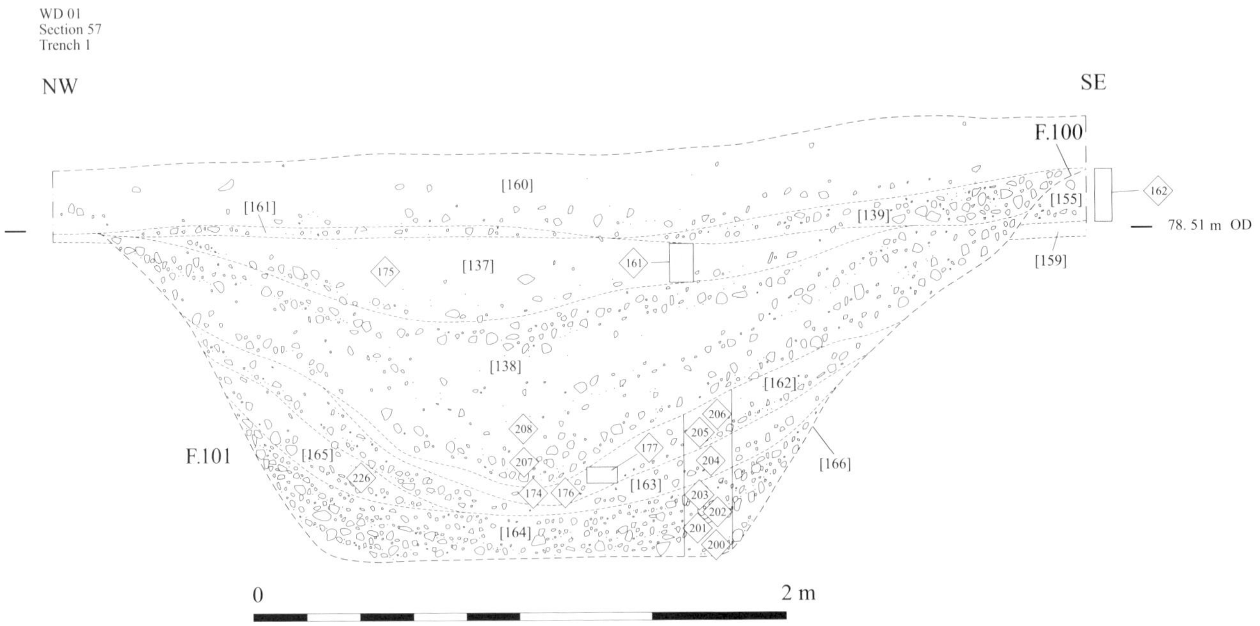

Figure 3.51. *The section of the north ditch of the Dorset Cursus in Trench 1 (M. Brudenell after C. French).*

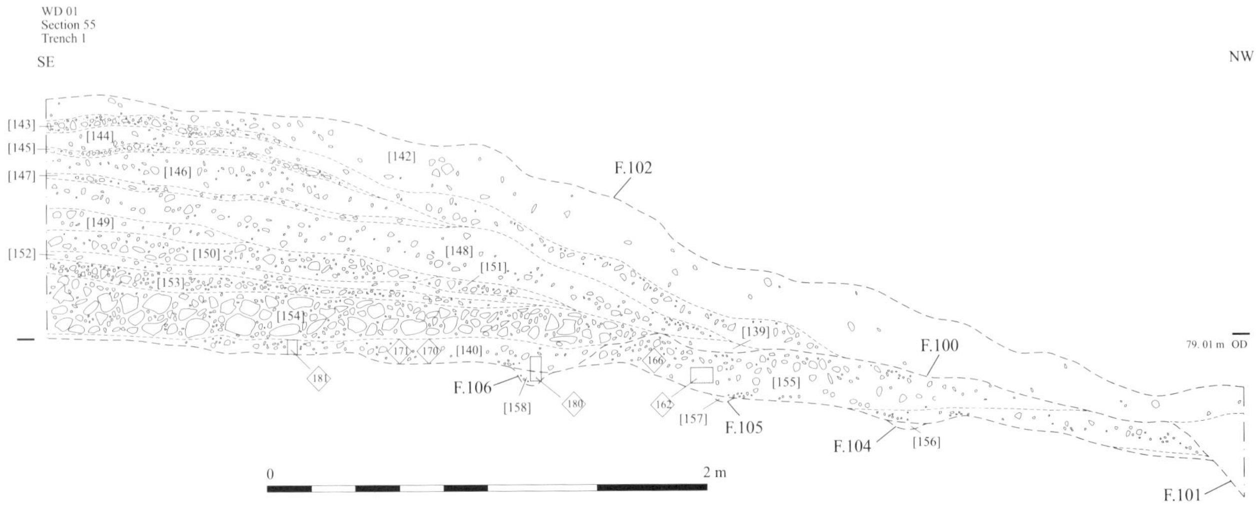

Figure 3.52. *The north ditch of the Dorset Cursus in Trench 1 continued, with the cursus bank, old land surface and the northern side of the agger of the Roman Ackling Dyke in Trench 1 (M. Brudenell after F. Sturt).*

monument on Wyke Down and in Fir Tree Field by Bradley (in Barrett *et al*. 1991a, 43–7) where homogeneous organic silt loam and fine chalk rubble fills suggested a natural infilling over a lengthy time period. Suggestions of relatively quick backfilling in this northern sector of what is believed to be the first major length of Dorset Cursus would suggest a complete change in the monument's importance, visibility and possibly function in the landscape by the later Neolithic. In addition, it may imply a complete re-alignment of any fields and boundaries in this part of the downland landscape. In contrast, on the other side of the valley to the south, the Gussage Cow Down Iron Age settlement boundaries and field systems clearly align themselves at right angles to the pre-existing Cursus which suggests long-term landscape organizational continuity (Fig. 2.13).

The buried soil present beneath the surviving Cursus banks was also already a thin rendzina by the end of the fourth millennium BC (Figs. 3.52–3.55). It appears to exhibit a slightly greater organic component, higher oriented clay component and better structure in comparison to the rendzinas observed

Figure 3.53. *View of Trench 1 across the Dorset Cursus north ditch and bank, and the Roman Ackling Dyke in the background (C. French).*

Figure 3.54. *The criss-cross ard marks preserved at the base of the buried rendzina/upper surface of the chalk substrate, sealed beneath the surviving chalk bank of the Cursus and the agger of the Ackling Dyke (C. French).*

beneath the subsequent Bronze Age barrows nearby (French *et al.* 2000, and see above). The presence of this soil type mirrors the palaeosol evidence from the near contemporary Knowlton South Circle henge further down-valley (Lewis 2001; forthcoming) and from beneath the Wyke Down barrows that were investigated in previous seasons of this project (French *et al.* 2000). Thus thin rendzina soils appear to be the common and ubiquitous downland soil type by the fourth millennium BC in the upper Allen valley area, and have remained so ever since.

From the two trenches across the northern Cursus ditch and bank and the two trenches across the southern Cursus ditch and bank in the Wyke Down to Bottlebush Down terminal length of Cursus (Figs. 3.33, 3.51 & 3.55–3.37) the following contexts were investigated using micromorphological analysis, as follows:

0 the pre-cursus bank buried soil from the north Cursus bank section in Trench 1 (3 samples);
1 the pre-Ackling Dyke buried soil and associated possible ploughmarks cut into the upper surface of the chalk subsoil in Trench 1 (1 sample);
2 the make-up of the Cursus bank in Trench 1 (1 sample);
3 probable standstill horizons in the primary, secondary and tertiary fills of the Cursus ditches in Trenches 1–4 (14 samples);
4 pre-Ackling Dyke ground preparation at the top of the infilled southern Cursus ditch in Trench 2 (3 samples).

The descriptions are summarized in Table 3.2 with detailed descriptions given in Appendix 4.7.

Cursus ditch fills

Probable standstill horizons were recognized at all fill levels from the primary, secondary and tertiary fill zones of the north and south Cursus ditch (Figs. 3.52 & 3.55–3.57). In the field these were described as composed of dark brown, small irregular aggregated, silt loam lenses.

In thin section, these lenses were composed of pale brown to pale greyish-brown, micritic (calcium carbonate) dominated fine sand which exhibited a fine micro-aggregated structure, with and without the inclusion of small irregular chalk fragments. These micro-aggregates are strongly reminiscent of a turf fabric that would be consistent with that formed on a rendzina. Indeed this fabric type has been commonly observed within the turf make-up of the Wyke Down barrow mounds investigated in this project (Lewis 1998a & Appendix 4.5).

Table 3.2. *Micromorphological descriptions for the palaeosols associated with the Dorset Cursus investigations by sample and context.*

Sample	Context	Description	Interpretation
Cursus ditch fills			
177	162; Tr. 1	weathered chalk rubble over even mix of chalk fragments and micritic fine sand with fragments of turf; in base of north Cursus ditch	chalk rubble-dumped material with turf fragments of rendsina in upper primary fill of north Cursus ditch
161	137; Tr. 1	amorphous iron-/organic-stained, chalk free, micritic sand with pellety to small, sub-angular blocky peds; in tertiary fill of north Cursus ditch	oxidized former turf in north Cursus ditch
179 (& 190)	182; Tr. 2	even mix of chalk fragments and micritic very fine sand with pellety to small, sub-angular blocky peds; in base of north Cursus ditch	turf and fine chalk rubble in upper primary/lower secondary level in north Cursus ditch
195 (& 193)	196; Tr. 3	mix of chalk fragments and pale brown micritic fine sand in small, irregular blocky peds and pellety void infills; in lower secondary fill of south Cursus ditch	mix of lower A horizon-like soil and weathered chalk as stabilisation lens in lower secondary fill of south Cursus ditch
194	197; Tr. 3	mix of micritic fabric with fine chalk fragments over dense chalk; in tertiary fill of south ditch of in tertiary fill of south Cursus ditch	lower A and A/C horizon material over redeposited chalk subsoil south Cursus ditch
192/1	194/upper; Tr. 3	brown micritic sand with a few medium fragments of chalk; in tertiary fill of south Cursus ditch	soil and minor chalk rubble accumulation in top of Cursus ditch; possible windblown fine soil component in south Cursus ditch
192/2	194/lower; Tr. 3	chalk-free, pale brown micritic fine sand in irregular small aggregates to pellety; in tertiary fill of south Cursus ditch	stabilization zone; lower A horizon-like material in top of Cursus ditch; possibly wind-sorted soil accumulation in south Cursus ditch
198	207/8; Tr. 4	alternating lenses of fine chalk rubble and pellety to small blocky micritic fine sand with punctuations of amorphous fine sand; in primary fill of south Cursus ditch	dumped/slumped chalk subsoil and turf-like material on the inner edge of the south Cursus ditch iron/organic stained micritic Cursus ditch
225/1	210/upper; Tr. 4	fine weathered chalk rubble over mix of fine weathered chalk and micritic fine sand in small, irregular to sub-rounded aggregates; in primary fill of south Cursus ditch	redeposited lower A horizon/turf material in south Cursus ditch
225/2	210/lower; Tr. 4	as above, except irregular impregnation with amorphous iron	as above; some oxidation
199	206; Tr. 4	even mix of fine chalk fragments and micritic fine sand in a porous, Cursus ditch pellety matrix; in secondary fill of south Cursus ditch	stabilization zone of turf in south
197	202; Tr. 4	weathered chalk, chalk fragments and micrite; in tertiary fill of south Cursus ditch	chalk subsoil backfill probably from associated bank in south Cursus ditch
Cursus Bank			
162	155; Tr. 1	porous mixture of chalk fragments and micritic fine sand with pellety to small, make-up of north Cursus bank	mixture of chalk subsoil and rendzina soil material in north Cursus bank sub-angular blocky peds
Pre-Ackling Dyke activity			
180	158; Tr. 1	mixture of 30% very fine chalk fragments and 70% micritic fine sand infilling open V-shaped cut into weathered chalk of C horizon; at base of buried soil	lower A horizon-type material infilling a pre-first century AD ard mark defining in the top of the chalk subsoil beneath Roman Ackling Dyke
163	177; Tr. 2	amorphous iron/organic stained, micritic fine sand with columnar/blocky peds; in tertiary fill of north Cursus ditch	oxidized former turf to level in top of north Cursus ditch prior to construction of Ackling Dyke in first century AD
178	141; Tr. 2	reddish-brown, micritic fine sand with columnar/blocky structure; in tertiary fill of north Cursus ditch	lower A-type material of a rendzina in top of north Cursus ditch prior to construction of Ackling Dyke in first century AD
188	177/9; Tr. 2	reddish/dark brown, chalk free, micritic fine sand as upper tertiary fill of south Cursus ditch	turf-like material formed in windblown material; used to finally infill/level north Cursus ditch prior to construction of Ackling Dyke in first century AD
Buried soil beneath Ackling Dyke			
185 (& 186)	170; Tr. 2	micritic fine sand with occasional fine chalk fragments occasional fine chalk fragments and reddish-brown/black micritic sand aggregates in small, irregular blocky peds; buried soil beneath Ackling Dyke	A horizon beneath Ackling Dyke
Ackling Dyke			
181	140; Tr. 1	brown micritic fine sand with scattered chalk over dense chalk; buried soil beneath Ackling Dyke	lower Ah horizon over A/C weathered chalk; truncated buried soil beneath first century AD Ackling Dyke

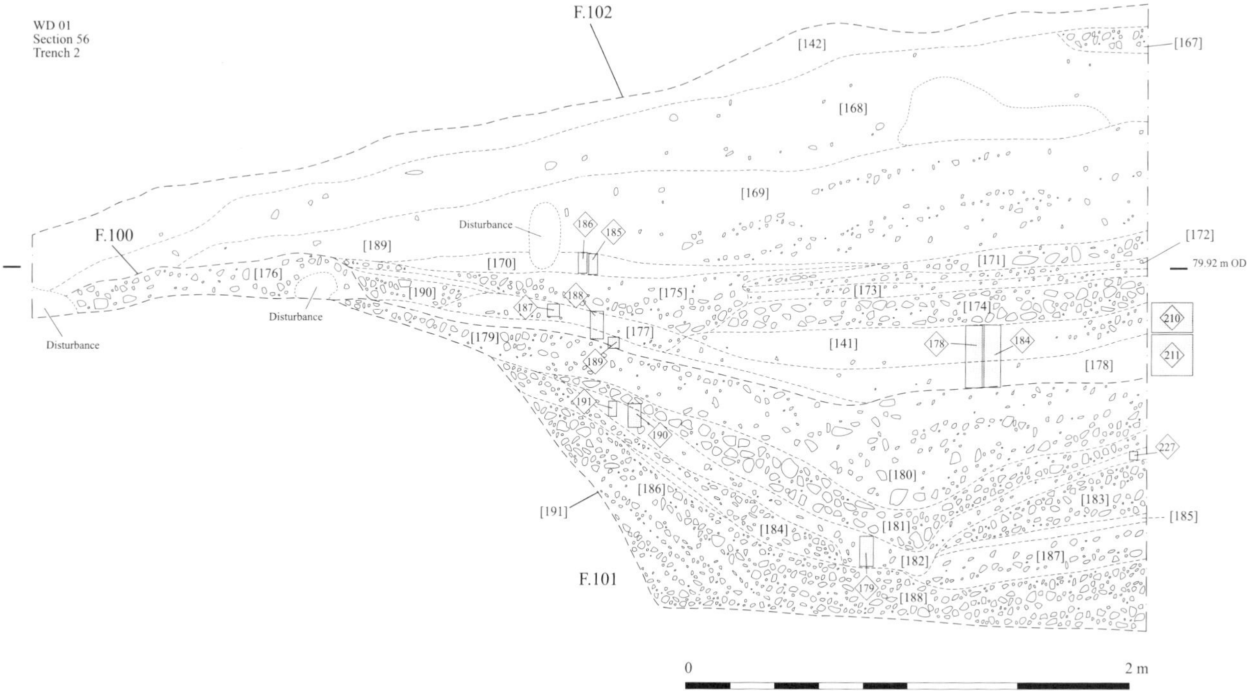

Figure 3.55. *Section of the north ditch of the Cursus and the southern side of the agger of the Roman Ackling Dyke in Trench 2 (M. Brudenell after D. Robinson & M. Green).*

Cursus bank make-up

In thin section, the north Cursus bank sampled (Fig. 3.52) was composed of a porous mixture of fine chalk fragments and micritic fine sand. The soil fabric was either pellety or small sub-angular blocky peds. This combination of material is suggestive of chalk subsoil upcast and turf and lower A horizon material of a rendzina, respectively.

Pre-Ackling Dyke soils and turf

Beneath the north and south Cursus banks there was a thin (*c.* 12–15 cm thick) rendzina buried soil surviving. There was also a thin (*c.* 10–12 cm thick) rendzina buried soil beneath the Ackling Dyke in Trench 1 (Fig. 3.52) at the base of which was a series of criss-cross ard marks (*c.* 5–8 cm deep & 5–12 cm in width; open V-shaped in section) (Fig. 3.54). The only ardmark sampled contained a juxtaposed mixture of *c.* 30 per cent very fine chalk fragments and *c.* 70 per cent micritic fine sand with a small, irregular blocky structure. This heterogeneous mixture of soil and subsoil is indicative of mechanical disturbance, and similar sized aggregates were certainly noted in ard ploughing experiments by Lewis (1998b). In addition, in the buried soil above the ard marks, there were irregular blocky aggregates of micrite and fine chalk fragments which could also have resulted from mechanical mixing processes associated with this other evidence for ploughing.

In Trench 2 (Fig. 3.55) through the Ackling Dyke and the north Cursus ditch, the tertiary fill of the Cursus ditch was marked by some very different deposits as compared to the other ditch sections. First there was a layer of reddish-brown, chalk-free, micritic fine to very fine sand (context 141) overlain by a distinctive and substantial layer of pellety micritic fine sand (contexts 177 & 179). Context 141 appears to be a turf-like fabric formed in or accumulating with well-sorted windblown material. This pale reddish-brown, fine, well-sorted, chalk-free fabric was also noted in the tertiary fill of Wyke Down barrow 34 and the Iron Age ditch on Gussage Cow Down (this volume) where both the molluscan and soil data suggest that this is wind-sorted and blown material derived from dry soils immediately adjacent. Contexts 177 and 179 are horizons of laid turves, which could also be seen in plan as rectilinear clods.

Unusually, the redeposited turves had a high amorphous organic content, as well as much amorphous iron replaced organic matter, and the fine groundmass contained a small but significant silty clay component. These features were not observed elsewhere in the soils associated either with the Cursus or the nearby barrow mounds make-up and buried soils. This might suggest that there were soils surviving in the vicinity of this part of the Cursus that were better developed and/or of this greater organic status

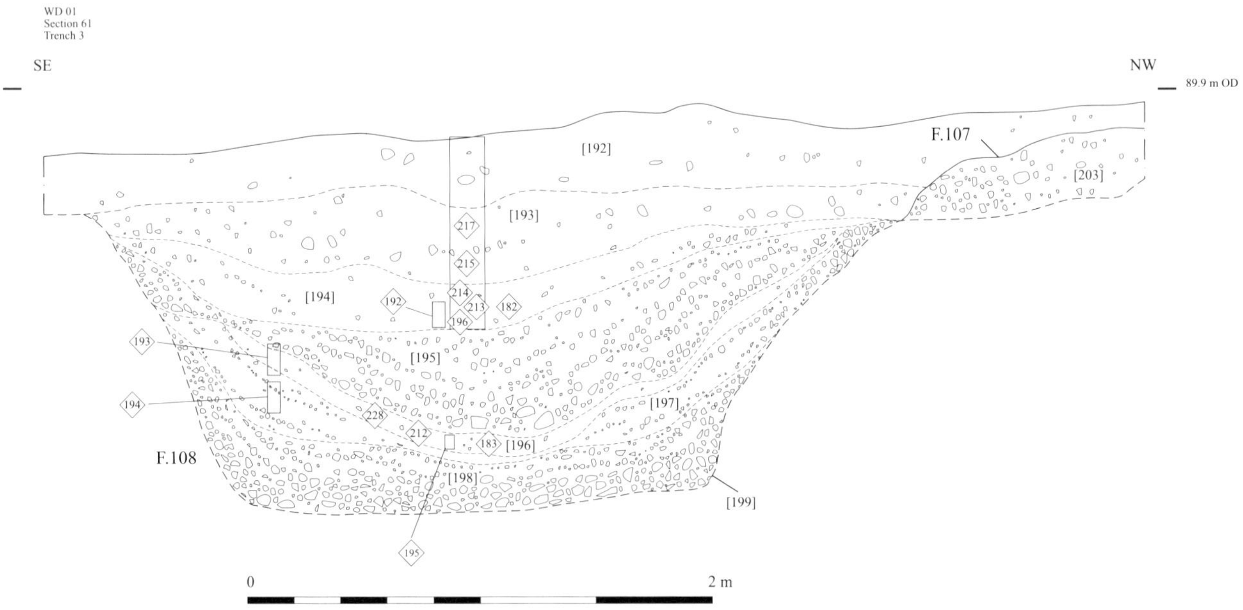

Figure 3.56. *Section through the south ditch of the Cursus at the Bottlebush Down terminal in Trench 3 (M. Brudenell after H. Lewis).*

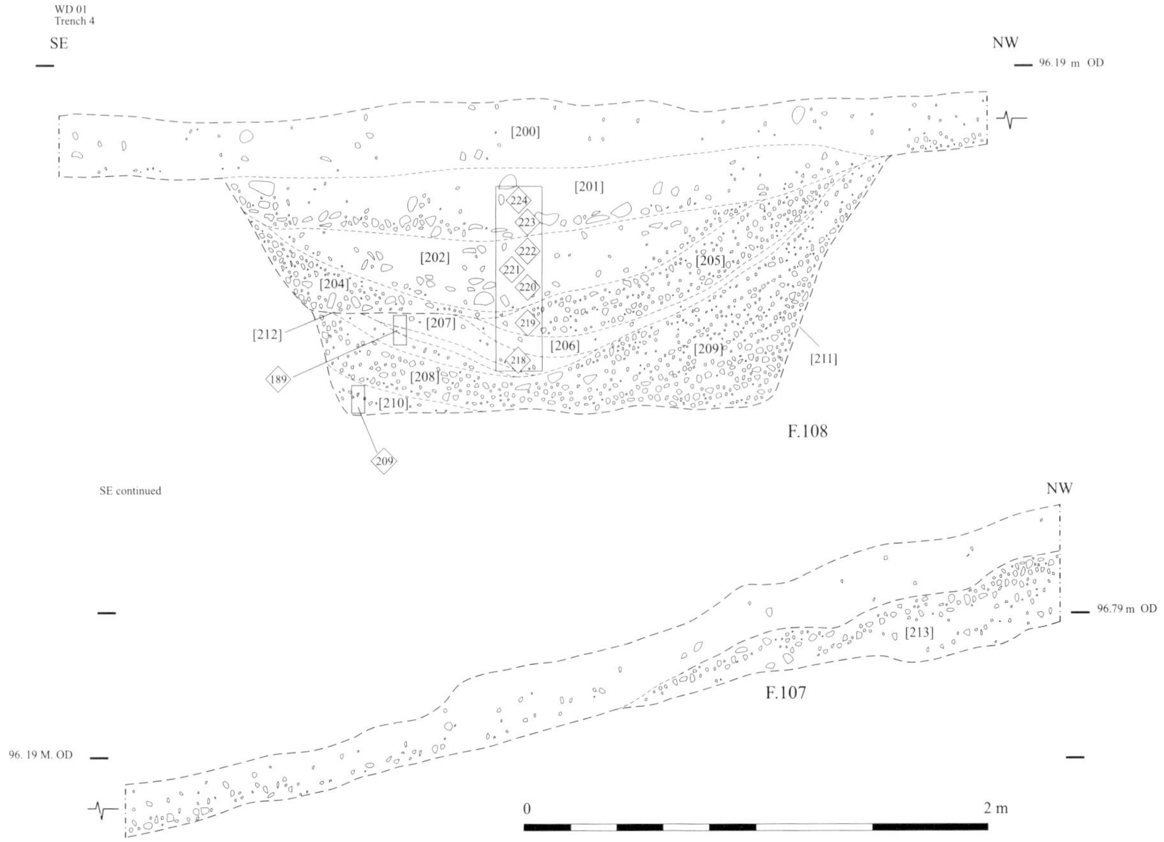

Figure 3.57. *Section through the south ditch of the Cursus and denuded bank at the Bottlebush Down terminal in Trench 4 (M. Brudenell after C. French).*

and structural integrity, pointing to the presence of former brown earth soils co-existing with depleted rendzinas in this part of the chalk downland in the upper Allen valley.

Discussion
The presence of thin, single horizon, highly calcareous, sandy soils beneath both Cursus banks and the Roman Ackling Dyke indicates that rendzina soils were already well established by the later Neolithic period and continued to be maintained throughout later prehistoric and into Roman times. But the turves used to level off the north Cursus ditch where it was to be diagonally crossed by the construction of the Ackling Dyke in the first century AD had a much greater organic status and structural integrity which indicate that there were possibly former brown earths still surviving in the immediate vicinity. This is a good example of related soil type variation occurring within short distances in this chalk downland landscape which hints at former soil types and routes of soil type change.

The Roman Road - Ackling Dyke
Charles French & Helen Lewis
Two half sections were cut through the first-century AD Roman Ackling Dyke on Bottlebush Down (Figs. 3.52 & 3.55). The road section exhibited a profile of alternating levels of chalk rubble make-up and either smooth chalk or metalled surfaces, similar to that excavated by Pitt Rivers at Bokerley Dyke to the north (Pitt Rivers 1892). Most importantly, the road overlay a thin (<10 cm) and compressed rendzina soil profile (Fig. 3.52).

In addition, one trench revealed a stratigraphic sequence which suggested that the line of the north Cursus ditch was still visible in Early Roman times. Prior to the construction of the Roman road, this remaining earthwork was back-filled and leveled off with laid turves. These are described above. Rather unexpectedly, micromorphological analysis of these turves indicated that they were clay-rich and well structured, with illuvial clay integral within the groundmass, unlike all the turf fabrics analysed from the other sites. This suggests that there may well have been areas in the near vicinity in this landscape of well-developed turf which were exploited at this time which had not been denuded previously. The use of these turves to backfill the ditch before road construction could also represent an engineering decision, in which case they may have been brought from further afield, perhaps from a site in the area, but probably not from the immediate surroundings, which would be predicted on the basis of this study to have shallow rendzinas well before the Roman period.

Monkton-up-Wimborne

On the opposite side of the study area on Bottlebush Down near Monkton-up-Wimborne important Neolithic sites were excavated by Green and sampled for palaeoenvironmental study. These included the Neolithic pit circle/shaft complex (MUW 97), the Neolithic enclosure (MUW 00), a pond barrow (MUW 02) and a later prehistoric enclosure (MUW 96) (Fig. 1.2).

Monkton-up-Wimborne Neolithic enclosure (MUW 00)
Charles French, Martin Green & Michael J. Allen
Near the top of Bottlebush Down at 94.65 m OD, a subrectangular enclosure (Figs. 1.2 & 3.58) was discovered adjacent to The Warren, Monkton-up-Wimborne (*c.* 450 m to the north of the Monkton-up-Wimborne, Late Neolithic, oval pit enclosure with shaft) (Green 2000). Enclosures of this type are well known in the Neolithic (e.g. Atkinson *et al.* 1951b) and were often termed mortuary enclosures. However, recent exhaustive analysis by Roy Loveday (2006; Loveday & Petchey 1982) has shown there is little, if any, evidence for this and he suggests the simple term long enclosures be used. This site was sampled by one northeast–southwest trench excavated across the southern end of the enclosure (Fig. 3.59). Although the northern end of the enclosure was not located (beyond the fence in the woodland), it is at least 20 m in length from the fence. This enclosure had an internal width of about 11 m and an external width of about 15 m. Two small postholes were found about 2.5 m in from the internal side of the eastern ditch section.

The ditch profiles (Figs. 3.59 & 3.60) were about 2–2.25 m in width, 1.2–1.25 m deep and flat bottomed, and much disturbed by subsequent rabbit burrowing. The primary fill comprised weathered chalk rubble interrupted in its upper levels by fine layers (1–2 cm thick) of black organic topsoil material. The secondary fill comprised a combination of these layers with a grey silt loam and fine chalk rubble, capped by a discontinuous horizon of black organic loam, presumed to be a turf stabilization layer, which was subsequently confirmed by micromorphological analysis. The orientation of the primary and secondary fills suggests that material was coming into the ditch from both sides through natural weathering processes and the redeposition by dumping of bank material. The tertiary fill, in contrast, was a brown, fine silt loam, with only a minor chalk rubble input. This inorganic silt and very fine sand is similar to the windblown material that was also found in the tertiary fill of Wyke Down Bronze Age barrow 38 ring-ditch, and of the Iron Age ditch on Gussage Cow Down (see above). A single small sherd of Beaker pottery and a

Figure 3.58. *Low-level oblique aerial view of the Monkton-up-Wimborne Neolithic enclosure (MUW 00) (M. Green).*

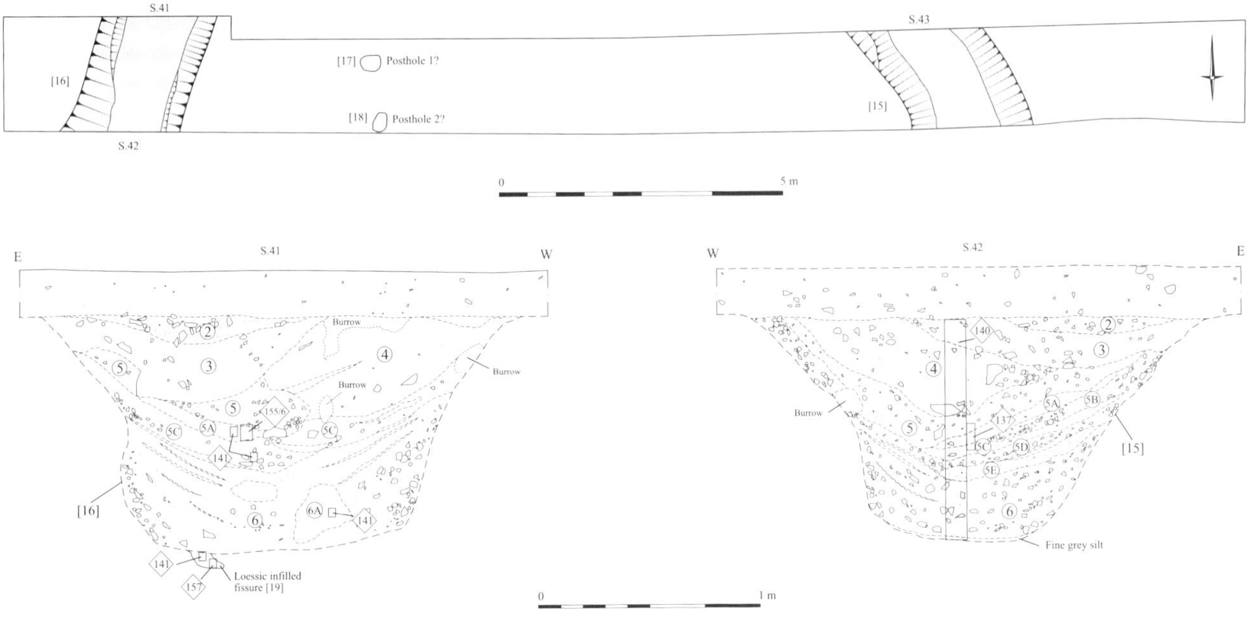

Figure 3.59. *Plan and ditch sections of the Neolithic enclosure at Monkton-up-Wimborne (MUW 00) showing the molluscan sample locations, and section S.42 showing the thin turf lines in the primary (6) and lower secondary fills (5E) (M. Brudenell after M. Green & H. Lewis).*

Figure 3.60. *Section S.41 of the Neolithic enclosure at Monkton-up-Wimborne (MUW 00) showing the thin turf lines (C. French).*

Figure 3.61. *Low-level oblique aerial view of the Monkton-up-Wimborne pit circle and shaft (MUW 97), looking northeast (M. Green).*

barbed and tanged arrowhead were recovered from the tertiary fill, but a number of flints more diagnostic of Neolithic times were recovered from all fill levels. There was also the suggestion that the upper and outer edge of the ditch infill had been subsequently re-cut by a narrow/shallow ditch.

Samples for molluscan analysis were taken from both ditch sections, and spot samples for micromorphological and palynological analyses were taken from the probable turf/standstill horizon which defined the upper surface of the secondary fill in the eastern ditch profile. This plough-damaged monument has produced molluscan data from the ditch suggesting a mixed open environment with some local woodland, rapidly giving way to assemblages reflecting very open habitats (see Allen below). This differs from the comparable Neolithic mortuary enclosure at nearby Handley Down (Allen 1997b), and suggests that this monument was located in an environment with larger clearings than at Handley Down. Despite the absence of any surviving buried soil associated with this monument, the redeposited soil and turf in the secondary fill of the enclosure ditch was more organic and clay-rich than the common later rendzinas, perhaps hinting at the former presence of a developed brown earth soil.

Monkton-up-Wimborne Late Neolithic pit circle/shaft complex (MUW 97) (see Appendix 4.9)

Martin Green

Summary

This remarkable monument was noticed on a number of aerial photographs taken on 26/9/1989 and held at the National Monument Record Centre (NMR4509/54, NMR4509/62 & NMR4509/65–6).

Located on a chalk ridge at 90 m OD overlooking the upper Allen Valley it forms part of the eastern edge of the monument cluster detailed in Figure 1.2. It consists of a ring of fourteen unevenly spaced pits with a diameter of 35 m which are broken by wide entrance gaps to the west of 14 m and to the east of 11 m (Figs. 3.61 & 3.62). Outside this latter entrance a fence line consisting of eight surviving postholes was traced to the edges of the excavation for a distance of 26 m with two larger postholes (F.16 & F.17) placed within the site axis some 7 m beyond the fence.

The pit circle enclosed a large vertically sided central pit some 11 m in diameter (F.1) which proved to be 1.5 m deep. The floor of this feature was remarkably smooth and even, which resulted from digging having ceased when a significant natural horizontal

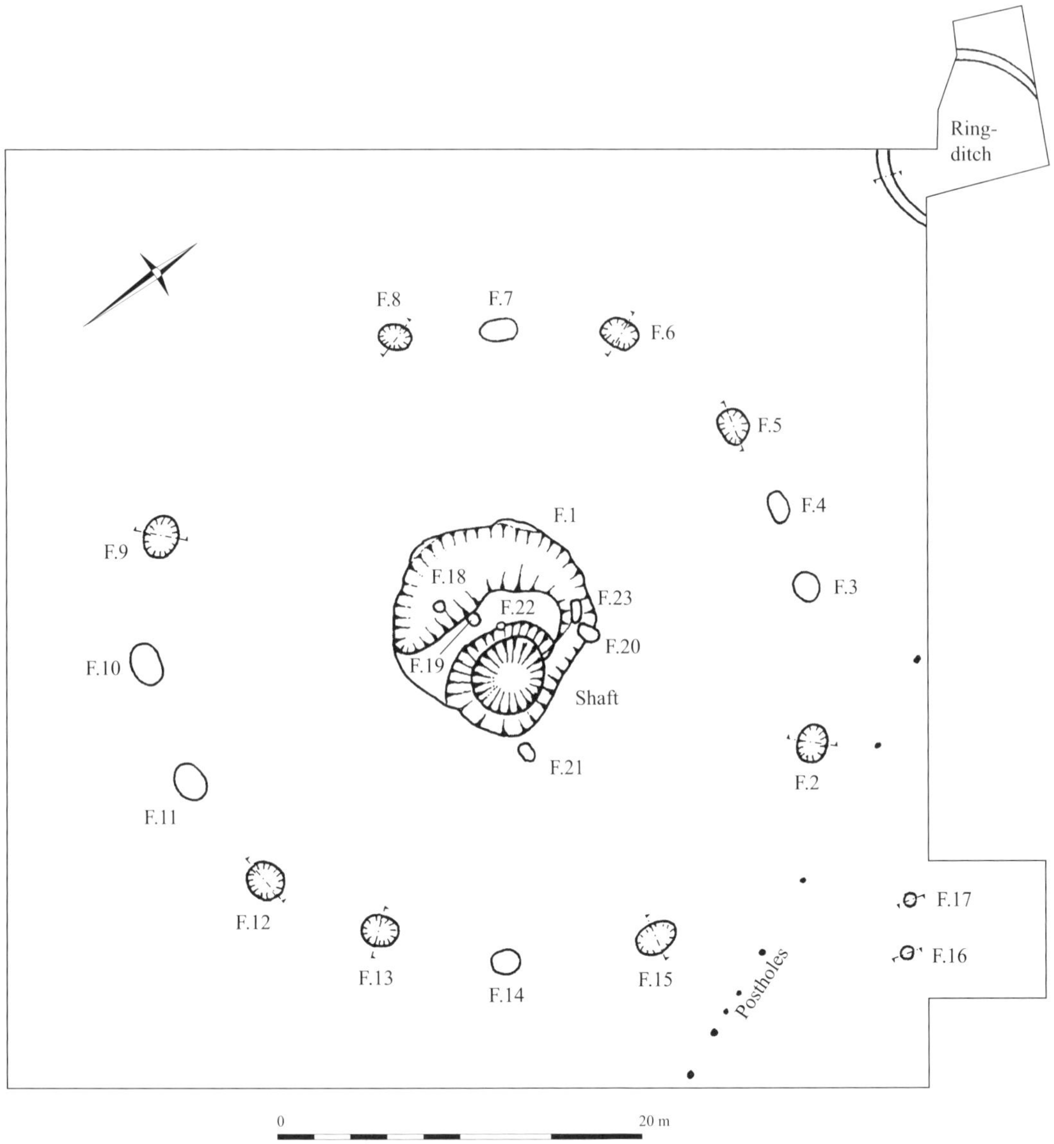

Figure 3.62. *Outline plan of the Monkton-up-Wimborne pit circle and shaft complex and associated features (N. Griffiths after M. Green).*

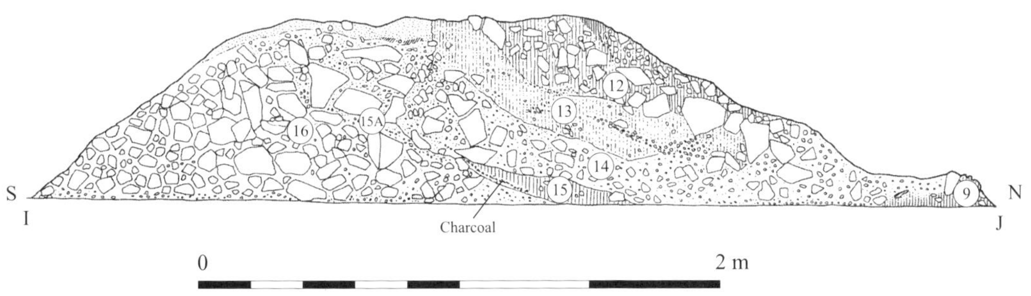

Figure 3.63. *Monkton-up-Wimborne pit circle and shaft (1997): section of the bank/platform within the F.1 pit (N. Griffiths after M. Green).*

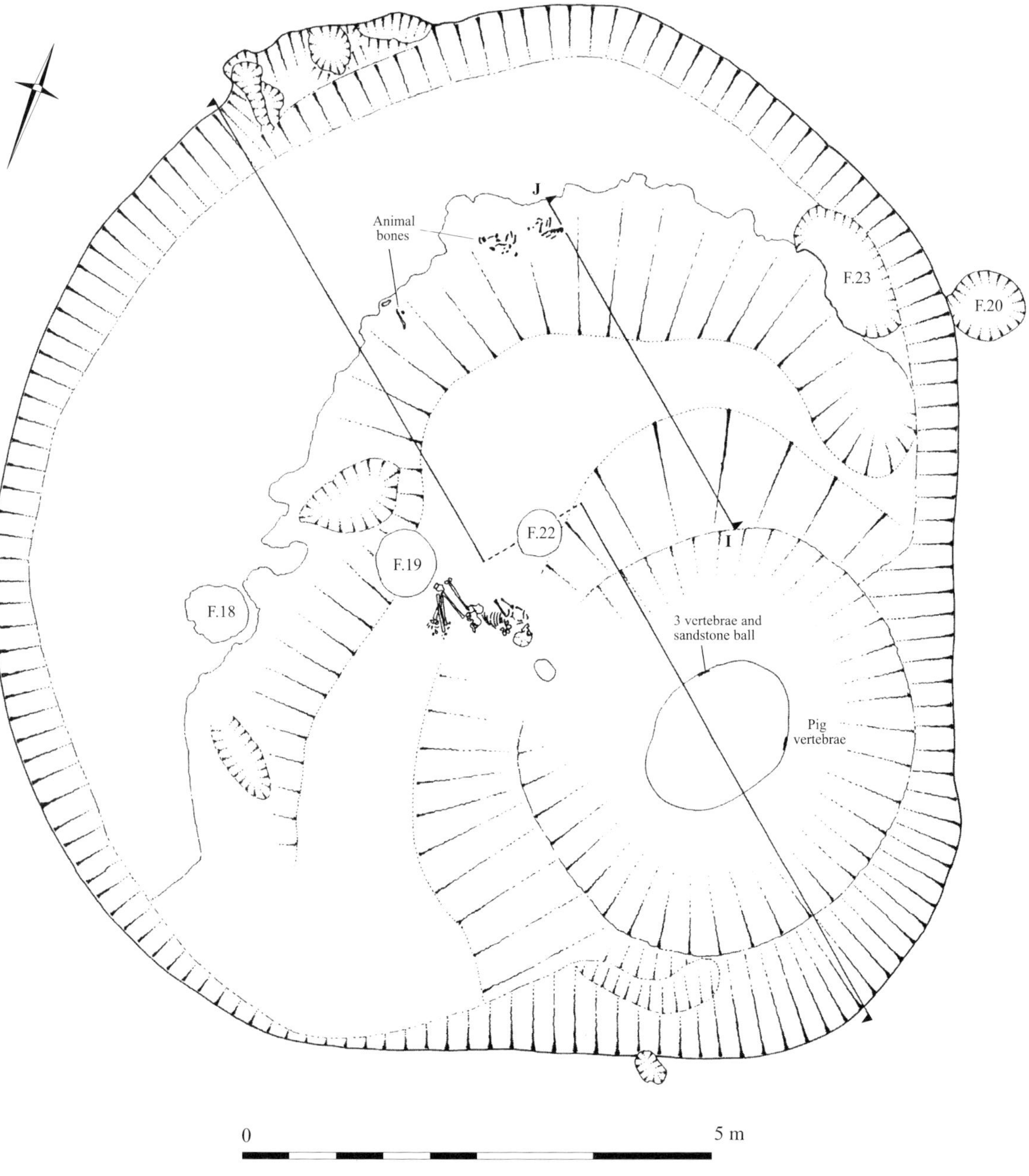

Figure 3.64. *Plan of the F.1 pit/shaft at Monkton-up-Wimborne (N. Griffiths after M. Green).*

joint within the chalk was reached. Within the secondary fills of the pit an Early Bronze Age crouched burial was uncovered (Fig. 3.64).

Dug through the floor of F.1 and partly undercutting its northern edge a grave was uncovered (F.23). It contained four tightly crouched individuals and had been carefully backfilled with neat chalk rubble. Much of the eastern floor and side of F.1 had been destroyed by the excavation of a tapering shaft some 4.5 m wide at the top and nearly 7 m deep. Some of the rubble from this latter excavation had been used to construct a sloping bank or platform providing an access point into F.1 from the south and bounding the shaft on its western side. Episodically constructed, the compacted surface of this platform and discovery of much animal bone on its western edge suggests it was a focus for feasting and other ceremonial activity.

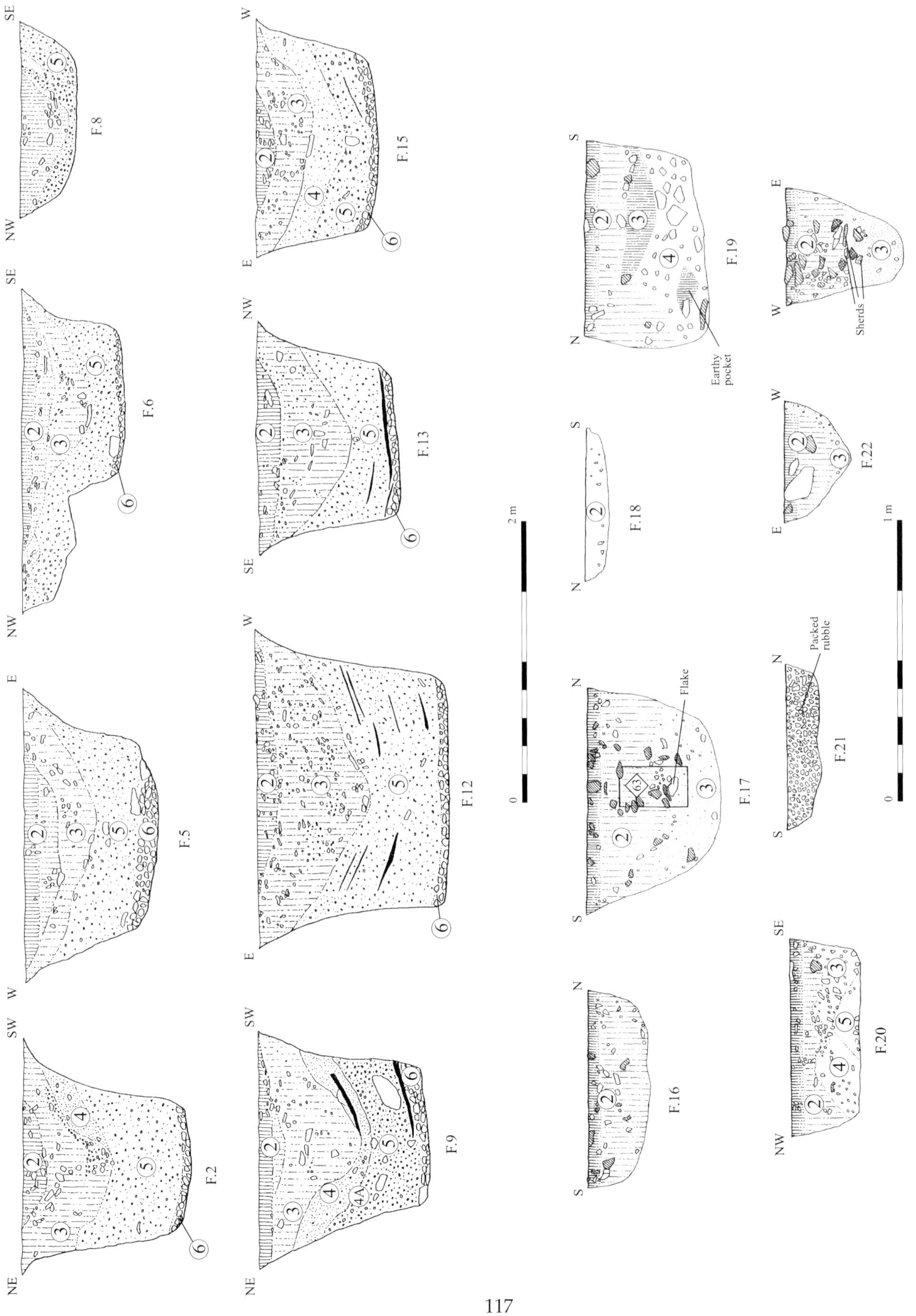

Figure 3.65. *MUW 97: sections of the pits comprising the outer pit circle (above), and other pit/posthole sections (below) (N. Griffiths after M. Green).*

A cemetery of three ring-ditches lies just to the north and west of the site, the nearest of which was incorporated by an extension to the north corner of the excavation (Fig. 3.62).

Outer pit circle

Of the fourteen pits forming the circle, four were excavated on the western side F.2, F.5, F.6 and F.8 and four on the eastern side F.9, F.12, F.13 and F.15 which included all the terminal pits. Not only were the pits unevenly spaced — excluding the entrances from between 2–6 m apart — but also varied considerably in depth from 0.38–1.4 m. No consistent pattern related to position or depth was apparent.

The pit fills were consistent (Fig. 3.65) and clearly had never held posts or stones but had been left to weather naturally after having been dug. The basal fill (layer 6) consisted of a concreted matrix of dark grey silt and chalk with a few scattered struck flints. At the interface between this layer and the overlying main chalk rubble (layer 5) occasional large blocks of chalk had been deposited, together with, in one case (F.9), a large cow bone. As we shall see later these blocks provide important evidence in understanding the site's phasing. Within layer 5 were occasional organic-rich lenses. Finer rubble (layer 4) completed the primary fills. The lower secondary fill (layer 3) consisted of a dark grey silt with much natural flint. A dark brown clayey decalcified soil (layer 2), also with much small natural flint, completed the sequence.

Finds within the pits were sparse and consisted of worked flint including part of a fabricator and a scraper in the primary fills, with a kite shaped arrowhead (Fig. A4.23, no. 8) and a flake from a polished flint axe from the secondary deposits. Small scraps of Peterborough Ware were present in the secondary fills of pits F.5 and F.12 with animal bone present in very small amounts in both primary and secondary levels.

Features outside the eastern entrance

Lying just outside the eastern entrance, a fence line consisting of eight postholes was located. The full extent of this was not uncovered as it continued to the edges of the excavated area in both directions. Where revealed, it ran for a distance of 26 m at a shallow oblique angle across the eastern entrance.

The southern end was more complete with five postholes covering a distance of 8 m at spacings between 1–2 m. The remaining three postholes covered 18 m with posthole 3 placed centrally in the eastern entrance. The shallowness of these postholes suggest others have been ploughed away. The purpose of the fence may have been to screen the activities taking place within the site. Further to the east, some 7 m beyond the fence line but still within the area covered by the site axis, two much larger postholes (F.16 & F.17) were uncovered. A well-stratified chisel arrowhead (Fig. A4.23, no. 9) found in F.17 suggests contemporaneity with the main building phase of the monument. These posts mark the position of the eastern entrance as they would have been the first visible elements to those approaching the monument from that direction. Their locations within the site axis suggest a possible additional astronomical function. Further excavation beyond these postholes showed they were not part of a larger structure lying beyond the confines of the original trench.

Central pit F.1, grave F.23, shaft and platform

The central pit proved to be remarkably regular with near vertical sides and a beautifully smooth and even surface 1.50 m below the ploughsoil (Figs. 3.64–3.66). This floor-like surface had been created by digging through an upper layer of small shattered chalk and then stopping at the interface with a lower layer of more massive blocky chalk thus producing the even 'natural' surface. A number of hazel twigs were recovered from the top of this surface which gave radiocarbon determinations of 3330–3210, 3190–3150 and 3130–2910 cal. BC (4427±42 BP; Wk-18753) (Table 1.1). However, a large part of this 'floor' had been subsequently destroyed by the digging of a shaft and much of the remainder covered by a rubble platform both shortly to be described.

At the northern end of F.1 a grave 1.5 m wide (F.23) had been cut to a depth of 80 cm below the floor and partly undercut the side where tool marks in the form of long vertical grooves were visible. Sealed by the primary fills of F.1 it must have been dug shortly after the former's completion. Carefully backfilled with packed chalk rubble it appears to have been deliberately concealed and was only revealed when careful cleaning of the chalk floor of F.1 took place. It contained a group of four inhumation burials (Figs. A4.27 & A4.28). Two were tightly crouched on their sides at the southern end and included an adult female and a child. At the northern end two further children were interred, one crouched on its side and the other on its back with legs drawn up partly overlying the former. Part of the subsequent analysis, sponsored by the BBC during the making of a television programme about the find (Richards 1999), provided a radiocarbon date for the adult of 3500–3100 cal. BC. A very small sherd of Peterborough Ware and three flakes were the only associated finds.

A shallow scoop containing dark grey silt (F.18) had also been cut into the floor of F.1. When first exposed the floor of F.1 had an almost polished appear-

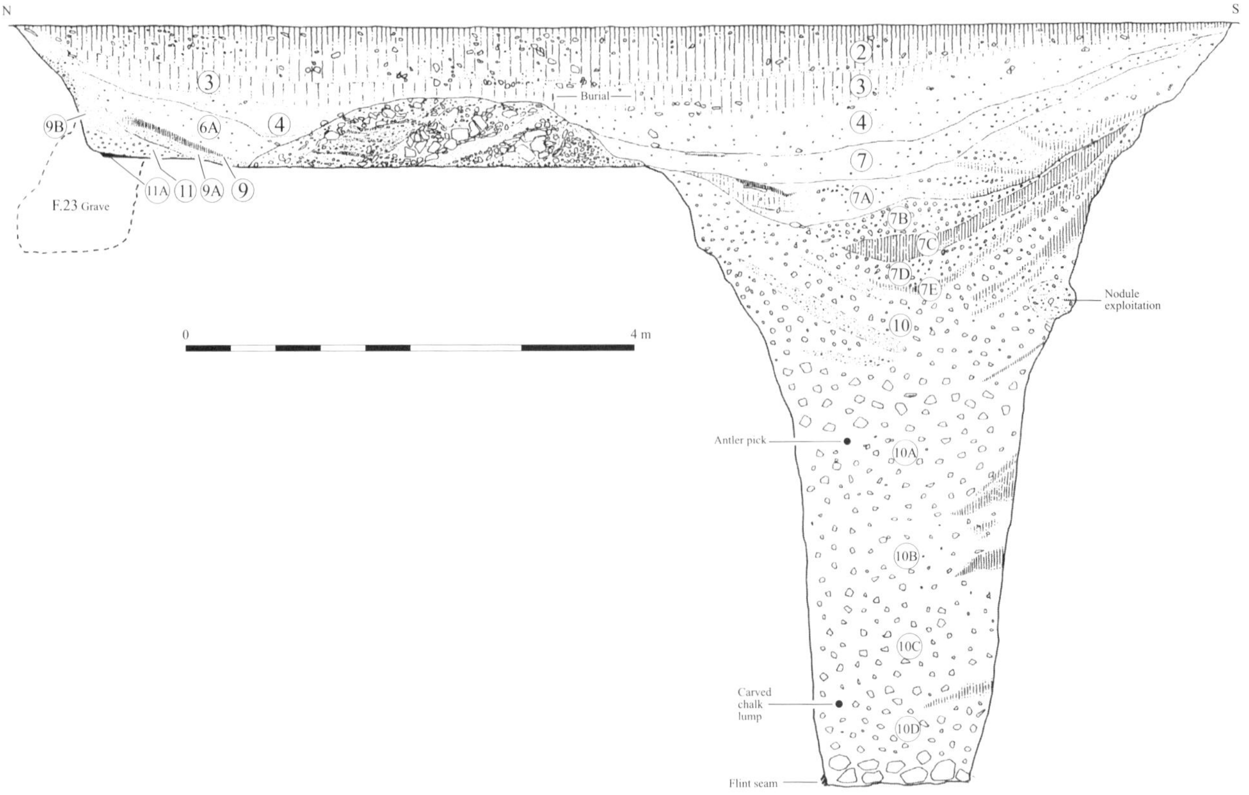

Figure 3.66. *Section of the F.1 pit/shaft at Monkton-up-Wimborne (N. Griffiths after M. Green).*

ance as though it had been worn smooth by the passage of feet. Resting partly upon it were the initial primary silts consisting of layers 9, 9a, 9b, 11 and 11a (Fig. 3.66). Within layer 9 were fresh sherds of two Mortlake style Peterborough Ware vessels (Fig. A4.25) providing a *terminus post quem* for the construction of F.1.

After this initial short period of weathering the floor area of F.1 was radically altered by the construction of a shaft in the southern half, slightly widening the edge of the feature at this point. The resulting bulge is clearly visible on the aerial photograph (Fig. 3.61). With a surface diameter of 4.5 m the shaft gradually tapered to a base diameter of 1.3 m some 6.9 m from the top of the ploughsoil. Tool marks were clearly visible in the last 2 m of the wall and consisted of shallow 'scallops' from a polished axe and long diagonal grooves perhaps produced by a pointed stake being hammered into the sides.

During the digging of this shaft some of the spoil was episodically used to build a platform some 0.7 m high which covered much of the remaining floor area of F.1 and provided access to the pit from its southern edge. This rubble was clearly observed to overlie the initial primary silts of F.1 suggesting only a short time interval, probably less than ten years, had elapsed between the two major digging episodes.

The base of the shaft had penetrated through a thin seam of flint which was visible in the lowest level of the wall. The final surface of the shaft floor was thus a 'natural' surface and bore the imprint of the removed flint, similar to the natural joint reached at the base of F.1 discussed earlier. Immediately above the floor were a few chalk blocks, one with a deep curved groove engraved into it, together with scattered animal bones. Most of these were derived from one animal, a very young pig, which bore clear butchery marks on at least two of the bones (Fig. A4.30a & b). Several bones had clearly been tucked into the angle between the base and wall including three vertebrae and a pecked sandstone 'ball' (Fig. A4.23, S1). Some 0.7 m above the floor, in the primary rubble of layer 10d, a remarkable decorated chalk block was discovered (Fig. A4.29). This block, approximately 15 × 30 cm, had a deep 'cup'-shaped depression cut into its flat side. When resting on this side its upper surfaces reveal a series of pecked lines, some parallel and others curvilinear forming two sets of nested arcs. Above this find a series of further rubble layers (10 & 10a–c), derived principally from the weathering of the sides, filled the shaft to within 1.3 m of the floor level of F.1. Within these rubble layers a few discreet deposits of a ritual nature were uncovered. In layer 10a a

Figure 3.67. *View of the central pit/shaft at Monkton-up-Wimborne, looking northeast (M. Green).*

portion of a pig skull was found with a lump of worked chalk and part of a large flint pebble with natural cavities derived from river gravel deposits. Slightly above this was a small pick made from the top of a red deer antler which had been cut and snapped (Fig. A4.30c). A few further single finds of worked chalk (Fig. A4.24) and introduced natural river gravel flints were encountered within these layers. At the level of the upper primary fills (layer 10) some minor opportunistic removal of a flint nodule seam had undercut the shaft wall (Fig. 3.66). Charcoal from this layer gave a radiocarbon determination of 3630–3590 and 3530–3360 cal. BC (4674±37 BP; Wk-18754) (Table 1.1). The remaining upper fills of the shaft, layers 7a–e were sealed by layers 7, 6a, 6, 4, 3 and 2. The former layers consisted of a series of alternating rubble and siltier accumulations probably partly deriving from activity upon the platform. Scattered charcoal, animal bone and antler, much partly burnt, was found throughout these levels. In addition an associated group consisting of a cow skull and portion of an antler beam was found in layer 7c close to human skull fragments (Fig. A4.30d). Nearby but slightly lower down a very unusual elongated chisel arrowhead was recovered (Fig. A4.23, no. 2). Within layer 7a an exceptionally fine leaf arrowhead was found (Fig. A4.23, no. 1) close to four disarticulated pieces of human bone including a portion of skull (see p. 377).

A few sherds of pottery were found from both the primary and secondary fills of the shaft and were exclusively of Peterborough Ware including two from the same vessel recovered at depths of 2 m and 5.95 m, respectively (Fig. A4.25). It seems likely this material too was derived from activity taking place on the platform which we shall now discuss.

The platform was constructed over much of the remaining floor area of F.1 principally around the northern half of the shaft. Its southern end was built against the near vertical side of F.1 where it was piled up to a level close to that of the natural chalk surface thus forming an access point to and from the mouth of the shaft. It sloped down from this point maintaining a fairly level the surface of about a metre in width which was highly compacted and puddled in nature. The sides gently sloped to an average overall width of about 3 m. The northern end rounded off to a bull-nosed profile before it reached the wall of F.1. Scattered along the platform's western edge a considerable concentration of animal bone, much burnt and with a high percentage of cow, was uncovered. The sections through the platform (Fig. 3.63) revealed a complex constructional sequence. Layer 16 consists of an original dump of coarse rubble derived from the shaft, just under 2 m wide. This was left for perhaps for a season allowing a thin weathering horizon (layer 15a) to develop. Further rubble was then piled up (layer 14) and both these dumps were then supported by a turf revetment (layer 15). This then stabilized before the predominantly earthy layer 13 was added. This layer may represent cleaning out of topsoil which had fallen into the shaft from around its southern and eastern sides. Similar lenses of material were encountered between the predominant rubble fills of the shaft, particularly on its southern side. These would have undoubtedly been derived from topsoil falling in from this direction. Finally a further layer of rubble (layer 12) was added which sealed the earthy layer 13 and increased the overall width to 3 metres.

Two small pits (F.19 & F.22) had been cut into the surface of the platform and although undated were sealed by layer 3, the lower secondary fill of F.1. Finds were fairly sparse within the platform layers but included much crudely struck flint including small knapping clusters, which from the appearance of the cortex was derived from nodules encountered during the digging of the shaft. Two sherds of Peterborough Ware were found in layer 15, odd scraps of animal bone from all layers and a carved chalk object from layer 12 (Fig. A4.24, no. C4). Charcoal was widespread and formed an almost continuous lens within layer 15a.

The upper primary fills within F.1 (layers 4, 6 & 6a) consisted of chalky washes and contained a characteristic later Neolithic flint industry which included chisel arrowheads (Fig. A4.22, no. 6), flakes from a polished flint axe and types particularly associated with the flint scatters located on the local clay-with-flints deposits (Fig. A4.22) (Green 2000). Following this phase a deep organic-rich soil horizon started to develop (layer 3). Within this lower secondary fill was a flint industry containing small scrapers and barbed and tanged arrowheads (Fig. A4.23, nos. 5, 6 & 7) together with a few scraps of Beaker pottery (Fig. A4.25) and sherds consistent with known Early Bronze Age urn fabrics found in the local area. Near the base of this layer at the centre of F.1 a small flint cairn had been constructed over a flexed male burial (Figs. A4.27 & A4.28). This individual had been placed within F.1 at a stage when it would have silted up to a pond-like profile with a maximum depth of 70 cm. No attempt had been made to dig a grave, with only the small flint cairn covering the body. The body had been carefully laid out to respect the axis of the monument with the head to the east and feet to the west. We will return to this fact during our discussion regarding the monument's sequence. The remaining final secondary fill (layer 2) consisted of a dark reddish-brown silt with much small flint, and contained pottery characteristic of the later Bronze Age in the area (Fig. A4.26). A few sherds of Romano-British date had penetrated 10 cm or so into this layer.

Two small rubble-filled pits (F.20 & F.21) lay on the northern and eastern perimeter of F.1 respectively and remain undated.

Ring-ditch

Only 14 m to the north of the monument one of a group of three ring-ditches known from aerial photography was substantially uncovered during the excavation. The ditch was very small being only 40 cm wide by 35–40 cm deep (Fig. 3.62) and enclosed an area 9 m in diameter.

The material originally dug from the ditch would have only produced a very small mound and it is likely that topsoil from the surrounding area would have been scraped up to enlarge the mound. The southern excavated segment produced substantial fragments of two well-stratified later Bronze Age vessels (Fig. A4.26) conclusively linking the construction to the final silting phase within F.1. No burials or features were found within the ring-ditch but are likely to have been removed through ploughing.

Sequence and discussion

The excavations revealed a unique combination of monuments — a massive unparalleled flat-bottomed pit, a shaft partially surrounded by a rubble-built platform and a pit circle which contained them. Evidence from the excavations has revealed the likely sequence of construction in three phases, which are likely to have happened over a relatively short space of time with only comparatively minor later embellishments, as follows:

Phase 1: the initial structure created at the site was the flat-bottomed pit F.1. The remarkably smooth and polished appearance of the floor suggests extensive wear caused by gatherings of people within. It is likely some form of external bank was created with the excavated spoil.

Phase 1A: the Phase 1 work appears to have been sanctified by a multiple foundation burial F.23, partly inserted into the northern edge of F.1.

Phase 2: shortly after the first phase, work began on the digging of a deep shaft and the concurrent construction of the platform. These were completed in an episodic manner on at least four separate occasions. Excess spoil would have probably gone to enlarge the small bank suggested in Phase 1.

Phase 2/3: at some time during or just after Phase 2 the outer pit circle would have been constructed.

The large chalk blocks, occasionally found deposited close to the floor of these pits, show that construction would relate closely to Phase 2. This blocky material could only have derived from the lower levels reached during the digging of the shaft. If this material had been left lying around for any length of time it would have disintegrated. Part of the pit circle, pits F.6–F.8 are built in a straight line as though they were avoiding a pre-existing structure — most likely the bank suggested in Phases 1–2. The spoil from these pits is likely to have made a further contribution to this bank. Supportive evidence for this putative bank is provided in the next phase.

Phase 4: an Early Bronze Age man is placed in the centre of the substantially naturally infilled pit F.1 and covered by a cairn of flint nodules. Radiocarbon dated to 1530–1400 cal. BC, at this stage the feature would have had the appearance of an overgrown pond with rampant vegetation. By this time, some 1800 years after is creation, the outer pit circle would have naturally backfilled. This conclusion is supported by the total lack of Early Bronze Age material in the upper fills of the pits although it is plentiful in layer 3 of F.1. The burial was carefully placed to respect the axis of the monument with the head to the east and feet to the west. With the outer pit circle backfilled by this stage it is likely the axis of the monument was clear to those depositing the burial by the entrances through the suggested bank. Much Early Bronze Age material continued to accumulate during the slow deposition of the organic rich horizon layer 3 (Fig. 3.66). Two of the three undated ring-ditches just to the west of the site are likely to have been built in this phase or the following.

Phase 5: the closest of the three ring-ditches was shown to be later Bronze Age in date and contemporary with the final silting phase of F.1 (layer 2).

Discussion

The form and date of the pit circle reveals it is at the transition between the demise of causewayed enclosures and the start of henge monuments. Sharing similar dates and transitional attributes, but built on a much larger scale, one can cite the enclosure at Flagstones near Dorchester (Smith *et al.* 1997) and the Aubrey Hole circle at Stonehenge (Cleal *et al.* 1995) The internal arrangements remain unique however; although the much smaller unexcavated site at Harlaxton 106 (Harding & Lee 1987) does appear superficially similar, it may be a timber monument closer in date to Sarn-Y-Bryn Caled (Gibson 1998). There appear to be no parallels for the F.1 pit or the shaft. Although sometimes claimed, conclusively dated Neolithic shafts other than flint mines are found to be wanting in the British record.

This complex represents a sophisticated open air temple of unique design. The enormous pit F.1 was dug down to a point where a natural joint was reached and exposed. Its polished surface indicates wear from the prolonged passage of human feet. Its construction was celebrated by the insertion of a foundation burial of four individuals, including a woman and her young daughter who were outsiders, and all were perhaps sacrificial victims. Detailed isotope analyses of the dental tissues of these individuals produced remarkable results which revealed their lifetime movements (Budd *et al.* 2000; Montgomery *et al.* 2000). Shortly afterwards the enormous pit was dramatically altered by the digging of a deep shaft in its southern half. The digging of this shaft stopped, like the F.1 pit, at a natural surface, this time provided by the removal of a thin layer of flint. Contemporaneous construction of the elaborate platform not only gave access to the F.1 pit and edge of the shaft but provided an elevated surface for ceremonial activity including feasting. From this platform material would have been selected, displayed and eventually symbolically placed deep within the shaft. That this site is just one component of a vast array of monuments in the immediate area (Fig. 1.2) testifies to the extraordinary power with which this landscape was imbued.

Subsequent developments included the burial of the man in the centre of F.1 some 1800 years later. Carefully laid out to respect the axis of the original monument this burial was covered by a small cairn of flint nodules. The appearance of the central feature by this stage would have been similar to a pond barrow and this may have been partly responsible for the burial. It seems significant that the pond barrow described in this volume only lies 300 m to the north and is of a similar size and profile to the F.1 pit at this time. The construction of two of the three barrows known to lie a few metres to the north of the site either took place at this time or, like the third, in the later Bronze Age. These barrows, probably built some 2000 years after the original monument, not only mark the end of significant ceremonial activity at the site but in the immediate study area.

Monkton-up-Wimborne pond barrow (MUW 02)

(see Appendix 4.10)

Martin Green

This monument, revealed as a dark 'blob' crop mark (Fig. 3.68), was also visible on some of the same aerial photographs which revealed the pit circle/shaft complex (MUW 97) previously described (NMR 4509/54, NMR 4509/62 and NMR 4509/65). It lies on the same ridge only 300 m north of this latter site and 150 m south of the sub-rectangular Neolithic enclosure (MUW 00), at a similar height, topographic situation and in the same current arable field. It forms part of a mini-complex within the greater monument cluster, with two previously unrecorded ring-ditches that lie a few metres to the west (Fig. 1.2). Visible on photograph NMR 15829/19, the closest ring-ditch (SU01621501) is about 30 m in diameter and cut by the present field boundary with the other just to the west (SU01601504) about 20 m. A 25 m square area was stripped to encompass the feature and allow for a generous area outside the monument to be examined (Figs. 3.69–3.73).

Features pre-dating the pond barrow

Two shallow pits F.3 and F.4 (Figs. 3.71 & 3.75) lay some 5–6 m to the southeast of the pond barrow and contained similar fills of a reddish-brown decalcified silt. Small amounts of worked flint were found in both, and a sherd of Beaker in F.3 (Fig. A4.25, P1). Amongst the flint assemblage was a fine discoidal knife/scraper (Fig. A4.31, no. 1) and a convex scraper. A few further residual Beaker sherds were uncovered within the fill of the pond barrow itself.

A small pit (F.2) lay less than a metre from the northern edge of the pond barrow (Figs. 3.71 & 3.75). The edge of the pond barrow deviates from its circular outline shortly before this point for a distance of about

Figure 3.68. *Low-level oblique aerial view of the Monkton-up-Wimborne pond barrow as first exposed (MUW 02), looking southeast (M. Green).*

3 m in order to come inside and clearly avoid this pre-existing feature. The pit contained a cremation within a collared urn (Fig. A4.32, no. P5) which had lost its base through ploughing.

The pond barrow

Besides the deviation mentioned, the barrow maintains a fairly even outer circumference with a diameter of about 11 m. It was excavated in quadrants revealing a mounded profile excavated into the chalk bedrock to a maximum depth of 34 cm (Fig. 3.73). Analogy with other pond barrows would suggest a low outer bank would have been constructed with the excavated rubble, although here extensive ploughing had obliterated any trace of this.

Cut into the barrow floor were two pits, a posthole and two stakeholes (Figs. 3.71–3.73). Pit F.7 (Fig. 3.75), 50 cm deep, was located just west of the centre of the pond hollow. It contained a cremation deposit associated with a burnt flint fabricator. Charcoal from this cremation gave a radiocarbon determination of 1890–1610 cal. BC (3423±50 BP; Wk-18749). Two clear stakeholes (S.1 & S.2), both 18 cm deep, lay just to the east of F.7 and may mark the position of a screen or the point at which the subsequent flint barrow platform was to terminate (see below). Pit F.8 (Fig. 3.75) was a funnel-shaped or double pit 0.8 m deep, consisting of an upper bowl-shaped pit with a narrow shaft, located close to the northern edge. Within the shaft was a cremation within an intact plain collared urn. Posthole 3 lay within the southeastern quadrant close to its southern edge and formed part of the entrance structure which lies principally outside the monument to the southeast.

Figure 3.69. *View of the pond barrow under excavation (D. Parry).*

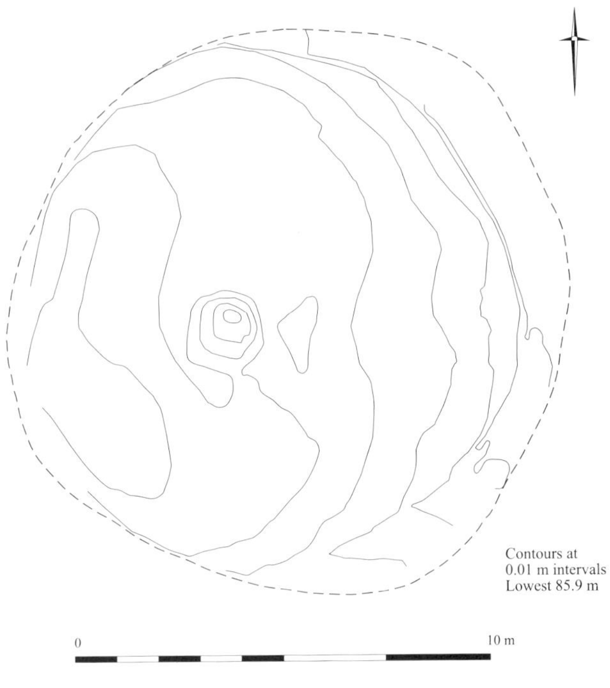

Figure 3.70. *Monkton-up-Wimborne pond barrow (MUW 02): contour plan, with contours at 10 cm intervals and the central/lowest contour at 85.9 m OD (D. Field).*

Within the barrow a little primary chalk rubble (layer 5) had initially accumulated around the perimeter before a more prolonged phase of weathering produced the chalky wash (layer 4) which filled in the edge of the hollow for a distance of about a metre before thinning out short of the central 'mound'.

This latter layer contained sherds of Early Bronze Age date (Fig. A4.32). Following these phases of weathering a flint nodule platform (layer 3) was carefully constructed over the incompletely filled

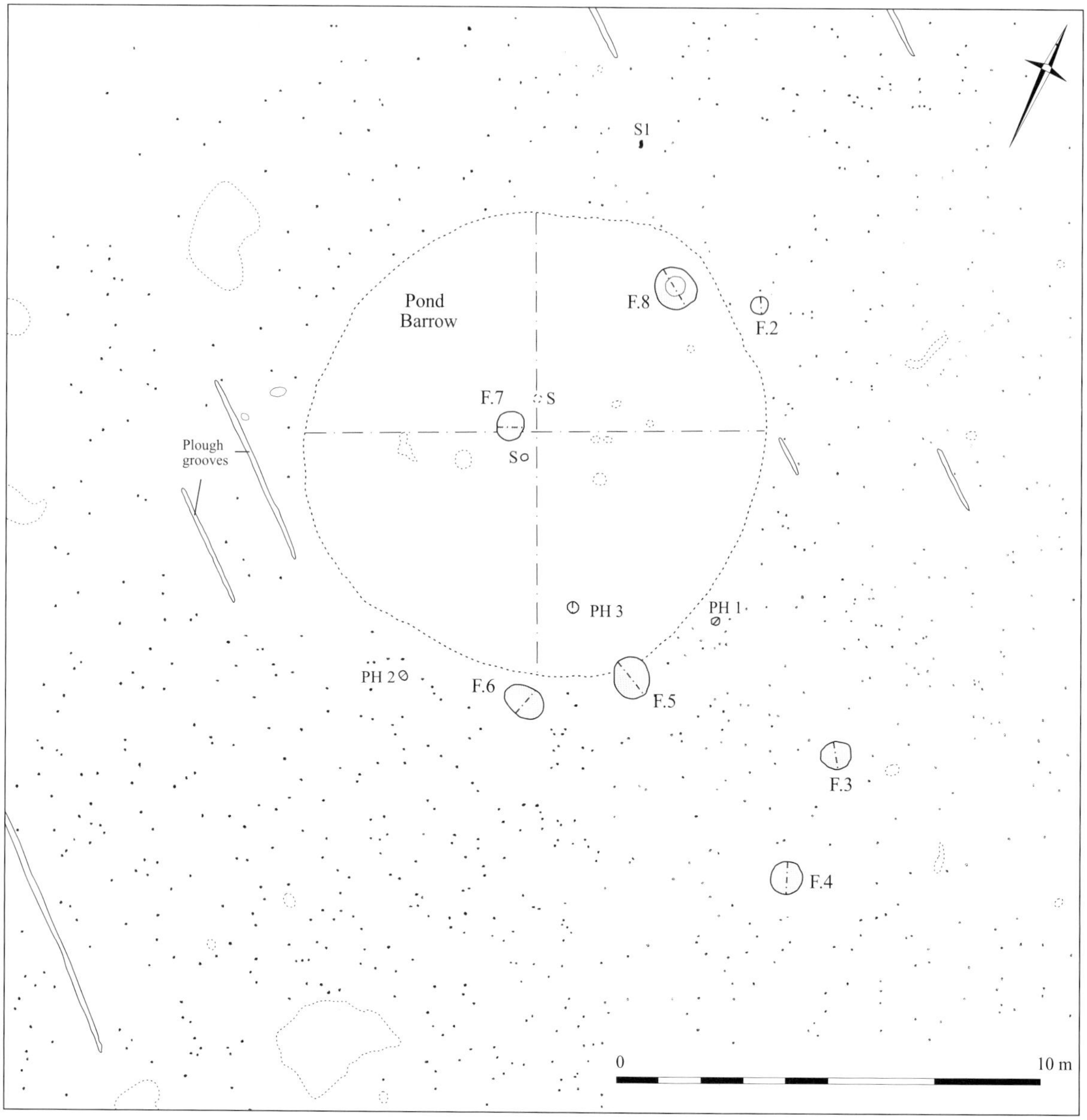

Figure 3.71. *Monkton-up-Wimborne pond barrow (MUW 02): overall plan (N. Griffiths after M. Green).*

remainder of the hollow, except in a 1.6 m diameter area surrounding the central cremation pit. From their frost-damaged and abraded appearance these nodules had clearly been collected from the surrounding field surface and were on average about 10 cm in size. Within this platform animal bones, worked flint, occasional lumps of heathstone and sherds of both Early and Middle Bronze Age date were carefully incorporated. A small area of the platform which directly overlay cremation burial F.8 had been heavily burnt and included lumps of heathstone. A few scraps of cremated bone from this area and in the adjoining northwestern quadrant suggest a pyre may have been fired at this point. Following this activity a final soil horizon (layer 2) containing exclusively Middle Bronze Age pottery accumulated.

Around the southeastern edge of the barrow an arrangement of postholes may form an entrance to the monument (Figs. 3.71 & 3.74). These were in an area which would suggest they were integrated into the suggested outer bank which is likely to have been penannular, with the gap containing the entrance. Two large postholes, F.5 and F.6 (Fig. 3.75), set 1.7 m apart may represent a gateway. F.6 was primarily rubble filled while F.5 contained packing stones including burnt heathstone. Within this material was a small burnt scraper. Flanking these two large postholes were two smaller examples (postholes 1

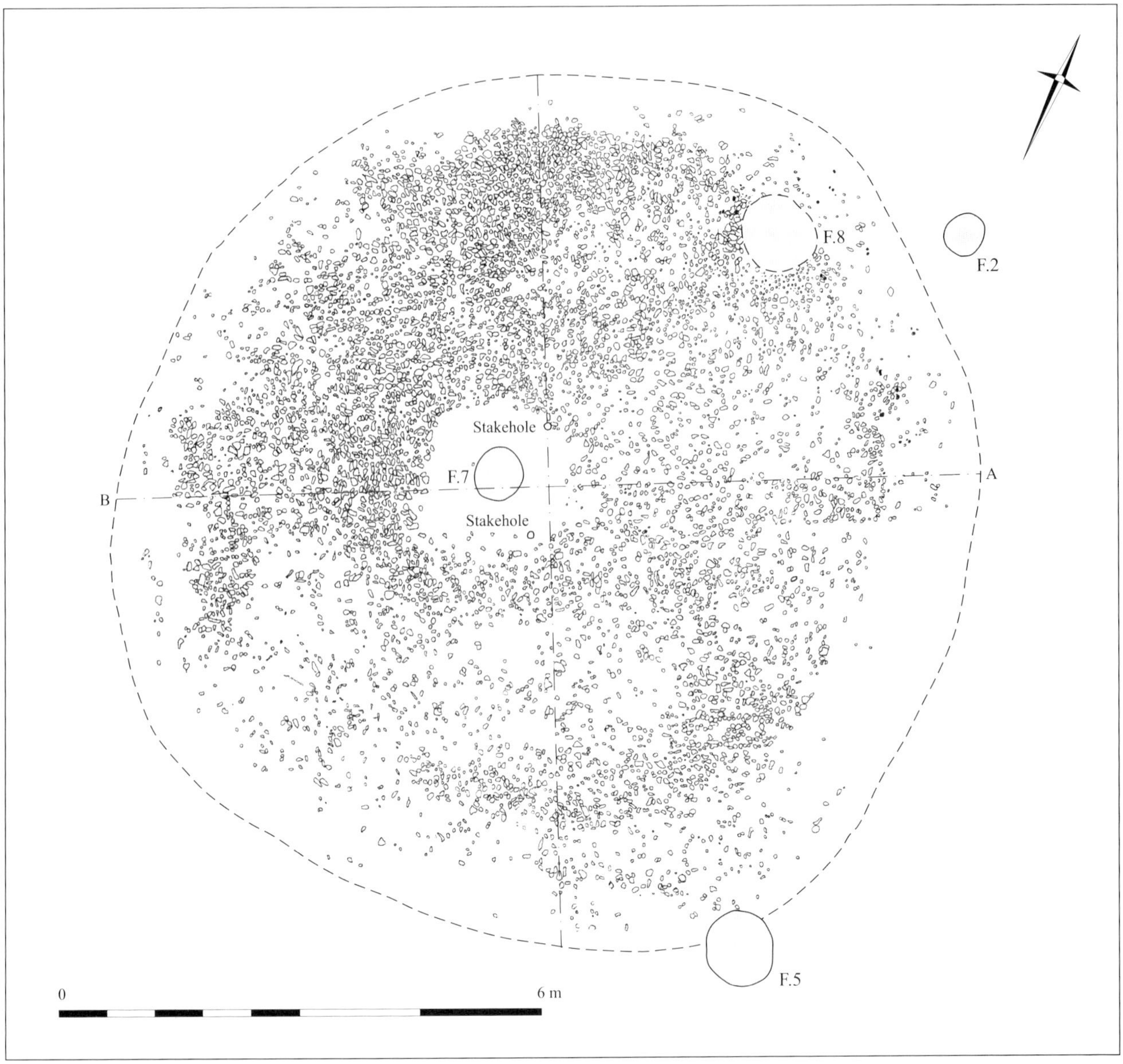

Figure 3.72. *Monkton-up-Wimborne pond barrow (MUW 02): as dug plan and the location of the main cross-section at A–B and outlying features (N. Griffiths after M. Green).*

& 2) at distances of 1.8 m and 2.3 m, respectively (Fig. 3.74). When combined with posthole 3 found within the pond barrow these postholes form a neat structure which can be interpreted as defining a formal entry point (Fig. 3.71). An extensive scatter of stakeholes was discovered cutting into the chalk outside the pond barrow. The similar nature of the upper fill of the pond barrow and its flint platform made it impossible to trace any stakeholes which may have penetrated into it. The essentially random stakehole distribution does not suggest any significant relationship to the barrow and the discovery of one stakehole (S.1: Fig. 3.74) with a rotted stake tip surviving argues for a later, perhaps recent, origin for the majority of these features.

Sequence and discussion

The observed sequence of features is as follows:

Phase 1 - the two pits F.3 and F.4 and a little residual pottery in the fill of the barrow attest to a Beaker presence in the vicinity;

Phase 2 - pit F.2 was dug to receive a cremation within an inverted collared urn;

Phase 2A - shortly afterwards the pond barrow was dug, with the excavated spoil likely to have been used to form an external bank, part of which would

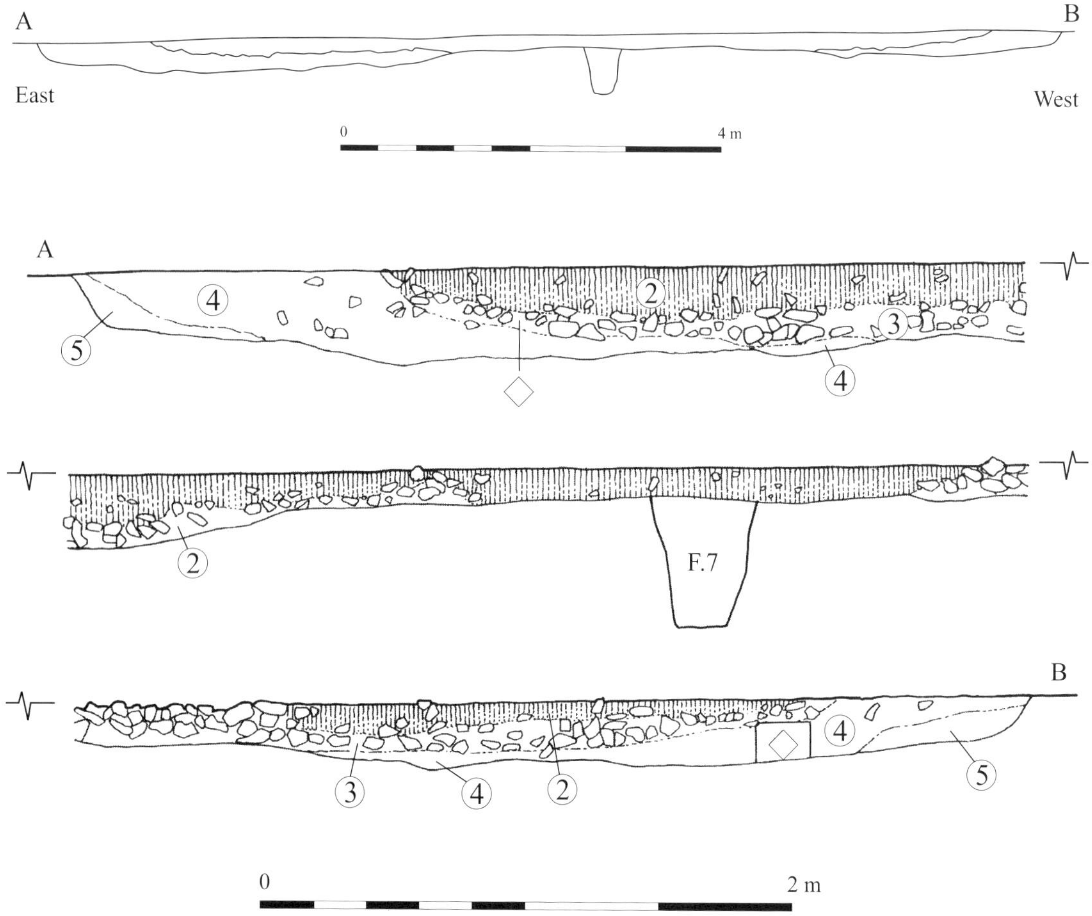

Figure 3.73. *Monkton-up-Wimborne pond barrow (MUW 02): cross-section A–B (N. Griffiths after M. Green).*

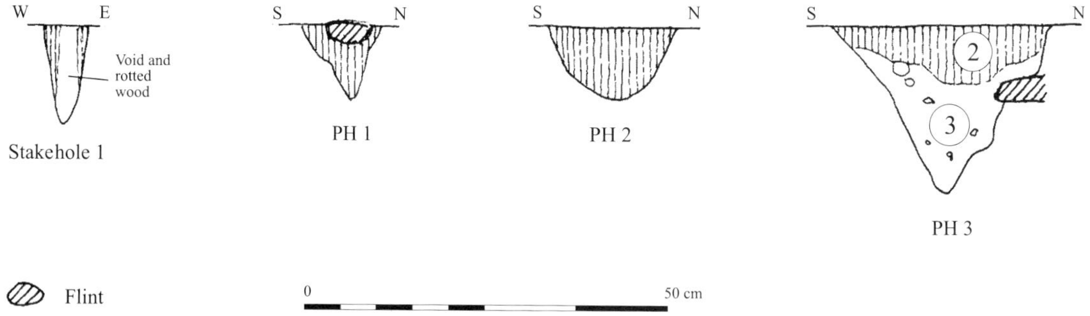

Figure 3.74. *MUW 02 pond barrow: post-holes associated with the 'entrance' structure to the south (N. Griffiths after M. Green).*

have covered the earlier cremation F.2. The two cremation pits F.7 and F.8 are likely to have been inserted shortly after the barrow's construction;

Phase 3 - there followed a period of weathering when the primary rubble layer 5 and the succeeding chalky wash layer 4 accumulated. Incorporated sherds date this phase to the Early Bronze Age;

Phase 4 - a flint platform was constructed over the remaining floor area of the barrow excluding the central area partly occupied by cremation pit F.7. Much debris is incorporated including pottery dating to the Middle Bronze Age. It is likely the two stakes S2 and S3 were driven in at this time to delimit part of the extent of the platform and

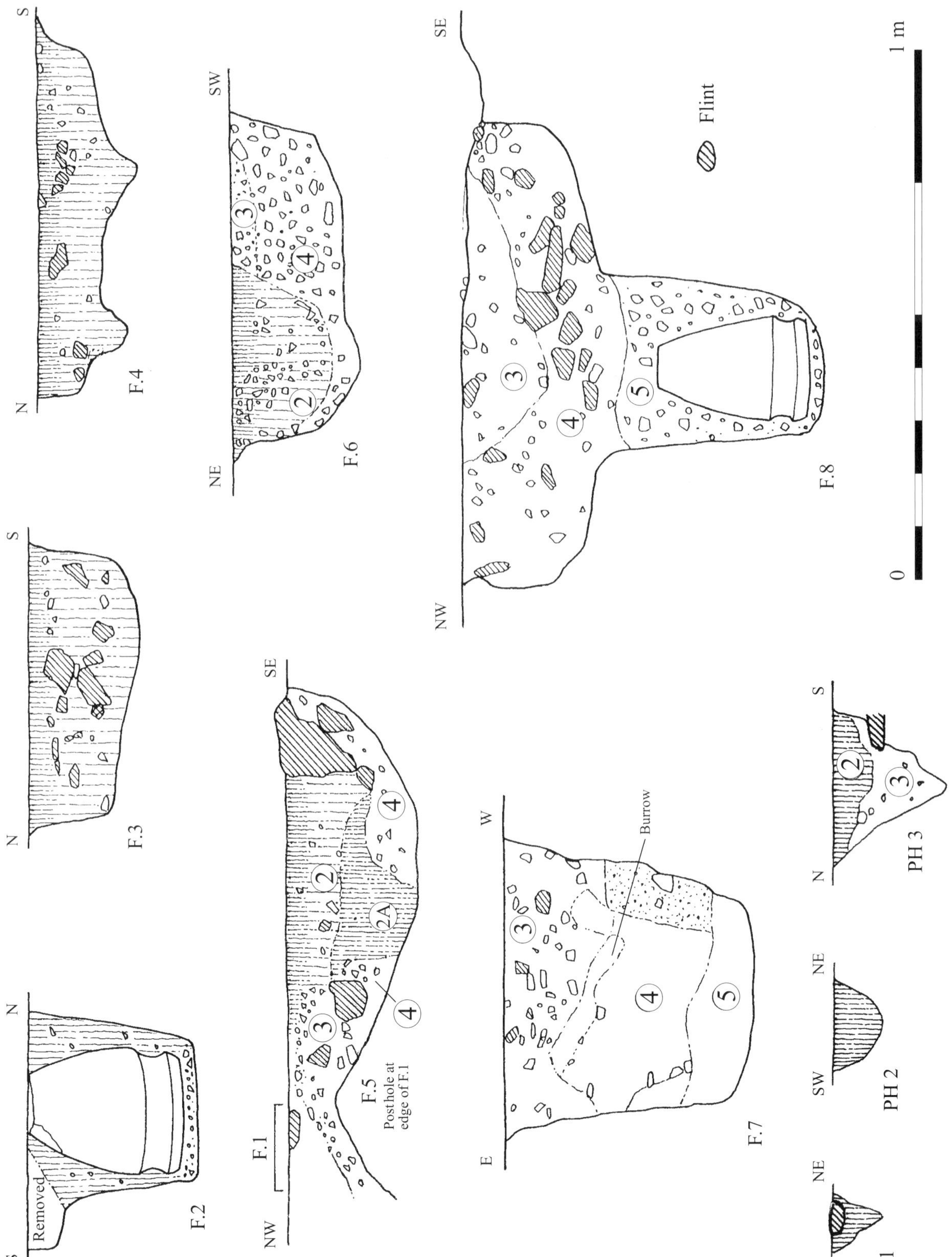

Figure 3.75. *MUW 02 pond barrow: F.3 & F.4 are Beaker pits; F.2, F.7 & F.8 are pits with cremations; F.5 and F.6 are large postholes forming the entrance to the southern side of the barrow (N. Griffiths after M. Green).*

to screen the central cremation pit. It is also likely the entrance structure on the southeast side was built within this phase to control access over the flint platform;

Phase 4A - a fire, perhaps representing a pyre, was lit close to the northern edge of the platform and immediately overlying the previously buried cremation pit F.8;

Phase 5 - a final soil horizon containing Middle Bronze Age sherds (layer 2) accumulated over the flint platform.

Pond barrows represent a rare form of Bronze Age burial monument which until relatively recently had received scant attention other than two examples excavated in the 1950s (Atkinson *et al.* 1951a; Thomas & Thomas 1955; Thomas 2005), although the former was not completely uncovered. The 'pond barrow' originally suspected at Wilsford and still claimed by some (Ashbee, in Ashbee *et al.* 1989) is not included here. The present author has no reason to doubt the excavator of the upper part of the shaft when she writes 'the interpretation of the site as a weathering cone around a shaft takes account of all the major physical features recorded and allows the present writer's thesis that there was no pond barrow at any stage' (Proudfoot 1989, 23).

However, the two now known from excavations on Cranborne Chase, are joined by others discovered during large-scale area excavations in Oxfordshire (Barclay & Halpin 1999), Cambridgeshire (Pollard 2002) and Essex (Germany 2007), all of which suggest they may be a more common phenomenon than previously recognized. Certainly surviving examples attest to the fragile nature of these monuments and their vulnerability to the affects of arable agriculture. Indeed, the extant bank at Winterborne Steepleton was only 15–18 cm in thickness. Those known from excavation range in the diameter of their excavated depressions from 19–6.5 m. Pollard (2002) described Pampisford as a ring-ditch. However, with its total area excavated below the surface of the natural, to this author, it clearly belongs to the pond barrow class. Although the few excavated examples display considerable variety (Fig. 3.76), there are a number of recurring elements, as follows:

- Cremation, with the exception of Radley 4866 (Barclay & Halpin 1999), is the dominant burial rite, and accompanying ceramics are collared urns with the exception of Down Farm (Barrett *et al.* 1991a), which also produced an enlarged food vessel.
- Inhumations often accompany the cremations and to date, identified human remains consist entirely of adult females and children with ceramic associations of food vessels, and in the case of Winterborne Steepleton (Atkinson *et al.* 1951a) transitional food vessel/collared urn forms.
- Monkton-up-Wimborne, Winterborne Steepleton and at least two unpublished excavated examples in the Poor Lot cemetery (Atkinson *et al.* 1951a, 14; Grinsell 1959) are known to possess flint platforms.
- Several examples have a number of burials around the periphery of the excavated hollows suggesting they would have been covered by an external bank.
- Monkton-up-Wimborne, Pampisford (Pollard 2002), two unpublished from Poor Lot (Grinsell 1959) and unexcavated examples quoted by Atkinson *et al.* (1951a, 15) have produced evidence of formal entrances. Down Farm also has a clear line of approach to the monument.
- Monkton-up-Wimborne and Pampisford are remarkably alike in size and form with both revealing gently mounded profiles excavated into the natural and possessing entrance structures.
- Down Farm, Snail Down and Monkton-up-Wimborne produced examples of cremations within the unusual funnel-shaped or double pits.
- Several sites produced numbers of 'empty' pits or pits containing ceramic finds with no accompanying human remains.
- Pond barrows are generally components of much larger Bronze Age burial complexes although Down Farm and Pampisford appear to be fairly isolated.

Since Atkinson's seminal discussion of these monuments (Atkinson *et al.* 1951a), we can increase their previously known distribution to include Oxfordshire, Essex and Cambridgeshire. He noted the similarity with smaller henge monuments of the previous millennium which also enclose a sacred circular space, are entered in a prescribed way, and whose special deposits can include human remains (Atkinson *et al.* 1951b). It is interesting to note that both Wyke Down henges (Barrett *et al.* 1991a; this volume) produced earlier Bronze Age pottery in the upper ditch fills and furthermore a pit was dug in the centre of Wyke Down 1 henge which was radiocarbon dated to 2030–1530 cal. BC, attesting to re-use during this phase.

A continuation in the practice of exposing significant natural surfaces, as exemplified at the Monkton pit circle/shaft complex (see above, this volume), is evident at Down Farm. Here within the 'floor' of the pond barrow a large ammonite fossil, an unusual occurrence at this zone of the chalk formation, was left deliberately exposed. Also one of the elaborate funnel shaped cremation pits (F.30) and two of the group of four central postholes had been dug down to a point where a continuous seam of sheet flint was exposed and left.

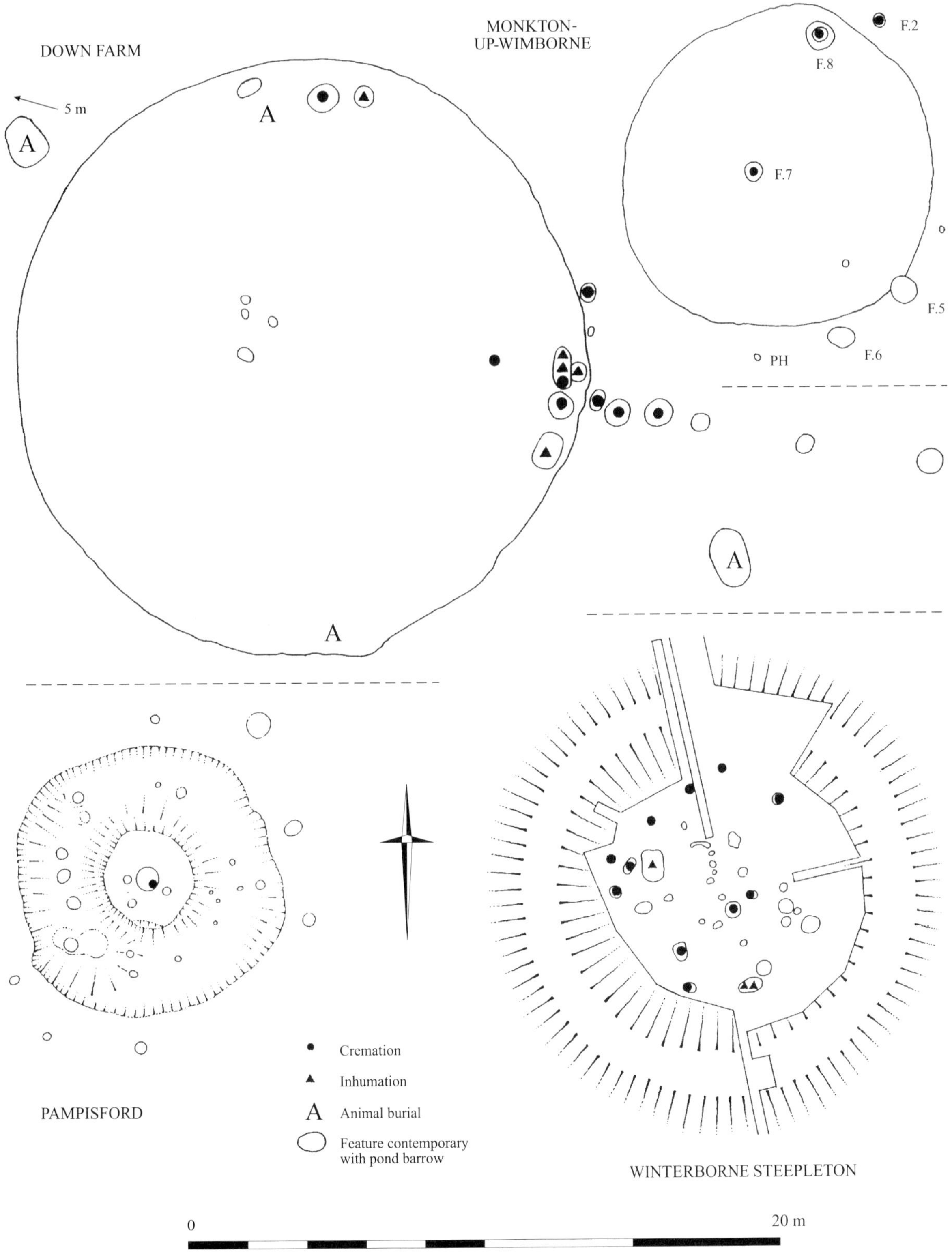

Figure 3.76. *Comparative plans of excavated pond barrows (M. Brudenell after M. Green).*

Although still somewhat enigmatic, pond barrows may owe their inspiration to natural and, in the case of Monkton-up-Wimborne, artificial phenomena encountered in the near vicinity. Certainly at Pampisford, Pollard (2002) argues a nearby natural hollow, which yielded significant amounts of worked flint, provided the inspiration. Furthermore, numbers of partly infilled 'sink-holes' lie close to Winterborne Steepleton, along the barrow rich South Dorset Ridgeway (Woodward 1991). Snail Down (Thomas & Thomas 1955) also lies in an area where Tertiary deposits overlie the chalk and which are likely to produce natural cavities into the ground. Down Farm is within 150 m of the Fir Tree Field shaft and Monkton-up-Wimborne (MUW 02) pond barrow lies 300 m from, what would have been by that stage, the substantially infilled central pit F.1 at the Monkton-up-Wimborne (MUW 97) Neolithic complex. The male burial placed at the centre of the F.1 pit and dated to 1530–1400 cal. BC clearly indicates the feature was still important at this date.

It seems the inspirations for these monuments are often the surviving visual elements of substantially infilled shafts which formerly descended deep underground. The surviving element, the pond-like depression, may have been seen as an entry point to the underworld. Creating visually similar monuments may have been a way of communicating with that world.

Figure 3.77. *The Monkton-up-Wimborne Late Bronze Age/Early Iron Age enclosure ditch (MUW 96) (with permission of English Heritage, Swindon Office, Crown Copyright).*

Monkton-up-Wimborne later Bronze Age to Early Iron Age enclosure (1996)

MARTIN GREEN

An oval enclosure is recorded on aerial photographs held at the National Monuments Record Centre. As the site straddles three current field boundaries, the full circuit of the enclosure was only revealed by piecing together numbers of photographs taken at different times (e.g. NMR 4509 SU0114/14). It is roughly 200 × 130 m, encloses about three hectares and lies on a south-facing slope immediately north of the hamlet of Monkton-up-Wimborne (Figs. 1.2 & 3.77). A clear entrance is known on the northern side. A remarkably similar enclosure lies only 1 km across the valley to the west (Fig. 1.2) and has produced surface finds ranging from Early Iron Age to Romano-British in date (see Corney, in Barrett *et al.* 1991a). As the present enclosure ditch crosses a piece of waste ground some 40 m west of the entrance, it provided an ideal opportunity to sample the ditch without impinging on the cropped areas and a 4 m wide section was excavated in 1996.

The excavation revealed a 2 m deep, V-shaped ditch (Fig. 3.78) which had partly cut an earlier and much smaller ditch on its northern side. The primary fills consisted of a series of chalk rubble bands (layers 7, 9 & 11) interspersed between two silty layers (layers 7 & 11). The secondary fills consisted of a series of silty loams (layers 2, 4 & 5) with layer 3 containing a high chalk rubble content. The remaining primary fill of the small ditch, layer 6, consisted of a grey silty clay with chalk rubble.

24 pottery sherds were found in primary fills. These reveal 12.5 per cent of Middle Bronze Age date, 50 per cent of Early to Middle Iron Age date, 21 per cent of Mid–Late Iron Age date and 8.5 per cent of the Late Iron Age to Romano-British periods. The small ditch to the north, partly removed by the digging of the large ditch, produced a single sherd of Early Iron Age date in its remaining primary fill, but by itself is insufficient to date this phase. The lithics tell a similar story, and it is suggested that this ditch is likely to date between the Middle Bronze Age and earlier Iron Age.

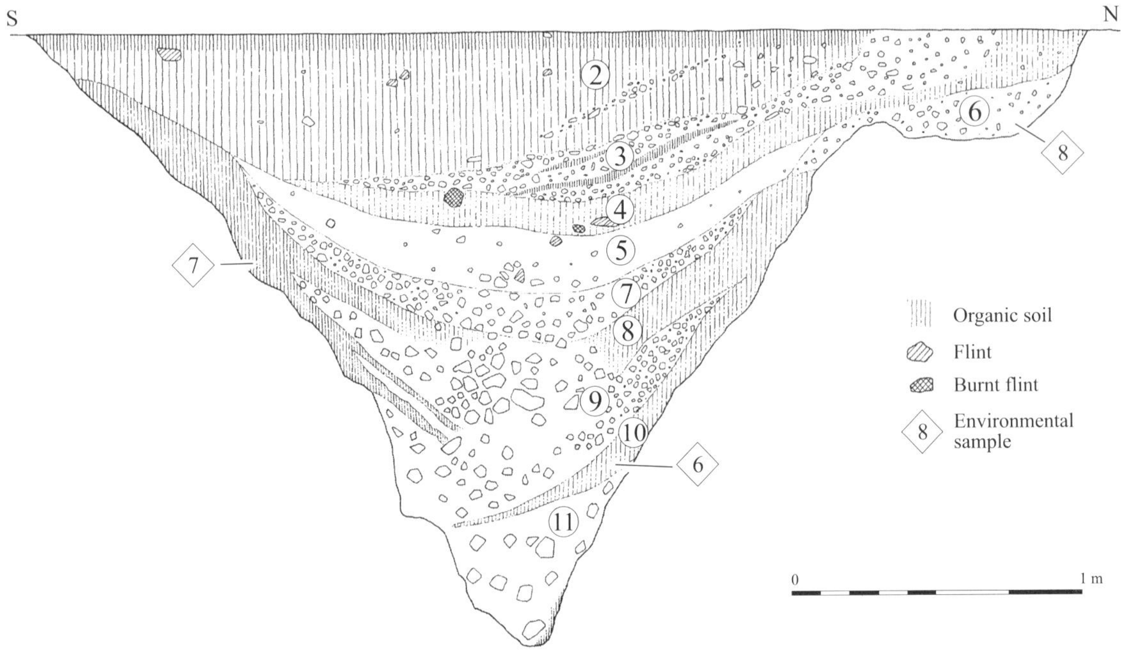

Figure 3.78. *Section of the Monkton-up-Wimborne enclosure ditch (MUW 96) (N. Griffiths after M. Green).*

Knowlton environs

Helen Lewis & Charles French

At the south end of the study area, a series of investigations was carried out around the Knowlton Circles monument complex (Figs. 1.1D, 2.1, 2.17, 3.79 & 3.80). This group of monuments is located just to the north of where the River Allen joins the River Stour. River confluences are known to have been significant locations regarding Neolithic henge construction, and at Knowlton a large henge (South Circle) was constructed, as well as three smaller henges including the well-preserved Church Henge. The history of this part of the landscape has been summarized above (Chapter 1). In addition to the important Neolithic complex, there are two substantial Bronze Age round barrow cemeteries, which have seen some recent investigation by Gale (in prep.). The site is also the location of an early church (Knowlton Church), being one of only a few known sites where early church buildings were placed within large henges. There are a few other comparable sites including Avebury, which is renowned for its importance. The significance of this arrangement at Knowlton has not been well investigated; with such comparisons it could be expected to be one of the most important landscapes in England. In any case, it was evidently a socially and presumably ritually important location in the Allen Valley landscape from at least the Middle Neolithic period on.

No attempt was made to excavate in the protected Knowlton Circles area itself. However, previous small investigations into the South Circle ditch and bank by Bournemouth University (Gale in prep.) provided some comparative material for the project from that site, and the landscape-wide auger survey (see Chapter 2) skirted the monument area. During that borehole survey, several flint scatters with worked tools (mostly scrapers) were noted on the low hills surrounding Knowlton, outside of the known barrow cemetery areas, in locations with no recorded archaeology. We decided to investigate these further as potential settlement and/or ploughed-out round barrow sites, and the resulting series of field walking activities followed by lithic analyses produced some significant scatters. A few of these, along with areas identified through new aerial photographs (see Chapter 2) were targeted for test excavations. The results of these studies, along with a new soil study at Deer Park Farm, a known final Upper Palaeolithic site in the area, are presented here.

Deer Park Farm (1999)

Charles French & Martin Green

Although this site has been fully reported on by Green *et al.* (1998), a brief summary will be given here as it is such an unusual site in the Wessex chalklands. The Deer Park Farm site (SU 05011109) was located a few hundred metres from the Edmondsham barrow (Proudfoot 1963) about 2 km southeast of Wimborne St

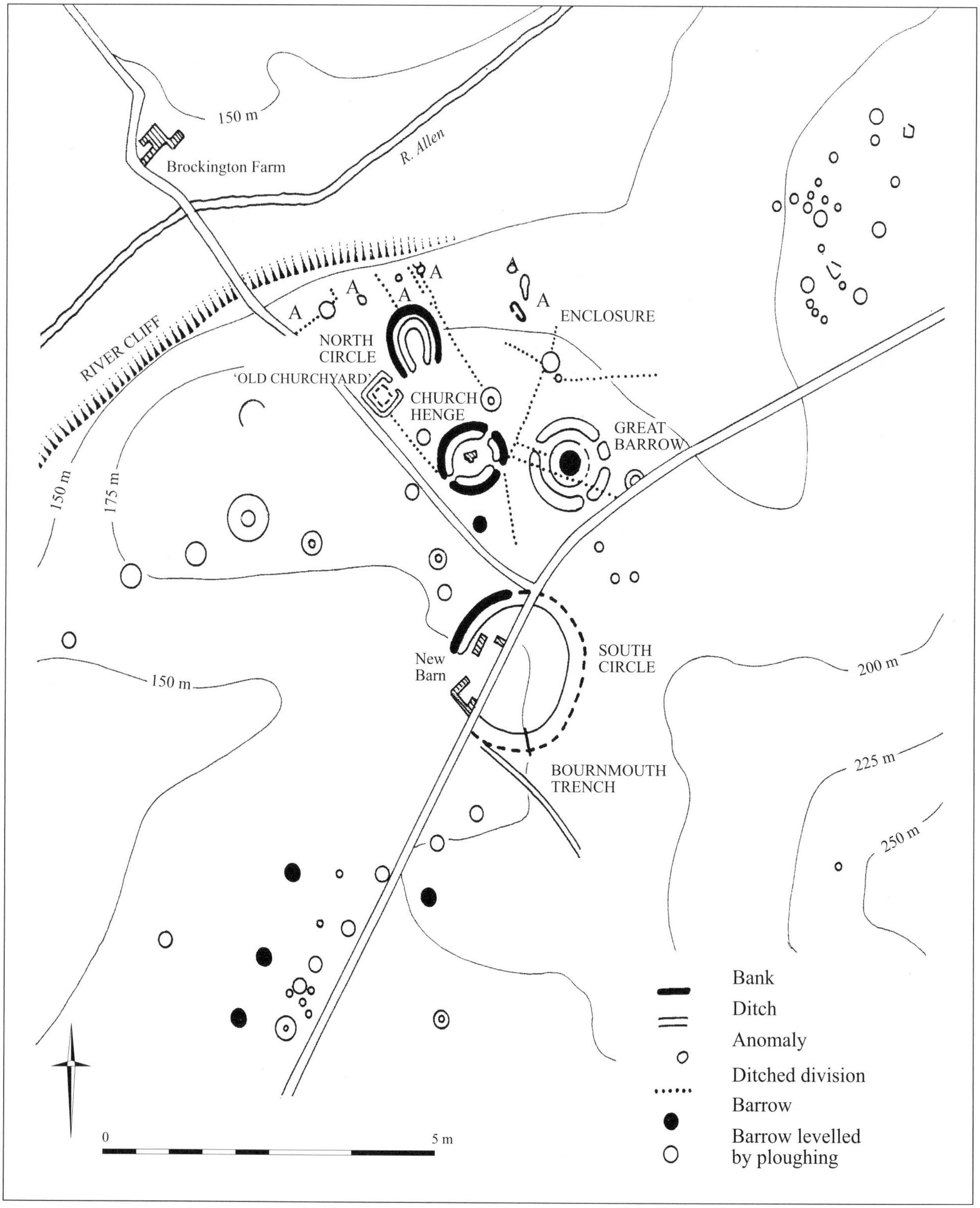

Figure 3.79. *The Knowlton complex (after RCHME 1975 with additions by M. Green; by permission of English Heritage, copyright reserved).*

Giles and *c.* 500 m south of the Avenue Lodge sites (see below) on the geological boundary between the chalklands of Cranborne Chase and the western margin of the Hampshire basin Tertiary deposits (i.e. gravels of Reading Beds and London Clay) (Fig. 2.1).

Detailed surface survey and test-pit excavation revealed two flint scatters, a discrete one some 30 m across and another more diffuse scatter some 100 m away, both similar but quite unlike the local Mesolithic flintwork (Arnold *et al.* 1988). These flint scatters were associated with a sandy clay loam ploughsoil overlying the lower half of an argillic brown earth palaeosol.

The patinated lithic assemblage, although largely devoid of retouched tools, shares characteristics of final Upper Palaeolithic type (*sensu* Barton 1992). The closest parallels appear to exist in assemblages of the Allerød or early Younger Dryas (12–10,700 BP) age from southern Britain and adjacent areas of mainland Europe (Barton 1992; Bodu & Valentin 1997; Coudret & Fagnart 1997). There is also a thermoluminescence date for the burnt flint in the associated palaeosol of 10,700 BP (68 per cent probability of the last heating occurring between 11,860 and 9620 calender years BP), which corroborates the late glacial/early Holocene age for this site.

A *c.* 22 cm long soil block was taken by Green from the deep sounding in the northeast corner of test pit 1. The field profile beneath the present day ploughsoil revealed *c.* 20 cm of orange clay overlying a grey clay layer, which became progressively sandier down-profile until a pure grey sand was reached at a depth of 70 cm below the ploughsoil (Fig. 2.6). The whole profile has developed on a geological transition zone between the chalklands and clays mixed with gravels of the Reading Bed series (Green *et al.* 1998). The orange clay was associated with minute charcoal flecks and a flint assemblage as mentioned above. The rare occurrence of such a well-dated soil with *in situ* artefacts made this worthy of micromorphological analysis.

Two large thin sections were made with the results summarized below. The detailed thin-section descriptions are found in Appendix 4.13.

Soil profile description

The base of the ploughsoil (Ap) was composed of grey, dense, very fine sandy clay with evidence of residual organic matter in the form of iron-replaced plant tissue, stem and root fragments. Up to about one-third of the soil fabric was affected by strong impregnation with amorphous sesquioxides.

This overlay a *c.* 6 cm thick horizon of yellowish brown, apedal, relatively dense, very fine sandy clay. The abundant clay component (35 per cent) was composed of non-laminated dusty or impure clay which exhibited irregular, strong zones of amorphous sesquioxide impregnation. These characteristics are suggestive of a weathered horizon subject to alternate wetting and drying, or a Bws horizon. Occasionally the voids were completely infilled with heavily iron impregnated fine groundmass material, suggestive of soil disturbance above, most probably associated with recent agricultural activities.

This horizon overlay mottled to striated orange/reddish-brown, very dense, fine to very fine sandy clay with very strong amorphous sesquioxide impregnation in irregular, vertical zones. The predominant, non-laminated, dusty clay component (>35 per cent) exhibited birefringent striations and cross striations. In addition, there were occasional non-laminated pure clay coatings of grains and in the groundmass (<10 per cent). In general, the dusty clay component was very heavily impregnated with amorphous sesquioxides. These textural pedofeatures are suggestive of an iron-rich, rather weakly developed, palaeo-argillic (Bst) horizon (Avery 1980; Limbrey 1975, 134–7).

Discussion

The surviving profile at Deer Park Farm is a thin, modern ploughsoil over a clay- and iron-rich B horizon. The lower horizon is characteristic of an argillic brown earth, but the middle to upper part of that type of profile is effectively missing, that is there is no evident depleted or Eb horizon under the topsoil. The loss of this horizon could have several explanations such as past agricultural practices leading to thorough horizon mixing, and/or erosion and truncation of the profile through water and wind agencies.

Although the micromorphological analyses of buried soils in this part of Cranborne Chase have only just begun, Fisher (1991, 17) previously suggested that palaeo-argillic soils were once much more prevalent in the area. These well developed, clay-enriched, woodland associated soils would have been naturally occurring where there were more clay-rich subsoil outcrops, and perhaps even on the chalk downland itself. Certainly, the micromorphological analyses completed of the mounds and old land surfaces of several Neolithic and Bronze Age barrows on Gussage and Wyke Downs and that associated with the Knowlton South Circle (see this chapter) suggests that there had once been some areas of better developed and clay-enriched (argillic) brown earths present on the chalk downlands themselves in the earlier Holocene. But how ubiquitous, how well developed and of what range of thickness is speculation at this stage.

Certainly the survival of the thick palaeo-argillic soil at Deer Park Farm may owe as much to subsoil variation, accident and time of burial, as to land use and erosion in the earlier Holocene. This model requires considerable further investigation and testing.

Conclusions
The 'orange clay' horizon beneath the ploughsoil which is associated with the very early Holocene lithic assemblage is in fact a very fine sandy clay loam. In effect the present-day ploughsoil is the upper part of the soil profile, but is now completely disturbed and altered by modern cultivation processes. The surviving lower half of the palaeosol is dominated by impure clay and is impregnated with amorphous iron oxides and hydroxides. These features are indicative of a palaeo-argillic horizon which has been much affected by variable groundwater conditions and consequent secondary iron formation.

The Knowlton prehistoric complex
Charles French & Helen Lewis

The complex of Neolithic henge monuments and associated round barrows at Knowlton are situated on a low chalk ridge overlooking the Allen valley to the northwest (Green 2000, fig. 60; Harding & Lee 1987) (Fig. 3.79). To date, they have eluded any intensive archaeological investigation, excepting Bournemouth University's sample excavation of the ditch of the South Circle. There are three certain Neolithic henge earthwork sites: the central Church henge (*c*. 95 × 105 m) with the twelfth-century church sited in its centre, the reversed 'D-shaped' North Circle (*c*. 60 m in diameter) about 150 m to the northwest, and the largest one (227 m in diameter), the South Circle, about 225 m to the south. There is also the Great Barrow about 75 m to the east, with an extensive barrow cemeteries about 500 m to the northeast and southwest, and a series of what look to be post-alignments running obliquely out from the Church henge in most directions (and see Charlton Dairy site 2 below).

Excavations of the ditch of the South Circle by John Gale and Steve Burrow of Bournemouth University (Burrow & Gale 1995) have suggested that this site is broadly contemporaneous with Avebury. The ditch was at least 5 m in width and 5 m deep, and the molluscan data from the fills indicated a stratified Late Neolithic sequence indicative of the gradual opening up of the surrounding landscape (Allen 1994). In addition, there was some survival of the associated ancient soil where the outer bank had once been (much eroded into the ditch in the past and now ploughed out). Here a 7 cm thick, dark brown and stone-free silty clay loam probable turf horizon was developed over an 11 cm thick, calcareous, silty clay stony layer or the weathered A/C horizon. This is indicative of a typical rendzina profile already being developed by the later Neolithic or later third millennium BC. Micromorphological analysis of this same profile suggests that this soil was once an argillic brown earth with hints of once having higher organic status and a clay-rich structure that could have been associated with a once wooded and subsequently more open woodland and meadow type of environment in the earlier–middle Neolithic, and which has suffered some pre-burial denudation and more recent disturbance through the impact of modern agriculture (see Lewis below).

Field walking and augering surveys of the fields to the northeast and southeast of the Knowlton complex (see Lewis below) have suggested that there is a dispersed settlement aspect, both contemporary with this monument complex and later in the Bronze Age. Also, in 2000 when we were asking Robert Antell of Knowle Hill Farm for permission to field walk on Shaftesbury Estate land that he farms, he informed us of two large stones that had been recovered from near the South Circle over 30 years ago. Both of these turned out to be substantial slabs of rectilinear heathstone, a local iron-rich sarsen stone from the adjacent Tertiary deposits, with one exhibiting four concentric rings pecked into its centre on one side (Fig. A4.33) (Lewis *et al*. 2000). Reminiscent of similar designs found on Grooved Ware pottery, such as found at Wyke Down Henge 1 and the Grooved Ware settlement adjacent to Wyke Down Henge 2 (Cleal 1991 & this volume; Green 2000), this type of carved motif on a megalith is very rare in the region and where it does occur it is found in association with barrows, such as Winterborne Came barrow 18b (Warne 1866; Grinsell 1959, 107) (see Bradley in Appendix 4.12). Thus, although it may be associated with the South Circle henge, it more probably originates from a cist grave in one of the ring-ditch, former barrow sites, just to the northeast of this site.

Knowlton South Circle soil micromorphology
Helen Lewis

Introduction
During the 1994 excavations of the southeast quadrant of Southern Circle at Knowlton (SU 023 102), samples were taken by M. Allen for soil micromorphological study. Two of these samples will be discussed here: sample 50 of buried soil under the chalk bank, and sample 40 of possible turf in the primary fill of the large henge ditch. Table 3.3 lists the samples and context summaries (after M. Allen pers. comm; Burrow & Gale 1994b, 17, 20–21; 1995, 131; Wessex Archaeology 1994, 5–6) and the detailed descriptions are in Appendix 4.12.

Table 3.3. *List of micromorphology samples from the Knowlton South Circle.*

Sample	Contexts	Thin sections
<50>	Taken through the base of the chalk bank (contexts 7, 8, 9), and the full buried soil profile (contexts 6, 26, 34).	KWC 50 A1–2
<40>	Turf slump (context 45) in primary fill (contexts 46 & 47) of henge ditch.	KWC 40 A1–2

Results

The buried soil and bank (sample 50): This profile represents the base of the bank (contexts 7/8/9) and the underlying buried soil (contexts 6, 26 & 34/bA/B). The soil profile as it stands today, and as it was when the monument was built, is the lower part of a rendzina soil (lower A horizon (topsoil) over chalk), under grassland, which has been strongly reworked through soil faunal action. It also has indicators, however, of a possible deeper soil profile once being present at this location, in the form of clay-rich soil material reflecting a tendency towards argillic brown earth development. There are no signs of cultural disturbance such as cultivation, with the possible exception of a small zone with charcoal and bone at the base of the topsoil. However, the upper part of the topsoil is not seen in the profile, suggesting the soil may have been deturfed or otherwise truncated as part of bank construction. A summary of the micromorphological characteristics is given in Table 3.4.

The bank material was described in the field as consisting of loosely packed chalk fragments in brown calcareous silty clay loam, with inclusions of modern rendzina cultivation soil (Burrow & Gale 1994b, 17, 20–21; Wessex Archaeology 1994, 5–6). In thin section, a compact and occasionally calcium carbonate (micrite) cemented deposit is seen. This micrite precipitation reflects oxidation of the profile, with calcium carbonate precipitating through exposure to air in this porous deposit. The small size of the crystals suggests relatively quick drying out of the profile after moistening (Kubiena 1938, 152–3). There is a great deal of mixing seen, with decalcified or less calcareous organic fragments frequently occurring within the primarily calcareous main fabric. This organic material appears to come from both the overlying topsoil and the underlying buried soil, and has been mixed in through bioturbation. Modern ploughing is mentioned as an influence on this profile but no clear indicators of this activity were observed in the thin sections.

The upper horizon of the buried soil (context 6) appears to be the lower part of the buried topsoil. There are no obvious signs of a surface horizon, whether vegetated or bare (e.g. a turf line, surface crust remains), and it seems that the uppermost part of the profile was removed before construction of the monument. The soil is very mixed regarding soil fabrics, with frequent inclusions of calcareous soil aggregates in the mainly organic and non-calcareous soil. These reflect soil fabrics found above and below context 6, and reflect earthworm activity and localized moisture regimes. Both the calcareous and non-calcareous fabrics are relatively clay rich, with fairly clean clay (although masked with amorphous organic matter), but no clearly alluvial or eluviated B horizon material is seen. This suggests a tendency toward argillic horizon development, but whether or not a more complex or deeper profile ever existed here is not evident. Micrite occurs as coatings and infills occasionally in the main fabric, reflecting oxidation in pore spaces.

In context 6 there are also areas with 'dusty' clay both in the groundmass and as coatings in modern pore space. This is clay with very fine inclusions of silt-sized material, and is held to represent disturbance that is strong enough to disrupt aggregates (thus releasing silt for translocation down-profile in suspension), unlike clean clay features, which represent the normal translocation of clay in solution (rainwater) in stable, vegetated situations (Greenland 1977; Thompson 1983; Holliday & Goldberg 1992, 251; Bullock *et al.* 1985, 66). The fact that this clay is present in the modern pore spaces suggests that its movement in the profile is an ongoing affair; perhaps it is dusty because of disruption related to the construction of overlying bank, or to the modern ploughing of the site. It is possible that some of these clay pedofeatures may have resulted from some pre-bank disturbance, but this is not evident from the thin sections. Regarding possible cultivation of the buried soil, there are no soil indicators specific to this type of activity (e.g. bare earth indicators or panning features). Rare fine charcoal was seen in this layer.

Context 26 is lower A horizon material, with slight iron and clay enrichment. It is generally calcareous, but the mixing of fabrics seen in the overlying horizon continues here. This layer may represent a remnant or incipient B horizon, but it has been strongly bioturbated with the organic material overlying and more calcareous material underneath, and is not clearly anything more than a lower A horizon.

At the base of the layer, in a 1 cm sized area, a few larger carbonized plant remains and a small fragment of bone are found. The location of these materials in such a small area is interesting; perhaps this small patch is derived from a nearby deposit representing domestic or funerary activity. These types of materials are commonly found in ancient cultivated soils as well, and this is also a possibility here, but it is difficult to support this interpretation on the basis of the micromorphological characteristics of this profile. Charred

Table 3.4. *Summary of micromorphological characteristics for profile 50, Knowlton South Circle.*

Context	Field description[1]	Microstructure	Porosity[2]	c:f ratio	Mineral components	Organic components[3]	Groundmass	Pedofeatures	Cultural components	Interpretation
7/8/9	Chalk fragments in brown calcareous silty clay loam	Mixed subangular blocky, crumb, channel	10–15%, channels and cracks	30–35: 65–70, sandy silt loam	Quartz, feldspar, micrite (in groundmass and as pedofeatures); chalk, flint, sandstone; earthworm granules	<5–5%, mostly 'punctuations'	Calcareous, crystallitic, with stipple speckled inclusions	Strong micrite coatings/infills, earthworm excrement	None evident	Bank material
6	Dark brown silty clay, stone free	Mixed channel, crumb, angular blocky	15%, channels and vughs	30:70, sandy clay loam-clay loam	As above, but very rare rock fragments; micrite as coatings; clay coatings in groundmass and as pedofeatures	5%, as in 7/8/9, plus rare fine charred remains and organic staining	Stipple speckled, non-calcareous, organic stained, with crystallitic, calcareous inclusions	Dusty clay coatings, micrite coatings, organic staining, earthworm excrement	Rare charcoal, clay disturbance indicators	Lower A horizon (topsoil) material (non-surface)
26	Brown silty loam, flint and chalk inclusions	Crumb	20%, channels, cracks, chambers	35–40: 60–65, sandy silt loam	As 7/8/9 iron staining, bone	5%, as in 6, but rare large charred remains	Crystallitic, calcareous, iron and organic stained	Micrite coatings, iron and organic staining, earthworm excrement	Rare charcoal, bone	Lower A horizon (topsoil) material; possibly some incipient B material
34/bA–B	Sandy silt loam, 50% chalk, over calcareous stony silty clay layer	Mixed channel, angular blocky, crumb	25–30%, channels, cracks, packing pores	30–35: 65–70, sandy silt loam	As 7/8/9	5%, as 7/8/9	As 7/8/9, with some iron staining	Micrite coatings, iron staining, earthworm excrement	None evident	A–C horizon

1 After Burrow & Gale 1994b, 17–21 and Wessex Archaeology 1994, 5–6.
2 Note that percentages represent visual estimate of area.
3 Note that percentages represent visual estimate with respect to groundmass (after Burrow & Gale 1994b, 17–21 and Wessex Archaeology 1994, 5–6).

Table 3.5. *Summary of micromorphological characteristics of profile 40, Knowlton South Circle.*

Context	Field description	Microstructure	Porosity	c:f ratio	Mineral components	Organic components	Groundmass	Pedofeatures	Cultural components	Interpretation
43	No description provided	Channel, vugh, blocky	15–25%, mostly channels, vughs	40:60, sandy silt loam	Quartz, micrite and feldspar, rare olivine. Chalk, sandstone, flint	<5%, 'punctuations', very fine carbonized remains	3 fabrics. 1) calcareous, crystallitic, iron stained; 2) calcareous, crystallitic, chalky 3) non-calcareous, stipple speckled, organic stained	Micrite coatings and infills, amorphous iron/organic/clay coatings, earthworm granules	Charcoal	A chalk-rich ditch fill
45	Dark stonefree soil lens, silty clay, organic, precipitated calcium carbonate on crumb surfaces, a large amount of charcoal	Channel, crumb, blocky	15–20%, channels	20–40: 60–80, sandy silt loam to clay loam	As above, but micrite mainly in calcareous aggregates and as pedofeatures. Chalk is only occasionally seen. Iron nodules with embedded sand grains	5–10%, 'punctuations', occasional fine carbonized remains	Fabrics 1) and 3) from above	Coatings as above, but decreasing in frequency with depth	Charcoal	Stabilization horizon – turf, with some possible disturbance at surface

grains were found in the old land surface under the bank during excavation/environmental processing (Wessex Archaeology 1994, 9–10).

The lowermost context studied (context 34/bA–B) is highly calcareous and chalk-rich, and appears to be the weathering zone at the surface of the chalk, where it interacts with the overlying soil profile. As such, this might be better labelled as a bA–C horizon. It has also seen bioturbation and slight iron enrichment.

The 'slump' and ditch deposits (sample 40): This sample was described as including contexts 45 ('slumped' turf or stabilization layer) and 46/47 (primary ditch fill). However, the field section drawing shows the upper part of the block to contain context 43 material, and this appears to be seen at the top of the thin section profile, while no obvious sign of the underlying primary fill is evident. A summary of the results from this profile is given in Table 3.5.

Context 43 is a mixed fill with angular chalk fragments 1–2 cm in size, and both calcareous and somewhat decalcified soil aggregates. Three types of soil material (based upon degree of calcareous influence) are present and intermixed, with the dominant soil fabric being an iron-stained but calcareous soil. This is mixed with a very calcareous, chalky fabric, and with non-calcareous organic soil material. The mixing seen appears to relate mainly to soil faunal activity and may also reflect localized moisture regimes. A further characteristic of this deposit is the presence of frequent micrite coatings and infills, suggesting that this layer has been relatively well aerated and loosely packed.

The possibility of several episodes of downward translocation of materials and development of micrite coatings is seen in the layering evident in these mineral coatings. In places these appear to be coated with, or overlie, amorphous organic/iron/clay coatings. Unfortunately, the micrite masks the main characteristics of these other lenses, but they appear to be generally slightly dusty, laminated and lightly stained with iron and/or amorphous organic matter. In some cases, there seem to be micrite-rich lenses in between clay lamina.

The turf horizon sampled is strongly influenced by rooting and earthworm activity, and its grassland type of structure suggests that it has at least partially stabilized during its history at this location. However, the horizon does not have a well-developed crumb structure, perhaps suggesting it did not have enough time to develop such a structure (i.e. it may have been buried relatively quickly). There are also disturbance indicators, including faunal mixing bringing aggregates of overlying and underlying material into the horizon, coatings and infills of soil material in pore spaces, coatings of amorphous iron/organic matter/clay associated with micrite coatings, and structural disruption in the upper 7–8 cm. This structural disruption takes the form of irregular blocky aggregates and less well-defined pore structure in the uppermost part of the context, with a more developed grassland type of structure evident with depth. The structural disturbance seen may relate to the deposition of the overlaying layer, or may simply reflect the exposed surface of this organic layer. The main fabric is organic material similar to the organic fabric in context 43 (see above), but the iron-stained calcareous fabric noted there is also well represented in context 45, and becomes the dominant fabric with depth (perhaps this fabric represents contexts 46/47). A very calcareous fabric seen in the overlying horizon is expressed as infills here, showing the continued influence of chalk weathering and micrite formation on exposed ped surfaces on this horizon.

Conclusion

In general, the micromorphological results support the mollusc and field interpretations in showing a grassland rendzina soil present before the monument was constructed. This was fairly shallow during the Neolithic, but there are some indicators that the profile may have had a tendency to develop a somewhat deeper, argillic soil in the past. The inclusion of charcoal and bone in the lower part of the buried topsoil horizon may represent the proximity of a cultural deposit (with the remains being intrusive to the soil profile), or cultivation of the soil during the Neolithic period, although no other indicators for the latter were seen.

The fabrics seen in the ditch profile are generally similar to those seen in the buried soil, although coatings and infills of micrite are more frequent and strongly expressed, presumably reflecting relatively loose and aerated materials, especially regarding the upper part of the profile. The organic soil material does appear to have been a topsoil/turf growing *in situ*, but this was not growing for the length of time needed to create a fully-fledged faunally reworked grassland soil. The upper part of this topsoil layer appears to have seen some physical disturbance.

All of the topsoil materials had relatively clay-rich fabrics, and this clay appears to have been fairly clean, suggesting undisturbed soils with stable vegetation. The clay in the groundmass of buried soil (contexts 6 & 26) might suggest that a deeper soil profile once existed, but there are no clear indicators of previous well-developed argillic or eluviated B horizons; this profile may always have been a fairly shallow, if clay-rich, rendzina soil. At the very least it was this type of soil by the later Neolithic.

These results fit with findings from Wyke Down to the north, where established rendzina soils have been revealed underneath Bronze Age barrows on chalk, showing that these shallow soils were well developed by the earlier Bronze Age. This contradicts ideas that Middle Bronze Age and later agricultural land use and clearance were primarily responsible for soil erosion and profile changes in the southern chalkland region (see e.g. Barrett 1994). If there was strong erosion of brown earth soils on the chalk, this appears to occur before the Late Neolithic–Early Bronze Age. These shallow soils may thus reflect changes related to pre-Bronze Age land use, and their presence indicates that the landscape has been a great deal more stable than is commonly believed. These findings underline the importance of site-specific research such as that carried out here in developing our understanding of the ancient landscape.

The Knowlton environs field survey (2001–2)

Helen Lewis

In the course of carrying out geoarchaeological survey of the upper Allen River valley for the Wyke Down Project (French *et al.* 2000), a number of previously unrecorded find-spots of worked flints were identified. These artefacts (mainly scrapers and flakes) were discovered in conditions of poor visibility, suggesting that potential flint scatters might be found in these locations. Since a major component of the initial project was to explore the relationships between monuments, land use and settlement in the area, it was decided that investigation of these potential settlement locations would be enlightening. In particular, the under-representation of Neolithic and Early Bronze Age settlement in the region is a vital issue in the underlying models being addressed, especially in terms of degree of sedentism or mobility, and how settlement patterns relate to the monumental landscape and to the land-use patterns the Wyke Down project set out to characterize.

On the basis of finds from the valley auger survey, five fields were field walked to get a general idea of the patterns and densities of artefact distributions (All Hallows or AH, Avenue Lodge or AVE/AL, Brockington Farm or BRO and BD, Farringdon Clump or FCL, and Knowle Hill or KH) (Figs. 3.80, 3.84 & 3.89), with funding from the McDonald Institute for Archaeological Research, University of Cambridge. With one exception (Field AH), the fields were walked along transects spaced at 20 m intervals, with material collected in *c.* 5 m wide strips along each transect. Artefacts were collected in bags for every 20 m distance along each transect. This method of non-intensive field survey aimed to simply identify if and where artefact scatters were present, and to provide enough material to enable a basic characterization of any scatters found. Field AH was walked using a 10 m transect and bag drop interval. This was the first field surveyed, and although it was known by hearsay to contain a flint scatter, the unanticipated large volume of flint collected using such a small interval led us to double our spacing in the other fields. In all fields, all artefacts not obviously modern were collected in these transects, covering all materials and all periods.

This field walking survey of the southern part of the study area around the Knowlton henge complex identified at least four possible zones or concentrations of Neolithic and Bronze Age lithic material. One interesting point about these scatters is that they are all on the opposite side of the river to the Knowlton monument complex itself. One field (All Hallows) included Mesolithic flints (see Gardiner below), revealing early use of an area further down the valley than expected, and on chalk geology. One field at Knowle Hill produced a large quantity of burnt flint, which Gardiner suggests could represent one or two (later) Bronze Age burnt mounds adjacent to the Neolithic monument complex. This hill is very prominent in the local landscape, which becomes less rolling and more plain-like where the river confluence occurs. A Roman period finds scatter is known from its lower slopes (outside of the survey area), and it would undoubtedly have been an important feature in the landscape in all periods. Unfortunately no survey was possible on the hill itself at this time.

Flint assemblages from the Knowlton area

Julie Gardiner

Transect field walking of seven fields (Fig. 3.80; Table A4.55; Appendix 4.12) produced 4221 worked flints. This discussion provides an assessment of the condition and raw materials present in this assemblage, a basic description of the composition of the assemblage as a whole and by field, and outline statements as to the probable date and possible activities represented where possible. Some indication of the relative density and distribution of the material is provided, as it is also for burnt flint, but detailed spatial analysis has not yet been undertaken.

Raw material and condition

All the material recovered, with the exception of one or two pieces of gravel flint and Portland Chert, is surface chalk flint with a medium thickness of buff/brown cortex, probably derived from the immediate area. The material is very typical of flint assemblages in Cranborne Chase generally (see sections by Gardiner in

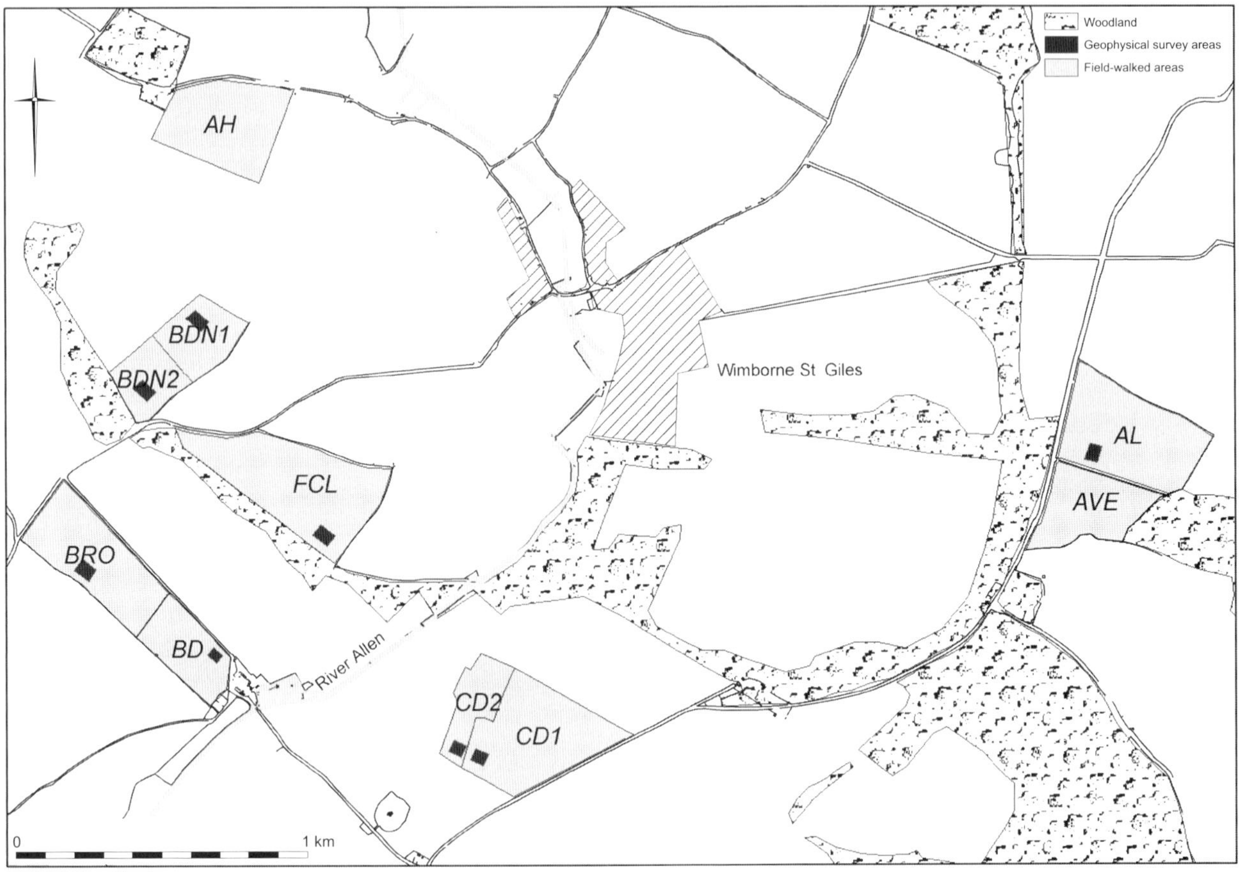

Figure 3.80. *Location map of the sites investigated through fieldwalking, geophysical and augering survey and test excavation in the Knowlton area (F. Sturt).*

Barrett *et al.* 1991a). Post-depositional edge damage is very common, especially on pieces recovered towards the ends of transects (i.e. towards the edges of the ploughed fields). Despite the occurrence in some fields of large quantities of burnt flint, few struck pieces are burnt.

Cortication varies considerable from almost none, through various stages of milky blue and white to dense white. In several of the fields the cortication of flint artefacts can be seen to vary across the field. The variation is often quite distinctive and more detailed examination may indicate some correlation between cortication and features such as break of slope or changes in soil composition.

All Hallows Farm (Field AH)

Field AH is the northernmost of the fields examined. It lies on a fairly gentle northeast facing slope overlooking the River Allen. A total of 818 struck flints was recovered (Table A4.55). Of this total 97 per cent is waste material with only 21 (3 per cent) retouched forms. Seventeen (81 per cent) of these are scrapers, with one retouched flake, a broken bifacially retouched object of unknown function, one small chisel and a fabricator.

Amongst the waste material from this field are seven cores or nodules/flakes exhibiting a few deliberate removals. Three of the cores produced small blades and it is noticeable that the blade:flake ratio for material from field AH is quite high (1:46) while the ratio of cores (and fragments) to all other waste material is very low by comparison with the other fields (1:99). Three tiny blades, unretouched but certainly suitable for use as microliths, were also recovered. While the majority of flints recovered from AH are generally undiagnostic there is some suggestion of a possible Mesolithic presence which is not seen in the assemblages from the other four fields. It is not possible to suggest any specific activities represented by the flint assemblage.

Initial indications on the distribution of material are that there is no obvious concentration of struck flint though totals are generally higher in the central transects (AH50–70).

Almost 36 kg of burnt flint were recovered. This material was widely spread across the field in generally fairly large quantities with no obvious concentrations.

Farringdon Clump (Field FCL)
Farringdon Clump was the largest field examined (Fig. 3.80). It straddles the top of a low, rounded hill with the slope falling away gently on the eastern side towards the River Allen. The top of the hill overlooks the Knowlton henge complex.

A total of 568 struck flints was recovered from this field (Table A4.55). Of these, 92 per cent are waste material including cores, core fragments, and two core rejuvenation flakes. There is a higher percentage of retouched forms here than in AH but 41 out of 46 of these (89.1 per cent) are scrapers with only a single crude chopping tool and four edge retouched pieces in addition. Two scrapers are manufactured on core fragments. The scrapers are quite diverse in type with end, side, hollow and multiple-use forms present as well as a familiar group of undiagnostic, rather roughly made, examples. The flint assemblage from FCL is typical of the generally undiagnostic, probably Late Neolithic–Early Bronze Age material that abounds on the surface of fields in Cranborne Chase. There is, however, a faily high proportion of bladelike flakes (7.2 per cent of the flake/blade assemblage) with at least two cores producing bladelike flakes/blades which may indicate an earlier Neolithic element.

The flint is fairly evenly distributed across the field with no major concentrations. Slightly higher figures are recorded in transects, or parts thereof, that also have a relatively high concentration of burnt flint (transects 140–180 and one corner of field). The overall quantity of burnt flint (*c.* 19.25 kg) is less than for AH, a field only one-sixth of its size.

Knowle Hill (Field KH)
Field KH lies immediately to the east of the largest of the Knowlton henges (Fig. 3.80), part of which lies within the northwestern corner of the field. KH slopes gently up to the top of Knowle Hill in the east, overlooking both the henges and the river. Field walking produced only 304 struck flints of which only eight (3 per cent) are retouched forms — five scrapers, two retouched flakes and one broken bifacially retouched piece (Table A4.55). The material is generally rather crude and roughly worked and appears to be later in date than the henges — probably mature Bronze Age.

The distribution of the worked flint does not, at first examination, suggest any obvious concentrations. Field walking, however, produced a great deal of burnt flint. Approximately 42 kg was recovered with concentrations of up to 5 kg per 20 m collection unit. Much of this material is heavily burnt and fragmented into small pieces. It is possible that one or more ploughed-out burnt mounds are present.

Brockington Farm (Fields BRO & BD)
Field BD was walked in 2001 and the adjacent field (BRO) to its northwest in 2002. These two fields lie on the southwest side of a very shallow dry valley that runs northwest to southeast, meeting the flood plain of the River Allen just beyond the boundary of field BD (Fig. 3.80). Field BRO is, therefore, upslope of BD in terms of the general topography but also encompasses part of the floor of the dry valley. The Knowlton henges lie just across the river.

Individually field BD was the most prolific in struck flint (1026) though it was by far the smallest of the fields examined (Table A4.55). Field BRO produced 761 struck flints from approximately twice the area (Table A4.55). There are differences in the composition of the flint assemblages from these two fields but, as discussed below, it seems probable that they are part of the same spread of material and should be considered together.

Together they produced more flintwork than the other three fields (1788 items, 51.4 per cent of field walking total). This combined assemblage comprises 95 per cent waste material and 5 per cent tools within which 19 out of 22 categories of recorded material are present and 11 out of 13 tool categories. A slightly higher percentage of tools was found in BRO than BD and noticeable differences between the two include the presence of heavy core tools (choppers, picks, a tranchet adze and a miniature axe) in BRO and that of diagnostic Late Neolithic arrowheads (one crude transverse and a broken oblique) in BD. There are more scrapers (of a greater variety of types) and borers in BRO but more simple retouched pieces in BD. There are also more cores/core fragments and pieces of irregular waste in BD than in BRO while BRO produced a high percentage of heavy flakes (27 per cent) and some bladelike flakes. Heavy flakes were also present in BD but as a much smaller percentage of the overall quantity of flakes and blades (10.7 per cent).

While none of this provides any clear evidence for specific activities, it suggests that the two fields represent broadly different activity areas of essentially Late Neolithic date within the dry valley. BRO seems to have seen some 'heavy-duty' flintworking and possibly some primary processing of animal hides. BD may have seen 'lighter' activities, perhaps more precise or 'finishing-off' flintworking and more domestic chores. Until more detailed work is undertaken on the distribution of the flintwork it is difficult to make any more meaningful comment and these observations may require ammendation. Transects producing flintwork in BRO suggest a fairly even spread though there are some blank areas. Within BD the quantity of material seems to increase gradually towards transect 80 and then gently tail off again but, again, there are some blank areas. On the other hand, BD produced the lowest quantity of burnt flint from the five fields (*c.* 12.25 kg) and the material seems to be fairly well spread. This perhaps mitigates against a strong argument for domestic activity here. BRO produced larger quantities (19.2 kg) with a notable concentration in transect 40.

It will be interesting to see if more detailed analysis provides further clues as to the use of this dry valley, or whether the distribution of the archaeological material in any way reflects that of colluvium or of the specific topography of the valley.

Avenue Lodge (Fields AL & AVE)

The two adjacent fields at Avenue Lodge, walked in 2003, lie immediately to the west of the B3078, just north of its junction with the B3081, about 0.6 km northeast of the Upper Palaeolithic flint scatter at Deer Park Farm and 2.5 km northeast of Knowlton Circles (Fig. 3.80). Field AL is to the north of field AVE. The fields are relatively flat with a gentle fall from east to west. Field walking of these two fields produced 744 worked flints (9625 g) and approximately 69.5 kg of burnt flint (Figs. 3.81–3.83; Table A4.55).

With the exception of one or two pieces of gravel flint, all the raw material is locally available surface flint with a white to buff cortex of varying thickness, some with an orangey-brown staining, probably from the clay-with-flints that is present as a thin deposit in parts of both fields. The colour of the flint itself is variable but most is corticated either to white or a milky blue, there being no spatial patterning to the variation. There is considerable post-depositional damage, particularly towards the ends of the runs (ie. close to the field boundaries).

Of the 744 worked pieces recovered, 707 (95 per cent) are debitage. Nineteen cores and core fragments were recorded, including one neat single platform blade core which is probably Mesolithic. The remainder is small and crudely worked with generally only a few removals (average weight 57 g). There is a wide range of shapes and sizes among the waste flakes, most of which appear to be hard hammer struck and there are a considerable number of hinge and a few step fractures. However, while there is only a single example of a well-struck blade, 12 per cent of the waste flakes have blade-like proportions (no formal metrical analysis has been undertaken). There appear to be several unretouched scraper blanks.

The range of tools is narrow (Fig. 3.83; Table A4.55). Formal tools consist of 25 (68 per cent) scrapers (including two hollow scrapers), two fabricators, two borers, a small tranchet adze and two bifacially retouched pieces that seem to be laurel leaves (one has failed in manufacture and the other is damaged). The scrapers come in a range of sizes and mostly have less than 50 per cent of the circumference steeply retouched. There are, however, two very shallow retouched examples and several end scrapers. On the whole the scrapers are quite carefully made. One of the two fabricators is roughly produced on the end of an elongated nodule but the other is a neatly made rod-like example. The tranchet adze is small and roughly made and the blade end is damaged.

In addition, there are two certainly edge retouched pieces and a broken, burnt fragment that seems to have come from a hammerstone.

The distribution of the worked flint shows no obvious spatial patterning (Fig. 3.83), other than a blank area *c.* 60 × 80 m approximately in the centre of AL and one *c.* 60 × 120 m in the western central area of AVE. In the latter case this corresponds roughly with a fall off in the weight of burnt flint but there is no such correspondence in AL. There is, perhaps a more consistent occurrence of worked flint towards the field boundary between the two fields but that could be a function of ploughing. The distribution of tools and cores appears random and there is no clear patterning either in the occurrence of blade-like flakes (see below).

A large amount of burnt flint was recovered but, again, there is no obvious spatial patterning (Figs. 3.81 & 3.82). There are a few 'hot spots' in field AL but even here collection units producing large amounts may be surrounded by units producing very little. In field AVE, transects 40 and 80 produced consistently large amounts along their lengths with much lower quantities in transects 20, 60, and 100. There is no evidence that this distribution corresponds with any linear features — archaeological or geological — within this field and the results could reflect collection bias.

There are some noticeable differences in the composition of the assemblages from the two fields which cannot easily be explained. Field AL has a much lower incidence of blade-like flakes than AVE (6.7 per cent against 16 per cent), though, in both fields, the percentages of blade-like flakes to other waste materials by collection unit shows considerable variation and no clear patterning. The tool component in AL is smaller than that in AVE (3.9 per cent/6.5 per cent) and scrapers form only 53 per cent of the tool assemblage in this field, compared with 80 per cent in AVE. Field AL also has a wider range of tool forms (5 in AL, 2 in AVE) (Fig. 3.83).

As there are no specific concentrations of flint material (worked and burnt) generally, or of cores or tools particularly, it is very difficult to determine if this

Figure 3.81. *Avenue Lodge flint distributions: number of burnt flints (F. Sturt).*

Figure 3.82. *Avenue Lodge flint distributions: weight of burnt flints (F. Sturt).*

Figure 3.83. *Avenue Lodge flint distributions: flint flake types (F. Sturt).*

indicates genuine differences in assemblage composition that may relate to either the relative date of material or to use areas within the scatter as a whole. One possibility is that the variation in flint assemblage at the surface relates in some way to variation in soil depth and/or to the presence of underlying features, or it could, at least in part, be due to collection bias.

Although the quality of the flintworking is generally fairly poor, taken together the characteristics of the assemblages from these two fields indicate an Early to Middle Neolithic date. This is suggested by the relatively high incidence of blade-like flakes, and the presence of a tranchet adze, end scrapers, laurel leaves and a rod-like fabricator.

Figure 3.84. *Oblique aerial view of the Brockingdon Down settlement features (M. Green).*

New test excavations in the Knowlton environs (2003)

Charles French

Introduction

With the exception of the Down Farm area in the northern part of the study area, the nature of the prehistoric settlement record of the valley and its relationship to the burial and ceremonial complexes are neither well studied nor understood. The Knowlton environs test evaluations were a first attempt at redressing this imbalance. Field walking and aerial photographic surveys of the southern part of the study area around the Knowlton henge complex identified at least four possible zones or concentrations of Neolithic and Bronze Age lithic material (see Żuk & Palmer and Gardiner above; Lewis & French 2002). In order to test whether these concentrations represented occupation or not, their date ranges and the level of preservation, this part of the project aimed to locate and elucidate the physical and palaeoenvironmental nature of these possible settlement sites through a field programme of targeted geophysical survey and test excavation (Fig. 3.80). This included aerial photographic searches by Rog Palmer and Martin Green, magnetometer survey by GSB Prospection of eight approximately half-hectare plots around each lithic concentration (see this volume) (Figs. 3.80 & A2.16); followed by test excavation pits/trenches (of 1 × 1 m, 2 × 2 m or 1 × 3 m in size) of eight locations with the test holes placed on the basis of the geophysical survey plots (Figs. 3.80, 3.85, 3.87 & 3.90). In addition to this, further transect field walking of the Avenue Lodge fields in the Knowlton environs took place concurrently with

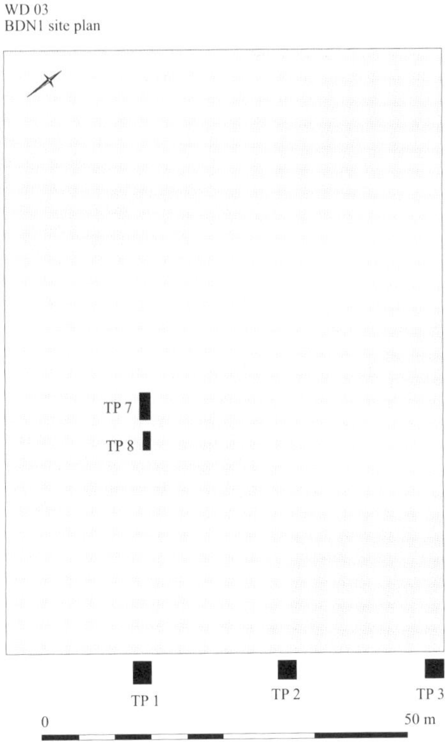

Figure 3.85. *Brockington Down 1 test-pit locations (M. Brudenell).*

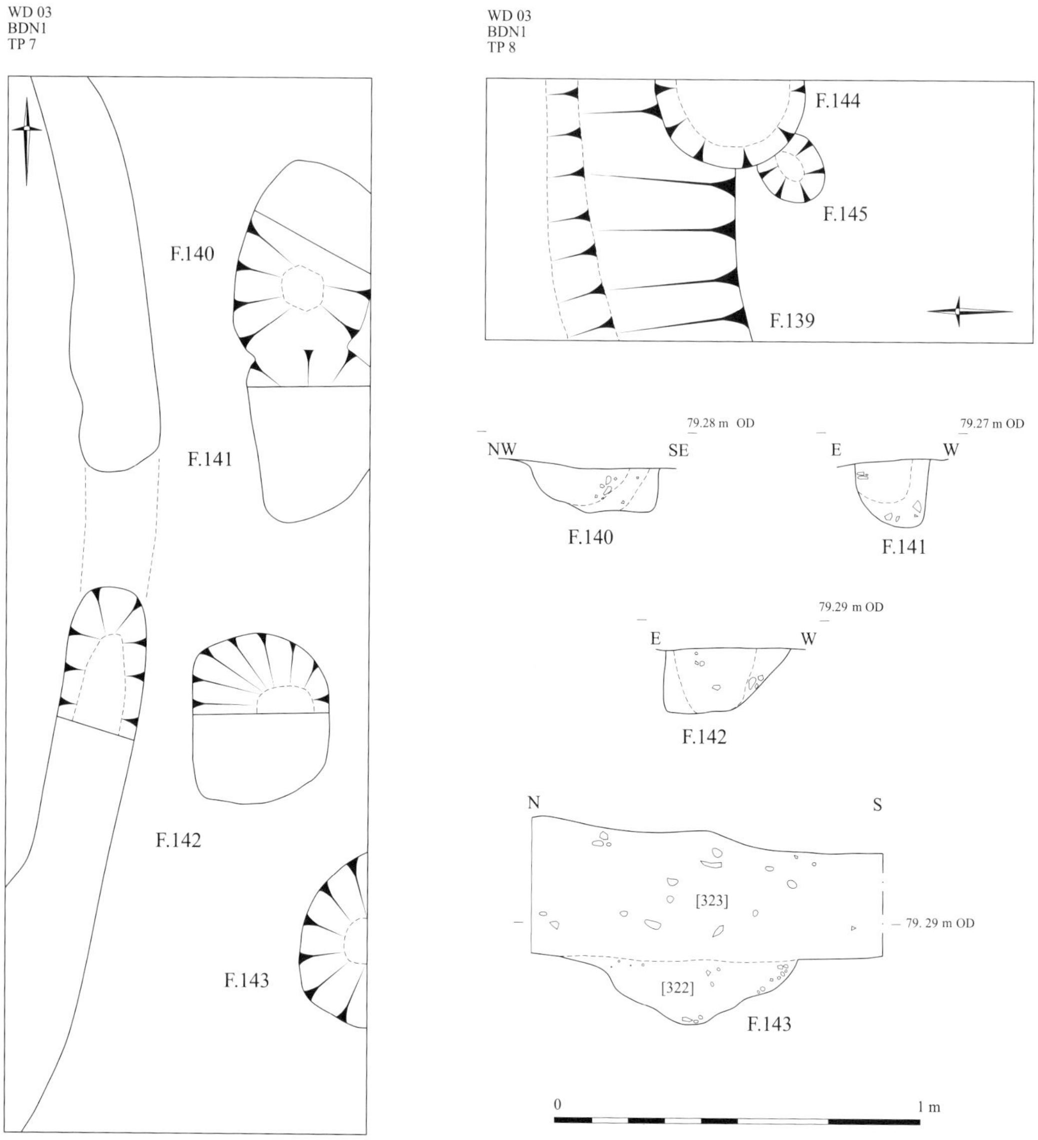

Figure 3.86. *Brockington Down 1 plans and sections of test pits 7 and 8 (M. Brudenell).*

the test excavation programme, building on the field walking survey completed in 2001–2 (Lewis & French 2002; see Gardiner above).

Field results by field and site

Brockington Down (BDN1 & 2)

Oblique aerial photographs taken by Martin Green in 1997 suggested that there were two areas of pits, small ring-ditches/gullies and linear features occurring just below the crest of Brockington Down, associated with slight ridges in the field's topography (Fig. 3.84). The subsequent geophysical survey (see Appendix 2) and test excavations suggested that area 1 (BDN1) to the northeast did contain structural and linear features (Figs. 3.84–3.86 & A2.16), whereas the second area (BDN2) to the southwest did not exhibit archaeological features (Figs. 3.84 & A2.16), but did have elements of the same possible field system that was evident in area BDN1.

Test excavation of area BDN1 at five locations (TP1–3, 7 & 8) revealed a number of postholes, possible stakeholes, postholes in post-trenches and one curvilinear gully (Figs. 3.85 & 3.86). These features were well defined against the chalk subsoil, and exhibited sufficient depth to have survived beneath the thin ploughsoil (<25 cm). About half of the features observed coincided exactly with the results of the magnetometer survey; the others were probably too small to have been picked out by this geophysical survey method. Lithic material in the ploughsoil was plentiful and was suggestive of a broad Neolithic to Bronze Age timescale being represented. Burnt flint was also common (see Gardiner below) and is normally ascribed to cooking/production processes of the later prehistoric periods. A few sherds of pottery were also encountered ranging from Romano-British New Forest Ware and Black Burnished Ware (M. Green pers. comm.) to more indeterminate but probably later Bronze Age coarse wares. Unfortunately, as is often the case, no diagnostic artefacts were found in the feature fills, which were 50 and 100 per cent excavated.

Test excavation of area BDN2 at three locations (TP4–6) did not reveal any definite cut features. Rather a series of amorphous, small pit-like features infilled with a reddish-brown silty clay did define in the upper surface of the chalk. It is suggested that these represent solution and tree-root holes in the chalk subsoil. Nonetheless, this does not mean that there are no archaeological features there if a greater area were to be opened up. Also, the geophysical plot was inconclusive, except for the presence of two linear features (Fig. A2.16).

Farringdon Clump (FCL)

Although this whole field had been previously field walked and had exhibited a good variety of Neolithic and Bronze Age lithic tool types and waste products, there were no specific concentrations. Nonetheless, a test excavation area was chosen on the slope overlooking the upper Allen river valley just to the south of Farringdon Clump (Fig. 3.80).

The magnetometer survey revealed no definite archaeological features (Fig. A2.16), and this was corroborated by the test-pit excavations. The four test pits (TP9–12) indicated that a thin rendzina soil over chalk by the Clump itself on the crest of the slope, soon gave way downslope to clay-with-flints deposits on the chalk. The resulting soil profile was much thicker with a thick ploughsoil containing an additional *c.* 15 cm depth of silty clay loam, most likely colluvially derived, developed over an *in situ* B horizon, in turn developed on a *c.* 50 cm thick clay-enriched or Bt horizon formed on about 70 cm of late glacial chalky silty clay wash. The B/Bt horizons most likely represent late glacial and early Holocene soil formation, such as observed at nearby Deer Park Farm where a similar soil profile was associated with a very late Upper Palaeolithic flint assemblage of about 11,000 BP (Green *et al.* 1998). Soil micromorphology (FCL, TPG-3) showed a possible old A horizon with several phases of clay illuviation (probably including modern), mostly of 'dusty', signifying disturbance of the upper horizons. The most recent illuviation events, however, have left relatively clean clay coatings in pore spaces, suggesting clay movement in a stable profile. The horizon is slightly oxidized, light reddish-brown sandy (clay) loam with low organic matter content, some of which is possibly charcoal (see Lewis Appendix 4.12).

In FCL pit profile A the sediment was more organic than in any of the soil profiles described (which were stained with amorphous iron oxides), and also contained more calcareous features. These characteristics could reflect a relatively organic pit fill, but more likely result from less disturbance (oxidation) than the soil profiles, and more earthworm activity is indicated by pedofeatures and mineral inclusions. The pit profile contains a metal fragment (haematite with alteration to something akin to goethite (Chris Doherty pers. comm.), occasional charcoal, and rare rounded clean clay fragments, presumably from a lower B/C horizon in the area (see Lewis Appendix 4.12).

Brockington Farm (BD & BRO)

Four test pits were placed in relation to both areas geophysically surveyed at lower (BD; TP 20 and 21) and upper (BRO; TP 22 and 23) slope locations (Fig. 3.80). The profiles at the lower slope test pits (BD) exhibited a colluvially thickened (*c.* 30–35 cm) ploughsoil overlying a thick (*c.* 1 m) B horizon of silty clay, developed on soliflucted coombe rock/flint rubble and chalk substrate deposits.

This occurrence of an *in situ* clay-enriched argillic brown earth soil associated with coombe rock or clay-with-flints deposits occurring towards the base of slope/dry valley is now being commonly observed at this southern end of the study area. This suggests that these areas of glacial drift deposits on the chalk weathered naturally to thick, silt and clay dominated soils, unlike the adjacent areas of just chalk substrate. These areas would have offered and supported a different soil and vegetation regime in the past that would have been slightly acidic, and were probably dominated by woodland and woodland soils.

The test pits placed upslope at BRO (Fig. 3.80) revealed no archaeological features, either on the chalk or on the thin horizon of coombe rock.

In both cases, the test pits would seem to confirm the magnetometer surveys (Fig. A2.16) and the absence of earth-fast archaeological features, despite the presence of good earlier prehistoric worked flint scatters in this field observed during field walking.

Charlton Dairy Farm, Knowlton (CD1 & 2)

This area was selected for geophysical and test-pit investigations purely on the basis of its location between the Knowlton henge complex and the barrow cemetery to the northeast in a small dry valley just to the southeast of the Allen valley flood plain (Fig. 3.80). Although there is a deserted medieval village immediately adjacent and to the north of this area and Charlton Dairy Farm, no medieval features or artefacts were found.

The geophysical survey suggests an absence of features in both areas tested (Fig. A2.16).

The test pits (TP 24–6) in the pasture field CD2 exhibited a turf and a thin rendzina soil profile developed on the chalk, with no evident archaeological features.

In contrast, the test pits in the arable field CD1 (TP 27–32) revealed a thick soil profile (*c*. 50–70 cm) of ploughsoil over an *in situ* B horizon of a brown earth developed on thin coombe rock deposits and weathered chalk as the slope dipped into the small dry valley (Figs. 3.87 & 3.88). Moreover, test pits 27 and 29–32 revealed a southwest–northeast aligned post-gully (F.17) (Fig. 3.88). This gully held an irregular line of almost contiguous postholes, traced over a distance of some 10 m and still continuing to the northeast, and had contained posts of different shapes and depths. The posts ranged from sharpened tipped, round to rectilinear to square to sub-square, suggesting that all were worked timbers, and most probably driven into the ground. The depths of the postholes, including the associated soil depth, ranged from about 60–120 cm. These features are indicative of either a post-built palisade or substantial wooden building.

In terms of artefacts, several small and abraded sherds of Romano-British Black Burnished Ware were recovered from upper fill of the postholes, although these may not necessarily be indicative of the age of the postholes. Moreover, there are a number of curious, linear pit line features leading up from/to the henges and barrows of the adjacent Knowlton complex (Green 2000, pl. 11) (Fig. 2.17), to which this feature may be similar. Indeed, similar lines of posts of later Bronze Age date have been identified at Barleycroft Farm, in the lower Great Ouse valley of Cambridgeshire (Evans & Knight 1996). Obviously, much more investigation is required.

Avenue Lodge (AL & AVE)

Both of the fields to the east of Avenue Lodge (Figs. 3.80, 3.89–3.91) were field walked with the help of

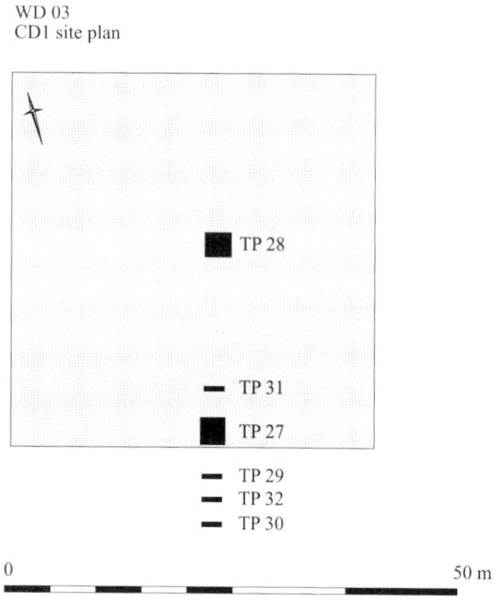

Figure 3.87. *Charlton Dairy 1 test-pit locations (M. Brudenell).*

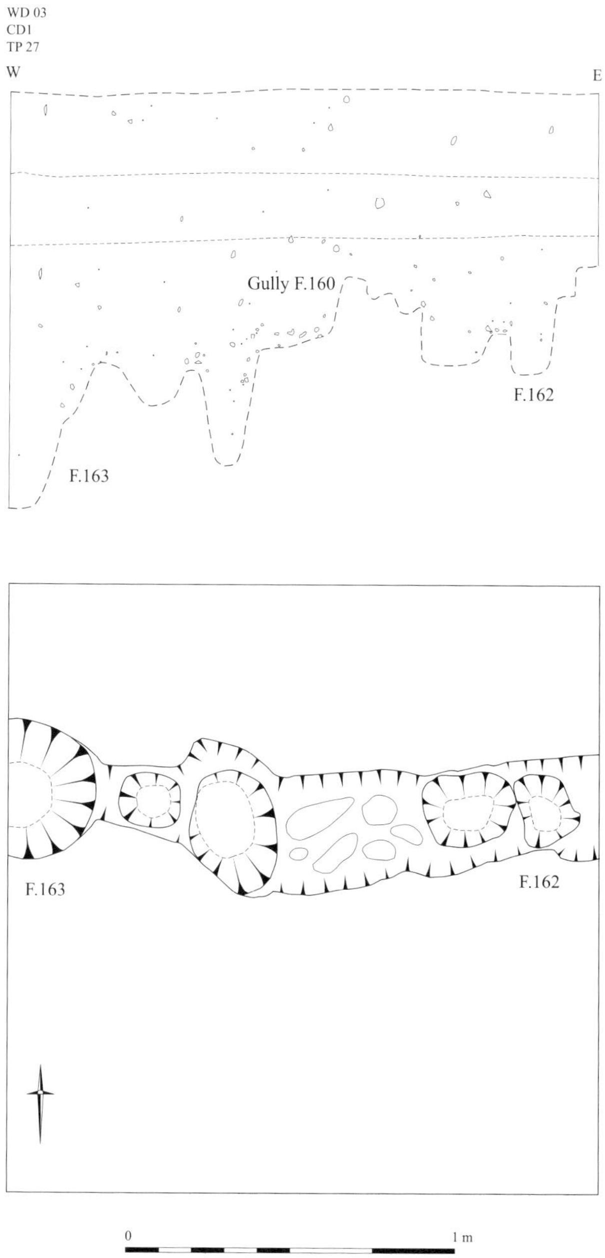

Figure 3.88. *Charlton Dairy 1 plan and section of test pit 27 (M. Brudenell after C. French).*

volunteers from the East Dorset Antiquarian Society and a sample area geophysically surveyed (Appendix 2; Fig. A2.16). First field impressions indicate that the area of pits showing on the 1997 NMR aerial photograph is matched by an extensive zone of burnt flint. Rather suprisingly, the magnetometer survey of the same tested area indicated very little sign of archaeological features (Fig. 3.89). This may have been due to the very dry conditions of the field, and it had just been drilled/harrowed/rolled. This zone of probable settlement is in a linear band just on and off the crest of the down overlooking the Crane valley and Cranborne village to the southeast, and is in effect on the watershed between the Allen and Crane valleys.

As the Avenue Lodge field was already drilled, a series of seven 1 m square test pits (TP 13–19) were hand excavated at 20 m intervals in a transect across the flint scatter and area of pits, 10 m out and parallel to the hedge line (Figs. 3.90 & 3.91). In all but three of these test pits, earth-fast archaeological features were identified including pits, postholes and gullies, cut both into the chalk and Reading Beds subsoils. These features indicate that a great density of settlement archaeology is surviving here. Both Middle Bronze Age and Romano-British pottery (K. Gdaniec pers. comm.), in small and abraded quantities, along with large quantities of burnt flint and moderate occurrences of rather poorly made worked flint were found associated with these features (see Gardiner below). These characteristics suggest that this is probably a later prehistoric settlement site with some superimposed Romano-British occupation.

The soil micromorphology profile sampled showed a bioturbated A horizon with some illuvial clay pedofeatures, overlying an iron-rich oxidized B horizon with clay illuviation. Both horizons consisted mainly of flint gravels with soil in between and therefore appears to have developed on a gravel fan. The clay coatings are mainly 'dusty', suggesting disturbance in the profile (see Lewis Appendix 4.12).

Lithic remains from the Knowlton environs test pits

Julie Gardiner

A series of test pits (TP 13–19) was dug at 20 m intervals along transect 0 in field AL (*c.* 10 m from the hedge-line separating AL and AVE) within collection units 0–20, 20–40 and 40–60 (Figs. 3.90 & 3.91). None of these units produced any worked flint from the surface. Fifteen struck flints were recorded from the test pits — a small, possibly Mesolithic core and 6 flakes from TP 14, and 1–3 flakes from each of the others. All test pits produced burnt flint (up to 260 g).

Figure 3.89. *Oblique aerial view of the Avenue Lodge settlement features (centre left) (with permission of English Heritage, Swindon Office, Crown Copyright).*

Other test pits in field walked areas

Farringdon Clump (FCL)

This field produced a surface assemblage of Late Neolithic–Early Bronze Age date. Test-pits dug close to the southern boundary of the field produced 120 worked flints, with a fair amount in each (24–36 pieces). Material recovered from the ploughsoil in each test pit was generally plough damaged, only that from the B horizon in TP 9 was in a fresh condition but the five flakes and a scraper from that level are not diagnostic. Flakes and irregular waste, cores and scrapers were recovered from all test pits.

Brockington Farm (BD/BRO)

Two test pits in BD produced 71 struck flints, all debitage apart from two rather roughly made scrapers. A neat blade core from the upper 'B' horizon in TP 20 is on a gravel nodule.

Two test pits in BRO produced 55 struck flints, again all debitage apart from a broken scraper. All test pits also produced burnt flint.

Brockington Down (BDN1 & 2)

Test pitting in two areas of field BDN (Figs. 3.85 & 3.86) produced 71 and 42 flints, respectively. There is insufficient material to provide any clear indication

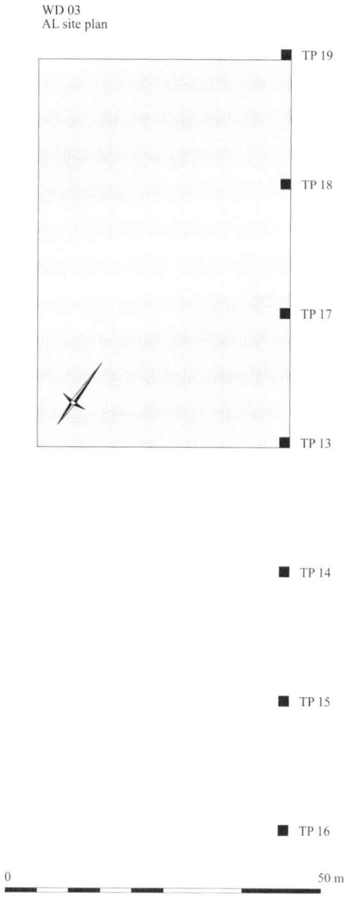

Figure 3.90. *Avenue Lodge test-pit locations. (M. Brudenell.)*

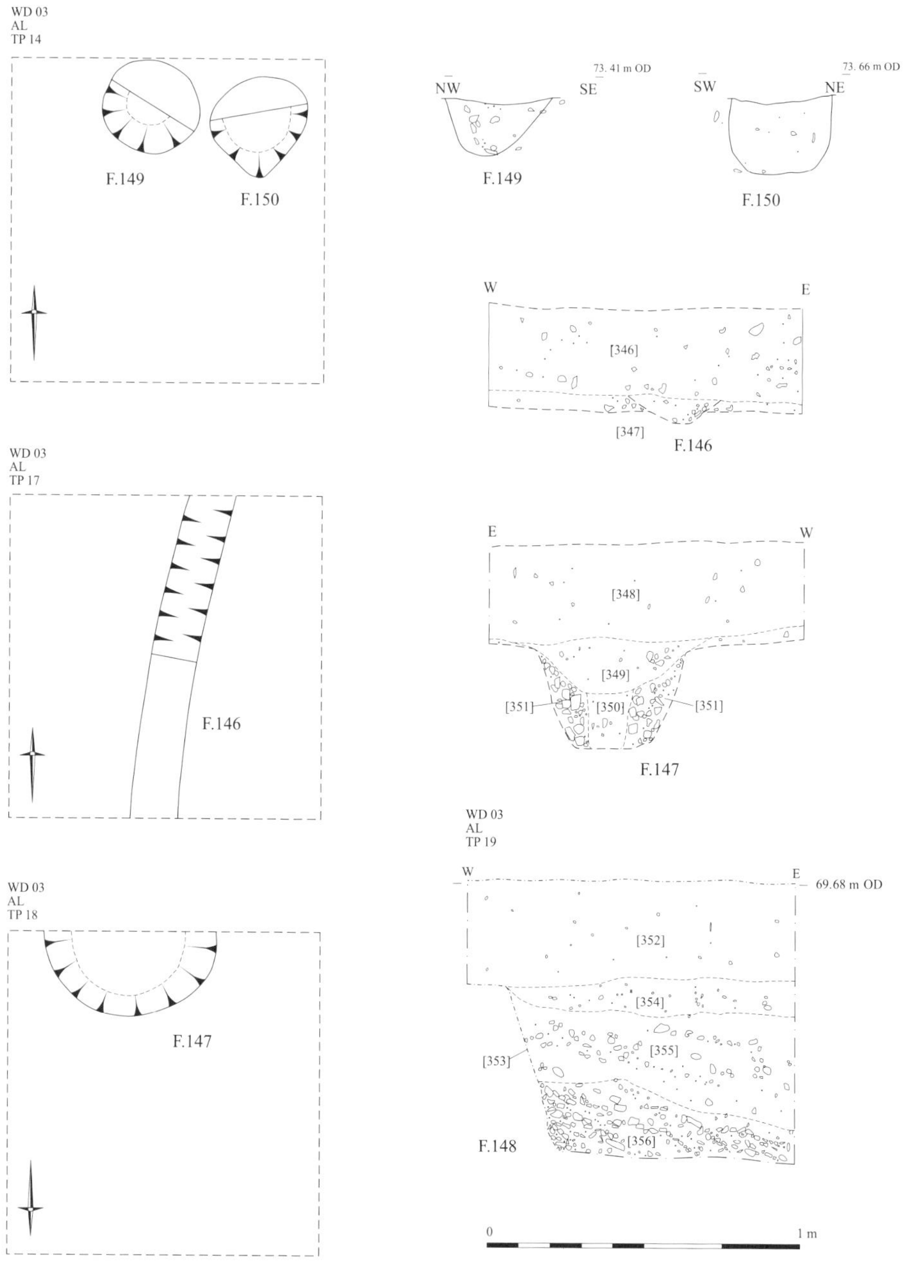

Figure 3.91. *Avenue Lodge plans and sections of test pits 14, 17 and 18 (M. Brudenell after H. Lewis & C. French).*

of date but there are relatively few hinge fractures, several blades and bladelike flakes and at least one neatly produced scraper which might suggest a Late Neolithic–Early Bronze Age rather than later Bronze Age date. Not surprisingly, given that the topsoil here is less than 25 cm in depth, much of the material is plough damaged. TP7 in BDN1 included several burnt worked flakes and 320 g of unworked burnt flint. TP4 in BDN2 produced a very small fragment of burnt clay or daub while TP5 also included 220 g of burnt flint.

Charlton Dairy Farm (CD1 & 2)
These fields lie just to the northeast of Knowlton Circles (Figs. 3.80 & 3.87). Test pits in area CD1 produced 98 worked flints, virtually all waste flakes but with one roughout transverse arrowhead from the Ap horizon of TP28. This test pit also produced over 1 kg of burnt flint, as did TP27.

Fifty-four pieces of debitage were recovered from test pits in area CD2 along with up to 650 g of burnt flint. Most of the material recovered is quite small and thin and includes several blades and blade-like flakes.

The assemblage is, again too small to provide clear dating evidence but seems more likely to be Late Neolithic–Early Bronze Age than later Bronze Age in date.

Knowlton environs soil micromorphology

Helen Lewis & Charles French

The test excavations at Farringdon Clump (Fig. 3.80) revealed a poorly structured to apedal, bioturbated, sandy clay loam to loam soil beneath a clay loam colluvial ploughsoil (Appendix 4.12). The whole profile exhibited frequent dusty clay additions in the voids, suggesting some downslope and down-profile colluvial movement of fine material. Unfortunately there were no good archaeological associations with this soil, and it really reflects the modern plough and hillwash regime more than anything else.

In contrast, the profile at the Charlton Dairy site, just downslope and to the northeast of the South Circle henge of the Knowlton complex (Fig. 3.80), was a thin, sandy clay loam, brown earth with some evident subangular blocky ped structure (Appendix 4.12). It exhibited an A and B horizon, the latter with clay coatings and infillings. This brown earth type of profile was more similar to the palaeosol found in the valley trench west of Monkton-up-Wimborne (see above; Fig. 2.7).

Although not examined micromorphologically, the profile at Avenue Lodge that was associated with the archaeological area of later prehistoric pits was more of a thickened and organic, silt loam, ploughsoil.

Considered together with all the other buried soil profiles examined within the study area, there is considerable variation in soil types now evident, not just blanket coverage of thin, calcareous silt loam rendzina profiles.

Land use and landscape development: the molluscan evidence

Michael J. Allen

A huge programme of land-snail analysis has been conducted over five seasons sampling ditch-fill sequences and buried soils from 11 separate monuments (Table 3.6) encompassing Neolithic to Iron Age sequences. The analytical programme has entailed the taking and processing of 154 samples (*c.* 270 kg) and the recovery of about 62,000 identified shells (excluding the burrowing, and thus palaeoecological insignificant, species *Cecilioides acicula*).

This data base provides a series of palaeoenvironmental sequences along a stretch (both north and south) of about 10 km^2 of the upper Allen Valley, from which to examine eleven changing local land-use histories. These, with previous work by Entwistle & Bowden (1991) which encompasses a further six sites, and research by this author (Allen 1998; 2002a) which includes a further 11 sites (252 samples; 11,520 snail identifications), provides probably the largest, and densest prehistoric land snail data base from any landscape (Table 3.7).

At present, however, this analysis is based upon the 154 samples analysed from this fieldwork set into the interpretive framework established by Entwistle & Bowden (1991), and that outlined by Allen (1994; 1998; 2000a; 2002a). Some of the molluscan analysis in this area conducted from sites excavated by Martin Green will be presented elsewhere (Allen in prep.) but are summarized in Appendices 4.2, 4.4 & 4.12.

Analysis

Analysis follows standard methods (Evans 1972), with molluscan groupings following Evans (1984). Statistical appraisal of the assemblages has been discussed (Allen, in French *et al.* 2003), but the application of these statistics is outlined below (see Appendix 3 & Tables A3.1–A3.13). Diversity indices are only employed and histograms constructed on assemblages with greater than at least 25 shells.

The use of the Shannon Index (*H'*) index which enables measurement of heterogeneity (Peet 1974) evaluates both species richness and evenness. It takes into account the number of species and proportions of those species to view evenness of the distribution, and is not affected by the total population. Hence it has been argued (Evans & Smith 1983), that this is ideal for sub-fossil assemblages. Indices are usually between 1.5 and 3.5, measurements above 2.3 in palaeo-mollusc assemblages are considered high.

The closely allied index, the Brillouin index (*HB*) takes account of total populations (unlike the Shannon index) as well as numbers of species and individuals (Magurran 1988). The significance of this is that the comparison of the two indices allows some measure of the completeness of the samples and thus indicates how representative the sub-fossil assemblage is likely to be of the original fauna from which it may be derived. Comparison of the Shannon and Brillouin indices, thus allow the suitability of sub-fossil assemblages to analysis by delta indices (Δ_2 & Δ_4) to measure the diversity of the ecological mosaic sampled, and thus examine the nature of the environment wider than that in the immediate proximity of the sampled location.

Thus the Shannon index provides an indication of sample richness. Comparison of Shannon and Brillouin indices allows us to evaluate the closeness of the sub-fossil assemblage to a likely real population. Then delta indices measure the complexity and maturity of the ecological mosaic. Diversity indexes are given in Appendix 3.

Table 3.6. *The excavations in the upper Allen valley where molluscan analyses have been conducted (* = reported in this volume; + = summarized in this volume).*

	Site	Period	Samples	Kg	Snails	Excavator
1	Monkton-up-Wimborne valley*	Iron Age and later	5	7.5	427	C. French, Univ. Cambridge
2	Gussage Cow Down 192/02*	Iron Age lynchet	9	14	2244	C. French, Univ. Cambridge
3	Monkton-up-Wimborne enclosure+	later Bronze Age to Iron Age	8	9.0	2018	M. Green, EDAS
4	South Lodge (Entwistle & Bowden 1991)	later Bronze Age	9	9.0	326	Barrett *et al.*
5	Home Field 1986 (Allen in prep.)	Bronze Age hollow	7	10.1	1111	M. Green, EDAS
6	Monkton-up-Wimborne 2002+	Bronze Age pond barrow	4	6.0	938	M. Green, EDAS
7	Home Field 1995 (Allen in prep.)	Middle Bronze Age lynchet	3	3.0	316	M. Green, EDAS
8	Wyke Down 1998 (Chalkpit Field barrow 34)*	Middle Bronze Age barrows	23	43.0	14,638	C. French, Univ. Cambridge
9	Wyke Down 1998 (Chalkpit Field barrow 41)*	Middle Bronze Age barrows	6	11.5	103	C. French, Univ. Cambridge
10	Wyke Down 2000 (barrow 36)*	Middle Bronze Age barrows	18	29.5	6416	C. French, Univ. Cambridge
11	Wyke Down 2000 (barrow 44)*	Middle Bronze Age barrows	13	21.1	6496	C. French, Univ. Cambridge
12	Down Farm ring ditch (Entwistle & Bowden 1991)	Early Bronze Age	6	6.0	667	Green/Barrett *et al.*
13	Down Farm pits and enclosure (Entwistle & Bowden 1991)	Grooved Ware & Middle Bronze Age	10	11.0	1513	Green/Barrett *et al.*
	Down Farm enclosure 2003+	Middle Bronze Age	9	14.3	1200	C. French, Univ. Cambridge
14	Fir Tree Field pits 1995+	Grooved Ware & Beaker	15	20.6	5188	M. Green, EDAS
15	Wyke Down 2 henge 1996+	Grooved Ware	28	44.3	4424	M. Green, EDAS
16	Wyke Down 2 pits\ph 1996+	Grooved Ware	8	9.2	1666	M. Green, EDAS
17	Wyke Down henge 1 (Entwistle & Bowden 1991)	Grooved Ware	4	5.5	1331	Green/Barrett *et al.*
18	Monkton-up-Wimborne enclosure*	later Neolithic	12	20.6	1090	C. French /M. Green
19	Monkton-up-Wimborne 2 1997+	?Peterborough–Mid Bronze Age	71	104.7	37,321	M. Green, EDAS
20	Knowlton Circles 1994*	Neolithic to Early Bronze Age	50	107	23,578	J. Gale, Bournemouth Univ.
21	Dorset Cursus 1982 & 1984 (Entwistle & Bowden 1991)	(Mesolithic) Mid to later Neo-lithic	16	25.8	595	Barrett *et al.*
	Dorset Cursus 2001*	Mid to later Neolithic	31	51.26	10,468	C. French, Univ. Cambridge
22	Handley Down mortuary enclosure 1996 (Allen in prep.)	earlier Neolithic–Early Bronze Age	27	29.8	21,224	S. Burrow, Bournemouth Univ
23	Gussage Cow Down 78/02*	Neolithic long barrow	6	11.75	612	C. French, Univ. Cambridge
24	Gussage Cow Down 294/02*	Neolithic long barrow	22	43.5	17,544	C. French, Univ. Cambridge
25	Fir Tree Field shaft 1992+	Mesolithic to Beaker	31	43.0	15,736	M. Green, EDAS
		Total	**451**	**712.01**	**179,190**	

Table 3.7. *Comparison of density of other molluscan data sets in chalkland landscapes and the calculation of a confidence factor.*

Study area	No. of data sets	km² study area	Density (data sets km²)	Confidence factor (density ×100)
Lewes	9	106	0.085	8.5
Isle of Wight	9	64	0.140	14.1
Strawberry Hill	1	10	0.100	10.0
Stonehenge 1	13	54	0.240	24.1
Stonehenge 2	19	80	0.238	23.8
Dorchester	12	35	0.343	34.3
Avebury	20	130	0.154	15.4
Winchester	3	16	0.187	18.7
Cranborne 1	22	150	0.147	14.7
Allen Valley/ Down Farm	62+	25	24.8	248

Interpretation

Many of the sequences discussed are from ditch fills that provide inherent difficulties in interpretation. In particular is the distinction between assemblages of the local microhabitat of the ditch and those reflecting the habitats and environment surrounding the ditch (Evans 1972, 321–8; cf. Bell 1990). This may be further complicated by the fact that ditches are often boundaries or edges between different micro-habitats. This is especially true of banks and ditches or barrows where the mound/bank may provide one habitat (e.g. grass) which is bound another (i.e. open tilled lands), and is separated by the habitat specific to the ditch and vegetation growing in it. The taphonomy of the ditch fills is, therefore, acutely important to the interpretation of the resultant assemblages. The combination of the taphonomy, application of species diversity indices, and recognition of these complexities all aid with providing more comprehensive local interpretation. An example is that of the Bronze Age round barrow at Round-the-Down, Lewes, East Sussex (Allen 1995a), where it was suggested that the mound was one of long grass set into an arable context, and here this interpretation was only possible though the factors discussed above, and the presence of local off-site data in the form of an analysed colluvial sequence (Allen 1995b; Southerham Grey Pit).

The tripartite division of ditch fills, as defined by Evans (1972, 321–8) and Limbrey (1975, 290–300), has proved to be applicable in archaeological situations (e.g. Stonehenge, see Allen 1995b). The complexities outlined above are exacerbated and particularly acute in secondary fills, and especially those where shady mollusc assemblages occur, such as will be seen in some key sequences described below. Shells from the primary fill largely originate from the old land surfaces through which the ditch (or pit) was cut and thus represent the pre-feature environment. Those from the tertiary fills represent the final phases of ditch infilling when the feature or monument was often subsumed into the broader landscape both physically and ecologically. In contrast, most acute variation occurs in the secondary fills, precisely those that accumulate during the duration of the monument use. So whilst being the most useful deposits for examining environments contemporaneous with the use of the site, they are undoubtedly the most complex to interpret. For instance, one lesson of the Experimental Earthworks Project (Bell *et al.* 1996) seems to be that an isolated shady ditch in a grassy landscape some hundreds of metres from the nearest real woodland does not tend to attract assemblages of shade-loving species, at least over a timescale of three decades (M. Bell pers. comm.). Nevertheless, shade-loving assemblages can migrate into rapidly open country rapidly over shorter distance of 50–100 m (Allen pers. observ.; in prep.). These two observations reinforce theoretical comments on ecological mosaics and scales of environmental resolution by Thomas (1985). Some groups of shade-loving species may represent local leaf-litter and vegetation communities within ditches, having migrated some time previously through open country from shady woodland conditions in the vicinity. Other groups such as those including *Clausilia* and *Ena montana*, for instance, tend to suggest the presence of more wooded conditions in the immediate proximity (less than 100 m) coeval with that of the ditch assemblages.

Aims

The aim of the analysis was simple: the development of high-resolution models (Allen 1997a; 2000b) of the land-use and landscape developments within the tight topographical constraints of the Allen Valley study area. In theory we would like to view the landscape though our ancestors eyes and be able to describe the landscape as if we were viewing it (Allen 2000a; 2002a), rather than reading about it (cf. Evans 1972). The data can be used to define the development of the ecosystem and the pedological history of the area. This can be used to compare with, and challenge existing system theories put forward for this area by Fisher (1991) and Entwistle & Bowden (1991) and the chalkland of southern England (Bell 1992; Allen 1992).

This independent palaeo-landscape data (Mollusca) can then be balanced by the wider spectra provided by the molluscan data from previous excavations in the study area, pollen, and interpretations tied to palaeosol data to develop interpretations of both the landscape mosaic and landscape changes. The action of human societies can be defined and modelled in the development of the prehistoric chalkland landscape.

The wider molluscan data base

The rich and diverse archaeological excavations conducted previously (Barrett *et al.* 1991a; Green 2000) in the upper Allen valley and as part of this project (French *et al.* 2000) have provided a wealth of well-dated Neolithic and Bronze Age contexts in the study area and the chalk downland region (Tables 3.6 & 3.7). The sampling of sequences through numerous sites starts to provide the potential to examine not just single site sequences, but to map the mosaic of land-use changes through the upper Allen valley (Allen 1998; French *et al.* 2000). This is only possible as a result of the large data set of samples, giving a greater density of data than seen in any other prehistoric landscape anywhere else in northwest Europe (Allen 1997a; 2000a,b).

The palaeo-molluscan results

Neolithic long barrows

Two profiles were examined. The first was the buried soils and turf mound of Gussage long barrow 78 which can be compared with that analysed from the Thickthorn long barrow (Kennard 1936; Entwistle 1985). The second longer sequence was from ditch fills of barrow 248.

Gussage Cow Down long barrow (GCD 78):
the pre-monument environment

A series of six spot samples from buried soils and turf horizons provide important evidence of the pre-monument, and Early Neolithic environment (Figs. 3.8 & 3.92; Table 3.8). Two short sequences were examined, and in both low shell numbers were present in the lower A horizon, with high shell numbers in the soil surface (upper A horizon). In both sequences the earliest assemblage from the lower A horizon (279, lower 278 & 291) contained small assemblages of only less than 30 shells (*c.* 12 kg), but significantly they all predominantly comprised open country and intermediate species. Very few shade-loving taxa were present. These small assemblages suggest the construction of this monument in a pre-existing open dry grassland, and the clearance of any former woodland at some considerable time previously.

The buried soil surface (context 278) and the turfline (291) in the bank are dominated by the *Vallonia* species, and suggest the establishment of open dry grassland with some bare soil (*Pupilla muscorum*/*Vertigo pygmaea*). A rare find of *Truncatellina cylindrica* in this landscape is typical of short dry, calcareous grassland (Kerney 1968) (and see also, for example, the Cursus). Apart from the rise in shell numbers which tends to confirm the presence of the soil surface, perhaps the most significant element of these assemblages is the relatively high proportions of *Pomatias elegans* (Fig. 3.92) which might be taken to suggest some local loose and bare soil. Here, however, these clearly do not indicate conditions created by woodland clearance, but might possibly suggest broken soil created by tillage or de-turfing. Low variance between Shannon and Brilliouin indices (Table A3.1) confirms autochthonous assemblages (cf. Allen, in French *et al.* 2003, 233–4) from these buried soils. Shannon function indices fluctuate around 1.2 to 1.43 and indicate low-diversity habitats such as grassland without significant scrub, and the uniformity of this local landscape is confirmed by the relatively low and constant Δ_2 and Δ_4 indices of *c.* 0.7 and 2.5 respectively (Table A3.1).

These assemblages superficially compare well with those from the buried soil surface beneath Thickthorn Down (Bradley & Entwistle 1985), indi-

Table 3.8. *Mollusca in the buried soil from beneath the Neolithic long barrows of Gussage Cow Down barrow 78 and Thickthorn Down (analysis by Kennard 1936; Entwistle 1985; and Bradley & Entwistle 1985).*

Feature	Long barrow 78, buried soils						Thickthorn			
Context	279	278	278	291	290	277	1 ols	3 ols	4 ols	OLS
Sample	248	246	245	256	255	243	1936	1936	1936	1985
Depth (cm)		sp 2	sp 1							
Wt (g)	2000	1750	2000	2000	2000	2000	hand	hand	hand	135
MOLLUSCA										
Pomatias elegans (Müller)	5	10	15	13	7	11	1	-	+	+
Cochlicopa lubrica (Müller)	-	-	-	-	-	-	1	-	-	2
Cochlicopa spp.	1	1	4	1	3	-	-	-	-	2
Truncatellina cylindaca (Férussac)	-	-	-	1	-	-	-	-	-	-
Vertigo pygmaea (Draparnaud)	-	1	-	-	4	5	-	-	-	3
Vertigo spp.	-	-	2	-	-	-	-	-	-	-
Pupilla muscorum (Linnaeus)	-	1	8	3	8	7	-	5	-	10
Vallonia costata (Müller)	7	2	84	5	110	46	5	6	6	7
Vallonia pulchella (Müller)	-	-	-	-	-	-	5	17	3	-
Vallonia excentrica Sterki	7	1	51	2	112	24	-	-	-	3
Vallonia spp.	-	-	-	-	-	-	-	-	-	5
Punctum pygmaeum (Draparnaud)	-	-	-	-	1	-	-	-	-	-
Discus rotundatus (Müller)	-	-	-	-	-	-	-	-	-	1
Vitrina pellucida (Müller)	-	-	-	-	1	-	-	-	-	-
Vitrea contracta (Westerlund)	1	-	-	-	-	-	-	-	-	-
Oxychilus cellarius (Müller)	-	-	1	-	-	-	-	-	-	-
Limacidae	1	-	-	-	-	-	-	-		-
Clausilia bidentata (Ström)	-	-	+	-	-	-	-	-	-	-
Helicella itala (Linnaeus)	-	-	2	-	2	3	-	-	-	4
Trichia hispida (Linnaeus)	1	-	12	1	10	4	-	-	-	-
Cepaea nemoralis (Linnaeus)	-	-	-	-	-	-	18	2	26	-
Cepaea spp.	2	4	3	+	+	1	-	-	-	1
Taxa	8	7	10	7	10	8	6	4	3	8
TOTAL	**25**	**20**	**182**	**26**	**258**	**101**	**30**	**30**	**35**	**36**

cating comparable 'rather dry grassland conditions' immediately prior to barrow construction. The lack of *P. elegans* at Thickthorn Down suggests perhaps a fuller grassy ground cover. We can see these monuments built in a stable, well-established short-sward (possibly grazed) dry grassland.

Gussage Cow Down long barrow (GCD 294): contemporaneous and post-monument environment

A series of 19 samples were taken through almost the entire northern ditch fill profile (F.111) (Figs. 3.4 & 3.92; Table 3.9). This sequence was augmented by a short sequence of three samples from a better-developed stasis horizon over the primary fill in the southern ditch (F.114) (Fig. 3.4).

Shell numbers are typically very low in the lower chalk rubble primary fills but soon pick up to moderate to high shell numbers in the main fill. The depauperate primary fill assemblages although dominated by open-country species, present a significantly higher number of shade-loving taxa than from barrow 78. Open-country (*Helicella itala* & *Vallonia excentrica*) and

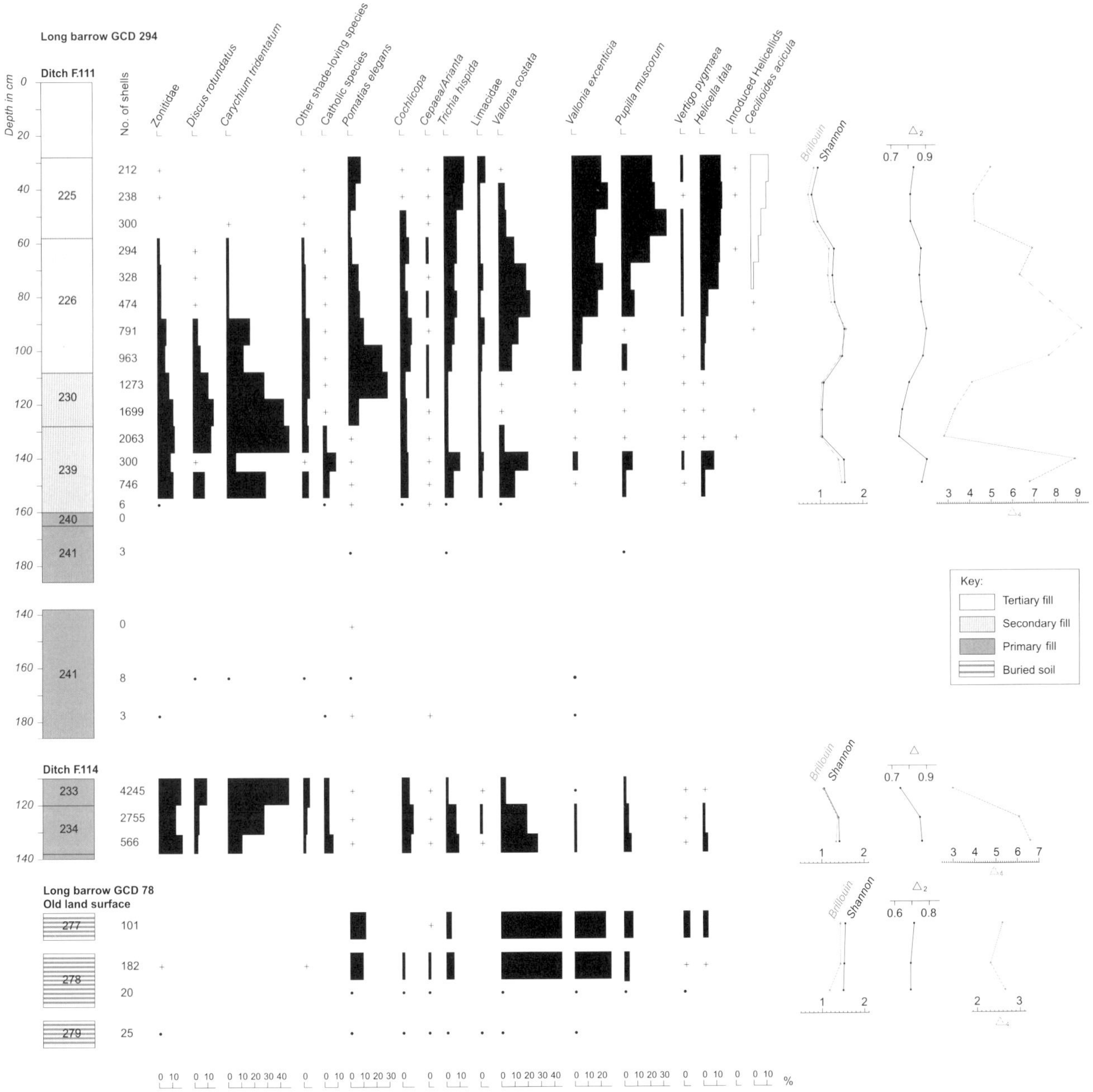

Figure 3.92. *Molluscan histogram of the ditch at Neolithic long barrow 294 and the old land surface beneath long barrow 78 on Gussage Cow Down (R. Goller, Wessex Archaeology).*

Table 3.9. *Mollusca from the ditch of Gussage Cow Down long barrow 294.*

Feature	south ditch F.114			north ditch F.111																		
Context	234		233	241				240	239			239/30	230		226					225		
Context	u primary soil			primary					secondary						tertiary							
Sample	3	2	1	1	2	3	4	5	6	7	8	9	10	11	12	13	14	15	16	17	18	19
Depth (cm)	131–138	120–131	110–120	170–185	157–170	138–150	165–186	160–165	155–160	145–155	138–145	128–138	118–128	108–118	98–108	88–98	78–88	68–78	58–68	48–58	38–48	28–38
Wt (g)	2000	2000	2000	2000	2000	2000	2000	1500	2000	2000	2000	2000	2000	2000	2000	2000	2000	2000	2000	2000	2000	2000
Pomatias elegans (Müller)	1	3	2	+	1	+	1	-	+	10	4	17	134	378	248	97	42	25	8	7	14	20
Carychium tridentatum (Risso)	63	764	2001	-	4	-	-	-	-	223	21	992	747	367	125	139	16	8	9	3	-	-
Cochlicopa lubrica (Müller)	12	8	28	-	-	-	-	-	1	5	13	10	4	4	20	17	-	2	4	4	-	-
Cochlicopa lubricella (Porro)	-	34	39	-	-	-	-	-	-	2	-	7	1	-	4	-	-	-	2	+	-	-
Cochlicopa spp.	26	198	172	-	-	-	-	-	-	48	18	96	72	49	53	55	28	10	15	12	2	3
Truncatellina cylindrica (Férussac)	-	-	-	-	-	-	-	-	-	-	-	-	-	-	-	-	-	-	-	-	-	-
Vertigo pygmaea (Draparnaud)	-	15	11	-	-	-	-	-	-	6	6	7	4	1	9	8	9	5	5	6	1	4
Vertigo pusilla (Müller)	-	2	2	-	-	-	-	-	-	-	-	-	-	-	-	-	-	-	-	-	-	-
Vertigo spp.	2	6	4	-	-	-	-	-	-	1	-	2	-	-	1	-	-	1	2	-	-	-
Pupilla muscorum (Linnaeus)	34	115	69	-	-	-	1	-	-	24	20	24	13	6	37	11	45	24	61	101	60	50
Vallonia costata (Müller)	160	557	197	-	-	-	-	-	4	93	65	59	18	9	92	118	112	68	33	18	13	2
Vallonia excentrica Sterki	9	43	17	1	1	-	-	-	-	5	12	6	4	8	72	63	93	76	53	56	65	49
Vallonia spp.	-	-	-	-	-	-	-	-	-	-	-	-	-	-	-	-	-	-	-	-	-	-
Acanthinula aculeata (Müller)	8	71	192	-	-	-	-	-	-	23	4	86	52	32	8	11	5	5	2	-	-	-
Ena montana (Draparnaud)	-	-	-	-	-	-	-	-	-	-	-	2	4	1	-	-	-	-	-	-	-	-
Ena obscura (Müller)	-	-	6	-	-	-	-	-	-	4	-	3	6	4	2	4	+	-	-	1	-	-
Punctum pygmaeum (Draparnaud)	23	72	112	-	-	-	-	-	-	22	8	51	14	9	-	2	3	1	2	-	-	-
Discus rotundatus (Müller)	21	131	441	-	1	-	-	-	-	65	2	282	290	166	61	29	3	1	1	-	-	-
Vitrina pellucida (Müller)	4	3	7	1	-	-	-	-	1	6	5	3	-	-	-	2	-	-	-	-	-	-
Vitrea crystallina (Müller)	5	6	13	-	-	-	-	-	1	13	4	17	3	8	1	4	-	1	-	-	-	-
Vitrea contracta (Westerlund)	76	154	215	-	-	-	-	-	2	23	23	54	27	11	3	5	1	2	1	-	-	-
Nesovitrea hammonis (Ström)	13	33	26	-	-	-	-	-	-	3	13	4	2	-	5	5	1	2	2	-	-	-
Aegopinella pura (Alder)	7	106	300	-	-	-	-	-	-	17	-	81	45	27	14	5	3	1	-	-	-	1
Aegopinella nitidula (Draparnaud)	12	79	185	1	-	-	-	-	-	28	2	113	95	60	33	39	10	5	4	-	1	-
Oxychilus cellarius (Müller)	-	14	13	-	-	-	-	-	-	5	-	12	28	7	-	2	-	-	-	-	-	-
Oxychilus alliarius (Miller)	2	-	-	-	-	-	-	-	-	3	1	-	-	-	-	-	-	-	-	-	-	-
Limacidae	8	46	23	-	-	-	-	-	-	24	12	21	42	32	28	42	10	11	5	7	5	11
Euconulus fulvus (Müller)	-	3	4	-	-	-	-	-	-	7	2	-	-	-	-	-	-	-	-	-	-	-
Cecilioides acicula (Müller)	-	-	-	-	-	1	-	-	-	-	-		1	-	-	2	1	7	16	24	29	28
Cochlodina laminata (Montagu)	3	-	1	-	-	-	-	-	-	-	-	2	8	12	14	4	1	1	-	1	-	-
Macrogastra riphii (Turton)	-	-	-	-	-	-	-	-	-	-	-	-	-	2	-	-	-	-	-	-	-	-
Clausilia bidentata (Ström)	1	14	18	-	1	-	-	-	-	8	+	23	16	17	29	22	9	4	3	2	1	1
Balea perversa (Linnaeus)	-	-	4	-	-	-	-	-	-	-	-	4	-	-	-	-	-	-	-	-	-	-
Candidula intersecta (Poiret)	-	-	-	-	-	-	-	-	-	-	-	1	-	-	-	-	-	-	-	-	-	-
Cernuella virgata (Da Costa)	-	-	-	-	-	-	-	-	-	-	-	-	-	-	-	-	-	-	1	-	-	3
Helicella itala (Linnaeus)	22	51	33	-	-	-	-	-	-	21	28	11	4	1	28	34	29	46	45	49	40	31
Trichia hispida (Linnaeus)	51	220	97	-	-	-	1	-	2	53	36	57	56	33	50	62	45	27	31	29	35	35
Arianta arbustorum (Linnaeus)	-	-	-	-	-	-	-	-	-	-	-	-	-	-	-	-	-	-	-	-	-	-
Helicigona lapicida (Linnaeus)	-	+	-	-	-	-	-	-	-	-	-	+	1	3	2	3	+	1	+	+	-	-
Cepaea nemoralis (Linnaeus)	-	-	2	-	-	-	-	-	-	-	-	-	-	-	-	-	-	-	-	-	-	-
Cepaea hortensis (Müller)	2	-	-	-	-	-	-	-	-	2	-	-	-	1	-	-	-	-	-	-	-	-
Cepaea spp.	-	-	-	+	-	-	-	-	-	-	-	-	-	25	24	-	-	-	-	-	-	-
Cepaea/Arianta spp.	1	7	11	-	-	-	-	-	+	2	1	16	9	-	-	8	9	1	5	4	1	2
Taxa	23	25	28	3	5	0	3	0	11	26	21	28	26	25	23	25	20	22	20	14	12	13
Shannon Index	2.42	2.40	2.06	1.20	1.38	-	1.1	-	1.64	2.58	2.56	2.05	2.05	2.09	2.53	2.58	2.35	2.30	2.34	1.96	1.82	1.97
Brillouin Index	2.34	2.38	2.04	0.60	0.93	-	0.60	-	1.18	2.50	2.43	2.02	2.02	2.05	2.48	2.62	2.27	2.19	2.22	1.87	1.74	1.87
Shannon Index - Brillouin Index	0.08	0.02	0.02	0.50	0.46	-	0.50	-	0.47	0.07	0.13	0.03	0.03	0.04	0.05	0.07	0.08	0.11	0.12	0.08	0.08	0.10
Δ2	0.87	0.86	0.75	0.67	0.69	-	0.67	-	0.78	0.87	0.90	0.74	0.76	0.80	0.88	0.90	0.87	0.86	0.87	0.81	0.81	0.83
Δ4	6.60	6.09	3.00	0.00	3.67	-	0.00	-	5.88	6.80	8.90	2.83	3.22	4.12	7.71	9.22	7.76	6.36	6.95	4.27	4.22	5.01
TOTAL	**566**	**2755**	**4245**	**3**	**8**	**0**	**3**	**0**	**6**	**746**	**300**	**2063**	**1699**	**1273**	**963**	**791**	**474**	**328**	**294**	**300**	**238**	**212**

intermediate species (*Trichia hispida, Pomatias elegans*) were also recorded. Too few shells were recovered to make any further comment.

The soil in the upper primary fill in ditch F.114 contained exceptionally high shell numbers, rising to 4245 (over 2000 kg), with very low values of *H' – HB* (Table A3.2) indicating autochthonous assemblages, possibly in an immature ranker soil forming on and over the final primary fill. The majority of this assemblage comprises shade-loving species; the zonitids comprising mainly *Ageopinella nitudula, A. pura* and *Vitrea contracta*), *Carychium tridentatum* (over 2000 in one sample), and *Discus rotundatus*. The other most common species are the open country species *Vallonia costata* and intermediate species *Trichia hispida*. This assemblage provides a conundrum in interpretation. The small difference between Shannon and Brilliuoin indices suggests that majority of the assemblage is probably generally autochthonous. The moderate delta indices (Δ_2 = *c.* 0.8 and Δ_4 *c.* 6 falling to 3) (Table A3.2) suggest limited ecological diversity but possibly changing environmental conditions. The assemblage contains the typical rock-rubble trio (*Discus rotundatus, Oxychilus* & *Vitrea*), but the deposit is soil and chalk-rich and not one of chalk rubble with vacuous interstices (Evans & Jones 1973). Further the predominance of *Aegopinella nitidula* and *A. pura* together with *Carychium tridentatum* are not indicative of rock-rubble habitats. Some shade is suggested by this assemblage, and certainly the predatory zonitids and *Carychium tridentatum* are particularly common in deciduous leaf-litter. Further, the Aegopinellas, *Carychium tridentatum* and *Trichia hispida* are more common together in stable, long-grass, swards and scrub. They are reminiscent of assemblages from Neolithic features on Hambledon Hill dating to 3700–3300 cal. BC, where the presence of accumulated leaf-litter has been argued to play an important role (Bell *et al.* forthcoming). The resultant assemblages, therefore tend to over-estimate the importance of the shade-loving elements. On balance this probably represents some soil formation on the top of the stable primary fill and long grasses and leaf litter accumulating (Allen in prep.) in the ditch producing a humic ranker. The absolute and relative progressive increase in *C. tridentatum* and comparable relative progressive decrease in *V. costata, Pupilla muscorum* and *Vertigo pygmaea* may suggest the establishments of a grassy vegetation mat.

A series of significant changes in the assemblages can be detected within the long sequence from the northern ditch F.111 (Fig. 3.4; Table 3.9). At the base of the secondary fill (239 at 145–150 cm), a broadly comparable assemblage is seen to that from the soil horizon at the top of the primary fill in the northern ditch F.114. The reduction in shell numbers reflects the demise of the buried soil and onset of gentle, but more rapid infill conditions. Increases in *V. costata* and other open country species concomitant with rising Δ_4 indices indicate stable and mature conditions with increasing habitat diversity (Table A3.2). The latter possibly reflecting both ditch and extra-ditch habitats. The shell-rich assemblage (746 shells) indicates a shady microenvironment, with accumulating leaf litter in the ditch and open shrub with stable ungrazed grassland around the monument. A lens of chalk inwash, indicated by the particle size data, contains very few *C. tridentatum* which tends to confirm the autochthonous nature of this species, probably in leaf litter and grass within the ditch, in contrast to the more open shorter grassland conditions surrounding the monument.

The main secondary fill (contexts 239 & 230) from 108–135 cm, contains very high shell numbers (*c.* 635 to 1030 kg) and high but falling proportions of *C. tridentatum*. Fewer open country species are present than the earlier secondary fills and very small differences in Shannon and Brillouin indices suggest the development of a long stable grass community within the ditch. Low delta diversity indices suggest stable environments, low habitat variation, but not ecological maturity. Falling proportions of *C. tridentatum* and rising significance of *Pomatias elegans* at the top of the secondary fill and base of the tertiary fill (context 226) mark a change in the wider landscape habitats. A range of open country species *Vallonia, Pupilla muscorum* and *Helicella itala* all occur, and shell numbers fall. Broken and open unstable ground is suggested, and rapidly increased Δ_4 indices, confirm habitat variation. This may suggest opening and preparing the landscape for pasture and tillage.

In the upper part of this tertiary fill (context 226, 68–88 cm) open dry downland conditions became well-established and many shade-loving species dwindle in their representation or disappear. High Shannon indices (Table A3.2) suggest a recently changed, but rapidly maturing environment. These are typical conditions for grazed chalk grassland, and the assemblage composition concurs with this. The final tertiary fill 225 (28–58 cm) sees the dominance of a typical xerophilous assemblage with *P. muscorum* and *V. excentrica, H. itala* is important but *V. costata* declines. The reduction in taxanomic range and all diversity indices suggest harsher environments, and probably the onset of tillage with bare and broken soil (*P. elegans, P. muscorum*). The presence of Introduced Helicellids (cf. Kerney 1966) may suggest a medieval or later date for this activity.

Discussion

The most important information from these long barrows is that all three (Thickthorn, Gussage Cow

Down long barrows 78 & 294) all seem to have been built in pre-existing well-established dry chalk grassland. Only GCD 294 on a south-facing slope seems to have any evidence of some shady conditions locally. No evidence of any former ancient woodland is indicated from any of the samples. The shady conditions indicated in Gussage 294 could be provided by light cover of bushes and shrubs in park woodland with grass. Enough deciduous species are suggested, however, to enable accumulations of leaf litter, with long grasses, in the ditch. Established and grazed grassland sward was established and maintained in the Neolithic period. Other changes have been documented, but when they occurred has yet to be discerned.

Dorset Cursus

Four sections were examined on Bottlebush Down (Figs. 3.33 & 3.51–3.57): two from the north(west) ditch (Tables 3.10, 3.12 & A3.3) and two from the south(east) ditch (Tables 3.11 & A3.4). These data are reviewed with the two published sections from the Cursus (Entwistle & Bowden 1991), both from the north(west) ditch. Excavations of the north(west) ditch beneath the agger of the Ackling Dyke Roman Road provided a rare opportunity to examine a buried soil not afforded previous workers. Detailed sampling of the ditches included the full sediment sequences as previously (Entwistle & Bowden 1991), but both sequences also included eroded soils not previously observed.

Previous work showed the similarity of two ditch sections of the north(west) ditch 600 m apart in Fir Tree Field (1982) and Chalkpit Field (1984) (Barrett *et al.* 1991a, fig. 2.13; Entwistle & Bowden 1991, fig. 3.2), but significant differences between the palaeo-molluscan sequences they contained. Considerable local variation in the vegetation mosaic over this 600 m was suggested. This has led to speculation about whether the palaeo-molluscan assemblages in the Chalkpit Field (1984) sequence were really contemporary with the Cursus (Allen 2000a; French *et al.* 2003, 228–9). In view of this, and tentative suggestions that the Cursus may represent a boundary and a division of two different habitats (i.e. woodland versus pasture), the data from the four sections are discussed in relation to the north or south ditch.

Excavations in 2001 produced four sections on Bottlebush Down that differed again. These were unlike those excavated previously in that they were dominated by clean chalk rubble (French 2003, 188), and may suggest rapid and deliberate re-interment of the bank material into the ditches. Samples were taken from all four sections, though the absence of a natural primary fill (*sensu* Evans 1972, 321–8; Limbrey 1975, 290–300) hinders recovery of assemblages truly contemporary with the Cursus.

Dorset Cursus, Bottlebush Down; north ditch sequences (F.101)

Buried soil (F.100): A thin relict buried soil was present beneath the bank (trench 1) (Fig. 3.52). Three spot samples were taken from the soil well sealed beneath the bank (context 140), that under the edge of the bank (context 155), and that over 155, and possibly over the bank (context 139). All three assemblages (Fig. 3.93; Table 3.10) are dominated by open-country species (especially *Vallonia excentrica*), with catholic species. No shade-loving species are present. If these really represent the soil contemporary with the construction of the Cursus, then they represent well-established open grass downland. All three assemblages are very similar with only minor fluctuations (Fig. 3.93), from which no palaeoecological significance can be determined. Species diversity indices are consistent in all of these samples (Table A3.3).

North ditch F.101: Fine deposits (context 162, trench 1) (Fig. 3.51) beneath the main primary fills (context 162) produced a mollusc assemblage dominated by open country (*V. costata, V. excentrica,* with, *H. itala, P. muscorum* & *Vertigo pygmaea*) and catholic species (*P. elegans, Cochlicopa* spp. & *Trichia hispida*). What is important here is the total absence of any shade-loving and rupestral species. If this assemblage is contemporary with the construction of the bank, then no woodland or shady refugia survived at this location; nor is there any evidence of any former or ancient woodland, not even in the form of the robust non-apical fragments of Clausiliidae. Few shells were present in the backfill/primary fills (Table 3.10), but the majority were open country taxa. The secondary fill (163), representing deposits soon after the backfill of the ditch, were almost identical to those from the lowest primary fill. Shannon diversity indices are only moderate (*c.* 1.7) and Δ_2 and Δ_4 both low (*c.* 0.74 & 3.8, respectively) (Table A3.3), suggesting restricted habitats, largely probably grassland, but with a moderate, though not high, sward height. We can suggest a lightly grazed, dense, generally dry grassland. Soil-rich deposits (context 162) over the secondary fill exhibit very similar characteristics, but the presence of *Truncatellina cyrlindica* is particularly noteworthy, indicating dry and very open grassland.

Soil material in the tertiary fill (contexts 178, 141 & 177, trench 2) (Fig. 3.51) contained open-country assemblages similar to those in the secondary fills. Again minor fluctuations are detected, and this represents dry chalk grassland. The increase in *P. muscorum* and *V. pygmaea* may represent bare soil, but the

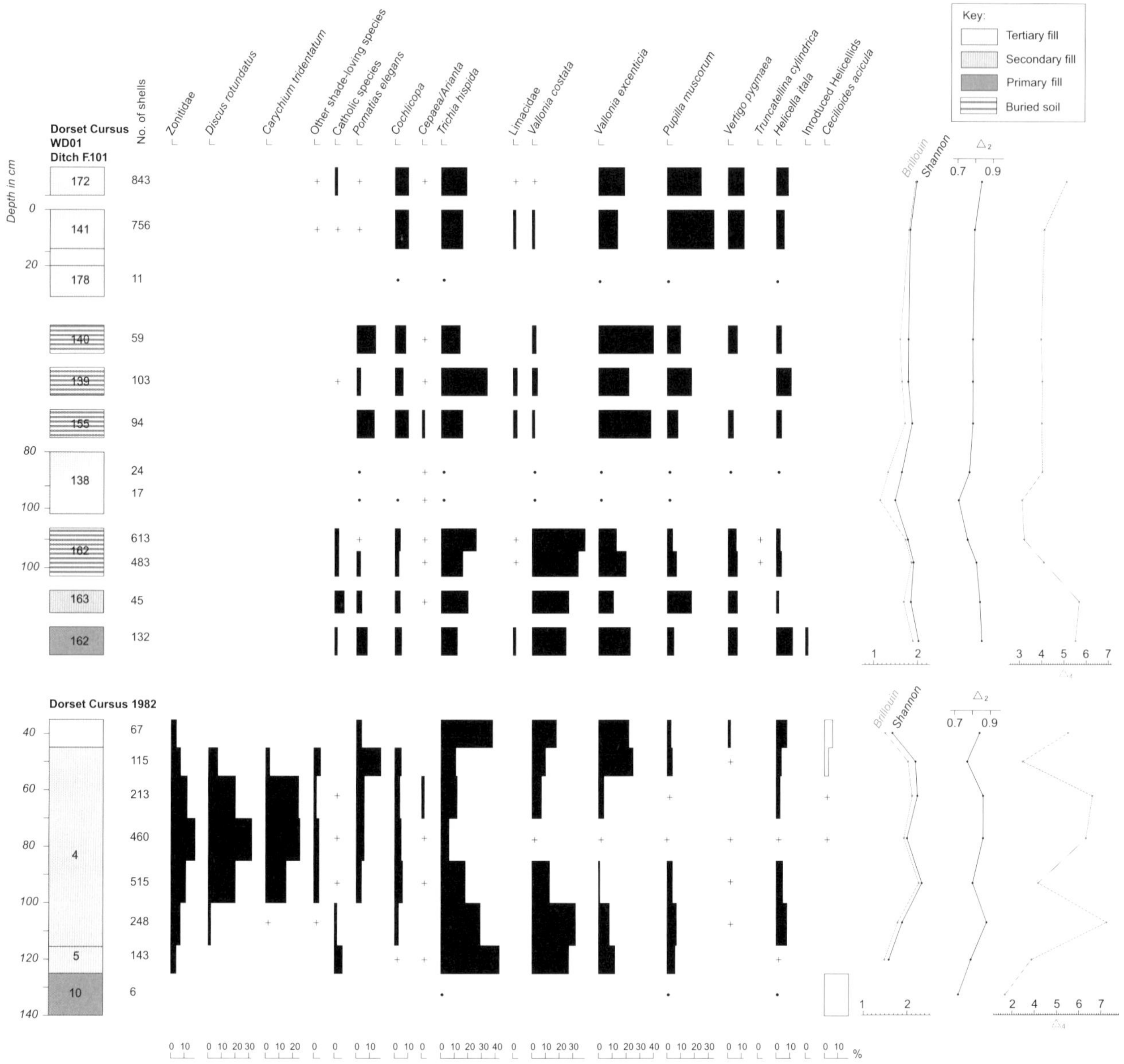

Figure 3.93. *Molluscan histogram of the north ditch of the Dorset Cursus (above) and the 1982 excavations (below) (R. Goller, Wessex Archaeology).*

reduction in *P. elegans* tends to suggest that this may be grazing and trampling pressure (cf. Cameron & Morgan-Huws 1975), rather than tillage. The tertiary fill in trench 21 (context 138) contained, surprisingly, very low shell numbers, though the assemblages look similar. Low shell numbers here may be due to locally more acidic soil conditions; certainly field records show slightly redder and stronger brown hues in this deposit.

What is surprising about the sequences described from the north ditch here is the lack of shade-loving species. One single specimen of *Oxychilus cellarius* was the only shade-loving specimen from the primary fills (context 164), and one apex of *Discus rotundatus* from eroded soil in the tertiary fill (context 162) were the only two shade-loving specimens among some 3199 identifications. This is all the more surprising in view of the heavily wooded environment and rich shade-loving assemblages, containing ancient woodland species such as *Ena montana*, from the Cursus ditch in Chalkpit Field (Entwistle & Bowden 1991). Such complete differences over less than 600 m are difficult to reconcile. The two parameters that require consideration here are the spatial and temporal components.

Dorset Cursus, Bottlebush Down: south ditch sequences (F.108)

The primary chalk rubble was almost devoid of shells (Table 3.11) and the earliest deposits containing shells were from a turf or soil stabilization horizon (context 196, trench 3) over the primary fill (Fig. 3.94, upper). Apart from *Vitrea contracta* there were no shade-loving species in this assemblage; and even this can be considered catholic and live in grassland. The assemblage is dominated by open-country species (70 per cent) with *V. costata, V. excentrica* and *H. itala* as the major components with *T. hispida.* This clearly represents long-established mature open country. Some broken ground may be reflected in the presence of *T. hispida,* but this might be the loose soil in and around the newly cut and weathering ditch and bank itself. Moderate Δ_2 and Δ_4 indices indicate mature habitats with moderate habitat diversity within the area, but a generally a well-established mature fauna reflected in the high Shannon index of 1.96 for grassland habitats with relatively low taxonomic range (12 taxa) (Table A3.4). This is sealed by dumped chalk over which the lower secondary fill (context 194) contains assemblages with completely different character: high numbers of *C. tridentatum* prevail with *D. rotundatus* and Zonitids. Open-country species represent a small proportion and the most significant of which, *V. costata,* can also inhabit open woodland. The assemblages contain a few rupestral species and high numbers of *D. rotundatus* that would not normally be expected in long grassland habitats that *C. tridentaum* and *P. pygmaeum* might inhabit. The invasion of a large number of new species indicate suitable refuges within the local landscape, and their rapid arrival is all the more surprising in view of the absence of most shade-loving taxa in earlier deposits. Very high shell numbers suggest that a number of these species resided locally in the ditch or on the bank, and that this horizon may represent some stabilization. A similar assemblage prevails in the base of the upper secondary fill (193), but this gives way to more open-country assemblages. A dramatic and almost instant loss of the shade-loving species at 50 cm hails the rise of *P. elegans* with *P. muscorum* and *H. itala.* This upper secondary fill suggests sudden change in the local environment, probably tillage, whilst some of the shade-loving and particular rupestral species may still have resided in the ditch on the vegetated adjacent bank. The presence of *Truncatellina cylindrica* in both the shade-loving and tillage assemblages suggests the presence of dry habitats in the vicinity.

The second sequence from the southern ditch, only 250 m to the north/south (Fig. 3.94, middle) superficially looks similar. It too shows a change from broadly open conditions at the base of the secondary fill to shadier habitats. The dumped chalk rubble (205) surprisingly contained 58 shells, mainly *V. costata,* but accompanied by other open country species and *T. hispida.* Although we must be cautious in interpreting this assemblage with relatively few shells, there are comparatively few shade-loving taxa and this probably represented an open long grassland habitat in, and surrounding, the ditch. The secondary fill (202), however, sees a rapid increase in *C. tridentatum* accompanied by other shade-loving species and a corresponding decline in open country taxa. The progressive increase in shade-loving taxa, ultimately including *D. rotundatus* and the Zonitids with high and Δ_2 and Δ_4 indices (0.88 & 7.3) and comparably high Shannon indices (>2.5) (Table A3.4) suggest mature and diverse local habitats. Certainly long damp grassland may be a component of this habitat, with long herbaceous vegetation and bushes. Unlike trench 3, there is a lack of rupestral species. Harsher conditions are indicated in the tertiary fill (201) by striking reductions in shell numbers. Changes which include increases in Limacidae and *P. elegans* might tentatively be suggested to represent the preparation of the ground prior to tillage as seen in trench 3. Here again the xerophile *T. cylindrica* is present in the secondary, more shade-loving assemblage.

The two profiles display broadly similar trends, but significant differences in the composition of each suggest local variation in habitats, probably predominantly on the bank and within the ditch rather than in the wider landscape.

Dorset Cursus discussion and summary

Firstly, there is no evidence of ancient mature woodland fauna in the base of any of the ditches with high diversity, or ancient woodland species (e.g. *Vertigo pusilla, Vertigo alpestris* & *Vertigo substriata*) although *Ena montana* has been reported from the Chalkpit Field section (Entwistle & Bowden 1991). We can be confident that the monument through Bottlebush and Wyke Down was constructed in a long-established well-developed open grass downland. The monument's history after that, however, is full of contradictions.

All of the northern ditch (trenches 1 & 2) shows very open conditions prevailing through its entire infill history; this contrasts with all the other sections including those recorded from Barrett and Bradley's excavations. Certainly these changes indicate temporal vegetation development, and although superficially they occur at different locations in the infill sequence, they may, in fact be concurrent. It is clear from the 2001 excavations that the ditches were deliberately infilled soon after construction. This contrasts with sections in

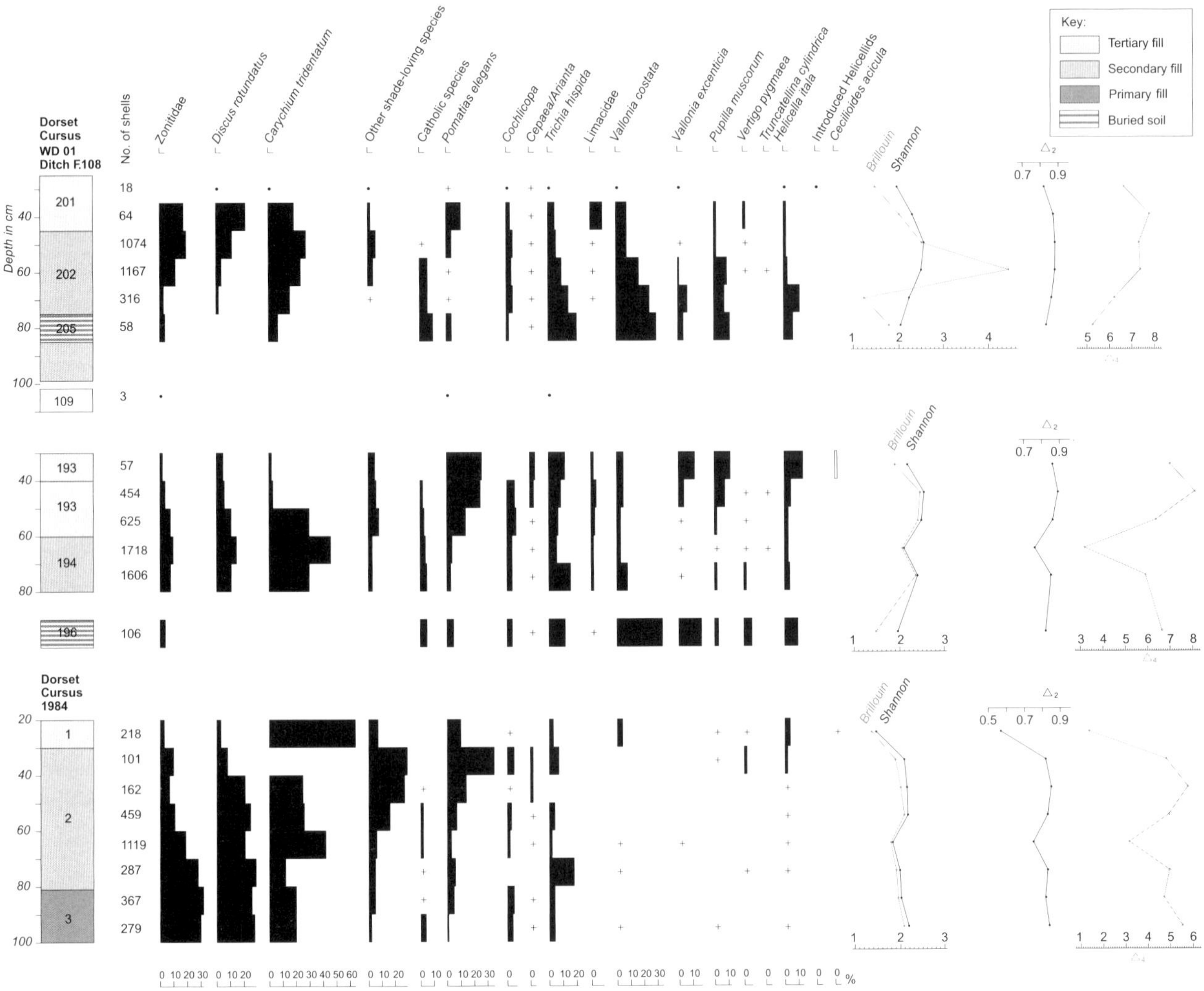

Figure 3.94. *Molluscan histogram of the south ditch of the Dorset Cursus (above) and the 1984 excavations (below). (R. Goller, Wessex Archaeology.)*

Chalkpit Field and Fir Tree Field in which very little primary fill occurring; a notable feature in this area was friable surface chalk giving rise to rapidly accumulating and deep primary fills (Pitt Rivers 1898, 24–5; author pers. observ.). Does this indicate not only that in some areas not only have the ditches been backfilled, but in others they may have been emptied?

The huge variations in the mollusc sequences from the same monument challenge even the bases of palaeo-molluscan studies (Tables 3.10–3.12, A3.3 & A3.4). Can one landscape contain such significantly different contemporaneous mollusc fauna over distances of 500 or 600 m? Perhaps it can. If woodland invasion was so rapid (cf. Entwistle & Bowden 1991), where did this refugium come from, especially in view of the now clear evidence of pre-existing open established grassland fauna? Perhaps we need to view this monument from two perspectives. First that the infill history differs considerably along the ditch, so substantial variation in the depth and nature of contemporaneous infills will occur. Second that as a linear monument traversing more than 10 km, crossing the Allen valley rising over both chalk ridges and rolling chalk uplands, its banks and ditches act as ecological corridors down which fauna and flora can migrate.

A concordance of both ditch fills and ecological interpretation allows us to suggest a history for the Cursus starting with open established mature chalk grassland. After some locations were backfilled with chalk rubble, possible slighting of the banks (south ditch section F.108 and north section F.101), while other locations such as in Chalkpit Field may have been emptied of their primary fill. Infilling of the north ditch (F.101) may have been almost complete, while that in the south only partial. Thus the northern profile contains little post-backfill sequence while that in the

Table 3.10. *Mollusca from Dorset Cursus 2001, north ditch, F.101.*

Feature	North ditch F.101										bank F.100		ditch F.101			
	Trench 1												Trench 2			
Context	164				163	162			138		155	139	140	178	141	177
Deposit type	primary				2ndry	base	eroded soil		tertiary		bank soil		ols	tertiary soil		turf
Sample	200	201	202	203	204	176	206	205	207	208	166	164	171	211	210	169
Depth (cm)	140–145	132–140	121–132	111–121	103–111		94–103	86–94	92–102	82–92				20–30	0–14	
Wt (g)	2000	2000	2000	2000	1500	2000	1500	1500	1500	2000	2000	2000	2000	1500	1500	2000
MOLLUSCA																
Pomatias elegans (Müller)	+	+	+	+	2	11	13	3	1	1	12	3	8	-	7	11
Carychium tridentatum (Risso)	-	-	-	-	-	-	-	-	-	-	[1]	-	-	-	-	-
Cochlicopa lubrica (Müller)	-	-	-	-	-	-	13	4	-	-	-	-	-	-	29	16
Cochlicopa lubricella (Porro)	-	-	-	-	1	1	-	6	-	-	2	2	-	-	-	17
Cochlicopa spp.	-	-	-	1	1	5	5	22	1	-	7	4	5	1	46	56
Truncatellina cylindrica (Férussac)	-	-	-	-	-	-	4	1	-	-	-	-	-	-	-	-
Vertigo pygmaea (Draparnaud)	-	-	-	-	3	9	34	39	-	1	4	-	4	-	81	102
Vertigo spp.	-	-	-	-	-	-	-	-	-	-	-	-	-	-	6	-
Pupilla muscorum (Linnaeus)	-	2	-	1	8	6	33	26	2	3	7	18	6	4	255	214
Vallonia costata (Müller)	1	-	-	2	12	33	164	239	3	4	2	4	2	-	14	7
Vallonia excentrica (Sterki)	1	2	-	2	5	30	99	79	8	9	36	22	23	2	127	162
Vallonia spp.	-	-	-	-	-	-	-	-	-	-	-	-	-	-	-	-
Acanthinula aculeata (Müller)	-	-	-	-	-	-	-	-	-	-	-	-	-	-	-	-
Ena montana (Draparnaud)	-	-	-	-	-	-	-	-	-	-	-	-	-	-	-	-
Ena obscura (Müller)	-	-	-	-	-	-	-	-	-	-	-	-	-	-	-	-
Punctum pygmaeum (Draparnaud)	-	-	-	1	3	2	14	18	-	-	-	1	-	-	7	14
Discus rotundatus (Müller)	-	-	-	-	-	-	-	-	-	-	-	-	-	-	-	-
Vitrina pellucida (Müller)	-	-	-	-	-	-	-	1	-	-	-	-	-	-	-	-
Vitrea crystallina (Müller)	-	-	-	-	-	-	-	-	-	-	-	-	-	-	-	-
Vitrea contracta (Westerlund)	-	-	-	-	-	-	-	-	-	-	-	-	-	-	-	-
Nesovitrea hammonis (Ström)	-	-	-	-	-	-	-	-	-	-	-	-	-	-	-	-
Aegopinella pura (Alder)	-	-	-	-	-	-	-	-	-	-	-	[1]	-	-	-	-
Aegopinella nitidula (Draparnaud)	-	-	-	-	-	-	-	-	-	-	-	-	-	-	-	-
Oxychilus cellarius (Müller)	1	-	-	-	-	-	-	-	-	-	-	-	-	-	-	-
Limacidae	-	-	-	-	-	3	2	2	-	-	3	3	-	-	13	12
Cecilioides acicula (Müller)	-	-	-	-	-	-	-	-	-	-	-	-	-	-	-	-
Cochlodina laminata (Montagu)	-	-	-	-	-	-	-	-	-	-	-	-	-	-	-	-
Macrogastra rolphii (Turton)	-	-	-	-	-	-	-	-	-	-	-	-	-	-	-	-
Clausilia bidentata (Ström)	-	-	-	-	-	-	-	-	-	-	-	-	-	-	+	-
Candidula interescta (Poiret)	-	-	-	-	-	-	-	-	-	-	-	-	-	-	-	-
Helicella itala (Linnaeus)	-	+	-	1	1	16	18	20	-	1	4	11	3	2	47	74
Trichia hispida (Linnaeus)	-	-	-	4	9	16	79	151	2	5	15	34	8	2	124	157
Arianta arbustorum (Linnaeus)	-	-	-	-	-	-	-	-	-	-	-	+	-	-	-	-
Helicigona lapicida (Linnaeus)	-	-	-	-	-	-	-	-	-	-	-	-	-	-	-	+
Cepaea nemoralis (Linnaeus)	-	-	-	-	-	-	-	-	-	-	-	-	-	-	-	-
Cepaea hortensis (Müller)	-	+	-	-	-	-	-	-	-	-	-	-	-	-	-	-
Cepaea spp.	-	-	-	-	-	-	-	-	+	+	-	1	+	-	-	-
Cepaea/Arianta spp.	-	-	-	-	+	+	5	2	-	-	2	-	-	-	-	1
Taxa	3	2	0	7	9	10	12	14	5	7	10	10	8	5	10	12
TOTAL	**3**	**4**	**0**	**12**	**45**	**132**	**483**	**613**	**17**	**24**	**94**	**103**	**59**	**11**	**756**	**843**

south (F.108) contains longer post backfill sequences. When no backfill occurred, and possibly even some emptying may have been conducted, then longer near-contemporary sequences may be present. The concordance enables contemporaneous environments to be defined, in which we can see relatively rapid colonization of the bank (and possibly the ditch) by tall herbaceous vegetation, and of bushes and shrubs. How long this lasted is unknown, but a dramatic change of land-use to a harsh (lowered shell numbers), and open probably arable landscape represents the tertiary fills of nearly every section.

Vegetation regeneration is demonstrable in the molluscan assemblages from the secondary fills in a

Table 3.11. *Mollusca from Dorset Cursus 2001, south ditch, F.108.*

Feature	F.109	South ditch, F.108													
		trench 4									trench 3				
Context	109	210	207	205	202			201		196	182		193		
Deposit type	perig	1 ry	lens	dump	secondary			tertiary		soil	secondary		tertiary		
Sample	230	209	218	219	220	221	222	223	224	212	213	214	215	216	217
Depth (cm)			99–110	75–85	65–75	55–65	45–55	35–45	25–35		70–80	60–70	50–60	40–50	30–40
Wt (g)	1938	1500	318	1500	1500	1500	1500	1500	1500	2000	1500	1500	1500	1500	1500
MOLLUSCA															
Pomatias elegans (Müller)	-	+	1	2	1	10	47	7	+	5	48	52	86	114	15
Carychium tridentatum (Risso)	-	-	-	4	50	277	298	12	1	-	488	772	188	57	1
Cochlicopa lubrica (Müller)	-	-	-	-	10	14	24	-	1	2	27	23	8	4	-
Cochlicopa lubricella (Porro)	-	-	-	1	5	6	6	-	-	-	-	-	1	-	-
Cochlicopa spp.	-	-	-	-	1	31	26	2	-	2	44	49	33	23	-
Truncatellina cylindrica (Férussac)	-	-	-	-	-	2	-	-	-	-	-	2	-	1	-
Vertigo pygmaea (Draparnaud)	-	-	-	-	-	3	3	1	-	6	21	5	4	4	-
Vertigo spp.	-	-	-	-	-	-	-	-	-	-	3	1	1	2	-
Pupilla muscorum (Linnaeus)	1	-	-	7	22	126	25	1	-	3	42	19	15	34	7
Vallonia costata (Müller)	-	-	-	18	79	197	87	4	1	36	128	52	19	24	3
Vallonia excentrica (Sterki)	-	-	-	2	21	22	5	-	2	18	17	4	8	13	7
Vallonia spp.	-	-	-	-	2	-	-	1	-	-	-	-	-	-	-
Acanthinula aculeata (Müller)	-	-	-	-	-	49	48	1	1	-	23	21	24	4	-
Ena montana (Draparnaud)	-	-	-	-	-	-	-	-	-	-	-	-	-	-	-
Ena obscura (Müller)	-	-	-	-	-	2	4	-	-	-	-	4	4	-	-
Punctum pygmaeum (Draparnaud)	-	-	-	-	13	49	30	-	-	4	69	50	10	6	-
Discus rotundatus (Müller)	-	-	-	-	9	50	136	14	1	-	177	259	66	27	3
Vitrina pellucida (Müller)	1	-	-	5	-	4	3	-	-	-	5	3	2	-	-
Vitrea crystallina (Müller)	-	-	1	1	-	12	14	1	-	-	5	1	1	1	1
Vitrea contracta (Westerlund)	-	-	-	1	6	57	55	1	-	4	26	33	3	1	-
Nesovitrea hammonis (Ström)	-	-	-	1	7	12	6	-	-	1	11	7	4	1	-
Aegopinella pura (Alder)	-	-	-	-	-	26	59	-	-	-	37	29	9	7	-
Aegopinella nitidula (Draparnaud)	-	-	-	-	2	31	71	7	-	-	46	81	28	10	-
Oxychilus cellarius (Müller)	-	1	-	-	-	3	10	2	-	-	13	28	6	1	-
Limacidae	-	-	-	-	2	9	12	6	-	1	26	31	17	16	1
Cecilioides acicula (Müller)	-	-	-	-	-	-	-		-	-	-	-	-	-	1
Cochlodina laminata (Montagu)	-	-	-	-	-	-	2	-	-	-	4	6	9	4	-
Macrogastra rolphii (Turton)	-	-	-	-	-	-	1	-	-	-	-	-	-	-	-
Clausilia bidentata (Ström)	-	+	-	-	1	6	12	-	-	-	20	24	14	23	3
Candidula intersecta (Poiret)	-	-	-	-	-	-	-	-	2	-	-	-	-	-	-
Helicella itala (Linnaeus)	-	-	-	4	35	37	17	1	3	11	70	43	16	22	7
Trichia hispida (Linnaeus)	-	-	1	12	49	126	61	3	6	13	251	109	42	42	7
Arianta arbustorum (Linnaeus)	-	-	-	-	-	-	-	-	-	-	+	+	-	-	-
Helicigona lapicida (Linnaeus)	-	-	-	-	-	+	1	-	-	-	-	1	-	1	+
Cepaea nemoralis (Linnaeus)	-	-	-	-	-	-	-	-	-	-	1	1	-	-	-
Cepaea hortensis (Müller)	-	-	-	-	-	-	-	-	-	-	-	-	-	1	-
Cepaea spp.	-	+	-	+	1	4	3	+	+	+	-	2	7	11	-
Cepaea/Arianta spp.	-	-	-	-	-	2	8	-	-	-	4	6	-	-	2
Taxa	2	1	3	12	17	25	27	15	9	12	23	26	25	24	12
TOTAL	**2**	**1**	**3**	**58**	**316**	**1167**	**1074**	**64**	**18**	**106**	**1606**	**1718**	**625**	**454**	**57**

number of Cursus sections. But, except perhaps for the assemblages from the Cursus ditches in Chalkpit Field to the south on Wyke Down, none can be seen to be of ancient woodland. Most assemblages appear to represent long grassland, and herbaceous vegetation, and perhaps bushes and shrubs creating a closed and stable environment but perhaps set within dry open grassland (*T. cylindrica*) landscape. This might enable us to conjure images of well-vegetated, almost 'hedged' banks of the Cursus. Bank and ditch provided stable corridors rich in species and perfect highways of ecological mobility.

It seems unlikely that ancient woodland existed, but with the absence of *Ena montana* from the Chalkpit

Table 3.12. *Mollusca from the Dorset Cursus excavations 1982 and 1984 (from Entwistle & Bowden 1991).*

	Mark Bowden								Marian Dalgleish and/or Jillian Parker							
	Dorset Cursus 1982 (Entwistle & Bowden 1991)								Dorset Cursus 1984 (Entwistle & Bowden 1991)							
Context	**10**	**5**	**4**	**4**	**4**	**4**	**4**	**4**								
Depth (cm)	**125–140**	**115–125**	**100–115**	**85–100**	**70–85**	**55–70**	**45–55**	**35–45**	**90–100**	**80–90**	**70–80**	**60–70**	**50–60**	**40–50**	**30–40**	**20–30**
Wt (g)	**1500**	**500**	**1000**	**1500**	**1000**	**1500**	**1500**	**1500**	**2300**	**2300**	**2100**	**2300**	**1400**	**1300**	**1900**	**1200**
MOLLUSCA																
Pomatias elegans (Müller)	-	-	-	18	28	12	21	3	7	19	17	49	31	23	35	22
Carychium tridentatum (Risso)	-	-	1	77	113	51	3	-	54	72	34	472	125	41	-	139
Cochlicopa lubrica (Müller)	-	-	-	28	20	-	6	-	2	9	-	5	2	1	-	2
Cochlicopa lubricella (Porro)	-	-	-	1	2	-	-	-	1	-	-	2	1	-	-	-
Cochlicopa spp.	-	2	8	-	-	8	-	-	9	9	-	14	12	-	5	1
Truncatellina cylindrica (Férussac)	-	-	-	-		-	-	-	-	-	-	-	-	-	-	-
Vertigo pygmaea (Draparnaud)	-	-	2	6	1	-	1	1	-	-	-	-	-	-	2	1
Vertigo spp.	-	-	-	-	-	-	-	-	-	-	2	-	-	-	-	-
Pupilla muscorum (Linnaeus)	2	8	18	18	2	3	4	2	1	-	-	-	-	-	1	1
Vallonia costata (Müller)	-	37	79	66	5	15	12	12	2	-	1	4	-	-	-	9
Vallonia excentrica Sterki	-	16	20	10	4	9	29	15	-	-	-	1	-	-	-	-
Vallonia spp.	-	-	-	-	-	-	-	-	-	-	-	-	-	-	-	-
Acanthinula aculeata (Müller)	-	-	-	15	10	4	-	-	-	10	11	42	16	10	8	2
Ena montana (Draparnaud)	-	-	-	-	-	-	-	-	-	-	-	1	5	5	2	-
Ena obscura (Müller)	-	-	-	1	-	-	-	-	-	-	-	1	1	-	-	-
Punctum pygmaeum (Draparnaud)	-	9	5	1	5	3	-	-	6	-	-	-	1	-	-	-
Discus rotundatus (Müller)	-	-	4	99	147	43	8	-	78	99	84	238	125	35	8	7
Vitrina pellucida (Müller)	-	-	1	-	-	-	-	-	4	3	3	17	6	1	-	-
Vitrea crystallina (Müller)	-	-	1	12	13	3	-	-	29	15	6	45	1	-	-	-
Vitrea contracta (Westerlund)	-	-	-	-	-	-	-	-	-	-	-	-	-	-	-	-
Nesovitrea hammonis (Ström)	-	-	-	-	-	-	-	-	-	-	-	-	-	-	-	-
Aegopinella pura (Alder)	-	-	-	-	-	-	-	-	37	19	31	10	19	3	2	3
Aegopinella nitidula (Draparnaud)	-	-	-	-	-	-	-	-	23	85	45	160	29	9	8	3
Oxychilus cellarius (Müller)	-	-	-	-	-	-	-	-	-	2	-	2	1	-	-	1
Zonitidae	-	6	16	43	68	23	8	3	-	-	-	-	-	-	-	-
Limacidae	-	1	-	-	-	-	-	-	-	-	-	-	-	-	-	-
Cecilioides acicula (Müller)	1	-	-	-	1	2	3	4	-	-	-	-	-	-	-	1
Cochlodina laminata (Montagu)	-	-	-	-	-	-	-	-	6	6	-	16	36	14	18	12
Macrogastra rolphii (Turton)	-	-	-	-	-	-	-	-	-	3	3	7	18	12	-	2
Clausilia bidentata (Ström)	-	-	1	4	-	-	-	-	-	-	-	-	-	-	-	-
Clausiliidae	-	-	-	-	8	1	6	-	-	-	-	-	-	-	-	-
Candidula interescta (Poiret)	-	-	-	-	-	-	-	-	-	-	-	-	-	-	-	-
Helicella itala (Linnaeus)	1	1	21	24	4	6	4	5	4	-	1	9	4	2	2	6
Heliicidae	-	-	-	-	-	-	-	1	-	-	-	-	-	-	-	-
Trichia striolata (C. Pfeiffer)	-	-	-	-	-	-	-	-	-	-	-	1	-	1	1	-
Trichia hispida (Linnaeus)	3	62	71	90	26	28	13	25	12	14	49	18	20	2	7	7
Arianta arbustorum (Linnaeus)	-	-	-	-	-	-	-	-	1	-	-	-	-	-	-	-
Helicigona lapicida (Linnaeus)	-	-	-	-	-	-	-	-	-	-	-	-	-	-	-	-
Cepaea nemoralis (Linnaeus)	-	-	-	-	-	-	-	-	2	-	-	-	-	-	-	-
Cepaea hortensis (Müller)	-	-	-	-	-	-	-	-	-	-	-	-	-	-	-	-
Cepaea spp.	-	1	-	2	4	4	-	-	1	2	-	5	6	3	2	-
Cepaea/Arianta spp.	-	-	-	-	-	-	-	-	-	-	-	-	-	-	-	-
Taxa	3	10	14	18	17	15	12	9	19	15	13	22	20	15	14	16
TOTAL	**6**	**142**	**248**	**515**	**460**	**213**	**115**	**67**	**279**	**367**	**287**	**1119**	**459**	**162**	**101**	**218**

Field assemblages, there is no evidence to suggest that such habitats were represented by these assemblages. The Cursus thus provides significant data from which to develop a hypothesis of the nature of the pre-Cursus landscape, of the nature of the woodland composition and of the nature of the woodland, or vegetation regeneration in the monument, and within the whole wider landscape.

Monkton-up-Wimborne Neolithic mortuary enclosure (MUW 00)

A series of 13 samples was takenfrom the Monkton-up-Wimborne Neolithic ditch fills for which no buried soil survived (Figs. 3.59 & 3.95).

Land snails seem to show that this monument was also built in an established open downland (Tables 3.13 & A3.5), but with a little shade-loving refugia locally (Fig. 3.95). No major vegetation re-generation phase is seen in the ditch of the surrounding landscape. Careful examination may enable this sequence to be chronologically related to the Monkton-up-Wimborne ceremonial site (MUW 97) from which a very large snail analysis programme (about 70 samples) has been completed.

Numbers of shells were moderate through the ditch fill sequence and are no way comparable with those from the Handley Down mortuary enclosure where shell numbers reached in excess of 2800 per sample and averaged over 775 per sample. The Monkton-up-Wimborne assemblages were only moderate in size and more in keeping with assemblages from the ceremonial site in the same field (Green 2000).

The assemblages indicate an open environment with little woodland locally which was maintained throughout the ditch-infill process. This differs from the comparable mortuary enclosure at Handley Down, and may suggest that the monument sampled here is later in date and may have resided in an environment with larger clearances than those at Handley Down (Allen 2000a). The evidence of some woodland or shade loving element indicates an environment perhaps not as open as that seen in the later fills of the ceremonial site.

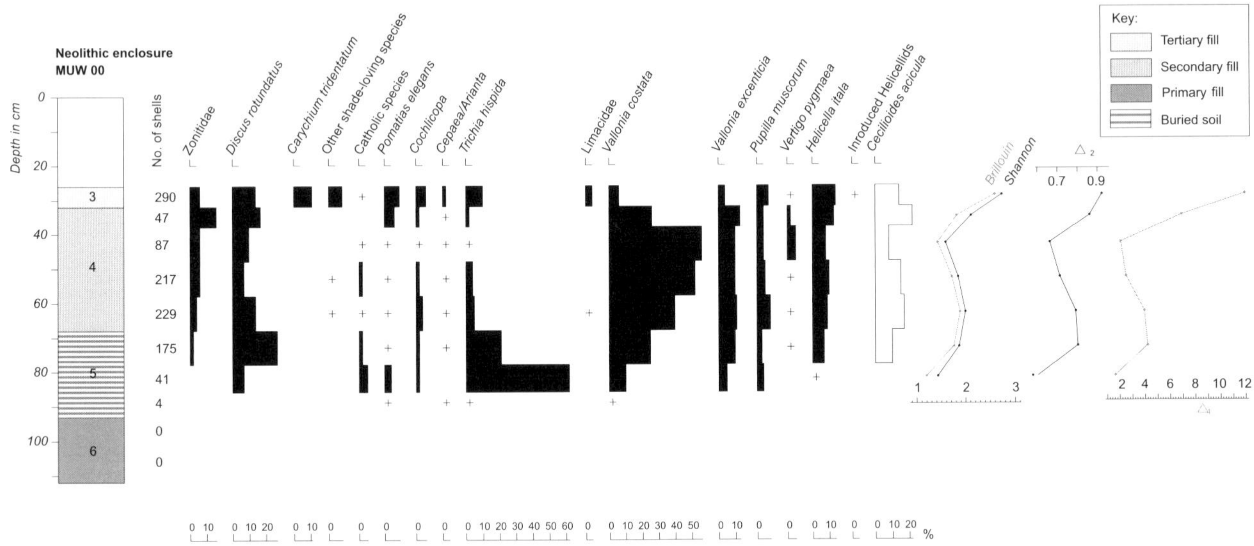

Figure 3.95. *Molluscan histogram of the Monkton-up-Wimborne Neolithic enclosure (R. Goller, Wessex Archaeology).*

Table 3.13. *Mollusca from Monkton-up-Wimborne mortuary enclosure (MUW 00).*

Feature	**section 44**		**section 42**									
Sample	**161**	**162**	**151**	**152**	**153**	**154**	**155**	**156**	**157**	**158**	**159**	**160**
Context	**14**	**13**	**6**		**5e**	**5d**	**5a**	**4**				**3**
Context	**l 1ry**	**u 1ry**	**primary**		**lenses**		**turf**	**secondary**				**3 ry**
Depth (cm)	**spot**	**spot**	**102–112**	**92–102**	**86–92**	**78–86**	**68–78**	**58–68**	**48–58**	**38–48**	**32–38**	**26–32**
Wt (g)	**2000**	**2000**	**2000**	**2000**	**2000**	**1500**	**1500**	**1500**	**1500**	**1500**	**1500**	**1500**
MOLLUSCA												
Pomatias elegans (Müller)	-	-	-	-	+	2	1	2	2	1	3	24
Carychium tridentatum (Risso)	-	-	-	-	-	-	-	-	-	-	-	31
Cochlicopa lubrica (Müller)	-	-	-	-	-	-	2	3	1	-	-	1
Cochlicopa lubricella (Porro)	-	-	-	-	-	-	-	1	1	-	-	2
Cochlicopa spp.	-	-	-	-	-	1	2	4	2	1	1	14
Vertigo pygmaea (Draparnaud)	-	-	-	-	-	-	2	3	2	4	1	3
Pupilla muscorum (Linnaeus)	-	-	-	-	-	2	5	20	10	3	2	19
Vallonia costata (Müller)	-	-	-	-	+	4	44	90	111	48	12	17
Vallonia excentrica Sterki	-	-	-	-	-	2	17	26	22	8	6	12
Acanthinula aculeata (Müller)	-	-	-	-	-	-	-	1	-	-	-	10
Ena obscura (Müller)	-	-	-	-	-	-	-	-	-	-	-	4
Punctum pygmaeum (Draparnaud)	-	-	-	-	-	1	-	-	2	1	-	4
Discus rotundatus (Müller)	-	-	-	-	-	3	48	31	16	9	8	39
Vitrina pellucida (Müller)	-	-	-	-	-	1	-	-	2	-	-	-
Vitrea crystallina (Müller)	-	-	-	-	-	-	-	-	2	-	-	-
Vitrea contracta (Westerlund)	-	-	-	-	-	-	1	8	4	5	4	7
Nesovitrea hammonis (Ström)	-	-	-	-	-	-	3	2	1	-	-	-
Aegopinella pura (Alder)	-	-	-	-	-	-	-	-	1	-	-	2
Aegopinella nitidula (Draparnaud)	-	-	-	-	-	-	2	2	7	-	3	4
Oxychilus cellarius (Müller)	-	-	-	-	-	-	1	-	-	-	-	4
Limacidae	-	-	-	-	-	-	-	2	-	-	-	10
Cecilioides acicula (Müller)	-	-	-	-	-	-	18	38	32	6	10	41
Clausilia bidentata (Ström)	-	-	-	-	-	-	-	-	-	-	-	8
Candidula intersecta (Poiret)	-	-	-	-	-	-	-	-	1	-	-	2
Helicella itala (Linnaeus)	-	-	-	-	-	+	10	21	21	6	6	39
Trichia hispida (Linnaeus)	-	-	-	-	4	25	37	11	7	1	1	29
Arianta arbustorum (Linnaeus)	-	-	-	-	-	-	-	-	-	+	-	-
Helicigona lapicida (Linnaeus)	-	-	-	-	-	-	-	-	-	-	-	+
Cepaea hortensis (Müller)	-	-	-	-	-	-	-	-	-	-	-	1
Cepaea spp.	-	+	-	-	+	-	+	2	2	-	+	4
Taxa	0	0	0	0	1	9	13	16	19	11	11	22
TOTAL	**0**	**0**	**0**	**0**	**4**	**41**	**175**	**229**	**217**	**87**	**47**	**290**

Wyke Down Bronze Age barrows

Wyke Down barrows 34 & 41 (WD 98)

A suite of 29 samples was analysed, primarily from the double-ditched and deeper buried soils of barrow 34 (Figs. 3.34, 3.40–3.43 & 3.96; Tables 3.14, A3.6 & A3.7). A good land snail sequence was obtained, particularly from the ditches of barrow 34 which had very good shell preservation. Analysis showed that these monuments were built in established open country and that subtle variations of land-use or barrow histories can be detected and related to a wider local pattern of activity (French 2000, fig. 12.7; French *et al.* 2003, fig. 18). These are summarized in French (2003, 179–84) and Allen in French *et al.* (2003, 225–8).

Wyke Down Bronze Age barrows 36 & 44 (WD 00)

As with barrows 41 and 34 from the 1998 season (Allen, in French *et al.* 2003), samples were removed from the old land surface and ditch fill sequences from the two barrows (18 samples from barrow 36; and 13 samples from barrow 44) (Figs. 3.34, 3.44, 68, 3.49 & 3.50) The land snail sequences (31 samples; Tables 3.15, 3.16, A3.8 & A3.9; Fig. 3.97) show that these barrows were also built in an established open downland. Very stable conditions become established in the secondary fills indicating grassland and arable conditions. Some shady and long grass refugia are still present, possibly on the mounds themselves.

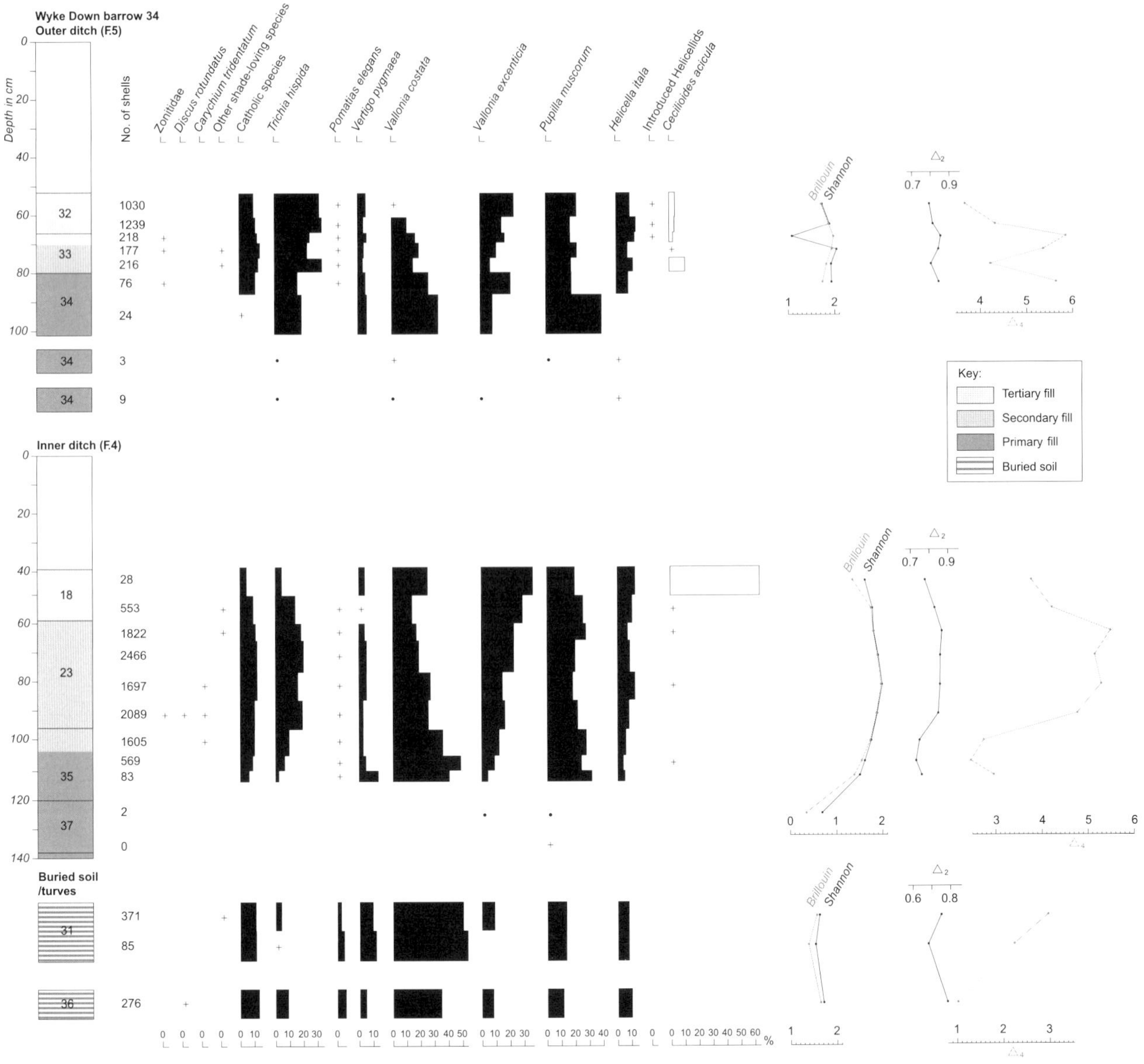

Figure 3.96. *Molluscan histograms for the inner (F.4) and outer (F.5) ditches of barrow 34, Wyke Down (R. Goller, Wessex Archaeology).*

Table 3.14. *Mollusca from Wyke Down barrows 34 and 41. (Figures in square brackets were hand picked and not included in totals.)*

Barrow	Barrow 41									Barrow 34																			
Feature	OLS		Ditch F.1				OLS\turves			Inner Ditch F4											Outer Ditch F.5								
Sample	52	53	35	36	37	38	17	4	2	41	42	43	44	45	46	47	48	49	50	51	25	26	27	28	29	30	31	32	33
Context	13	11	10	9	5	36	31	37	35	23	18	34	33	32/3	32														
	ols\glacial		prim	secondary		3ry	ols	turf		prim	primary and lenses			secondary fills					tertiary		primary fills				secondary		U 2nd	tertiary	
Depth (cm)	spot	spot	spot	89–98	80–89	50–58	spot	10–20	0–10	129–141	119–129	109–112	104–109	95–104	85–95	75–85	65–75	59–65	49–59	39–49	spot	spot	88–101	80–88	75–80	70–75	66–70	61–66	53–61
Wt (g)	1500	2000	2000	2000	2000	2000	2000	2000	2000	2000	2000	2000	2000	2000	2000	2000	2000	2000	2000	2000	2000	2000	2000	2000	1500	1500	1000	1500	1500
LAND																													
Pomatias elegans (Müller)	-	-	-	-	+	-	15	3	7	-	-	+	4	1	+	1	2	3	5	-	-	-	-	+	+	+	+	2	+
Carychium tridentatum (Risso)	-	-	-	-	1	-	-	-	-	-	-	-	-	2	1	2	-	-	-	-	-	-	-	-	-	-	-	-	-
Cochlicopa lubrica (Müller)	-	-	-	1	-	-	6	1	13	-	-	-	5	45	65	32	65	29	4	-	-	-	-	-	4	2	8	40	29
Cochlicopa lubricella (Porro))	-	-	-	-	-	-	-	-	-	-	-	1	3	8	3	6	4	2	-	-	-	-	-	-	+	6	3	6	3
Cochlicopa spp.	-	-	-	5	1	-	17	4	22	-	-	5	22	77	78	71	159	112	39	1	-	-	+	1	11	9	6	59	39
Vertigo pygmaea (Draparnaud)	1	-	-	3	2	1	13	9	31	-	-	10	17	337	50	66	95	42	7	1	-	-	1	2	4	5	11	41	45
Vertigo spp.	-	-	-	-	-	-	-	-	-	-	-	-	-	-	1	-	-	-	-	-	-	-	-	-	-	-	-	-	-
Pupilla muscorum (Linnaeus)	-	-	3	3	5	+	30	11	47	+	1	25	126	419	408	288	433	496	132	5	-	1	9	12	34	35	32	183	208
Vallonia costata (Müller)	-	3	1	15	24	1	119	45	174	-	-	32	270	580	537	443	436	312	78	7	5	+	8	19	32	30	34	117	7
Vallonia excentrica Sterki	-	-	-	3	2	-	14	-	7	-	1	3	50	184	337	242	537	403	160	10	3	-	2	16	16	20	37	184	248
Vallonia spp.	-	-	-	-	-	-	7	1	-	-	-	2	5	7	15	16	8	2	-	-	-	-	-	2	-	-	-	-	-
Punctum pygmaeum (Draparnaud)	-	-	-	2	3	-	-	2	1	-	-	-	13	25	44	58	17	1	1	-	-	-	-	4	8	4	3	4	5
Discus rotundatus (Müller)	-	-	-	-	-	-	+	-	-	-	-	-	-	-	1	-	-	-	-	-	-	-	-	-	-	-	-	-	-
Vitrina pellucida (Müller)	-	-	-	-	3	-	-	-	-	-	-	-	-	-	6	5	20	1	-	-	-	-	-	-	1	-	1	-	-
Nesovitrea hammonis (Ström)	-	-	-	-	-	-	-	-	-	-	-	-	-	-	1	8	-	-	-	-	-	-	-	-	1	1	-	1	-
Aegopinella nitidula (Draparnaud)	-	-	-	-	-	-	-	-	-	-	-	-	-	-	1	-	-	-	-	-	-	-	-	1	-	-	1	-	-
Oxychilus cellarius (Müller)	-	-	-	-	-	-	-	-	-	-	-	-	-	-	1	-	-	-	-	-	-	-	-	-	-	2	-	-	-
Limacidae	-	-	-	-	3	-	3	2	-	-	-	-	-	9	3	5	2	12	-	-	-	-	-	3	2	2	3	24	20
Cecilioides acicula (Müller)	-	-	-	3	2	5	-	-	-	-	-	-	1	-	-	1	-	6	2	18	-	-	-	-	22	2	5	27	27
Cochlodina laminata (Montagu)	-	-	-	-	-	-	-	-	-	-	-	-	-	-	-	-	-	-	-	-	-	-	-	-	2	1	-	-	-
Clausilia bidentata (Ström)	-	-	-	-	-	-	-	-	1	-	-	-	-	-	-	-	-	1	1	-	-	-	-	-	-	-	-	-	-
Candidula intersecta (Poiret)	-	-	-	-	-	-	-	-	-	-	-	-	-	-	-	-	-	-	-	-	-	-	-	-	-	-	1	1	-
Candidula gigaxii (L. Pfeiffer)	-	-	-	-	-	-	-	-	-	-	-	-	-	-	-	-	-	-	-	-	-	-	-	-	-	-	-	-	1
Helicella itala (Linnaeus)	-	-	+	-	3	2	25	6	26	-	-	3	17	65	151	183	189	104	48	3	+	+	-	5	23	13	25	155	90
Trichia hispida (Linnaeus)	-	-	-	3	8	-	21	1	6	-	-	2	35	146	386	271	499	300	77	1	1	2	4	11	77	47	52	421	332
Arianta arbustorum (Linnaeus)	-	-	-	-	-	-	-	-	+	-	-	-	-	-	+	-	-	-	-	-	-	-	-	-	-	-	-	-	-
Helicigona lapicida (Linnaeus)	-	-	-	-	-	-	-	-	+	-	-	-	-	-	-	-	-	-	-	-	-	-	-	-	-	-	-	-	-
Cepaea hortensis (Müller)	-	-	-	-	-	-	[9]	-	-	-	-	-	-	-	-	-	-	-	-	-	-	-	-	-	-	-	-	-	-
Cepaea spp.	-	-	-	-	-	-	[5]	-	-	-	-	+	2	-	+	-	+	2	-	+	-	-	-	-	1	+	+	1	+
Cepaea/Arianta spp.	-	-	-	-	-	-	6	-	4	-	-	-	-	-	-	-	-	-	+	-	-	-	-	-	-	-	-	-	-
FRESH-/BRACKISH-WATER																													
Pisidium cf. *casertanum*	-	-	-	-	1	-	-	-	-	-	-	-	-	-	-	-	-	-	-	-	-	-	-	-	-	-	-	-	-
Taxa	1	1	2	7	12	3	10	9	11	0	2	7	11	12	16	14	12	14	10	7	3	2	5	9	13	13	14	14	11
TOTAL	**1**	**3**	**4**	**35**	**56**	**4**	**276**	**85**	**371**	**0**	**2**	**83**	**569**	**1605**	**2089**	**1697**	**2466**	**1822**	**553**	**28**	**9**	**3**	**24**	**76**	**216**	**177**	**218**	**1239**	**1030**
Magnetic susceptibility x 10-8SI\kg	47	38	4	31	31	49	91	96	95	8	20	51	66	53	39	33	41	63	56	51	7	6	10	18	34	45	39	45	53

Table 3.15. *Mollusca from Wyke Down barrow 36.*

Feature	treehole			OLS	ditch													
Context	89	88	66/88	72	78	77	76	75	68	67								
Context				OLS	primary fill				U primary		secondary			stabilization			teritary	
Sample	129	128	127	102	103	105	106	107	108	109	110	111	112	113	114	115	116	117
Depth (cm)					spot	118–128	108–118	98–108	88–98	82–88	77–82	72–77	66–72	60–66	50–60	46–50	38–46	32–38
Wt (g)	2000	2000	2000	2000	2000	1500	1500	1500	1500	1500	1500	1500	1500	1500	1500	1500	1500	1500
MOLLUSCA																		
Pomatias elegans (Müller)	-	-	+	-	-	-	-	-	-	-	+	+	-	2	1	1	-	-
Carychium tridentatum (Risso)	-	-	-	-	-	-	-	-	-	-	-	-	-	-	-	-	-	-
Cochlicopa lubrica (Müller)	-	-	-	-	-	-	-	1	-	-	-	5	16	83	24	14	14	9
Cochlicopa lubricella (Porro)	-	-	-	-	-	-	-	-	-	-	-	1	+	-	-	1	-	-
Cochlicopa spp.	-	-	1	-	1	-	-	-	-	1	1	13	11	69	104	40	17	8
Vertigo pygmaea (Draparnaud)	-	1	2	-	-	-	-	-	-	1	3	8	16	101	64	31	15	6
Vertigo spp.	-	-	-	-	-	-	-	-	-	-	-	1	3	14	10	7	1	-
Pupilla muscorum (Linnaeus)	-	3	3	-	1	-	-	-	-	1	12	83	99	117	153	198	218	70
Vallonia costata (Müller)	-	1	1	-	-	-	-	-	-	1	4	51	122	351	187	29	10	5
Vallonia excentrica Sterki	-	-	4	-	1	-	-	-	-	2	8	56	149	568	318	118	97	72
Acanthinula aculeata (Müller)	-	-	-	-	-	-	-	-	-	-	-	-	-	-	-	-	-	-
Ena obscura (Müller)	-	-	-	-	-	-	-	-	-	-	-	-	-	-	-	-	-	-
Punctum pygmaeum (Draparnaud)	-	-	-	-	-	-	-	-	-	-	4	21	10	51	3	5	4	-
Discus rotundatus (Müller)	-	1	2	1	-	-	-	-	-	1	-	-	-	-	-	-	-	-
Vitrina pellucida (Müller)	-	-	-	-	-	-	-	-	-	-	1	1	4	3	2	-	1	-
Vitrea crystallina (Müller)	-	-	-	-	-	-	-	-	-	-	-	-	-	-	-	-	-	-
Vitrea contracta (Westerlund)	-	-	-	-	-	-	-	-	-	-	-	-	-	-	-	-	-	-
Nesovitrea hammonis (Ström)	-	-	1	-	-	-	-	-	-	-	2	11	17	40	13	-	1	-
Aegopinella pura (Alder)	-	-	-	-	-	-	-	-	-	-	-	-	-	-	-	-	-	-
Aegopinella nitidula (Draparnaud)	-	1	1	-	-	-	-	-	-	-	-	-	-	-	-	-	-	-
Oxychilus cellarius (Müller)	-	-	-	-	-	-	-	-	-	-	-	-	-	-	-	-	-	-
Limacidae	-	-	-	-	-	-	-	-	-	-	-	2	12	65	46	9	5	1
Cecilioides acicula (Müller)	-	-	-	-	-	-	-	-	-	-	-	-	-	-	2	-	-	-
Cochlodina laminata (Montagu)	-	-	-	-	-	-	-	-	-	-	-	-	-	-	-	-	-	-
Clausilia bidentata (Ström)	-	+	-	-	-	-	-	-	-	-	-	-	-	-	-	-	-	-
Helicella itala (Linnaeus)	-	7	4	-	2	+	-	3	-	2	4	13	33	147	62	27	25	27
Trichia hispida (Linnaeus)	-	-	1	-	3	-	-	1	1	3	6	92	328	739	451	121	79	56
Helicigona lapicida (Linnaeus)	-	-	-	-	-	-	-	-	-	-	-	-	-	-	-	-	-	-
Cepaea spp.	-	-	-	-	-	-	-	-	-	-	-	-	-	+	-	-	-	-
Cepaea/Arianta spp.	-	-	+	-	-	-	-	-	-	-	-	-	-	-	-	-	-	-
Taxa	0	6	10	1	5	0	0	3	1	8	10	12	11	12	13	11	11	8
TOTAL	**0**	**14**	**20**	**1**	**8**	**0**	**0**	**5**	**1**	**12**	**45**	**358**	**820**	**2350**	**1438**	**601**	**487**	**254**

Table 3.16. *Mollusca from Wyke Down barrow 44.*

Feature	turf	turf	ols	Ditch									
Context	99	100	107	111		110		109		108		105	
Context				primary fill				u 1ry	secondary			tertiary	
Sample	134	133	140	141	142	143	144	145	146	147	148	149	150
Depth (cm)				125–135	115–125	105–115	95–105	85–95	75–85	65–75	55–65	45–55	35–45
Wt (g)	2000	2000	2000	2000	1300	1300	1500	1500	1500	1500	1500	1500	1500
MOLLUSCA													
Pomatias elegans (Müller)	1	+	1	-	+	+	-	+	+	3	+	+	-
Carychium tridentatum (Risso)	-	-	-	1	-	-	3	2	11	7	2	1	-
Cochlicopa lubrica (Müller)	-	-	-	-	-	-	-	1	44	43	48	18	6
Cochlicopa lubricella (Porro)	-	-	-	-	-	1	-	-	-	-	-	-	-
Cochlicopa spp.	+	+	-	-	-	1	2	4	21	29	52	21	10
Vertigo pygmaea (Draparnaud)	1	-	2	-	-	1	-	2	14	50	102	37	7
Vertigo spp.	-	-	-	-	-	-	-	-	2	7	14	5	2
Pupilla muscorum (Linnaeus)	8	6	-	4	2	4	6	14	306	342	579	237	129
Vallonia costata (Müller)	5	2	-	-	-	2	10	13	196	246	234	44	6
Vallonia excentrica (Sterki)	14	2	6	2	-	1	9	8	120	277	654	235	71
Acanthinula aculeata (Müller)	-	-	-	-	-	-	1	2	13	12	1	-	-
Ena obscura (Müller)	-	-	-	-	-	-	-	-	-	2	1	6	5
Punctum pygmaeum (Draparnaud)	-	-	-	-	-	1	-	8	35	34	59	-	-
Discus rotundatus (Müller)	-	-	-	-	-	1	-	-	5	8	1	-	-
Vitrina pellucida (Müller)	-	-	-	-	-	-	-	-	3	-	-	1	-
Vitrea contracta (Westerlund)	-	-	-	-	-	3	4	3	7	6	-	-	-
Nesovitrea hammonis (Ström)	-	-	-	1	-	1	3	1	38	29	3	-	-
Aegopinella pura (Alder)	-	-	-	-	-	-	-	-	7	12	-	-	-
Aegopinella nitidula (Draparnaud)	-	-	-	-	-	-	2	2	13	15	-	-	-
Oxychilus cellarius (Müller)	-	-	-	-	-	-	-	-	-	-	-	-	-
Limacidae	-	1	-	-	1	-	2	-	49	20	38	4	8
Cecilioides acicula (Müller)	-	-	-	-	-	-	-	-	-	-	1	4	6
Cochlodina laminata (Montagu)	-	-	-	-	-	-	-	-	-	4	-	-	-
Clausilia bidentata (Ström)	-	-	-	-	-	-	1	-	9	12	2	-	1
Helicella itala (Linnaeus)	8	1	1	3	2	2	3	4	65	55	80	44	28
Trichia hispida (Linnaeus)	4	2	1	1	-	2	8	15	308	420	426	97	36
Helicigona lapicida (Linnaeus)	-	-	-	-	-	-	-	-	-	+	-	-	-
Cepaea spp.	1	-	+	-	+	-	-	-	-	-	-	-	-
Cepaea/Arianta spp.	-	-	-	-	-	-	-	-	2	2	+	-	-
Taxa	8	6	5	6	3	11	13	13	19	21	15	11	10
TOTAL	42	14	11	12	5	20	54	79	1268	1635	2296	750	309

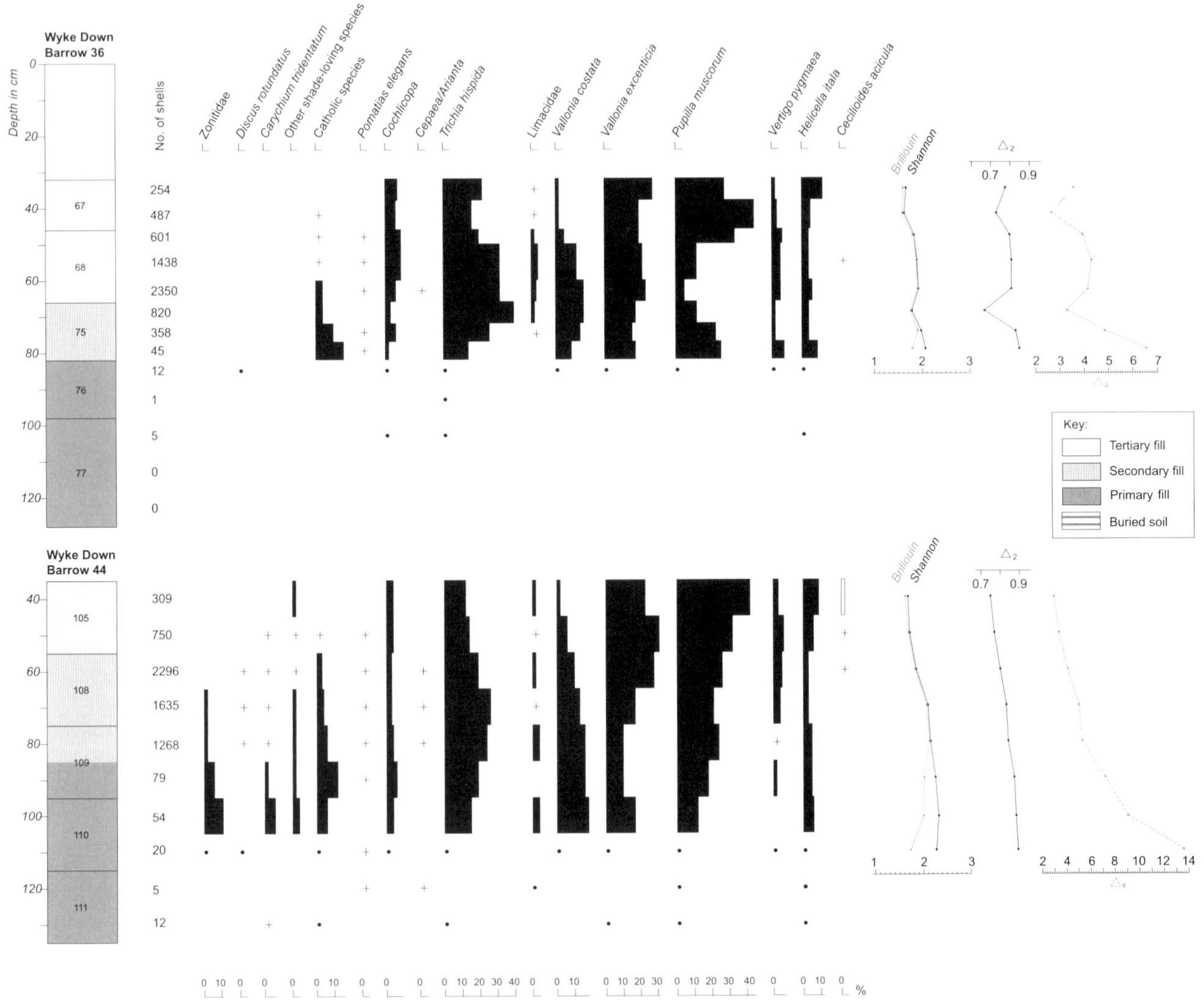

Figure 3.97. *Molluscan histogram of the Wyke Down barrows 36 and 44 (R. Goller, Wessex Archaeology).*

Pre-barrow contexts were few, and shell preservation in the treehollow (barrow 36) is poor, but shells seem to be largely open country species (*Helicella itala, Pupilla muscorum* and *Vallonia excentrica*) which may call into question the interpretation of this feature (Table 3.15).

The old land surfaces (and turf core) contexts contained typically depauperate assemblages. Those species present, however, were *Trichia hispida, Helicella itala, Vallonia* spp. and *Pupilla muscorum* indicate established open-country conditions and dry, grazed grassland.

The ditch fills typically contain low numbers in primary fills but upper secondary, tertiary and especially stabilization horizons, contained high to exceptional numbers of shells in the flots. Numbers decreased in the tertiary fills. The assemblages were predominately open country and no clear well-defined zones of major vegetation regeneration can be detected in either ditch sequence.

Very similar sequences to those already recorded from barrows 41 and 34 seem to be present. This indicates that in areas of coarse vegetation mosaics (i.e. with low spatial biotic variation), land snail sequences may be referable over very large areas, perhaps of 800 m^2.

The sequences from the four barrows may allow detailed local chronological mapping from these in conjunction with the two henges on Wyke Down (henge 1; Entwistle & Bowden 1991; WD96 unpublished analysis by M.J. Allen; see Appendix 4.4).

The land-snail sequence shows that these barrows were also built in an established open downland. Very stable conditions become established in the secondary fills indicating grassland and arable conditions. Some shady and long grass refugia were still present, possibly on the mounds themselves.

Fir Tree Field, Down Farm, Middle Bronze Age enclosure (WD 03)
A column of nine samples was taken through the ditch fills and into the overlying old Ap (context 376) and the modern turf (Ah, context 375) from a section cut through the southwestern unexcavated portion (Fig. 3.98; Tables 3.16 & A3.10). This was within 5 m of the section sampled and analysed for snails by Entwistle (Entwistle & Boden 1991; Barrett *et al.* 1991, fig. 5.27).

The ditch was 0.9 m deep from the chalk surface and was similar in profile to sections recorded (Barrett *et al.* 1991a, 184–6, fig. 5.32) and sampled previously (Entwistle & Bowden 1991, 32–6, fig. 3.11). The nature and origin of the fills are important to the taphonomy and interpretation of the assemblage (cf. Evans 1972, 321–8; Limbrey 1975, 290–300). The combination of this data, together with the species diversity indices so deftly employed by Entwistle (Entwistle & Bowden 1991), enable a better index of the scale and extent of the vegetation refugia reflected in the snail assemblages. The main fills were dated previously by radiocarbon to about 1500–750 cal. BC.

The enclosure ditch section here, as previously excavated in 1978, contained a shallow primary fill (380–379), surmounted by secondary fills (377) and deposits classed as 'midden' (378; Barrett *et al.* 1991, 185–6). The tertiary fills above (372) were 25 cm thick and similar to those from the closest, and previously sampled section (Barrett *et al.* 1991a, fig. 5.32; Entwistle 1990a, fig. 3.11). However, unlike the previous excavations the full profile including the old ploughsoil (376) and the modern grassland turf (Ah, 375) were also sampled. Thus the sequence here has longer record than that reported by Entwistle, excepting the lack of the basal 5 cm which were not sampled in section 381, DF03.

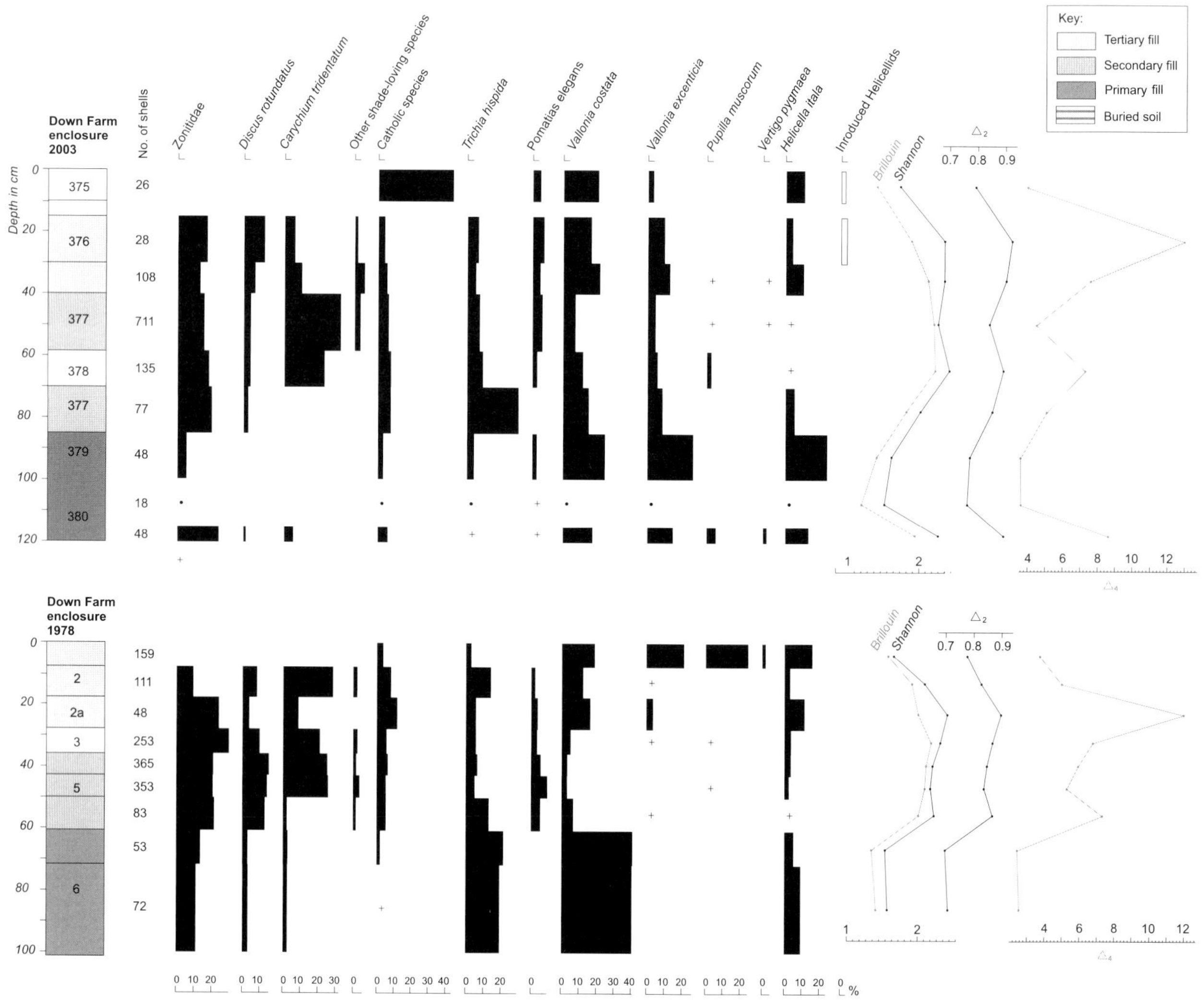

Figure 3.98. *Molluscan histogram of the Middle Bronze Age enclosure ditch in Fir Tree Field in the 2003 and 1978 excavations (R. Goller, Wessex Archaeology).*

The sequence

Overall the assemblages are remarkably similar to those reported by Entwistle & Bowden (1991). The histograms of the two sequences (Fig. 3.98) show great parity, with only differences in the occurrence of *Vallonia excentrica* being significantly different. Whether this represents spatial variation in the *Vallonia* populations along the ditch, or identification differences cannot be confirmed.

Assemblages from the chalk rubble primary fill confirm that this enclosure ditch was built in a pre-existing and established open downland. The basal sample of the primary rock-rubble, although depauperate, does contain some shade-loving species *A. nitidula, O.* cellarius and *D. rotundatus* (Fig. 3.98) as also recorded by Entwistle & Bowden (1991, 34), but probably represents the presence of long ungrazed grassland refugia. The secondary fills and dumped deposits ('midden') indicate flourishing grassland conditions, possibly representing longer grasses and herbaceous vegetation on the adjacent bank, set in to a generally open landscape, possibly with some woody vegetation. In the tertiary deposits, increases in *V. costata* and *H. itala* might represent the onset of arable and tilled conditions in the later Bronze Age.

The modern turf was shell-poor and has quite a different molluscan composition. It contains slug plates (Limacidae) predominantly, which are typical of lush, species rich grassland with an acidic turf horizon.

Table 3.17. *Mollusca from the Down Farm Bronze Age enclosure ditch; samples taken in 2003 (this project) and 1978 (Entwisttle & Bowden 1991) (numbers in [ ] are specimens with periostricum and considered modern, and not included in the totals).*

Feature	**Down Farm Enclosure 2003**									**Down Farm Enclosure 1978**								
Fill	**primary**			**2ndry**	**dump**	**tertiary**	**old Ap**		**Ah**	**primary**		**secondary**		**dump**		**tertiary**		
Context	**380**	**379/80**	**379**	**377**	**378**	**377**	**376/7**	**376**	**375**	**6**		**5**	**-**	**3**	**2a**	**2**		
Sample	**8**	**7**	**6**	**5**	**4**	**3**	**2**	**1**	**9**	**9**	**8**	**7**	**6**	**5**	**4**	**3**	**2**	**1**
Depth (cm)	**115–120**	**100–115**	**85–100**	**70–85**	**58–70**	**40–58**	**30–40**	**15–30**	**0–10**	**72–102**	**61–72**	**50–61**	**43–50**	**36–43**	**28–36**	**18–28**	**8–18**	**0–8**
Wt (g)	**2000**	**2000**	**2000**	**1500**	**1500**	**1500**	**1500**	**1500**	**800**	**1000**	**1000**	**1000**	**1000**	**1000**	**1000**	**1000**	**1000**	**1000**
MOLLUSCA																		
Pomatias elegans (Müller)	+	+	1	-	3	40	4	2	1	-	-	4	35	14	5	1	2	-
Carychium tridentatum (Risso)	3	-	-	-	37	276	13	2	-	2	1	20	118	76	61	4	38	-
Cochlicopa lubrica (Müller)	-	-	-	-	2	-	2	-	[1]	-	-	1	9	3	1	1	2	-
Cochlicopa lubricella (Porro)	-	-	-	-	1	2	-	+	-	-	-	-	-	2	-	-	1	5
Cochlicopa spp.	2	-	1	5	6	28	-	-	-	1	1	4	9	6	7	1	3	-
Vertigo pygmaea (Draparnaud)	1	-	-	-	-	4	1	-	-	-	-	-	-	-	-	-	-	3
Pupilla muscorum (Linnaeus)	3	-	-	-	3	4	1	-	-	-	-	-	1	-	3	-	-	43
Vallonia costata (Müller)	9	6	13	12	16	42	26	5	[6]	34	27	5	9	6	10	7	15	34
Vallonia excentrica Sterki	8	6	15	8	8	38	16	3	1	-	-	-	2	-	1	2	1	40
Acanthinula aculeata (Müller)	-	-	-	-	3	23	1	-	-	-	-	2	9	6	7	-	1	-
Ena obscura (Müller)	-	-	-	-	3	-	1	-	-	-	-	-	-	-	1	-	1	-
Punctum pygmaeum (Draparnaud)	-	-	-	-	1	6	1	-	[2]	-	-	-	-	-	4	1	2	-
Discus rotundatus (Müller)	1	-	-	3	8	42	9	4	-	3	2	13	68	48	30	2	12	-
Vitrina pellucida (Müller)	1	1	-	-	-	-	-	-	-	-	-	-	-	-	-	-	-	-
Vitrea contracta (Westerlund)	5	-	1	5	16	49	7	1	-	1	1	6	59	19	40	4	6	-
Nesovitrea hammonis (Ström)	-	-	-	-	-	3	-	-	-	-	-	-	1	-	2	2	3	1
Aegopinella pura (Alder)	1	-	-	-	5	46	7	2	-	-	-	3	4	22	31	4	7	-
Aegopinella nitidula (Draparnaud)	1	1	1	5	4	27	1	2	-	2	2	5	22	16	10	1	-	-
Oxychilus cellarius (Müller)	6	-	1	7	2	-	-	-	-	6	6	7	8	7	7	2	-	-
Limacidae	-	-	-	-	-	11	1	1	10	-	-	-	-	-	-	-	-	-
Clausilia bidentata (Ström)	-	-	-	-	-	7	-	-	-	-	-	-	4	-	-	-	-	-
Candidula gigaxii (L. Pfeiffer)	-	-	-	-	-	-	-	[3]	[2]	-	-	-	-	-	-	-	-	-
Helicella itala (Linnaeus)	7	3	13	4	2	6	12	1	[3]	7	3	1	8	9	7	5	3	28
Trichia hispida (Linnaeus)	+	1	2	26	13	48	5	2	-	16	14	12	19	24	21	3	19	5
Helicigona lapicida (Linnaeus)	-	-	-	-	-	2	-	-	-	-	-	-	3	-	-	-	-	-
Cepaea hortensis (Müller)	-	-	-	-	-	-	-	-	-	-	-	-	1	-	-	-	-	-
Cepaea spp.	+	-	-	2	2	7	1	+	+	-	-	-	3	8	2	-	-	-
Taxa	13	6	9	10	18	20	18	12	9	9	9	12	18	14	18	14	15	8
TOTAL	**48**	**18**	**48**	**77**	**135**	**711**	**109**	**28**	**26**	**72**	**57**	**83**	**392**	**266**	**250**	**40**	**116**	**159**

Gussage Cow Down Iron Age bank and ditch (GCD 192)
A sequence of 8 samples through the ditch augmented by a single spot sample from the buried soil under the bank were examined (Figs. 3.11 & 3.99; Tables 3.18 & A3.11). Shell numbers from the ditch and buried soil are moderate to high. High numbers of shells are present in the old ground surface (context 253) and dominated by *Pomatias elegans* indicating loose soil created by clearance or cultivation (Fig. 3.99). Very open dry and established open downland existed in the main (secondary fills) with high numbers of *Vallonia* and *H. itala* with *T. hispida* suggesting stable grass or arable landscapes.

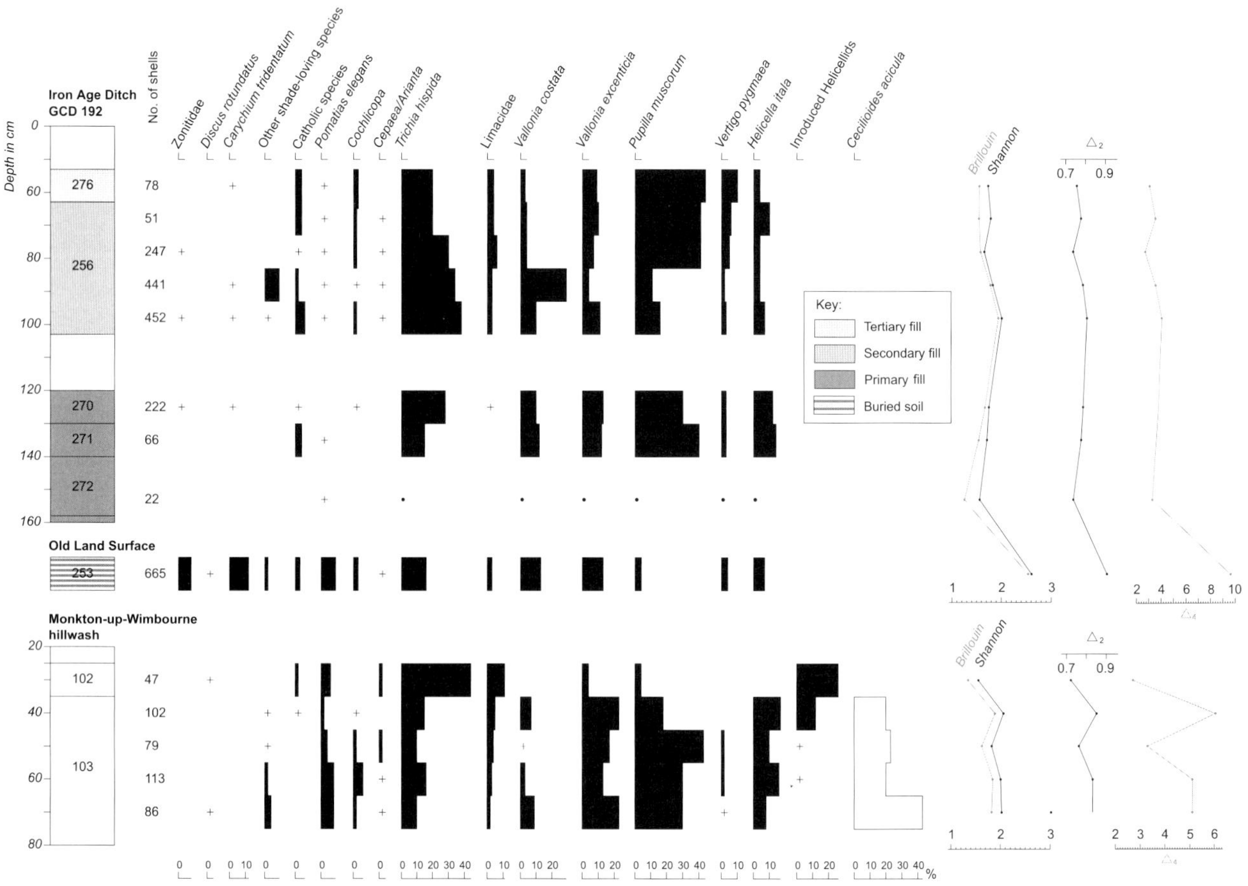

Figure 3.99. *Molluscan histograms of the Gussage Cow Down site 192 Iron Age ditch and dry valley/hillwash section near Monkton-up-Wimborne (R. Goller, Wessex Archaeology).*

Table 3.18. *Mollusca from the Iron Age ditch and buried soil section at Gussage Cow Down 192 (numbers in [ ] are specimens with periostricum and considered modern, and not included in the totals).*

Feature	**Buried soil**		**Ditch**							
Context	**253**		**272**	**271**	**270**	**256**	**256**	**256**	**256**	**276**
Context	**soil**			**primary**		**secondary - windblown**				**3ry**
Sample	**237**		**1**	**2**	**3**	**4**	**5**	**6**	**7**	**8**
Depth (cm)	**spot**		**140–148**	**130–140**	**120–130**	**93–103**	**83–93**	**73–83**	**63–73**	**53–63**
Wt (g)	**2000**		**1500**	**1500**	**1500**	**1500**	**1500**	**1500**	**1500**	**1500**
MOLLUSCA										
Pomatias elegans (Müller)	58	-	+	+	-	1	2	1	+	+
Carychium tridentatum (Risso)	80	-	-	-	1	6	4	-	-	1
Cochlicopa lubrica (Müller)	4	-	-	-	-	4	2	1	-	-
Cochlicopa lubricella (Porro)	-	-	-	-	-	-	-	1	-	-
Cochlicopa spp.	18	-	-	-	1	6	9	2	1	2
Vertigo pygmaea (Draparnaud)	25	+ [1]	2	2	6	14	48	13	3	7
Pupilla muscorum (Linnaeus)	30	+ [6]	10	26	66	71	127	102	21	34
Vallonia costata (Müller)	87	+ [4]	2	8	23	42	18	9	2	2
Vallonia excentrica Sterki	84	+ [7]	2	8	28	50	40	18	5	7
Acanthinula aculeata (Müller)	8	-	-	-	-	2	-	-	-	-
Punctum pygmaeum (Draparnaud)	8	+ [1]	-	2	2	8	6	2	1	3
Discus rotundatus (Müller)	8	-	-	-	-	-	-	-	-	-
Vitrina pellucida (Müller)	1	-	-	-	-	-	-	-	-	-
Vitrea crystallina (Müller)	15	-	-	-	-	1	-	-	-	-
Vitrea contracta (Westerlund)	5	+ [1]	-	-	-	2	-	1	-	-
Nesovitrea hammonis (Ström)	12	-	-	1	-	17	5	1	1	-
Aegopinella pura (Alder)	17	-	-	-	-	-	-	-	-	-
Aegopinella nitidula (Draparnaud)	17	+ [2]	-	-	1	1	-	-	[1]	-
Oxychilus cellarius (Müller)	1	-	-	-	-	1	-	-	-	-
Limacidae	20	-	-	-	3	13	12	12	2	3
Cochlodina laminata (Montagu)	3	-	-	-	-	-	-	-	-	-
Macrogastra rolphii (Turton)	1	-	-	-	-	-	-	-	-	-
Clausilia bidentata (Ström)	1	-		-	-	2	-	-	-	-
Helicella itala (Linnaeus)	46	-	3	9	28	41	20	11	5	3
Trichia hispida (Linnaeus)	110	+ [6]	3	10	63	170	148+[1]	73	10	16
Helicigona lapicida (Linnaeus)	1	-	-	-	-	+	-	-	-	-
Cepaea spp.	5	-	-	-	-	-	-	-		-
Cepaea/Arianta spp.	-	-	-	-	-	+	+	+	+	-
Helix aspersa (Müller)	-	-	-	-	-	+	-	-	-	-
Taxa	25	[9]	6	8	11	18	13	13	10	10
Shannon Index	2.61		1.56	1.71	1.75	2.01	1.83	1.66	1.79	1.74
Brillouin Index	2.54		1.26	1.54	1.67	1.94	1.78	1.58	1.55	1.56
Shannon Index - Brillouin Index	0.07		0.30	0.17	0.80	0.07	0.06	0.08	0.24	0.18
Δ2	0.90		0.73	0.77	0.78	0.80	0.78	0.73	0.77	0.75
Δ4	9.63		3.28	3.62	3.79	4.06	3.56	2.72	3.55	3.09
TOTAL	**665**	**+[28]**	**22**	**66**	**222**	**452**	**441+[1]**	**247**	**51+[1]**	**78**

Monkton-up-Wimborne valley trench (Iron Age colluvium)
Five samples were taken from the shallow (0.75 m deep) colluvial sequence (Figs. 2.8 & 3.99; Tables 3.19 & A3.12). The undated colluvial sequence contained lower shell numbers than other sites in the Allen Valley and Down Farm (Allen 1998; 2000a). Nevertheless, sufficient shells were present in the lower four samples (average 95 shells per sample, but less than 76 kg) to make some interpretation (Table 3.19). The assemblages were dominated by open-country species typically found in hillwash (cf. Bell 1983): *Pupilla muscorum, Vallonia excentrica, Helicella itala* and *Trichia hispida* (Fig. 3.99). Assemblages are typical of colluvium, excepting the relatively high proportions of *Pomatias elegans* which may be a function of taphonomy or preservation. In the basal four samples covering the bBw and bA1 horizons no major change can be detected in the sequence and thus no detailed interpretation can be gained from this short sequence. It probably represents arable and mixed pasture.

Table 3.19. *Mollusca from Monkton-up-Wimborne valley colluvium.*

Sample	170	171	172	173	174
Context		103			102
	bBw		bA1	bAh	A1
Depth (cm)	65–75	55–65	45–55	35–45	25–35
Wt (g)	1500	1500	1500	1500	1500
MOLLUSCA					
Pomatias elegans (Müller)	7	9	3	2	3
Cochlicopa lubricella (Porro)	1	-	-	-	-
Cochlicopa spp.	1	7	2	1	-
Vertigo pygmaea (Draparnaud)	1	2	2	-	-
Pupilla muscorum (Linnaeus)	25	34	34	18	2
Vallonia costata (Müller)	8	3	1	7	-
Vallonia excentrica (Sterki)	20	15	14	23	2
Punctum pygmaeum (Draparnaud)	-	-	-	1	1
Discus rotundatus (Müller)	1	-	-	-	+
Limacidae	2	3	3	5	5
Cecilioides acicula (Müller)	37	22	18	20	-
Cochlodina laminata (Montagu)	1	-	-	-	-
Clausilia bidentata (Ström)	2	2	1	1	-
Candidula gigaxii (Poiret)	-	1	-	-	-
Candidula intersecta (L. Pfeiffer)	-	-	-	9	12
Cernuella virgata (Da Costa)	-	-	1	2	-
Helicella itala (Linnaeus)	7	18	8	17	-
Trichia hispida (Linnaeus)	9	18	8	16	21
Cepaea spp.	1	1	2	-	-
Cepaea/Arianta spp.	-	-	-	-	1
Helix aspersa (Müller)	-	-	-	+	+
Taxa	13	12	12	12	8
TOTAL	**86**	**113**	**79**	**102**	**47**

The upper portion (above *c.* 55 cm) from the base of the A1 horizon, however shows a reduction in shell numbers and increase in *T. hispida* and concomitant decrease in the open country species *V. excentrica* and *Pupilla muscorum*. Reduction in taxa and species diversity may indicate a more uniform constant, harsher, arable habitat: this is medieval or later in date in view of the presence of the Introduced Helicellids (*Cernuella virgata, Candidula intersecta* & *Candidula gigaxii* are all present). There are, however, probably intrusive (modern or more recent) shells in this profile. The lower part of the profile (75–65 cm) looks, however, from the land-snail composition to be later prehistoric (?Iron Age onwards) (cf. Allen 1994; 1997a). The particle size texture (hand texturing) and high percentage of sample less than 0.5 mm tends to indicate a strong aeolian component typical of an open arable landscape or a landscape stripped of turf.

What is surprising here is that the whole sequence post-dates the main, mid to Late Neolithic monuments (Cursus, Monkton-up-Wimborne shaft and pit circle, Wyke Down henges, Knowlton South Circle) and the Bronze Age barrows (on Wyke Down).

The South Circle at Knowlton
The exposure of 4.5 m of the 5.5 m deep henge ditch profile of the South Circle at Knowlton as excavated by Bournemouth University, in combination with surprisingly good preservation of the old ground surface, enabled comprehensive sampling of these contexts to examine the local land-use history of this Knowlton monument (Allen 1995a).

A total of 47 samples (Tables 3.20–3.23) was taken as columns of contiguous samples, augmented with occasional spot samples, through the old ground surface, the henge ditch, and both the shallow inner and outer gully fills. The samples through the henge ditch did not include the basal *c.* 1 m primary fills, nor chalk wash expected at the base of the ditch, as these contexts were not excavated. However, soil material derived from the old land surface and the buried soil under the relict bank, were sampled. All samples were processed following standard methods (Evans 1972) and the nomenclature follows Kerney (1999). The results are presented in the species groupings following Evans (1984) and Entwistle & Bowden (1991). A number of the shade-loving species (*sensu* Evans 1972), are considered to have more catholic preferences occurring in longer grass swards and open shrubs where damp mesic habitats exist at the base of the vegetation (Cameron & Morgan-Huws 1975; Evans 1984). In the accompanying histograms (Fig. 3.40), the species are grouped as follows:

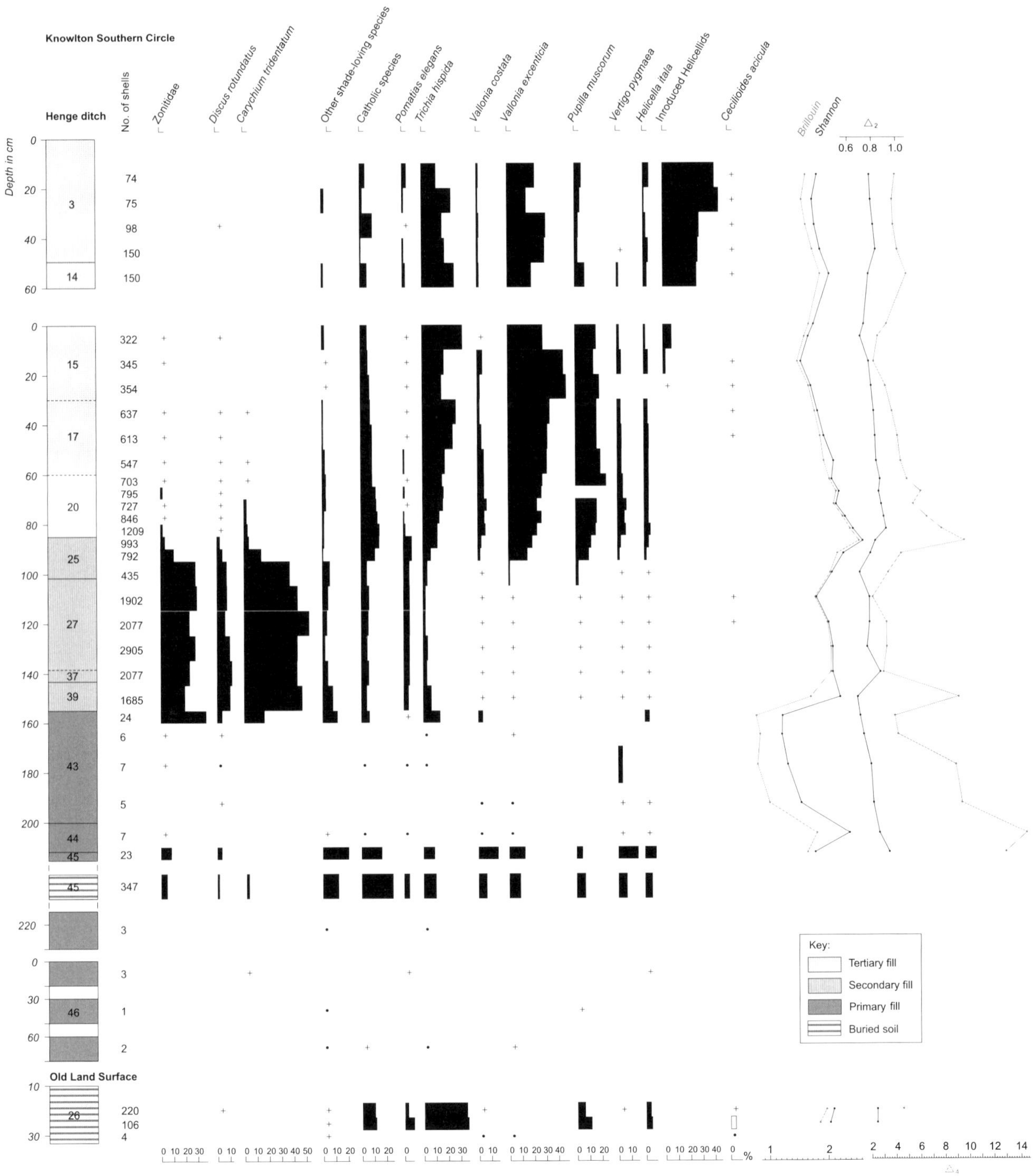

Figure 3.100. *Molluscan histogram of the South Circle henge ditch at Knowlton (R. Goller, Wessex Archaeology).*

Shade-loving species:

Zonitidae: *Vitrea crystallina, V. contracta, Aegopinella pura, A. nitidula, Oxychilus cellarius, O. allarius;*
Discus rotundatus: plotted separately in the histograms, but included with shade-loving species;
Carychium tridentatum: plotted separately in the histograms, but included with shade-loving species;
Other shade-loving species: Clausiliidea (*Cochlodina laminata, Macrogastra rolphii, Clausilia bidentata, Balaea perversa*); *Ena obscura, E. montana, Acanthinula aculeata, Lauria cylindracea.*

Catholic species:

Cochlicopa spp., *Punctum pygmaeum, Euconulus fulvus, Cepaea* spp., *Vitrina pellucida, Nesovitrea hammonis*, Limacidae;
Trichia hispida: plotted separately in the histograms, but included with catholic species;
Pomatias elegans: plotted separately in the histograms, but included with catholic species.

Open-country species:

Vertigo pygmaea, Abida secale, Pyramidula rupestris, Vallonia costata, Vallonia excentrica, Pupilla muscorum, Helicella itala, Introduced Helicellids, *Helix aspersa.*

Species diversity calculations (Table A.13) have been employed to aid in interpretation of the sequences of molluscan assemblages. The analysis of (delta) species diversity indices (Hurlbert 1971) allows us to attempt to examine how 'uniform' the wider landscape is (cf. Entwistle, in Allen *et al.* 1990; Entwistle & Bowden 1991), and the application of species diversity indices clearly allow us to examine generally the nature of the environment (cf. Bell *et al.* forthcoming):

i) closeness of sampled assemblage to a likely real population; Shannon index (*H'*) - Brillouin index (HB) (see Evans 1979; Evans & Smith 1983; see Appendix 1 in French *et al.* 2003));
ii) the diversity of the ecological mosaic sampled;
iii) the complexity or maturity of the assemblages and thus the environment they inhabited.

To complement land-snail analysis and the basic interpretation from the sediment descriptions (Appendix 4.12), magnetic susceptibility was recorded on 10 g air dried samples <2 mm to aid in the characterization of the deposits.

Results

A total of 23,578 identifications were made from the 107 kg of soil processed for land snails (Tables 3.20–3.23). Although preservation was variable between context groups (i.e. old land surface, primary fills, secondary fills and tertiary fills), huge and dramatic changes in the species composition were recorded, providing evidence for defining events and of several significantly different land-use episodes.

Old ground surface

The old ground surface (26) survived as 0.16 m thick dark brown (10YR 3/3) stone free silty clay with some fine sand or coarse silt, with weak small to medium blocky structure. This was seen as a typical brown earth or argillic brown earth (Avery 1980), with an intermittent turf and worm-worked horizon (6). Three samples were taken through this horizon adjacent to the Kubiena sample for soil micromorphology. All samples from the old ground surface were low in shell numbers (Table 3.20), but those present were all typical of very open conditions and broken ground. No shade-loving species were present, not even fragments of the more robust taxa Clausiliidea. From this sparse assemblage we can, however, very tentatively suggest a pre-monument (i.e. pre-2570–2190 cal. BC; 3890±60 BP; Beta-141096) environment of long-established open downland, and we can postulate the absence of woodland in close proximity, in contrast to the local woodland clearance demonstrated from earlier Neolithic Handley Down mortuary enclosure (Allen 1997c; 2000a, 45). Where snails were recovered, from two large 10 litre samples taken by the excavators from the turf-line and old ground surface, inevitably sampling was not as accurate and this is reflected in the number of modern specimens (shells which still retain their periostricum) and Introduced Helicellids, of medieval or later date, that were present (Table 3.20). The assemblages are again typical of very open, well-established chalk grassland; an almost complete, though not total, lack of shade-loving species again testifies to the distance, both in time and space, of the proximity of woodland. The three shade-loving species present (*Discus rotundatus, Clausilia bidentata* & *Nesovitrea hammonis*) occur as no more than two specimens, and we cannot be sure if these are modern or ancient in view of the shallow burial of the old land surface and the character of the sample; all of these species were recorded in 1997 and 1999 living in hedges and field boundaries within 50 m of the sample location.

Henge ditch

The henge ditch was shown to be 5.5 m deep and contained a typical tripartite ditch infill sequence (cf. Evans 1972, 321–8; Limbrey 1975, 290–301). Full descriptions of the sampled ditch section were made in the field specifically to provide information about the formation processes of each context from which shells were derived. In summary the sampled infill sequence is as follows. The vacuous coarse chalk rubble primary fill accounted for nearly half of the profile (contexts 46 & 43) and within which was a mass of fallen soil material (45). The secondary fills (27, 37 & 39) were finer, very calcareous and silty and sealed by a darker more humic stabilization horizon (20). The tertiary fills were typically colluvial with the upper of these (14 & 3) being probably medieval or later ploughwash.

The significant contexts of the lower 1 m of primary ditch fill were not excavated and could not be sampled. The poor molluscan assemblages from the Neolithic buried soil (Table 3.20), however, provide some indication of the pre-monument environment, moreover as primary rock rubble fills accumulate rapidly (cf. Overton Down: Bell *et al.* 1996), material from the lowest fills excavated probably do not sample significantly different environments to those that exist on the base of the henge ditch. Indeed large (2 kg) samples of the lowest chalk rubble (46) yielded very few shells (Table 3.21). Those that were present largely reflected the assemblages recovered from the old land surface. Fallen soil (context 45) was the earliest deposit to provide suitable assemblages for palaeo-environmental interpretation. Two samples and a large (12 kg) bulk sample provided assemblages that indicate very open dry grassland, with loose and broken soil perhaps represented by *Pomatias elegans.* Some relict shade-loving species, notably the Zonitidae, occur in very low numbers, and indicate some shady refugia or leaf litter habitats within the vicinity. These, however, cannot have formed a significant component of the local landscape. The loose rubbly upper primary fill (context 43) contained only sparse shells (Table 3.20); although predominately open-country species (*Helicella itala, Vallonia* species), an increase in the taxonomic range and numbers of shade-loving species may reflect the local shady and rock-rubble habitats (Evans & Jones 1973).

Secondary fills (contexts 39, 37 & 27) are not only clearly characterized by the presence of a poorly sorted soil matrix with many small rounded chalk pieces but also by very high shell numbers (1685 to 2905) (see Table 3.22). High numbers of shells indicate significant stasis in ditch infill rates and the stabilization of the chalk sides typical of secondary fills. Assemblages dominated by Zonitidae, *Carychium tridentatum* with *Discus rotundatus* and with high ($H' = 2.10$) and rising Shannon and Brillouin diversity indices (Table A3.12) indicate a lush mesic vegetation sward. Closeness of these two indices indicates samples of autochthonous populations probably living in the ditch. Although the assemblages are species-rich (20 to 30 taxa), and suggestive of highly favourable environments, the low (delta) diversity indices tend to indicate a mature or wide habitat variation.

During a period of more stable ditch infill conditions there is clear indication of vegetation colonization in the ditch. Although the numbers of shade-loving species increase significantly, many of the species present are common in longer grassland succession habitats (Cameron & Morgan-Huws 1975) and communities typical of the early stages of ditch colonization by plants (Evans 1972; 1984). The main Zonitidae present are not those typical of rock-rubble habitats (Evans & Jones 1973) and significant components of the assemblage are taxa which are not found in rock rubble such as *Aegopinella* sp., *Carychium tridentatum,* and *Trichia hispida.*

The ditch itself, at *c.* 1.0 to 1.5 m deep at this point, itself provides shady micro-habitats, but we cannot attribute this assemblage entirely to this dicth mircro-environment and long grass. Although some of the dominant taxa of these rich assemblages are perhaps typical of tall grassland habitats (*Carychium tridentatum, Vitrea contracta, Aegopinella pura, Punctum pygmaeum*) a number are not: *Discus rotundatus, Aegopinella nitidula, Oxychilus cellarius* and the five Clausiliidea species present. *Discus rotundatus* which occurs in very high numbers is clearly is not common in such environments. It seems to require leaf litter and shady habitats that could not be provided solely by long grasses. A number of species, including *Discus,* are common in leaf litter and have been recorded in both archaeological (e.g. Hambledon; Bell *et al.* forthcoming) and modern ditch fills (Offham; Balksbury; Allen pers. observ.). We can be sure in both contexts that *Oxychilus cellarius* and *Vitrea contracta* are not associated with rock-rubble habitats, and that they probably represent leaf-litter in the ditch; clearly mature vegetation cannot have been in the ditch itself.

The presence of a large number of other shade-loving taxa and rupestral species (Clausiliidea & *Acanthinula aculeata*) suggest that, despite the wholly open-country assemblages recovered from the primary fills and old ground surface, shady refugia must have existed within the vicinity. The occurrence of the *Ena montana,* a species of ancient woodland, also indicates proximity of this environment to the henge. If these assemblages were wholly a result of local conditions within the ditch within an open landscape, then the assemblages should show a progression from open to more mesic (damper/shadier) habitats, and some open-country species from the local landscape would be present in a significant proportion (cf. Thomas 1985). We can, therefore, suggest that this might reflect both the local ditch environment (colonizing plants and seasonally accumulating leaf litter), and

Table 3.20. *Mollusca in the old ground surface (OGS), lower primary fills and fallen turf, Knowlton South Circle (Note: figures in square parenthesis are probably recent specimens).*

Feature	OGS		turf & OGS			lower primary fill				fallen soil		
Sample	**10**	**11**	**12**	**201**	**200**	**41**	**42**	**43**	**13**	**39**	**203**	**14**
Context	**26**	**26**	**6**	**26**	**6**	**46**	**46**	**46**	**46**	**45**	**45**	**45**
Depth (cm)	**28–33**	**23–28**	**17–23**	**spot**	**spot**	**60–70**	**30–40**	**0–10**	**215–230**	**spot**	**spot**	**210–215**
Wt (g)	**1500**	**1500**	**1500**	**12,000**	**10000**	**2000**	**2000**	**2000**	**2000**	**1500**	**12,000**	**2000**
MOLLUSCA												
Pomatias elegans (Müller)	+	2	6	8	7	+	-	+	-	-	16	-
Carychium tridentatum (Risso)	-	-	-	-	-	-	-	1	-	-	9	-
Cochlicopa lubrica (Müller)	-	-	-	-	-	1	-	-	-	-	3	-
Cochlicopa lubricella (Porro)	-	-	-	-	-	-	-	-	-	-	4	-
Cochlicopa spp.	1	1	1	2	2	-	-	-	-	1	51	3
Pyramidula rupestris (Draparnaud)	-	-	-	-	-	-	-	-	-	-	-	-
Vertigo pygmaea (Draparnaud)	-	-	-	-	1[+1]	-	-	-	-	1	9	4
Vertigo spp.	-	-	-	-	-	-	-	-	-	-	16	-
Abida secale (Draparnaud)	-	-	-	-	-	-	-	-	-	-	-	-
Pupilla muscorum (Linnaeus)	-	-	2	12	13	-	1	-	-	-	20	1
Lauria cylindracea (da Costa)	-	-	-	-	-	-	-	-	-	-	-	-
Vallonia costata (Müller)	1	-	-	9	23[+8]	-	-	-	-	1	40	-
Vallonia excentrica Sterki	1	3	2	2[+8]	14[+12]	1	-	-	-	4	30	2
Vallonia spp.	-	-	-	-	-	-	-	-	-	-	5	1
Acanthinula aculeata (Müller)	-	-	-	-	-	-	-	-	-	-	1	-
Ena montana (Draparnaud)	-	-	-	-	-	-	-	-	-	-	-	-
Ena obscura (Müller)	-	-	-	-	-	-	-	-	-	-	-	
Punctum pygmaeum (Draparnaud)	-	-	-	-	2	-	-	-	1	3	32	2
Discus rotundatus (Müller)	-	-	-	-	2	-	-	-	-	-	6	1
Vitrina pellucida (Müller)	-	-	-	-	-	-	-	-	-	1	5	-
Vitrea crystallina (Müller)	-	-	-	-	-	-	-	+	-	-	2	-
Vitrea contracta (Westerlund)	-	-	-	-	-	-	-	-	-	-	3	1
Nesovitrea hammonis (Ström)	-	-	-	-	1	-	-	-	-	-	4	-
Aegopinella pura (Alder)	-	-	-	-	-	-	-	-	-	-	3	-
Aegopinella nitidula (Draparnaud)	-	-	-	-	-	+	-	-	-	1	4	-
Oxychilus allarias (Müller)	-	-	-	-	-	-	-	-	-	-	2	-
Oxychilus cellarius (Müller)	-	-	-	-	-	-	-	-	-	-	2	2
Limacidae	-	-	1	4	9	-	-	-	-	-	14	-
Euconulus fulvus (Müller)	-	-	-	-	-	-	-	-	-	-	1	-
Cecilioides acicula (Müller)	23	20	51	127	166	-	-	-	-	-	8	-
Cochlodina laminata (Montagu)	-	-	-	-	+	-	-	-	-	-	-	-
Macrogastra rolphii (Turton)	-	-	-	-	-	-	-	-	-	-	-	-
Clausilia bidentata (Ström)	-	-	-	1	1	-	-	-	-	-	7	1
Balea perversa (Linnaeus)	-	-	-	-	-	-	-	-	-	-	-	1
Candidula intersecta (Poiret)	-	-	-	-	cf. 29	-	-	-	-	-	-	-
Candidula gigaxii (L. Peiffer)	-	-	-	-	-	-	-	-	-	-	-	-
Cernuella virgata (Da Costa)	-	-	-	4	4	-	-	-	-	-	-	-
Helicella itala (Linnaeus)	1	-	2	5	9	-	-	2	-	2	20	2
Hellicellids	-	-	-	9	-	-	-	-	-	-	-	-
Trichia hispida (Linnaeus)	-	-	-	38	77	+	-	-	2	4	33	2
Arianta arbustorum (Linnaeus)	-	-	-	-	-	-	-	-	-	-	-	-
Helicigona lapicida (Linnaeus)	-	-	-	-	-	-	-	-	-	-	-	-
Cepaea nemoralis (Linnaeus)	-	-	-	-	-	-	-	-	-	-	-	-
Cepaea hortensis (Müller)	-	-	-	-	-	-	-	-	-	-	-	-
Cepaea spp.	-	+	+	-	1	+	-	-	-	-	-	-
Cepaea/Arianta spp.	-	-	-	-	-	-	-	-	-	-	5	-
Helix aspersa (Müller)	-	-	-	4	4	-	-	-	-	-	-	-
Valvata cristata (Müller)	-	-	-	-	-	-	-	-	-	-	-	-
Bitynia tentaculata (Linnaeus)	-	-	-	-	-	-	-	-	-	-	-	-
Taxa	4	3	6	11	17	2	1	2	2	9	25	12
TOTAL	**4**	**6**	**14**	**106**	**220**	**2**	**1**	**3**	**3**	**18**	**347**	**23**
Shannon Index	-	-	-	2.03	2.09	-	-	-	-	2.01	1.80	2.37
Brillouin Index	-	-	-	1.86	1.97	-	-	-	-	1.53	1.67	1.83
Δ2	-	-	-	0.82	0.82	-	-	-	-	0.85	0.93	0.80
Δ4	-	-	-	4.77	4.59	-	-	-	-	8.56	13.11	14.81
Magnetic susceptibility	43	61	105	-	-	2	5	5	-	98	-	41

Table 3.21. *Mollusca in the henge ditch, upper primary fill, Knowlton South Circle.*

Feature	**upper primary fill**					**secondary fill**							
Sample	**15**	**16**	**17**	**18**	**19**	**20**	**21**	**22**	**23**	**24**	**25**	**26**	**27**
Context	**43**	**43**	**43**	**43**	**43**	**29**	**37**	**27**	**27**	**27**	**25**	**25**	**25**
Depth (cm)	**200–210**	**185–200**	**170–185**	**160–170**	**155–160**	**145–155**	**135–145**	**125–135**	**115–125**	**105–115**	**95–105**	**90-95**	**85-90**
Wt (g)	**2000**	**2000**	**2000**	**2000**	**2000**	**1500**	**1500**	**1500**	**1500**	**1500**	**1500**	**1500**	**1500**
Pomatias elegans (Müller)	+	-	+	-	+	26	90	140	97	100	20	57	72
Carychium tridentatum (Risso)	-	-	-	-	4	787	900	1272	1118	829	167	114	39
Cochlicopa lubrica (Müller)	-	-	-	-	-	14	9	11	4	16	4	15	30
Cochlicopa lubricella (Porro)	-	-	-	-	-	3	3	4	-	2	-	3	-
Cochlicopa spp.	+	-	-	-	-	32	26	51	37	50	6	35	63
Pyramidula rupestris (Draparnaud)	-	-	-	-	-	-	-	-	-	-	-	-	1
Vertigo pygmaea (Draparnaud)	-	1	-	-	1	2	-	1	1	2	1	19	37
Vertigo spp.	1	-	-	-	-	-	3	1	-	1	-	-	-
Abida secale (Draparnaud)	-	-	-	-	-	-	1	-	-	-	-	-	-
Pupilla muscorum (Linnaeus)	-	-	-	-	-	14	12	18	10	9	6	82	131
Lauria cylindracea (da Costa)	-	-	-	-	-	2	-	-	-	-	-	-	-
Vallonia costata (Müller)	-	-	2	-	1	40	14	16	6	9	15	109	149
Vallonia excentrica Sterki	-	-	-	1	-	3	2	6	2	3	6	119	201
Vallonia spp.	1	1	-	-	-	-	-	-	-	-	-	3	-
Acanthinula aculeata (Müller)	-	-	-	-	1	57	37	43	35	31	11	6	2
Ena montana (Draparnaud)	-	-	-	-	-	-	-	3	-	-	-	-	-
Ena obscura (Müller)	-	-	-	-	-	2	6	6	2	7	2	+	2
Punctum pygmaeum (Draparnaud)	-	-	-	-	-	4	22	7	15	11	6	2	3
Discus rotundatus (Müller)	-	1	3	1	1	182	273	317	139	160	33	35	24
Vitrina pellucida (Müller)	-	-	-	-	1	2	-	1	6	4	-	-	1
Vitrea crystallina (Müller)	-	-	1	1	6	83	64	75	37	79	9	5	-
Vitrea contracta (Westerlund)	-	-	-	-	3	78	110	279	191	166	31	23	9
Nesovitrea hammonis (Ström)	-	-	-	-	-	18	4	5	8	2	-	1	4
Aegopinella pura (Alder)	-	-	-	-	-	81	154	258	120	130	42	27	11
Aegopinella nitidula (Draparnaud)	-	-	-	-	-	75	142	211	126	184	39	33	14
Oxychilus allarias (Müller)	-	-	-	-	-	-	-	-	-	-	-	-	-
Oxychilus cellarius (Müller)	1	-	-	-	-	12	15	22	19	14	4	2	2
Limacidae	-	-	-	-	-	25	36	19	29	31	7	15	33
Euconulus fulvus (Müller)	-	-	-	-	-	2	5	3	3	2	-	-	-
Cecilioides acicula (Müller)	-	-	-	2	-	6	1	2	2	2	5	7	12
Cochlodina laminata (Montagu)	-	-	-	-	1	4	9	-	4	3	1	1	2
Macrogastra rolphii (Turton)	-	-	-	-	-	2	-	3	-	-	-	-	-
Clausilia bidentata (Ström)	2	-	-	-	-	8	11	11	5	7	5	3	7
Balea perversa (Linnaeus)	-	-	-	-	-	-	-	1	1	1	-	-	-
Candidula intersecta (Poiret)	-	-	-	-	-	-	-	-	-	-	-	-	-
Candidula gigaxii (L. Peiffer)	-	-	-	-	-	-	-	-	-	-	-	-	-
Cernuella virgata (Da Costa)	-	-	-	-	-	-	-	-	-	-	-	-	-
Helicella itala (Linnaeus)	2	2	-	-	1	13	7	3	2	3	4	20	39
Hellicellids	-	-	-	-	-	-	-	-	-	-	-	-	-
Trichia hispida (Linnaeus)	-	-	1	3	3	101	66	101	49	38	15	52	110
Arianta arbustorum (Linnaeus)	-	-	-	-	-	+	-	-	-	-	-	-	-
Helicigona lapicida (Linnaeus)	-	-	-	-	-	5	7	5	1	1	+	1	1
Cepaea nemoralis (Linnaeus)	-	-	-	-	-	-	2	1	-	-	-	1	-
Cepaea hortensis (Müller)	-	-	-	-	-	3	-	1	1	-	-	-	1
Cepaea spp.	+	-	+	-	-	5	14	10	9	7	1	9	5
Cepaea/Arianta spp.	-	-	-	-	1	-	-	-	-	-	-	-	-
Helix aspersa (Müller)	-	-	-	-	-	-	-	-	-	-	-	-	-
Valvata cristata (Müller)	-	-	-	-	-	-	-	-	-	2	-	-	-
Bitynia tentaculata (Linnaeus)	-	-	-	-	-				1	3	-	-	-
Taxa	5	4	4	4	12	29	27	20	29	30	22	24	25
TOTAL	**7**	**5**	**7**	**6**	**24**	**1685**	**2044**	**2905**	**2078**	**1902**	**435**	**792**	**993**
Shannon Index	1.56	1.33	1.23	1.24	2.22	2.10	2.10	2.03	1.83	2.09	2.28	2.60	2.44
Brillouin Index	1.02	0.82	0.86	0.80	1.73	2.07	2.07	2.01	1.81	2.06	2.18	2.54	2.39
$\Delta 2$	0.78	0.72	0.69	0.67	0.86	0.75	0.77	0.77	0.69	0.78	0.82	0.91	0.89
$\Delta 4$	9.50	9.00	4.25	4.00	9.22	3.08	3.35	3.33	2.19	3.47	4.53	9.73	7.85
Magnetic susceptibility	12	11	13	-	-	37	42	42	49	30	30	40	64

Table 3.22. *Mollusca in the henge ditch, stabilisation horizon, upper secondary and tertiary fills, Knowlton South Circle.*

Feature	**Stabilization horizon**					**Lower tertiary fill**						**Ploughwash**				
Sample	**28**	**29**	**30**	**31**	**32**	**33**	**34**	**35**	**36**	**37**	**38**	**44**	**45**	**46**	**47**	**48**
Context	**20**	**20**	**20**	**20**	**20**	**17**	**17**	**17**	**15**	**15**	**15**	**14**	**3**	**3**	**3**	**3**
Depth (cm)	**80–85**	**75–80**	**70–75**	**65–70**	**60–65**	**50–60**	**40–50**	**30–40**	**20–30**	**10–20**	**0–20**	**50–60**	**40–50**	**30–40**	**20–30**	**10–20**
Wt (g)	**1500**	**1500**	**1500**	**1500**	**1500**	**1500**	**1500**	**1500**	**1500**	**1500**	**1500**	**1000**	**1000**	**1000**	**1000**	**1000**
MOLLUSCA																
Pomatias elegans (Müller)	29	14	5	11	5	7	7	5	2	2	+	4	2	+	1	3
Carychium tridentatum (Risso)	38	16	13	9	5	1	-	1	-	-	6	-	-	-	-	-
Cochlicopa lubrica (Müller)	37	12	13	8	11	8	8	11	1	1	1	-	-	-	-	-
Cochlicopa lubricella (Porro)	2	1	+	-	-	-	-	-	-	-	-	-	-	-	-	-
Cochlicopa spp.	86	45	36	35	31	21	26	14	10	7	4	2	1	1	-	-
Pyramidula rupestris (Draparnaud)	-	-	-	-	-	-	-	-	-	-	-	-	-	-	-	-
Vertigo pygmaea (Draparnaud)	86	57	59	43	38	25	29	21	10	13	7	2	1	1	-	-
Vertigo spp.	-	-	-	-	-	-	-	-	-	-	-	-	-	-	-	-
Abida secale (Draparnaud)	-	-	-	-	-	-	-	-	-	-	-	-	-	-	-	-
Pupilla muscorum (Linnaeus)	193	148	123	169	174	109	107	110	71	51	54	13	4	3	3	4
Lauria cylindracea (da Costa)	-	-	-	-	-	-	-	-	-	-	-	-	-	-	-	-
Vallonia costata (Müller)	138	83	74	61	42	33	24	15	8	16	3	4	3	2	1	1
Vallonia excentrica Sterki	291	238	176	221	201	177	201	222	171	160	95	29	47	32	12	17
Vallonia spp.	-	-	-	-	-	-	-	-	-	-	-	-	-	-	-	-
Acanthinula aculeata (Müller)	3	3	3	1	3	-	-	1	-	-	-	-	-	-	-	-
Ena montana (Draparnaud)	-	-	-	-	-	-	-	-	-	-	-	-	-	-	-	-
Ena obscura (Müller)	3	1	1	-	-	-	1	5	-	-	-	-	-	-	-	-
Punctum pygmaeum (Draparnaud)	9	9	22	16	19	16	9	-	1	3	1	1	-	-	-	-
Discus rotundatus (Müller)	13	4	3	3	+	2	+	+	-	-	1	-	-	1	-	-
Vitrina pellucida (Müller)	-	-	-	-	-	-	-	-	-	-	-	-	-	-	-	-
Vitrea crystallina (Müller)	2	-	-	-	-	-	-	-	-	-	-	-	-	-	-	-
Vitrea contracta (Westerlund)	1	4	1	1	2	1	-	-	-	-	1	-	-	-	1	-
Nesovitrea hammonis (Ström)	3	-	1	-	-	-	-	-	-	-	-	-	-	-	-	-
Aegopinella pura (Alder)	10	-	4	5	2	-	-	-	-	-	1	-	-	-	-	-
Aegopinella nitidula (Draparnaud)	7	3	4	6	-	2	1	2	-	2	1	-	-	-	-	-
Oxychilus alliarius (Miller)	-	-	-	-	-	-	-	-	-	-	-	-	-	-	-	-
Oxychilus cellarius (Müller)	-	-	-	-	1	-	-	-	-	-	-	-	-	-	-	-
Limacidae	39	45	33	33	16	16	18	20	11	12	11	6	12	10	2	4
Euconulus fulvus (Müller)	-	-	-	-	-	-	-	-	-	-	-	-	-	-	-	-
Cecilioides acicula (Müller)	13	41	51	11	17	30	47	18	29	33	20	52	70	69.	77	52
Cochlodina laminata (Montagu)	1	-	-	-	-	-	-	-	-	-	-	-	-	-	-	-
Macrogastra rolphii (Turton)	-	-	-	-	-	-	-	-	-	-	-	-	-	-	-	-
Clausilia bidentata (Ström)	5	3	3	3	6	1	2	2	2	1	+	1	-	-	-	-
Balea perversa (Linnaeus)	-	-	-	-	-	-	-	-	-	-	-	-	-	-	-	-
Candidula intersecta (Poiret)	-	-	-	-	-	-	-	-	-	-	1	34	32	24	25	26
Candidula gigaxii (L. Peiffer)	-	-	-	-	-	-	-	-	-	-	-	-	1	-	-	1
Cernuella virgata (Da Costa)	-	-	-	-	-	-	-	-	1	6	22	8	11	5	10	5
Helicella itala (Linnaeus)	64	38	33	35	33	25	29	31	10	11	7	5	7	3	1	4
Hellicellids	-	-	-	-	-	-	-	-	-	-	-	-	-	-	-	-
Trichia hispida (Linnaeus)	146	121	119	134	112	102	150	176	54	60	105	39	28	16	18	9
Arianta arbustorum (Linnaeus)	-	-	-	-	-	-	-	-	-	-	-	-	-	-	-	-
Helicigona lapicida (Linnaeus)	+	-	+	-	+	-	+	-	-	-	-	-	-	-	-	-
Cepaea nemoralis (Linnaeus)	-	-	-	-	-	-	-	-	-	-	-	-	-	-	-	-
Cepaea hortensis (Müller)	-	-	-	-	-	-	-	-	-	-	-	-	-	-	-	-
Cepaea spp.	3	-	1	1	2	1	1	-	-	-	1	1	-	-	-	-
Cepaea/Arianta spp.	-	1	-	-	-	-	-	1	2	+	-	-	-	-	1	-
Helix aspersa (Müller)	-	-	-	-	-	-	+	+	+	+	-	1	1	+	+	+
Valvata cristata (Müller)	-	-	-	-	-	-	-	-	-	-	-	-	-	-	-	-
Bitynia tentaculata (Linnaeus)	-	-	-	-	-	-	-	-	-	-	-	-	-	-	-	-
Taxa	23	19	20	18	17	16	14	15	13	13	17	15	13	11	11	10
TOTAL	**1209**	**846**	**727**	**795**	**703**	**547**	**613**	**637**	**354**	**345**	**322**	**150**	**150**	**98**	**75**	**74**
Shannon Index	2.31	2.16	2.21	2.09	2.12	1.96	1.86	1.75	1.58	1.71	1.80	2.06	1.91	1.82	1.78	1.86
Brillouin Index	2.27	2.12	2.16	2.05	1.96	1.90	1.81	1.71	1.52	1.64	1.72	1.91	1.78	1.67	1.60	1.67
Δ2	0.87	0.85	0.86	0.83	0.82	0.81	0.79	0.77	0.70	0.73	0.77	0.83	0.81	0.79	0.78	0.79
Δ4	6.64	5.53	6.15	5.02	4.53	4.26	3.83	3.29	2.34	2.68	3.41	5.03	4.30	3.98	3.89	4.12
Magnetic susceptibility	70	74	69	65	59	65	56	59	62	64	62	56	48	51	48	55

Table 3.23. *Mollusca in the north and south gullies, Knowlton South Circle.*

Feature	South gully				North gully				
Sample	6	7	8	9	1	2	3	4	5
Context	6.1	32	32	2	10	24	24	24	12
Depth (cm)	spot	50–62	36–50	24–36	spot	50–60	40–50	27–40	20–27
Wt (g)	1500	1500	1500	1500	1500	1500	1500	1500	1500
MOLLUSCA									
Pomatias elegans (Müller)	+	19	17	11	6	10	12	7	4
Carychium tridentatum (Risso)	1	-	2	6	2	1	3	-	-
Cochlicopa lubrica (Müller)	-	2	9	15	9	-	6	6	6
Cochlicopa lubricella (Porro)	-	2	2	2	2	6	1	1	3
Cochlicopa spp.	5	16	6	19	25	6	9	13	16
Vertigo pygmaea (Draparnaud)	-	4	7	15	10	5	9	10	17
Vertigo spp.	-	1	2	7	3	5	-	4	2
Pupilla muscorum (Linnaeus)	40	51	55	79	54	29	18	39	69
Vallonia costata (Müller)	3	10	41	129	11	7	4	8	30
Vallonia excentrica Sterki	9	25	37	108	82	59	47	24	57
Acanthinula aculeata (Müller)	-	1	1	1	-	-	-	-	-
Punctum pygmaeum (Draparnaud)	1	2	5	11	10	3	4	6	6
Discus rotundatus (Müller)	-	2	1	-	+	2	+	+	-
Vitrea contracta (Westerlund)	-	1	1	-	-	1	-	-	-
Nesovitrea hammonis (Ström)	-	-	1	3	2	-	1	1	1
Aegopinella pura (Alder)	-	1	1	-	-	-	-	-	-
Aegopinella nitidula (Draparnaud)	1	3	1	1	-	-	-	-	4
Oxychilus cellarius (Müller)	1	2	3		1	-	-	-	-
Limacidae	1	11	7	14	1	2	2	5	7
Cecilioides acicula (Müller)	4	16	43	224	7	8	1	45	149
Cochlodina laminata (Montagu)	-	1	-	-	-	1	-	-	-
Clausilia bidentata (Ström)	+	4	2	1	1	-	-	-	3
Candidula intersecta (Poiret)	-	-	-	7	-	-	-	1	-
Cernuella virgata (Da Costa)	-	-	5	3	-	-	-	-	3
Helicella itala (Linnaeus)	11	32	33	28	17	8	15	17	46
Trichia hispida (Linnaeus)	1	20	32	118	20	28	17	49	116
Helicigona lapicida (Linnaeus)	-	+	+	-	1	+	-	-	-
Cepaea spp.	-	+	2	3	+	+	1	+	-
Helix aspersa (Müller)	-	-	1	2	-	-	-	-	1
Mytilus edulis	-	-	-	-	-	+	-	+	-
Taxa	11	19	23	20	16	14	14	13	16
TOTAL	**74**	**210**	**274**	**583**	**257**	**173**	**149**	**191**	**391**
Shannon Index	1.53	2.37	2.41	2.18	2.08	2.01	2.15	2.14	2.09
Brillouin Index	1.36	2.23	2.28	2.11	1.97	1.88	2.00	2.02	2.02
Δ2	0.66	0.88	0.88	0.85	0.82	0.81	0.84	0.85	0.83
Δ4	2.06	7.30	7.68	5.69	4.77	4.50	5.54	5.89	5.11

shrubby species such as hawthorn invading the adjacent, ungrazed bank.

In addition to these assemblages the secondary fills contain several examples of freshwater taxa; *Valvata cristata* and *Bithynia tentaculata*. In total six specimens of these two taxa were recovered from two samples and we cannot invoke the hypothesis that they were all brought here from riverine or river margin environments by a bird (cf. Thomas 1977). Their occurrence as species from large bodies of slow moving water (*B. tentaculata*) and of slowly flowing water with rich vegetation and muddy substrates (*V. cristata*) are more likely to indicate the human exploitation of resources in local River Allen nearly 1 km to the west, perhaps for water, reeds for bedding or thatching, or clay/mud, as suggested for similar finds at Balksbury, Hampshire (Allen 1995b, 92–100). What is surprising about these secondary fill assemblages is the strong parallels with the secondary (Beaker) fills from the Coneybury Henge, Wiltshire (Bell & Jones 1990), even down to the presence of an aquatic taxa.

Major, important ecological changes can be seen within the secondary fills where the nature of the assemblages changes rapidly and dramatically. Nearly all of the major shade-loving components abundant in the previous six samples (contexts 39, 37 & 27) are reduced to just a presence, and rises in nearly all open country taxa occur in the top of the secondary fills, washed from the bank (25). Assemblages are dominated by *Vallonia excentrica, Pupilla muscorum* and *Trichia hispida*, with *Helicella itala* being present in significant numbers for the first time and Shannon and Brillouin indices increase (Table A3.12) indicating some maturity of this community. During this change there is a peak in *Pomatias elegans*, a species common in disturbed and broken ground and typical of clearance episodes. The complete change in the molluscan fauna indicates changes both within the ditch and beyond; the lack of a number of the shade-loving taxa previously so significant indicates quite widespread changes in the local environment that can only be accounted for by some extensive human activity. This indicates the removal of the shady environment and the establishment of stable open county grassland.

At the top of the secondary fill a stone-free stabilization horizon was recorded (20), though shell numbers, although high at the base, were not markedly increased (Table 3.22). The assemblages, however, are not significantly different from those established at the top of the previous deposits and suggest a mature, and establishing grass in the ditch, and probably beyond. This indicates active use of the local area, and of the monument; there is no indication of neglect, abandonment or vegetation regeneration and this suggests renewed use of the monument.

Unfortunately the lack of artefacts from the excavations does not allow us to date either the rapid occurrence of the shady environments or its demise and the establishment of stable open grassland conditions. The removal of shady conditions and founding of grassland is, however, coincident with the construction of a fence or revetment within the ditch, and thus the major ecological change in the assemblages can be seen to be a direct result of renewed but different human activity at the henge.

Deposits sealing the stabilization horizon are typically colluvial in nature and the molluscan assemblages reflect this. They continue to be dominated by open-country species, especially *V. excentrica* and *T. hispida,* and show a gradual decrease in shell numbers (Table 3.22). These are assemblages typical of hillwash and lynchet deposits (cf. Bell 1983; Allen 1995d), but the higher proportions of *Pupilla muscorum* may indicate some grassland. Subtle and significant ecological changes occur within the lower tertiary fill (contexts 17 & 15). Although *T. hispida* rises steadily through context 17, a dramatic reduction in relative and absolute abundance concomitant with a large rise in *V. excentrica* occurs precisely at the rather indistinct boundary between these two layers. What this means in terms of land-use and local environment is difficult to define, except that it is interesting to note that it does occur when the presence of *Helix aspersa* fragments occur in all samples from context 17. As this is an introduced species (Kerney 1966) this might indicate changing land-use practices in the Roman or post-Roman periods.

The assemblages from the upper tertiary fills (contexts 14 & 3) show a reduction in *P. muscorum* and increase in *Helicella itala* and Introduced Helicellids (Table 3.22). This change is accompanied by a reduction in all species diversities (Table A3.13) and probably reflects environments typical of drier open calcareous and intensive arable conditions, while the strong presence of Introduced Helicellids (*Candidula intersecta* & *Cernuella virgata*) indicates a medieval or post-medieval date for these deposits.

Inner and outer gullies

The fills of both the inner and outer shallow gullies were sampled (Table 3.23). The assemblages from both are similar and show little major changes through the sequences. They are dominated by open-country species throughout, especially *Pupilla muscorum, Vallonia* species and *Helicella itala* with *Trichia hispida.* These are typical of open dry grassland conditions, and the lack of shade-loving elements in these fills indicates that such conditions were already established when these gullies were cut.

Discussion of Knowlton

The well-preserved molluscan sequence provides the opportunity to examine the pre-/during/post-monument land-use and activity. It also provides an opportunity for a closer phasing between the main henge infill sequence and the gullies defining the bank.

What was the pre-monument environment at Knowlton; is there evidence for previous activity?

Evidence from both the old land surface and from the lowest primary fills available for sampling, indicates an open landscape around the henge. It is not surprising that the henge was built in a clearing; the composition of the species and diversity indices indicate that woodland clearance was not performed for the construction of the main henge monuments. Clearance on a wider scale had preceded this construction and must have been associated with some earlier episode of activity.

What was the environment and land use at the time of construction?

Immediately prior to construction the area of the Knowlton South Circle, at least, was under cleared open grassland. Grassland had been in existence long enough for the demise of any former woodland molluscan assemblage and the colonization of a wholly open-country one.

How quickly did vegetation regeneration in the ditch occur?, and how far away was the woodland?
It is not possible to be precise about the extent of clearance and the distance to the woodland, but some indication is provided by the composition and speed of colonization of the ditch and local environment by both longer vegetation and woodland mollusc species. The primary main fills do not indicate any recolonization, but the finer and poorly sorted secondary fills immediately show established vegetation and shade-loving molluscan communities. By analogy with the Overton Down experiment (Bell *et al.* 1996) we can assume that the primary fills, although comprising nearly 50 per cent of the vertical depth of the ditch, represent rapid accumulation, probably in the order of 50 years rather than several centuries to a millennium. The appearance of a large number of taxa not present in previous deposits tends to suggest proximity of suitable habitats. Further the diverse range of taxa (five Clausiliidea, *Ena Montana,* etc.) suggests that woodland cannot have been very far away. Topographically we can postulate that woodland existed on the higher ground to the east, but beyond that we can only indicate that some ancient species-rich deciduous woodland was probably within hundreds of metres.

Does vegetation regeneration in the secondary fill represent monument abandonment?
Vegetation regeneration typically occurs in ditches when sedimentation rates slow down in the secondary fills, and comprise finer deposits (Evans 1972, 323–5; Limbrey 1975, 294–8). Typical specialized molluscan fauna often inhabit the juvenile vegetation sere, comprising open country species and catholic species typical of longer grassy vegetation. These have sometimes been taken, erroneously, to indicate re-vegetation and abandonment of the monument as a whole rather than stabilization and vegetation colonization of the ditch. This fauna, however, is not present at Knowlton in the secondary fills. Instead species typical of leaf litter, logs and tree trunks are present indicating the existence of these habitats in close proximity, if not in the ditch itself. The assemblage could just indicate a combination of leaf litter accumulating in the 1.5 m deep ditch amongst the tall grassy vegetation and small shrubs, and the revegetation of the henge bank. If the monument was still in use under these conditions, we would expect a significant component of open-country species relating to the open grassland interior. There is little evidence of the presence of this habitat during the main secondary fills from which we can infer that this indicates a wider landscape phenomenon as indicated by the low (delta) diversity indices that indicate a mature or wide habitat variation. Thus we can infer some phase of monument abandonment, lack of maintenance or change of its function within society. In this respect the sequence closely parallels the land-use history of the slightly earlier, Late Neolithic monument of Coneybury Henge, Wiltshire (*c.* 3050–3500 cal. BC), within the Stonehenge landscape (Bell & Jones 1990).

What was the reason for the major ecological change at the top of the secondary fills?
Without any dating evidence we have no idea how long the re-vegetated state of the henge lasted. However, before the completion of the secondary infills there is evidence of some chalk bank wash (context 25) and of rapid ecological change to open-country grassland with some bare earth and disturbed conditions which can only have been facilitated by vegetation clearance on site. This seems to have occurred coincident with the insertion of posts for a palisade or revetment within the ditch. This open dry grassland seems to have been maintained as a grassland soil formed within the ditch and probably extended across much of the site. This indicates a phase of re-use of the henge, probably in the later Neolithic to earlier Bronze Age.

How do the assemblages from the inner and outer gullies relate to the henge ditch sequence?
The assemblages from both gullies are very similar as might be expected from contemporaneous features, though it does confirm a uniformity of land use either side of the bank. These assemblages indicate that the gullies were dug, and existed in, dry grassland conditions. With a complete lack of direct dating evidence from these features we can do no more than attempt to compare the environment and land use with that from the main henge ditch. Comparison with the henge ditch indicates two places in the prehistoric sequence where open grassland existed. The first is that associated with the pre-henge and primary henge ditch fills, and the second is that associated with the stabilization horizon (context 20) at the top of the secondary fill. In terms of species composition (moderate representation of *P. muscorum* and *V. excentrica*), much better analogues are seen in the assemblages associated with the second clearance phase (context 25) and the stabilization in the ditch (context 20). Ironically, this is precisely where construction of a palisade or revetment is seen within the ditch, and the gullies may represent similar activity to redefine the bank.

The wider 'Knowlton' landscape

The molluscan analysis clearly provides detailed information about the nature of the local land use, and phases of abandonment and re-use associated with the Knowlton South Circle. It has enabled some suggestion of phasing within otherwise poorly dated archaeological events. This long sequence has the potential to provide information on the land-use of the wider landscape (Allen 1998; 2000a), and may be compared with sequences at other Late Neolithic monuments in the Allen Valley such as the Wyke Down henges and Fir Tree Field Grooved Ware pits (Allen 2000a) and the long sequence from the Fir Tree Field Shaft (Allen & Green 1998; Green & Allen 1997). The lack of dating for the phases of the Knowlton monument makes this difficult, especially as this cannot even provide a chronological framework for the environmental events which we wish to compare within the Allen valley landscape (see Allen 1997a, 139; 2000b). Nevertheless some comparisons may be made with respect to the nature of the changing wider landscape.

The Knowlton henge complex is situated on the eastern edge of the Cranborne Chase landscape as studied by Barrett *et al.* (1991a), and detailed and comprehensive molluscan studies by Entwistle & Bowden (1991) focused largely on the Down Farm landscape *c.* 5 km to the northwest. Indeed this same area is also the focus of resent research by the current author (Allen 1998; 2000a; in prep.) and others (Gale in prep.). Nevertheless some distinct trends and contrasts can be discerned. Although deciduous woodland existed throughout the Allen Valley in early post-glacial and earlier Neolithic periods, there are indications of some human clearances as early as the later Mesolithic (4200–3800 cal. BC: Allen & Green 1998, 34), and certainly locally around earlier Neolithic mortuary monuments (e.g. Handley Mortuary Enclosure, Thickthorn Down etc. *c.* 3800–3400 cal. BC). Much more extensive woodland clearance existed for the construction and use of the Dorset Cursus (3500–3200 cal. BC) and Monkton-up-Wimborne ceremonial site. In the later Neolithic, the evidence from both the Wyke Down henges, and Grooved Ware pits at Wyke Down and Fir Tree Field indicates an open landscape. We can start to infer that quite extensive tracts of the Allen Valley were cleared as well as swathes crossing it (Dorset Cursus) by the time Knowlton South Circle was constructed at 2900–2300 cal. BC (Allen 2000a, 46–7). This concurs with the detailed analysis presented above, but contrasts markedly with the picture from the Hambledon Hill landscape. At this hilltop location, only the hilltop itself seems to have been cleared and the wider landscape local to the hill was only cleared in later prehistory (Bell *et al.* forthcoming).

The most significant evidence, however, is the 'abandonment' phase at Knowlton, occurring probably in the later Neolithic and possibly earlier Bronze Age. No significant effect such as this is detected in either of the Wyke Down henges, or any of the Grooved Ware pits, though minor local and subtle vegetation regeneration phases are detected (Entwistle & Bowden 1991; Allen 1999b). At present we cannot consider this to be a landscape phenomenon; i.e. a Late Neolithic abandonment phase, and this was clearly closely looked for by Entwistle & Bowden (1991). Nevertheless, this does leave us with clear evidence of vegetation colonization, which we assume to represent neglect, at the Knowlton South Circle. Its re-use might relate to the Beaker phase re-use of other sites in the Allen Valley and Cranborne Chase landscape, but the determination of this awaits further critical radiocarbon dating of environmental and monumental events.

Discussion of molluscan data

Palaeo-molluscan interpretation and the wider palaeoenvironmental discussion

The data presented here from a variety of sites allow exploration of a range of themes varying from challenging the presence of ancient woodland, and asking whether we have in fact accurately detected it in many other Neolithic monuments in Wessex. These concepts range from those of palaeo-molluscan interpretation to the development of new landscape models in conjunction with the soil and pollen data. This outline discussion touches on a number of these issues.

Pollen evidence suggests that open environments of bracken and fern may have existed. Although palaeo-molluscan analysis can suggest 'open' landscapes, can we define environments of bracken and ferns by palaeo-molluscan analysis? Unfortunately to date, there has been insufficient ecological work done on the snail faunas in these kinds of vegetational habitats to even begin to be precise about this kind of environment from a molluscan perspective.

Can we really recognize 'ancient' woodland in the molluscan assemblages, and how often have we found it?

The analyses from these sites and from another eleven sites in this landscape (Tables 3.6 & 3.7; Allen unpubl.) cannot define an ancient Atlantic or Sub-Boreal woodland comprising rich molluscan fauna with anthropophobes and species rare in the present British fauna (Evans 1972). It is assumed that such assemblages would represent deciduous woodland with rich sub-canopy flora, high leaf litter and moder humus over brown earth profiles.

On these assumptions we can indicate that we have yet to find this type of woodland anywhere throughout the Cranborne Chase landscape, an area more intensively studied than any other chalkland in northwest Europe. The snail data base alone comprises 27 sample locations, over 450 samples and in excess of 180,000 identified snails (Tables 3.6 & 3.7) in one of the smallest archaeological study landscapes (Allen 2000b, table 2.2). Thus the resolution of our interpretation should be high (Allen 1997a; 2002a), and the fact that this environment has not been detected questions some fundamental assumptions about whether it ever existed (see below).

Can we use our data purely chronologically?

Creating a chronological model has often denied spatial variation. With more data sets, and the higher resolution of interpretation now required within archaeology, it is inevitable that generalisms that were ground breaking in the 1970s (Evans 1972; Evans & Jones 1979) now seem unhelpful.

Purely chronological models tend not to account for spatial variation. Yet these data allow us to detect contemporaneous different environments within even the relatively small landscape we are studying. There are a several distinct topographic, if not catenary components to the study area, and in general terms these can be divided into the upper slopes of steep chalklands of Gussage Cow Down to the south; valley bottom and lower slopes of the broad Allen Valley from Gussage Cow Down to Handley and Monkton-up-Wimborne embracing Down Farm and Wyke Down; and thirdly the lower and upper slopes of the rolling chalk uplands to north including Handley and Monkton-up-Wimborne. Any model must first be developed on a chronological and general scale, and then interpretation applied to a much more specific and local scale to understand individual site and catenary histories. The multiple and spatial location of the snail analyses make it clear that there are real spatial patterns of the vegetation and habitats within the landscape mosaic (i.e. open grassland and wooded patches), and that change and variation are not purely temporal. This significantly increased level of land-snail analysis enables us to look at spatial resolution and mapping of past environments *within* landscapes rather than considering single sites, or sample points, to be wholly representative of their landscapes.

Is Cranborne Chase a 'classic' example of chalk downland in southern England?

If we are to suggest that the accepted vegetation history for southern England (Evans & Jones 1979, 209; cf. Allen 2000a, table 2) did not apply to Cranborne Chase, then is this area typical of chalk downland in southern England? If Cranborne was atypical, this would invalidate the application of any model to other areas of southern England. If however Cranborne was typical it would suggest revision of the model was required (French *et al.* 2003).

Challenging fundamental assumptions of landscape and soil development

The accepted model of post-glacial vegetation history follows Tansley's ecological developmental progression (1939) through vegetation seres to climax woodland. But the nature of the early post-glacial and Neolithic chalk landscape is one currently under great review and speculation. Firstly the perceived history of post-glacial forestation has been challenged on ecological grounds (e.g. Bush 1988; 1989; Bush & Flenley 1987; Thomas 1989), and via interpretation of archaeological landscapes from molluscan and other data (Allen 1997b, 278; 2002a; French *et al.* 2003). The lack of any detailed chronological control within which to relate the mollusc sequences seriously hinders discussion of the Neolithic landscape development versus patchiness in the mosaic of contemporaneous Neolithic environments (cf. Allen 2002b; 2005a). Although it is quite clear that large areas of the chalk downlands of southern England were forested by the Neolithic period, it has commonly been assumed to have been a continuous blanket of woodland. What is evident is that this woodland development was retarded in some locations and it seems increasingly likely that early post-glacial woodland did not exist across much of the Dorchester chalkland environment (Allen 1997b, 279), with isolated exceptions. Maiden Castle is one such exception, probably existing in the Early Neolithic period as a densely covered wooded hill, superficially akin to Danebury today.

The reasons for this apparent lack of woodland evidence are not clear, but it has also recently been detected in Cranborne Chase (Allen 1998; 2002b; French *et al.* 2003), and has been postulated for the Stonehenge environs (Allen 1997a). We may speculate that in part this may be natural ecological variation and patchiness in a large and extensive forest, with natural glades almost the size of parishes. These would invite vegetation diversity including fruiting trees and bushes, with the nuts and berries attracting wild animals (deer and cattle). Obviously these fruits provide food for human populations (Jones 2000; Moffett *et al.* 1989; Robinson 2000). Such areas provide ideal opportunities for human populations to exploit this ecological diversity (Wiens 1976), and in which to locate and hunt deer and other animals feeding on the fringes of these open spaces.

Evidently woodland development was retarded in some locations. It seems increasingly likely that natural clearings or glades comprising grassland, scrub and some trees formed a part of the natural ecological variation and patchiness in a large and extensive forest. Such a location would also invite Mesolithic communities to hunt and gather, and possibly even maintain these glades artificially and perpetuate them. It is perhaps not too far-fetched to suggest that these areas attracted some of the first large human populations and that in consequence they survive today as zones of more concentrated monuments and sites. It is no coincidence that three of the apparently most densely populated parts of the prehistoric downland, all containing major earlier Neolithic monuments, all have significant evidence of Mesolithic activity. And it is probably also no coincidence that these areas (Dorchester, Cranborne and Stonehenge) all seem to have been open, not closed wooded downland. This concept has been explored further for the southern chalkland (Allen 2000b; Allen & Scaife in press), and is discussed in more detailed elsewhere (see below).

Ironically, these ideas of open park woodland propagated initially by Bush (1988) and Bush & Flenley (1986) for the Yorkshire Wolds were independently postulated for areas of the Dorset Downs from 1988 to 1997 (see Allen 1997b, 278). Reassuringly these ideas conform directly to ecological models developed by Vera (2000), that landscape mosaics existed in which 'half-open' and park-like landscape existed in lowland areas, and were maintained by wild herbivores. We postulate that precisely these areas were exploited by Mesolithic populations because of their openness, wild fruit and berries, and the concentration of herbivores (cf. Allen 2002a).

It is perhaps not too far fetched to suggest that it is no coincidence that these areas (Dorchester, Cranborne and Stonehenge) attracted some the first large human populations, and as a consequence they became areas of denser and more concentrated Neolithic monuments and sites.

How robust is the accepted model for southern England?
This original model for the southern English chalkland was presented by the seminal work of Evans (1967; 1972). It essentially confirmed ecological ideas presented by Tansley (1939), and is corroborated by other data presented by Godwin (1975), Dimbleby (1985) and Scaife (1988; 1995). The presence of shade-loving molluscan fauna in numerous Neolithic monuments has also been taken to confirm the presence of the expected Neolithic post-glacial woodland. But how many of these really indicate dense forest, and how many represent local vegetation regeneration specific to individual monuments and how many represent park woodland or 'secondary' woodland?

In the Dorchester area for instance, extensive analysis has been conducted of Neolithic monuments (Allen 1997b; 2000b; Evans & Rouse 1991). All Neolithic contexts lack diagnostic ancient woodland species, such as *Ena montana, Vertigo pusilla, Columella* and *Acuicla fusca,* and the low taxanomic ranges tend to suggest that they do not represent an ancient mature woodland like that defined at Maiden Castle (Evans & Rouse 1991). The site mollusc zones (SMA 'a' & 'b') at Maiden Castle pre-date 3090–2910 cal. BC, and probably date to about 3900–3500 cal. BC. All the assemblages elsewhere in the Dorchester landscape seem to indicate a more open scrubby woodland in view of the presence of open-country species including *Helicella itala, Pupilla muscourm, Vertigo* cf. *pygmaea* and *Valliona* spp. Mixed assemblages from Neolithic pits at Thomas Hardeye School (Allen 2004) seem to be similar to those from the basal fill of pit 00221 at Flagstones dated to 3970–3540 cal. BC and interpreted as representing open and secondary woodland (Allen 1997a, 167). Thus in the Dorchester environs the only closed ancient woodland seems to have been on Maiden Castle itself; a heavily wooded hill top within a wider a park woodland landscape (cf. Vera 2000).

A number of the classic sequences from ditches of long barrows such as Cherhill (Evans & Smith 1983), South Street, Horslip, and Beckhampton (Ashbee *et al.* 1979; Evans 1972), Easton Down (Whittle *et al.* 1993), Corton (Allen & Gardiner 2004) indicate the presence of woodland prior to the barrow, but we have been not been very astute at always defining the precise nature of that wood. Although the causewayed enclosures at Windmill Hill (Fishpool 1999), Whitesheet Hill (Allen, in Rawlings *et al.* 2004) and Offham (Thomas 1977), contain clear evidence of shady or even woodland environments (Allen 2005a) in the immediate vicinity of the ditch circuits, we can be relatively sure this is a dense deciduous wood. Elsewhere however, e.g. Kingsborough, Kent (Allen & Lievers in press), the causewayed enclosures were established in a pre-existing open downland. Indeed when we examine the presence of some Mesolithic and Neolithic woodland in more detail, many have clear evidence of open woodland canopies and of open habitats (e.g. Stonehenge: Allen 1995b). Many could represent a park woodland and local ditch vegetation. Thus the concept of major forest across the chalk of southern England needs re-evaluating (Allen & Scaife 2007). The molluscan, soil and pollen data provide the opportunity to define the chalkland ecology and the nature of the palaeoecological assemblages associated with it. This then provides the basis of re-examination of other data sets and interpretations such as those mentioned above.

Back to the Allen valley

Taking these wider issues of chalkland ecology into account we can briefly re-examine some of the data from Cranborne Chase, and suggest some implications in this landscape. We note that Entwistle & Bowden interpret the molluscan palaeoecology of the two Dorset Cursus ditch sections on the premise and assumption of the presence of a post-glacial woodland maximum (cf. Entwistle & Bowden 1991, 23 & table 2; Evans & Jones 1979, table 26). They have to argue for very rapid re-colonization and suggest very limited woodland clearing despite the immense size of the monument (10 km long by 90 m wide). If this post-glacial woodland assumption was removed and had to be proven, then interpretation for both the Cursus as a single monument, and the local landscape becomes less contradictory with the archaeological evidence and postulated presence of open park woodland (cf. Vera 2000).

The level of presence of Mesolithic predecessors to this Neolithic landscape was significant. Although obviously not monumental, this area contains some of the densest distributions of Mesolithic artefacts on the chalk anywhere in the country (Gardiner 1988; Gardiner 1991; Arnold *et al.* 1988). We postulate that this high level of activity here, like the significant Mesolithic activity (timber poles/totems) at Stonehenge (Allen 1995b), was encouraged by the presence of open-park woodland habitats and the diverse flora (including edible fruits) and fauna (especially herbivores) they supported. Notably cattle, which before domestication were natural browsers rather than grazers, would have been ideally suited to this landscape, providing a very attractive ecology for human exploitation.

It is no coincidence that once this area contained significant established Mesolithic communities, ultimately too it became the centre and concentration of Neolithic monuments, as indeed many Neolithic monuments were constructed in places that had already acquired special significance (Bradley 1991a; 1993, 45) and which already had a long and established human history (see Allen & Gardiner 2002). Thus, environmental archaeological enquiry can provide the basis of clear interpretations of cultural and community action as postulated for Stonehenge (e.g. Allen 1977a) demonstrating that palaeo-environmental evidence and interpertation is crucial to archaeological interpretations of social, cultural and perceptual actions.

The soil micromorphological data

Helen Lewis

This section sets out the main soil types and features as characterized through soil micromorphological analysis, and then examines the local and regional implications of the soils data. These include buried soils under monuments, monument construction soils and ditch fills, and soil profiles representing typical modern profiles, as well as profiles from non-monumental sites. In general, most soils studied represented rendzina profiles, often simple and shallow soils with relatively few disturbance indicators. These types of soils are commonly found today over ploughed and grazed fields on chalk in the Allen Valley. The second most frequent type of soil profile seen was thicker and slightly more clay-rich rendzinas and rankers, as found today over gravel fans and coombe rock deposits. More rarely seen were argillic brown earths, found only at the Mesolithic Fir Tree Field site over clay-with-flints, and seen in one location under woodland over chalk, near a stream within the woodland ride of the St Giles's Park estate, and in the same estate at Deer Park Farm. The St Giles profile was revealed in an auger hole and could not be sampled for thin-section analysis. All archaeological soils and sediments sampled included small amounts of finely comminuted charcoal, which is to be expected given their location in or adjacent to archaeological monuments and settlement sites; as such charcoal is discussed here only where more frequently found, or found in an organized manner, despite its ubiquitous low level presence. For all tables the coarse:fine limit is 100 μm, PPL is plane polarized light, XPL is crossed polarized light, and all percentages represent visual estimates of area. Full descriptions of slides and contexts are presented in Appendix 4, with a simple summary in Table 3.24.

Gussage Cow Down long barrows and bank

Soil micromorphology samples were collected from two Gussage Cow Down long barrows: upstanding long barrow 78 and the ploughed-out long barrow 294 with its surviving ditches. In addition, samples were taken from soils and sediments associated with an Iron Age bank and ditch related to the Gussage Cow Down field system (SAM 192) (Table 3.24; Appendix 4.1).

At upstanding long barrow 78, contexts [278–9] are the buried soil profile underneath the mound. They show four main layers/horizons: <1 cm of chalk rock fragments, representing the base of the chalk mound; 5 cm of dark brown buried A horizon (top of buried soil), with partially sorted rock fragments (0.3–1 cm; developing stone line); 4–5 cm of medium brown A horizon (lower A), with unsorted chalk rock fragments (0.3–2 cm); and 1.5 cm of chalk weathering horizon mixed with some lower A zones (top of chalk rock). Context [277] also represented turf at the top of this sequence, directly overlying chalk rock. These soils all show standard shallow rendzina characteristics, revealing developed grassland immediately predating the long barrow. There were no clear indicators of any previous soil or land-use types, suggesting that a grassy clearing was present at this location for an extended period of time before the barrow was built. Context [278], being the upper part of the buried A horizon, is interesting in showing some effects of compaction, with elongated sub-angular blocky structure seen. This appears to relate to the pressure of the mound on the old land surface.

The upper ditch fills from long barrow 294 comprise three horizons: contexts [220], 3–4 cm of angular blocky soil (0.5–1 cm), well-sorted rock fragments (chalk, 0.5 cm); [221], 1–1.5 cm of dense orange apedal soil; and [222], 3–4 cm of crumb and sub-angular blocky (<0.5 cm) soil with stone line in the middle (chalk rock fragments 0.5—0.7 cm in size). The fill of the plough marks comprises two main horizons: soil over a 1 cm thick very dense orange layer, with charcoal visible and some iron mottling. This zone could be the base of the plough zone (following Jongerius 1983; see Lewis 1998b).

The buried soil (context 273) under the Iron Age bank (F.122) is very gravelly (chalk fragments make up 20–25 per cent of the area) and poorly sorted, well-developed sub-angular blocky clay loam. The soil contains some textural pedofeatures related to leaching (clay coatings), as well as pottery fragments (including possible grog fragments), and one aggregate with 30 per cent organic components (mainly charcoal) which appear to have originated from a combustion feature or hearth rake-out context. Although relatively clayey, the soil again appears to show a rendzina history before burial.

Dorset Cursus contexts (Table 3.24; Appendix 4.6)
Overall, the various Cursus ditch fills are silt loams with little organic matter (occasionally including some charcoal), frequent to occasional chalk rock fragments and frequent micrite. Sand grains are predominantly quartz. Porosity varies from 10 per cent up to 50 per cent, often in relation to frequency of chalk rock fragments (gravels) or to structural variation, with pellety and subangular blocky structures being seen in samples of lower porosity; pores are normally vughs and channels. Amorphous sesquioxide staining and impregnation occurs both as a general feature of some fabrics, and as pedofeatures in most thin sections. Some fills show mixing or infills with more micritic or chalky fabrics. The ditch fills have been interpreted by French as representing the following types of materials:

- oxidized turf material dumped into ditch; context [137];
- turf, sometimes with chalk rubble, dumped into ditch; primary/lower secondary fill; contexts [182] and [210];
- stabilization layer of turf, with chalk rubble; secondary fill contexts [196] and [206];
- lower A & A/C material over tertiary chalk fill; context [197];
- soil and chalk rubble dump, some possible windblown fine soil; uppermost fill context [194] upper;
- stabilization layer of turf, with some possible windblown fine soil; uppermost fill context [194] lower;
- dumped & slumped chalk subsoil and turf; contexts [207/8] and [202].

The Cursus bank material [155] is very similar to some of the more gravel-rich ditch fills, showing mixed rendzina topsoil and chalk fragments, the only major variation noted being a very micritic nature. This could reflect increased oxidation of the bank in comparison to the ditch fills and the increased influence of frequent chalk gravel on the soil; since the bank was heavily ploughed out, this observation is unsurprising.

Ackling Dyke contexts (Table 3.24; Appendix 4.7)
In comparison to many of the buried soils, contexts [177] and [141], which represent the uppermost leveling or fill materials under the Ackling Dyke, are slightly more clay-rich; although they are still classified as silt loams, they contained visible clean clay in their main fabrics, and the increased amount of clay is reflected in their structures, which are well-developed sub-angular blocky to columnar. Although [141] appears to be a rendzina growing on top of the cursus ditch, [177] is interpreted as a leveling deposit of turf, possibly from a soil on top of gravels or coombe rock in the vicinity (but not found in this immediate location). Buried soil context [177/9] is different, being more sandy (quartz sand) and more organic-stained in nature, suggesting different origins and/or post-depositional activities. French (App. 4.7) suggests that there is a windblown component.

Ard marks were seen cutting into chalk underneath the Ackling Dyke in Trench 1. These were V-shaped, 3 cm deep and 6 cm wide at top of cut, and 1 cm wide at base. They were filled with the same type of soil as the main buried soil context [140], which appears to be a shallow rendzina A horizon, similar to those found underneath many of the monuments discussed here.

Wyke Down round barrows (Table 3.24; Appendix 4.5) Four earthen round barrows were sampled for the soils of their mounds and ditches, as well as for their buried soil profiles. Turves are frequently marked by very fine iron and manganese mottles, reflecting fine rooting channels. The round barrow soils also more frequently contained a suite of mineral components noted only rarely in other parts of the study, including occasional to rare feldspars, amphibole, epidote and possibly garnet and zircon (C. Doherty pers. comm.), in addition to the more standard fare of micrite, quartz, calcite and occasional glauconite. Barrow 41 lies on a deposit of flint gravels which is clay-rich, and these mineralogical indicators suggest these represent a glacial outwash fan (C. Doherty pers. comm.). The other nearby barrows, while overlying chalk rock, may also be influenced by this deposit.

Barrow 34 is a large upstanding mound which is a double barrow (with two mounds and ditches), and comprises a large turf stack over a chalk mound, over a turf deposit, over a primary chalk core, over turf, over chalk. The original ditch is covered by the second large turf mound, which is bounded by the second ditch. The following contexts were studied:

- [21] a single turf with edges marked by iron and manganese mottling in the turf stack;
- [31] turves over the main chalk core; this context was relatively organic-stained compared to the other soils sampled, and appears to also represent some stabilization period with soil growth on an exposed chalk core, suggesting some time gap before the second mound was erected, and that the original mound was left as an exposed white chalk dome. Earthworm sorting was also noted;
- [36] A/A–C of buried rendzina soil (Fig. 3.101d & e). The latter comprises strongly mixed calcareous and non-calcareous fabrics. Some of this mixing is definitely related to soil faunal activity. However, such mixing was also typical of tilled rendzinas from Butser Ancient Farm in Hampshire, where it was not a major feature of untilled comparatives (Lewis 1998b). This is a physically disrupted soil, but whether that relates to monument construction, cultivation or other influences is uncertain.

Barrow 36 is a large upstanding barrow mound comprising a turf stack on chalk rock. The following contexts were sampled:

- [72] turf mound material comprising a blocky clay loam (Fig. 3.101f). A line of flattened angular peds (3000 × 7000 μm) is seen at the base of the mound, similar to those seen in the long barrow mound discussed above; similar features have been found associated with compaction zones at the edges of ard marks (Lewis 1998b), but there are no clear indicators of cultivation here; these flattened aggregates relate to compaction of the mound.
- [83] is a buried soil. The apparent boundary between contexts [72] and [83] is marked by a horizontal line of manganese mottles. This appears to be the rooting line of a turf layer, suggesting the barrow was built on top of a grassed surface. Contexts [91] and [87] represent the base of the turf mound/top of buried soil. No visible difference is seen between the two sampled contexts, but under the upper 1.5 cm there is a 'line' of manganese & iron mottles which again appears to represent the top of the old, grassed land surface beneath the barrow mound. In the upper part of this is also rounded flint gravel (0.5 cm). The soil material is otherwise the same; it appears to be turf, but with disturbance indicators ('dusty' to 'dirty' clay coatings).

Barrow 41 is a partially ploughed-out and probably originally small barrow comprising a turf stack over flinty drift or fan. This barrow had no clear buried soil; indeed the mound appeared to have been erected directly on a layer of flint, suggesting truncation of the previous soil profile. Context 6 represents the lower mound which comprises a silty clay loam with frequent 'clean' and 'dusty' to 'dirty' clay coatings, the latter suggesting strong disturbance (possibly the ploughing over of the monument in recent times). Context 13 is a buried lower A(-Bt), strongly influenced by and growing on a clay-rich drift or fan deposit. In the field this was reddish-brown (5YR4/4) to dark brown (10YR3/3) silty clay to silty clay loam with flint cobbles and chalk rubble. In thin section, some areas look like a 'palaeosol' with rubified clay in pores and in the groundmass, including some rounded fragments of the same, mixed in with what appears to be topsoil material. The fabric has a reticulate and granostriated groundmass, frequent 'clean' and 'dusty' clay coatings, and general amorphous iron staining. This profile is very different from all others examined in the project, and represents variation based primarily on the influences of parent material. Unfortunately there was no buried soil profile; in this area we could expect a soil growing over this type of deposit to be an argillic brown earth. The original profile was truncated in prehistory (through stripping or erosion) as well as being disturbed historically through deep ploughing.

Barrow 44 is a large upstanding barrow mound comprising a turf stack on chalk rock. Context [107] mound turf has a stone line of chalk and flint rock fragments, revealing earthworm sorting in the base of the mound. The soil is blocky clay loam, with clay coatings and iron and manganese mottling as described above;

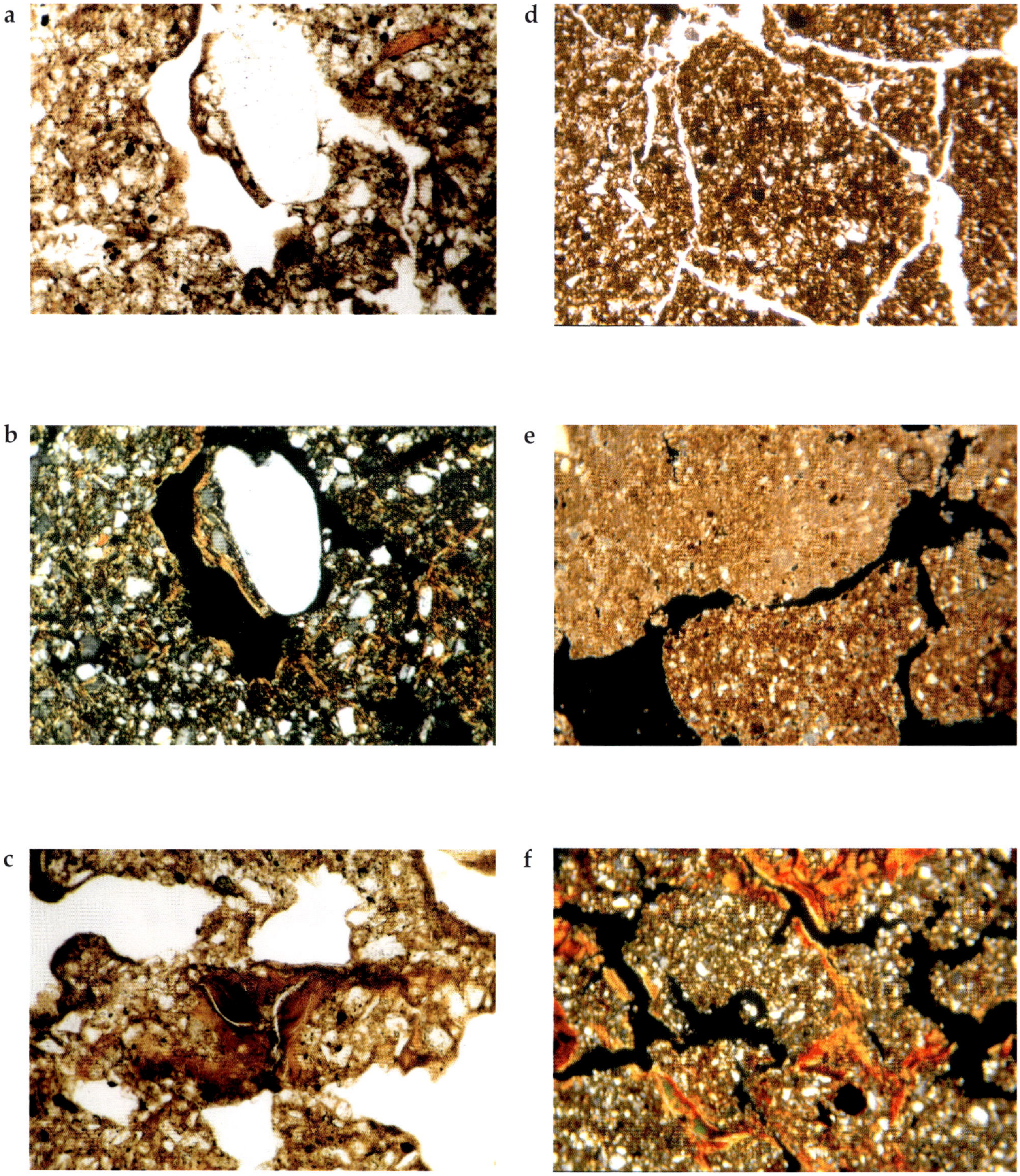

Figure 3.101. *Thin-section photomicrographs of: (a) the lower Ah fabric of a buried argillic brown earth palaeosol at Farrington Clump, test pit 9 (plane-polarized light; frame width 2.25 mm); (b) the Bt fabric of a buried argillic brown earth at Farringdon Clump, test pit 9 (cross-polarized light; frame width 2.25 mm); (c) a sequence of clay infills in the lower part of this same soil at Farringdon Clump (plane-polarized light; frame width of 2.25 mm); (d) the well-structured lower A horizon of a rendzina beneath Wyke Down barrow 34 (plane-polarized light; frame width of 4.25 mm); (e) the A/C horizon of the rendzina beneath barrow 34 (cross-polarized light; frame width 4.25 mm); (f) the illuvial clay infills in the turf mound of barrow 36 (cross-polarized light; frame width 4.25 mm) (C. French & H. Lewis).*

these are also the main characteristics of context 100 basal turves/upper buried soil. The ring-ditch fills (Table 3.24) contexts [102] and [103]) also show similar soil materials.

Knowlton environs

Farringdon Clump test pits (Table 3.24; Appendix 4.12)
The test pits in the Farringdon Clump field showed mixed structure and loamy texture, with frequent and varying rock fragments (chalk, sandstone, chert/flint), low organic matter (occasionally charcoal), and frequent 'dusty' and 'clean' clay pedofeatures, some of which reflect leaching of both undisturbed and disturbed soil material (Fig. 3.101a–c). The latter occur in the same profiles, and suggest a variation probably related to land use (e.g. stable wood- or grassland to ploughing); this could be a relatively recent event, but the history of the field was not confirmed. An iron metal fragment was found in thin section, but could not be further identified. In general, most features appear to reflect the modern land use (ploughing), but the clean clay pedofeatures seen suggest that a stable vegetated profile was once present. Possibly these could also reflect thicker profiles related to pockets of surficial gravel deposits in parts of the field, although such deposits were not revealed in the test pits.

Charlton Dairy House Field & Fir Tree Cottage profiles (Table 3.24; Appendices 4.12 & 4.3)
These profiles revealed the typical variation in modern soil profiles seen in the upper part of the Allen valley through auger survey: rendzinas over both chalk and gravels, the latter with thin oxidized B horizons. The Charlton Dairy profile (Appendix 4.12) showed a typical modern plough soil over gravel deposits; similar profiles were frequently revealed over gravel deposits in the Allen Valley augering survey, and these tend to relate to flint gravels found at the head of the valley. The profile shows a well-developed disturbed A horizon, with indicators of bioturbation as well as physical disruption ('dusty' clay features). The B horizon is typically marked by its iron stained and slightly more clayey nature; clay is more frequently seen both in the groundmass, weathering off sand grains and rock fragments, and as disturbance-related infills and coatings.

Fir Tree Cottage samples showed a typical very organic upper topsoil layer over a more oxidized lower A horizon with some clay leaching and weathering (App. 4.3). This profile is typical of a local modern grassland profile, although some previous disturbance here or upslope is evident in the presence of chalk gravels in the upper horizon.

Knowlton South Circle bank and ditch (Table 3.24; Appendix 4.12)
The Knowlton South Circle samples were taken by Michael Allen through exposures in the bank and ditch of this large henge monument. The profiles are summarized here by context.

Context 7/8/9 is the base of a chalk bank, with bioturbated topsoil inclusions (earthworms and rooting). This layer is compact and somewhat cemented; it is calcareous, with frequent micrite in the groundmass and micrite coatings in most modern pore spaces. In some areas a sort of chalk 'slurry' appears to have formed, seen surrounding aggregates. This is especially frequent at the base of the layer, where the bank context meets the underlying soil horizon.

The groundmass fabric of context 6 (buried soil under bank) suggests a topsoil layer, and structure and fabric pedofeatures suggest mixing by earthworms and other soil fauna both during growth and post-burial. The rooting structure seen is sometimes marked by infills (mostly calcium carbonate precipitation), and shows a network of fine channels evenly distributed, with general vertical orientation. This suggests the soil was once a grass/turf-covered soil, as suggested by the mollusc interpretation (Wessex Archaeology 1994, 5–6). 'Dusty' to 'dirty' clay pedofeatures are seen throughout. These characteristics might relate to plough intrusion mentioned by Burrow & Gale (1994b, 17 & 20–21), although could indicate other types of disturbance, such as clearance or disruption related to bank construction.

Context 26 underlies context 6, and comprises lower A horizon material, showing slight amorphous iron staining. The base of this horizon or layer sees a small accumulation of cultural material (charcoal, bone) in a 1 cm sized zone. Such materials are seen at no other location in the profile; this small patch appears to have been introduced from a deposit relating to cooking, middening, or possibly burial, activities.

Context 34 represents the base of the buried soil profile, said in the field to be bA–B, but perhaps better named as bA–C. It appears to become more calcareous with depth, and presumably overlies the chalk geology, representing the top of the weathering zone at the surface of the rock, where it interacts with the overlying soil profile. It has been influenced by earthworm and rooting activity, and sees some light amorphous iron staining.

In the ditch, the upper primary fill context 43 comprises calcareous and organic soil fabrics mixed through bioturbation, and with zones of cementing of included chalk fragments. Two fabrics, varying in calcareous or organic nature, occur as aggregates, and as mammilated excrements and old infills within pore

Table 3.24. *Summary of the soil micromorphology data for the study area (H. Lewis).*

Site	Soil description	Interpretation
Gussage Cow Down		
Long barrow 78	pre-mound buried soil of very dark brown turf line over mixed silt loam and chalk developed on chalk with hints of sub-angular blocky ped structure	poorly developed, but organic and structured brown earth transformed to turf grassland rendzina
Long barrow 294	upper ditch fill of chalk fragments and silt loam with chalk pebble line and ploughmarks	no buried soil present, but upper ditch fill is base of plough zone
Iron Age ditch/bank (192)	pre-bank buried soil of well-developed sub-angular blocky clay loam with charcoal and pottery fragment inclusions and clay coatings with abundant chalk fragments	? former brown earth to rendzina soil
Fir Tree Field MBA enclosure	organic silt loam over leached/weathered lower A horizon on chalk	modern grassland rendzina profile
Dorset Cursus		
Ditch fills	silt loams with variable amounts of chalk rubble inclusions with turf-like standstill horizons	combination of ditch edge and bank erosion with episodic organic accumulation
Bank make-up	mixture of soil-derived silt loam material and chalk rubble with neo-formed calcite; on very thin silt loam buried soil with excremental fabric	mixture of soil and chalk substrate from line of ditch; on thin turf line of poorly developed rendzina
Ard-mark fill	similar to buried soil above	ard mark defined as shallow (<3 cm) V-shaped cut infilled with rendzina soil material
Wyke Down barrows		
Buried soils	bioturbated calcareous/non-calcareous mix of silt to clay loam fabrics with some oxidation mottling	physically disturbed, thin rendzina soils and former clay-enriched brown earths
Mound make-ups	oxidation mottled turf fabrics, sometimes containing clean to dusty clay in voids	redeposited turves of rendzina soils, possibly subject to fine colluvial inwash inclusions prior to turf inclusion in mound
Ackling Dyke	buried soil of more clay-rich silt loam with clean clay in groundmass and a well-developed sub-angular blocky ped structure; with top of ditch containing rendzina-like soil material and strong very fine sand/coarse silt component	brown earth soil type with clear evidence of clay illuviation; but by the time the ditch became infilled there was rendzina soil formation and a possible windblown component
Knowlton South Ring henge pre-bank buried soil	dusty/dirty clay- and calcite-enriched and bioturbated turf/silt loam fabric over lower A on chalk with some pot/bone fragments at base	buried grassland rendzina with some included cultural material, possibly ploughed in the past
Knowlton environs		
Faringdon Clump	dusty and clean clay-enriched silt loam soil beneath hillwash	may reflect combination of past land use of stable wood/grassland to ploughing under later hillwash, but these events are undated
Charlton Dairy	thin silt loam on chalk with disturbed/bioturbated A horizon	ploughed rendzina
Deer Park Farm	thick, clay-enriched sandy clay loam on Reading Beds associated with very late Upper Palaeolithic flint assemblage, but eluviated E horizon above, only the modern ploughsoil	lower half of *in situ* argillic brown earth on gravel dominated geology; upper half of palaeosol truncated/incorporated in modern ploughsoil

spaces. Context 45 is a humic silty clay loam which also shows very mixed fabrics. This does not appear to represent *in situ* soil growth (i.e. it is not a stabilization layer), but fits the interpretation of a dumped or slumped soil layer. One very calcareous fabric, frequent in overlying context 43, is here seen as infills, showing the continued influence of overlying chalk weathering and leaching on this horizon. Although earthworms and rooting clearly account for most of the mixing, some more substantial disturbance may also be indicated, probably related to its location within a ditch, and from deposition of overlying materials.

Deer Park Farm (Table 3.24; Appendix 4.13)

This profile revealed an argillic brown earth soil, but lacking an E horizon, possibly through ploughing truncation and mixing of that horizon into the Ap. This profile comprised Ap over Bws over Bts horizons, the latter being a sandy clay associated with Mesolithic lithics, and with a thermoluminescence date of 10,700 BP (see Appendix 4.13).

The implications of the soil studies

From a soils perspective, the aim of this project was to explore the geoarchaeological evidence for land use and landscape change in later prehistory, and in particular to address the issue of change in soil types and sedimentation history related to proposed major changes in agricultural and settlement practices in the Bronze Age.

In southern England soil thickness and fertility are perceived to have decreased throughout the Neolithic and Bronze Age, to the point at which regional soil types changed, and this is associated with erosion episodes and perceived settlement pattern alterations. A combination of regional pollen, molluscan, soil profile and erosion evidence has led to the previous suggestion that pre-Neolithic thick brown forest soils evolved (or devolved, depending on perspective) to thinner rendzina soils through the combined influences of prehistoric forest clearance and cultivation on the Wessex chalklands (e.g. Ball 1975; Catt 1977; 1979; Moffat & Cope 1984). In the Wyke Down studies we have found no evidence of widespread thicker brown forest soils. Rather our work has revealed that thicker, more clay- and organic-rich soils existed where they continue to exist — on locations with flinty drift, coombe rock or fan deposits in the head of the Allen valley, such as under round barrow 41, and possibly represented in the clayey rendzina turves (with similar mineralogy) of round barrows 44 and 36. Other locations where thicker soils (argillic brown earths) were found are also those with similar modern soils — on the clay-with-flints at the top of the valley, and represented in sites such as Deer Park Farm. On the chalk rock proper we have no evidence of Late Mesolithic, Neolithic, Bronze Age or later thick brown forest soils, with shallow well-developed rendzinas being seen in all locations and from all periods. This suggests that much of the Wyke Down area was covered in grassland; indeed the fact that even the clayey turves in the round barrows are rendzinas suggests that even those clay-rich locations which could be expected to have seen tree cover in the past were already grasslands by the Early Bronze Age if not earlier.

In the regional model, changes in soil type are thought to come about through forest clearance, which leads to oxidation of the topsoil and a reduction in surface organic matter accumulation, thus lowering organic matter content. Clearance and, later, cultivation result in the physical break-up of soil aggregates, and the disruption of soil aggregation properties, allowing clay, some minerals, and fine organic particles to be washed down through the soil under the action of rainwater. These are either lost to the profile through slope erosion, or deposited in the lower horizons of the soil as coatings on mineral grains and in voids through the processes of leaching, eluviation and illuviation. We do have some evidence of such disruption in the soils from the round barrows in particular, but relating this to processes additional to the disturbance inherent in monument creation is difficult. And the soils were clearly grassland at the time of barrow construction, as shown by their turf lines and general nature.

Other evidence for such soil disturbance has been suggested regionally through the erosion record. Erosion and soil loss occur more easily when soils are disrupted, especially if they are left bare, and especially where rainwater, slope processes and wind action have a strong impact. Cultivation is often a major factor in downslope erosion due to its role in loosening the topsoil, creating barriers to water infiltration and (often) reducing organic matter content. Deposition of such eroded material occurs in rivers or in valley bottoms (as colluvium) (Bell 1983; Allen 1988; Boardman 1990). As discussed in Chapter 2, our valley study suggests that any major erosion episodes are either situated in historic times (colluvium possibly dating to the medieval period from dry valleys adjacent to the main Allen River valley), or occurred much farther back in prehistory than envisioned by current archaeological models (i.e. before the Neolithic, and certainly before the Bronze Age). The round barrow soils discussed above do show some minor evidence for down-profile and/or slopewash deposition of clays, which could be related to clearance and agriculture, but we have no large-scale erosion evidence in the system. In general, the Wyke Down soil histories do not fit the general model which suggests the truncation of thick brown forest soils and the development of rendzinas accelerated on the Wessex chalklands with both the introduction and intensification of cultivation (Limbrey 1975, 185–6, 191; Macphail *et al.* 1987). Indeed, the evidence of the Early Neolithic long barrows and the other, later Neolithic monuments studied suggests that stable and well-developed (i.e. old) rendzinas existed in many places in the Wyke Down area by the Neolithic, and that these soils persisted into later times.

The extant model of soil change related to land use in the chalklands is itself fraught with points of contention regarding timing, cause and impact of the proposed changes in the physical landscape, and meaning of these for the cultural landscape. Archaeological evidence for deposition of windblown soil materials in some areas has been related to the presence of bare (possibly tilled) soils (Cornwall 1958; Mercer 1981, xv); in Wyke Down this type of evidence is seen in the Iron Age Gussage Cow Down ditch fills, and in the final fills of later prehistoric monuments. The relationship of colluvium to cultivation is also problematic if interpreted to have regional significance: in some locations, chalky colluvium buries earlier brown earth colluvium, suggesting continuing exposure of the calcareous subsoil to erosion (Macphail *et al.* 1987; Boardman 1992). Dry valley colluvium, sometimes up to 3 metres thick (Bell 1992), covers some Neolithic and Bronze Age sites (e.g. Durrington Walls) (Evans 1971, 23), and is normally dated to the Middle Bronze Age or later. But it also covers earlier colluvial deposits (as at Strawberry Hill) and earlier soils which have been truncated, indicating that erosion was occurring before later Bronze Age material was deposited (Allen 1992, 46, 48; Mercer 1981, xii). Other studies suggest that pre-Bronze Age colluvium has been lost 'down valley' (Bell 1983). The erosion evidence is complex, and suggests that the timing and location of soil loss and profile changes vary greatly across the larger region (Bell 1992).

Many authors suggest that the progressive change in soil properties represented in buried soils under monuments and erosion evidence in southern England and other regions is related to clearance and/or cultivation in the Early Neolithic (e.g. Evans 1972, 274–7; Limbrey 1975, 181, 185–7, 191; Kerney *et al.* 1964). Others suggest that the erosion seen in some places reflects the onset of extensive arable practices following an intensive arable phase during the earlier Neolithic (e.g. Bradley 1978; Gebhardt 1992, 20). A rare few believe that it relates solely or mainly to an intensification of arable agriculture during the Bronze Age (e.g. Barrett 1994, 148). Moffatt *et al.* (1989) have suggested that major soil losses mainly occur later, through Iron Age and Roman arable intensification, and modern cultivation. None of these models fits the evidence of the Wyke Down soils and sediments, which suggests a stable long-term grassland landscape (with patches of trees), dating from at least the Neolithic on.

We know that cultivation occurred in the region during the Neolithic from the presence of ard marks underneath Neolithic and Bronze Age monuments. The presence of ard marks cut into chalk under barrows is said to indicate that tilled soils were already shallow in the region by the Bronze Age (as ards only cut to a certain depth) (Fowler & Evans 1967), thus implying pre-Bronze Age 'degradation' in soil thickness and fertility. While no early ard marks were found under Wyke Down monuments (with only probable Iron Age or Roman period marks seen under the Ackling Dyke), the suggestion that soils were already shallow rendzinas by the Early Bronze Age is supported by the evidence from the Wyke Down barrows, whose soils also show disturbance indicators (even if these cannot be tied down to a type of disturbance). Even this generalization is too general, however: the evidence from South Street long barrow, where ard marks underlie up to 30cm of later soil development before the Neolithic monument was built (Ashbee *et al.* 1979; Fowler & Evans 1967, 290), suggests that in some localities Neolithic soils were not simply becoming shallower in the region, but that they were occasionally building up after early tillage episodes.

At present, based on the soil profile, sedimentary and micromorphological studies, it seems clear that *if* erosion and major profile altering processes were influencing the soils in the Allen valley, then these must have been active mainly during the Mesolithic and/or the Iron Age and later. The former has been proposed elsewhere in southern England (Catt 1979; Macphail & Scaife 1987; Macphail *et al.* 1987; Scaife & Macphail 1983; Scaife & Burrin 1983). There are comparable examples of buried rendzinas under Neolithic monuments (e.g. Ashbee *et al.* 1979), in addition to decalcified brown earths and argillic brown earths (e.g. Evans *et al.* 1993; see Macphail 1987 for a review of the evidence).

Allen (1992) describes colluvial deposits that support the view that soil loss began during the Neolithic, and he relates this to arable activity. He suggests that grassland can also be identified and fitted into a pattern of arable-pasture rotation, which is supported by pollen (Scaife 1984) and molluscan evidence (Allen 1994). These analyses suggest two different land-use types coexisting in the region from an early time — pasture in valley bottoms with arable land on adjacent slopes. Settlement is proposed to have been located at the base of slopes or in valley bottoms. Later prehistoric erosion due to extensive soil tillage, however, is said to have led to the production of thin rendzinas on the valley sides and thicker (colluvial) soils on the bottoms. It is suggested that settlement then moved to the slopes and plateaux, while tillage became concentrated on the valley bottoms (Allen 1992, 47). In this model, the settlement and agricultural systems changed in tandem with changes in soils. In our study of Wyke Down we see no such progression or alteration in sedimentation or soil types. Unlike most other parts of Wessex, early settlement (Neolithic) is seen, and this occurs in the head of the valley (Green 2000).

To observe accurately the pedological effects of agricultural land use in thin section through soil micromorphology is not always easy or straightforward. Comparison with the pollen (Chapter 2) and, especially, molluscan (above) methods of interpreting land use has provided both corroboration and a more complex picture of land-use history than any one method alone can give. In addition, issues relating to the interpretation of buried soils and monumental soil evidence cannot be understated; in particular it is important to remember that monuments are special places in many ways, and it is not appropriate to simply assume that environmental evidence from monuments is representative of the landscape as a whole. However, through comparison to wider landscape data, such as that produced through pollen and sedimentary studies, it is possible to investigate the level of representation seen in monumental soils. This comparison at Wyke Down shows that monumental soils do represent the general landscape during the later Neolithic and Bronze Age, but that early monument locations may have already been distinguishable 'places' in the landscape in the earliest Neolithic, and even into the Mesolithic. These issues are discussed further in Chapter 4.

Towards a dynamic ecosystem model for the Neolithic of the upper Allen valley

Edgar Samarasundera

Introduction

This section summarizes the development of a GIS-based simulation model designed to test the plausibility of a pastoral-dominated Neolithic economy in the upper Allen valley. Such types of dynamic spatial model are commonly employed in landscape ecology but have received little attention so far within palaeoecology. The relevant variables and parameters are discussed in depth; these include landscape structure, ungulate feeding impacts on vegetation and direct human clearance of vegetation. This discussion is followed by a summary of the model and an algorithm description. The preliminary results from the simulations run to date tentatively support the hypothesis that livestock feeding could have caused forest recession in the Neolithic Allen landscape but further model runs and post-modelling analyses are needed before firmer conclusions can be drawn.

The palaeoenvironmental evidence for the study area has suggested that the upper Allen valley was predominantly woodland in the southern part with a gradual transition to a more complex vegetation mosaic in the vicinity of Down Farm further to the north during the Late Mesolithic. The collective picture presented by the molluscan, palaeosol and pollen evidence has also indicated that the Neolithic witnessed the opening up of the landscape such that open grassland dominated by the Early Bronze Age. The limited evidence to date of colluviation suggests that arable agriculture was not the dominant economic activity of the Neolithic population and hints that livestock grazing may have been central. This section summarizes the development of a dynamic ecosystem model for the Neolithic upper Allen valley and presents some preliminary simulation results.

Dynamic spatial models are commonly used in landscape ecology to try to predict the future effects of both current and proposed human activities upon ecosystems, as well as to inform conceptual understanding of ecosystem behaviour (e.g. Ares *et al.* 2003; Dale & Pearson 1999; Ludeke *et al.* 1999; Mottet *et al.* in press) but are only starting to receive attention in palaeoecology (e.g. Heiri *et al.* 2006). This approach offers the opportunity to test the plausibility of the pastoral economy hypothesis for the Neolithic of the study area under different Late Mesolithic landscape scenarios and different grazing-ecology models, with model validation in part involving comparison of simulation outcomes against the palaeoenvironmental record. Two notable strengths of this research framework are that both the dynamics of ecological processes and the effects of spatial adjacency can be represented together (Costanza & Voinov 2004, 3–20; Mladenoff & Baker 1999, 1–4; Turner *et al.* 2001, 47–70). In combination these two strengths, in the specific context of the early Holocene of the upper Allen valley, enable the interplay between the tendency for livestock grazing and anthropogenic vegetation clearance to fragment woodland versus the propensity for woodland to regenerate to be explored, using palaeoenvironmental data.

Model-development stages

There are three broad stages in developing any landscape (palaeo)ecological model: developing an algorithm, producing initial conditions input maps and validating the model. In the context of a dynamic spatial model for the Neolithic of the upper Allen valley, the first of these stages involves considering ecological dynamics issues such as livestock grazing behavior, direct anthropogenic clearance of vegetation, the tendency towards woodland expansion and vegetation succession within an ecological patch. The second stage includes producing plausible alternative landscape structure maps for the terminal Mesolithic. The last phase is that of validating the model in terms of assessing its ecological realism, comparing simulation outcomes against the data (i.e. the palaeoenvironmental record) and sensitivity to changes in input parameters.

Developing a model for the upper Allen valley

Livestock grazing

There are a number of factors that need to be evaluated in developing a grazing sub-model for the upper Allen valley model. These are the quantitative relationship between grazing and its impacts upon vegetation, the appropriate spatial and temporal scales for modelling grazing palaeoecology, growth in livestock numbers over time, ecosystem carrying capacity for ungulates, livestock grazing preferences and Neolithic livestock management practices. These issues are discussed here and a grazing sub-model is proposed.

Fundamental to modelling grazing impacts is the ability to quantitatively estimate the relationship between livestock density and forest fragmentation. Suggested figures vary widely in the published literature, differences primarily attributable to the biogeographical and landscape contexts in which the relevant studies were conducted. The literature pertaining to temperate woodland and wood pasture landscapes suggests that even at cattle (or horse) densities of 20 individuals per 100 ha, forest will return in less than fifty years (Vera 2000, 147–8). Rackham (1980) found that, after initial grazing impacts, forest recovery took place within probably less than fifty years during the seventeenth century in Epping Forest in a context of 1500 ungulates (deer, cattle and sheep) feeding inside 2000 ha of woodland (75 animals per 100 ha). By contrast, a study of grazing in late eighteenth-century Prussia (in modern Poland) shows that herbivore densities of 250-300 per 100 ha result in the complete transition from arboreal-dominated to grassland landscapes within decades (Vera 2000, 172–3). The outcomes of a field study carried out over a timeframe of almost thirty years in the New Forest has been summarized by Ratcliffe (1992, 239–41) and yield further insight. Ratcliffe's research suggests that ungulate densities of 100 animals per 100 ha would be sufficient to result in the dominance of open vegetation communities but with the retention of some arboreal cover, so long as grazing pressure continues. Such medium-term studies and historical accounts, combined with research into present-day, short-term effects of browsing and grazing in woodlands and wood pastures (e.g. Best 1998; Teich *et al.* 2005) yield a body of literature that provides us with the crude quantification relationship and livestock stocking rate upon woodland fragmentation given below, assuming a timeframe of thirty to seventy years:

- less than 0.1 livestock unit per ha: no impact on closed canopy forest;
- 0.3 livestock unit per ha: shift from forest to open canopy woodland;
- 0.8 livestock unit per ha: forest fragments to produce a complex mosaic;
- 1 livestock unit per ha: closed canopy forest becomes scrub;
- 2.5 livestock units per ha: complete woodland destruction, grassland forms.

Therefore five categories of livestock grazing intensity and five classes of vegetation cover are pertinent for modelling long-term grazing-vegetation dynamics; this contrasts with short-term field studies for which ample evidence of species-level interactions exists (e.g. Bartolome *et al.* 2000; Ratcliffe 1992; Vallentine 2001, 261–302).

The next consideration is that of the temporal resolution appropriate for the upper Allen valley. It is widely agreed that major transitions in landscape structure in temperate woodland and wood pasture ecosystems typically take a few decades to complete, yet both observed and inferred timeframes vary widely (Delcourt & Delcourt 2004, 90–131; Frelich 2002; Kirby 2004) hence two strategies have been pursued in the model developed here to acknowledge temporal uncertainty and avoid false precision:

- run scenarios with an 'average' figure of fifty years for each vegetation state change to take place;
- run scenarios with twenty five years for half-state changes to be modelled;
- run scenarios with transition periods varying between stage types, for example having a longer period for woodland regeneration than for woodland break-up; this approach has been attempted by Kirby (2004) in a non-spatial model for a hypothetical landscape.

In terms of spatial scale, modelling the effects of grazing at the resolution of 1 ha is appropriate considering comparison with the stocking rate literature is necessary. The input initial conditions vegetation maps were therefore produced with 1 ha resolution cells (see previous section on initial conditions scenarios).

Another necessary variable to reflect on are the ecological preferences of Neolithic European livestock. Modern cattle, horses, pigs and sheep have significantly different positions along the browsing-grazing continuum. Cattle and sheep, though having a preference for grasses due to their ease of digest, and high nitrogen and phosphorous levels (Mayer *et al.* 2003; Vandenberghe *et al.* 2007), are nevertheless known to browse seedlings and in the case of cattle young saplings too (Hester *et al.* 1996; Mayer *et al.* 2006; Papanastasis *et al.* 1995; Teich *et al.* 2005; Vandenberghe *et al.* 2007), thus retarding forest regeneration under sufficient ungulate density levels over decadal timeframes. The ancestor of modern cattle, the aurochs, is thought to have had a preference for large clearings within open canopy woodlands (Kirby 2004; Vera 2000), though it is unknown what the balance between any browsing and grazing behaviour may have been, adding to the uncertainty over when modern cattle may have evolved their current grazing behaviour. Horses are known to be both browsers and grazers, whilst pigs are typically foragers and thus horses and pigs are considered here to have a preference for areas composed of complex vegetation mosaics. The dietary preferences of sheep in Neolithic Britain are probably the most difficult to ascertain with considerable variation in the feeding behaviour between the earliest-domesticated breeds, for example Pryor (1998b) has discovered that Soay sheep have a particular interest in young saplings. These considerations, combined with a lack of information on the composition of Neolithic livestock communities in southern Britain, mean that we have to consider two possibilities for overall grazing behaviour within the upper Allen valley model: a general livestock preference for (a) open grassland/herbaceous communities and (b) complex vegetation mosaics. Therefore simulation runs were performed for both grazing sub-models.

Two further variables need to be factored into account in order to produce an ecologically realistic grazing sub-model: the inevitable change in livestock numbers over time and the issue of ecosystem carrying capacity. The former can be assumed to have involved an increase in livestock over the course of the Neolithic and into the Early Bronze Age, though the total numbers of livestock and the rates of increase in their populations can only be speculated. Hence five alternative scenarios have been included: a slow linear increase with a low final (i.e. Late Neolithic/Early Bronze Age) population, a moderate linear increase with a moderate final population, a rapid linear increase with a high final population, an initially rapid but later asymptotic increase with a high final population and an initially slow but later exponential increase with a high final population. The growth and final population numbers issue is closely related to both ecosystem carrying capacity and archaeological plausibility questions: how many livestock could the Neolithic human population of the area have had? From the perspective of landscape carrying capacity, following on from the quantitative relationship between livestock density and woodland recession given above, whilst ungulate densities in excess of 2.5 livestock units (LUs) per ha (1 cow is typically considered by ecologists to be equivalent to 0.8 livestock unit, sheep to be 0.1 units each) are within the carrying capacity of temperate north and central European grasslands and are common in modern contexts (e.g. Hester *et al.* 1996; Palmer *et al.* 2004), a realistic uppermost figure for the total livestock population for a Neolithic pastoral economy is more problematic to estimate. The study area in the upper Allen valley model is 8 km by 8 km (6400 ha) and since a carrying capacity of three livestock units per ha is ecologically plausible, a sustainable livestock population of over 19,000 is possible. However, it is patently unlikely that the Neolithic human populace of the Cranborne Chase area had livestock populations even approaching this value. Modern, non-industrialized pastoral communities have widely differing herd sizes with a community of 150 households in Tanzania reported as having over 45,600 livestock units (Coppolillo 2000), whilst another group of 77 households elsewhere in the same country had less than 1000 livestock units (Sellen 2003) and a study of an Andean community of nine households found a collective herd of 225 livestock units (Kuznar 1991). Even though the biogeographical and socio-cultural contexts of these studies are radically different from that of the English chalklands, they do demonstrate that herd size can vary widely even once the size of the accompanying human population is taken into account. Likewise, historical European herd sizes show marked dissimilarity between cases, for example one area of seventeenth-century France had an average of 548 livestock units per community whilst a community in nineteenth-century Switzerland had a mean of 196 units in an area of 250 ha (Vera 2000, 172).

Although both modern and historical case studies yield some insight into potential Neolithic upper Allen valley livestock population sizes, the added complications caused by the possibility of the size of livestock herds being an expression of social status during the Neolithic in all or parts of southern Britain place extra uncertainty in estimating plausible limits on the range of values for the herd-size model variable. This situation is compounded further: even though it is reasonable to speculate that Early Neolithic farmers could have built up the mooted livestock population gradually in the upper Allen valley, migration(s) of pastoralists with well-established animals from other parts of southern England cannot be discounted (although it should be noted that the simulation modelling assumes an early introduction to restrict the number of runs to a manageable number). Therefore simulation runs exploring different values for Late Neolithic/Early Bronze Age herd sizes in the upper Allen valley are necessary and in the research presented here, informed by estimates of the number of households in the landscape. Fowler (1983, 32–6) and Roberts (1990, 11–12) considered an average population density for Britain of 11 people per sq. km for the Middle Bronze Age, here interpreted as 3–5 households, to be possible. Using this as a starting point and considering the study period is Neolithic, 1–3 households per sq. km for the Neolithic upper Allen valley is a reasonable suggestion considering the uncertainties over geographic and temporal patterns and fluctuations in population levels during the transition to established agricultural economies. Since the archaeological evidence from the Allen area points towards settlement activity having been concentrated in the north of the study area, perhaps in an area of less than 8–10 sq. km, a range of 8–20 for the number of households in the Neolithic landscape is in line with the thoughts of Fowler and Roberts. This array of values gives some indication of conceivable herd sizes. Following on from the above discussion on modern and historical pastoralists and community/household to herd size ratios, a theoretical range of 100–400 livestock units for a community of 8–20 households is appropriate, consequently seven alternative population maxima have been used for different simulation scenarios for the Late Neolithic/Late Bronze Age upper Allen valley over this range of values increasing in increments of 50 livestock units between runs. One final deliberation is necessary before a plausible grazing sub-model can be derived: the spatial management of livestock. Herding at some level must clearly have been performed particularly in relation to the protection of arable areas, though it is implausible to suggest that Neolithic farmers practiced woodland protection especially considering the free-ranging approach to pastoralism that occurred in Europe well into the nineteenth century. Thus within the grazing sub-model a mechanism for preventing simulated grazing of GIS cells allocated as being arable was included but beyond this qualification livestock have been assumed to be free-roaming. This issue is closely tied, from a modelling building perspective, to that of the carrying capacity of each cell (patch), so again, a mechanism has been incorporated to prevent livestock numbers exceeding an average of three livestock units in a cell per timestep.

Vegetation dynamics

In addition to vegetation responses to long-term grazing, other aspects of vegetation dynamics need to be incorporated into the upper Allen valley model. These variables are the tendency for vegetation succession to occur within patches, for forest patches to expand and for wind-throw to create canopy openings. The transitional stages for the first two variables operate over periods thirty to seventy year years and can be modelled as discussed in the section on modelling grazing impacts. For each timestep, within-patch succession can be easily expressed by changing the state of a GIS cell from one vegetation class to the next class up the hierarchy of grassland → scrub → complex mosaic → woodland → forest. The expansion of forest/woodland patches into adjacent less wooded areas can be modelled in the same way in terms of ecological theory, though the actual algorithm implementation within a GIS is necessarily different as within-patch changes do not involve spatial expansion (see model summary section below). The effects of storms and wind damage upon forests in the form of wind-throw must be incorporated into the model to maintain ecological realism. There are three key factors to consider in order to effect this: the approximate temporal frequency of wind-throw events, the typical size of canopy clearings caused by such incidents and any spatial patterning to be expected. Research by Quine & Bell (1998) into wind-throw in eight present-day wooded areas in the west of Britain has revealed that typically 1–3 per cent of a landscape would be affected by the results of wind-throw at any given time, with 80 per cent of wind-throw caused canopy gaps being less than 1 sq. km in extent. The mean figure for gap size was around 0.85 sq. km, the mode 0.1 sq. km and the median 0.3 sq. km; over 95 per cent of clearing sizes were 0.3 sq. km or less. Before these modern day analogues can be used as a basis for estimating the impacts of wind-throw in the Neolithic upper Allen valley, the fact that the west of Britain is more susceptible to strong Atlantic winds needs to be taken into account. As a result a conservative estimate of 1 per cent landscape damage has been made for the wind-throw sub-model to incorporate the reduced frequency of wind-throw events in Dorset, which translates to 64 cells in the upper Allen valley model (1 cell = 0.1 sq. km = 1 ha). The next stage is to consider the ratios between canopy clearing sizes. Following the relationships derived by Quine & Bell, ratios which other researchers have also found to hold for other parts of the world (e.g. Liu & Hytteborn 1991), the distribution that holds for the upper Allen valley model is: 51 wind-throw cells of 1 ha, 10 cells of 2 ha and 3 cells of 3 ha. To simulate wind-throw effects in the upper Allen valley wind-throw sub-model, a pseudo-random number generator was used for each time step using the ratios stated above to mimic a random spatial pattern of storm damage in the landscape.

Anthropogenic vegetation clearance

The last aspect of the upper Allen valley model-building process, the role of direct anthropogenic clearance, is somewhat more complex in empirical terms due to the lack of even semi-quantitative evidence. The archaeological and palaeoenvironmental records point toward that Neolithic settlement activity having been primarily located in the north of the study area in the vicinity of the Dorset Cursus (and around Knowlton) and that arable agriculture played a minor role during this period. Using these indicators of the spatial foci and extents of direct vegetation clearance, four alternative scenarios have been envisaged for the upper Allen valley model:

- linear increase and early permanent settlement: slow rate of increase with permanent clearances by lower quartile of model timeframe;
- linear increase and late permanent settlement: slow rate of increase with shifting clearances until upper quartile of model timeframe;
- fluctuating and early permanent settlement: erratic changes in rate, foci and extent of increasing vegetation clearance; permanent clearances by lower quartile of model timeframe;
- fluctuating and late permanent settlement: erratic changes in rate, foci and extent of increasing vegetation clearance; permanent clearances by upper quartile of model timeframe.

Initial conditions input maps

Whilst the sub-fossil record suggests a landscape that becomes increasingly dominated by arboreal cover the further one moves south (see Scaife in Chapter 2), the spatial structure of the landscape in terms of parameters such as the percentage of each community (patch) type, the spatial arrangement of different communities and the level of connectivity between like patches cannot be determined from the palaeoenvironmental data alone. As a result it is necessary to create a range of virtual maps in a geographic information system (GIS) to provide quantifiable, alternative initial conditions scenarios. Consequently methods for quantifying landscape structures need to be considered and currently there are two potential avenues for quantifying past landscape patterns: pollen transport modelling and landscape metrics.

Simulation models of pollen transport have been increasingly used to reconstruct past landscapes and quantitatively estimate human impacts upon them (e.g. Caseldine & Fyfe 2006; Sugita 1993; Sugita *et al.* 1999). Such approaches offer much potential in landscape studies, however due to the pioneering, exploratory state of such pollen-modelling research, it has been decided to use the tried and tested suite of methods offered by landscape metrics to produce the reconstructions of the Late Mesolithic vegetation patterns in the upper Allen valley model instead. This latter suite of techniques includes measures of spatial clustering of patch types, community diversity scores, the proportion of a landscape covered by a community type, fractal indices and connectivity between similar patch types (Gergel & Turner 2002, 85–146; Turner *et al.* 2001, 93–134). Landscape ecologists experience a very different research context from landscape palaeoecologists with respect to quantifying vegetation patterns. The former can use satellite imagery, aerial photography and/or take field samples to produce accurate or near-accurate input maps for simulation models, whereas the latter are obliged to create a range of plausible initial conditions maps that can be quantified for the purposes of model assessment.

In this study hypothetical initial conditions raster maps were created in ASCII text format, each representing an 8 km by 8 km grid with an origin at 397,000, 110,000 cell size of 100 m by 100 m (1 ha). Each of these grids/maps was in turn imported into FragStats 3.1, a landscape ecology software package used for quantifying vegetation patterns. Landscape metrics were calculated for composition and spatial arrangement were performed to assess the percentage of arboreal cover, degree of patch connectivity and the extent of woodland clustering in each initial conditions map. The quantified input vegetation maps were then imported into ArcGIS 9.0 for the simulation runs, Figure 3.102 provides an example initial conditions vegetation map. The classification of cells was approached in two ways, the first being the grassland → scrub → complex mosaic → woodland → forest, providing five categories of vegetation class; the second aimed at incorporating half-stages in vegetation structure and represented as percentages of arboreal cover; to illustrate this concept class 2 (11–20 per cent) in Figure 3.102 represents a patch transitional between scrub and open grassland. The landscape structure variations and two vegetation classification systems provided a total of forty hypothetical initial conditions vegetation maps for the simulation runs. For each of these runs, an initial conditions grazing map was produced for each of the two theoretical grazing models (preference for open areas and complex mosaics, respectively). Figure 3.103 is the accompanying grazing intensity map, with a grazing preference for complex mosaics, to the hypothetical vegetation map in Figure 3.102. For each of these initial conditions vegetation maps, alternative wind-throw and anthropogenic clearance input maps were assigned, each being updated as each simulation timestep passes. Figure 3.104 shows the pattern and extent of direct anthropogenic vegetation clearance after ten fifty-year timesteps for a scenario of early permanent settlement with erratic changes in rate, foci and extent of increasing vegetation clearance, from the same simulation run as Figures 3.102 and 3.103.

Model summary and implementation

So far the model building process has been discussed; the reader's attention is now turned to a summary of the model algorithm, number of runs made and the range of scenarios included, before the implementation of the model in a GIS is briefly addressed.

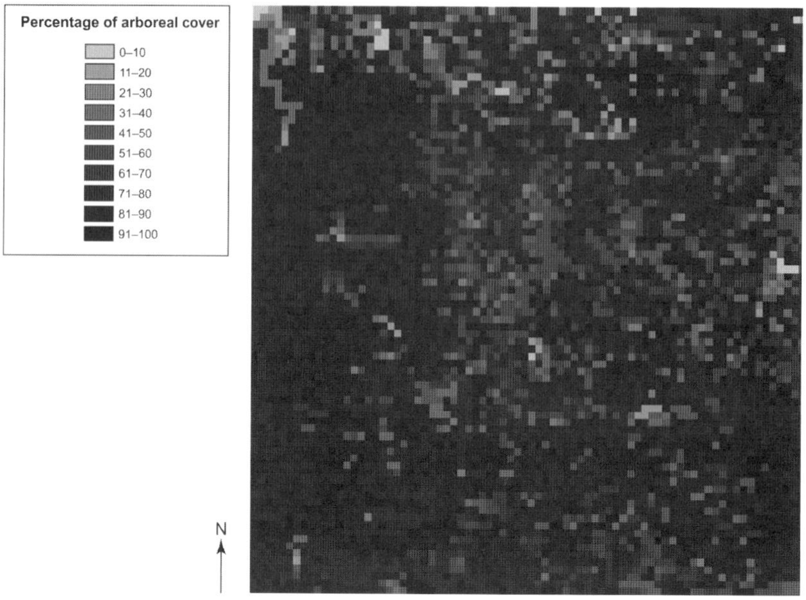

Figure 3.102. *Initial landscape conditions for scenario 3 showing the percentage of arboreal cover per model cell (E. Samarasundera).*

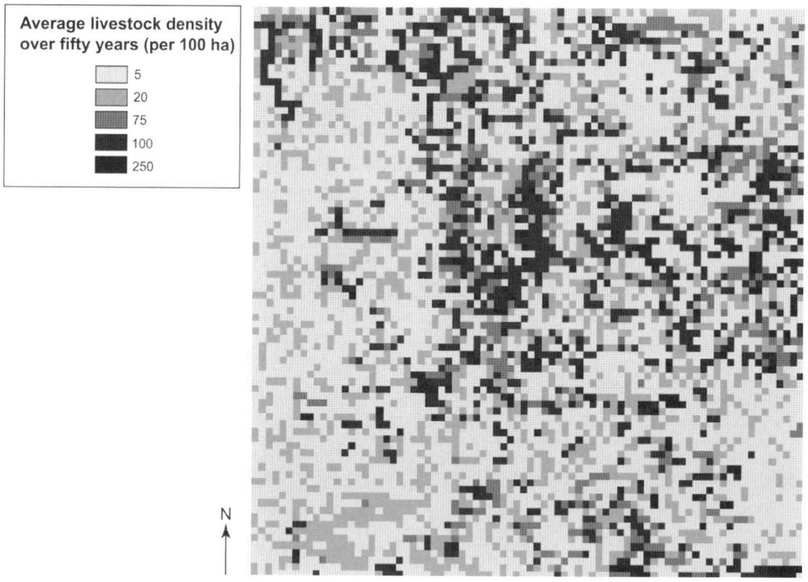

Figure 3.103. *Associated grazing intensity over a 50-year period per 100 ha (E. Samarasundera).*

In total, 3024 simulation runs formed the core of the model framework with variations between runs in the form of different combinations of the following variables:

- initial conditions landscape structure (9 scenarios);
- livestock population growth and final population (7 scenarios);
- timestep variations (3 scenarios);
- grazing behaviour (2 scenarios);
- anthropogenic vegetation clearance spatio-temporal patterns (4 scenarios);
- vegetation classification system (2 scenarios).

Each simulation scenario was performed using the same model structure, the high-level algorithm for which is given below:

- set initial conditions (t_0) vegetation map
- set initial conditions number and spatial pattern of livestock

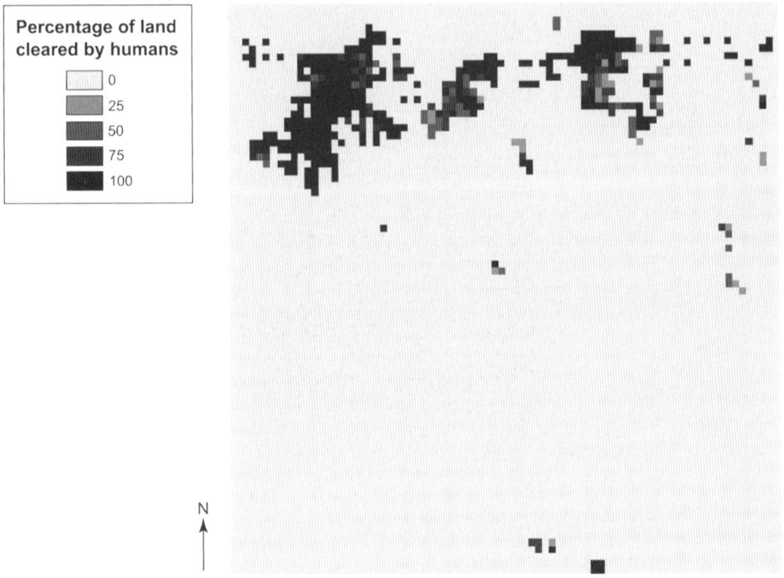

Figure 3.104. *Extent of direct anthropogenic impacts after ten fifty-year timesteps (or timestep 10) for scenario 3 (E. Samarasundera).*

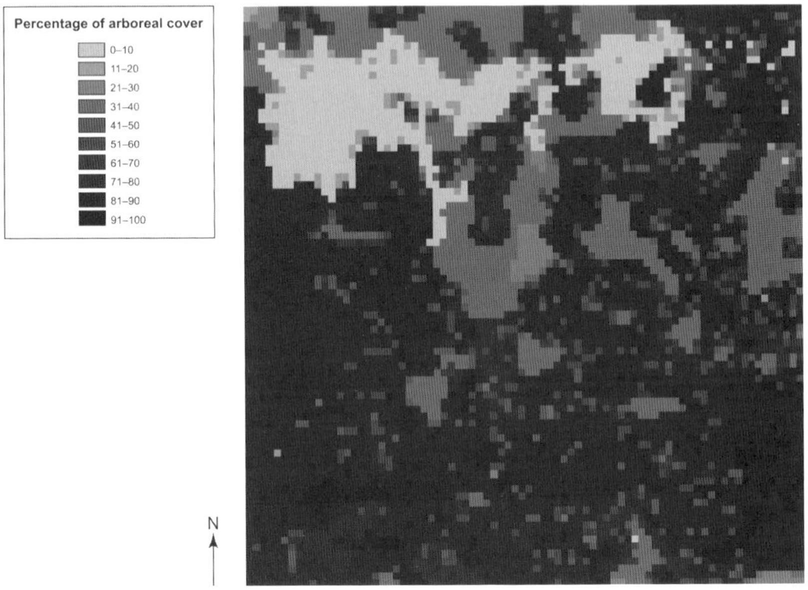

Figure 3.105. *Landscape structure showing the percentage of arboreal cover at timestep 16 for scenario 3 (E. Samarasundera).*

- check whether carrying capacity has been reached for a cell
- if so → reassign excess LUs to cells of sub-optimal vegetation classes
- create initial conditions grazing map
- run anthropogenic clearance sub-model
- run grazing sub-model
- model within-patch vegetation succession
- model woodland patch expansion into neighbouring patches
- model wind-throw
- combine effects of sub-models
- create timestep 1 (t_1) vegetation map and use as new input vegetation map
- repeat process until timestep 30 (t_{30})

The model was implemented using the Spatial Analyst extension within ArcGIS 9.0 in SQL syntax; for details of the low-level algorithm and SQL script see Samarasundera (forthcoming). Figure 3.105 shows the landscape structure after 16 timesteps for scenario 8 in Table 3.25.

Table 3.25. *Landscape structure after 1500 years (30 timesteps).*

Scenario	Scenario description	% open	% mosaic	% woodland
1	Livestock preference for grassland, slow increase in livestock numbers, late introduction of arable	16	49	35
2	Livestock preference for grassland, slow increase in livestock numbers, early introduction of arable	17	50	33
3	Livestock preference for grassland, rapid increase in livestock numbers, late introduction of arable	64	25	11
4	Livestock preference for grassland, rapid increase in livestock numbers, early introduction of arable	65	27	8
5	Livestock preference for mosaics, slow increase in livestock numbers, late introduction of arable	15	65	20
6	Livestock preference for mosaic, slow increase in livestock numbers, early introduction of arable	15	65	20
7	Livestock preference for mosaic, rapid increase in livestock numbers, late introduction of arable	17	68	15
8	Livestock preference for mosaic, rapid increase in livestock numbers, early introduction of arable	20	71	9

(These results are a representative selection of results but they do not give any indication of model sensitivity to parameters (e.g. spatial arrangement of vegetation communities.) In terms of issues such as connectivity between like patches of landscape, the modelling runs that yield a largely open or open-dominated mosaic by the Late Neolithic/Early Bronze Age had the underlying assumption of clustered canopy openings during the latest Mesolithic (rather than fragmented canopy gaps). In addition, all simulation runs assume that direct anthropogenic vegetation clearance starts and largely remains in the north and to a lesser extent the centre part of the upper Allen valley landscape, with no clearance in the southern part of the study area. In all simulation runs, the southern part of the study area is where the woodland and tree-dominated mosaics are to be found after 1500 years of pastoral activity up-valley. The landscape only becomes predominantly open if a rapid rise in livestock numbers is assumed to occur early in the Neolithic. Also, it was assumed that the north of the landscape is largely for the purposes of arable and non-farming activities (e.g. ceremonial, burial and settlement purposes) with grazing shifting gradually towards the centre of the study area over time.

Results

Some preliminary results are presented for regular timesteps of fifty years (Table 3.25), fluctuating rates of direct anthropogenic clearance and a vegetation classification system comprised of half-stage communities. The definition used in the table for a 'slow' increase in livestock numbers refers to a final (i.e. Late Neolithic/Early Bronze Age) population of 200 livestock units, whilst that for a 'rapid' rise in numbers refers to 400 livestock units by timestep 30 of a simulation run. The 'early' introduction of arable farming concept involves small-scale shifting agriculture from the first timestep of the simulation followed by a fluctuating, erratic transition to permanent, larger arable plots; the 'late' version introduces arable farming at the fifteenth timestep though the overall number of cells directly cleared by humans is the same in all scenarios presented here. With respect to the landscape structure, all of the scenarios in Table 3.25 assume the same initial conditions: moderate connectivity between and moderate clustering of like patches. For a quantitative definition of these terms in the context of the upper Allen valley model see Samarasundera (forthcoming).

The results, albeit representing a limited range of scenarios and only a small number of runs, suggest that livestock numbers are influential on model outcomes but that the date of introduction of arable farming is not. Scenarios 3 and 4 correlate most closely with the palaeoenvironmental record. The two grazing models tested provide a more complex set of results to interpret.

Discussion

Modelling human impacts on past landscapes, particularly in prehistoric contexts, is different in some respects from simulating anthropogenic activity in present-day and near-future landscapes. The former is in its infancy and this fact combined with the taphonomic biases inherent in the sub-fossil record means that 'off-the-shelf' models from the landscape ecology literature will not do and that relatively crude models such as the one presented here have to be the starting point for environmental archaeology. Thus a generalized model has been built here, one which requires further simulation runs and post-modelling analyses. Whilst the scenario outcomes presented are a representative selection of the preliminary results, it must be stressed that they do not give any indication of model sensitivity to a wider set of parameters and variables (e.g. degree of clustering of vegetation communities and the final number of cells directly cleared by people). Additionally, analyses of potential critical thresholds in the ecosystem model have yet to be carried out. Such issues are currently being explored. It should also be noted that in all scenarios, including those not published yet, it is assumed that the north of the Neolithic landscape near the Dorset Cursus was largely for the purposes of arable and non-farming activities (e.g. dwellings, 'ritual' purposes, etc.) with grazing shifting gradually towards the centre of the study area over time as the north became increasingly devoted to arable, settlement and cultural activities.

The small selection of simulation runs presented here tentatively hints that model outcomes are sensitive to livestock stocking density. This is to be expected and bodes well for the future development and testing of the upper Allen valley model. Nevertheless great caution should be exercised when interpreting the results of the eight scenarios presented here, for a number of reasons. One reason is that the livestock population growth sub-models included are linear, whereas in reality non-linear scenarios (e.g. asymptotic increases in livestock units) are more likely. Further research will no doubt aid in assessing whether model outcomes are sensitive to the growth rate parameter. A second foundation for caution is that the results discussed all assume the same initial conditions landscape structure and it is plausible that other vegetation patterns may have a significant influence on the impacts of grazing within the model. Other reasons include the assumptions of regular, fifty-year timesteps and the extent of arable activity in the Neolithic Allen landscape. With respect to the timing of the introduction of arable farming, whether it is introduced at the beginning of a simulation run or at the halfway stage appears not to have a significant difference on the final landscape structure. Future work may, however, demonstrate that under certain initial conditions and ecological assumptions this may not always hold true within the model.

The variation between outcomes with respect to the grazing model employed show that under an ungulate preference for open vegetation communities, feeding can cause major forest fragmentation if livestock densities are sufficient. Regardless of livestock densities, a theoretical livestock niche within vegetation mosaics does not produce outcomes that correlate with the palaeoenvironmental record in scenarios 5–8. However, it must be stressed again that these early model results consider only one hypothesized spatial arrangement of vegetation communities for the Late Mesolithic landscape. Therefore the assumption of a livestock preference for complex mosaics cannot be dismissed at this stage, though it should be noted that the eventual dominance of scrub over arboreal communities in scenarios 5–8 has clearly been influenced by the initial conditions landscape which is characterized by a marked presence of complex mosaics in the centre and southeast of the study area (Figs. 3.102 & 3.105). The choice of a grazing sub-model with a preference for meadow/pasture communities, scenarios 3 and 4 do produce a landscape predominantly open under the initial conditions assumed but only if a rapid rise in livestock population and final stock of 400 livestock units is specified.

The preliminary simulation results presented in this volume provide a tentative corroboration of the theory of a pastoral-dominated Neolithic economy in the Allen landscape. As already stated other scenarios are being investigated at the time of writing. The current, ongoing research is starting to put the results presented here into a wider context. It is becoming apparent, from the yet-to-be-published simulation runs, that the upper Allen valley model outcomes are sensitive to initial conditions of landscape structure, for example the degree of clustering or conversely fragmentation of Late Mesolithic canopy openings during the latest Mesolithic appearing to be particularly significant. Both current and future modelling work, along with the associated post-modelling analyses, will no doubt provide a more thorough picture of both the utility of the model and the plausibility of the mooted archaeological theory. A specific point to note in the context of model corroboration with the palaeoenvironmental data, is that in all simulation runs made to date for the upper Allen area, even those scenarios not included in the material here, the south of the landscape is where the woodland and tree-dominated mosaics are to be found after 30 timesteps (1500 years) involving pastoral activity.

As a postscript, the reader is asked to consider the value of linking dynamic ecological models with archaeological and palaeoenvironmental data beyond its merits of testing archaeological hypotheses. Such long-term palaeoecological modelling can potentially aid in a better understanding of long-term human impacts upon landscapes and has practical applications in woodland management and environmental impact assessment. Whilst landscape ecologists can draw upon superior quality data and can validate their models more robustly and reliably, they deal in short-term changes which are effectively the 'noise' in the longer-term trends in the palaeoenvironmental record. Linking the past, present and future environmental datasets in simulation models opens up a range of possibilities and pitfalls in assessing current and potential future human impacts upon nature. Over the next few decades this is a likely major area of research, one in which archaeologists can play a significant role.

Another potential avenue for simulation modelling is in archaeological prospection. It may be the case that a well-tested, robust and reliable model can play a role in predicting 'hotspots' within a landscape meriting further fieldwork, thus supplementing the role that satellite remote sensing and aerial photography already play.

Conclusions

The key conclusions are:

1. An early introduction of livestock appears necessary along with a rapid increase in the importance of pastoral farming, for an open landscape to develop in the model by the Late Neolithic. This assumes a livestock preference for open vegetation communities.
2. The date of introduction of arable farming has only a negligible effect on model outcomes.
3. The scenarios presented in Table 3.25 and Figures 3.102–3.105 hide significant differences in the nature of the mosaics, that is park-type communities and scrub are distinguished in the model but not in the table. Considerable caution must therefore be exercised when interpreting scenarios 5–8.

Chapter 4

Prehistoric Land Use and Landscape Development in the Upper Allen Valley

Charles French & Helen Lewis

Comparisons with Wessex and revisions to the general palaeoenvironmental models for the region
CHARLES FRENCH

The evidence for closed versus open woodland
A commonly held view is that the majority of evidence elsewhere in the Wessex chalk region and beyond in much of Britain strongly suggests that closed woodland was almost ubiquitous by the late Mesolithic (Pennington 1974; Bennett 1988; Edwards 1982). This woodland exhibited considerable variation between regions and within different landscape zones in many study areas. Throughout the subsequent Neolithic there were a variety of intensities of inroads into that woodland (Smith 1970; Edwards 1982; 1989; 1993), but it must be seen as a greater or lesser backdrop to Neolithic activity on the chalk downlands. The palaeoenvironmental data for southern England was very well summarized by Whittle *et al.* 1993, 226–33, over a decade ago. Woodland on the lower slopes and plateaux of the Upper and Middle Chalk is well attested, for example at South Street, Avebury and Windmill Hill (Ashbee *et al.* 1979; Evans 1972; Evans *et al.* 1993; Fishpool 1999), Knap Hill (Connah 1965; Evans 1972) and Fyfield Down (Evans 1972), in dry valleys such as at Beckhampton Road where there was an early Holocene forested phase (Evans 1972, 250–51; Ashbee *et al.* 1979), and in river valleys as at Avebury Trusloe (Evans *et al.* 1988; 1993) to the foot of the chalk scarp at Cherhill (Evans *et al.* 1978; Evans & Smith 1983). Evidence for woodland is also attested from elsewhere in the greater region and beyond, for example in the Chilterns (Evans 1966), the Cotswolds of Oxfordshire (Evans 1971), the South Downs in Sussex (Thomas 1977), in flint mines (Evans & Jones 1991), Maiden Castle in south Dorset (Sharples 1991), the southern part of the Lincolnshire Wolds (Evans & Simpson 1991), the North Downs of Kent (Kerney *et al.* 1964), and the lower Exe and Exebridge valleys of Devon (Fyfe *et al.* 2003).

There are also significant geographical gaps in terms of the coverage of the evidence for southern England, such as for the Salisbury Plain area. Here there is sparse data from molluscs, namely from Durrington Walls (Wainwright & Longworth 1971), Woodhenge (Wainwright 1979), and from Stonehenge itself (Allen 1997b; Evans 1984; Cleal *et al.* 1995) (Fig. 4.1). But pollen data from the Avon valley just to the east of Durrington Walls (Scaife 1994b; 2004) hint that a more mosaic landscape was present when these monuments were in use during the Late Neolithic and Early Bronze Age, comprising marsh, grassland, arable fields, scrubland and woodland, just a few kilometres away from the predominantly open short sward grassland around Stonehenge on Salisbury Plain (Allen 1997b; Evans 1984).

These conclusions are primarily based on molluscan data, with pollen data only available from a handful of sites such as the River Avon near Durrington Walls, and even less sub-regional data is available from off-site records such as in dry valleys and/or river valley systems. Moreover, when pollen data are obtained, they often suggests a much greater degree of variability and openness than the molluscan data would suggest at first glance. For example, rather poor pollen data were recovered from Easton Down, Wiltshire, which suggested that there was a mosaic of open-canopy plant communities of grassland, waste ground, bracken and hazel immediately around the long barrow site, with woods nearby (Cruise 1993). Also, the recent studies of Fyfe *et al.* (2003) in the Lower Exe and Exebridge reaches suggest a partly wooded and open valley floor in the Mesolithic which developed into almost closed woodland by about 7600 cal. BC. Additional evidence from further afield in the Pannel valley, east Sussex (Waller 1993), Lismore Fields in Derbyshire (Wiltshire & Edwards 1993), and the Etton landscape in the lower Welland valley, Cambridgeshire (Scaife 2005), suggests river valley and riverine areas may have been much

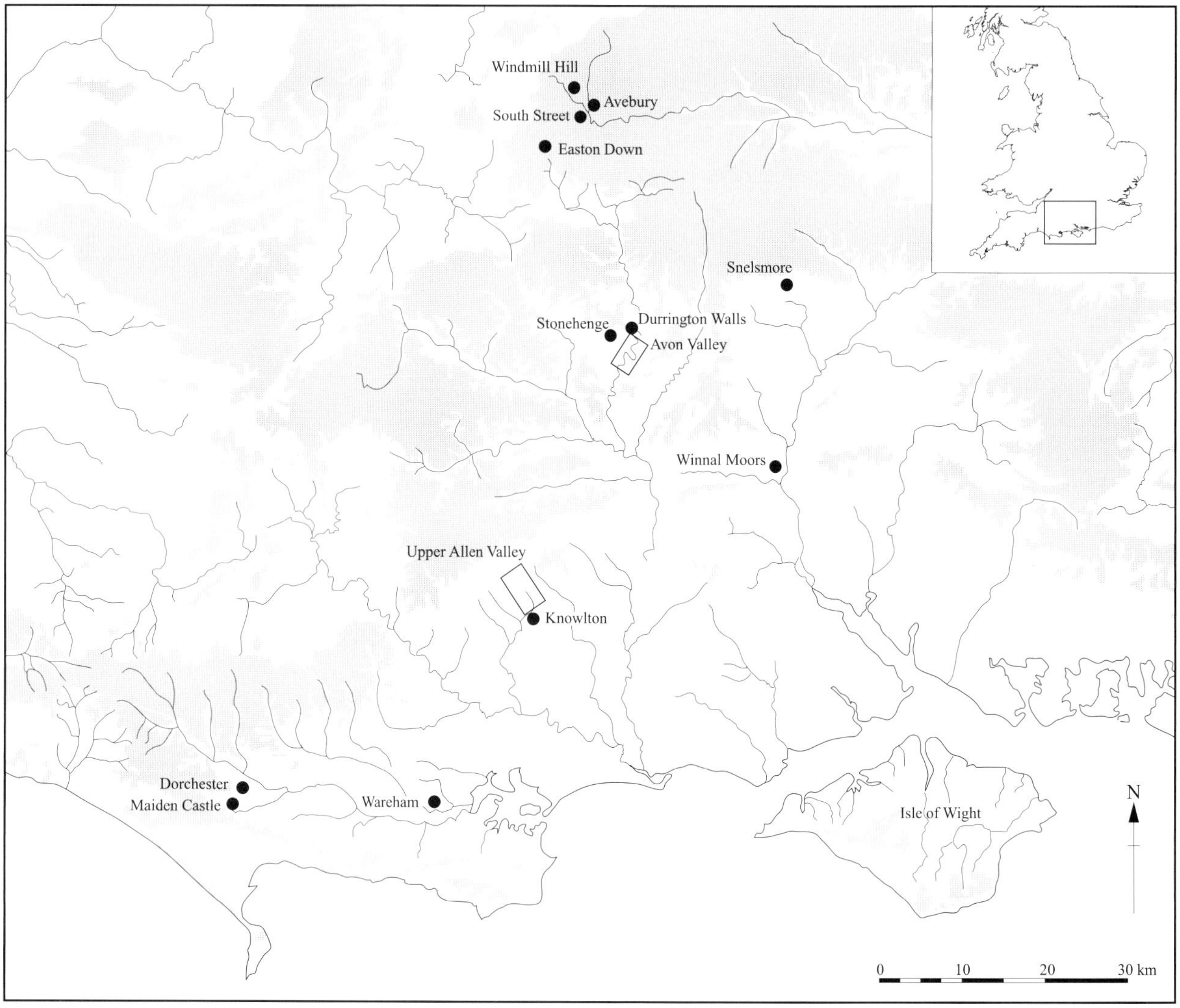

Figure 4.1. *Location map of important sites and survey areas for molluscan and palynological data in Wessex (C. Begg).*

more open than adjacent upland areas. The reasons for this are not absolutely clear, but may relate to a variety of factors such as the distribution of woodland in the Mesolithic, variation in woodland composition, windthrow, natural fires and the impact of grazing animals (Brown 1997; Fyfe *et al.* 2003). In sum, these factors may have helped to create areas of landscape that were more open and disturbed in the Mesolithic and therefore more susceptible to exploitation within the Neolithic.

The earlier studies of Thorley (1981) and Waton (1982a,b; Waton & Barber 1987) do, however, suggest the presence of woodland on the chalk from the Mesolithic period, as does Scaife's work at Wimborne Minster in the later Mesolithic (see above). Waton (1982a,b; Waton & Barber 1987) investigated a series of ten different valley systems (with 70 cores) for new pollen data on the chalk of southern England, but only one core contained sufficient pollen to count, at the site of Winnall Moors in the Itchen valley, just north of Winchester (Fig. 4.1). He then tried searching just off the chalk substrate by up to about 500 metres, and found five sites with sufficiently good preservation to count for pollen. The Winnall Moors site produced a full sequence from the Boreal of open herb/grassland communities to birch/pine woodland, through a very early elm decline at 4720–4430 cal. BC (5630±90 BP; HAR-4342) followed by extensive woodland clearance with arable and pastoral farming from the Neolithic onwards until alluviation and the development of water meadows from the seventeenth century AD. Significantly, pasture tended to predominate over arable until about 1000 BC when cereal cropping increased for a brief period, then pastoral conditions again dominated.

The other sites showed considerable variation in their woodland and clearance histories (Waton

1982a,b). The pollen evidence from Snelsmore, *c.* 40 km north of Winchester in the Berkshire Downs (Fig. 4.1), suggested that there was minimal disturbance of the forest cover until the Early Iron Age. At Rimsmoor, near Wareham in south Dorset (Fig. 4.1), an elm decline was evident at the end of the Early Neolithic, but major clearance associated with cereal cultivation did not occur until later Bronze Age and Iron Age times (Waton & Barber 1987). There was also heathland and podzol formation in the Wareham area from the Bronze Age (Piggott & Dimbleby 1953). Similarly, at Amberley Wild Brooks, a site off the downs near Arundel in Sussex, three major waves of clearance in the first millennium BC led to largely open chalk downland development. Thus it appears that some areas of the chalk downlands were exploited earlier than others from the Early Neolithic, but many areas were not subject to widespread clearance activities until much later in prehistoric times. Nevertheless, there was an underlying tendency for grassland for pasture to be the most common land use once the downland had been cleared.

The palynological studies conducted by Scaife in this project (see above) have suggested that both the valley bottom and chalk downland slopes remained much more substantially open throughout most of the Mesolithic period than there was hitherto evidence for. This period of partially open downland has been dated from *c.* 11,000 to 7000 BP, at least in the upper reaches of the valley to the west of Wimborne St Giles, but whether this predominantly open landscape extended up to and into the Neolithic is more questionable. Certainly substantial woodland cover (of *c.* 45–65%) of the lower reaches of the study area is only attested at the Allenbourne car park sequence, Wimborne Minster, some 12 km south of Wimborne St Giles, towards the end of the Mesolithic period around 7000 BP. But this partly wooded element to the upper Allen valley is ostensibly on the wane by the Early Bronze Age at *c.* 4000 BP, and open grass-dominated downland has begun to predominate.

The modelling study by Samasundera (see Chapter 3 above) has attempted to visualize the palynological and molluscan data from the upper Allen valley over a series of timestep scenarios (Table 3.25; Figs. 3.102–3.105) and simulated the development of an open landscape by the Late Neolithic in the Gussage/Wyke/Bottlebush Downs area. This assumed clustered canopy openings already existed during the latest Mesolithic (rather than fragmented canopy gaps). An early introduction of livestock appears necessary along with a rapid increase in the importance of pastoral farming. The date of introduction of arable farming has only a negligible effect on model outcomes. In contrast, in the southern part of the study area, woodland and tree-dominated mosaics continued to be found well into the Bronze Age after some 1500 years of pastoral activity.

The work of Waton (1982a) at Winnall Moors, Hampshire, Waller (1993) in Sussex and Wiltshire, Edwards (1993) in Derbyshire and Fyfe *et al.* (2003) in Devon suggests that there were degrees of openness in river valley landscapes in the Mesolithic, and especially when compared to their adjacent uplands. But our investigations in the upper Allen valley, the work of Allen around Dorchester (1997a) and Stonehenge (Allen 1997b), and the work of Bush & Flenley (1986) and Bush (1988) in the Yorkshire Wolds (in the Great Wold valley and around Kilham long barrow) suggests that more substantial open areas may have been the case for some considerable tracts of chalk downland in the Mesolithic. Indeed, it has been suggested that grassland communities could have persisted in an otherwise forested landscape throughout the earlier Holocene in isolated refugia (Pigot & Walters 1954). A greater degree of open grassland may also have been a consequence of the long-term management of game, which then allowed and/or encouraged the construction of major monument clusters, such as Neolithic long barrows and cursus monuments. The creation and management of open ground associated with deliberate use of fire in the Mesolithic has often been suggested as a way of improving productivity of plants and shrubs suitable for grazing and therefore advantageous in attracting game (Mellars 1976; Mellars & Dark 1998; Law 1998; Dinnin & van der Noort 1999; Mithen 2000). Despite the Thomas (1989) and Bush debate (1989) about the nature and degree of openness of the chalk downland, the new sub-regional pollen data would suggest that there were different trajectories of soil and vegetational development on the chalk, and that certain areas may well have been much more open, and kept open, since Mesolithic times. Thus the Mesolithic population may well have deflected and altered the post-glacial vegetational succession in some parts of the British Isles.

Another feature of this project is that no valley sedimentation is really evident until historic times in the upper Allen valley, and even then it is very limited in extent and thickness (Figs. 2.2–2.10). Almost no colluvial accumulations are evident in the dry valleys or flood-plain edge in the upper Allen valley, and those that have been found are thin, generally in the range of *c.* 15–30 cm in thickness, and occur in a very localized manner. Similarly, alluvial deposits are few and far between, and only begin to occur as the flood plain widens out downstream from Knowlton and around Wimborne Minster, but as shallow deposits (*c.* 30–50 cm) with very variable extent. Similarly in the Kennet valley around Avebury, Evans *et al.* (1993) suggested that sedimentation in the valley floor was a rare event. There, in the early Holocene the valley was open and marshy, but by about 8400 BP, this gave way to dense woodland. Woodland clearance, paludification and some limited soil erosion began after 3800 BP with open grassland valley slopes predominant by the Bronze Age. Limited sedimentation in the valley floor occurred only from the later Neolithic to Iron Age and again in the post-medieval period. Importantly, the Kennet valley floor was cleared early on in the Neolithic, with no evidence of later Neolithic woodland regeneration. For much of the Mesolithic and Neolithic, as well as the Middle Ages, the valley floor environment was little different from the valley slopes.

Returning to the molluscan data, there are a number of sites from the southern chalklands that suggest varying degrees of well-developed grassland prior to construction of the first monuments, usually Neolithic long barrows. John Evans's seminal study of some 20 Neolithic sites investigated on the Wessex chalk (1972), suggested that only three of twenty sites investigated had a woodland fauna in the buried soil, with the majority having an open-country fauna in the surface layers of the buried soil beneath the mounds. For example, the long barrows at Beckhampton Road (Ashbee *et al.* 1979; Evans 1972), Horslip (Evans 1972), Ascott-under-Wychwood (Evans 1971) and Thickthorn Down, just to the south of our study area at the southern end of the Dorset Cursus (Kennard 1936; Entwistle 1985), exhibited predominantly open-country faunas associated with rendzina grassland soils directly beneath the long mounds, as did the two long barrows that our project investigated on Gussage Cow Down (GCD 78 & 294) (Fig. 3.1). Moreover, all of the ditch and buried soil faunas associated with the Wyke/Bottlebush Down parts of the Dorset Cursus exhibited dry and very open grassland with few vestiges of any woodland habitats surviving by the later Neolithic (see Allen above). At Gussage Cow Down site 78we are able to say not only that the barrow was built in a grassy clearing, but also that this was a clearing for a substantial time before monument construction. In addition, there are no indicators of a forest soil within the thin sections studied. This clearing could be a feature of the later Mesolithic landscape, as well as the early Neolithic, and pre-Neolithic landscape history could be important to monument siting decisions; i.e. the location of this and other long barrows in the landscape could relate as much to, and tell as much about, Mesolithic decisions as Early Neolithic ones. Remarkably, two of Evans's sites which were exceptional in being associated with a woodland molluscan assemblage were the causewayed enclosures at Windmill Hill and Knap Hill, which appear to have been built in clearings within hazel woodland. As Evans (1972, 247) suggested, perhaps these sitings were 'in some way special' to begin with.

Certainly, the vast majority of buried soil profiles beneath Neolithic long barrows in the Wessex region are evidently rendzina soils associated with and indicative of calcareous grassland soils. Although the turves of some mounds showed a history of deriving ultimately from argillic brown earths (relict indicators of clay-rich and/or eluviated Eb horizons), intact buried soils under all monuments investigated here, with the exception of round barrow 41, were well-developed shallow rendzinas associated with the establishment of turf and the maintenance of grassland. This is true for all periods, from the Early Neolithic through to Roman times, regardless of position in the valley. This is discussed further by Lewis below. Moreover, all standstill horizons in ditches, where they occurred, were also indicative of turf development. It was only very occasionally, such as in the Monkton-up-Wimborne valley trench, that a brown earth was evident, and was perhaps associated with a degree of woodland cover. Well-developed woodland soils or argillic brown earths, were only exceptionally found. This soil type was only found on the southeastern margins of the study area, associated with either clay-with-flints or Reading Bed geology and on the highest parts of the Chase to the north and outside the study area, again associated with clay-with-flints outcrops (Fisher 1991; Barrett *et al.* 1991a). Only one (modern) argillic brown earth was found on chalk, in a borehole in the St Giles Park woodland ride, suggesting that this area of woodland has been there for a long time. However, it was not possible to sample this profile for land use history; i.e. we cannot say whether that location was always woodland.

Evidence for arable agriculture

In both the Beckhampton Road and Ascott-under-Wychwood cases, it was suggested that a rendzina soil type had succeeded a brown earth. Also, at several other sites there are indications of disturbance prior to the building of the long barrow such as at Skendleby (Evans & Simpson 1991), Horslip, South Street and West Kennet, as well as beneath the mound at Silbury Hill and beneath the bank at Avebury (Evans 1972). At Hazelton North long barrow, the soil micromorphological data showed that the brown earth had seen the addition of midden-like material, soil homogenization and illuvial dusty clay coatings in voids, derived from bare ground through cultivation (Macphail 1986; 1990). At Easton Down, there were strong hints in the pollen, soil micromorphological and molluscan evidence for soil horizon mixing caused by ploughing (Cruise 1993; Macphail 1993). Most importantly, several sites show successions of land-use change in the snails and soil data, such as South Street where a sequence is suggested of ard ploughing, grassland, renewed tillage and then a return to grassland (Evans 1972). At Gussage Cow Down long barrow site 78, the molluscs hint at some bare ground in the vicinity, just possibly ploughed, associated with pre-existing open and dry grassland, and the soil evidence suggests a change from a brown earth to a rendzina profile prior to burial. Unfortunately, other sites with good soil data are few and far between, and similar molluscan data is only corroborated by soil micromorphological data at Easton Down and Hazelton North, and at the sites investigated in the upper Allen valley (see above).

In addition, there are sometimes molluscan data to suggest some woodland regeneration was occurring later on in the sequence of the infilling of the ditches. This occurs at the Easton Down long barrow (Rouse & Evans 1993), and the Knap Hill (Connah 1965) and Windmill Hill (Whittle *et al.* 1999) causewayed enclosures. In each of these cases, slight woodland regeneration is followed by expanding areas of grazed grassland conditions, reflecting a more widespread opening up of the upper downland zone in the later Neolithic. At first this was also thought to be happening at the Gussage Cow Down long barrow 294 (see above), as well as at the Handley Down Neolithic mortuary structure (Allen 1997c), both on the high ground to either side of the upper Allen valley in our study area, and also in the secondary fills of the Dorset Cursus (see Allen below; Entwistle & Bowden 1991). But Allen now suggests (above) that what we may be seeing is locally shaded conditions produced by a light cover of bushes and shrubs in a park woodland type of environment which is otherwise an established, grazed grassland sward.

Molluscan data showing regeneration of woodland, often in secondary, later Neolithic contexts, can be problematic. As the resolution of snail data is probably best measured in terms of small fields, or over a distance of no more than 50–100 m, the snail populations probably primarily reflect the nature of the ditch, its particular micro-environments and its immediate environs only. Also, the amount of chalk rubble that often acts as much of the secondary fill of the ditches associated with large monuments can create moist and shaded micro-environments that could just as easily be on the woodland floor. Also, different land-use treatment immediately around a monument like a long barrow may lead to local regeneration of trees and shrubs, and one needs to think about the diversity of the assemblage, and whether key woodland species associations really exist to ascribe the immediate conditions to woodland as opposed to locally scrubby conditions or indeed refugia. Thus, ascribing wider woodland regeneration on the basis of molluscan data alone can be a difficult interpretative call, and may well only be reflecting very localized conditions.

Evidence for colluviation and alluviation

As in the upper Allen valley, finding thick colluvial and alluvial deposits that are ascribable to early prehistoric soil erosion in the dry valleys and valley bottoms is not as common in the chalk downlands of Wessex as one would have expected. Good examples of colluvial profiles with erosion beginning in the later Neolithic such as Strawberry Hill and Durrington Walls, Wiltshire, Redcliff and Newbarn Coombe, Isle of Wight (Allen 1992) are often viewed as normative. But as Evans *et al.*'s work (Evans *et al.* 1993) in the Kennet valley and our own in the upper Allen valley have shown, there may have only been a slow trickle of sedimentation downslope during later prehistoric and post-medieval times. It is hard to replicate the prehistoric, deep and complex erosion sequences found for example at Kiln Combe in the dry valleys of East Sussex by Bell (1981; 1983; 1992). Very possibly, this has much to do with reaching erosion thresholds dependent on the degree and length of slope and the particular combination of geological substrate, slope angle, soil and vegetation cover at each locale (after Selby 1993), as well as the particular circumstances of human impacts and land-use history in each part of each valley. Generic and sweeping models of 'deforestation equals colluviation' in the dry valleys of the chalk downlands of Wessex are now much too simplistic. Each valley system, and indeed different reaches of the same valley system will have different developmental histories. A similar conclusion was reached by Wilkinson (2003), who discusses colluvial sequences in dry valleys in the southern chalkland region, and suggests that these should be related to palaeoenvironmental and land-use changes on a local scale only.

What does seem clear is that vegetation type and slope are major controlling factors in creating the colluvial sequences in the dry valleys within our study area. The wide and shallow slopes of the upper part of the Allen valley around Down Farm (Fig. 1.9), generally with less than 15–20 degrees of slope, would appear to preclude more than minor amounts of valley sedimentation, especially as a long-term grassland environment seems to have been the norm since at least Neolithic times. But where lynchet systems have been established, presumably in later prehistoric times and associated with cultivation, such as on the north-facing slope of Gussage Down, just

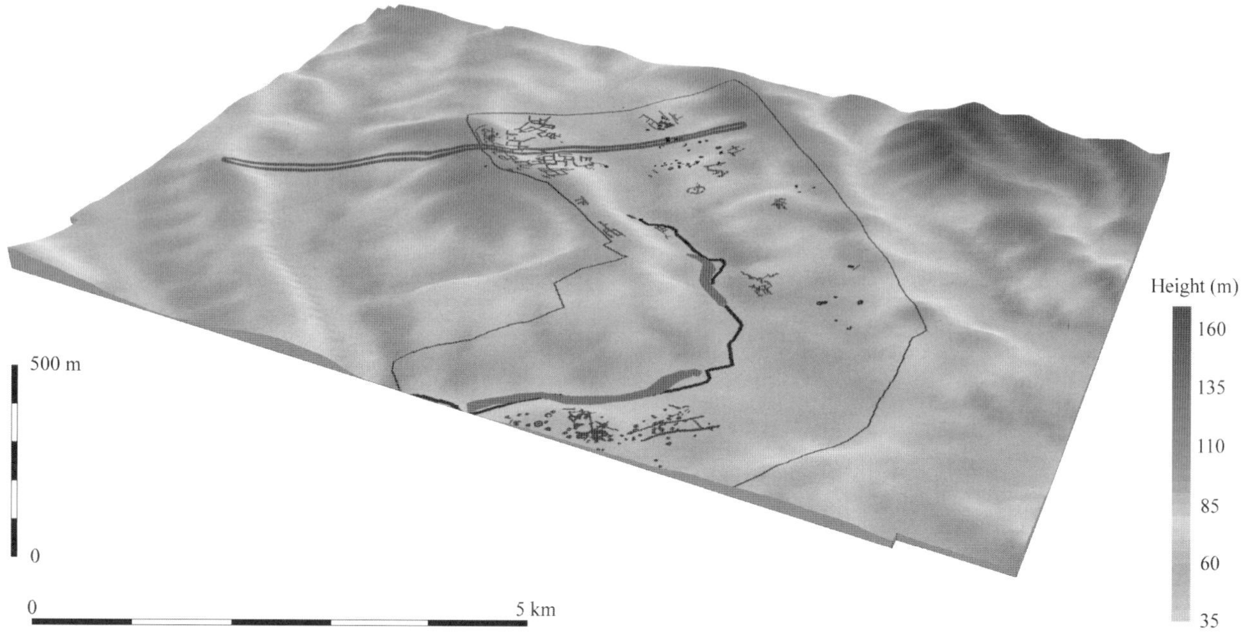

Figure 4.2. *A digital terrain model of the upper Allen valley looking northwestwards from the Knowlton complex upstream and up-valley to Wyke/Bottlebush Downs and the Dorset Cursus (C. Begg).*

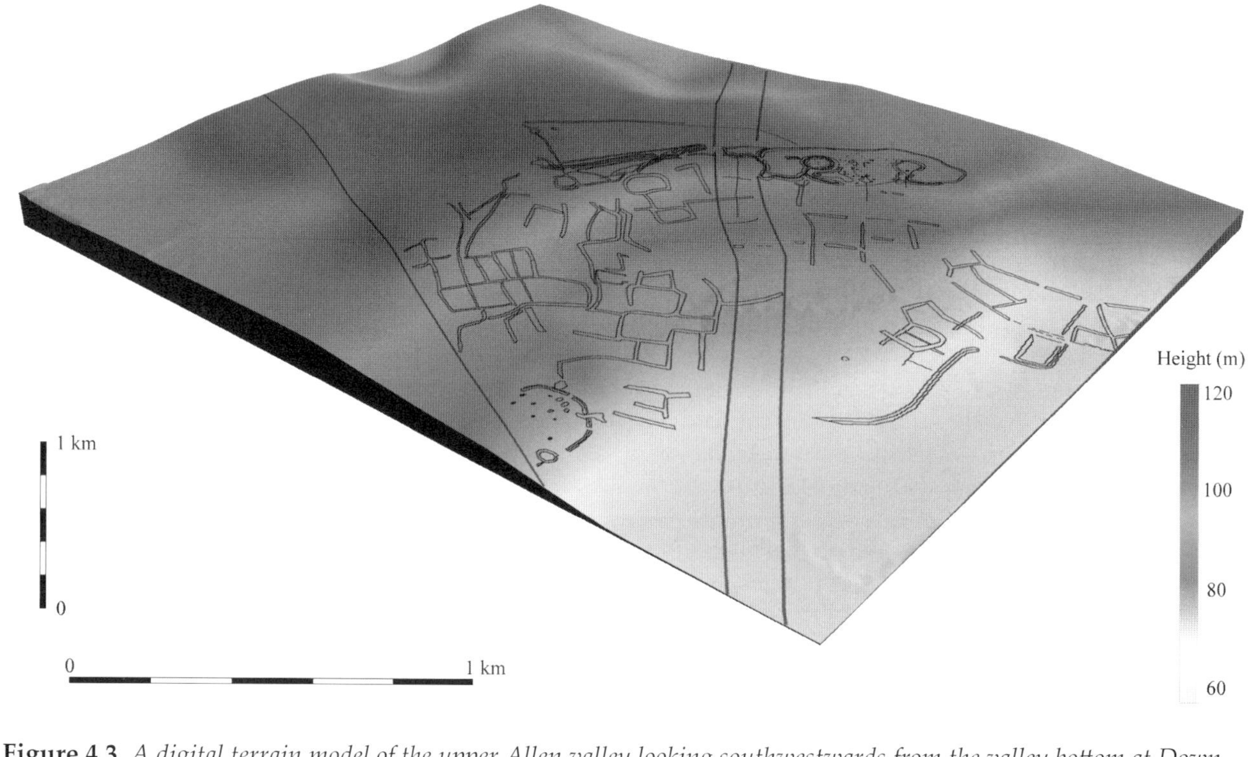

Figure 4.3. *A digital terrain model of the upper Allen valley looking southwestwards from the valley bottom at Down Farm up to the top of Gussage Cow Down which shows the arrangement of the later prehistoric field systems against the line of the Neolithic Dorset Cursus (C. Begg).*

behind Down Farm, and where the slope is slightly greater in incline (*c.* 25–30 degrees), spatially discrete colluvium is found on the uphill side of the lynchets (*c.* 15–30 cm thick) and in the dry valley bottom at the southern edge of Fir Tree Field (up to 1.2 m thick). Similarly on the opposite side of the valley on Bottlebush Down, a 50 cm depth of colluvium was located at the base of a steeper-sloped dry valley (Figs. 2.7 & 2.8). It should be pointed out that these small zones of colluvial deposits were the exception, not the rule (as first expected), and they appear to be of the later historic period in date. Figures 2.2 and 2.9 show the maximum potential area of land susceptible to erosion in the study area, conservatively less than one-eighth of the total landscape available, and immediately one is struck by what a small area that is. Similarly, the only colluvial and alluvial deposits in the river flood plain that have been found are to the west of St Giles House and Knowlton, and these are thin (<50 cm), generally located on the margins of the flood plain, and very late in the sequence, that is above prehistoric peat formation in cut-off channel meanders.

Conclusions

Through the combination of the extensive molluscan studies, palynological analyses in four palaeochannels between Wimborne St Giles and Wimborne Minster, and the on- and off-site soil and geoarchaeological studies, a detailed picture of landscape development has emerged for this part of Cranborne Chase. The combination of several new pollen sequences and the dense number of molluscan analyses, as well as the use of diversity indices for the first time in this landscape to help interpret the molluscan results, has given a very time and space extensive and more detailed impression of the environmental and land-use change in the study area. Woodland development in the early Holocene was slower and more patchy in the uppermost part of the Allen valley, contrary to most pre-existing models of vegetational development for southern England. The relatively more open/woodland mosaic of this part of the chalk downs was probably soon exploited by Mesolithic hunters, and it may be the encouragement of game through the use of fire and the creation of grassland for grazing which maintained the area as quite open, essentially inhibiting the development of full closed canopy woodland here. This 'parkland' landscape was most marked around the uppermost part of the valley from Gussage Cow Down to Wyke and Bottlebush Downs (Figs. 1.2 & 2.1), thus perhaps pre-adapting that area to the later strong human focus with the Neolithic long barrows, the Dorset Cursus and the Early Bronze Age round barrows (Figs. 2.1 & 4.3), and similarly around the Knowlton complex of Neolithic henges and Bronze Age barrows (Figs. 2.17 & 4.2). This degree of openness is seen particularly well in the soils where long established grassland appears to have already existed by the Neolithic in this same upper part of the Allen valley. Soil change from argillic brown earth to brown earths and rendzinas has occurred, and turf grassland had became well developed after some degree of soil disturbance. This pre-supposes the long-term exploitation of parts of the high parts of the downs and downland slopes, ostensibly as grassland, in the later Mesolithic and Early Neolithic. Moreover, this pre-existing relative openness in the Mesolithic appears to have been capitalized on by the construction of the lengthy Dorset Cursus in the fourth millennium BC. Perhaps this was along long-established routeways of people and game from the coast to the south to the Salisbury Plain and Stonehenge area to the north, as well as areas frequented for flint resource procurement from the clay-with-flints areas on the higher parts of the downland as identified by Barrett *et al.* (1991a, fig. 2.3).

Perhaps something different occurred in the upper part of the Allen valley, possibly associated with the absence of full or patchy woodland development in the earlier Holocene. But perhaps it is better not to think of it as different, rather as representing just one extreme in a mosaic and variably vegetated landscape; at the other extreme would be a firmly closed woodland in the later Mesolithic/earlier Neolithic. We are not saying that everywhere in the study area or elsewhere in the Wessex chalklands was as open as the Gussage to Wyke to Bottlebush Downs area. Indeed, the relatively blank archaeological zone in the central part of our study area (Figs. 2.1 & 4.2) suggests either different land use or activities that do not leave the same archaeological signal existed in prehistory, and/or was possibly associated with better-developed and longer-lived wooded conditions. Moving southwards, there is the explosion of earlier prehistoric activity in the Knowlton area, especially in the later Neolithic, but here we know that there is woodland soil cover as a backdrop on the slopes to the east and southeast associated with Reading Beds and clay-with-flints geology. Further south towards Wimborne Minster there is also significant development of mixed deciduous woodland by the later Mesolithic and into the Neolithic period. These differences in vegetation complex testify to the very localized variation in the density and development of woodland cover in the Mesolithic over the upper Allen valley over a distance of about 8–15 km.

There is little doubt that there was ubiquitous and well-maintained grassland throughout the study area by the beginning of the Early Bronze Age. The whole valley from Down Farm to Wimborne St Giles to Knowlton became fully open, with grassland predominating, as indicated by the pollen profiles in infilled channel meanders in the flood plain at St Giles House and just northwest of Knowlton. This is corroborated by the ubiquitous open-country snail faunas associated with numerous Bronze Age sites throughout the study area.

The landscape at this time appears to be stable and without any real valley sedimentation going on. It at first seemed remarkable that so few deposits of hillwash (colluvium) or alluvium (overbank flood deposits in the valley bottom) existed in the study area despite intensive and extensive field investigations, and contrary to the picture derived from many other areas of the chalk downlands. But the absence of well-developed brown forest soils, except in the southeastern part of the study area, implies relatively thin soils. Coupled with the early and long-lived establishment of grassland in much of the study area and the relatively minor impact of arable agriculture, there is no need to invoke wholesale soil erosion in later prehistoric times. What little erosion there is, usually less than 50 cm in the dry valleys and valley flood plain, and very limited and variable in its spatial coverage, suggests that there was little sedimentation in the valley system and most of that accumulated in historic to more recent times.

Strong signatures for arable land and cultivation are rare and where they do occur, slight, suggesting that arable agriculture was probably of secondary importance in the area to grazing in terms of overall agricultural land use. Turf creation and maintenance was obviously also very important for constructing burial monuments, and there is evidence that turf was sometimes chosen from particular local parts of the landscape for this purpose (see Lewis below). The main period of field system creation in this landscape is in the Iron Age, in particular on Gussage Cow Down. As this extensive system respects and is aligned parallel and at right angles to the Dorset Cursus, however, it suggests a considerable longevity to either the Cursus monument itself, or the land division seen in this part of the upper Allen valley. Finally, the rare and late occurrence of soil erosion acting to infill dry valleys and the flood-plain margins, suggests that arable agriculture was not the driving force in this landscape until relatively recent times.

Holocene soil change in the upper Allen valley

Charles French

What is striking about the bulk of the study area is the ubiquity of thin rendzina type soils, both as the modern-day soil profile and beneath various types of prehistoric monuments. This has been a regular occurrence at many other monuments in the Wessex region (see above), and has given rise to interpretations involving soil degradation and erosion associated with human use of this landscape since Neolithic times, but especially occurring in the later Bronze Age and Iron Age times. Moreover, this soil type is associated with both grassland and as the present-day ploughsoil under an intensive arable farming regime. It is generally only where the geology changes in the extreme southeastern part of the study area, beyond the Knowlton henge complex, to an area of Pleistocene soliflucted, head deposits that the present day soils become more clay-rich, more moisture retentive and better developed as argillic brown earths and brown earths.

Exceptions to the rendzina soil type, past and present, are few and far between. These exceptions include the occurrence of an argillic brown earth associated with the late Upper Palaeolithic palaeosol at the excavated site of Deer Park Farm which is developed on gravelly Reading Beds, and strong hints of former brown earths being present beneath the bank of the South Circle at Knowlton and beneath the Neolithic long barrow SAM 78 on Gussage Down, both on chalk, and some of the round barrow mounds. In the short augering transects on the western side of Wimborne St Giles, stony brown earths also appeared to be present beneath thin colluvial deposits on the valley slope (Fig. 1.11, transects 17 and 18 & Fig. 2.5). Nevertheless, at Gussage long barrow site 78 on the top of the down, the brown earth had already been transformed to a turf and rendzina profile by the time the first barrow mound had been built, at some point in the fourth millennium BC. On the downland slopes at the head of the upper Allen valley, both the Dorset Cursus banks and Wyke Down round barrows were built on short-turved rendzina soil profiles. By comparison with the profile beneath Gussage long barrow 78, these later Neolithic and earlier Bronze Age soils contain less organic matter and exhibit a less well-developed Ah horizon or organic matt, and are more of a dark reddish-brown than a greasy, very dark grey to black in colour. Although these observations are only qualitative and relatively minor in nature, they do suggest that further degradation of the organic and structural status of the rendzina soil had occurred during the latter part of the Neolithic period in this downland area.

How does the brown earth soil type become a rendzina? What is essential to maintaining a rendzina profile is being under turf with a very active soil fauna, all on a calcareous substrate. Although there is no easy, absolute and straight-forward answer, the 'degradation' sequence of forest/argillic brown earth to brown earth to rendzina is primarily dependent on the vegetation complex, physical disruption of the soil profile and the movement of soil bases or nutrients, and the establishment of turf grassland. One possible route of soil change is as follows: under a deciduous woodland, the soil is well nourished, and receives rainfall, organic litter and a high rate of microbial attack on the plant tissue, allowing the organic litter to become finely comminuted and mixed with the mineral soil, which in turn forms a nutrient rich organic A horizon and gives the soil a strong and well-aerated crumb structure. Clay minerals form 2:1 layer lattices which give the soil structural integrity and retain nutrients derived from the plant complex; the physical/chemical alteration of the parent material (limestone) leads to the formation of a weathered Bw (or cambic horizon), which is characteristic of any brown earth. Where the parent material is base-rich on a calcareous substrate, as on the chalk downlands of Wessex, calcium carbonate is precipitated and may form a diffuse whitening or discrete concentration as a horizon in the base of the B horizon, even though the A and upper B horizon may be relatively decalcified, to become a brown calcareous earth (Avery 1980) or a calcareous brown forest soil (Limbrey 1975, 133). To become a *sol lessivé* or argillic brown earth, the high base status of the soil ceases to be maintained. This leads to the dispersion of colloidal fine material in the upper part of the soil profile, its slow migration down-profile and immobilization towards the base of the profile, often at a textural or chemical boundary, and/or just above the ambient groundwater table, and where the base concentration increases, an illuvial accumulation is produced of fine material, generally clay, leading to the formation of an argillic (or Bt) horizon (Limbrey 1975, 134–7).

For an argillic brown earth to develop into a rendzina, this sequence of dispersal, mobilization (or leaching) and immobilization processes must be disrupted. This mainly occurs through events such as deforestation and arable agriculture, and phenomena such as climate change with increased rainfall, alteration of vegetative cover, and associated mechanical disturbance, and the failure to replenish nutrients lost through the mobilization process through either manuring and/or insufficient lay or fallow periods. In order to disrupt this *sol lessivé* sufficiently, it is generally considered that either severe climatic change and/or disruptive human activity must persist to allow sufficient leaching to remove most of the fine components of the soil (i.e. silt, clay, organic matter, bases/nutrients). As rendzinas develop and are maintained largely through earthworm activity, a soil's chemical and nutritional status must become amenable to increased earthworm influence which will eventually homogenize the brown earth horizons into one main A horizon. Physical breaking up of the soil, say through tree-throw and ploughing, and the reduction of input of organic matter into the soil to retain structural integrity, (if ploughing) without manuring or fallow/lay periods, are seen as key factors in setting this degradation cycle in motion. Once degraded, the trajectory appears to be the development of turf grassland, this then allows stability and constant earthworm mixing that are essential to maintaining the rendzina soil type. But what controls and sets in motion the change in land use is harder to say — it could be driven by the recognition of the degradation process itself as much as social, economic or climatic factors as well as changes in population dynamics.

To produce these effects on the chalk downland of the upper Allen valley, the following more specific scenarios could perhaps be envisaged. If woodland cover occurred later in the Mesolithic and was variable in its formation and extent of coverage, as the palynological data from this project would suggest, this would imply that both horizon development and structural integrity of the brown forest soil may have been compromised and less well developed in the earlier Holocene than would have been expected. Indeed, these soils may also have not had sufficient formation time to have become very thick and well developed, that is they were in the 30–35 cm range rather than twice that thickness. As the soils were naturally dominated by silt-sized material and had a low clay component, the soils were very permeable, and once tree/vegetative cover had been removed and the organic binding qualities had become depleted, they were readily susceptible to leaching and the mobilization of fine material and nutrients both down-profile and into the soil water system downslope. When these soils were either ploughed for the first time, and/or were not receiving new organic input through manuring or lengthy fallow periods, they soon became structurally weak and nutrient/base-poor. Thus both the A and B horizons of the brown earths were easily mixed by soil fauna, along with any tillage, and at the same time easily became prone to both wind-blow and downslope erosion processes when bare. As there is little evidence of significant amounts of colluvial hillwash in the dry valleys and/or alluvial sedimentation in the river valley bottom, significant leaching, decalcification and a lowering of base status must be seen as the real causes of soil change to mono-horizontal soil cover with thin brown earths, i.e. not the erosion of bare, tilled soils. At a stroke, this suggested scenario makes it easier to accept the apparent lack of earlier prehistoric signatures of arable agriculture, the large open areas of grassland available to the Cursus, henge and barrow builders in the Neolithic and Early Bronze Age, the formation of rendzina soils on an extensive basis on the chalk downland, and the lack of evidence for any great erosion history in the area during prehistoric times.

This goes hand in hand with envisaging a largely grassland landscape utilized extensively for pasture on the chalk downs and slopes, especially throughout the Neolithic and Bronze Age periods. This is not to say that there was no arable agriculture in these peoples' lives, but it was a relatively minor activity as opposed to the more ubiquitous livestock husbandry. Although there is no specific physical archaeological evidence to support the following suggestion, it is possible that arable crops, which are consistently present in the pollen record from the middle of the third millennium BC (see Scaife above), were produced in fields set adjacent to the flood-plain edge where the flood plain is wider and the valley edges have low relief, and where in some areas flinty drift/fan soils are found. Here, it is very possible that moisture retention in the brown earth and rendzina soils would not have been an issue, with winter and spring-time replenishment through natural winterborne flooding, and any arable soils would not have been subject to the erosive forces expected on the downland slopes, and therefore not have produced obvious colluvial or alluvial signatures. As these flood-plain edge areas would have been effectively acting as receiving zones for the replenishment of soil bases, nutrients and groundwater, they would have been ideal, self-replenishing areas that were much more suitable for arable crop (and fodder grass) production than the adjacent, thinly soil covered and depleted downland slopes.

Problems of land-use recognition

Charles French

So, why are we as archaeologists not able to detect these different areas of land use easily, especially in this open, rural landscape? In this particular study area, these same flood-plain edge areas are today often occupied by farms, farmyards, houses, villages and roads, and their archaeological visibility is poor. Also, these areas are often today used as homefield pasture, paddocks for horses, as garden areas, or are covered in concrete as farmyards. In addition, thin and variable colluvial and alluvial cover may in combination with these other present-day factors be obscuring the archaeological recognition of field boundaries. Moreover, the boundaries that we are looking for may not take the form of banked and ditched enclosures as on the adjacent downland, but have been organic boundaries formed mainly of fences or hedges formed out of water's edge plants such as sedges, and stands of alder, hazel or willow, for example, which may have served a twin function as coppice and field boundary. All of these vegetation components are present in the pollen record, but it is much harder to ascribe actual spatial locations to these components.

This situation appears to begin to change in later prehistoric times. Specifically on Gussage Cow Down an extensive series of sub-square to sub-rectangular enclosed fields (Figs. 2.1, 2.13, 4.2 & 4.3) is constructed (incorporating the much earlier Dorset Cursus). There are also hints of other less well-developed field systems to the south in the Brockington Down area, the north and northeast at the head of the upper Allen valley, as well as extensive systems on Oakley Down and the next valley to the northeast, and the Knowlton area to the south (Figs. 2.1, 2.17 & 4.2), although the dating of all of these systems is open to question through lack of investigation. The formation of lynchets on the slope beneath Gussage Hill, now just surviving as darker soil scars on the slope, would suggest their use as arable fields in the Iron Age or later.

What is distinctive about the bulk of the remainder of the study area is that such field systems are not evident on the chalk downland slopes, and they certainly should be, given the high archaeological visibility of the area and the number of aerial surveys that have been undertaken (see Żuk & Palmer in Chapter 2; Bowen 1990). Coincidently, this largely blank area in the aerial photographic record more or less coincides with the present-day extent of the lands of the Shaftesbury Estate. Established in the mid-seventeenth century, substantial woodland rides were planted around most of the boundaries of the estate lands, the Allen River was effectively canalized and straightened in the Wimborne St Giles stretch, an ornamental lake made just downstream of St Giles House, and a substantial parkland planted between Wimborne St Giles and the Cranborne road, as the garden of St Giles House itself. Despite these landscape modifications, our augering survey has demonstrated that these areas are also dominated by chalk geology and thin rendzina soils (with only one argillic brown earth profile on the chalk), so the good archaeological visibility should be the same whether one is outside of this area in the Down Farm or Knowlton areas or inside the estate area. Indeed, earlier monuments (barrows, enclosures) are found inside the estate. Thus the lack of visible ditched field and lynchet systems is probably a true reflection of the archaeological record, so maybe this large area of downland really is substantially unenclosed in prehistoric times. Having said that, the Monkton-up-Wimborne Neolithic pit circle and enclosure sites (see above), both within the estate boundary, have only shown up through persistent fieldwork presence and low-level aerial work by Martin Green in recent years, and would suggest that more sites may emerge through rigorous prospection, and that aerial detection is not fool-proof even in this chalk downland landscape.

The influence of concepts of soil fertility and sustainability on land use and settlement interpretations

Helen Lewis

Connotations of land use play an important role in interpretation in landscape archaeology. Discussions of prehistoric farming land use practices frequently refer to soil or land fertility as a controlling, or at least prominent, factor in decision making, for instance relating land fertility and ease of tillage based on soil-type distribution to settlement location and technology at the time in question. It is assumed that people practising farming in any given region would have settled soils the most fertile for their particular crops. In southern England, for example, Neolithic settlement was once presumed to have concentrated on the chalk, rather than the sands (Clark 1952, 91). Soil properties, and the technological ability to deal with these, are thus seen to be major limiting factors in settlement location.

Soils are not, however, inherently fertile or degraded; they are fertile or degraded only when perceived in terms of the production of some desired item or the ability to sustain some desired ecosystem, land-use activity or landscape. Fertility and degradation are subjective terms — one must always ask 'fertile for what?' or 'degraded in what sense?' The word 'fertile' is generally used in its connotation meaning 'reproductive'. From the perspective of crop production, land fertility is normally classified in a hierarchical scheme related to farming system, and farming systems are related to social organization through evolutionary perspectives such as those presented by Boserup (1965) and Goody (1976).

Neolithic and Early Bronze Age southern England is a vital region regarding the integration of soil studies with general landscape perspectives because it is an area where land use evidence appears to be in contradiction with some recent landscape archaeology models. Some studies have tended to gloss over or ignore the evidence of land-use studies, relying instead on an approach based on architectural monumental and settlement landscapes to deal with land use issues (e.g. Barrett 1994; Brück 1997a,b), even when discussing proposed agricultural 'revolutions'. Since in this region we have mainly monuments, with few domestic sites, ideas of what people were doing off-site have been strongly influenced by models developed from ethnographic analogy of how settlement (or lack thereof) relates to land use. This approach, however, is not appropriate for studying how settlement and land use interacted in past cultural landscapes. Indeed, developing general archaeological approaches to land use and landscape based on soil evidence might offset the poverty of the domestic settlement record in Wessex during these periods.

As discussed above, in southern England soil thickness and fertility are perceived to have decreased throughout the Neolithic and Bronze Age, to the point at which regional soil types changed, and this is associated with erosion episodes linked to perceived land use changes and settlement pattern alterations (Middle Bronze Age land division and interpreted sedentism).

That cultivation occurred in the region during the Neolithic at least is clear from the presence of ard marks underneath Neolithic and earlier Bronze Age monuments. In some parts of Wyke Down it seems clear that soils were mainly thin rendzinas long before any proposed Middle Bronze Age land use and settlement changes, and even before any cultivation occurred — during the Mesolithic, probably associated with forest clearance or localized decline (see above).

The barrows and valley deposits examined to date in the project show that predicted patterns of soil change often associated with intensification of agricultural land use and settlement during the middle–later Bronze Age, appear to be seen at earlier dates (pre-barrow construction, i.e. during the Neolithic–earlier Bronze Age or earlier). In the valley as a whole, it appears that a stable soil, slope and valley system had been established well before the second millennium BC. This lack of change contradicts the archaeological model for Middle Bronze Age landscape changes in the region, and has major implications for the interpretation of land-use practices in the Neolithic and Bronze Age, as well as for assumed relationships of land-use practices with later field systems and associated settlement evidence. While we have only minimal evidence regarding specific agricultural land uses, it appears that any intensification of arable land use in the Bronze Age is not reflected in the local soil or erosion evidence. As this is exactly where such land-use changes should be visible, the perception of the impact of middle–later Bronze Age agricultural practices on the physical and cultural landscape of the downland area as a whole may need revision on a localized scale.

The land evaluation approach holds that variation in land type and 'the values' attached to this influence land uses, and that this variation is partially systematic, such that it can be mapped and the influence of land uses on a particular, given piece of land can be predicted (Coppock 1964; Rossiter 1996, 166). Ethnographic studies suggest that 'traditional' farmers classify and value their soils differently than we do, although they do view certain types of soils in their landscapes as being appropriate for specific land uses, not all of which are related to subsistence (Malinowski 1935; Ollier *et al.* 1971; Wilshusen & Stone 1990). It could be possible to examine ancient land use with the issue of *ancient* perceptions of land and soil suitability for specific land uses in mind. This approach requires the recognition that the values attached to land or land use in prehistory may not coincide with our own.

For instance, the commonly-held belief that both prehistoric and recent agrarian practices can be inherently damaging to the earth may be partially related to the modern landscape preference for a wild 'Arcadia' (*sensu* Schama 1995), and is reflected in recent agricultural policies regarding environmental impact, with a focus on loss of diversity. More recently the 'sustainability' concept has emerged in agricultural planning, research and practice (Whatmore 1993, 88; Várallyay 1989), defined by Goldman (1995; in Bogucki 1996, 290) as being '... indicated by the persistence of an identified quality, activity or system over a given period of time', and this term will be used in this sense here. For arable activities, sustainability relates to the retention of soil fertility, the reduction of erosion, and the ability of the farming system to continue in the face of soil degradation and many other factors, most of which are cultural. However, for other land uses, such as grazing or browsing, settlement, constructing burial mounds etc., we should expect that sustainability will relate to other factors: the ability to produce sweet grasses, the ability to produce well-developed turves, the quality of being able to see along the entire length of the Cursus in the Allen Valley.

Soil degradation implies a decline in fertility in relation to the perceived appropriate land use for a given area of earth. It is a concept centred around the ability or inability of a soil to produce desired agricultural and 'natural' products (such as wood) or to sustain a preferred ecosystem or landscape, and implies that there was a more fertile starting point in this regard at some time in the history of the soil. Inappropriate land use is held to lead to soil degradation (Rossiter 1996, 165). In respect to agrarian farming, degradation is often linked to practices which lead to a decline in soil organic matter and nutrient content (fertility), loss of suitable structure or hydrological conditions, and to soil loss (erosion), as discussed above (Stone 1993, 74; Várallyay 1989, 171; Lal *et al.* 1989). With such a perspective, changes in soil type, for instance from thick brown forest soils to thin rendzinas, are perceived to have been negatively perceived in the past, representing the damaging effect of human activities on the ecosystem, marking the loss of the great wild forests, etc.

However, if the people of Neolithic and Early Bronze Age Wessex were more interested in pastoral than in arable activities, as is suggested by some models (e.g. Fowler 1981; Fleming 1989; Barrett 1994), a proposed soil change from forest brown earths to rendzinas may be better viewed not as a degradation, but as sustaining or even improving desirable soil qualities for grassland production. If monument visibility was as important as some believe, the vegetation cover represented by rendzinas is also a positive land-use factor. Indeed, monument construction in many regions suggests grassland soils were vital for many cultural activities beyond subsistence — such as building barrows. In addition, in Wyke Down at least, it seems that rendzina growth reflects Mesolithic activities and perspectives at least as much as Neolithic and later ones. Could rendzinas and other shallow brown earths have been selected for, and consciously developed and maintained for these purposes? If we equate rendzinas to grasslands (although they are sometimes found under forests in other parts of the world) then it is quite likely, given what we know about Neolithic monumental and burial landscapes, and what we know about Bronze Age pastoralism, that these soils were the soil of choice, and that they were consciously sustained and maintained, at least in some parts of the landscape, from the Mesolithic on.

In addition, rendzinas are not infertile soils for crop production. On the contrary, they are highly organic and relatively stable, with a favourable structure (Courty *et al.* 1989, 148). Assuming that manuring and/or rotation were practised from at least the Bronze Age, if not earlier in Britain (Allen 1992; Bell 1981), there is no reason to think that 'less fertile' rendzinas would be detrimental to arable farming at the low level envisioned in the Neolithic and Early Bronze Age. Modern perceptions of ideal soil depths and chemical characteristics for crop growth may be largely irrelevant to the sustainability of arable practices in prehistory.

The proposed 'evolution' of rendzinas may be perceived to be a major disaster if one is concerned with the maintenance of a landscape covered by deciduous forests. It seems possible from the evidence for long-term clearance activities that this was not the overriding concern of later prehistoric peoples in all parts of southern England. There is also no reason to believe that Neolithic or Bronze Age farmers were concerned with maximizing arable production on *any* type of soil. Indeed, there are suggestions that arable activity was limited to 'subsistence' levels of production, if even that (e.g. Robinson & Williams 1987, 72), and such levels would not necessitate a concentration on the most fertile soils or the most productive practices, even if such practices were known. Soil suitability for arable farming does not equate to cultural requirements — '... (other) aspects of land evaluation concern social habits and prejudices' (Burrough 1986, 193; Blaikie & Brookfield 1987, 35–6; Bailey 1989).

Specific sorts of land use carry certain landscape connotations, and modern preferences are attributed to the past with little exploration of how land use is perceptually linked to landscape. There is a great deal of variation seen in ethnographic and historical studies of agricultural practices, suggesting that the prediction of patterns of land use, the relationship of land use to social organization, and the stability of landscape types may be more complicated than many authors appear to believe. Studying site-specific ancient land-use practices is one means of assessing our pre-conceptions about the sustainability of ancient farming practices, ecosystems, social systems and landscapes. Research in the upper Allen valley suggests that it is vital to address these issues, and that archaeological approaches using soil data can provide a means to explore both modern and ancient landscape perceptions.

Updating our understanding of pre-Neolithic Cranborne Chase

Charles French

From Bradley's work of over two decades ago, the Mesolithic focus of activity was found to be located on the highest parts of the downland of Cranborne Chase, most often associated with outcrops of clay-with-flints (Barrett *et al.* 1991a, fig. 2.3). This was equated with smaller-scale clearances in the woodland cover around those areas of possible flint resource outcrops, and possibly associated with the interception of the seasonal movement of game (cf. Mithen 2000). Since then, a few more sites have been discovered in the lower parts of the downland slopes, notably the extensive scatter of Mesolithic flint along the river cliff in Chalk Pit Field in the base of the upper Allen valley (Green 2000, figs. 11 & 35). Again, this natural geological phenomenon could have been seen to have produced a break in the woodland cover which made its exploitation easier, and it was just above an area in the base of the valley that would have flooded regularly as a winterborne (as it still does occasionally today). The coincidence of a natural outcrop in the chalk, creating a break in the woodland cover, and a water source would make this an obvious choice for both people and game to frequent. Nonetheless, the occurrence of Mesolithic flint artefacts and radiocarbon dates from faunal material more indicative of the Late Mesolithic period in the north Cursus ditch in a lower slope position on Wyke Down, as well as a scattering of occasional Mesolithic type flints found elsewhere throughout the valley during this project, would indicate some Mesolithic use of a much wider part of this chalk downland landscape, not just on the tops of the downs in and around outcrops of clay-with-flints as concluded by Barrett and Bradley (Barrett *et al.* 1991a).

Rather unexpectedly, the late glacial to Mesolithic palynological sequences recovered from peat filled relict channels about 2.5 km downstream (Fig. 1.11:1 & 2) produced evidence of a much more open downland landscape than had hitherto been expected. In essence, deciduous woodland does not appear to become fully established until after 6500 cal. BC (see Scaife above), with the first two and a half millennia of the Mesolithic witnessing only about one-half to two-thirds of the landscape being tree covered. The pollen data in the Wimborne St Giles sequences for the earlier Holocene testify to a lengthy persistence of pine, the development of hazel, the slow colonization of the downland with deciduous woodland species, especially oak and elm, and persistence of a widespread and diverse herb community on the downland, indicative of a considerable and unusual degree of openness with calcareous grassland occurring right into the Late Mesolithic. Immediately, movement of game and people can be envisaged as much easier; 'occupation sites' in this semi-open early Holocene landscape may have been that much more transient and consequently much harder to detect as discrete scatters of activity.

Although only hinted at by the palynological and flint artefact data, different parts of the study area may have had slightly different trajectories in terms of how vegetation cover developed in the earlier Holocene/Mesolithic period. In contrast to the upper part of the valley, down-valley around Wimborne Minister mixed deciduous woodland was already well developed and predominant by about 6000 cal. BC, as if following the more conventional trajectory of the development of climax woodland in the early Holocene. Also, much better and more complete deciduous woodland development appears to have occurred along the flood-plain edge of the upper Allen valley, with willow, sedges, reeds and areas of open water suggestive of a willow fringed sedge swamp in the flood-plain zone itself.

This implies that some parts of the downland maintained a degree of openness throughout the Mesolithic. Unusually, there is also sufficient pine remaining in the pollen record to suggest the survival of stands of pine in places on the downland into the Neolithic period. In contrast, the flood-plain edge and interfluves may have acted as a refuge for woodland throughout the Neolithic with much more marshy to alder/willow carr conditions pertaining in the flood plain, and the first indicators of cereal type pollen and the occurrence of ribwort plantain are seen. But, by the Early Bronze Age all areas were certainly substantially open and remained so until the present day. This certainly conforms to recent ideas of a much more variable sub-regional complexity in the development of the vegetation of the chalk downland.

Site-specific indicators of land use in the Neolithic and Bronze Age

Charles French

At each of the Neolithic and Bronze Age sites investigated in this project and those previously investigated by Martin Green (Green 2000; this volume), the ancient soil and molluscan data is indicative of long and well-established grassland. This grassland environment occurs in all areas from the top of the chalk downs, across the slopes to the edge of the flood plain. In contrast, the pollen data from the relict channel near the Knowlton henge complex both corroborates this openness, with woodland cover gradually giving way to grassland conditions, and gives a picture of a more variable vegetation mosaic persisting. Despite this substantially open and gradually opening up landscape, mixed deciduous woodland including ash, and rather unusually pine, continued to persist in the Neolithic. Exactly where these woodland elements were located in this downland system is, however, not clear. There is also a noticeable decline in elm around 3500–3000 cal. BC, and this is coincident with the first indications of cereal pollen and associated weeds. But there appears to be only very localized evidence of cereal cultivation, and indeed small amounts of minerogenic sediment appearing in the peat which does indicate some minor soil erosion getting into the drainage system which may be associated with this. It is only within the flood plain and along the flood-plain edge that the vegetation is substantially different with alder carr fringing more marshy areas of grass-sedge fen.

In only one instance beneath an earthwork has the physical evidence of ploughing been observed in this study. Several criss-cross ard plough marks were found defined in the base of the rendzina beneath the southern bank of the Dorset Cursus and agger of the Ackling Dyke in Trench 1 (Figs. 3.33 & 3.55), just cutting into the weathered surface of the chalk at mid-slope. Significantly, there were no hints in the micromorphological evidence of ploughing occurring in this palaeosol nor any of the buried soils analysed (see Lewis above), although it is hard to say how typical these areas of downland may have been. But, the dominance of calcareous grassland and the relatively very minor indications of cereal cultivation in the pollen record for both the Neolithic and Bronze Age periods (see Scaife above) would seem to substantiate this lack of evidence in the palaeosol record for arable cultivation in this landscape. Nonetheless, as discussed above, the mechanical disturbance of brown earths caused by ploughing in earlier Neolithic times, conceivably associated with clearance activities, could have been a contributory factor in the creation of the ubiquitous rendzina soil type in this downland landscape.

In many respects there is an apparent, slight mis-match in emphasis between the palynological data from the palaeochannels in the valley bottom and the palaeosol records present beneath the prehistoric monuments that were investigated on the downland slopes in this project. Some of this may be due to locational factors, and some possibly to taphonomic factors. The evidence from the pre-monument soils is unequivocal; they are thin rendzina soils characteristic of short-turved grassland that are well established under the Neolithic long barrows on Gussage Down, the late Neolithic Dorset Cursus on Wyke and Bottlebush Downs, and under several of the Bronze Age round barrows of the Wyke Down barrow group. This may reflect the different contexts of the data. The downland slopes had less woodland and long-term forest cover than previously envisaged, and they were cleared earlier and more comprehensively in prehistoric times than the valley bottom and flood-plain areas. Moreover, these areas may have been primarily kept and managed as grassland due to the type and use of the monuments concerned (i.e. for burial and ceremony). In contrast, the valley bottom and flood-plain areas retained a fringing woodland of varying density and species composition and more marshy areas throughout most of prehistoric and historic times, thus putting some taphonomic constraints on the inputs of pollen rain into the palaeochannels. As most of the fills of the palaeochannels were peats, rather than entrained river-borne sediments, this should not have had a significant effect on the nature of the pollen assemblage, but the dominance of different species assemblages close by the context of deposition is undoubtedly more reflective of the immediate vegetation conditions than the surrounding downland vegetation. But on balance, the palynological sequence corroborates and complements the archaeological, molluscan and soil data, essentially in a valley catena sequence.

There are also indications at several sites from the molluscan data such as at the Wyke Down henge site 2 (see Allen in Chapter 3 above) and in the pollen record from the Allenbourne car park site in Wimborne Minster (see Scaife above) for some possible regeneration of woodland. The molluscan data suggest this may have been very localized, and perhaps even monument specific in the sense of colonizing an abandoned site. In the pollen data, both ash and then lime appear to have become more important for a time, and this is often seen in the immediate post-elm decline period as a general trait. In fact, lime continues to be important right up to and through the later Bronze Age, although this may be related to taphonomic factors. Again the scale and extensiveness of this woodland regeneration is not known, and much could be due to the management of surviving copses and adjoining grassland. Nevertheless, peat formation itself is suggestive of an overall trend towards a reduction of woodland cover in the landscape and increased surface run-off into low-lying areas which were conducive to peat formation under high groundwater table and anaerobic conditions.

The evidence from the round barrow soils

Helen Lewis

As was noted above, three of the four upstanding round barrows investigated were constructed of sods which were relatively clay-rich in comparison to the modern rendzina soils around them, and to the buried soils under many of the monuments investigated. One of these, barrow 41, was standing on a flinty drift or fan deposit at the head of the Allen Valley, with clay-rich subsoil found in hollows in the chalk. This, it was suggested, was a location in which brown earth, and probably argillic brown earth soils would have developed in the past, based on surficial geological properties, and it was proposed that the relatively clay-rich barrow turves could have been cut at the very spot on which the monument was built. However, the other round barrows all stand on chalk, with rendzina buried soils, and there was no evidence of early argillic horizons or clay enhancement at the spots on which they stand. This could suggest that sods in these barrows came from a slightly different part of the landscape than that on which they stand, one with clay-rich soils, possibly nearby.

Soil micromorphological analysis has further shown that these turves are mineralogically similar to each other, and to the soils at barrow 41 on the flinty drift or fan deposit. All samples are dominated by quartz grains, but unlike the long barrows or other monuments studied, as well as unlike all of the buried rendzinas-on-chalk underneath the round barrows themselves, round barrow turves contained relatively frequent numbers of minerals found in Pleistocene deposits in the region: angular plagioclase and possibly potassium feldspars, amphibole, epidote, and possibly garnet and zircon (C. Doherty pers. comm.); these were also found frequently in all of the barrow 41 deposits on the flinty drift or fan. The frequency of these minerals requires quantification, but there is a clear visible difference in thin section between the drift/fan-round barrow mound samples and the buried soils on chalk. The turves used to build barrows 34 and 36, for example, appear to have originated from a deposit similar in nature to the drift or fan deposit at the top of the valley. This may explain their relatively clay-rich nature — they come from soils developed on flinty clay deposits over chalk. They may not reflect changing soil types growing on chalk rock *per se*.

While buried soils do reflect the landscape as a whole, they mostly tell us about the history of a specific place. Because buried soils are mainly found to survive under upstanding monuments in the chalk downlands of Wessex, unlike in other regions where they are more widespread (e.g. under alluvium in the East Anglian Fens), methods such as soil micromorphology inform mostly on particular, and special, places in the landscape. These places are the places where people decided to build monuments. Can we conclude that these locations are actually representative of the landscape as a whole? The evidence of the round barrow turves discussed above shows that monument soils and monument locations can reflect the larger landscape, but that we must interrogate these remains with an understanding that goes well beyond any individual spot in that landscape.

In order to move beyond the monument itself and into the landscape, site-specific analyses are compared with each other, and with the other sedimentary, soil, excavation and environmental evidence as discussed previously. Through such comparisons it is possible to discern whether the locations chosen for monument construction were substantially different from the landscape as a whole, and which parts of the physical surroundings were chosen for specific constructions and functions. The presence of mature rendzina soils under Early Bronze Age barrows shows that in some specific locations very long-term grassland clearings were present by this time, but that not all locales in the vicinity had seen the same history, with some local Early Bronze Age grassland soils being more clay-rich brown earths (and their topsoils chosen for turves used in barrow construction). These variations do still fit into a larger model of progressive change in soil profile through erosion, from possible forest soils to grassland soils, but only in the limited locations where brown earths and argillic brown earths would necessarily have developed given the underlying geology. This history fits the regional model, but it also demonstrates that a regionalized model is insufficient for landscape archaeological studies, as different parts of the landscape are changing at different rates, have differing starting points, and, of course are used and selected for different cultural activities.

Choices of where and how to erect a monument might be related to these variations. Why, for instance, were the barrows not built (with one exception) on degraded (argillic) brown earths? And why were they not all constructed using mature rendzina sods? In other regions, such as south Jutland in Denmark, well preserved Early Bronze Age barrows on sands show intricate construction details, revealing systematic and knowledgeable choice and organization of a mixture of types of soil materials during barrow construction (Lewis & Hart 2003; Breuning-Madsen & Holst 1992; 1998; Breuning-Madsen *et al.* 2001). Deliberate choice of mound construction materials is also known from Britain (e.g. Cocksbarrow, Cornwall: Miles & Miles 1971). It is possible that the choice of more clayey turf soils for round barrow construction at Wyke Down was consciously made, part of the engineering and design of these barrows, not reflecting the immediate location, but revealing the variety of materials in the local landscape. In addition to choosing materials for specific engineering issues (such as the creation of anaerobic barrow cores in the Danish barrows), the soil materials come from and reflect various parts of the local landscape, various local land uses (often including wet pasture, dry pasture, once-arable soils, etc.), and possibly also various local social groups. These monuments are the sum of the local soil types.

Perhaps the Wyke Down barrow group was built in this part of the valley not only because of the location of earlier monuments, or the location of springs and river cliffs, or the location of a routeway. Perhaps the cluster was also built here because of the presence of these soils, and the inherent quality of the turf, which was evidently consciously chosen building material for these round barrows. Field (1998) noted a relationship between round barrow distributions and 'simplified' geology and geography in the Weald, and discussed how landscape features were important to barrow clustering. In the Wyke Down barrow group, at least as far as it has been systematically investigated, we have a classic situation of clusters arranged in relation to topographic features, the river, physical and possibly spiritual 'boundaries' (including soil-type boundaries), earlier monuments (memory), and routes. These are obviously all integral to barrow 'cemetery' creation, and are also intertwined with each other.

In addition, we can suggest that proximity of construction materials may have been very important, and in more than a simply pragmatic manner. The flinty clay deposits are located near the location of the upper river springhead, near to the ancient river cliff, and its seasonal lake, and near the area of naleds in the valley bottom at the head of the upper Allen valley (Fig. 1.2). If these phenomena were special for ancient people, then possibly the different soil found in those locations represented the special character of the upper part of the valley. Or maybe clay-rich turves just make for better barrows. One thing is clear, however 'simplified' geology is not subtle enough for us to access ancient understanding of these very specific locations in the landscape. Given the findings of the few Wyke Down round barrows investigated, we would suggest that further systematic work on the issue of relationships between barrow clusters and geology or soils should be carried out through the combination of site-specific and landscape methods applied here. We cannot substantially address these issues in our landscape study, despite mapping soils and geology, because we have only sampled from four barrow mounds. It would be relatively easy to extend this study in both the northern and southern barrow groups of the upper Allen River valley. Moreover, it would also be appropriate to attempt to calculate the shear acreage volume of turves necessary to construct not just the round barrow mounds, but earlier monuments as well such as the bank of the Dorset Cursus and the mounds of the long barrows on Gussage Cow Down.

The valley survey and monument ditch-fill studies suggest that there was little erosion from the Neolithic through until the Iron Age, or even into historic times in the case of colluviation, and suggest a relatively stable soil and slope system during most of later prehistory. This has been interpreted, in light of the evidence of the buried soils discussed here, as revealing a stable grassland landscape from the Neolithic on. This grassland appears to extend to areas like the flinty deposits at the top of the valley by at least the Early Bronze Age, suggesting that even on 'less favourable' soils (where one might expect woodland maintenance) turf was being grown. As discussed above, there are indications of herbaceous clearings in the early Holocene, of arable farming in a largely grassland environment in the Neolithic and Bronze Age, and of stability through the supposedly revolutionary land-use changes associated with first field system construction in the Middle Bronze Age — in short there is a mosaic of vegetation and land-use types across the landscape in all periods, and, apparently, on all soil types. Landscape interpretations based heavily on large-scale soil or geology distribution mapping are not sufficient in and of themselves to understand land-use history.

The combination of detailed site-specific soil and molluscan studies with detailed landscape-scale investigations of palaeoenvironment and land use has revealed both problems with interpreting ancient landscape from either regional-scale or site-based information, as well as solutions to these methodological issues. Through the approaches described above it is possible to assess regional models of landscape change, and to understand how individual sites and locations are part of both change and stability in the prehistoric landscape.

Where is the Middle Bronze Age transition to arable farming and colluviation?
Charles French

On the face of it, there does not appear to be a visible transition to arable farming on a wide scale in the Middle Bronze Age in the upper Allen valley as is often postulated for the rest of Cranborne Chase and most of Wessex (cf. Barrett 1994). As set out above, field systems and lynchets on any scale do not appear until the Iron Age with the Gussage Down system, despite the earlier existence of the Down Farm settlement enclosure being tied into substantial field enclosures in the Middle Bronze Age. Similar sites are singularly lacking elsewhere in the study area, although other examples occur to the south and north in adjacent valley systems (Green 2000). There are also two large oval enclosures, one investigated at Monkton-up-Wimborne (MUW 96) by Martin Green of later Bronze Age to Iron Age date and another one on the southwestern slope of Wyke Down (Fig. 1.2), which has not been investigated through trenching, only fieldwalking, but is also thought to be of Iron Age date. Importantly also, the soil erosion record for these prehistoric periods is just not in evidence, as discussed above. Moreover, the pollen record for later prehistoric times is indicative of relatively small numbers of trees and shrubs with oak and hazel on the drier slopes, alder in the valley bottom and fringing the river, and a predominantly grassland/pasture habitat with only a small amount of cereal pollen present.

Indeed, the bulk of the Wyke and Bottlebush Downs landscape does not seem to have visible evidence of activity past the period of barrow building in the earlier Bronze Age. Despite intensive archaeological prospection by Martin Green in these areas, there is no hint of settlement, or indeed burial or ceremonial monuments from the Middle Bronze Age onwards. This does not mean that it is not present, but the settlement and monumental focus does seem to have shifted elsewhere to the southern side of the valley around Down Farm and Gussage Down, and to new loci further downstream and on the upper parts of the downs.

However, hints of new settlement activity, probably of later Bronze Age date, have been discovered through this project. In particular, new sites have been discovered and assessed through a combination of fieldwalking, aerial and geophysical prospection and test excavation at Brockington Down in the central southern part of the study area (Figs. 3.80 & 3.84) and at Avenue Lodge in the southeastern part of the study area, in both cases just off the top of the down (Figs. 3.80 & 3.89). At the former site, several round structures defined by eavesdrip gullies with pits and ditches are present (Figs. 3.84–3.86), and at the latter site, pits, postholes, gullies and copious quantities of burnt flint (Figs. 3.81–3.86). In both cases, there are sufficient dispersed concentrations of Neolithic and Early Bronze Age flint tools (Fig. 3.83; Table A4.55) to suggest that more, as well as earlier, settlement activity may be going on at these locations (see Gardiner in Appendix 4.12). These sites, beyond the areas thoroughly covered by Martin Green in the northern half of the study area, hold out the prospect of a more fulsome settlement record of later prehistoric times being forthcoming.

The historic landscape and conservation issues
Charles French

The land-use record
As indicated by the pollen evidence for the later Bronze and Iron Ages in the profiles at St Giles House and Knowlton, the landscape remains primarily grassland and pasture with only minor evidence of arable land through into Roman times. Nonetheless, by medieval and post-medieval times in the Pippins Bridge sequence, Wimborne Minster, there are strong suggestions of significantly increased areas of arable cultivation, waste and disturbed ground being present alongside areas of chalk grassland, which still remains dominant, as indicated by the predominantly herb taxa assemblage (e.g. including cereals, weeds, segetals and ruderals). There is some limited woodland with oak and hazel, perhaps as managed coppice, lime and ash, and alder and sedges continuing to mark the flood-plain edges, with bodies of slow moving water present from time to time. Although this Pippins Bridge sequence remains undated, the absence of exotic and planted species suggests that it is pre-AD 1750 in date.

Monument erosion and conservation
As a result of the test interventions in the upper Allen valley, it has been possible to record exact amounts and timings of erosion of several upstanding monu-

ments in different topographical positions in the study area. In particular, good evidence of past and present erosion has been observed for the Dorset Cursus on Bottlebush Down (SAM 310), for four Bronze Age barrows on Wyke Down (SAM 299 c, f, g & i), and for the two long barrows investigated on Gussage Down (GCD 78 & 294) (Figs. 3.1, 3.2 & 3.5). Although this has been reported on as part of a larger DEFRA study (French 2002) and a summary of the main findings is included in Appendix 4.6, some salient points are brought out here.

Both the northern and southern banks of the Dorset Cursus on Bottlebush Down were affected by the deliberate back-filling of the ditch with bank material as well as pre-Roman ploughing (Figs. 3.51–3.54). The Cursus banks occasionally survive to a thickness of about 25–30 cm where they are protected either by other monuments or later field boundaries. Where they are unprotected and have been ploughed for the past 45 years, no earthwork survives and there is only a dissipated linear scar visible in the field today. There is almost no survival of the southern Cursus bank at the Bottlebush Down terminal despite the presence of an apparent earthwork. In this case, what is scheduled as an earthwork is not the actual south bank of the Cursus, but a contour change in the chalk geology that is at a very slightly oblique angle to the actual line of the Cursus ditch and bank.

The Wyke Down barrow group consists of only seven upstanding barrow mounds out of a total of at least 29 ring-ditch sites (Fig. 3.34). The ploughed-out barrow mounds are just visible in the field under the right light conditions as very slight rises in the field, but are not otherwise detectable. Ploughing since the latter 1950s of the scheduled and upstanding barrows has resulted in the sculpting of the mounds resulting in ovoid to sub-rectangular to sub-square shapes to the once circular mounds. In addition, this same ploughing has created truncation steps around all four of the upstanding barrows investigated, with up to *c.* 25–78 cm of chalk subsoil having been scoured away over about 45 years. This has particularly affected the ring-ditch, berm and outer edges of the mounds, and is particularly marked in barrows 36 and 44 (Figs. 3.44 & 3.49). In some cases the upstanding barrows are standing on plinths of up to 20 cm of chalk rock, created through soil loss from repeated ploughing around the edge of the protected zones. This makes them appear much larger in the landscape than they actually were, and is a significant finding for monument visibility and prominence studies.

In one case at barrow 34, the conversion of former arable land to pasture by Martin Green for the last 20 years has dramatically slowed this truncation process on its southern side (Fig. 3.39). It has resulted in a less abrupt and more gently sloping talus edge to the surviving barrow mound, which better reflects the natural collapse and erosion of the mound over time.

Both the long barrows on Gussage Down have also been affected by recent human activities. Despite long barrow 78 surviving as high as 4 m above the modern ground surface (Fig. 3.8), most of the north-eastern end of the mound and ditch has had a *c.* 15 × 15 m area destroyed either to the level of the basal horizon of the buried soil under the mound or the ditch being two-thirds back-filled through quarrying for chalk in the early twentieth century. Subsequently this has facilitated much destruction by rabbit burrowing.

In contrast, long barrow 294 is now only visible as a slight contour effect of *c.* 25–35 cm, but no actual mound was observed to still be surviving in our intervention of 2003. With about a metre's denudation of the long mound since Grinsell recorded the dimensions of this barrow in the early 1950s (Grinsell 1959), ploughing has effectively destroyed this monument in 45–50 years. Similarly, the lynchets and field systems occurring on the Allen valley side of the Gussage Down 192 complex (Figs. 3.1 & 3.10) where they are under a modern ploughing and cropping regime only show now as linear spreads of darker earth, with no actual earthwork survival at all. In contrast, in Martin Green's youth (pers. comm.), they were still upstanding low (*c.* 50–75 cm in height), linear banks of mixed ploughsoil and chalk, with a pronounced dip on the downhill side.

Although some destruction of upstanding mounds and earthworks can sometimes be proven to have occurred in antiquity, much severe sculpting and smoothing of the shape of earthworks and considerable reductions in the size and height of the earthwork monuments has occurred as a result of modern ploughing and other human activities over very short time periods of between 45 and 70 years. This has demonstrated that the scheduled status of these prehistoric monuments is insufficient to really protect the monuments where modern arable agriculture is practised. Without more frequent monitoring of these sites and/or management agreements in place to slow the destruction of these upstanding earthworks, these sites may be all but destroyed in half a century under an arable regime of land use, irrespective of their protected status. Finally, in at least two cases, severe encroachment on the extent and survival of a barrow mound and berm area has occurred in just the past two decades.

A new model of landscape change for the upper Allen valley and the rest of Wessex

Charles French, Helen Lewis, Rob Scaife & Michael J. Allen

As a result of five seasons of fieldwork and analyses, a picture has emerged of greater variety and complexity within the upper Allen valley, with landscape change occurring on a variety of scales at an earlier period than hitherto envisaged. The model that is put forward looks like this: a stable soil, slope and valley system had been established well before the second millennium BC. Certain parts of this downland environment may have been slow to become wooded in the first place, that is in the earlier part of the Holocene or Mesolithic period, and may even have never been as uniformly wooded as conventional models have previously always maintained. Perhaps some patches of the downland remained as open grassland from the very beginning of the Holocene, and this might in part be associated with the extensive Mesolithic exploitation of the downland ridges and upper Allen valley in this area (see Barrett *et al.* 1991a, fig. 2.3; Arnold *et al.* 1988). Nonetheless, by the third millennium BC or later Neolithic period, this landscape was substantially cleared, a process which appears to have gathered pace and extent of impact over at least the two previous millennia. In turn, this implies that some parts of this downland landscape may have developed to greater or lesser extents in terms of the vegetation and soil complex that it supported, and this could have been controlled by the nature of the subsoil variability and the amounts of late glacial loessic accumulation. In particular, thick and well-developed argillic brown forest earths do not appear to have formed uniformily across the downland prior to the Neolithic period as suggested by previous authors (i.e. Fisher 1991). Rather, some areas of downland slope most probably supported thin calcareous brown earths for a very long time, and were therefore pre-disposed to becoming grassland and developing thin, rendzina-type soils. This then removes much of the necessity for envisaging large-scale and extensive erosion episodes associated with arable land use and bare soil-scapes as has occurred in many other areas of the southern English chalk downlands (Allen 1992; Bell 1992). Localized soil erosion certainly did occur, especially on upper margins of the dry valleys which dissect the downland slopes to either side of the valley. But this is on a relatively minor scale, and certainly not catastrophic.

The most probable scenario is that major soil, vegetation and hydrological change took place during the Mesolithic and Neolithic periods rather than during the middle–later Bronze Age as has been suggested previously. This landscape was much more of a mosaic-composite characterized by different vegetation-soil complexes. Certainly the predominant soil type was already largely well established and exploited, that is thin rendzinas being more or less ubiquitous by the main period of barrow and Cursus building in the Neolithic and earlier Bronze Age. Whatever the scale, intensity and frequency of deforestation, human agricultural, burial and settlement activities, and wind and water erosion, they have not generated thick aggraded deposits in the any of the valley bottoms investigated, unlike what has been either postulated and/or found elsewhere in the region. There was also a greater through-put of water through the upper Allen valley in the past as demonstrated by the discovery of earlier and later Holocene palaeochannel systems. These former river systems may have been responsible for carrying some eroded soils through this part of the valley system to beyond our study area, but this has yet to be proven. The only reasonable answer is that limited erosion of variably developed brown earth soils had largely occurred much earlier in the Holocene than in the Bronze Age. Subsequent and intensifying land use has kept the soils exposed, thin and mono-horizonal from then until the present day.

Needless to say, the research should not end here. The data so far is only suggesting this new model for one small area of the greater chalk downlands of Wessex. Further investigation is required at a number of levels, for example of further Neolithic and Bronze Age monuments and the prehistoric settlement record of the upper Allen valley, of other river valleys elsewhere in the region such as in the Avon, Kennet and Wylye valleys, as well as finding appropriate data to decipher the periods and periodicity of erosion versus stability in the ostensibly cleared later prehistoric and historic landscapes, with the emphasis probably veering towards historic and recent times, and locating further data to test whether the vegetation canopy was ever closed or that early Holocene soils were of variable thickness and development status because of existing variable geological substrates. Finding new repositories of waterlogged deposits, or at any rate loci providing sufficiently good conditions of preservation for the survival of pollen, will never be easy in these alkaline and relatively dry valleys, such that more of the middle Holocene vegetational record may remain elusive for awhile longer.

Although we have deliberately downplayed the wide-ranging significance of these findings, as one of the prominent conclusions of this project is that change should be understood locally and at a landscape scale, there are potential implications for the broader-based understanding of later prehistory in the Wessex chalk downlands. For one thing, if there is no evident land-use change in the Middle Bronze Age, and if this is significant beyond the Wyke Down landscape, an important question to follow up would be: how does land use relate to land division, proposed tenurial changes and sedentism during this period? Many examples from the anthropological and ethnographic record suggest that the construction of field systems and 'permanent' settlements should be related to (and thus reflected in) land-use patterns and tenurial organization. However, we do not see evident land-use changes associated with these activities at this time. Is Wyke Down simply an exception … or the exception that proves the rule? As there has been so little research into land-use change itself at the Middle Bronze Age transition in Wessex, mainly because of a lack of comparative settlement remains for the earlier part of the Bronze Age, it is difficult to say at this time. It is also, however, difficult to support the view that a 'true agricultural revolution' happened at this time in Wessex, associated with land division. The soils record of barrows should see revision and much more investigation for evidence of settlement and land use during the Early Bronze Age in particular, in order to generate the type of information needed to address this issue.

Other regional issues explored above include the huge variability in extensiveness of woodland cover in Mesolithic. The evidence of possibly extensive clearings dominated by grass and herbaceous plant communities locally in the Mesolithic has implications for our understanding of Mesolithic settlement patterns, of woodland regeneration in the earlier Holocene, and is important for understanding the landscape in a historical sense. That is to say, what happens in the Neolithic is undoubtedly precipitated by the importance of the presence of some extensive and spatially variable gaps in the wooded landscape during the Mesolithic. While it is clear that the river cliff and lake, the springs, the river course and the hills were of great significance to Neolithic monument builders, it is increasingly important to recognize that the construction of monuments may not represent completely new meanings in the landscape. There were evidently maintained and substantial clearings in these locations well before long barrows or cursus monuments were built.

However, one cannot dismiss the significance of creating such enormous, prominent markers and burial sites in the landscape, even if many of these were built on already important places. One of the less-discussed aspects of this project is the great blessing of being allowed to excavate small holes into these incredible structures (to get at the soils underneath), even more so since they are located in such an archaeologically important region. Each monument, even in such small slices, has proven to be complex and fascinating in its history. It is difficult to interpret and understand large monuments from very small holes, so we have not attempted to extrapolate our information too far in terms of monument construction histories or the significance of individual monuments and clusters. However, the project has generated a good deal of extremely detailed information on the individual monuments studied, and we hope that future work will expand and incorporate this data set into the vast amount of information and the numerous models in the study of monuments on the chalk downs. In the upper Allen valley, one group of monuments in particular requires significant investigation: the Knowlton Circles complex of henges and barrows is obviously an extremely important place in this landscape, and it will be impossible to understand the ancient river and its relationship to the Dorset Cursus and later monuments, ancient settlement patterns, or the clustering of round barrows without significant work at Knowlton.

Finally, the data and story from the upper Allen valley is just one set of scenarios set amongst a huge range of variability in terms of data coverage, range of results, the time frames covered and the geographical areas studied within the Wessex chalklands. The mosaic of different landscape change sequences evident suggest that it is very dangerous to apply blanket models of landscape development. Rather, just as in the East Anglian Fens, each valley system must be studied, system by system, landform zone by zone, in relation to the course of the archaeological record.

References

Adams, J.C., 1986. *Outline of Orthopaedics*. London: Churchill Livingstone.

Adams, J.C., 1987. *Outline of Fractures*. London: Churchill Livingstone.

Albarella, U. & S. Payne, 2005. Neolithic pigs from Durrington Walls, Wiltshire: a biometrical database. *Journal of Archaeological Science* 32, 589–99.

Albarella, U. & D. Serjeantson, 2002. A passion for pork: meat consumption at the British Late Neolithic site of Durrington Walls, in *Consuming Passions and Patterns of Consumption*, eds. P. Miracle & N. Milner. (McDonald Institute Monographs.) Cambridge: McDonald Institute for Archaeological Research, 33–50.

Allen, M.J., 1988. Archaeological and environmental aspects of colluviation in southeast England, in *Man-Made Soils*, eds. W. Groenman-van Waateringe & M. Robinson. (British Archaeological Reports British Series 410). Oxford: BAR, 67–92.

Allen, M.J., 1992. Products of erosion and the prehistoric land-use of the Wessex Chalk, in *Past and Present Soil Erosion*, eds. M. Bell & J. Boardman. (Oxbow Monograph 22.) Oxford: Oxbow Books, 37–52.

Allen, M.J., 1994. The Land-use History of the Southern English Chalklands with an Evaluation of the Beaker Period using Colluvial Data: Colluvial Deposits as Environmental and Cultural Indicators. Unpublished PhD thesis, University of Southampton.

Allen, M.J., 1995a. Knowlton Circles, palaeo-environmental assessment. *Proceedings of the Dorset Natural History and Archaeological Society* 116, 119.

Allen, M.J., 1995b. Before Stonehenge: the environment sections, in *Stonehenge in its Landscape*, eds. R. Cleal, K. Walker & R. Montague. (Archaeological Report 10.) London: English Heritage, 4–6, 34, 41–65, 116–17, 168–9 & 332–4.

Allen, M.J., 1995c. The prehistoric land-use and human ecology of the Malling-Caburn Downs. *Sussex Archaeological Collections* 133, 19–43.

Allen, M.J., 1995d. 'Land-use history of Round-the-Down: the molluscan evidence', in The excavation of a Bronze Age round barrow at Round-the-Down, near Lewes, East Sussex, by C. Butler. *Sussex Archaeological Collections* 133, 13–16.

Allen, M.J., 1997a. Environment and land-use: the economic development of the communities who built Stonehenge; an economy to support the stones, in *Science and Stonehenge*, eds. B. Cunliffe & C. Renfrew. Oxford: Proceedings of the British Academy, 115–44.

Allen, M.J., 1997b. Land-use history: land molluscan evidence and landscape, land-use and farming, in *Excavations Along the Route of the Dorchester by-pass, Dorset, 1986-8*, by R.J.C. Smith, F. Healy, M.J. Allen, E.L. Morris, I. Barnes & P.J. Woodward. (Wessex Archaeology Report 11.) Salisbury: Wessex Archaeology, 166–83 & 277–83.

Allen, M.J., 1997c. Handley Down Mortuary Enclosure: Mollusc Assessment, Recommendation and Potential. Unpublished report to School of Conservation Sciences, University of Bournemouth.

Allen, M.J., 1998. A note on reconstructing the prehistoric landscape environment in Cranborne Chase: the Allen Valley. *Proceedings of the Dorset Natural History and Archaeological Society* 120, 39–44.

Allen, M.J., 2000a. Soils, pollen and lots of snails, in *A Landscape Revealed: 10,000 Years on a Chalkland Farm*, by M. Green. Stroud: Tempus, 36–49.

Allen, M.J., 2000b. High resolution mapping of Neolithic and Bronze Age landscapes and land-use; the combination of multiple palaeo-environmental analysis and topographic modelling, in *Plants in Neolithic Britain and Beyond: Landscape and Environment, Economy and Society*, ed. A.S. Fairbairn. (Neolithic Studies Group 5.) Oxford: Oxbow Books, 9–26.

Allen, M.J., 2002a. The chalkland landscape of Cranborne Chase: a prehistoric human ecology. *Landscapes* 3, 55–69.

Allen, M.J., 2002b. The land molluscs and summary of landscape evolution, in *Excavations at Alington Avenue, Fordington, Dorchester, Dorset, 1984–87*, eds. S.M. Davies, P.S. Bellamy, M.J. Heaton & P.J. Woodward. (Dorset Natural History and Archaeological Society Monograph 15.) Dorchester: Dorset Natural History and Archaeological Society, 48–53, 108–11, 182, 185–6.

Allen, M.J., 2003. 'Molluscan analysis', in Archaeological and palaeo-environmental investigations of the Upper Allen Valley, Cranborne Chase, Dorset (1998–2000): a new model of earlier Holocene landscape development, eds. C. French, H. Lewis, M.J. Allen, R.G. Scaife & M. Green. *Proceedings of the Prehistoric Society* 69, 201–34.

Allen, M.J., 2004. 'Thomas Hardeye School; land snails and aspects of the Neolithic environment', in Excavations at Thomas Hardeye School, Coburg Road, Dorchester. Unpublished Client Report, Wessex Archaeology.

Allen, M.J., 2005a. 'Evidence of local Neolithic landscape and land-use; the land snail evidence', in Excavations along the Old Sarum Water Pipeline, north of Salisbury, by A.B. Powell, M.J. Allen, J. Chapman, R. Gale, P. Harding, S. Knight, J.I. McKinley & C. Stevens. *Wiltshire Archaeology and Natural History Magazine* 98, 250–80.

Allen, M.J., 2005b. Considering prehistoric environmental changes on the Marlborough Downs, in *The Avebury*

Landscape: Aspects of Field Archaeology of the Marlborough Downs, eds. G. Brown, D. Field & D. McOmish. Oxford: Oxbow Books, 77–86.

Allen, M.J., 2006. Professor John Gwynne Evans, 1941–2005. Aka 'Snails' Evans. *Journal of Conchology* 39, 111–17.

Allen, M.J., in prep. Open leaf litter and rock-rubble habitats; modern analogues for ancient fauna No 1. *Journal of Conchology*.

Allen, M.J. & L. Allen, 1999. Advances in the Application of Diversity Indices to Archaeological Snail Assemblages. Unpublished manuscript, Codford, St Peter.

Allen, M.J. & J. Gardiner, 2002. A sense of time — cultural markers in the Mesolithic of southern England?, in *Inscriber Landscapes: Marking and Making Place*. Honolulu (HI): University of Hawaii Press, 139–53.

Allen, M.J. & J. Gardiner, 2004. Neolithic of the Wylye Valley 1: millennium re-investigation of the Corton long barrow, ST 9030 4034. *Wiltshire Archaeology and Natural History Magazine* 96, 63–77.

Allen, M.J. & M. Green, 1998. The Fir Tree Field shaft: the archaeological potential of solution features in chalk. *Proceedings of the Dorset Natural History and Archaeological Society* 120, 25–38.

Allen, M.J. & M. Lievers, in press. Neolithic causewayed enclosures and later prehistoric farming: duality, imposition and predecessors at Kingsborough, Isle of Sheppey, Kent. *Proceedings of the Prehistoric Society* 74.

Allen, M.J. & R. Scaife, 2007. A new downland prehistory: long-term environmental change on the southern English chalklands, in *Landscape History after Hoskins: Prehistoric and Roman Landscapes*, eds. A. Fleming & R. Hingley. Macclesfield: Windgather Press, 16–32.

Allen, M.J., R. Entwistle & J. Richards, 1990. Molluscan studies, in *The Stonehenge Environs Project*, by J. Richards. (English Heritage Archaeological Report 16.) London: English Heritage, 253–9.

Allen, M.J., C. Bloomfield, R. Macphail & S. Wyles, 1993. A record of soil and molluscan changes resulting from landuse management at Butser Archaeological Farm: some preliminary results. *Butser Archaeological Farm Yearbook 1992/3*.

Allen, M.J., M. Morris & R.H. Clark, 1995. Food for the living: re-assessment of a Bronze barrow at Buckskin, Basingstoke, Hampshire. *Proceedings of the Prehistoric Society* 61, 157–89.

Andersen, S.T., 1970. The relative pollen productivity and pollen representation of North America trees, and correction factors for tree pollen spectra. *Geological Survey of Denmark* RII, 96, 1.

Andersen, S.T., 1973. The differential pollen productivity of trees and its significance for the interpretation of a pollen diagram from a forested region, in *Quaternary Plant Ecology*, eds. H.J.B. Birks & R.G. West. Oxford: Blackwell, 109–15.

Ares, J., M. Bertiller & A. Bisigato, 2003. Modeling and measurement of structural changes at a landscape scale in dryland areas. *Environmental Modeling and Assessment* 8, 1–13.

Armour-Chelu, M., 1998. The animal bone, in *Etton: Neolithic Causewayed Enclosure, near Maxey, Cambs. – Excavations 1982–7*, by F. Pryor. (English Heritage Archaeological Report 18.) London: English Heritage, 273-88.

Arnold, J., M. Green, B. Lewis & R. Bradley, 1988. The Mesolithic of Cranborne Chase. *Proceedings of the Dorset Natural History and Archaeological Society* 110, 117–25.

Ashbee, P., 1970. *The Earthen Long Barrow in Britain*. London: Dent & Sons Ltd.

Ashbee, P., 1979. 'The Horslip (or Windmill Hill) long barrow', in Excavation of three long barrows near Avebury, Wiltshire, by P. Ashbee, I.F. Smith & J.G. Evans. *Proceedings of the Prehistoric Society* 45, 207–25.

Ashbee, P., I.F. Smith & J.G. Evans, 1979. Excavation of three long barrows near Avebury, Wiltshire. *Proceedings of the Prehistoric Society* 45, 207–300.

Ashbee, P., M.G. Bell & E. Proudfoot, 1989. *Wilsford Shaft: Excavations 1960–62*. (English Heritage Archaeological Report 11.) London: English Heritage.

Atkinson, R.J.C., 1955. The Dorset Cursus. *Antiquity* 29, 4–9.

Atkinson, R.J.C., J.W. Brailsford & H.G. Wakefield, 1951a. A pond barrow at Winterbourne Steepleton, Dorset. *Archaeological Journal* 108, 1–24.

Atkinson, R.J.C., S. Piggott & A. Sanders, 1951b. *Excavations at Dorchester, Oxon*. Oxford: Ashmolean Museum.

Avery, B.W., 1980. *Soil Classification for England and Wales*. (Soil Survey Technical Monograph 14.) Harpenden: Lawes Agricultural Trust.

Ayala, G., 2003. Landscape and Land Use Change in North-central Sicily: a Geoarchaeological Approach. Unpublished PhD, Department of Archaeology, University of Cambridge.

Ayala, G. & C. French, 2003. Holocene landscape dynamics in a Sicilian upland river valley, in *Alluvial Archaeology in Europe*, eds. A. Howard, M. Macklin & D. Passmore. Lisse: Swets and Zeitlinger, 81–92.

Bailey, G.N., 1989. Concepts of time in quaternary prehistory. *Annual Review of Anthropology* 12, 165–92.

Baker, C.A., P.A. Moxey & P.A.M. Oxford, 1978. Woodland continuity and change in Epping Forest. *Field Studies* 4, 645–69.

Ball, D.F., 1975. Processes of soil degradation: a pedological point of view, in *The Effect of Man on the Landscape: the Highland Zone*, eds. J.G. Evans, S. Limbrey & H. Cleere. (Council for British Archaeology Research Report 11.) York: Council for British Archaeology, 20–27.

Barber, K.E. & M.J. Clarke, 1987. Cranes Moor, New Forest: palynology and macro-fossil stratigraphy, in *Quaternary Research Association Field Guide for Wessex and the Isle of Wight*, ed. K.E. Barber. London: Quaternary Research Association, 33–44.

Barclay, A. & C. Halpin, 1999. *Excavations at Barrow Hills, Radley, Oxfordshire*, vol. 1: *The Neolithic and Bronze Age Monument Complex*. (Thames Valley Landscape Series 11.) Oxford: Oxford Archaeology.

Barrett, J.C., 1994. *Fragments from Antiquity: an Archaeology of Social Life in Britain 2900–1200 BC*. Oxford: Blackwell.

Barrett, J., R. Bradley & M. Green, 1991a. *Landscape, Monuments and Society: the Prehistory of Cranborne Chase*. Cambridge: Cambridge University Press.

Barrett, J., R. Bradley & M. Hall, 1991b. *Papers on the Prehistoric Archaeology of Cranborne Chase.* (Oxbow Monograph 11.) Oxford: Oxbow Books.

Bartolome, J., J. Franch, J. Plaixats & N.G. Seligman, 2000. Grazing alone is not enough to maintain landscape diversity in the Montseny Biosphere Reserve. *Agriculture, Ecosystems and Environment* 77, 267–73.

Barton, N. & A. Roberts, 2004. The Mesolithic period in England: current perspectives and new research, in *Mesolithic Scotland and its Neighbours*, eds. A. Saville. Edinburgh: Society of Antiquaries of Scotland, 339–58.

Barton, R.N.E., 1992. Long blade technology and the question of British Late Pleistocene/Early Holocene lithic assemblages, in *Stone Age Archaeology: Essays in honour of John Wymer*, eds. N. Ashton, F. Healy & P. Pettitt. (Oxbow Monograph 102.) Oxford: Oxbow Books, 158–64.

Bass, W.M., 1987. *Human Osteology.* Columbia: Missouri Archaeological Society.

Beek, G.C. van, 1983. *Dental Morphology: an Illustrated Guide.* Bristol: Wright PSG.

Bell, M., 1981. Valley sediments and environmental change, in *The Environment of Man: the Iron Age to Anglo-Saxon Periods*, eds. M. Jones & G.W. Dimbleby. (British Archaeological Reports British Series 87.) Oxford: BAR, 75–91.

Bell, M., 1982. The effect of land use and climate change on valley sedimentation, in *Climatic Change in Later Pre-history*, ed. A.F. Harding. Edinburgh: Edinburgh University Press, 114–26.

Bell, M., 1983. Valley sediments as evidence of prehistoric land-use on the South Downs. *Proceedings of the Prehistoric Society* 49, 119–50.

Bell, M.G., 1990. Sedimentation rates in the primary fills of chalk-cut features, in *Experimentation and Reconstruction in Environmental Archaeology*, ed. D.E. Robinson. Oxford: Oxbow Books, 237–48.

Bell, M.G., 1992. The prehistory of soil erosion, in *Past and Present Soil Erosion*, eds. M. Bell & J. Boardman. (Oxbow Monograph 22.) Oxford: Oxbow Books, 21–36.

Bell, M.G. & S. Johnson, 1996. Land molluscs, in *The Experimental Earthwork Project 1960–1992*, eds. M.G. Bell, P.J. Fowler & S.W. Hillson. (Research Report 100.) York: Council for British Archaeology, 140–42.

Bell, M.G. & J. Jones, 1990. Land mollusca (Coneybury Henge), in *The Stonehenge Environs Project*, by J.C. Richards. (English Heritage Archaeological Report 16.) London: English Heritage, 154–8.

Bell, M.G., P.J. Fowler & S.W. Hillson (eds.), 1996. *The Experimental Earthwork Project 1960–1992.* (Research Report 100.) York: Council for British Archaeology.

Bell, M.G., M.J. Allen, R.W. Smith & S. Johnson, forthcoming. Mollusc and sedimentary evidence for the palaeo-environmental history of Hambledon Hill and its surroundings, in *Hambledon Hill*, by R. Mercer & F. Healy. (English Heritage Archaeological Report.) London: English Heritage.

Bennett, K.D., 1988. A provisional map of forest types for the British Isles 5000 years ago. *Journal of Quaternary Science* 4, 141–4.

Bennett, K.D., G. Whittington & K.J. Edwards, 1994. Recent plant nomenclatural changes and pollen morphology in the British Isles. *Quaternary Newsletter* 73, 1–6.

Berry, A.C. & R.J. Berry, 1967. Epigenetic variation in the human cranium. *Journal of Anatomy* 101, 261–379.

Best, J.A., 1998. Coppice grazing in Rockingham Forest, in *The Ecological History of European Forests*, eds. K.J. Kirby & C. Watkins. Wallingford: CABI, 63–72.

Birks, H.J.B., 1989. Holocene isochrone maps and patterns of tree-spreading in the British Isles. *Journal of Biogeography* 16, 503–40.

Blaikie, P. & H. Brookfield, 1987. *Land Degradation and Society.* London: Methuen.

Boardman, J., 1990. Soil erosion on the South Downs: a review, in *Soil Erosion on Agricultural Land*, eds. J. Boardman, I.D.L. Foster & J.A. Dearing. Wiley: Chichester, 87–105.

Boardman, J., 1992. Current erosion on the South Downs: implications for the past, in *Past and Present Soil Erosion*, eds. M. Bell & J. Boardman. (Oxbow Monograph 22.) Oxford: Oxbow Books, 9–19.

Bodu, P. & B. Valentin, 1997. Groupes a Federmesser ou Aziliens dans le sud et l'ouest du Bassin Parisien. Propositions pour un nouveau modele d'evolution. *Bulletin de la Société Préhistorique Française* 94, 341–7.

Bogucki, P., 1988. *Forest Farmers and Stockbreeders. Early Agriculture and its Consequences in North-central Europe.* Cambridge: Cambridge University Press.

Bogucki, P., 1996. Sustainable and unsustainable adaptations by early farming communities of northern Poland. *Journal of Anthropological Archaeology* 15, 289–311.

Boserup, E., 1965. *The Conditions of Agricultural Growth.* London: Allen & Unwin.

Bowen, H.C., 1990. *The Archaeology of Bokerley Dyke.* London: HMSO.

Bradley, R., 1978. *The Prehistoric Settlement of Britain.* London: Routledge & Kegan Paul.

Bradley, R.J., 1984. Studying monuments, in *Neolithic Studies*, eds. R.J. Bradley & J. Gardiner. (British Archaeological Reports British Series 133.) Oxford: BAR, 61–6.

Bradley, R.J., 1986. *The Dorset Cursus: the Archaeology of the Enigmatic.* (Wessex Lecture III. CBA Group 12.) London: Council for British Archaeology.

Bradley, R., 1991a. The evidence of earthwork monuments, in *Landscape, Monuments and Society: the Prehistory of Cranborne Chase*, eds. J. Barrett, R. Bradley & M. Green. Cambridge: Cambridge University Press, 35–58.

Bradley, R., 1991b. Monuments and places, in *Sacred and Profane*, eds. P. Garwood, D. Jennings, R. Skeates & J. Toms. (Oxbow Monograph 32.) Oxford: Oxbow Books, 135–41.

Bradley, R., 1992. The gravels and British prehistory from the Neolithic to the Early Iron Age, in *Developing Landscapes of Lowland Britain: the Archaeology of the British Gravels. A Review*, eds. M. Fulford & E. Nichols. (Occasional Paper 14.) London: Society of Antiquaries, 15–22.

Bradley, R., 1993. *Altering the Earth: the Origins of Monuments in Britain and Continental Europe.* (Monograph 8.) Edinburgh: Society of Antiquaries, Scotland.

Bradley, R., 1997. *Rock Art and the Prehistory of Atlantic Europe*. London: Routledge.

Bradley, R. & R. Entwistle, 1985. Thickthorn Down long barrow — a new assessment. *Proceedings of the Dorset Natural History and Archaeological Society* 107, 174–6.

Bradley, R. & J. Thomas, 1984. Some new information on the henge monument and Maumbury Rings, Dorchester. *Proceedings of the Dorset Natural History and Archaeological Society* 106, 132–4.

Bradley, R., R. Cleal, J. Gardiner, A. Legge, F. Raymond & J. Thomas, 1984. Sample excavation of the Dorset cursus, 1984: preliminary report. *Proceedings of the Dorset Natural History and Archaeological Society* 106, 128–32.

Breuning-Madsen, H. & M.K. Holst, 1992. Genesis of iron pans in Bronze Age mounds in Denmark. *Journal of Danish Archaeology* 11, 80–86.

Breuning-Madsen, H. & M.K. Holst, 1998. Recent studies on the formation of iron pans around the oaken coffins of the Bronze Age burial mounds of Denmark. *Journal of Archaeological Science* 25, 1103–10.

Breuning-Madsen, H., M.K. Holst & M. Rasmussen, 2001. The chemical environment in a burial mound shortly after construction — an archaeological-pedalogical experiment. *Journal of Archaeological Science* 28, 691–7.

Brewster, T.C.M., 1982. *The Excavation of Garton and Wetwang Slacks*. London: National Monuments Record.

Brewster, T.C.M., 1984. *The Excavation of Whitegrounds Barrow 1, Burythorpe, North Yorkshire*. Malton: East Riding Research Committee Publications.

Bridges, E.M., 1978. *World Soils*. 2nd edition. Cambridge: Cambridge University Press.

British Geological Survey, 1979. *Geological Survey Ten Mile Map South Sheet*. 3rd edition. London: HMSO.

Brooks, S.T., 1955. Skeletal age at death: the reliability of cranial and pubic age indicators. *American Journal of Physical Anthropology* 13, 567–97.

Brothwell, D.R., 1971. Palaeodemography and earlier British populations. *World Archaeology* 4, 75–87.

Brothwell, D.R., 1972. *Digging Up Bones*. London: British Museum Natural History.

Brothwell, D. & S. Zakrzewski, 2004. Metric and non-metric studies of archaeological human remains, in *Guidelines to the Standards for Recording Human Remains*, eds. M. Brickley & J.I. McKinley. British Association for Biological Anthropology and Osteoarchaeology and Institute for Field Archaeology, 24–30.

Brown, A., 1991. Structured deposition and technological change among the flaked stone artefacts from Cranborne Chase, in *Papers on the Prehistoric Archaeology of Cranborne Chase*, eds. J. Barrett, R. Bradley & M. Hall. (Oxbow Monograph 11.) Oxford: BAR, 101–33.

Brown, T., 1997. Clearances and clearings: deforestation in Mesolithic/Neolithic Britain. *Oxford Journal of Archaeology* 16, 133–46.

Brück, J., 1997a. The Early–Middle Bronze Age Transition in Wessex, Sussex and the Thames Valley. Unpublished PhD Thesis, University of Cambridge.

Brück, J., 1997b. Body metaphors and technologies of transformation in the English Middle and Late Bronze Age, in *Bronze Age Landscapes: Tradition and Transformation*, ed. J. Brück. Oxford: Oxbow Books, 149–60.

Budd, P., J. Montgomery, J.A. Evans & B. Barreiro, 2000. Human tooth enamel as a record of the comparative lead exposure of prehistoric and modern people. *The Science of the Total Environment* 263, 1–10.

Buehler Ltd, 1998. *Omnimet Enterprise Handbook*. Lake Bluff: Buehler Ltd.

Buikstra, J.E. & D.H. Ubelaker, 1994. *Standards for Data Collection from Human Skeletal Remains*. (Research Series 44.) Fayetteville (AR): Arkansas Archaeological Survey.

Bull, G. & S. Payne, 1982. Tooth eruption and epiphysial fusion in pigs and wild boar, in *Ageing and Sexing Animal Bones from Archaeological Sites*, eds. B. Wilson, C. Grigson & S. Payne. (British Archaeological Reports British Series 109.) Oxford: BAR, 55–71.

Bullock, P., N. Fedoroff, A. Jongerius, G. Stoops & T. Tursina, 1985. *Handbook for Soil Thin Section Description*. Wolverhampton: Waine Research.

Burrough, P.A., 1986. *Principles of Geographical Information Systems for Land Resources Assessment*. (Monographs on Soil and Resources Survey 12.) Oxford: Blackwell Scientific.

Burrow, S., 1996. *Survey and Excavations on Enclosures in Cranborne Chase*. Electronic publication at http://csweb.bournemouth.ac.uk/consci/proj_cran. Accessed 03/08/00.

Burrow, S.P. & J.D. Gale, 1994a. *Knowlton: Southern Circle. A Project Design for an Archaeological Investigation*. Poole: Department of Conservation Sciences, Bournemouth University.

Burrow, S.P. & J.D. Gale, 1994b. *Knowlton: Southern Circle. Post Excavation Assessment Report*. Poole: Department of Conservation Sciences, Bournemouth University.

Burrow, S.P. & J.D. Gale, 1995. Survey and excavation at Knowlton Rings, Woodland Parish, Dorset 1993–5. *Proceedings of the Dorset Natural History and Archaeology Society* 117, 131–2.

Bush, M.B., 1988. Early Mesolithic disturbance: a force on the landscape. *Journal of Archaeological Science* 15, 453–62.

Bush, M.B., 1989. On the antiquity of British grasslands: a response to Thomas. *Journal of Archaeological Science* 16, 555–60.

Bush, M.B. & J.R. Flenley, 1987. The age of the British chalk grasslands. *Nature* 395, 484–6.

Buxton, L.H.D., 1921. Excavations at Frilford. *Antiquaries Journal* I, 87–97.

Cameron, R.A.D., 1978. Interpreting buried land-snail assemblages from archaeological sites: problems and progress, in *Research Problems in Zooarchaeology*, eds. D.R. Brothwell, K.D. Thomas & A. Clutton-Brock. London: Institute of Archaeology Occasional Publication 3, 19–23.

Cameron, R.A.D. & D.I. Morgan-Huws, 1975. Snail faunas in the early stages of a chalk grassland succession. *Biological Journal of the Linnean Society* 7, 215–29.

Cameron, N. & R.G. Scaife, 1988. 'The pollen record', in Excavation and survey on Furzey Island, Poole Harbour, Dorset, 1985, by P. Cox. *Proceedings of the Dorset Natural History and Archaeological Society* 110, 65–70.

Carter, S.P. & D.A. Davidson, 1998. An evaluation of the contribution soil micromorphology to the study of ancient agriculture. *Geoarchaeology* 13, 535–47.

Caseldine, C. & R. Fyfe, 2006. A modelling approach to locating and characterising elm decline/landnam landscapes. *Quaternary Science Reviews* 25, 632–44.

Catt, J.A., 1977. Loess and coversands, in *British Quaternary Studies: Recent Advances*, ed. F.W. Shotton. Oxford: Clarendon Press, 132–6.

Catt, J.A., 1978. The contribution of loess to soils in lowland Britain, in *The Effect of Man on the Landscape: the Lowland Zone*, eds. S. Limbrey & J.G. Evans. (Research Report 21.) York: Council for British Archaeology 12–20.

Catt, J.A., 1979. Soils and Quaternary geology in Britain. *Journal of Soil Science* 30, 607–42.

Catt, J.A., M. Green & N.J. Arnold, 1980. Naleds in a Wessex Downland Valley. *Proceedings of the Dorset Natural History and Archaeological Society* 102, 69–75.

Clark, J.G.D., 1932a. The curved flint sickle blades of Britain. *Proceedings of the Prehistoric Society of East Anglia* VII, 67–81.

Clark, J.G.D., 1932b. *The Mesolithic Age in Britain*. Cambridge: Cambridge University Press.

Clark, J.G.D., 1934. Derivative forms of the *petit tranché* in Britain. *Archaeological Journal* 91, 32–58.

Clark, J.G.D., 1952. *Prehistoric Europe: the Economic Basis*. London: Methuen.

Clarke, D.L., 1970. *Beaker Pottery of Great Britain and Ireland*. Cambridge: Cambridge University Press.

Cleal, R.M.J., 1988. The occurrence of drilled holes in later Neolithic pottery. *Oxford Journal of Archaeology* 7, 139–44.

Cleal, R.M.J., 1991. Cranborne Chase: the earlier prehistoric pottery, in *Papers on the Prehistoric Archaeology of Cranborne Chase*, eds. J. Barrett, R. Bradley & M. Hall. (Oxbow Monograph 11.) Oxford: Oxbow Books, 134–200.

Cleal, R.M.J., 1995. Pottery fabrics in Wessex in the fourth to second millennia BC, in *Unbaked Urns of Rudely Shape. Essays on British and Irish Pottery for Ian Longworth*, eds. I. Kinnes & G. Varndell. (Oxbow Monograph 55.) Oxford: Oxbow Books, 185–94.

Cleal, R.M.J., 2004. The dating and diversity of the earliest ceramics of Wessex and south-west England, in *Monuments and Material Culture. Papers in Honour of an Avebury Archaeologist: Isobel Smith*, eds. R.M.J. Cleal & J. Pollard. Salisbury: Hobnob Press, 164–92.

Cleal, R.M.J. & M.J. Allen, 1994. Investigation of tree-damaged barrows on King Barrow Ridge and Luxenborough Plantation, Amesbury. *The Wiltshire Archaeological and Natural History Magazine* 87, 54–84.

Cleal, R.M.J., J. Cooper & D. Williams, 1994. Shells and sherds: identification of inclusions in Grooved Ware, with associated radiocarbon dates, from Amesbury, Wiltshire. *Proceedings of the Prehistoric Society* 60, 445–8.

Cleal, R.M.J., K.E. Walker & R. Montague, 1995. *Stonehenge in its Landscape*. (English Heritage Archaeological Report 10.) London: English Heritage.

Cleal, R.M.J., M.J. Allen & C. Newman, 2004. An archaeological and environmental study of the Neolithic and later prehistoric landscape of the Avon valley and Durrington Walls environs. *Wiltshire Archaeological and Natural History Magazine* 97, 218–48.

Clutton-Brock, J., 1984. *Neolithic Antler Picks from Grimes Graves, Norfolk and Durrington Walls, Wiltshire: a Biometrical Analysis*. (Excavations at Grimes Graves, Norfolk 1972–1976 Fascicule 1.) London: British Museum,

Colt Hoare, R., 1812. *The Ancient History of Wiltshire*. London: William Millar.

Colt Hoare, R., 1819. *The Ancient History of North Wiltshire*. London: Lackington, Hughes, Harding, Mayor and Jones.

Connah, G., 1965. Excavations at Knap Hill, Alton Priors, 1961. *Wiltshire Archaeology and Natural History Magazine* 60, 1–23.

Coppock, J.T. ,1964. *An Agricultural Atlas of England and Wales*. London: Faber.

Coppolillo, P.B., 2000. The landscape ecology of pastoral herding: spatial analysis of land use and livestock production in East Africa. *Human Ecology* 28, 527–60.

Corney, M., 1991. Later first millennium settlement morphology, in *Landscape, Monuments and Society: the Prehistory of Cranborne Chase*, eds. J. Barrett, R. Bradley & M. Green. Cambridge: Cambridge University Press, 228–36.

Cornwall, I.W., 1953. Soil science and archaeology. *Proceedings of the Prehistoric Society* 2, 129–47.

Cornwall, I.W., 1958. *Soils for the Archaeologist*. London: Phoenix House.

Costanza, R. & A. Voinov, 2004. Introduction: spatially explicit landscape simulation models, in *Landscape Simulation Modeling: a Spatially Explicit, Dynamic Approach*, eds. R. Costanza & A. Voinov. New York (NY): Springer-Verlag, 3–20.

Coudret, P. & J.-P. Fagnart, 1997. Les industries a Federmesser dans le basin de la Somme: chronologie et identité des groupes culturels. *Bulletin de la Société Préhistorique Française*, 349–59.

Courty, M.-A., P. Goldberg & R.I. Macphail, 1989. *Soils and Micromorphology in Archaeology*. Cambridge: Cambridge University Press.

Cox, P. & C. Hearne, 1991. *Redeemed from the Heath: the Archaeology of the Wytch Farm Oilfield, 1987–90*. (Monograph 9.) Dorchester: Dorset Natural History and Archaeological Society

Crabtree, K., 1996. Geomorphology of the ditch infill, in *The Experimental Earthwork Project 1960–1992*, eds. M. Bell, P.J. Fowler & S.W. Hillson. (Research Report 100.) York: Council for British Archaeology, 39–42.

Crawford, O.G.S. & A. Keiller, 1928. *Wessex from the Air*. Oxford: Oxford University Press.

Cruise, G.M., 1993. 'Soil pollen', in A Neolithic downland monument in its environment: excavations at the Easton Down long barrow, Bishops Cannings, North Wiltshire, by A. Whittle, A.J. Rouse & J.G. Evans. *Proceedings of the Prehistoric Society* 59, 219–21.

Cunnington, M.E., 1929. *Woodhenge.* Devizes: George Simpson.

Dale, V.H. & S.M. Pearson, 1999. Modeling the driving factors and ecological consequences of deforestation in the Brazilian Amazon, in *Spatial Modeling of Forest Landscape Change*, eds. D.J. Mladenoff & W.L. Baker. Cambridge: Cambridge University Press, 256–76.

David, A., 1995. *Geophysical Survey in Archaeological Field Evaluation.* (English Heritage Research and Professional Services Guideline No. 1.) London: English Heritage.

Dayrell-Reed, T., 1931. Notes on the outer bounds of Cranborne Chase. *Proceedings of the Dorset Natural History and Archaeological Society* 53, 215–27.

Delcourt, P.A. & H.R. Delcourt, 2004. *Prehistoric Native Americans and Ecological Change.* Cambridge: Cambridge University Press.

Degerbøl, M. & B. Fredskild, 1970. *The Ursus (*Bos primigenius *Bojanus) and Neolithic Domesticated Cattle (*Bos taurus domesticus *Linné) in Denmark.* (Det Kongelige Danske Videnskabernes Selskab Biologiske Skrifter 17, 1.) Copenhagen: Munksgaard.

Dimbleby, G.W., 1985. *The Palynology of Archaeological Sites.* London: Academic Press.

Dimbleby, G.W. & J.G. Evans, 1974. Pollen analysis and land-snail of calcareous soils. *Journal of Archaeological Science* 1, 117–33.

Dinnin, M. & R. van der Noort, 1999. Wetland habitats, their resource potential and exploitation: a case study from the Humber wetlands, in *Bog Bodies, Sacred Sites and Wetland Archaeology*, eds. B. Coles, J. Coles & M.S. Jorgensen. Exeter: WARP, 69–78.

Drew, C. & S. Piggott, 1936. The excavation of long barrow 163a on Thickthorn Down, Dorset. *Proceedings of the Prehistoric Society* 2, 77–96.

von den Driesch, A., 1976. *A Guide to the Measurement of Animal Bones from Archaeological Sites.* (Monograph 1.) Cambridge (MA): Peabody Museum, Harvard University.

Edmonds, M., 1995. *Stone Tools and Society*. London: Batsford.

Edwards, K.J., 1982. Man, space and the woodland edge-speculation on the detection and interpretation of human impact in pollen profiles, in *Archaeological Aspects of Woodland Ecology*, eds. S. Limbrey & M. Bell. (Symposia of the Association for Environmental Archaeology No. 2.) (British Archaeological Reports International Series 146.) Oxford: BAR, 5–22.

Edwards, K.J., 1983. Quaternary palynology: multiple profile studies and pollen variability. *Progress in Physical Geography* 7, 587–609.

Edwards, K.J., 1989. The cereal pollen record and early agriculture, in *The Beginnings of Agriculture*, eds. A. Milles, D. Williams & N. Gardner. (British Archaeological Reports International Series 496.) Oxford: BAR, 113–35.

Edwards, K.J., 1993. Models of mid-Holocene forest farming for north-west Europe, in *Climate Change and Human Impact*, ed. F.M. Chambers. London: Chapman & Hall, 133–45.

Ellis, C., 1985. Flandrian molluscan biostratigraphy and its application to dry valley deposits in East Sussex, in *Palaeobiological Investigations: Research Design, Methods and Data Analysis*, eds. N. Fieller, D. Gilbertson & N.G.A. Ralph. (British Archaeological Reports International Series 266.) Oxford: BAR, 157–65.

Ellis, C., 1986. The postglacial molluscan succession of the South Downs dry valleys, in *The Scientific Study of Flint and Chert*, eds. G. de G. Sieveking & M.B. Hart. Cambridge: Cambridge University Press, 175–94.

Entwistle, R., 1985. 'Molluscan analysis', in Thickthorn Down long barrow — a new assessment, by R. Bradley & R. Entwistle. *Proceedings of the Dorset Natural History and Archaeological Society* 107, 175–6.

Entwistle, R., 1989. Relativism and Interpretation in Prehistoric Archaeology: Some Thoughts on the Formulation of Archaeological Evidence, with Special Reference to the Use of Palaeoenvironmental Data. Unpublished PhD thesis, University of Reading.

Entwistle, R., 1990a. 'Statistical appraisal', in. *The Stonehenge Environs Project*, by J.C. Richards. (Archaeological Reports 16.) London: English Heritage, 254.

Entwistle, R., 1990b. Land mollusca (Lesser Cursus, Amesbury long barrow), in *The Stonehenge Environs Project*, by J.C. Richards. (English Heritage Archaeological Reports 16.) London: English Heritage, 88–93 & 105–9.

Entwistle, R., 1994. The environmental setting of the linear ditches system, in *Prehistoric Land Divisions on Salisbury Plain: the Work of the Wessex Linear Ditches Project*, eds. R. Bradley, R. Entwistle & F. Raymond. (English Heritage Archaeological Report 2.) London: English Heritage, 101–21.

Entwistle, R. & M. Bowden, 1991. Cranborne Chase; the molluscan evidence, in *Papers on the Prehistoric Archaeology of Cranborne Chase*, eds. J. Barrett, R. Bradley & M. Hall. (Oxbow Monograph 11.) Oxford: Oxbow Books, 20–48.

Eogan, G. & H. Roche, 1999. Grooved Ware from Brugh na Bóinne and its wider context, in *Grooved Ware in Britain and Ireland*, eds. R.M.J. Cleal & A. MacSween. Oxford: Oxbow Books, 98–111.

Evans, C. & M. Knight, 1996. Barleycroft Farm, Bluntisham. Unpublished report, Cambridge Archaeological Unit, University of Cambridge.

Evans, J.G., 1966. Late Glacial and Post-glacial subaerial deposits at Pitstone, Buckinghamshire. *Proceedings of the Geological Association* 77(3), 347–64.

Evans, J.G., 1967. The Stratification of Mollusca in Chalk Soils and their Relation to Archaeology. Unpublished PhD thesis, Institute of Archaeology, University of London.

Evans, J.G., 1971. Habitat change on the calcareous soils of Britain: the impact of Neolithic Man, in *Economy and Settlement in Neolithic and Early Bronze Age Britain and Europe,* ed. D.D.A. Simpson. Leicester: University Press, 27–73.

Evans, J.G., 1972. *Land Snails in Archaeology*. London: Seminar Press.

Evans, J.G., 1979. 'The South Street long barrow', in Excavation of three long barrows near Avebury, Wiltshire, by

P. Ashbee, I.F. Smith & J.G. Evans. *Proceedings of the Prehistoric Society* 45, 250–300.

Evans, J.G., 1984. New Things to Do with Snails and Slugs: a Study in Diversity and Association. Unpublished manuscript, Cardiff.

Evans, J.G., 1991. An approach to the interpretation of dry-ground and wet-ground molluscan taxocenes from central-southern England, in *Modelling Ecological Change: Perspectives from Neoecology, Palaeoecology and Environmental Archaeology*, eds. D.R. Harris & K.D. Thomas. London: Institute of Archaeology, 75–89.

Evans, J.G. & H. Jones, 1973. Sub-fossil and modern land-snail faunas from rock rubble habitats. *Journal of Conchology* 28, 103–29.

Evans, J.G. & H. Jones, 1979. Mount Pleasant and Woodhenge: The land mollusca, in *Mount Pleasant, Dorset: Excavations 1970–71*, by G.J. Wainwright. (Research Report 37.) London: Society of Antiquaries, 190–213.

Evans, J.G. & A. Rouse, 1991. The land mollusca, in *Maiden Castle: Excavations and Field Survey 1985–6*, by N.M. Sharples. (Archaeological Report 19.) London: English Heritage, 118–25.

Evans, J.G. & D.D.A. Simpson, 1991. Giant's Hills 2 long barrow, Skendleby, Lincolnshire. *Archaeologia* 109, 1–45.

Evans, J.G. & I.F. Smith, 1983. Excavations at Cherhill, North Wiltshire, 1967. *Proceedings of the Prehistoric Society* 49, 101–9.

Evans, J.G. & D. Williams, 1991. Land mollusca from the M3 archaeological sites — a review, in *Archaeology and the M3: the Watching Brief, the Anglo-Saxon Settlement at Abbots Worthy and Retrospective Sections*, by P.J. Fasham & R.J.B. Whinney. (Monograph 7.) Winchester: Hampshire Field Club and Archaeological Society, 113–42.

Evans, J.G., C. French & D. Leighton, 1978. Habitat change in two Late-glacial and Post-glacial sites in southern Britain: the molluscan evidence, in *The Effect of Man on the Landscape: the Lowland Zone*, eds. J.G. Evans & S. Limbrey. (Research Report 21.) London: Council for British Archaeology, 63–74.

Evans, J.G., M.W. Pitts & D. Williams, 1985. An excavation at Avebury, Wiltshire, 1982. *Proceedings of the Prehistoric Society* 51, 305–20.

Evans, J.G., S. Limbrey, I. Maté & R. Mount, 1988. Environmental change and land-use history in a Wiltshire river valley in the last 14,000 years, in *The Archaeology of Context in the Neolithic and Bronze Age: Recent Trends*, eds. J.C. Barrett & I. Kinnes. Sheffield: University of Sheffield, 97–103

Evans, J.G., S. Limbrey, I. Mate & R. Mount, 1993. An environmental history of the upper Kennet valley, Wiltshire, for the last 10,000 years. *Proceedings of the Prehistoric Society* 59, 197–240.

Field, D., 1998. Round barrows and the harmonious landscape: placing Early Bronze Age burial monuments in southeast England. *Oxford Journal of Archaeology* 17, 309–26.

Field, N.H., C.L. Matthews & I.F. Smith, 1964. New Neolithic sites in Dorset and Bedfordshire, with a note on the distribution of Neolithic storage pots in Britain. *Proceedings of the Prehistoric Society* 15, 352–81.

Finnegan, M., 1978. Non-metric variations of the infracranial skeleton. *Journal of Anatomy* 125, 23–37.

Fisher, P.F., 1982. A review of lessivage and Neolithic cultivation in southern England. *Journal of Archaeological Science* 9, 299–304.

Fisher, P.F., 1991. The physical environment of Cranborne Chase, in *Papers on the Prehistoric Archaeology of Cranborne Chase*, eds. J. Barrett, R. Bradley & M. Hall. (Oxbow Monograph 11.) Oxford: Oxbow Books, 11–19.

Fishpool, M., 1999. Land mollusca, in *The Harmony of Symbols: the Windmill Hill Causewayed Enclosure*, by A. Whittle, J. Pollard & C. Grigson. Oxford: Oxbow Books, 127–38.

Fitzpatrick, E.A., 1993. *Soil Microscopy and Micromorphology*. Chichester: John Wiley and Sons.

Fleming, A., 1989. *The Dartmoor Reaves: Investigating Prehistoric Land Division*. London: Batsford.

Folk, R.L., 1959. Practical petrographic classification of limestones. *American Association of Petrological Geology Bulletin* 43, 1–38.

Fowler, P., 1981. Wildscape to landscape, in *Farming Practice in British Prehistory*, ed. R. Mercer. Edinburgh: Edinburgh University Press, 9–54.

Fowler, P.J., 1983. *The Farming of Prehistoric Britain*. Cambridge: Cambridge University Press.

Fowler, P.J. & J.G. Evans, 1967. Plough-marks, lynchets and early fields. *Antiquity* 41(164), 289–301.

Frelich, L.E., 2002. *Forest Dynamics and Disturbance Regimes*. Cambridge: Cambridge University Press.

French, C.A.I., 1983. An Environmental Study of the Soil, Sediment and Molluscan Evidence Associated with Prehistoric Monuments on River Terrace Gravels in North-west Cambridgeshire. Unpublished PhD thesis, University of London.

French, C., 1985. Soil/sediments and molluscan analyses at Barnack/Bainton and Barnack/Pilsgate, in *The Fenland Project*, no. 1: *Archaeology and Environment in the Lower Welland Valley*, by F. Pryor & C. French. (Report 27.) Cambridge: East Anglian Archaeology, 284–96.

French, C., 2002. Effects of past and present ploughing on the prehistoric monuments in Wyke Down, Cranborne Chase, Dorset, in *The Management of Archaeological Sites in Arable Landscapes*, appendix Jiii, ed. Oxford Archaeology. York: DEFRA.

French, C., 2003. *Geoarchaeology in Action: Studies in Soil Micromorphology and Landscape Evolution*. London: Routledge.

French, C. & H. Lewis, with contributions by R. Scaife & M. Allen, 2005. New perspectives on Holocene landscape development in the southern English chalklands: The upper Allen valley, Cranborne Chase, Dorset, England. *Geoarchaeology* 20, 109–34.

French, C., H. Lewis, M. Allen & R. Scaife, 2000. Palaeoenvironmental and archaeological investigations on Wyke Down and in the upper Allen valley, Cranborne Chase, Dorset, England: interim summary report for 1998–9. *Proceedings of the Dorset Natural History and Archaeological Society* 122, 53–71.

French, C., H. Lewis, M. Allen & R. Scaife, 2003. Archaeological and palaeoenvironmental investigations of the upper Allen valley, Cranborne Chase, Dorset (1998–2000): a new model of earlier Holocene landscape development. *Proceedings of the Prehistoric Society* 69, 210–34.

Fyfe, R.M., A.G. Brown & B.J. Coles, 2003. Mesolithic to Bronze Age vegetation change and human activity in the Exe valley, Devon. *Proceedings of the Prehistoric Society* 69, 161–82.

Gaffney, C. & J. Gater, 2003. *Revealing the Buried Past*. Stroud: Tempus.

Gaffney, C., J. Gater & S. Ovenden, 2002. *The Use of Geophysical Techniques in Archaeological Evaluations*. (Paper 6.) Reading: Institute of Field Archaeologists.

Gale, J. (ed.), in prep. *Archaeological Investigations in the Environs of Knowlton Henge Complex, Dorset 1993–7*. Oxford: British Archaeological Reports.

Gardiner, J., 1985. Intra-site patterning in the flint assemblage from the Dorset cursus. *Proceedings of the Dorset Natural History and Archaeological Society* 105, 87–93.

Gardiner, J.P., 1988. The Composition and Distribution of Neolithic Surface Flint Assemblages in Central Southern England. Unpublished PhD, University of Reading.

Gardiner, J., 1991. Spatial analysis: the lithic scatter and the Cursus, in *Landscape, Monuments and Society: the Prehistory of Cranborne Chase*, by J. Barrett, R. Bradley & M. Green. Cambridge: Cambridge University Press, 73–5.

Gebhardt, A., 1990. Evolution du paléopaysage agricole dans le nord-ouest de la France, Aport de la micromorphologie. Unpublished PhD thesis, Université de Rennes I.

Gebhardt, A., 1992. Micromorphological analysis of soil structure modifications caused by different cultivation implements, in *Prehistoire de l'agriculture: Nouvelles approaches experimentales et ethnographiques*. (Monograph 6.) Valbonnes: C.R.A., 373–81.

Gejvall, N.G., 1981. Determination of burnt bones from Prehistoric graves. *OSSA LETTERS* 2, 1–13.

Gergel, S.E. & M.G. Turner, 2002. *Learning Landscape Ecology: A Practical Guide to Concepts and Techniques*. New York (NY): Springer-Verlag.

Germany, M., 2007. Neolithic and Bronze Age monuments and Middle Iron Age settlements at Lodge Farm, St Osyth, Essex. *East Anglian Archaeology* 117, 36–8 & 109–12.

Gibson, A., 1998. *Stonehenge and Timber Circles*. Stroud: Tempus.

Gillings, M., J. Pollard & D. Wheatley, 2002. Excavations at the Beckhampton Enclosure, Avenue and Cove, Avebury: an interim report on the 2000 season. *Wiltshire Archaeological and Natural History Magazine* 95, 249–58.

Girling, M.A., 1988. The bark beetle *Scolytus scolytus* (Fabricius) and the possible role of elm disease in the early Neolithic, in *Archaeology and the Flora of the British Isles*, ed. M. Jones. Oxford: Oxford University Committee for Archaeology, 34–8.

Godwin, H., 1962. Vegetational history of the Kentish Chalkdowns as seen at Wingham and Frogholt. *Veroff Geobot. Inst. Rubel*, Zurich 37, 83–99.

Godwin, H., 1975. History of the natural forests of Britain: establishment, dominance and destruction. *Philosophical Transactions of the Royal Society* B 271, 47–67.

Goldman, A., 1995. Threats to sustainability in African agriculture: searching for appropriate paradigms. *Human Ecology* 23, 291–334.

Goodman, D., 1975. The theory of diversity-stability relations in ecology. *The Quarterly Review of Biology* 50, 237–66.

Goody, J., 1976. *Production to Reproduction*. Cambridge: Cambridge University Press.

Gordon, D. & C. Ellis, 1985. Species composition parameters and life tables: their application to detect environmental change in fossil land molluscan assemblages, in *Palaeoenvironmental Investigations: Research Design, Methods and Data Analysis*, eds. N.R.J. Fieller, D.D. Gilbertson & N.G.A. Ralph. (British Archaeological Reports International Series 258.) Oxford: BAR, 157–74.

Gosden, C. & G. Lock, 2003. Frilford: Romano-British ritual pool. *Current Archaeology* 16(184), 156.

Grant, A., 1982. The use of toothwear as a guide to the age of domestic ungulates, in *Ageing and Sexing Animal Bones from Archaeological Sites*, eds. B. Wilson, C. Grigson, & S. Payne. (British Archaeological Reports British Series 109.) Oxford: BAR, 91–108.

Gray, H., 1977. *Anatomy*. New York (NY): Bounty Books.

Green, M., 1985. Excavations in Home Field, Down Farm. *Proceedings of the Dorset Natural History and Archaeological Society* 108, 171–3.

Green, M., 2000. *A Landscape Revealed: 10,000 Years on a Chalk Downland Farm*. Stroud: Tempus.

Green, M., forthcoming. Excavation of Endless Pit, Sixpenny Handley. *Proceedings of the Dorset Natural History and Archaeological Society*.

Green, M. & M.J. Allen, 1997. An early prehistoric shaft on Cranborne Chase. *Oxford Journal of Archaeology* 16, 121–32.

Green, M., R.N.E. Barton, N. Debenham & C.A.I. French, 1998. A new late-glacial open-air site at Deer Park Farm, Wimborne St Giles. *Proceedings of the Dorset Natural History and Archaeological Society* 120, 85–8.

Greenland, D.J., 1977. Soil damage by intensive arable cultivation: temporary or permanent? *Philosophical Transactions of the Royal Society of London* B 281, 193–208.

Greig, J.R.A., 1982. Past and present lime woods of Europe, in *Archaeological Aspects of Woodland Ecology*, eds. M. Bell & S. Limbrey. (Assoc. Environ. Arch. Symposia vol. 2.) (British Archaeological Reports International Series 146.) Oxford: BAR, 23–55.

Grigson, C., 1965. Measurements of bones, horncores, antlers and teeth, in *Windmill Hill and Avebury: Excavations by Alexander Keiller 1925–1939*, I. Smith. Oxford: Oxford University Press, 145–67.

Grigson, C., 1982a. Sexing Neolithic domestic cattle skulls and horn cores, in *Ageing and Sexing Animal Bones from Archaeological Sites*, eds. B. Wilson, C. Grigson & S. Payne. (British Archaeological Reports British Series 109.) Oxford: BAR, 25–35.

Grigson, C., 1982b. Porridge and pannage: pig husbandry in Neolithic England, in *Archaeological Aspects of*

Woodland Ecology, eds. M. Bell & S. Limbrey. (British Archaeological Reports British Series 146.) Oxford: BAR, 297–314.

Grigson, C., 1999. The mammalian remains, in *The Harmony of Symbols: the Windmill Hill Causewayed Enclosure, Wiltshire*, by A. Whittle, J. Pollard & C. Grigson. Oxford: Oxbow, 164–252.

Grimm, E.C., 1991. *TILIA and TILIA-GRAPH*. Springfield (IL): Illinois State Museum.

Grinsell, L.V., 1936. *The Ancient Burial-Mounds of England*. London: Methuen & Co. Ltd.

Grinsell, L.V., 1959. *Dorset Barrows*. Dorchester: Dorset Natural History and Archaeological Society.

Grinsell, L., 1982. *Dorset Barrows Supplement*. Dorchester: Dorset Natural History and Archaeology Society.

GSB, 1998. *Wyke Down, Gussage St Michael and Handley, Dorset*. (GSB Report 98/65.) Bradford.

GSB, 1999. *Wyke Down II, Gussage St Michael and Handley, Dorset*. (GSB Report 99/99.) Bradford.

GSB, 2001. *Bottlebush Down, Sixpenny Handley, Dorset*. (GSB Report 2001/87.) Bradford.

GSB, 2002a. *Gussage Cow Down, Dorset*. (GSB Report 2002/05.) Bradford.

GSB, 2002b. *Gussage Cow Down II*. (GSB Report 2002/85.) Bradford.

GSB, 2003. *Upper Allen Valley, Dorset*. (GSB Report 2003/70.) Bradford.

Haigh, J.G.B., 1989. Rectification of aerial photographs by means of desk-top systems, in *Computer Applications and Quantitative Methods in Archaeology*, eds. S. Rahtz & J. Richards. (British Archaeological Reports International Series 548.). Oxford: BAR, 111–19.

Halstead, P., 1985. A study of mandibular teeth from Romano-British contexts at Maxey, in *Archaeology and Environment on the Lower Welland Valley*, vol. 1, by F. Pryor & C. French. (East Anglian Archaeological Report 27.) Cambridge: East Anglian Archaeology, 219–24.

Hambleton, E., 1999. *Animal Husbandry Regimes in Iron Age Britain: a Comparative Study of Faunal Assemblages from British Iron Age Sites*. (British Archaeological Reports British Series 282.) Oxford: BAR.

Hamilton, M.A., 1997. Pottery, in *Sacred Mound, Holy Rings. Silbury Hill and the West Kennet Palisade Enclosures: a Later Neolithic Complex in North Wiltshire*, by A. Whittle. Oxford: Oxbow Books, 93–117.

Harcourt, R.A., 1971. Animal bones from Durrington Walls, in *Durrington Walls: Excavations 1966–1968*, by G.J. Wainwright & I.H. Longworth. (Research Committee Report 29.) London: Society of Antiquaries of London, 338–50.

Harcourt, R.A., 1974. The dog in prehistoric and early historic Britain. *Journal of Archaeological Science* 127, 56–63.

Harcourt, R.A., 1979. Animal bones from Mount Pleasant, in *Mount Pleasant, Dorset: Excavations 1970–71*, by G.J. Wainwright. (Research Committee Report 27.) London: Society of Antiquaries of London, 214–23.

Harding, D.W. & G.E. Lee, 1987. *Henge Monuments and Related Sites of Great Britain*. (British Archaeological Report British Series 175.) Oxford: BAR.

Harris, D.R. & K.D. Thomas (eds.), 1991. *Modelling Ecological Change: Perspectives from Neoecology, Palaeoecology and Environmental Archaeology*. London: Institute of Archaeology.

Haskins, L.E., 1978. The Vegetational History of South-east Dorset. Unpublished PhD thesis, Department of Geography, University of Southampton.

Hawkes, C.F.C., 1959. The ABC of the British Iron Age. *Antiquity* 33, 170–82.

Hawkes, C.F.C. & S. Piggott, 1947. Britons, Romans and Saxons round Salisbury and in Cranborne Chase. *Archaeological Journal* 104, 27–81.

Heiri, C., H. Bugmann, W. Tinner, O. Heiri & H. Lischke, 2006. A model-based reconstruction of treeline dynamics in the central Swiss Alps. *Journal of Ecology* 94, 206–16.

Hester, A.J., F.J.G. Mitchell & K.J. Kirby, 1996. Effects of season and intensity of sheep grazing on tree regeneration in a British upland woodland. *Forest Ecology and Management* 88, 99–106.

Higgs, E.S., 1959. The excavation of a late Mesolithic site at Downton, near Salisbury, Wiltshire. *Proceedings of the Prehistoric Society* 25, 209–32.

Hillson, S.W., 1990. *Teeth*. (Cambridge Manuals in Archaeology.) Cambridge: Cambridge University Press.

Hingley, R., 1985. Location, function and status: a Romano-British 'religious complex' at the Noah's Ark Inn, Frilford (Oxfordshire). *Oxford Journal of Archaeology* 4, 201–14.

Holden, J.L., P.P. Phakley & J.G. Clement, 1995a. Scanning electron microscope observations of incinerated human femoral bone: a case study. *Forensic Science International* 74, 17–28.

Holden, J.L., P.P. Phakley & J.G. Clement, 1995b. Scanning electron microscope observations of heat-treated human bone. *Forensic Science International* 74, 29–45.

Holliday, V.T. & P. Goldberg, 1992. Glossary of selected soil science terms, in *Soils in Archaeology: Landscape Evolution and Human Occupation*, ed. V.T. Holliday. Washington (DC) & London: Smithsonian Institution Press, 247–54.

Hurlbert, S.H., 1971. The nonconcept of species diversity: a critique and alternative parameters. *Ecology* 52, 577–86.

Iscan, M.Y., S.R. Loth & R.K. Wright, 1984. Age estimation from the ribs by phase analysis: White males. *Journal of Forensic Science* 29, 1094–104.

Iscan, M.Y., S.R. Loth & R.K. Wright, 1985. Age estimation from the ribs by phase analysis: White females. *Journal of Forensic Science* 30, 853–63.

Iversen, J., 1949. The influence of prehistoric man on vegetation. *Geological Survey of Denmark* RIV 3 (6), 1–25.

Janssen, C.R., 1969. *Alnus* as a disturbing factor in pollen diagrams. *Acta Botanica Neere* 8, 55–8.

Jones, G., 2000. Evaluating the importance of cultivation and collecting in Neolithic Britain, in *Plants in Neolithic Britain and Beyond*, ed. A.S. Fairbairn. (Neolithic Studies Group Seminar Papers 5.) Oxford: Oxbow Books, 79–84.

Jongerius, A., 1983. The role of micromorphology in agricultural research, in *Soil Micromorphology*, eds. P.

Bullock & C.P. Murphy. Berkhamsted: A B Academic, 111–38.
Kennard, A.S., 1936. 'Report in the non-marine mollusca', in The excavation of long barrow 163a on Thickthorn Down, Dorset, by C.D. Drew & S. Piggott. *Proceedings of the Prehistoric Society* 2, 94–5.
Kenward, H.K., 1978. The analysis of archaeological insect assemblages: a new approach, in *The Archaeology of York*, Fascicle 19.1. London: CBA/York Archaeological Trust.
Kerney, M.P., 1966. Snails and man in Britain. *Journal of Conchology* 26, 3–14.
Kerney, M.P., 1968. Britain's fauna of land mollusca and its relation to the Post-glacial thermal optimum. *Symposia Zoological Society London* 22, 273–91.
Kerney, M., 1999. *Atlas of Land and Freshwater Molluscs of Britain and Ireland*. Colchester: Harley Books.
Kerney, M.P., E.H. Brown & T.J. Chandler, 1964. The late-glacial and post-glacial history of the chalk escarpment near Brook, Kent. *Philosophical Transactions of the Royal Society* B 248, 135–204.
Kerney, M.P., R.C. Preece & C. Turner, 1980. Molluscan and plant biostratigraphy of some Late Devensian and Flandrian deposits in Kent. *Philosophical Transactions of the Royal Society of London* B291, 1–43.
Kinnes, I., 1979. *Round Barrows and Ring-Ditches in the British Neolithic.* (Occasional Paper 7.) London: British Museum.
Kinnes, I., 1999. Longtemps ignorées: Passy-Rots, linear monuments in northern France, in *Pathways and Ceremonies: the Cursus Monuments of Britain and Ireland*, eds. A. Barclay & J. Harding. (Neolithic Studies Group Seminar Papers 4.) Oxford: Oxbow, 148–54.
Kirby, K.J., 2004. A model of a natural wooded landscape in Britain as influenced by large herbivore activity. *Forestry* 77, 405–20.
Knüsel, C.J. & A.K. Outram, 2004. Fragmentation: the Zonation method applied to fragmented human remains from archaeological and forensic contexts. *Environmental Archaeology* 9, 85–97.
Kubiena, W.L., 1938. *Micropedology*. Ames: Collegiate Press.
Kuznar, L.A., 1991. Herd composition in an Aymara community of the Peruvian Altiplano: a linear programming problem. *Human Ecology* 19, 369–87.
Lal, R., G.F. Hall & F.P. Miller, 1989. Soil degradation, I: basic processes. *Land Degradation and Rehabilitation* 1, 51–69.
Law, C., 1998. The uses and fire-ecology of reedswamp vegetation, in *Star Carr in Context*, eds. P. Mellars & P. Dark. (McDonald Institute Monographs.) Cambridge: McDonald Institute for Archaeological Research, 197–208.
Legge, A.T., 1981. Aspects of cattle husbandry, in *Farming Practice in British Prehistory*, ed. R. Mercer. Edinburgh: Edinburgh University Press, 169–81.
Legge, A.T., 1991. The animal remains from six sites at Down Farm, Woodcutts, in *Papers in the Prehistoric Archaeology of Cranborne Chase*, by J. Barrett, R. Bradley & M. Hall. Oxford: Oxbow, 54–100.
Legge, A.T. & P.A. Rowley-Conwy, 1988. *Star Carr Revisited: a Re-analysis of the Large Mammals*. London: Centre for Extra Mural Studies, University of London.
Lewis, B. & R. Coleman, 1982. Pentridge Hill, Dorset: trial excavation. *Proceedings of the Dorset Natural History and Archaeological Society* 104, 59–65.
Lewis, H., 1998a. Summary Report on the Buried Soils and Turf Mounds from Wyke Down Barrows 34 and 41. Report for the Wyke Down Project, University of Cambridge.
Lewis, H.A., 1998b. The Characterisation and Interpretation of Ancient Tillage Practices through Soil Micromorphology: a Methodological Study. Unpublished PhD thesis, University of Cambridge.
Lewis, H., 2001. Knowlton 1994 (KWC 94) Southern Circle Excavation Soil Micromorphology Report. Unpublished report, Bournemouth University.
Lewis, H., 2003. Marcham/Frilford 2003: the 'Amphitheatre', Buried Soils, General Notes on the Site, and Report of Soil Micromorphology Sampling. Unpublished report, Oxford, OUDCE.
Lewis, H., forthcoming. Knowlton South Circle soil micromorphology, in *Knowlton Circles Excavations*, ed. J. Gale. (Excavation Monographs.) Bournemouth: Bournemouth University.
Lewis, H. & C. French, 1999. Preliminary Report on Fieldwork Carried Out in Wyke Down and the Allen Valley in July 1999. Report for English Heritage, Dorset County Council and AHRB.
Lewis, H. & C. French, 2002. Looking for prehistoric settlement in the Allen Valley: report on the Knowlton Environs Project 2002. Report for the McDonald Institute of Archaeological Research, University of Cambridge.
Lewis, H. & A.M. Hart, 2003. Soil Micromorphology at Skelhøj and Tobøl I, Ribe Amt., South Jutland, Denmark (NMI 8282/02): Initial Sampling of the Early Bronze Age Barrow Mounds. Report for Aarhus University & National Museum of Denmark
Lewis, H., C. French & M. Green, 2000. A decorated megalith from Knowlton, Dorset. *PAST* 35, 1–3.
Limbrey, S., 1975. *Soils and Archaeology*. London: Academic Press.
Liu, Q. & H. Hytteborn, 1991. Gap structure, disturbance and regeneration in a primeval *Picea abies* forest. *Journal of Vegetation Science* 2, 391–402.
Long, A.J., R.G. Scaife & R.J. Edwards, 1999. Pine pollen in intertidal sediments from Poole Harbour, UK: implications for late Holocene sediment accretion rates and sea-level rise. *Quaternary International* 55, 3–16.
Longworth, I.H., 1971. The Rinyo-Clacton culture reconsidered, in *Durrington Walls: Excavations 1966–1968*, eds. G.J. Wainwright & I.H. Longworth. London: Society of Antiquaries, 235–306.
Longworth, I.H., 1979. The Neolithic and Bronze Age pottery, in *Mount Pleasant, Dorset: Excavations 1970–1971*, by G.J. Wainwright. London: Society of Antiquaries, 75–124.
Longworth, I.H., 1984. *Collared Urns of the Bronze Age in Great Britain and Ireland*. Cambridge: Cambridge University Press.

Longworth, I.H. & C. Haith, 1992. Henry Durden and his collection. *Proceedings of the Dorset Natural History and Archaeological Society* 114, 151–60.

Loveday, R., 2006. *Inscribed Across the Landscape — the Cursus Enigma*. Stroud: Tempus.

Loveday, R. & M. Petchey, 1982. Oblong ditches: a discussion and some new evidence. *Aerial Archaeology* 8, 17–24.

Ludeke, M.K.B., O. Moldenhauer & G. Petschel-Held, 1999. Rural poverty driven soil degradation under climate change: the sensitivity of the disposition towards the Sahel syndrome with respect to climate. *Environmental Modeling and Assessment* 4, 315–26.

Macphail, R.I., 1986. Palaeosols in archaeology, in *Paleosols*, ed. V.P. Wright. Oxford: Blackwell Scientific, 263–90.

Macphail, R.I., 1987. A review of soil science in archaeology in England, in *Environmental Archaeology of England: A Regional Review*, ed. H.M.C. Keeley. London: HBMC. 332–77.

Macphail, R.I., 1990. The soils, in *Hazelton North, Gloucestershire, 1979–82: the Excavation of a Neolithic Long Cairn of the Cotswold-Severn Group*, by A. Saville. (English Heritage Archaeological Report 13.) London: English Heritage, 223–7.

Macphail, R.I., 1992. Soil micromorphological evidence of ancient soil erosion, in *Past and Present Soil Erosion: Archaeological and Geological Perspectives*, eds. M. Bell, & J. Boardman. (Oxbow Monograph 22.) Oxford: Oxbow Books, 197–215.

Macphail, R.I., 1993. 'Soil micromorphology', in Neolithic downland monument in its environment: excavations at the Easton Down long barrow, Bishops Cannings, North Wiltshire, by A. Whittle, A.J. Rouse & J.G. Evans. *Proceedings of the Prehistoric Society* 59, 197–239.

Macphail, R.I., 1998. A reply to S. Carter and D. Davidson's 'An evaluation of the contribution of soil micromorphology to the study of ancient agriculture'. *Geoarchaeology* 13, 549–64.

Macphail, R.I., 1999. Soils, in *The Harmony of Symbols: the Windmill Hill Causewayed Enclosure, Wiltshire, by A.* Whittle, J. Pollard & C. Grigson. (Cardiff Studies in Archaeology.) Oxford: Oxbow Books, 121–7.

Macphail, R.I. & R.G. Scaife, 1987. The geographical and environmental background, in *The Archaeology of Surrey to 1540*, eds. J. Bird & D.G. Bird. Guildford: Surrey Archaeological Society, 31–51.

Macphail, R.I., J.C.C. Romans & L. Robertson, 1987. The application of micromorphology to the understanding of Holocene soil development in the British Isles; with special reference to early cultivation, in *Soil Micromorphology. Proceedings of the 7th International Working Meeting on Soil Micromorphology, Paris - July 1985*, eds. N. Federoff, L.M. Bresson & M.A. Courty. Plaisir: L'Association Français pour l'Étude du Sol, 647–56.

Macphail, R.I., M.A. Courty & A. Gebhardt, 1990. Soil micromorphological evidence of early agriculture in north-west Europe. *World Archaeology* 22, 53–69.

Magurran, A.E., 1988. *Ecological Diversity and its Measurement*. London: Chapman and Hall.

Malim, T., 1999. Cursuses and related monuments of the Cambridgeshire Ouse, in *Pathways and Ceremonies: the Cursus Monuments of Britain and Ireland*, eds. A. Barclay & J. Harding. (Neolithic Studies Group Seminar Papers 4.). Oxford: Oxbow Books, 77–85.

Malinowski, B., 1935. *Coral Gardens and their Magic: a Study of the Methods of Tilling the Soil and of Agricultural Rites in the Trobriand Islands*. London: Allen and Unwin.

Maltby, M., 1990a. Animal bones, in *The Stonehenge Environs Project*, by J. Richards. (Monograph 16.) London: English Heritage, 150–54.

Maltby, M., 1990b. The exploitation of animals in the Stonehenge environs in the Neolithic and Bronze Age, in *The Stonehenge Environs Project*, by J. Richards. (Monograph 16.) London: English Heritage, 247–9.

Maltby, M., 2004. 'Animal bones', in Investigation of the Whitesheet Down Environs 1989–90: Neolithic causewayed enclosure and Iron Age settlement, by M. Rawlings, M.J. Allen & F. Healy. *Wiltshire Archaeological and Natural History Magazine (Wiltshire Studies)* 97, 167–71.

Manby, T., 1980. Bronze Age settlement in eastern Yorkshire, in *Settlement and Society in the British Later Bronze Age*, eds. J.C. Barrett & R. Bradley. (British Archaeological Reports British Series 83.) Oxford: BAR, 307–70.

Manby, T., 1988. *Archaeology in Eastern Yorkshire. Essays in Honour of T.C.M. Brewster.* Sheffield: Department of Archaeology and Prehistory, University of Sheffield.

Manchester, K., 1983. *The Archaeology of Disease*. Bradford: University of Bradford Press.

Mayer, A.C., V. Stockli, C. Huovinen, W. Konold, B.L. Estermann & M. Kreuzer, 2003. Herbage selection by cattle on sub-alpine wood pastures. *Forest Ecology and Management* 181, 39–50.

Mayer, A.C., V. Stockli, W. Konold & M. Kreuzer, 2006. Influence of cattle stocking rate on browsing of Norway spruce in subalpine wood pastures. *Agroforestry Systems* 66, 143–9.

McKinley, J.I., 1993. Bone fragment size and weights of bone from modern British cremations and its implications for the interpretation of archaeological cremations. *International Journal of Osteoarchaeology* 3, 283–7.

McKinley, J.I., 1994a. *The Anglo-Saxon Cemetery at Spong Hill, North Elmham*, part VIII: *The Cremations*. (East Anglian Archaeology 69.) Dereham: East Anglian Archaeology.

McKinley, J.I., 1994b. Bone fragment size in British cremation burials and its implications for pyre technology and ritual. *Journal of Archaeological Science* 21, 339–42.

McKinley, J.I., 1997a. The cremated human bone from burial and cremation-related contexts, in *Archaeological Excavations on the Route of the A27 Westhampnett Bypass, West Sussex, 1992*, vol. 2, by A.P. Fitzpatrick. (Report 12.) Salisbury: Wessex Archaeology, 55–72.

McKinley, J.I., 1997b. Bronze Age 'barrows' and the funerary rites and rituals of cremation. *Proceedings of the Prehistoric Society* 63, 129–45.

McKinley, J.I., 1998. Archaeological manifestations of cremation. *The Archaeologist* 33, 8–20.

McKinley, J.I., 2000a. The analysis of cremated bone, in *Human Osteology*, eds. M. Cox & S. Mays. London: Greenwich Medical Media, 403–21.

McKinley, J.I., 2000b. Human bone and funerary deposits, in *Twyford Down, Hampshire. Archaeological Investigations the M3 Motorway from Bar End to Compton, 1990–93*, by K.E. Walker & D.F. Farwell. (Monograph 9.) Winchester: Hampshire Field Club, 85–119.

McKinley, J.I., 2004a. The human remains and aspects of pyre technology and cremation rituals, in *The Roman Cemetery at Brougham, Cumbria*, by H. Cool. (Britannia Monograph Series 21.) London: Roman Society, 283–309.

McKinley, J.I., 2004b. Compiling a skeletal inventory: disarticulated and co-mingled remains, in *Guidelines to the Standards for Recording Human Remains*, eds. M. Brickley & J.I. McKinley. Reading: British Association for Biological Anthropology and Osteoarchaeology and Institute for Field Archaeology, 13–17.

McMinn, R.M.H. & R.T. Hutchings, 1985. *A Colour Atlas of Human Anatomy*. London: Wolfe Medical Publications.

Mellars, P., 1976. Fire ecology, animal populations and man: a study of some ecological relationships in prehistory. *Proceedings of the Prehistoric Society* 42, 15–45.

Mellars, P. & P. Dark, 1998. *Star Carr in Context: New Archaeological and Palaeoecological Investigations at the Early Mesolithic Site of Star Carr, North Yorkshire*. (McDonald Institute Monographs.) Cambridge: McDonald Institute for Archaeological Research.

Melville, R.V. & E.C. Freshney, 1982. *The Hampshire Basin and Adjoining Areas*. British Regional Geology. London: HMSO.

Mercer, R.J. (ed.), 1981. *Farming Practice in British Prehistory*. Edinburgh: Edinburgh University Press.

Miles, H. & T.J. Miles, 1971. Excavations on Longstone Downs, St Stephen-in-Brannel and St Mewan. *Cornish Archaeology* 10, 5–28.

Mithen, S. (ed.), 2000. *Hunter-gatherer Landscape Archaeology: The Southern Hebrides Mesolithic Project 1988–1998*. (McDonald Institute Monographs.) Cambridge: McDonald Institute for Archaeological Research.

Mladenoff, D.J. & W.L. Baker, 1999. Development of forest and landscape modeling approaches, in *Spatial Modeling of Forest Landscape Change*, eds. D.J. Mladenoff & W.L. Baker. Cambridge: Cambridge University Press, 1–13.

Moffat, A.J. & D.W. Cope, 1984. The Hampshire chalklands, in *Soils of the Southampton District*, M.G. Jarvis & D.C. Findlay. London: British Society of Soil Science, 82–116.

Moffett, L., M.A. Robinson & V. Straker, 1989. Cereals, fruit and nuts: charred plant remains from Neolithic sites in England and Wales and the Neolithic economy, in A. Milles, D. Williams & N. Gardner, *The Beginnings of Agriculture*. (British Archaeological Reports International Series 496.) Oxford: BAR, 243–61.

Molleson, T.I., 1993. The human remains, in *Poundbury* vol. 2: *The Cemeteries*, by D.E. Farwell & T.I. Molleson. (Monograph 11.) Dorset Natural History and Archaeological Society, 142–214.

Montgomery, J., P. Budd & J.A. Evans, 2000. Reconstructing the lifetime movements of ancient people: a Neolithic case study from southern England. *European Journal of Archaeology* 3, 370–85.

Moore, P.D., 1977. Ancient distribution of lime trees in Britain. *Nature* 268, 13–14.

Moore, P.D. & J.A. Webb, 1978. *An Illustrated Guide to Pollen Analysis*. London: Hodder and Stoughton.

Moore, P.D. & A. Wilmott, 1973. Prehistoric forest clearance and the development of peatlands in the Uplands and lowlands of Britain. *Proceedings of the VI International Peat Congress*, Poznan, Poland 1976, 1–15.

Moore, P.D., J.A. Webb & M.E. Collinson, 1991. *Pollen Analysis*. 2nd edition. Oxford: Blackwell Scientific.

Mottet, A., M.P. Julien, G. Balent & A. Gibon, in press. *Environmental Modeling and Assessment*.

Mukherjee, A.J., R.P. Evershed & A.M. Gibson, 2004. Compound-specific Stable Isotope Analysis of Organic Residues from Cranborne Chase. Unpublished report, University of Bristol.

Murphy, C.P., 1986. *Thin Section Preparation of Soils and Sediments*. Berkhamsted: AB Academic.

Needham, S., 2005. Transforming Beaker culture in northwest Europe: processes of fusion and fission. *Proceedings of the Prehistoric Society* 71, 171–217.

Nielsen-Marsh, C., A. Gernaey, G. Turner-Walker, R. Hedges, A. Pike & M. Collins, 2000. The chemical degradation of bone, in *Human Osteology in Archaeology and Forensic Science*, eds. M. Cox. & S. Mays. London: GMM, 439–54.

Ollier, C.D., D.P. Drover & M. Godelier, 1971. Soil knowledge amongst the Baruya of Wonenara, New Guinea. *Oceania* 42, 33–41.

Ortner, D.J. & W.G.J. Putschar, 1985. *Identification of Pathological Conditions in Human Skeletal Remains*. Washington (DC): Smithsonian Institution Press.

OS, 2006. http://www.ordnancesurvey.gov.uk/productpages/landline/positional-background.htm (accessed August 2007).

O'Sullivan, M., 2006. La Boyne et autre sites: une revue de l'art megalithique d'Irlande, in *Origine et developpment de megalithisme de l'ouest de l'Europe*, eds. R. Joussaume, L. Laporte & C. Scarre. Bougon: Musée des tumulus de Bougon, 645–86.

Palmer, R., 2000. Wyke Down Project: Air Photo Input. Unpublished report, Air Photo Services, Cambridge.

Palmer, R., 2003. The Archaeology of Wyke Down, Dorset: Supplementary Aerial Photographic Survey. Unpublished report, Air Photo Services, Cambridge.

Palmer, R. & A. Oswald, forthcoming. The field survey, in *Hambledon Hill Causewayed Enclosure Complex, Dorset. Excavations 1974–84*, eds. R. Mercer & F. Healy. London: English Heritage.

Palmer, S.C.F., I.J. Gordon, A.J. Hester & R.J. Pakeman, 2004. Introducing spatial grazing impacts into the prediction of moorland vegetation dynamics. *Landscape Ecology* 19, 817–27.

Papanastasis, V., Z. Koukoura, D. Alifragis & I. Makedos, 1995. The effects of thinning, fertilisation and sheep grazing on the understory vegetation of *Pinus pinaster* populations. *Forest Ecology and Management* 77, 181–9.

Peet, R.K., 1974. The measurement of species diversity. *Annual Review of Ecological Systems* 5, 285–307.

Peglar, S.M., 1993. The mid-Holocene *Ulmus* decline at Diss

Mere, Norfolk, UK: a year-by-year pollen stratigraphy from annual laminations. *The Holocene* 3, 1–13.
Peglar, S.M., 2006. The Ouse channel Flandrian sequence, in *A Woodland Archaeology: Neolithic Sites at Haddenham*, by C. Evans & I. Hodder. (McDonald Institute Monographs.) Cambridge: McDonald Institute for Archaeological Research, 26–9.
Pennington, W., 1974. *The History of British Vegetation*. London: The English Universities Press Ltd.
Pielou, E.C., 1975. *Ecological Diversity*. New York (NY): Wiley.
Pigot, C.D. & S.M. Walters, 1954. On the interpretation of the discontinuous distribution shown by certain British species of open habitats. *Journal of Ecology* 42, 95–117.
Piggott, S., 1954. *The Neolithic Cultures of the British Isles*. Cambridge: Cambridge University Press.
Piggott, S. & G.W. Dimbleby, 1953. A Bronze Age barrow on Turner's Puddle Heath. *Proceedings of the Dorset Natural History and Archaeology Society* 75, 34–5.
Piggott, S. & C.M. Piggott, 1944. Excavation of barrows on Crichel and Launceston Downs, Dorset. *Archaeologia* 90, 47–80.
Pitt Rivers, A.H.L.F., 1887. *Excavations in Cranborne Chase, near Rushmore*, vol. 1: *Excavations in the Romano-British Village on Woodcutts Common, and Romano-British Antiquities in Rushmore Park*. Privately printed.
Pitt Rivers, A.H.L.F., 1888. *Excavations in Cranborne Chase, near Rushmore*, vol. II.: *Excavations in Barrows near Rushmore; Excavations in Romano-British Village, Rotherley; Excavations in Winklebury Camp; Excavations in British Barrows and Anglo Saxon Cemetery, Winklebury Hill*. Privately printed.
Pitt Rivers, A.H.L.F., 1892. *Excavations in Cranborne Chase, near Rushmore*, vol. III. Privately printed.
Pitt Rivers, A.H.L.F., 1898. *Excavations in Cranborne Chase, near Rushmore on the Borders of Dorset and Wiltshire, 1893–1896*, vol. IV. Privately printed.
Pollard, J., 1993. Traditions of Deposition in Neolithic Wessex. Unpublished PhD, University of Wales, Cardiff.
Pollard, J., 1995. The Durrington 68 timber circle: a forgotten Late Neolithic monument. *The Wiltshire Archaeological and Natural History Magazine* 88, 122–6.
Pollard, J., 2002. The ring-ditch and the hollow: excavation of a Bronze Age 'shrine' and associated features at Pampisford, Cambridgeshire. *Proceedings of the Cambridge Antiquarian Society* XCI, 5–22.
Poole, R.W., 1974. *An Introduction to Quantitive Ecology*. Tokyo: McGraw Hill.
Preece, R.C. & D. Bridgland, 1999. Holywell Coombe, Folkestone: a 13,000 year history of an English Chalkland Valley. *Quaternary Science Reviews* 18, 1075–125.
Proudfoot, E., 1963. Report of the excavation of a bell barrow in the parish of Edmondsham, Dorset, England, 1959. *Proceedings of the Prehistoric Society* 49, 395–425.
Proudfoot, E., 1989. The pond barrow and the weathering cone, in *Wilsford Shaft: Excavations 1960–62*, by P. Ashbee, M.G. Bell & E. Proudfoot. (Archaeological Report 11.) London: English Heritage, 15–25.
Pryor, F., 1984. *Excavation at Fengate, Peterborough, England: the Fourth Report*. (Royal Ontario Museum Archaeological Monograph 7/Northamptonshire Archaeological Society Monograph 2.) Toronto/Northampton: Royal Ontario Museum/Northamptonshire Archaeological Society.
Pryor, F., 1998a. *Etton: Neolithic Causewayed Enclosure, Near Maxey, Cambs. – Excavations 1982–7*. (Archaeology Report 18.) London: English Heritage.
Pryor, F., 1998b. *Farmers in Prehistoric Britain*. Stroud: Tempus.
Quine, C.P. & P.D. Bell, 1998. Monitoring of windthrow occurrence and progression in spruce forests in Britain. *Forestry* 71, 87–97.
Rackham, O., 1980. *Ancient Woodland: its History, Vegetation and Uses in England*. London: Edward Arnold.
Ratcliffe, P.R., 1992. The interaction of deer and vegetation in coppice woods, in *Ecology and Management of Coppice Woodlands*, ed. G.P. Buckley. London: Chapman and Hall, 233–46.
Rawlings, M.N., M.J. Allen, R.M.J. Cleal & F. Healy, 2004. Investigations of the Whitesheet Down environs 1989–90; Neolithic causewayed enclosure and Iron Age settlement. *Wiltshire Archaeology & Natural History Magazine* 97, 144–96.
RCHME, 1975. *Inventory of Dorset*, vol. V. London: RCHME, 113–16.
Richards, C., 1996. Henges and water: towards an elemental understanding of monumentality and landscape in Late Neolithic Britain. *Journal of Material Culture* 1, 313–36.
Richards, C. (ed.), 2005. *Dwelling Among the Monuments: the Neolithic Village of Barnhouse, Maeshowe Passage grave and Surrounding Monuments at Stenness, Orkney*. (McDonald Institute Monographs.) Cambridge: McDonald Institute for Archaeological Research.
Richards, J., 1990. *The Stonehenge Environs Project*. (Archaeological Report 16.) London: English Heritage.
Richards, J., 1999. *Meet the Ancestors*. London: BBC.
Richmond, I., 1968. *Hod Hill*, vol. 2: *Excavations Carried Out Between 1951–58*. London: British Museum.
Roberts, B.K., 1990. Perspectives on prehistory, in *An Historical Geography of England and Wales*, eds. R. Dodgshon & R.A. Butlin. London: Academic Press, 1–24.
Roberts, C. & M. Cox, 2003. *Health and Disease in Britain from Prehistory to the Present Day*. Stroud: Sutton.
Roberts, C. & K. Manchester, 1997. *The Archaeology of Disease*. 2nd edition. Stroud: Sutton.
Robinson, D.A. & R.B.G. Williams, 1983. The soils and vegetation of Sussex, in *Sussex: Environment, Landscape and Society*, ed. Geography Editorial Committee, University of Sussex. Gloucester: Sutton, 109–26.
Robinson, M.A., 2000. Further considerations of Neolithic charred cereals, fruits and nuts, in *Plants in Neolithic Britain and Beyond*, ed. A.S. Fairbairn. (Neolithic Studies Group Seminar Papers 5.) Oxford: Oxbow, 85–90.
Robinson, M. & B. Williams, 1987. A survey of the environmental archaeology in the south midlands, in *Environmental Archaeology: a Regional Review*, vol. II, ed. H.C.M. Keeley. London: HBMC, 16–100.
Robledo, B., G.J. Trancho & D. Brothwell, 1995. *Cribra orbitalia*: health indicator in the late Roman population of

Cannington (Sommerset [*sic.*], Great Britain). *Journal of Palaeopathology* 7, 185–93.

Rogers, J. & T. Waldron, 1995. *A Field Guide to Joint Disease in Archaeology*. Chichester: Wiley.

Rogers, J., T. Waldron, P. Dieppe & I. Watt, 1987. Arthropathies in palaeopathology: the basis of classification according to most probable cause. *Journal of Archaeological Science* 14, 179–93.

Rossiter, D.G., 1996. A theoretical framework for land evaluation (with discussion). *Geooderma* 75, 165–202.

Rothwell, A.M., 2002. An Analysis of the Animal Bone Assemblages from a Henge at Wyke Down, Dorset, and a Comparison with other Neolithic henges of the British Isles. Unpublished BSc dissertation, Bournemouth University, Department of Archaeology.

Rouse, A.J. & J.G. Evans, 1993. 'Land molluscs', in A Neolithic downland monument in its environment: excavations at the Easton Down long barrow, Bishops Cannings, North Wiltshire, by A. Whittle, A.J. Rouse, & J.G. Evans, 1993. *Proceedings of the Prehistoric Society* 59, 211–17.

Samarasundera, E., forthcoming. Modelling the Dynamics of Holocene Landscapes: a Methodological Investigation using Case Studies from the English Chalklands. Unpublished PhD thesis, University of London.

Scaife, R.G., 1980. Late Devensian and Flandrian Palaeoecological Studies in the Isle of Wight. Unpublished PhD thesis, Department of Geography, King's College, University of London.

Scaife, R.G., 1982. Late Devensian and early Flandrian vegetational changes in southern England, in *Archaeological Aspects of Woodland Ecology* eds. S. Limbrey & M. Bell. (Symposia of the Association for Environmental Archaeology 2/British Archaeological Reports International Series 146.) Oxford: BAR, 57–74.

Scaife, R.G., 1984. Gallibury Down, Isle of Wight: pollen analysis of a Bronze Age downland palaeosol. *Ancient Monuments Laboratory Report* 4240.

Scaife, R.G., 1987a. The Late Devensian and Holocene vegetation and environment of the Isle of Wight, in *Quaternary Research Association Field Guide for Wessex and the Isle of Wight*, ed. K. Barber. London: Quaternary Research Association, 156–80.

Scaife, R.G., 1987b. A review of later Quaternary plant microfossils and macrofossil research in southern England; with special reference to environmental archaeological evidence, in *Regional Environmental Archaeological Reviews*, ed. H.C.M. Keely. London: English Heritage Commission, 125–203.

Scaife, R.G., 1988. The ulmus decline in the pollen record of southeast England and its relationships to early agriculture, in *Archaeology and the Flora of the British Isles*, ed. M. Anes. Oxford: Oxbow, 21–33.

Scaife, R.G., 1990. Pollen analysis of the old land surface, in *Hazelton North, the Excavation of a Neolithic Long Cairn of the Cotswold-Severn Group*, by A. Saville. (Archaeological Report 13.) London: English Heritage, 218–19.

Scaife, R.G., 1991. Summary of the earlier Holocene vegetation: pollen investigation and vegetational history, in *Redeemed from the Heath: the Archaeology of the Wytch Farm Oilfield, 1987–90*, by P. Cox & C. Hearne. (Monograph 9.) Dorchester: Dorset Natural History and Archaeological Society 180–97.

Scaife, R.G., 1994a. Pollen Analysis of the River Allen Valley Floodplain, Wimborne Town Centre, Dorset. Unpublished report for Southern Archaeological Services.

Scaife, R.G., 1994b. 'Avon valley floodplain sediments: the pre-Roman vegetational history', in An Archaeological and Environmental Study of the Neolithic and Later Prehistoric Landscape of the Avon Valley between Durrington Walls and Earl's Farm Down, by R.M.J. Cleal, M.J. Allen & C. Newman. Unpublished report for Wessex Archaeology.

Scaife, R.G., 1995. Boreal and sub-boreal chalk landscape: pollen evidence and pollen from the buried Neolithic soil, in *Stonehenge in its Landscape: Twentieth-century Excavations*, by R.M.J. Cleal, K.E. Walker & R. Montague. (Archaeological Report 10.) London: English Heritage, 51–5 & 61–8.

Scaife, R G., 1999. Pollen analysis of the Churn river channel, in *Excavations Alongside Roman Ermine Street, Gloucestershire/Wiltshire: Archaeology of the A419/A417 Swindon-Gloucester Road Acheme*, vol. 2: *Medieval and Post-medieval Activity, Finds and Archaeological Evidence*, by A. Mudd, R.J. Williams & A. Lumpton. Oxford: Oxford Archaeological Unit, 502–12.

Scaife, R.G., 2000. Pollen analysis, in *The Holocene Evolution of the London Thames*. by J. Sidell, K. Wilkinson & N. Cameron. (Monograph 5.) London: Museum of London.

Scaife, R.G., 2001. 'Pollen analysis of valley fen peats of Romano-British date', in Clatterford Roman Villa. An assessment report for the excavations at Clatterford Roman Villa, Isle of Wight (CAS Project 550), by P. Busby. *Hampshire Field Club and Archaeological Society* 56, 119–24.

Scaife, R.G., 2003a. 'Palynological analyses', in Archaeological and palaeo-environmental investigations of the Upper Allen Valley, Cranborne Chase, Dorset (1998–2000): a new model of earlier Holocene landscape development, by C. French, H. Lewis, M.J. Allen, R.G. Scaife & M. Green. *Proceedings of the Prehistoric Society* 69, 217–22.

Scaife, R.G., 2003b. The palaeoecological background, in *The Isle of Wight Flora*, by C. Pope, L. Snow & D. Allen. Wimborne: Dovecote Press, 19–31.

Scaife, R.G., 2004. 'Pollen analysis', in An archaeological and environmental study of the Neolithic and later prehistoric landscape of the Avon valley and Durrington Walls environs, by R.M.J. Cleal, M. Allen & C. Newman. *Wiltshire Archaeological and Natural History Magazine* 97, 218–48.

Scaife, R.G., 2005. Palynological analyses of the relict river courses, in *Archaeology and Environment of the Etton Landscape*, by C. French & F. Pryor. (East Anglian Archaeology 109.) Peterborough: East Anglian Archaeology.

Scaife, R.G. & M.J. Allen, 1994. 'Pollen', in Investigation of tree-damaged barrows on King Barrow Ridge and Luxenborough Plantation, Amesbury, by R.M.J. Cleal

& M.J. Allen. *The Wiltshire Archaeological and Natural History Magazine* 87, 54–84.

Scaife, R.G. & P.G. Burrin, 1983. Floodplain development in, and the vegetational history of, the Sussex High Weald and some archaeological implications. *Sussex Archaeological Collections* 121, 1–10.

Scaife, R.G. & R.I. Macphail, 1983. 'The post-Devensian development of heathland soils and vegetation', in Soils of the heathlands and chalklands, ed. C.P. Burnam. *Proceedings of the South-east Soils Discussion Group, Seesoil* 1, 70–99.

Schama, S., 1995. *Landscape and Memory*. London: Harper-Collins.

Scheuer, L. & S. Black, 2000. *Developmental Juvenile Osteology*. London: Academic Press.

Schulting, R.J., A. Tresset & C. Dupont, 2004. From harvesting the sea to stock rearing along the Atlantic façade of north-west Europe. *Environmental Archaeology* 9, 143–54.

Schutkowski, H., 1993. Sex determination of infant and juvenile skeletons, I: Morphological features. *American Journal of Physical Anthropology* 90, 199–205.

Scollar, I., 2002. Making things look vertical, in *Aerial Archaeology: Developing Future Practice*, eds. R.H. Bewley & W. Rączkowski. (NATO Science Series 337.) Amsterdam: NATO, 166–72.

Seagrief, S.C., 1959. Pollen diagrams from southern England: Wareham, Dorset and Nursling, Hampshire. *New Phytologist* 58, 316–25.

Seagrief, S.C., 1960. Pollen diagrams from southern England: Cranes Moor, Hampshire. *New Phytologist* 59, 73–83.

Seagrief, S.C. & H. Godwin, 1960. Pollen diagrams from southern England: Elstead, Surrey. *New Phytologist* 59, 84–91.

Selby, M.J., 1993. *Hillslope Materials and Processes*. 2nd edition. Oxford: Oxford University Press.

Sellen, D., 2003. Nutritional consequences of wealth differentials in East African pastoralists: the case of the Datoga of northern Tanzania. *Human Ecology* 31, 529–70.

Serjeantson, D. & J. Gardiner, 1995. Antler implements and ox scapulae shovels, in *Stonehenge: a Monument in its Landscape*, by R.M.J. Cleal, K.E. Walker & R. Montague. (Archaeological Report 10.) London: English Heritage, 414–30.

Sharples, N.M., 1991. *Maiden Castle: Excavations and Field Survey 1985–6*. London: Historic Buildings and Monuments Commission for England.

Shee Twohig, E., 1981. *The Megalithic Art of Western Europe*. Oxford: Clarendon Press.

Shennan, S.J., 1988. *Quantifying Archaeology*. Edinburgh: Edinburgh University Press

Sheridan, A., 1999. Grooved Ware from the Links of Noltland, Westray, Orkney, in *Grooved Ware in Britain and Ireland*, eds. R.M.J. Cleal & A. MacSween. Oxford: Oxbow Books, 112–24.

Simmons, I.G., 1969. Evidence for vegetation changes associated with Mesolithic man in Britain, in *The Domestication and Exploitation of Plants and Animals*, eds. P.J. Ucko & G.W. Dimbleby. London: Duckworth, 110–119.

Smith, A.G., 1970. The influence of Mesolithic and Neolithic man on British vegetation: a discussion, in *Studies in the Vegetational History of the British Isles*, eds. D. Walker & R.G. West. Cambridge: Cambridge University Press, 81–96.

Smith, A.G., 1981. The Neolithic, in *The Environment in British Prehistory*, by I.G. Simmons & M. Tooley. London: Duckworth, 125–209.

Smith, A.G. & J.R. Pilcher, 1973. Radiocarbon dates and the vegetational history of the British Isles. *New Phytologist* 72, 903–14.

Smith, I.F., 1965. *Windmill Hill and Avebury: Excavations by Alexander Keiller 1925–1939*. Oxford: Clarendon Press.

Smith, I.F., 1985. 'The pottery', in The excavation of a ring-ditch at Tye Field, Lawford, Essex. *Archaeological Journal* 142, 165–77.

Smith, I.F., 1991. Round barrows Wilsford cum Lake G51–G54: excavations by Ernest Greenfield in 1958. *Wiltshire Archaeological Magazine* 84, 11–39.

Smith, R.J., F. Healy, M.J. Allen, E.L. Morris, I. Barnes & P.J.K. Woodward, 1997. *Excavations Along the Route of the Dorchester By-pass, Dorset, 1986–8*. (Report 11.) Salisbury: Wessex Archaeology.

Southwood, T.R.E., 1978. *Ecological Methods*. London: Chapman and Hall.

Sparey-Green, C., 1997. Excavations at the Town Farm House site, now The Orchard, Dean Lane, Sixpenny Handley. *Proceedings of the Dorset Natural History and Archaeological Society* 119, 87–102.

Sparey-Green, C., 1998. Interim report on excavations at Myncen Farm, Sixpenny Handley, Dorset, 1998. *Proceedings of the Dorset Natural History and Archaeological Society* 120, 91–4.

Stace, C., 1991. *New Flora of the British Isles*. Cambridge: Cambridge University Press.

Stead, I.M., 1991. *Iron Age Cemeteries in East Yorkshire*. (Archaeological Reports 22.) London: English Heritage.

Stoertz, C., 1997. *Ancient Landscapes of the Yorkshire Wolds*. London: RCHME.

Stoertz, C., 2005. *Knowlton Circles Project, Dorset*. (Internal Report AER/6/2005.) Swindon: English Heritage.

Stoertz, C. & J. Gale, in prep. Aerial photographic transcription of features in the vicinity of the prehistoric complex at Knowlton Circles, Dorset, in *Archaeological Investigations in the Environs of Knowlton Henge Complex, Dorset 1993–7*, ed. J. Gale. (British Archaeological Reports British Series.) Oxford: BAR.

Stoops, G., 2003. *Guidelines for Analysis and Description of Soil and Regolith Thin Sections*. Madison (WI): American Society of Soil Science.

Stone, G.D., 1993. Agricultural abandonment: a comparative study in historical ecology, in *The Abandonment of Settlements and Regions: Ethnoarchaeological and Archaeological Approaches*, eds. C.M. Cameron & S.A. Tomka. Cambridge: Cambridge University Press, 74–81.

Stuiver, M. & J. van der Plicht, 1998. Editorial comment. *Radiocarbon* 40(3), xii–xiii.

Stuiver, M., P.J. Reimer, E. Bard, J.W. Beck, G.S. Burr, K.A. Hughen, B. Kromer, G. McCormac, J. van der Plicht & M. Spurk, 1998. INTCAL98 radiocarbon age calibra-

tion 24,000–0 cal BP. *Radiocarbon* 40(3), 1041–83.
Sugita, S., 1993. A model of pollen source area for an entire lake surface. *Quaternary Research* 39, 239–44.
Sugita, S., M.-J. Gaillard & A. Brostrom, 1999. Landscape openness and pollen records: a simulation approach. *Holocene* 9, 409–21.
Summers, P.G., 1941. A Mesolithic site, near Iwerne Minster, Dorset. *Proceedings of the Prehistoric Society* 7, 145–64.
Talma, A.S. & J.C. Vogel, 1993. A simplified approach to calibrating C14 dates. *Radiocarbon* 35, 317–22.
Tansley, A.G., 1939. *The British Islands and their Vegetation.* Cambridge: Cambridge University Press.
Tauber, H., 1965. Differential pollen dispersion and the interpretation of pollen diagrams. *Danm. Geol. Unders.* II 89, 1–69.
Tauber, H., 1967. Investigation of the mode of pollen transfer in forested areas. *Review Palaeobotany Palynology* 3, 277–87.
Teich, I., A.M. Cingolani, D. Renison, I. Hensen & M.A. Giorgis, 2005. Do domestic herbivores retard *Polylepis australis* Bitt. woodland recovery in the mountains of Cordoba, Argentina? *Forest Ecology and Management* 219, 229–41.
Thomas, J., 1991. *Rethinking the Neolithic.* Cambridge: Cambridge University Press.
Thomas, K.D., 1977. 'The land mollusca from the enclosure on Offham Hill', in The excavation of a Neolithic causewayed enclosure on Offham Hill, East Sussex, 1976, by P.L. Drewett. *Proceedings of the Prehistoric Society* 43, 234–9.
Thomas, K.D., 1978. Population studies on molluscs in relation to environmental archaeology, in *Research Problems in Zooarchaeology*, eds. D. Brothwell, K.D. Thomas & J. Clutton-Brock. (Institute of Archaeology Paper 3.) London: Institute of Archaeology, 9–19.
Thomas, K.D., 1985. Land snail analysis in archaeology: theory and practice, in *Palaeoenvironmental Investigations: Research Design, Methods and Data Analysis,* eds. N.R.J. Fieller, D.D. Gilbertson & N.G.A. Ralph. (British Archaeological Reports International Series 266.) Oxford: BAR, 131–57.
Thomas, K.D., 1989. Vegetation of the British Chalklands in the Flandrian period: a response to Bush. *Journal of Archaeological Science* 16, 549–53.
Thomas, N., 2005. *Snail Down, Wiltshire.* (Monograph 3.) Salisbury: Wiltshire Archaeological and Natural History Society.
Thomas, N. & C. Thomas, 1955. Excavations at Snail Down, Evereleigh 1953 and 1955: an interim report. *Wiltshire Archaeological and Natural History Magazine* 56, 127–48.
Thomas, W.R. & T.C. Foin, 1982. Neutral hypotheses and patterns of species diversity: fact or fiction? *Palaeobiology* 8, 44–55.
Thompson, T.R.E., 1983. Translocation of fine earth in some soils from an area in mid Wales, in *Soil Micromorphology II. Soil Genesis. Proceedings of the 6th International Working Group Meeting on Soil Micromorphology, London 1981*, eds. P. Bullock & C.P. Murphy. Berkhamsted: A. B. Academic, 531–40.
Thorley, A., 1971a. An Investigation into the History of Native Tree Species in Southeast England Using the Pollen Analysis Technique. Unpublished PhD thesis, University of London King's College.
Thorley, A., 1971b. Vegetational history of the Vale of Brooks. *Institute of British Geographers Conference* 5, 47–50.
Thorley, A., 1981. Pollen analytical evidence relating to the vegetation history of the chalk. *Journal of Biogeography* 8, 93–106.
Troels-Smith, J., 1960. Ivy, mistletoe and elm: climatic indicators — fodder plants. *Geological Survey of Denmark* RIV 4(4), 1–32.
Trotter, M. & G.C. Gleser, 1952. Estimation of stature from long bones of American whites and Negroes. *American Journal of Physical Anthropology* 10, 463–514.
Trotter, M. & G.C. Gleser, 1958. A re-evaluation of estimation of stature bases on measurements of stature taken during life and of long bones after death. *American Journal of Physical Anthropology* 16, 79–123.
Turner, J., 1962. The *Tilia* decline: an anthropogenic interpretation. *New Phytologist* 61, 328–41.
Turner, J., 1970. Post-Neolithic disturbance of British vegetation, in *Studies in the Vegetational History of the British Isles*, eds. D. Walker & R.G. West. Cambridge: Cambridge University Press, 97–116.
Turner, M.G., R.H. Gardner & R.V. O'Neill, 2001. *Landscape Ecology in Theory and Practice*. New York (NY): Springer-Verlag.
Usai, M.R., 2001. Textural pedofeatures and pre-Hadrian's Wall ploughed palaeosols at Stanwix, Carlisle, UK. *Journal of Archaeological Science* 28, 541–53.
Vallentine, J.F., 2001. *Grazing Management.* San Diego (CA): Academic Press.
Vandenberghe, C., F. Frelechoux, M.A. Moravie, F. Gadallah, F. & A. Buttler, 2007. Short-term effects of cattle browsing on tree sapling growth in mountain wood pastures. *Plant Ecology* 188, 253–64.
Várallyay, G., 1989. Soil degradation processes and their control in Hungary. *Land Degradation and Rehabilitation* 1, 171–88.
Vera, F.W.M., 2000. *Grazing Ecology and Forest History*. Wallingford: CABI.
Wainwright, G.J., 1968. The excavation of a Durotrigian farmstead near Tollard Royal in Cranborne Chase. *Proceedings of the Prehistoric Society* 34, 102–47.
Wainwright, G.J., 1979. *Gussage All Saints: an Iron Age Settlement in Dorset*. London: HMSO.
Wainwright, G.J. & I.H. Longworth, 1971. *Durrington Walls: Excavations 1966–1968*. London: Society of Antiquaries.
Walker, K.E. & D.E. Frawell, 2000. *Twyford Down, Hampshire: Archaeological Investigations on the M3 Motorway from Bar End to Compton, 1990–93*. (Monograph 9.) Salisbury: Trust for Wessex Archaeology/Hampshire Field Club.
Waller, M., 1993. Flandrian vegetational history of south-eastern England: pollen data from Pannel Bridge, East Sussex. *New Phytologist* 124, 345–69.
Waller, M., 1994. The *Tilia* decline and paludification in southern England. *Holocene Studies* 4, 430–33.
Waller, M.P. & S.D. Hamilton, 1998. The vegetational history of the South Downs: Mount Caburn, in *The Quaternary*

of Kent and Sussex. Field Guide, by J.B. Murton, *et al.* London: Quaternary Research Association, 115–20.

Waller, M. & S.D. Hamilton, 2000. Vegetational history of the English Chalklands: a mid-Holocene pollen sequence from Caburn, East Sussex. *Journal of the Quaternary Science* 15(3), 253–73.

Warne, C., 1866. *The Celtic Tumuli of Dorset.* London: John Russell Smith.

Waton, P.V., 1982a. Man's impact on the chalklands: some new pollen evidence, in *Archaeological Aspects of Woodland Ecology,* eds. M. Bell & S. Limbrey. (British Archaeological Reports International Series 146.) Oxford: BAR, 75–91.

Waton, P.V., 1982b. A Palynological Study of the Impact of Man on the Landscape of Central Southern England, with Special Reference to the Chalklands. Unpublished PhD thesis, University of Southampton, Department of Geography.

Waton, P.V. & K.E. Barber, 1987. Rimsmoor, Dorset: biostratigraphy and chronology of an infilled doline, in *Wessex and the Isle of Wight, Field Guide,* ed. K. Barker. Cambridge: Quaternary Research Association, 75–80.

Webb, P.A.O. & J.M. Suchey, 1985. Epiphyseal union of the anterior iliac crest and medial clavicle in a modern multiracial sample of American males and females. *American Journal of Physical Anthropology* 68, 457–66.

Wessex Archaeology, 1994. Knowlton Circles: Palaeoenvironmental Assessment: Land Snail, Charred Plant Remains and Charcoal Assessment, Pollen and Soils Potential. Report for Department of Conservation Sciences, University of Bournemouth. Trust for Wessex Archaeology.

Wessex Archaeology, 2006. Practical Archaeology Training Course 2005, Down Farm, Sixpenny Handley, Dorset, Excavation Report. Unpublished.

Whatmore, S., 1993. Agricultural geography. *Progress in Human Geography* 17, 84–91.

Whittle, A., A.J. Rouse & J.G. Evans, 1993. A Neolithic downland monument in its environment: excavations at the Easton Down long barrow, Bishops Cannings, North Wiltshire. *Proceedings of the Prehistoric Society* 59, 197–239.

Whittle, A., J. Pollard & C. Grigson, 1999. *The Harmony of Symbols: the Windmill Hill Causewayed Enclosure, Wiltshire.* (Cardiff Studies in Archaeology.) Oxford: Oxbow Books.

Wiens, H.A., 1976. Population responses to patchy environments. *Annual Review of Ecological Systems* 7, 81–120.

Wilkinson, K.N., 2003. Colluvial deposits in dry valleys of southern England as proxy indicators of palaeoenvironmental and land-use change. *Geoarchaeology* 18, 725–55.

Williams, D. & A. Miles, 1987. Snails by numbers. *Circaea* 5, 18–19.

Wilshusen, R.H. & G.D. Stone, 1990. An ethnoarchaeological perspective on soils. *World Archaeology* 22, 104–14.

Wiltshire, P. & K. Edwards, 1993. Mesolithic, early Neolithic, and later prehistoric impacts on vegetation at a riverine site in Derbyshire, England, in *Climate Change and Human Impact on the Landscape,* ed. F.M. Chambers. London: Chapman & Hall, 157–68.

Woodward, P., 1988. Pictures of the Neolithic: discoveries from the Flagstone House excavations, Dorchester, Dorset. *Antiquity* 62, 266–74.

Woodward, P., 1991. *The South Dorset Ridgeway Survey and Excavations 1977–1984.* (Monograph 8.) Dorchester: Dorset Natural History and Archaeological Society.

Yalden, D., 1999. *The History of British Mammals.* London: Poyser.

Yeates, S., 2005. Religion, Community, and Territory: Defining Religion in the Severn Valley and Adjacent Gills from the Iron Age to the Early Medieval Period. Unpublished PhD thesis, University of Oxford.

Żuk, L. & R. Palmer, 2001a. The Archaeology of Wyke Down, Dorset: Aerial Photographic Survey. Unpublished report, Air Photo Services, Cambridge.

Żuk, L. & R. Palmer, 2001b. The Archaeology of Wyke Down, Dorset: Supplementary Aerial Photographic Survey. Unpublished report, Air Photo Services, Cambridge.

Appendices

Appendix 1

Excavation and Recording Methods

Charles French & Helen Lewis

This project was initially designed to be a feasibility study investigating former land surfaces beneath Early Bronze Age barrows on chalk downland. Subsequently, this was expanded to include the test excavation of monuments of most prehistoric periods, in combination with a valley wide augering survey, and fieldwalking survey in the southeastern part of the study area (as it had not received such archaeological thorough coverage as Martin Green had given to the northwestern half of the study area). Initially, former antiquarian trenches were used to facilitate the archaeological investigation of buried land surfaces, combined with the application of molluscan, soil micromorphological and palynological analyses.

As a first step to a larger project, our approach was tested on a small scale at the Wyke Down barrow group at Down Farm, Dorset, in 1998 and 1999 (French *et al.* 2000). Here there were four potential barrow candidates for initial assessment, Dorset SMR numbers 34, 36, 41 and 44. Each of these is upstanding about 1–1.5 m above the present ground surface, and each exhibits linear depressions suggestive of the position of former antiquarian trenches, except barrow 41, which was <1 m high and did not show any clear 'old' trenches.

The importance of investigating these particular scheduled barrows stems from the following:

1. the wealth of past and current archaeological investigations in this immediate area, including a new dating programme and environmental investigations (Green 2000);
2. the existing land-use models on the Cranborne Chase area to test (Barrett *et al.* 1991a);
3. the fact that the barrows were well enough preserved to seal a relatively intact buried land surface, and that previous 'damage' to the barrows would allow easier access to this for sampling purposes without the need for great disturbance to the mounds themselves;
4. there are many barrows in the same area which may have the potential to provide the environmental detail that will be necessary to discern land-use pattern on an intensive and extensive scale;
5. also, the present-day soils are believed by Fisher (1991) to be not as calcareous and biologically mixed as elsewhere in the Wessex chalk downlands, and therefore hold better prospects for the preservation of pollen and the preservation of indicative soil micromorphological features.

Field methodology

The approach in the field was as follows:

1. initial hand auger survey to confirm the presence/absence and thickness of buried soil survival beneath the mound and on the inner berm of the ring-ditch, and to ascertain the depth of infill deposits in the ring-ditch;
2. contour survey and plan of each monument at a scale of 1:20;
3. magnetometer and resistivity survey to delimit the disturbed/former trenched area of each barrow mound, and the position of the ring-ditch(es);
4. full photographic record of the barrows as they survive today;
5. only then did excavation begin, with the amount of new ground disturbance kept to a minimum at all times; this involved first the removal of the turf, and then cleaning out the backfill of the former trench by starting along one section face, working from the outer edge of the mound inwards; initial field inspection suggested that the size of trench required was no more than about 2 m in width and 5–6 m in length; all archaeological finds were recovered through excavation and dry sieving techniques (10 mm mesh). If *in situ* burials were encountered, advice from relevant authorities was to have been taken as to whether it would be preferable to leave them *in situ* or to excavate them. In the event, no burials were found;
6. depending on the depth and extent of disturbance

of the former trench, there was further excavation to modern standards down to the subsoil surface;
7. the sections thus created were thoroughly described and recorded at 1:10 according to Wessex Archaeology conventions with schematic infill representations of M. Green (Fig. A4.35), and photographed;
8. then two–three series of samples were taken for molluscan, soil micromorphological and palynological analyses from the buried soil and mound turves, and if possible, samples for radiocarbon assay;
9. and a 2 m wide section was excavated through the ring-ditch(es), with appropriate recovery of artefacts and recording of sections, and sampling for molluscs and pollen;
10. finally, reinstatement involved the soil as removed being backfilled into the excavation trenches, and the turves replaced, and with every effort made to follow the pre-existing contours. The barrows were re-seeded with local wild grass mixture to ensure quick reinstatement of vegetation.

During the course of the fieldwork and post-excavation analyses it was possible to acquire detailed information on the present condition of the monuments and the sequence of past land-use history of the sites with respect to their current scheduled status. Indeed, this information was submitted to a DEFRA study on the management of archaeological sites in arable landscapes compiled by Oxford Archaeology (French 2002), and is summarized in Appendix 4.5 and Table A4.30. This new information could aid in future management decisions taken by both national and county curators.

Once this first phase of work in 1998 and 1999 had been deemed successful in terms of information, return value for time and money spent on the excavations, the project was expanded to take in a selection of other scheduled monuments in the area dating to other periods, namely the Dorset Cursus, the Ackling Dyke, two of the Gussage Cow Down long barrows, and the Gussage Cow Down enclosure and lynchet system. This work particularly targeted Neolithic and Mesolithic land surfaces and features, and post-Bronze Age features. The same methodological approach as itemized above was used throughout.

This was further augmented by the contemporaneous excavations of Martin Green at several unscheduled prehistoric sites within the study area, mainly between Bottlebush Down and Monkton-up-Wimborne. Finally, a series of six sites in the south-eastern part of the study area that had been identified as potential prehistoric settlement sites during the course of the valley geoarchaeological survey and the Knowlton environs field survey were also treated to similar geophysical survey and test excavation methods as set out above.

The valley geoarchaeological survey

Charles French & Helen Lewis

Off-site investigations entailed obtaining landowner's permission for access, and then carrying out augering transects (set at approximately 500 m intervals) across the valley system, with about 100 m between auger holes (Fig. 1.11). A full aerial photographic mapping of the study area was conducted, delimiting zones of thin and thick soil cover with respect to the known archaeological record (Figs. 2.1 & 2.2). The results of these procedures allowed an erosion assessment map for the region to be produced (Fig. 2.9), on the basis of which we could target zones of colluvial aggradation or lynchet systems for sample excavation using small trial trenches (1 × 2 m).

The Knowlton environs field survey

Helen Lewis

This field survey is discussed in more detail in Chapter 3. On the basis of finds from the valley auger survey, five fields were fieldwalked to get a general idea of the patterns and densities of artefact distributions (AH, AL/AVE, BRO/BD, FCL, KH) (Figs. 3.80–3.83). With one exception (Field AH), the fields were walked along transects spaced at 20 m intervals, with material collected in *c.* 5 m wide strips along each transect. Artefacts were collected in bags for every 20 m distance along each transect. This method of non-intensive field survey aimed to simply identify if and where artefact scatters were present, and to provide enough material to enable a basic characterization of any scatters found. Field AH was walked using a 10 m transect and bag drop interval. This was the first field surveyed, and although it was known by hear-say to contain a flint scatter, the unanticipated large volume of flint collected using such a small interval led us to double our spacing in the other fields. In all fields, all artefacts not obviously modern were collected in these transects, covering all materials and all periods. These were analysed as discussed in Chapter 3 and tabulated in Table A4.55.

This fieldwalking survey of the southern part of the study area around the Knowlton henge complex identified at least four possible zones or concentrations of Neolithic and Bronze Age lithic material. In order to then test whether these concentrations represent occupation or not, their date ranges and the level of preservation, the next phase of the project involved a field programme of targeted geophysical survey and test excavation, which took place in the late summer of 2003 (Figs. 3.80 & 3.84–3.91). This included aerial photographic searches by Rog Palmer and Martin Green (Figs. 3.79, 3.84 & 3.89), magnetometer survey of eight half-hectare plots around each lithic concentration by GSB Prospection (Figs. 3.80 & A2.16), followed by test excavation pits/trenches (of 1 × 1 m, 2 × 2 m or 1 × 3 m in size) of eight locations with the test holes placed on the basis of the geophysical survey plots (Figs. 3.85, 3.87 & 3.90). In addition to this, further systematic transect fieldwalking of the Avenue Lodge fields (Figs. 3.81–3.83) took place concurrently with the test excavation programme. This fieldwalking survey was conducted by Dr Helen Lewis with volunteers from the East Dorset Antiquarian Society and several Cambridge PhD students, with the distribution plots drawn by Dr Fraser Sturt on the basis of the flint analysis by Dr Julie Gardiner.

Appendix 2

Geophysical Survey in the Upper Allen Valley, Cranborne Chase

GSB Prospection & Chris Gaffney

Magnetic and resistance surveys were carried out in six phases as part of this project. The first survey was undertaken in July 1998 and the final one in September 2003. In general, the surveys were conducted at the front end of the project's summer fieldwork, although the survey of long barrow (SAM No. 294) was undertaken during January 2002.

The surveys were all comparatively small scale and are probably best described as 'surgical' in that they have been undertaken largely to investigate known archaeological sites or monuments. In the main, surveys were carried out not to find new sites but to identify appropriate areas to excavate samples for environmental evaluation within or around sites. In total about 6.5 ha was subjected to magnetic investigation, although this was spread over 16 areas. Additionally, six of these areas were partially covered using the resistance technique and two very small resistance data sets were collected without the benefit of any additional method. The total area covered using resistance was only *c.* 1.3 ha. The position of the surveys can be seen in Figures 3.1, 3.33, 3.80 & A2.8.

All too often it is assumed that magnetic surveys, in particular, over chalk lands are straightforward. However, that belies the complexity of the archaeology, especially when ploughing and interventions, both antiquarian and modern, are considered. The surveyors undertaking the work have benefited from the correlation of geophysical data with subsequent excavation.

Methodology

The survey methodology that was used at all sites adhered to the standards that were common at the time of the surveys (David 1995; Gaffney *et al.* 2002). In all cases the area of interest was divided into 20 m blocks and C. French and colleagues tied this grid into local map features. The method utilized in the two techniques is as follows and a wider discussion of the topics can be found in Gaffney & Gater (2003).

Magnetometer survey

Survey measurements were collected using Geoscan Research Fluxgate Gradiometer instruments (FM36/256). While all the data were captured along traverses that were spaced 1 m apart, the sample interval along the line was varied. In the earlier surveys two samples were collected each metre, while in the final phase of work four samples were collected; this change is in line with advances in good practice in the profession. The data were captured in the internal memory of the instrument and then downloaded to portable computers in the field. The individual grids were then matched together to produce an overall plan of the variation in the vertical component of the Earth's magnetic field within the surveyed area.

Resistance survey

The data were captured in the internal memory of a Geoscan Research RM15 Resistance Meter. In all cases the instrument was set up using the standard Twin-Probe configuration with 0.5 m separation between the mobile probes. The variation in the earth resistance was measured at 1 m intervals in both directions. At regular intervals the data were downloaded to portable computers and individual grids were then matched together to produce an overall plan of the resistance of the ground within the surveyed area.

The data from both techniques were analysed using a variety of in-house and commercial software. A full report, including text, diagrams and interpretation based on a CAD system, was prepared after each site visit. The following is a summary of the results of the surveys arranged broadly by monument type.

Results

Images were reproduced of all the significant results from the geophysical work and these are discussed in the relevant sections in Chapter 3 and in this Appendix. The summary section of each survey episode relates to the overall data collected, which were reported in full elsewhere and are referenced accordingly.

Long barrows

Gussage Cow Down (SAM No. 294) (GSB 2002a)

NGR: ST 992 131

The long barrow is a Scheduled Ancient Monument that lies approximately 1.5 km to the east of North Farm and 1.5 km to the south of Canada Farm in the parish of Gussage St Michael (Fig. 3.1). It is one of a dense concentration of prehistoric monuments occupying the chalk downland at this point. The aims of the survey were to locate and identify the nature and extent of the surviving archaeological remains associated with the long barrow. As part of the investigation a portion of a trackway adjacent to the barrow was surveyed with a gradiometer in order to locate any remains of the barrow entrance.

Gradiometer survey (Figs. A2.1 & A2.3)

Two parallel linear anomalies clearly indicate the long barrow boundary ditches. They terminate 65 m to the west of the boundary that divides the two survey areas. The eastern limit of the ditches is unclear. There is a suggestion of a ditch end in the survey carried out on the track. However, the interpretation is cautious; it is possible that this anomaly has been produced by ferrous debris.

A number of linear trends aligned northeast to southwest were recorded. They are assumed to be due to ploughing disturbance and are aligned with the boundary to the east. However, one anomaly is well defined and could represent part of a ditch feature.

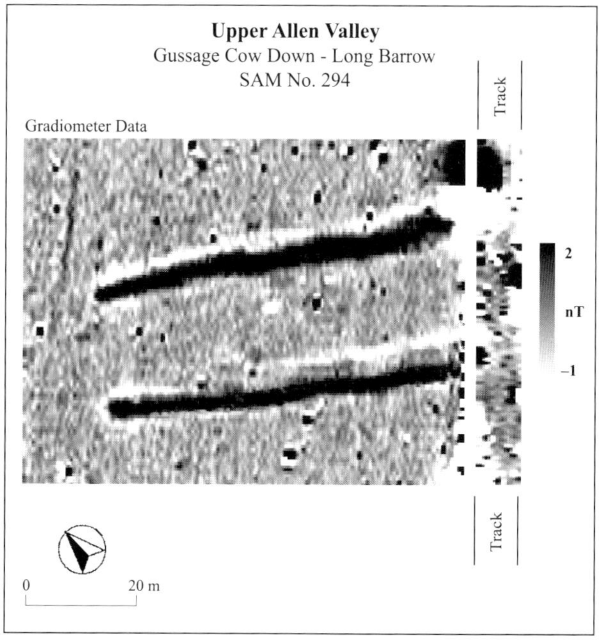

Figure A2.1. *The gradiometer data for Gussage Cow Down long barrow 294 (GSB).*

A number of small-scale responses were recorded. Many of them have been attributed to ferrous debris in the topsoil that is likely to be modern in origin, although some are considered to have archaeological potential as they may represent pit features. It should be noted that deeply buried ferrous objects and natural variations in the underlying bedrock could also have produced these anomalies.

Resistance survey (Figs. A2.2 & A2.3)

The remains of the long barrow mound are well defined as a band of high resistance. Low-resistance anomalies on each side are partly

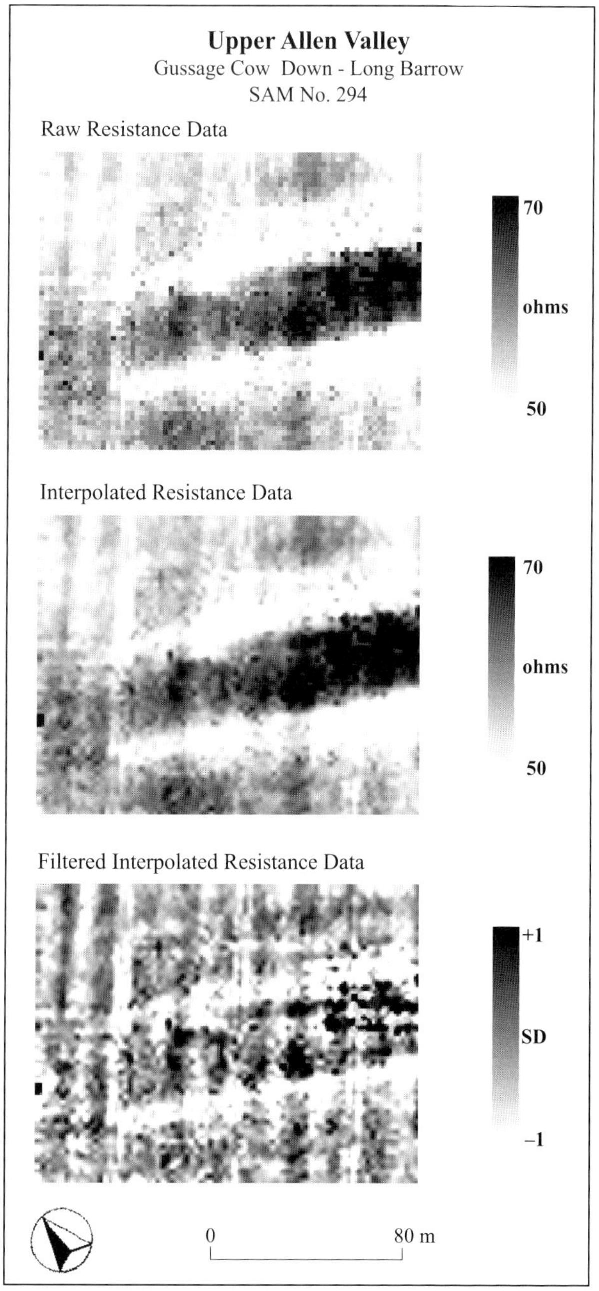

Figure A2.2. *The raw resistance, interpolated resistance and filtered interpolated resistance data for Gussage Cow Down long barrow 294 (GSB).*

a response from the underlying ditch features and a topographic effect caused by moisture trapped within surface undulations.

Other regions of increased high resistance are present in the data. These are considered to be natural in origin and/or due to ground surface changes and are not thought to be of archaeological interest.

A series of linear anomalies running northeast to southwest across and parallel to the boundary are due to ploughing. The effect of these variations in the soil can best be seen in the plot of the filtered and interpolated data. Bands of varying resistance cut the mound of the long barrow that are the result of ploughing.

The data plots show an increase in resistance values toward the boundary at the eastern edge. This could relate to the quantity of chalk stone in the composition of the mound. It is likely that the eastern end of the barrow was more substantial when constructed. In addition, it is likely that the proximity of the adjacent boundary may have helped in the preservation of the mound from plough damage. In Figure A2.3 the magnetic data are draped over a 3D surface plot of the resistance survey; the correlation between the two sets of results is very high.

Summary

The geophysical survey has successfully identified the location and extent of the boundary ditches of the long barrow. The results confirm that the eastern edge of the barrow is likely to have been severely damaged or destroyed by a trackway that runs north–south across the presumed eastern end.

Other possible archaeological type responses have been identified in the gradiometer data, including a ditch type anomaly. A by-product of the resistance survey has been an indication of the effect of plough damage on the mound of the barrow. Subsequent excavation (see Chapter 3) revealed that the mound did not survive and that the resistance survey had recorded variations in the chalk bedrock sculpted by ploughing.

Gussage Cow Down II (SAM No. 78) (GSB 2002b)

NGR: ST 9945 1136

Geophysical survey was carried out over a long barrow protected as a Scheduled Ancient Monument that is located on Gussage Cow Down, approximately 2 km NNE of the village of Gussage St Michael, Dorset (Fig. 3.1). Quarrying has damaged the northeastern end of the barrow. Geophysical survey was aimed at investigating the immediate vicinity of the monument.

Gradiometer survey (Fig. A2.4)

On top of the mound, a group of positive anomalies was detected, forming the shape of an elongated letter 'E' and aligned with the east–west axis of the barrow mound. For the most part these responses are weak and ill defined, although a few stronger discrete anomalies can be discerned. The overall pattern of the results clearly suggests that the anomalies relate to features within the barrow, though their precise nature remains uncertain. Positive magnetic anomalies of the magnitude recorded here usually indicate archaeological ditches or pits, but only anomaly (1) has a corresponding low-resistance response that would corroborate the interpretation. The magnetic anomalies lie *adjacent* to high-resistance responses and this could indicate that the instruments are seeing a build-up of magnetic soil against possible stone features.

Several negative linear anomalies were noted; these too follow a broadly rectilinear arrangement and all lie outside the main positive anomalies. Negative magnetic responses can be associated with stone features or earthen banks. With the exception of anomalies (2) the negatives recorded here have no parallels in the resistance data. The possibility that some of the responses are produced by topography cannot be discounted.

Ferrous anomalies recorded at the eastern end of the barrow coincide with the area of disturbance/damage arising from modern quarrying. They are attributed to ferrous material — either buried or

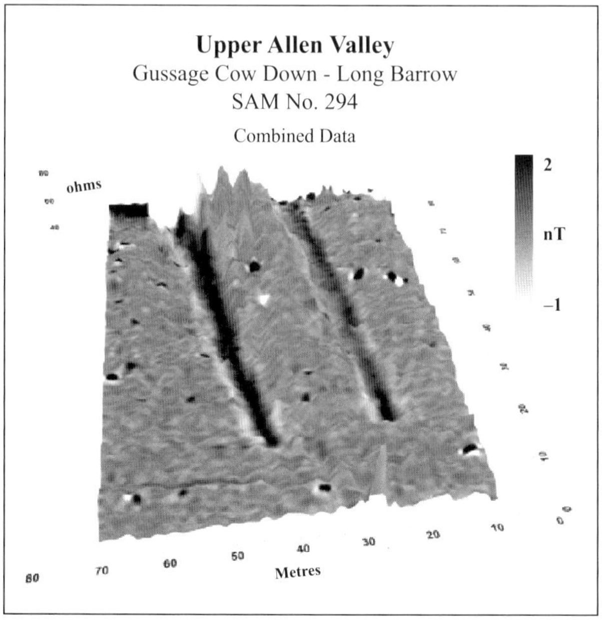

Figure A2.3. *The combined geophysical data for Gussage Cow Down long barrow 294 (GSB).*

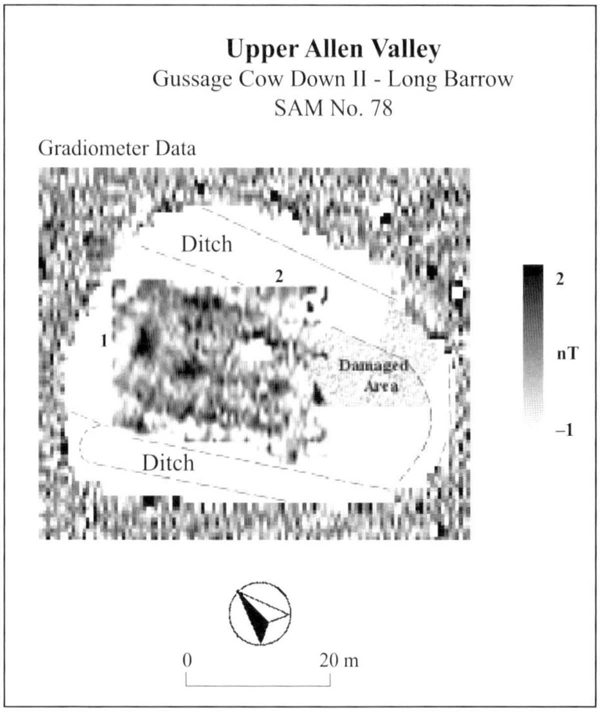

Figure A2.4. *The gradiometer data for Gussage Cow Down long barrow 78 (GSB).*

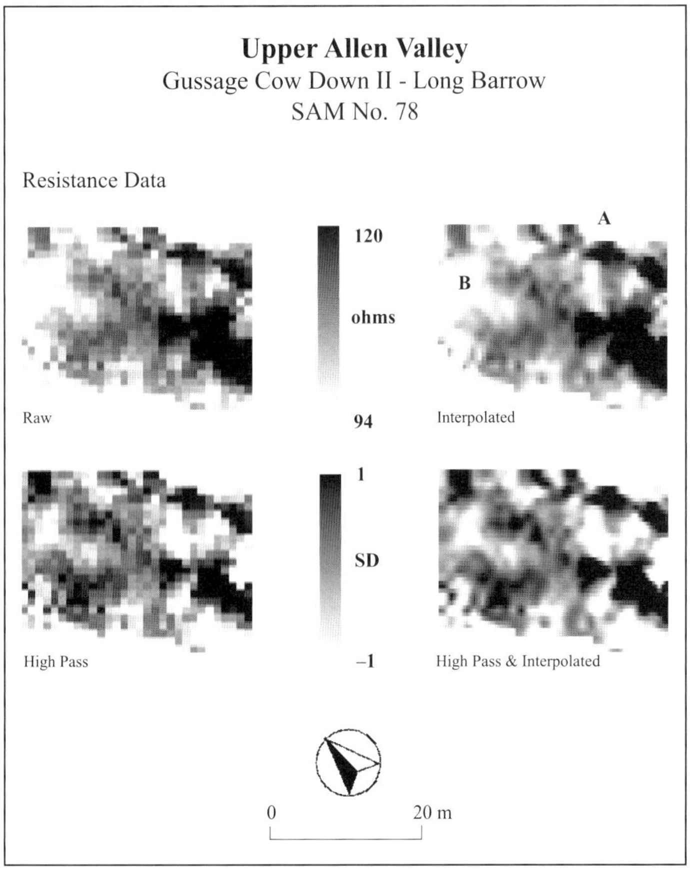

Figure A2.5. *The resistance data for Gussage Cow Down long barrow 78 (GSB).*

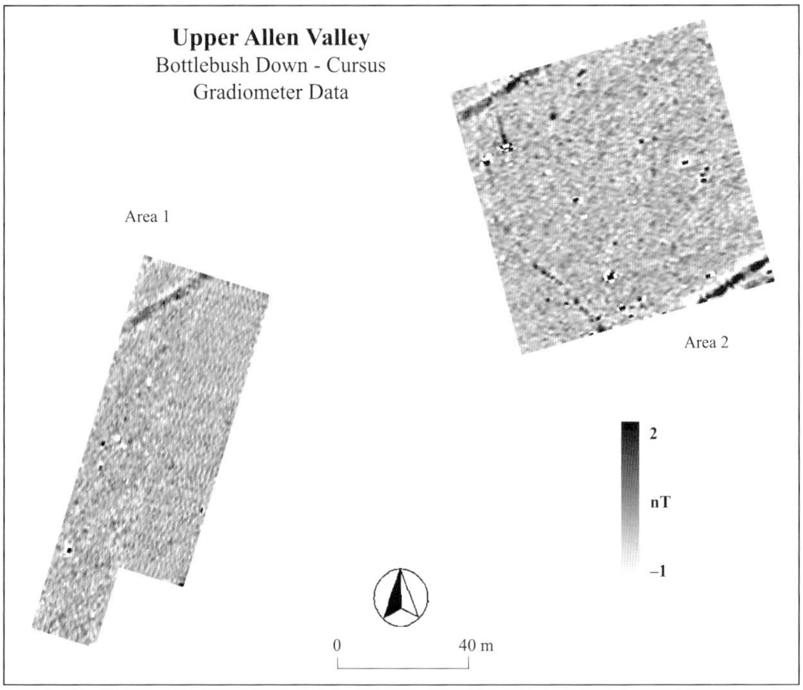

Figure A2.6. *The gradiometer data for the Dorset Cursus on Bottlebush Down (GSB).*

caught in the undergrowth — and assumed to be associated with this recent activity.

No definitive archaeological anomalies were detected in the immediate vicinity of the barrow.

Resistance survey (Fig. A2.5)

Three discrete low-resistance anomalies are tentatively highlighted as archaeological, suggesting pits/sections of ditch or a build-up of soil. Of these anomaly (A) is perhaps the most convincing, not least because it coincides with magnetic response (1).

Several linear high resistance anomalies, varying in strength and definition, were recorded by the resistance survey. They are broadly aligned with the long axis of the barrow and could indicate the remains of construction features. The anomalies at (B) coincide in part with negative magnetic responses (2) which could support this interpretation. In general, however, the limited size of the survey area makes it difficult to ascertain their precise nature.

Areas of high and low resistance along the eastern and southern edges of the grid are thought to be the product of modern disturbance, topography and surface features (trees and bushes).

Summary

Despite the hindrances caused by ground conditions and the limited size of survey area, the geophysical survey has provided some tentative evidence for features within the barrow. Although there are only two instances where magnetic and resistance anomalies can be directly related, the two data sets are generally complementary. The most prominent anomalies detected by the gradiometer suggest possible pits/short lengths of ditch or a build-up of soil. These anomalies lie *adjacent to* high-resistance responses suggestive of structural remains. Subsequent excavation (Chapter 3) has confirmed the resistance interpretation. A primary barrow with a turf core and chalk revetments was found to have been buried by the construction of a larger second tumulus of chalk and further chalk revetments. The interpretation of the gradiometer results, however, remains uncertain.

Dorset Cursus

Bottlebush Down, Sixpenny Handley (GSB 2001)

NGR: SU 0120 1535

Geophysical survey was undertaken within four areas on and around the intersection of the Ackling Dyke Roman road and the Dorset Cursus in order to establish the exact position and courses of the Cursus ditches (Fig. 3.33).

Gradiometer survey (Fig. A2.6)

Area 1

Gradiometry located a single low-magnitude linear ditch-type response; this coincides with the line of the northernmost ditch of the Cursus as shown on the Ordnance Survey map.

A number of trends can be discerned within the data. Where these are parallel they are attributed to recent ploughing. Those on a differing orientation may be archaeological, but these are close to the limits of detection and any firm interpretation would be speculative.

Area 2

Two broadly parallel linear responses, approximately 90 metres apart, are evident in the data and form part of the Cursus monument. The intervening area is largely devoid of anomalies, although a small number of pit-type responses have been noted which may be of archaeological interest. A discontinuous linear anomaly, comprising pit-type and short ditch-type responses, whilst categorized as possible archaeology, coincides with a former field boundary.

Numerous trends can be seen in the data; those that share a parallel orientation coincide with the current ploughing trend and are discounted as modern. Those that remain may be archaeological, although alternative explanations, such as a periglacial origin, are equally plausible.

Resistance survey (Fig. A2.7)

Area 3

The data for this area display disturbance associated with a modern trackway cutting through Ackling Dyke; this cut is clearly visible in the high-pass filtered data. A number of amorphous low-resistance anomalies and trends were attributed to this trackway.

A band of high-resistance readings, with clearly defined edges, coincides with the top and upper slope of Ackling Dyke and is thought to represent the compacted rubble of the Roman road. Resistance values decrease to background levels as one descends the slope of the dyke. Two poorly defined bands of intermediate resistance values lie either side of, and parallel with, the high-resistance anomaly of the road, and may reflect road-side ditches that have become infilled by the slumping of coarse materials from the road and its embankment. Their interpretation, however, remains cautious as the sample size is very small.

High-resistance responses away from the road may be archaeological, possibly reflecting debris that was redistributed when the trackway was cut through the dyke.

Area 4

Owing to the presence of thick scrub, it was only possible to survey the western flank of the Ackling Dyke. Nevertheless, the data clearly show three bands of resistance values indicating the metalling of the road, a putative side ditch, and low-resistance responses presumed to reflect the natural background. A faint edge or trend seen is likely to represent the line of one of the ditches of the Cursus.

Summary

Gradiometer survey located two parallel ditch-type anomalies that indicate the Dorset Cursus. These are approximately 90 m apart and, in the two areas surveyed, few intervening anomalies of archaeological potential were detected.

Resistance survey over the Ackling Dyke suggests a metalled road with possible roadside ditches that have become infilled with rubble from the road. A faint low-resistance response is cautiously interpreted as part of the cursus that underlies the Ackling Dyke.

Tumuli

Wyke Down, Gussage St Michael and Handley (GSB 1998)

NGR: SU 0076 1520

The survey investigated Tumuli 34 and 41 that are part of a series of such monuments known as the Wyke Down barrow group, as well as the land in between the two (Figs. 3.33 & A2.8). Both gradiometer and electrical resistance surveys were utilized to detect any ditches or other features associated with the two barrows and to locate any previous excavation trenches. The northeastern limit of the survey area abutted a field boundary marked by a partially removed wire fence. This produced an area of magnetic disturbance and spurious low-resistance readings.

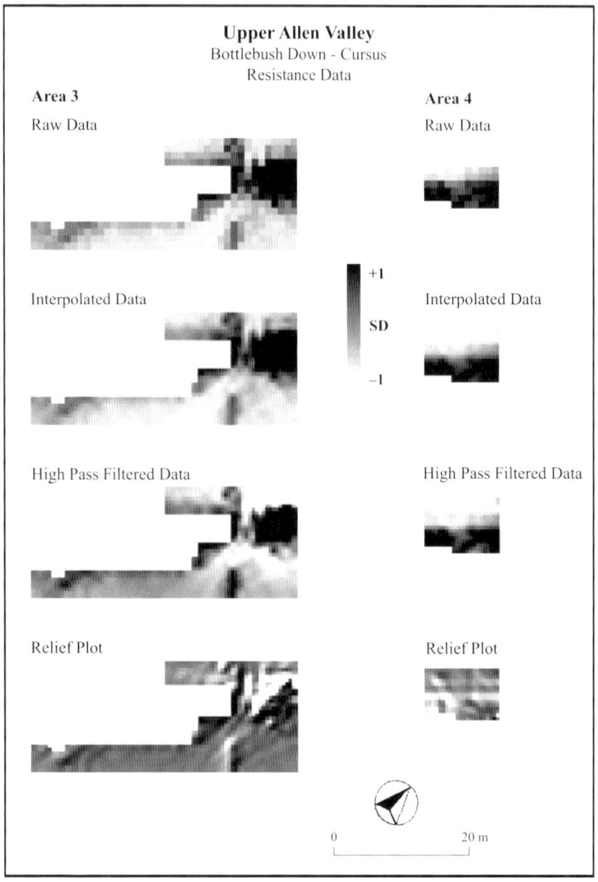

Figure A2.7. *The resistance data for the Dorset Cursus on Bottlebush Down (GSB).*

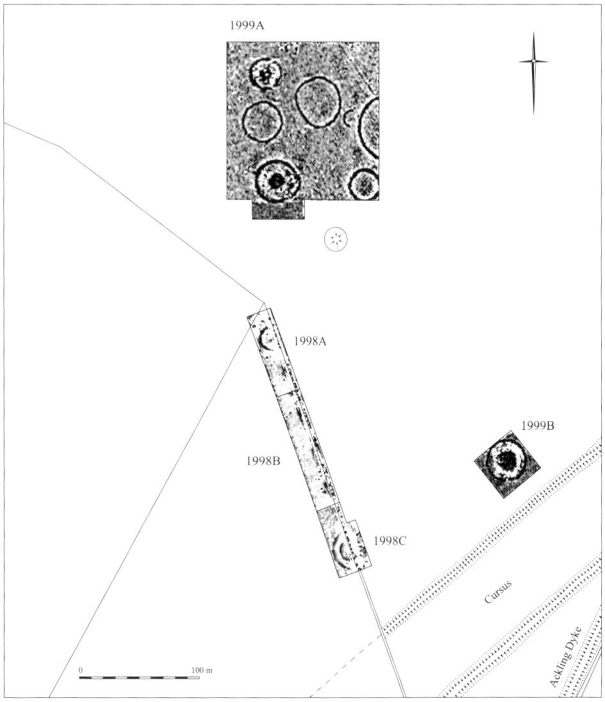

Figure A2.8. *The geophysical survey areas of the Wyke Down barrows (C. Begg after GSB).*

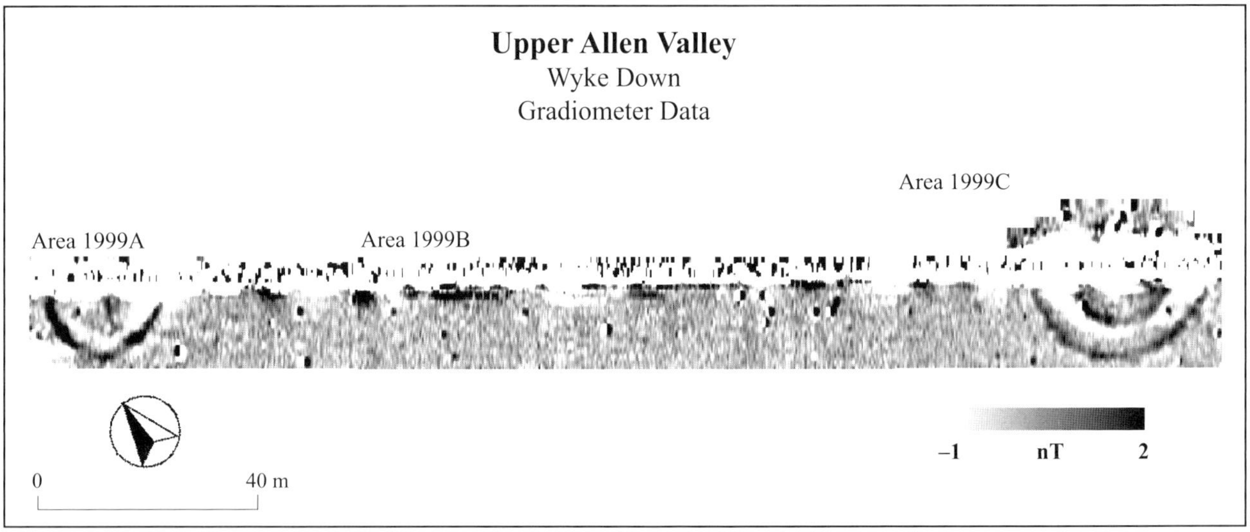

Figure A2.9. *The gradiometer data for Wyke Down barrows 41 and 34 and the intervening dry valley (GSB).*

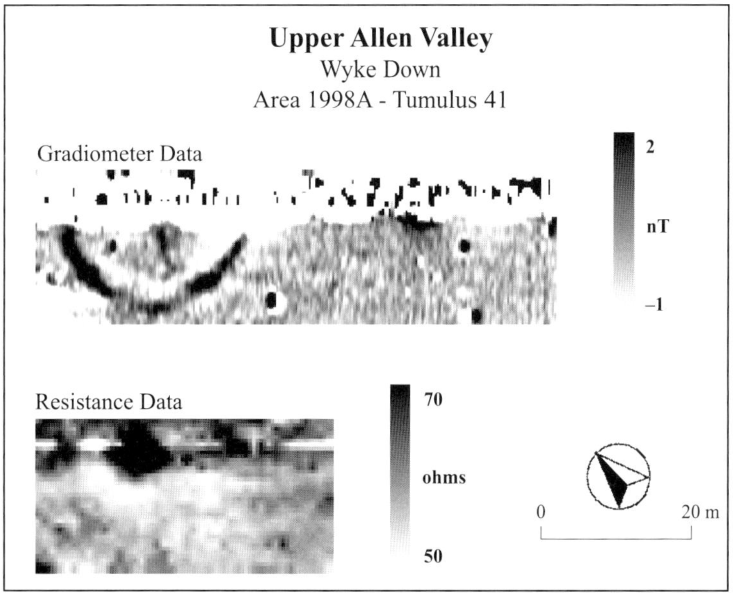

Figure A2.10. *The gradiometer and resistance data for Wyke Down barrow 41 (GSB).*

Area 1998A - Tumulus 41 (Figs. A2.9 & A2.10)
Gradiometer survey
Gradiometry detected a semi-circular anomaly that is believed to represent part of the former ditch surrounding the extant barrow. A short curvilinear anomaly detected may or may not be of archaeological interest, as a modern agency, such as recent ground disturbance or contamination, might account for it.

Resistance survey
Resistance survey detected an amorphous anomaly that corresponds approximately with the extant mound while immediately to the southwest a low-resistance anomaly was also detected. They lie within the area enclosed by the ring-ditch detected by gradiometry and it is probable, therefore, that the low-resistance anomaly does not represent part of the ring-ditch *per se*; it is likely that the image effectively reflects topographic variation rather than the buried archaeology.

Area 1998C - Tumulus 34 (Figs. A2.9 & A2.11)
Gradiometer survey
Two concentric semi-circular anomalies were detected that represent the two ditches of the barrow feature. It is the case that any continuation of the semi-circular anomalies will have been masked by the magnetic disturbance associated with the wire fence and the field boundary.

Resistance survey
A high-resistance anomaly was detected within the circular low-resistance anomaly. Its nature is unclear, though, given its proximity to an obvious depression (perhaps a backfilled trench), it may represent upcast spoil or more generally disturbed ground. Survey detected a circular low-resistance anomaly; this appe ars to correspond with the inner ditch-type anomaly as detected by gradiometry.

Summary
Gradiometer survey successfully detected the ditches associated with the two tumuli, although there were problems resulting from the fenceline that crosses the monuments. There were no internal features identified with any certainty, although there is some potential in Tumulus 41. No anomalies of archaeological interest were noted in the area between the monuments.

Resistance survey of Tumulus 41 was disappointing. A high-resistance anomaly with an adjacent low-resistance anomaly to its southwest was detected, although their exact nature is unclear. The high-resistance anomaly coincides with the core of the tumulus; it is not clear whether this simply relates to construction material or more significant archaeological deposits. The presence of the modern fenceline has certainly hindered the interpretation of these data. Resistance survey over Tumulus 34 detected a circular low resistance anomaly, which correlates with the inner-ditch anomaly detected by gradiometry. The nature of a high-resistance anomaly within the circular low-resistance anomaly is unclear, though, given its context, may represent upcast spoil. Such an interpretation, however, remains tentative. Both sets of resistance data contain a

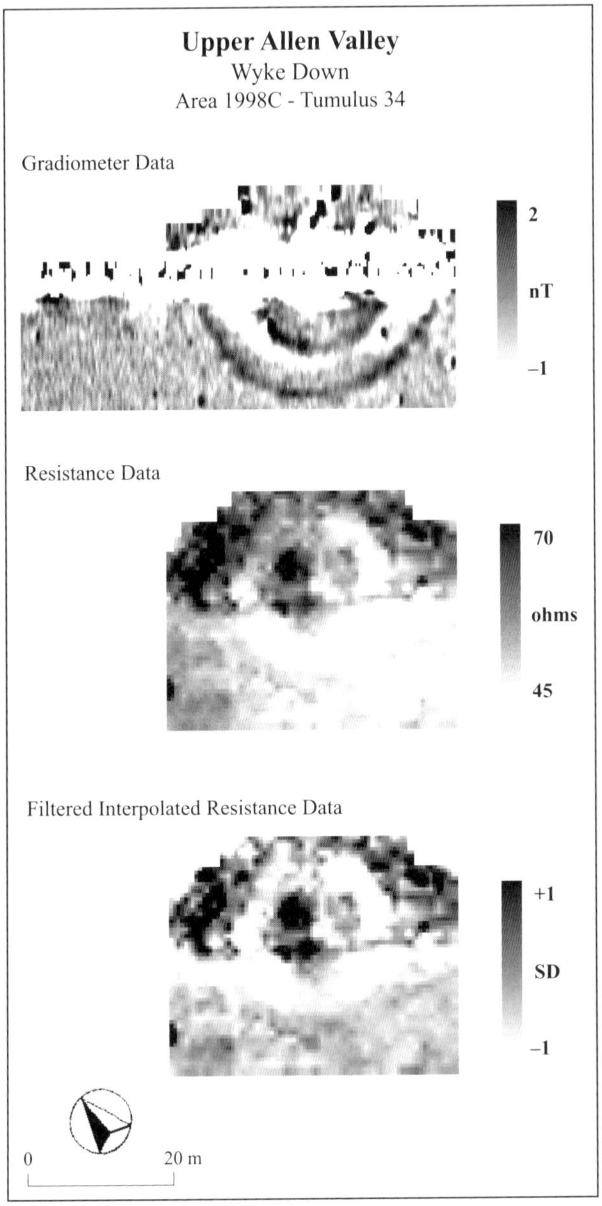

Figure A2.11. *The gradiometer, resistance and filtered interpolated resistance data for Wyke Down barrow 34 (GSB).*

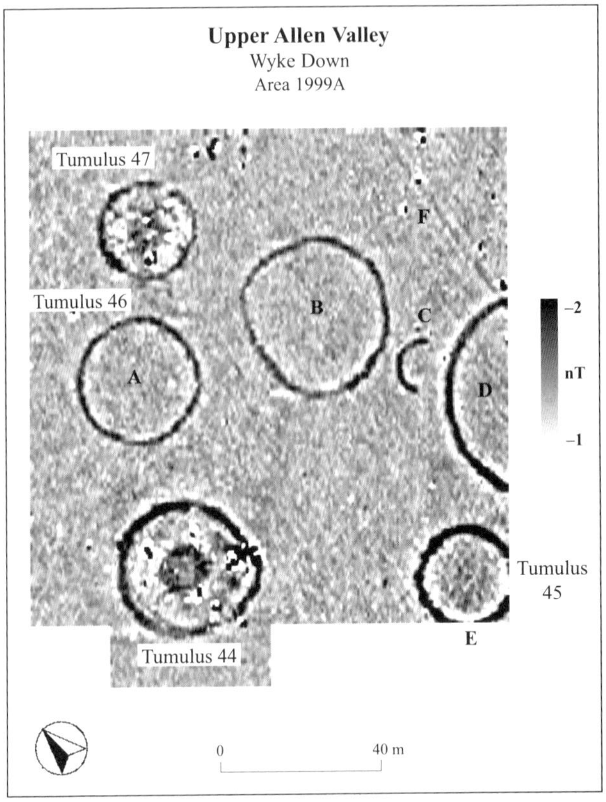

Figure A2.12. *The gradiometer survey of the main area of the Wyke Down barrow group, including barrows 45–7 (GSB).*

distinct edge, which divides an area of generally higher-resistance values from an area of lower-resistance values; it is likely that this edge represents the limit of recent ploughing which has encroached upon the barrows.

Gussage St Michael and Handley (GSB 1999)

NGR: SU 0071 1548

The second survey of the Wyke Down barrow group (SMR number 299) investigated the area of land occupied by tumuli 44, 45, 46 and 47 while another looked at tumulus 36 (Figs. A2.8 & A2.12–A2.14). The first and largest area lies to the north of the first survey undertaken at Wyke Down (Areas 1998A–C), while the second area is due east. In addition, resistance survey was conducted over Tumuli 44 and 36. The Dorset Cursus runs to the southeast of the survey area, as does the Ackling Dyke Roman road. The survey aimed to locate any features that might be associated with the standing barrows, and to confirm the aerial photographic evidence of barrow cropmarks.

Gradiometer survey

Area 1999A - Tumuli 44, 45, 46 and 47 (Fig. A2.12)

Seven ring-ditches and curvilinear anomalies were located in this area, four of which contained anomalies suggestive of internal features.

Survey over Tumulus 44 detected an outer ring-ditch which encircles the upstanding mound; at its centre an ill-defined anomaly, or group of anomalies, was detected. These anomalies may reflect buried pits, an inner ditch, or some aspect of the construction of the mound — such as a turf core. Between the outer ring-ditch and the inner group of anomalies, several other amorphous pit-type anomalies can be seen.

Survey over Tumulus 47 also detected an outer ring-ditch that encircles the standing mound of the monument. Within the circle of the ring-ditch, i.e. below the present mound, an area of increased response was noted. This is probably mainly archaeological in origin but a firm conclusion cannot be drawn owing to its lack of form. This area of increased response may represent a collection of pits, or it may reflect disturbed features and deposits.

Two other ring-ditches (A) and (B) were detected which display no apparent internal anomalies. This is most probably explained by ploughing which is likely to have destroyed and dispersed any other features that may have existed. Anomaly (A) appears to correspond with Tumulus 46 which is visible on AP data. A semi-circular

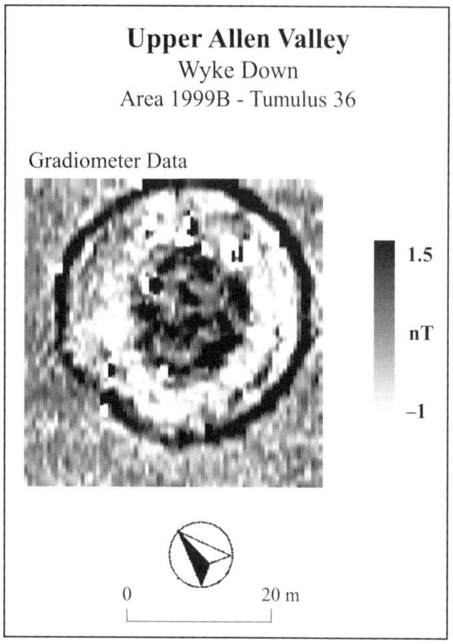

Figure A2.13. *The gradiometer data for Wyke Down barrow 36 (GSB).*

anomaly (C) is also interpreted as being archaeological and may represent a lesser ring-ditch that had been subsequently truncated by the (presumed) ring-ditch visible as anomaly (D). Anomaly (E) spatially is likely to represent tumulus 45, though anomaly (D) cannot be discounted entirely in this identification.

Anomaly (E) and, to a lesser extent, anomaly (D) contain internal areas of increased magnetic response. However, at the time of survey (E) had no surface manifestation and (D) was barely perceptibly as a low rise in the ground surface. The most credible explanation for these areas of increased magnetic responses, therefore, is that they reflect magnetically enhanced materials disordered by ploughing. The prevailing direction of ploughing can be clearly perceived as parallel linear trends orientated northwest to southeast. A linear trend (F) is noted in the interpretation because of its signal strength relative to the responses attributed to ploughing. Whilst an archaeological explanation cannot be wholly dismissed, it is more supportable that (F) is due to the 'smearing' of underlying magnetically enhanced materials by ploughing.

Area 1999B - Tumulus 36 (Fig. A2.13)

The data contain a ring-ditch that encompasses the mound of the standing monument. A curvilinear anomaly can be seen to lie just within the circle of the ring-ditch and is strongest in its southern part. The anomaly appears to lie parallel with the main ring and may represent the remnant of some lesser inner ditch. A central group of amorphous, pit-type anomalies is visible within the ring-ditch; these may relate to an inner ditch, pits, or some element of the barrow's construction.

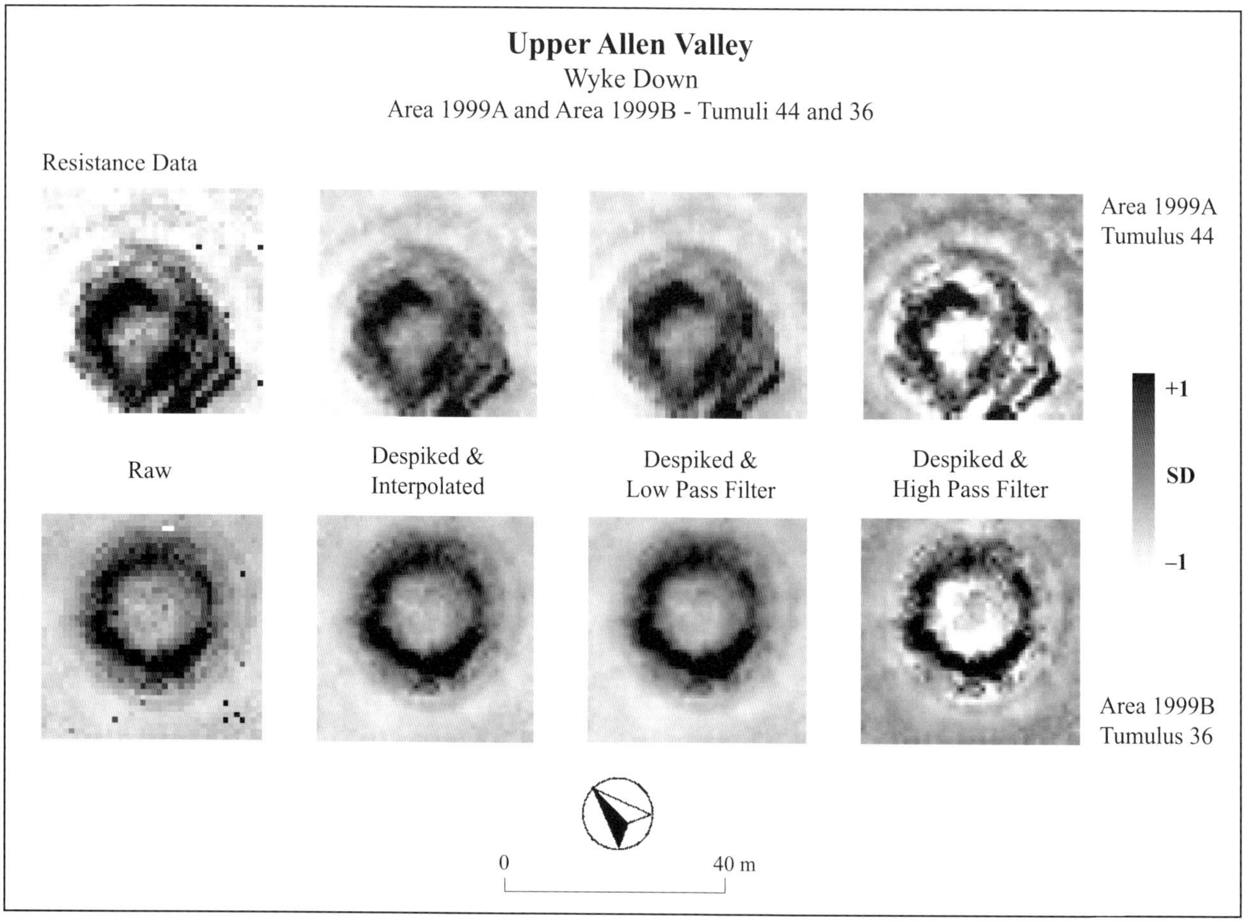

Figure A2.14. *The raw, despiked and interpolated, despiked and low pass filter, and despiked and high pass filter resistance data for Wyke Down barrows 36 and 44 (GSB).*

Resistance survey

Area 1999A - Tumulus 44 (Fig. A2.14)

The results from this survey add detail to those from the gradiometry. A central core of low resistance is visible that coincides with the central concentration of positive magnetic anomalies. This is surrounded by an irregular band of high resistance that generally corresponds with the circular band of low-susceptibility material evident in the gradiometer data between the ring-ditch and the inner group of anomalies. These results are commensurate with a feature constructed with a central turf core surrounded by chalk rubble, although a central 'robber pit' that was backfilled with soil might also produce such results.

The barrow also appears to have suffered disturbance in its southeastern quarter that corresponds to a concentration of ferrous-type responses recorded during the gradiometer survey. This adds weight to the possibility of ground disturbance. Although these anomalies may be archaeologically significant, they may relate to more recent causes such as disturbance from trees or boundaries that may have been recently removed, agricultural practice or even excavation of this part of the mound.

Close comparison of the gradiometry and resistance data indicates that the ring-ditch visible in the magnetic results does not correlate consistently with any one of the curvilinear high- or low-resistance anomalies that can be seen about the mound. On the western and much of the eastern flanks of the mound, the ring-ditch (as seen in the gradiometer data) does correspond with an area of low resistance. The northern and southeastern section of the ring-ditch correlates with an area of high resistance. This may reflect variations in the nature of the ditch fill, or it may be that the method is detecting edges within the ditch. It may, however, relate to seasonal peculiarities in micro-scale drainage within the ditch that can produce ephemeral resistance anomalies.

The resistance data suggest the existence of at least one, and possibly two, concentric high-resistance anomalies with an intervening low-resistance band. These may relate, therefore, to two banks and a ditch respectively, outside the ring-ditch described by gradiometry. However, it may be the case that it is the edges within this possible ditch that are being detected.

Area 1999B - Tumulus 36 (Fig. A2.14)

The resistance data for Tumulus 36 indicate a central area of low resistance (which coincides with a group of magnetic anomalies) encircled by a band of high resistance. It is suggestive of a central body of finer textured soil surrounded by coarser materials. This could suggest either a turf core or a 'robber pit' backfilled with soil.

There is an apparent spatial discrepancy between the ring-ditch detected by gradiometry and the band of low resistance, which is clearest to the southwest of the mound. This is explicable in terms of the slumping of coarse (high-resistance) material from the mound over the top of the ditch. The ring-ditch would still appear as a magnetic anomaly but a region of low resistance would occur away from the mound where this 'skirt' of high-resistance debris is thinnest.

Summary

Gradiometer survey over the standing tumuli 36, 44 and 47 detected outer ring-ditches as well as concentrations of anomalies suggestive of internal ditches and pits within the barrow mounds. Gradiometry also detected a further five circular or curvilinear features of the Wyke Down barrow cemetery, two of which contain areas of increased magnetic response. This is consistent with ploughing having destroyed any internal features and dispersing any magnetically enhanced materials they enclosed. The data also contained several linear trends, the bulk of which are believed to reflect the prevailing direction of ploughing, or enhanced materials having been disturbed and 'smeared' by ploughing. One weak curvilinear anomaly is no more than a trend but, given its form, an archaeological explanation cannot be excluded. Several of the ring-ditches exhibit concentric negative anomalies and it is possible that these represent the remains of concentric banks constructed from the chalk soil upcast from the ditch. A number of ferrous anomalies have been recorded at some of the tumuli sites that are likely to be due to discarded modern debris. The ferrous anomalies present within the ring-ditches of some of the barrows could relate to the remains of fence posts.

Resistance survey over Tumulus 44 suggests the mound to be constructed of a core of low-resistance materials encircled by coarser high-resistance material; this is consistent with a barrow with a turf core surrounded by chalky rubble. In addition to the ring-ditch described by gradiometry, the data also contain two possible outer concentric high-resistance anomalies, with an intervening band of low resistance, suggestive of banks and a ditch. A group of anomalies in the southeastern quarter of the barrow are noteworthy as they may be archaeological or be due to more recent disturbance. The data for survey over Tumulus 36 again suggest a mound constructed of low resistance material surrounded by coarser chalk rubble.

Gussage Cow Down II (SAM NO. 192) (GSB 2002b)

NGR: ST 9955 11375

Geophysical survey was carried out over an area of lynchets, preserved as a Scheduled Ancient Monument, located on Gussage Cow Down, approximately 2 km north-northeast of the village of Gussage St Michael, Dorset (Figs. 3.1 & A2.15). They are part of a series of earthworks associated with an Iron Age settlement and field systems, and those in the southeastern end of this site are the subject of the current investigation.

The bank and ditch pair at the northern end of the grid is the best preserved of the earthworks in this area. A strong magnetic anomaly (1) follows the line of the ditch and most probably represents an accumulation of strongly enhanced material in the ditch. That some of this material may be modern debris is suggested by the presence of several ferrous anomalies forming part of (1). Anomaly (2) could indicate a continuation of the ditch that has been disturbed or damaged by a modern farm track that dissects the survey area.

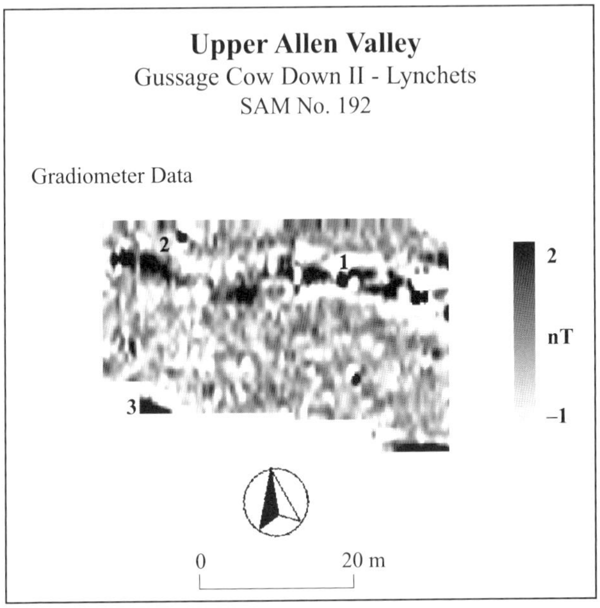

Figure A2.15. *The gradiometer data for the Gussage Cow Down bank and ditch complex SAM 192.*

Negative magnetic anomalies were recorded on either side of (1) and (2), which could reflect parts of the banks. However, they are considerably narrower than the upstanding banks and do not follow the line of the earthworks precisely. They could simply be a 'shadow' produced by the strongly enhanced magnetic material in the ditch.

The less well-preserved earthworks did not produce anomalies comparable in strength and coherence to those at (1) and (2). A few trends were highlighted, but two of these follow the line of the modern track, not the earthworks. Strong positive and negative anomalies (3) could indicate the remains of ditch and bank features, but are perhaps more likely to be caused by modern ferrous debris caught in the undergrowth adjacent to the fence.

Summary

The anomalies recorded by the gradiometer in this area seem to relate directly to the extant nature of the earthworks. The well-preserved ditch in the northern half of the grid produced a strong well defined magnetic anomaly while the denuded earthworks to the south appear only as faint trends. None of the banks produced definitive anomalies; although a few negative magnetic linears were recorded, they could simply be magnetic 'shadow'.

Knowlton environs (GSB 2003)

NGR: SU 030 115 (centre)

A total of eight small areas were examined near the village of Wimborne St Giles and close to the Knowlton Circles (Fig. 3.80). Numerous burial mounds and the Knowlton henge complex lie within the locality. The survey areas were either identified as cropmarks on aerial photographs and/or chosen as a result of field walking. There are suggestions that settlement remains dating to the Neolithic and Bronze Age may be present in the target areas and it was hoped that gradiometer survey would locate and identify anomalies of archaeological significance associated with this activity.

Gradiometer survey (Fig. A2.16)

Area BDN1 (Brockington Down)

The two linear anomalies (1) and (2) are likely to relate to buried ditches. The results indicate that they continue outside the survey area, a fact that is not apparent in the corresponding cropmarks recorded on aerial photographs. The arrangement of these features and their irregular nature is curious. The possibility that these anomalies are due to cracks in the underlying bedrock cannot be discounted.

Anomaly (3) coincides with a ring-ditch that may relate to a gully around a large pit, while anomaly (4) corresponds to a probable large pit. Both are visible as cropmarks on aerial photographs.

The survey has identified other possible ring-ditches (5) that may represent the sites of dwellings. These responses are barely visible above background levels and the interpretation is cautious. However, the features, if they are gullies, may not be very substantial especially if ploughing, which is apparent in the data, has damaged them.

Other pit-type anomalies were recorded that could be associated with former occupation. In addition, an area of increased magnetic response (6) was detected between the ditch anomalies (1) and (2). This could represent a cluster of anomalies and/or a spread of magnetically enhanced material; there is also the suggestion of part of a ring-ditch within (6). A series of parallel linear trends, aligned broadly north–south, indicate the direction of recent ploughing and are not considered to be of archaeological interest.

Area BDN2 (Brockington Down)

Two well-defined linear responses (7) and (8) were recorded. Anomaly (8) appears as two linear responses at the western edge of the survey area. These are interpreted as former field ditches.

A scatter of small-scale pit type anomalies was also recorded that may indicate remains associated with settlement activity, though no clear pattern emerges from the results.

Parallel linear trends similar to those recorded in the previous survey area, but magnetically stronger, are due to disturbance from recent ploughing.

Area BRO (Brockington Farm)

Areas BRO and BD lie to the northwest of Brockington Farm about 1.5 km to the southwest of Wimborne St Giles.

A linear anomaly (9) and magnetically very weak and ill-defined trend hint at part of a possible enclosure. Elsewhere, a scatter of small-scale anomalies is highlighted on the interpretation diagram. Some are obviously ferrous in origin and are due to modern debris in the topsoil. However, some of the pit responses could be of archaeological interest.

Anomalies recorded along the western edge of the survey area could relate to a pipe running down the adjacent hedge.

Area BD (Brockington Farm)

Well-defined linear anomalies (10) and (11) were detected in this area, and appear to be of archaeological significance. They may represent the remains of settlement features and/or parts of ancient field systems. Anomaly (12) is an isolated pit-type response that could indicate a feature with a strongly enhanced fill, perhaps incorporating burnt material.

A series of parallel trends aligned approximately northwest–southeast in the centre of the survey area is due to recent ploughing disturbance.

Area FCL (Farringdon Clump)

Area FCL lies immediately to the southeast of Farrington Clump approximately 1 km southwest of Wimborne St Giles.

Compared to the results of survey in the previous areas there is a noticeably lower level of background response in Area FCL. A number of pit type anomalies were detected though they appear to be isolated. Anomaly (13) is the most substantial and offers the greater archaeological potential. It should be noted that modern debris or a natural hollow in the near surface bedrock could account for this response.

A series of weak linear responses and trends are shown on the interpretation diagram. Although an archaeological origin cannot be dismissed it is likely that they are natural or agricultural in origin.

Area CD1 (Charlton Dairy)

Areas CD1 and CD2 lie immediately to the south of Charlton Dairy Farm, to the east of Knowlton Circles and about 1.5 km to the south-southwest of Wimborne St Giles.

The results from this survey area are dominated by a series of strong parallel linear anomalies aligned broadly northeast–southwest. They are due to recent ploughing and could indicate that magnetically enhanced material from disturbed archaeological deposits is present in this area.

Two pit-type anomalies and a magnetically very weak trend describing a circle could be of interest. However, disturbance produced by ploughing mask, or perhaps even form, these anomalies and the interpretation is doubtful.

Area CD2 (Charlton Dairy)

A band of increased magnetic response is present in the western part of the survey area within which there is a cluster of small ferrous-type responses. Although an archaeological origin cannot be ruled out it is possible that this increase in magnetic activity is associated with a former quarry site a short distance to the west.

Two pit-type anomalies were recorded in the western corner of the survey area. The potential of these responses is low given the likelihood of disturbance associated with quarrying. A series of linear trends visible in the eastern part of the survey area is

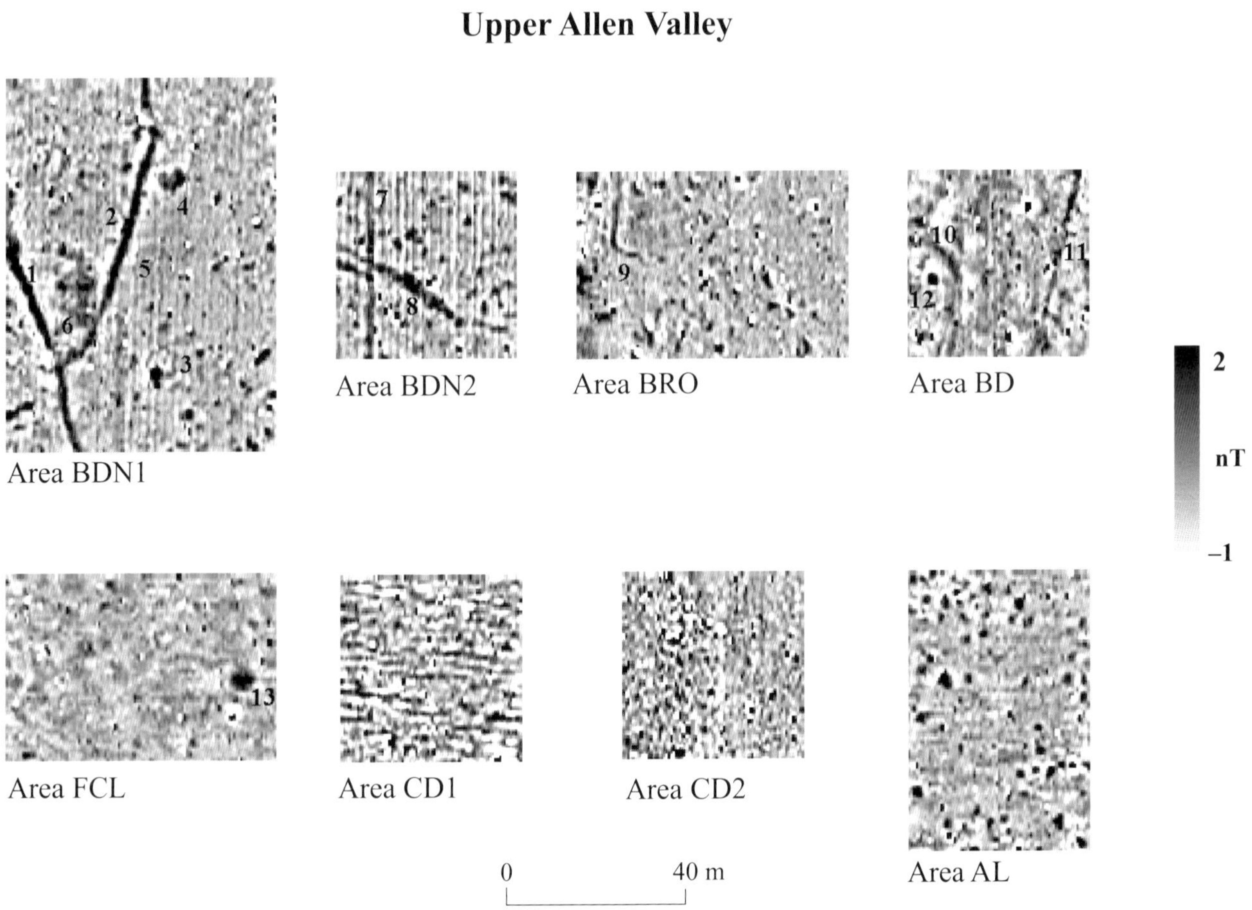

Figure A2.16. *The gradiometer data for the sample survey areas in the Knowlton environs: Brockington Down areas 1 and 2, Brockington Farm areas BRO and BD, Farringdon Clump, Charlton Dairy areas 1 and 2, and Avenue Lodge.*

due to past ploughing, despite the fact that the field is presently pasture.

Area AL (Avenue Lodge)

Area AL lies to the east of Avenue Lodge, near the B3078 road about 1.5 km east of Wimborne St Giles.

A number of large pit-type anomalies were recorded that coincide with a cluster of cropmarks recorded on aerial photographs. It is possible that these anomalies are associated with past un-enclosed settlement as no enclosing ditch anomalies were recorded; evidence for such features is absent on the photographs also.

A number of parallel linear trends have been identified. However, the archaeological importance of these is low; they are likely to be due to recent ploughing disturbance.

Summary

Samples of detailed gradiometer survey were conducted in eight locations within the study area of the Upper Allen Valley. Archaeological-type anomalies indicative of possible occupation activity were successfully identified in four locations. Within these areas, suggestions of enclosure ditches, possible ring-ditches and pit clusters were encountered.

Within the remaining four sites responses considered to be archaeological in nature were recorded but the interpretation is less certain.

Owing to the limited extent of the survey areas a wider context for the archaeological responses could not be determined, for example the relationship between the anomalies recorded in Areas BDN1 and BDN2 is unknown. The results suggest that more features are present than appear as cropmarks and that the remains of a large occupation site could be present in this area.

A second complication of limited survey size is that an assessment of the variation in background magnetic response could not be made. It is possible that many of the pit type anomalies relate to natural and widespread variations in the underlying bedrock and/or hollows left from the removal of tree stumps.

There is a noticeable change in the level of background response between these areas, despite the similar soil depths and conditions known to be present within the study area. Linear responses produced during recent ploughing vary in magnetic strength from area to area. It is likely that they are strong where plough damage to archaeological deposits has occurred. It is believed that in Area CD1, where such anomalies are at their strongest, archaeological remains have been destroyed and magnetically enhanced material from these features is present in the topsoil.

Conclusions to the geophysical work in the upper Allen valley

The resistance and magnetic surveys that form part of the investigations in the upper Allen valley have been targeted on sites or features of archaeology known from cropmark, earthwork or fieldwalking evidence. The interpretations have turned out to be exceptionally accurate, particularly for the long and round barrows where relatively fine details of the construction of the monuments have been identified, verified by excavation and so placed in a wider context.

Additionally, evidence has been obtained on the correlation between geophysical evidence and excavation but also between the two geophysical techniques that were used. While it is often assumed that geophysical survey over chalk is relatively straightforward, the correlation between the resistance and magnetic data sets was often discordant at the feature level. This is particularly true on some of the earthwork sites where small-scale variation in the moisture content of the soil is highly influenced by topographic variation.

The project team comprised Dr C. Gaffney, Dr J. Gater, L. Harvey, E. Heapy, J. Leigh, C. Martinez, D. Shiel, A. Shields, C. Stephens, B. Urmston and Dr D. Weston.

Appendix 3

Molluscan Diversity

3.1. Definitions of the Shannon, Brillouin and delta indices

Michael J. Allen

Shannon index (H')

The Shannon index assumes that individuals are randomly sampled from an 'indefinitely large' population (Pielou 1975). It is ideal for molluscan assemblages as it estimates the diversity of the unsampled, as well as the sampled, portion of the community. It takes into account the number of species and proportions of those species and views the evenness of the distribution of species present. It is not however, affected by total population which is why it is particularly applicable to sub-fossil assemblages where completeness of the assemblage cannot be guaranteed.

$$H' = -p_i \ln p_i$$

where
H' = index of diversity
p_i = proportion of individuals found in the ith species
ln = log to the base e

Brillouin index (HB)

$$HB = \frac{1n\ N! - \sum 1n - n_i!}{N}$$

The Brillouin index is closely related to the Shannon index and as it takes account of total populations (unlike the Shannon function) as well as the numbers of species and individuals. The significance of the combination of the two indices is that they provide some measure of the completeness of the sample and therefore it is possible to say something of the assemblage taphonomy. The closer *HB* is to *H'*, the more representative the subfossil assemblage is of the fauna from which it derives, providing that sample size is constant to within a few kilograms (Evans & Smith 1983, 102). In some cases the *value* of the difference between *H'* and *HB* is of ecological significance, particularly when it is applied to long continuous sequences of mollusc assemblages (e.g. colluvium) where it is a sensitive indicator of taphonomic change (Evans & Williams 1991). This is particularly important in colluvial sequences. If assemblages can be seen to be close to 'complete' or can be paralleled to those from ditch or buried soils then it is possible, if not likely, that they are more or less representative of death assemblages. Because of the problems of using diversity indices in the interpretation of palaeo-molluscan fauna (*H'* and *HB*) the results often plotted as curves (Evans 1979; Evans & Smith 1983) to provide an additional commentary on the data and for examining taphonomic factors in how the sampled assemblage varies from the original population, that is *H'* – HB (Allen 1994).

Delta (Δ) indexes

The delta indexes have been utilized by Gordon & Ellis (1985), and deployed in Cranborne Chase by Entwistle & Bowden (1991) to provide some discrimination of the wider landscape in which the sampled populations reside. Gordon & Ellis (1985, 153–6), however, have argued that measures of the probability of non-lethal interspecific encounters (Δ_2), and the ratio of interspecific to intraspecific competition (Δ_4) (Hurlbert 1971; Goodman 1975), are more applicable to biological populations than those derived from Information Theory, such as the Shannon function, and therefore are more sensitive indicators of environmental change. In view of the persuasive arguments put forward by Gordon & Ellis, and specifically when applied to sub-fossil molluscan assemblages, Entwistle (1989) adopted the Hurlbert indices throughout his studies and particularly in studies of the Stonehenge Environs (Entwistle 1990b), Cranborne Chase (Entwistle & Bowden 1991) and the Wessex Linear Ditches (Entwistle 1994).

This can be mathematically simplified following arguments presented by Allen & Allen (1999) as follows:

$$\Delta_2 = \frac{N^2 - \sum_{i=1}^{s} N_i}{N^2}$$

and

$$\Delta_4 = \frac{N^2 - \sum_{i=1}^{s} N_i^2}{\sum_{i=1}^{s} N_i^2 - N}$$

A number of other statistical techniques have been applied to land mollusca including both arithmetic and logarithmic correlation coefficients, allometric coefficients and sample *t*-tests (Thomas 1978; 1985). Williams & Milles have demonstrated the potential of rank order curves (1987) while Evans & Williams (1991) used a whole series of statistical techniques including the Simpson-Yule index (*D*), Berger- Parker index (*d*) and Fisher's Alpha index (Southwood 1978; Poole 1974; Pielou 1975) to provide data to create CLUSTAN groups for the entire M3 mollusc data base (Evans & Williams 1991). Other indexes have been advocated by Gordon & Ellis (1985), namely the probability of non-lethal interspecific encounters (*D*2) and the ratio of interspecific to intraspecific competition (*D*4) (Hurlbert 1971; Goodman 1975) which have been employed by Entwistle (1990a). While these analytical and statistical approaches are appreciated, the main thrust of this work was one of land-use history and landscape appraisal, and thus only the more established mathematical techniques were employed. However, the mollusc identifications are held on a single data base and can be interrogated using other statistical appraisals (cf. Shennan 1988) at some future date.

3.2. Molluscan species diversity data

Michael J. Allen

Species diversity data tables (Tables A3.1–A3.13).

Table A3.1. *Species diversity data from the buried soils at Thickthorn Down and Gussage Cow Down long barrow 78.*

Sample	Depth (cm)	Context	Fill-type	Molls	mpkg	Shannon	Brillouin	H' – HB	Δ_2	Δ_4
Thickthorn long barrow										
1985	spot	-	ols	36	267	1.83	1.56	0.27	0.81	5.00
Gussage Cow Down long barrow GCD 78										
243	spot	277	ols	101	51	1.55	1.43	0.12	0.71	2.59
245	spot	278	ols	182	91	1.52	1.43	0.09	0.69	2.31
246	spot	278	ols	20	11	1.50	1.17	0.33	0.69	2.65

Table A3.2. *Species diversity data from the ditch of Gussage Cow Down long barrow 294.*

Sample	Depth (cm)	Context	Fill-type	Molls	mpkg	Shannon	Brillouin	H' – HB	Δ_2	Δ_4
Gussage Cow Down long barrow GCD 294										
North ditch F.111										
19	28–38	225	tertiary	212	106	1.97	1.87	0.10	0.83	5.01
18	38–48	225	tertiary	238	119	1.82	1.74	0.08	0.81	4.22
17	48–58	225	tertiary	300	150	1.96	1.87	0.08	0.81	4.27
16	58–68	226	tertiary	294	147	2.34	2.22	0.12	0.87	6.95
15	68–78	226	tertiary	328	164	2.30	2.19	0.11	0.86	6.36
14	78–88	226	tertiary	474	237	2.35	2.27	0.08	0.87	7.76
13	88–98	226	tertiary	791	396	2.58	2.62	0.07	0.90	9.22
12	98–108	226	tertiary	963	482	2.53	2.48	0.05	0.88	7.71
11	108–118	230	secondary	1273	637	2.09	2.05	0.04	0.80	4.12
10	118–128	230	secondary	1699	850	2.05	2.02	0.03	0.76	3.22
9	128–138	239/230	secondary	2063	1032	2.05	2.02	0.03	0.74	2.83
8	138–145	239	secondary	300	150	2.56	2.43	0.13	0.90	8.90
7	145–155	239	secondary	746	373	2.58	2.50	0.08	0.87	6.80
6	155–160	239	secondary	6	3	-	-	-	-	-
5	160–165	240	primary	0	0	-	-	-	-	-
4	165–185	241	primary	3	2	-	-	-	-	-
3	138–150	241	primary	0	0	-	-	-	-	-
2	157–170	241	primary	8	4	-	-	-	-	-
1	170–185	241	primary	3	2	-	-	-	-	-
South ditch F.114										
1	110–120	233	u primary soil	4245	2123	2.06	2.04	0.02	0.75	3.00
2	120–131	234	u primary soil	2755	1378	2.40	2.38	0.02	0.86	6.09
3	131–138	234	u primary soil	566	283	2.42	2.34	0.08	0.87	6.60

Table A3.3. *Species diversity data from the north ditch of the Dorset Cursus, 2001, 1984 and 1982 excavations.*

Sample	Depth (cm)	Context	Fill-type	Molls	mpkg	Shannon	Brillouin	H' – HB	Δ_2	Δ_4
Cursus 2001: North(west) ditch F.101										
Ditch F.101; trench 2										
169	spot	172	turf	843	422	1.99	1.96	0.03	0.84	5.11
210	0–14	141	tertiary soil	756	504	1.84	1.81	0.03	0.80	4.11
211	20–30	178	tertiary soil	11	7	-	-	-	-	-
171	spot	140	ols	59	30	1.80	1.61	0.19	0.79	3.97
Bank F.100; trench 1										
164	spot	139	bank soil	103	52	1.79	1.65	0.14	0.79	4.02
166	spot	155	bank soil	94	47	1.88	1.72	0.16	0.79	4.00
Ditch F.101; trench 1										
208	82–92	138	tertiary	24	12	1.65	1.34	0.31	0.77	4.02
207	92–102	138	tertiary	17	11	1.50	1.16	0.34	0.71	3.12
205	86–94	162	eroded soil	613	409	1.78	1.73	0.04	0.76	3.21
206	94–103	162	eroded soil	483	322	1.91	1.86	0.05	0.81	4.09
204	103–111	163	secondary	45	30	1.85	1.69	0.26	0.83	5.69
203	111–121	164	primary	12	6	-	-	-	-	-
202	121–132	164	primary	0	0	-	-	-	-	-
201	132–140	164	primary	0	0	-	-	-	-	-
200	140–145	164	primary	3	2	-	-	-	-	-
176	spot	162	basal	132	66	2.02	1.89	0.113	0.84	5.51
Cursus 1982										
-	35–45	5		67	45	1.67	1.51	0.17	0.77	3.48
-	45–55	4		115	78	2.19	2.02	0.17	0.86	6.61
-	55–70	4		213	142	2.23	2.11	0.12	0.86	6.33
-	70–85	4		460	307	2.00	1.93	0.07	0.80	4.17
-	85–100	4		515	343	2.33	2.26	0.07	0.88	7.26
-	100–115	4		248	165	1.89	1.79	0.09	0.79	3.87
-	115–125	4		143	95	1.59	1.49	0.10	0.72	2.68
-	125–140	10		6	4	-	-	-	-	-
Cursus 1984										
-	20–30	-		218	182	1.47	1.37	0.10	0.57	1.37
-	30–40	-		101	53	2.09	1.89	0.20	0.82	4.82
-	40–50	-		162	125	2.15	2.01	0.14	0.85	5.78
-	50–60	-		439	328	2.17	2.09	0.08	0.83	4.94
-	60–70	-		1119	487	1.83	1.79	0.04	0.75	3.16
-	70–80	-		287	136	1.99	1.91	0.08	0.83	4.96
-	80–90	-		367	160	2.02	1.95	0.07	0.82	4.72
-	90–100	-		279	121	2.19	2.08	0.11	0.84	5.54

Table A3.4. *Species diversity data from the south ditch of the Dorset Cursus, 2001 excavations.*

Sample	Depth (cm)	Context	Fill-type	Molls	mpkg	Shannon	Brillouin	H' – HB	Δ_2	Δ_4
Cursus 2001: South(east) ditch F.108										
Ditch F.108; trench 4										
224	25–35	201	tertiary	18	12	1.96	1.48	0.48	0.82	6.65
223	35–45	201	tertiary	64	43	2.30	2.01	0.30	0.87	7.80
222	45–55	202	secondary	1074	716	2.56	2.51	0.05	0.88	7.33
221	55–65	202	secondary	1167	778	2.50	4.45	0.05	0.88	7.39
220	65–75	202	secondary	316	211	2.23	1.24	0.10	0.86	6.22
219	75–85	205	dump	58	39	2.04	1.78	0.26	0.83	5.26
218	91–99	209	lens	3	2	-	-	-	-	-
Ditch F.108; trench 3										
217	30–40	193	tertiary	57	38	2.18	1.90	0.28	0.86	7.02
216	40–50	193	tertiary	454	303	2.55	2.46	0.10	0.89	8.11
215	50–60	193	tertiary	524	417	2.49	2.41	0.08	0.86	6.39
214	60–70	194	secondary	1718	1145	2.10	2.06	0.03	0.76	3.21
213	70–80	194	secondary	1606	1071	2.40	2.36	0.04	0.85	5.91
212		196	soil	18	9	1.96	1.48	0.48	0.82	6.65

Table A3.5. *Species diversity data from the Monkton-up-Wimborne 2000 Neolithic enclosure.*

Sample	Depth (cm)	Context	Fill-type	Molls	mpkg	Shannon	Brillouin	H' – HB	Δ_2	Δ_4
Monkton-up-Wimborne Neolithic enclosure MUW 00										
160	26–32	3	tertiary	290	193	2.74	2.60	0.14	0.92	11.89
159	32–38	4	secondary	47	31	2.12	1.82	0.30	0.86	6.89
158	38–48			87	58	1.59	1.43	0.17	0.66	2.05
157	48–58			217	145	1.85	1.72	0.13	0.71	2.46
156	58–68	4	secondary	229	153	2.00	1.89	0.11	0.79	3.92
155	68–78	5a	turf	175	117	1.87	1.76	0.11	0.80	4.18
154	78–86	5d	soil lens	41	27	1.43	1.20	0.24	0.60	1.63
153	86–92	5e	soil lens	4	2	-	-	-	-	-
152	92–102	6	primary	0	0	-	-	-	-	-
151	102–112	6	primary	0	0	-	-	-	-	-
162		13	L primary	0	0	-	-	-	-	-
161		14	U primary	0	0	-	-	-	-	-

Table A3.6. *Species diversity data from barrow 34, Wyke Down.*

Sample	Depth (cm)	Context	Fill-type	Molls	mpkg	Shannon	Brillouin	*H'* – HB	Δ_2	Δ_4
Wyke Down Barrow 34 (Chalkpit Field)										
Outer ditch: F.5										
33	53–61	32	tertiary	1030	687	1.73	1.71	0.02	0.79	3.67
32	61–66	32	tertiary	1239	826	1.89	1.87	0.02	0.81	4.32
31	66–70	32/3		218	145	2.08	1.97	0.11	0.85	5.85
30	70–75	33	secondary	177	118	2.04	1.94	0.10	0.84	5.36
29	75–80	33	secondary	216	143	1.92	1.82	0.10	0.80	4.21
28	80–88	34	primary	76	38	1.93	1.74	0.19	0.84	5.64
27	88–101	34	primary	24	12	-	-	-	-	-
26	spot	34	primary	3	2	-	-	-	-	-
25	spot	34	primary	9	5	-	-	-	-	-
Inner ditch: F.4										
51	39–49	18	tertiary	28	14	1.62	1.35	0.27	0.76	3.78
50	49–59	18	tertiary	553	277	1.78	1.75	0.03	0.81	4.23
49	59–65	23	secondary	1822	911	1.82	1.80	0.02	0.85	5.50
48	65–75	23	secondary	2466	1233	1.92	1.90	0.02	0.84	5.16
47	75–85	23	secondary	1697	849	1.99	1.97	0.02	0.84	5.30
46	85–95	23	secondary	2089	1045	1.88	1.87	0.01	0.83	4.78
45	95–104	23	secondary	1605	803	1.75	1.73	0.02	0.73	2.75
44	104–109	35	primary	569	285	1.61	1.54	0.07	0.71	2.47
43	109–102	35	primary	83	42	1.50	1.38	0.12	0.74	2.96
42	119–129	35	primary	2	1	0.69	0.35	0.34	-	-
41	129–141	37	primary	0	0	-	-	-	-	-
2	0–10	31	turf	371	186	1.63	1.57	0.06	0.75	2.97
4	10–20	31	turf	85	43	1.54	1.39	0.15	0.68	2.24
17	spot	36	ols	276	138	1.72	1.65	0.07	0.78	1.02

Table A3.7. *Species diversity data from barrow 41, Wyke Down.*

Sample	Depth (cm)	Context	Fill-type	Molls	mpkg	Shannon	Brillouin	*H'* – HB	Δ_2	Δ_4
Wyke Down Barrow 41 (Chalkpit Field)										
38	50–58	5	tertiary	4	2	-	-	-	-	-
37	80–90	9	secondary	56	28	1.94	1.67	-	0.88	8.54
36	89–98	10	secondary	35	18	1.67	1.42	-	0.76	3.65
35	spot	11	primary	4	2	-	-	-	-	-
53	spot	13	ols	3	2	-	-	-	-	-
52	spot	13	ols	1	1	-	-	-	-	-

Table A3.8. *Species diversity data from barrow 36, Wyke Down.*

Sample	Depth (cm)	Context	Fill-type	Molls	mpkg	Shannon	Brillouin	H' – HB	Δ_2	Δ_4
Wyke Down Barrow 36										
117	32–38	67	tertiary	254	169	1.65	1.59	0.06	0.78	3.58
116	38–46	67	tertiary	487	325	1.61	1.57	0.04	0.73	2.66
115	46–50	68	stabilization	601	4001	1.82	1.79	0.04	0.80	3.95
114	50–60	68	stabilization	1438	957	1.88	1.87	0.02	0.81	4.31
113	60–66	68	stabilization	2350	1567	1.91	1.89	0.14	0.81	4.16
112	66–72	75	secondary	820	410	1.77	1.74	0.03	0.77	3.29
111	72–77	75	secondary	358	239	1.97	1.90	0.07	0.83	4.86
110	77–82	75	secondary	45	30	2.06	1.78	0.29	0.85	6.56
109	82–88	76	U primary	12	8	-	-	-	-	-
108	88–98	76	U primary	1	1	-	-	-	-	-
107	98–108	77	primary	5	3	-	-	-	-	-
106	108–118	77	primary	0	0	-	-	-	-	-
105	118–128	77	primary	0	0	-	-	-	-	-
103	spot	78	primary	8	4	-	-	-	-	-
102	spot	72	ols	1	1	-	-	-	-	-
127	spot	66	treehole	20	10	2.14	1.64	0.50	0.87	10.18

Table A3.9. *Species diversity data from barrow 44, Wyke Down.*

Sample	Depth (cm)	Context	Fill-type	Molls	mpkg	Shannon	Brillouin	H' – HB	Δ_2	Δ_4
Wyke Down Barrow 44										
150	35–45	105	tertiary	309	206	1.68	1.62	0.06	0.75	2.98
149	45–55	105	tertiary	750	500	1.72	1.69	0.03	0.77	3.41
148	55–65	108	secondary	2296	1531	1.85	1.83	0.02	0.80	4.11
147	65–75	108	secondary	1635	1090	2.10	2.07	0.03	0.83	5.02
146	75–85	109	secondary	1268	845	2.16	2.13	0.03	0.84	5.31
145	85–95	109	U primary	79	53	2.26	2.02	0.24	0.87	7.23
144	95–105	110	primary	54	36	2.33	2.01	0.32	0.88	9.08
143	105–115	110	primary	20	15	2.28	1.73	0.55	0.89	13.62
142	115–125	111	primary	5	4	-	-	-	-	-
141	125–135	111	primary	12	6	-	-	-	-	-
140	spot	107	ols	11	6	-	-	-	-	-
132	spot	100	turf	14	7	-	-	-	-	-
134	spot	99	turf	42	21	1.74	1.51	0.23	0.79	4.28

Table A3.10. *Species diversity data from the Down Farm enclosure excavations in 1978 and 2003.*

Sample	Depth (cm)	Context	Fill-type	Molls	mpkg	Shannon	Brillouin	H' – HB	Δ_2	Δ_4
Down Farm Bronze Age enclosure (2003)										
9	0–10	375	turf	26	33	1.73	1.40	0.33	0.77	4.00
1	15–30	376	tertiary old Ap	28	19	2.36	1.89	0.47	0.90	13.00
2	30–40	376	tertiary old Ap	109	73	2.36	2.13	0.23	0.88	7.61
3	40–58	377	secondary	711	474	2.27	2.21	0.06	0.82	4.52
4	58–70	378	'midden'	135	90	2.43	2.23	0.20	0.87	7.31
5	70–85	377	secondary	77	51	2.02	1.82	0.20	0.83	5.10
6	85–100	379	U primary	48	24	1.61	1.40	0.21	0.75	3.31
7	100–115	239/80	primary	18	9	1.51	1.19	0.32	0.74	3.63
8	115–120	380	primary	48	24	2.27	1.94	0.33	0.88	8.64

Sample	Depth (cm)	Context	Fill-type	Molls	mpkg	Shannon	Brillouin	H' – HB	Δ_2	Δ_4
Down Farm Bronze Age enclosure (1978)										
-	0–8	2	tertiary	72	72	1.66	1.58	0.08	0.78	3.75
-	8–18	2	tertiary	57	57	2.10	1.92	0.18	0.83	4.99
-	18–28	2a	tertiary	83	83	2.42	2.01	0.31	0.90	12.00
-	28–36	3	dump	392	392	2.32	2.19	0.13	0.87	6.76
-	36–43		secondary	266	266	2.21	2.12	0.09	0.85	5.92
-	43–50	5	secondary	250	250	2.18	2.10	0.08	0.84	5.29
-	50–61	5	secondary	40	40	2.23	2.01	0.22	0.87	7.30
-	61–72	6	primary	116	116	1.54	1.35	0.19	0.70	2.45
-	72–102	6	primary	159	159	1.57	1.41	0.16	0.71	2.54

Table A3.11. *Species diversity data from the Gussage Cow Down 192 Iron Age ditches and banks.*

Sample	Depth (cm)	Context	Fill-type	Molls	mpkg	Shannon	Brillouin	H' – HB	Δ_2	Δ_4
Gussage Cow Down Iron Age bank and ditch GCD 192										
8	53–63	276	tertiary	78	52	1.74	1.56	0.18	0.75	3.09
7	63–73	256	secondary	51	34	1.79	1.55	0.24	0.77	3.55
6	73–83	256	secondary	247	165	1.66	1.58	0.08	0.73	2.72
5	83–93	256	secondary	441	294	1.83	1.78	0.06	0.78	3.56
4	93–103	256	secondary	452	301	2.01	1.94	0.07	0.80	4.06
3	120–130	270	primary	222	148	1.75	1.67	0.80	0.78	3.79
2	130–140	271	primary	66	44	1.71	1.54	0.17	0.77	3.62
1	140–148	272	primary	22	15	1.56	1.25	0.30	0.73	3.28
237		253	ols	665	333	2.61	2.54	0.07	0.90	9.63

Table A3.12. *Species diversity data from the dry valley trench at Monkton-up-Wimborne.*

Sample	Depth (cm)	Context	Fill-type	Molls	mpkg	Shannon	Brillouin	H' – HB	Δ_2	Δ_4
Monkton-up-Wimborne valley trench (hillwash)										
174	25–35	102	A1	47	31	1.56	1.35	0.21	0.72	2.71
173	35–45	103	bAh	102	68	2.07	1.90	0.17	0.85	6.05
172	45–55	103	bA1	79	53	1.83	1.63	0.20	0.76	3.30
171	55–65	103	bBw	113	75	2.01	1.85	0.16	0.83	5.10
170	65–75	103	bBw	86	57	2.03	1.83	0.20	0.83	5.10

Table A3.13. *Species diversity data from the Knowlton South Circle.*

Sample	Depth (cm)	Context	Fill-type	Molls	mpkg	Shannon	Brillouin	H' – HB	Δ_2	Δ_4
Southern Circle henge ditch and OLS										
48	10–20	3	tertiary	74	74	1.86	1.67	0.19	0.79	4.12
47	20–30	3	tertiary	75	75	1.78	1.60	0.18	0.78	3.89
46	30–40	3	tertiary	98	98	1.82	1.67	0.15	0.79	3.98
45	40–50	3	tertiary	150	150	1.91	1.78	0.13	0.81	4.30
44	50–60	14	tertiary	150	150	2.06	1.91	0.15	0.83	5.03
38	0–10	15	tertiary	322	215	1.80	1.72	0.08	0.77	3.41
37	10–20	15	tertiary	345	230	1.71	1.64	0.07	0.73	2.68
36	20–30	15	tertiary	354	236	1.58	1.52	0.06	0.70	2.34
35	30–40	17	tertiary	637	425	1.75	1.71	0.04	0.77	3.29
34	40–50	17	tertiary	613	409	1.86	1.81	0.05	0.79	3.83
33	50–60	17	tertiary	547	365	1.96	1.90	0.06	0.81	4.26
32	60–65	20	stabilization	703	469	2.12	1.96	0.16	0.82	4.53
31	65–70	20	stabilization	795	530	2.09	2.05	0.04	0.83	5.02
30	70–75	20	stabilization	727	485	2.21	2.16	0.05	0.86	6.15
29	75–80	20	stabilization	846	564	2.16	2.12	0.04	0.85	5.53
28	80–85	20	stabilization	1209	806	2.31	2.27	0.04	0.87	6.64
27	85–90	25	secondary	993	662	2.44	2.39	0.05	0.89	7.85
26	90–95	25	secondary	792	528	2.60	2.54	0.06	0.91	9.73
25	95–105	25	secondary	435	290	2.28	2.18	0.10	0.82	4.53
24	105–115	27	secondary	1902	1268	2.09	2.06	0.03	0.78	3.47
23	115–125	27	secondary	2078	1385	1.83	1.81	0.02	0.69	2.19
22	125–135	27	secondary	2905	1937	2.03	2.01	0.02	0.77	3.33
21	135–145	37	secondary	2044	1363	2.10	2.07	0.03	0.77	3.35
20	145–155	39	secondary	1685	1123	2.10	2.07	0.03	0.75	3.08
19	155–160	43	primary	24	12	2.22	1.73	0.49	0.86	9.22
18	160–170	43	primary	6	3	1.24	0.80	0.44	0.67	4.00
17	170–185	43	primary	7	4	1.23	0.86	0.37	0.69	4.25
16	185–200	43	primary	5	3	1.33	0.82	0.51	0.72	9.00
15	200–215	43	primary	7	4	1.56	1.02	0.52	0.78	9.50
14	210–215	45	soil	23	12	2.37	1.83	0.54	0.80	14.81
39	spot	45	soil	18	26	2.01	1.53	0.48	0.85	8.56
203	spot	45	soil	347	29	1.80	1.67	0.13	0.93	13.11
13	215–230	46	primary	3	2	-	-	-	-	-
43	0–10	46	primary	3	2	-	-	-	-	-
42	30–40	46	primary	1	+	-	-	-	-	-
41	60–70	46	primary	2	1	-	-	-	-	-
200	spot	6	turf	220	22	2.09	1.97	0.12	0.82	4.59
201	spot	26	buried soil	106	9	2.03	1.86	0.07	0.82	4.77
12	17–23	6	turf	14	9	-	-	-	-	-
11	23–28	26	buried soil	6	4	-	-	-	-	-
10	28–33	26		4	3	-	-	-	-	-
Knowlton Circle inner (north) gully										
5	20–27	12	tertiary	391	261	2.09	2.02	0.07	0.83	5.11
4	27–40	24	secondary	191	127	2.14	2.02	0.12	0.85	5.89
3	40–50	24	secondary	149	99	2.15	2.00	0.15	0.84	5.54
2	50–60	24	secondary	173	115	2.01	1.88	0.13	0.81	4.50
1	Spot	1	primary	257	171	2.08	1.97	0.11	0.82	4.77
Knowlton Circle outer (south) gully										
9	24–36	2	tertiary	583	389	2.18	2.11	0.07	0.85	5.69
8	35–50	32	secondary	274	183	2.41	2.28	0.13	0.88	7.68
7	50–62	32	secondary	210	140	2.37	2.23	0.14	0.88	7.30
6	spot	6.1	primary	74	608	1.53	1.36	0.17	0.66	2.06

Appendix 4

Excavation Results by Site

This appendix aims to present all of the available field-derived descriptive data which is either not specifically presented in the main text, and/or augments that text. It is organized on a site-by-site basis from Gussage Cow Down across Down Farm to Wyke and Bottlebush Downs, Monkton-up-Wimborne and finally Knowlton and its environs.

4.1. The Gussage Cow Down monuments (2002)

Long barrow 78

Lithics

Julie Gardiner

One hundred and eighteen flints were recovered, all but 13 of them from layer 290, the buried soil beneath the mound (Table A4.1). This context produced a cache of very fresh flakes and blades that appeared to represent a discrete episode of flint deposition. The material seemed, on discovery, to represent one or more broken up nodules with a further group of flint flakes close by. The former group was collected in one bag with the remaining material from the buried soil between the two phases of mound revetment being individually marked on the trench plan and then being bagged separately. These groups appeared to the excavators to resemble the flint clusters, or 'nests' described from Thickthorn Down long barrows by (Grinsell 1959) (see Chapter 3 above). For purposes of this assessment the material has not been recombined though it is clear that refits are present.

This group of struck flints represents the deliberate deposition of a quantity of knapping debris. The flint is very good-quality, very fresh chalk flint, mottled grey-white in colour with a thick brown rindlike cortex and patinated densely white. It was presumably encountered during construction of the monument. Three or four nodules are probably present and it is clear that the full knapping sequence is not represented. Primary flakes are underrepresented and there are no residual cores. Rather more tertiary flakes would be expected if the full sequence were present and there is a clear absence of possible tool blanks.

No metrical analysis or detailed study of technical attributes has yet been undertaken but simple visual examination indicates that more than 50 per cent of the pieces are blades or of bladelike proportion. The assemblage is certainly Early Neolithic in date and undoubtedly exactly contemporary with construction of the monument. It appears that any useful pieces and cores were removed from the site and a proportion of the resulting waste dumped immediately or very soon after knapping, possibly as a foundation deposit or symbolic act. Similar dumps of fresh knapping debris have been recently recorded from the ditches of the Dorset Cursus excavated in 1984 (Gardiner 1985; 1991, 72–5).

Table A4.1. *The lithic types present at Gussage Cow Down long barrow 78.*

Context	**290**	**290 (nodule)**	**283**	**Unlabelled**	**Total**
Primary	1	3	0	0	4
Secondary	23	24	3	1	51
Tertiary	16	37	7	2	62
Sub-total	40	64	10	3	117
Irregular waste	1	0	0	—	1
Total	41	64	10	3	118
Chip	11	98			109
Weight					
Bladelike	20 (50%)				
Blades	5 (12.5%)				

Soil micromorphology

Helen Lewis

The micromorphology samples taken from long barrows 78 and 294, and the lynchet site 192 on Gussage Cow Down are as follows:

Sample no.	Site	Context no.	Description
249	78	278	turf beneath first mound
250	78	279	lower half of buried soil beneath first mound
253/A–B	78	278/279	complete buried soil profile beneath first mound (second set)
257	78	277	turf at base of first mound just above buried soil
259	294	220/1/2	fill of medieval/later ploughmarks in top of/inner edge of long barrow ditch F.111
260	294	294/5/6	turf lens from primary/secondary fill transition in north ditch (F.111)
267	294	234	turf lens in ditch F.111 at base of context 234
273	192	275	Buried soil under Iron Age bank F.122 (2 sections)
275/1–3	192	253/260/263	transition from bank (F.120) to buried soil beneath (253) (sections 275/1, 275/3 upper & lower)

The micromorphological descriptions of the samples taken from Gussage Cow Down long barrow 78 are as follows (Table A4.2):

Samples 253 A & B: Complete buried soil profile under long barrow 78

Samples 253 A and B are the buried soil profile underneath the mound. They show four main layers/horizons: <1 cm of chalk rock fragments (<1 cm), representing the base of the chalk mound; 5 cm of dark brown buried A horizon (top of buried soil), with partially sorted rock fragments (0.3–1 cm; developing stone line); 4–5 cm of medium brown A horizon (lower A), with unsorted chalk rock fragments (15%, 0.3–2 cm); and 1.5 cm of chalk weathering horizon mixed with some lower A zones (top of chalk rock). The soil horizons are described below.

Upper A horizon

Structure: Horizontally elongated sub-angular blocky (0.3–0.8 × 0.5–1.3 cm) with crumb (<1 cm).

Porosity: 5–10% channels and interpedal cracks (300–1200 × 1000–25,000 μm, accommodated to partially accommodated, straight and zigzag).

Mineral components: Chalk rock fragments; quartz sand; occasional flint fragments, shells, calcite grains and calcitic earthworm granules. Clays are masked by strong organic and sequioxide staining. Groundmass is very dark brown (PPL), very dark reddish-brown (XPL), stipple speckled (where clay zones are visible), occasionally crystallitic.

C:f ratio: 15:85 (very coarse sand <2%, coarse sand 5%, medium sand 5%, fine sand 5%, very fine sand 20%, silt 30%, clay 35%).

Organic components: <5%, black angular and rounded fragments 20–200 μm, rarely >1000 μm, occasionally with structure visible (charcoal).

Table A4.2. *Gussage Cow Down long barrow 78: buried soil micromorphological descriptions (H. Lewis).*

Context	Structure	Porosity	Mineral components	C:f ratio	Organic components	Groundmass	Pedofeatures
B78 [277] soil	Crumb	10–15% channels & cracks	Frequent chalk fragments; flint; shell; quartz sand; earthworm granules; possible bone	20–25:75–80 Silt loam	5–15%, roots & fine black fragments, some charcoal	Medium-dark reddish-brown (PPL & XPL), stipple speckled to crystallitic	Occasional crystallitic fabric as pore infills & ped coatings; micrite coatings and infills; light iron-oxide staining
B78 [277] chalk	Apedal, with crumb microaggregates in pores	5–10% packing pores, channels & cracks	As above; frequent micrite	10–15:85–90 Silt (loam)	<5%, as above	Medium-light reddish-brown (PPL), greyish-yellow (XPL), crystallitic	As above
B78 [278]	Horizontally elongated subangular blocky	As above	As [277] soil; earthworm calcitic granules	15:85 Loam	As above	Very dark brown (PPL), very dark reddish-brown (XPL), stipple speckled	Occasional calcareous inclusions
B78 [279]	Sub-angular blocky & crumb	As above	As above	15:85 Silt loam	5%, as above	Light reddish-brown (PPL), light brown (XPL), mixed stipple speckled & crystallitic	As above; coatings of calcareous fabric; excremental infills

Pedofeatures: Inclusions (1000–2000 μm) of greyish-brown (PPL), light brown (XPL) crystallitic groundmass, with c:f ratio of 5:95 (fine sand 5%, very fine sand 10%, silt 40%, clay 45%). Sand and silt are quartz; there are sand-sized chalk rock inclusions. Organic components are <5%, black rounded fragments <200 μm.

Lower A horizon
Structure: sub-angular blocky and crumb (<1 cm); zones of apedal channel (coalesced peds).
Porosity: 5–10%, channels (100–300 × 300–1000 μm, partially accommodated, straight & curved).
Mineral components Coarse as in upper A. Fine fabric is mixed light reddish-brown and greyish-brown (PPL), light reddish-brown (PPL) and light brown (XPL), stipple speckled and crystallitic groundmass. The darker reddish-brown zones are lightly stained with amorphous organics and iron oxide; the lighter brown zones are lightly stained with amorphous organics, and appear to be slightly more calcareous.
C:f ratio: For both fabrics: 15:85 (very coarse sand <5%, coarse sand 5%, medium sand <5%, fine sand 5%, very fine sand 5%, silt 45% clay 35%).
Organic components: For both fabrics 5%, mostly 'punctuations' and very small (20–150 μm) black angular & rounded fragments.
Pedofeatures: as in upper A; zones (<300 μm) of dark organic fabric infilling channels (earthworm); micrite in pores; rare chambers partially filled with micro-excrements; occasional peds have coatings (<200 μm thick) of calcareous fabric material (2nd main fabric).

Chalk rock surface
Apedal channel (channels 100–300 μm diameter) grey (PPL & XPL) crystallitic groundmass with frequent chalk rock fragments (3000–25,000 μm); soil fabric is primarily silty clay (micrite dominates) with 5–10% very fine sand, representing dissolving chalk rock mixing with soil. Organic components are as in lower A. There are rare iron oxide mottles (<200 μm).

Sample 257: Turf at base of long barrow 78, above buried soil
Four horizons were evident, all with three mixed fabrics. Turf: 4.5–5 cm organic fabric with well-sorted chalk rock fragments (3000–5000 μm, rarely 12,000 μm) making up 10% of the area. Sharp boundary with chalk mound horizon: 2.5 cm, mainly chalk fragments (95% of area), mainly 4000–10,000 μm. Under this is a one cm organic horizon, diagonally oriented, overlying 0.5 cm of further chalk (chalk rock). The two organic horizons comprise similar fabrics, and will be described under 'turf' below, with variations noted therein.

Turf & buried soil
Structure & porosity: sub-angular and angular blocky (4000–8000 μm), 20% porosity, mainly packing pores and channels (100–1000 × 500–5000 μm, straight and curved, partially accommodated and unaccommodated).
Mineral components: mainly quartz grains, chalk, micrite; also flint, shell, calcite grains & earthworm granules.
C:f ratio: 25:75: very coarse sand <5%, coarse sand 5%, medium sand <5%, fine sand 10%, very fine sand 20%, silt 40%, clay 15%. Sandy silt loam.
Organic components: 5%, 'punctuations', small angular black fragments and charcoal (<300 μm).
Groundmass: crystallitic with zones of stipple speckled, and zones of undifferentiated (strongly organic stained) fabrics; main fabric is light reddish-brown (PPL, XPL & RL).
Pedofeatures: 1) silty clay infills (100 μm thick), 80% silt (mainly micrite, occasional quartz), 20% clay; 2) very micritic/chalky fabric in zones <2000 μm; 3) depleted/amorphous organic zones <1000 μm, depleted of micrite, stained with amorphous organic matter (bioturbation).

Chalk mound & chalk rock surface
Structure & porosity: sub-angular blocky and angular blocky peds (2000–4000 μm) between rock fragments. 5–20% porosity, mainly packing pores & channels as in turf horizons.
Mineral components: mainly chalk; other components as in turf horizons, but rare earthworm calcitic granules.
C:f ratio: soil fragments have texture as in turf horizons; rest is 5–10:90–95: very coarse, coarse & medium sand 0%, fine sand 5–10%, very fine sand 10–15%, silt 55%, clay 15%. Silt loam.
Organic components: <5%, small black fragments (<150 μm) and 'punctuations', except in soil peds, where as in turf horizons.
Groundmass: crystallitic (chalky), light grey-brown (PPL & XPL), white-yellow (RL), with zones of stipple speckled and undifferentiated.
Pedofeatures: 1) frequent micrite and clay infills (as in turf horizons), up to 500 μm thick; 2) zone of soil similar to turf horizons, but with frequent chalk rock fragments of medium sand size, frequent dark organic-stained zones, and frequent iron 'mottles' (100–200 μm); porosity 5–10%, channels (100 × 1000–2000 μm, straight, partially accommodated; 2) frequent micrite; 3) zones with strong amorphous organic staining; 4) iron 'mottles' (100–200 μm).

Long barrow 294

Lithics
JULIE GARDINER

Fifteen hundred and one struck flints were recovered from the excavation of long barrow 294 with flint occurring in contexts throughout the depth of the barrow ditch (Table A4.3). However, nearly 77 per cent of the material was recovered either from the primary silts or from a feature (F.118) cut into the primary fills in ditch section F.111.

In terms of raw material, all the flint is very fresh, good-quality chalk flint with a medium thickness of fairly smooth, rind-like white-buff cortex. It was presumably collected from the chalk during construction of the monument. Most of the pieces have a dense white cortication. Flint from the primary silts is mostly extremely fresh and sharp and many pieces carry a concretion of calcium carbonate resulting from the percolation of water through the rubbly chalk fill. Higher up the profile, there is some evidence of slightly rolled and damaged edges and a generally less 'fresh' appearance to the flintwork.

Over *c.* 43% of the total assemblage was recovered from the primary silts ([237], [240] [241]) and consists entirely of waste material from knapping, including many small chips and spalls. No metrical or technological analysis has yet been undertaken but it is clear that a few relatively large nodules are represented. Blades and bladelike flakes occur in considerable numbers and as a high proportion of the total (*c.* 45%) indicating that the assemblage is earlier Neolithic in date. Rapid examination of flake butts suggests the use of hard-hammer technique throughout, with little or no obvious platform preparation. Three cores, all producing bladelike flakes, and a tested nodule are the only core fragments from these contexts and the material appears to be a series of deliberate dumps of knapping debris made either during or very shortly after construction of the monument, a feature seen also in the ditches of the Dorset cursus and long barrow 78 (above).

Feature F.118 [244] was a small pit cut into the primary silts close to the centre of the ditch. The fill of this feature [244] produced 508 struck flints including five cores. The flint is not quite as fresh in appearance as that from the primary silts and an initial impression is that a slightly smaller proportion of the pieces is blades or of bladelike proportions (a rough estimate suggests *c.* 38 per cent). The cores resemble little more than smashed nodule fragments with no systematic platform working, and vary in weight from 156–273 g (average 195 g) suggesting that the original nodules were of considerable size. This group again seems to be a deliberate dump of knapping debris and the position of the feature within the ditch indicates that it was made within a few years of construction of the monument.

Far fewer flints were found in the secondary and tertiary fills of the ditch. Layer [239], immediately above the primary fills in F.111, produced 135 flints (partly from sieving), all waste material including three core fragments, though with some edge damage and calcium carbonate concretion which suggests that some of it, at least, may have originated in the upper part of feature F.118. The core fragments are little more than unsystematically smashed nodules.

Layer [228], the modern ploughsoil horizon, produced 73 struck flints including four cores. Although the raw material is identical to that found in the lower ditch silts, the material is rolled and edge damaged reflecting its position within the ploughsoil. The pieces are generally heavy with rather irregular flakes, with just one example having bladelike proportions, and the cores are small and crudely worked. A single, roughly made, small end-scraper was recovered. Overall this material gives the appearance of being considerably later in date than the monument although the raw material was probably quarried from the bank or ditch.

Table A4.3. *The lithic types present at Gussage Cow Down long barrow 294.*

	F.114					Plough-soil	F.111									
	231	**232**	**233**	**236**	**237**	**228**	**223**	**225**	**226**	**229**	**230**	**239**	**244**	**240**	**241**	**Total**
Primary	0	1	3	15	60	9	3	1	0	0	3	24	104	1	31	255
Secondary	0	14	1	17	62	27	14	4	0	0	2	51	132	2	65	391
Tertiary	5	8	7	24	163	29	6	5	0	1	3	44	224	8	217	744
Sub-total	*5*	*23*	*11*	*56*	*285*	*65*	*23*	*10*	*0*	*1*	*8*	*119*	*460*	*11*	*313*	*1390*
Irreg.	0	0	0	3	12	3	0	0	1	0	0	13	43	0	17	92
Core	0	0	0	0	3	4	0	0	0	0	0	0	5	0	0	12
Core frag.	0	0	0	0	0	0	0	0	0	0	0	3	0	0	0	3
Tested nodule	0	1	0	2	1	0	0	0	0	0	0	0	0	0	0	4
Sub-total	*0*	*1*	*0*	*5*	*16*	*7*	*0*	*0*	*1*	*0*	*0*	*16*	*48*	*0*	*17*	*111*
Ret	0	0	0	0	0	1 scraper	0	0	0	0	0	0	0	0	0	1
Total	**5**	**24**	**11**	**61**	**301**	**73**	**23**	**10**	**1**	**1**	**8**	**135**	**508**	**8**	**330**	**1499**
Chips	0	0	0	0	42	1	2	0	0	0	0	75+	150+	0	160+	430+
Weight (g)	1215	890	218	950	4835	1336	210	172	28	27	30	1074	4784	64	2153	17,986
Burnt (g)	3	64				231					30					
	231	**232**	**233**	**236**	**237**	**228**	**223**	**225**	**226**	**229**	**230**	**239**	**244**	**240**	**241**	
Total flakes/ blades	**5**	**23**	**11**	**56**	**285**	**65**	**23**	**10**	**1**	**8**	**0**	**119**	**460**	**11**	**313**	**1271**
Bladelike	2	4/17.4%	0	12/21.4%	79/27.7%	4/6.2%	5/21.7%	6	1	2	0	30/25.2%	101/22.0%	6	69/22.0%	321/25.3%
Blade	0	2/8.3%	2	3/5.4%	60/21.1%	1/ 1.5%	0	0	0	0	0	17/14.3%	76/16.5%	0	63/20.1%	224/17.6%
																545/42.9%

Soil micromorphology

Helen Lewis

The micromorphological descriptions of the samples taken from Gussage Cow Down long barrow 294 are as follows (Table A4.4):

Sample 259: Fill of ploughmarks on inner edge of long barrow ditch F.111

The fill of the ploughmarks comprises two main horizons: soil over a 1 cm thick very dense orange layer, with charcoal visible and some iron mottling. This zone could be the base of ploughzone (following Jongerius 1983; Lewis 1998b).

Structure: Upper: angular to sub-angular blocky (5000–10,000 μm); lower: dense, coalesced crumb and granular (100 μm).
Porosity: 25%; 5% vughs, in peds, irregular to sub-rounded, <250 um and 3–5 mm; 15-0% channels, <500 μm wide, <3 cm long, accommodated, serrated to weakly serrated, in all orientations.
Mineral components: rounded chalk fragments (10–15%, <10,000 μm), micrite and microsparite in pores, quartz, earthworm granules.
C:f ratio: coarse/fine ratio: <20/80; coarse fraction: 3% coarse, 7% medium and 10% fine quartz sand, sub-rounded to sub-angular, 100–750 μm; fine fraction: 20–30% very fine quartz sand, 50–100 μm, sub-rounded; <10% silt; 30% micrite; <10% dusty clay; pale golden brown (CPL), brown (PPL).
Organic components: 5–10+%, mostly charcoal.
Groundmass: calcareous, crystallitic groundmass, stipple speckled in some soil zones; reddish-brown to orange (PPL), reddish-brown (XPL), orange (RL).
Pedofeatures: 1) iron mottles <300 μm); 2) organic staining in some zones; 3) micrite and microsparite growth in pores; 4) mite excrements.

Sample 260 (turf lens in long barrow ditch F.111)

Two units: 7.5 cm of turf and chalk rock fragments, overlying 0.5 cm of chalk fill. Both show frequent chalk rock fragments, with sorting and diagonal bedding in the turf horizon. In the upper 1–1.5 cm chalk (3000–10,000 μm sized fragments) comprises 70% of the area; in the following 1–1.5 cm chalk fragments <3000 μm in size make up 60% of the area; in the rest of the turf horizon chalk fragments are mainly <5000 μm, but with occasional 10,000 μm sized fragments, and make up 40–50% of the area. In the chalk fill layer, chalk rock fragments are 95% of the area; these are <5000 μm, and appear to be forming a stone line.

Turf

Structure & porosity: crumb and sub-angular blocky (3000–5000 μm); 20–25% porosity in upper part (50–300 × 700–1500 μm, rarely up to 5000 μm, straight & curved, partially accommodated), becoming 15% with depth (channels as above).
Mineral components: chalk, quartz and micrite frequent, occasional shells, calcitic earthworm granules, calcite grains and flint rock fragments.
C:f ratio: 15–20:80–85: very coarse sand 5%, coarse sand 5%, medium sand <5%, fine sand 5%, very fine sand 5–10%, silt 30%, clay 45% (silty) clay.
Organic components: 5%, <20 μm angular & rounded black fragments and 'punctuations', rare <100 μm charcoal, rare <200 μm brown (PPL) roots.
Groundmass: greyish-brown (PPL & XPL), light yellow (RL), crystallitic fine groundmass; open porphyric-related distribution.
Pedofeatures: 1) frequent clay coatings, light greyish-brown (PPL), light brown (XPL), yellow (RL), slightly 'dusty', micrite rich, comprising 80% clay: 20% silt (micrite & 'punctuations'); 2) chalky zones (as in 275/3), occasionally with 5–10% charcoal, linear, 50 × 200 μm; 3) micrite coatings and infills; 4) iron-stained 'mottles' <250 μm; 5) channels infilled with mite excrements (ovoid & round, 100 μm); 6) <250 μm zones of brown to very dark brown (PPL & XPL) earthworm sorted fabric, with only very rare very fine sand, rest strongly organic-stained clay and silt.

Chalk fill

Structure & porosity: crumb (<500 μm), 10–15% porosity, mostly packing pores and cracks & channels (50–300 × 1000–3000 μm, accommodated and partially accommodated, straight).

Table A4.4. *Gussage Cow Down long barrow 294: micromorphological descriptions of the ditch fills (H. Lewis).*

Context	Structure	Porosity	Mineral components	C:f ratio	Organic components	Groundmass	Pedofeatures
[220] plough marks in upper fill	Apedal crack	10–20% cracks & channels	Chalk, micrite & microsparite, quartz, calcite	10:90 (Silty) clay	5–10%, charcoal	Reddish-brown to orange (PPL), reddish-brown (XPL), orange (RL); crystallitic	Iron mottles; organic staining; micrite & microsparite; mite excrements
[221] & [222] (as above)	Crumb & sub-angular blocky	As above	Quartz, chalk, flint, micrite, calcite, shell	As above	5%, charcoal, roots, sclerotia	Red (PPL), reddish-orange (XPL), orange (RL); stipple speckled	Zones of crystallitic & calcareous fabric; micrite coatings; iron nodules; iron staining; 'dusty' clay coatings
[294]/[295]/[296] turf lens	Crumb & sub-angular blocky	15–25% channels	Chalk, quartz, micrite, shell, calcite, flint	15–20:80–85 (Silty) clay	5% 'punctuations', rare charcoal & roots	Greyish-brown (PPL & XPL), light yellow (RL); crystallitic	'Dusty' clay coatings; calcareous zones; micrite coatings; iron staining; mite excrements
[294]/[295]/[296] chalk fill	Crumb	10–15% packing pores & cracks	As above	15:85 (Sandy) silt loam	<2% 'punctuations'	Grey (PPL & XPL), white-yellow (RL), crystallitic	Micritic clay coatings & infills; micrite coatings; iron staining

Mineral components: As in turf layer, but predominantly chalk and micrite, with occasional flint and shell.
C:f ratio: 15:85: very coarse sand 5%, coarse sand <5%, medium sand <5%, fine sand 5%, very fine sand 5–10%, silt 50–55%, clay 20%. (Sandy) silt loam.
Organic components: <2%, mainly 'punctuations in crumb peds and occasional <100 μm angular black fragments throughout.
Groundmass: grey (PPL & XPL), white-yellow (RL), crystallitic.
Pedofeatures: 1) infills and coatings of <100 μm thick micrite-rich clay; 2) micrite coatings on pores; 3) iron oxide on some peds and rock fragments.

Sample 267 (turf lens in ditch F.111 at base of context 234)

One horizon, 30–40% rock fragments (chalk and flint, 3000–15,000 μm, angular and sub-angular), unsorted.

Structure & porosity: crumb (2000–3000 μm), occasional sub-angular blocky (3000–4000 μm); 10–30% porosity, mainly packing pores, cracks and channels (50 × 1000 and 100–500 × 1000–5000 μm, unaccommodated and partially accommodated, straight).
Mineral components: chalk and flint rock fragments, quartz grains, calcite grains, micrite, shell.
C:f ratio: 20:80: very coarse sand <5%, coarse sand 5%, medium sand 5%, fine sand 5–10%, very fine sand 10%, silt 40%, clay 30%. Clay loam.
Organic components: 5%, mainly 'punctuations' and occasional charcoal (<500 μm, rarely 1000 μm); rare brown root fragments (100 × 1500 μm) in pores.
Groundmass: crystallitic, with some stipple speckled zones; medium greyish-brown (PPL & XPL), yellow to orange-brown (RL). Open porphyric-related distribution.
Pedofeatures: 1) occasional micritic silty clay coatings (50–250 μm thick) on peds, with 5% quartz very fine sand and coarse silt; 50% fine silt, 45% clay; 2) very rare tiny orange-yellow (PPL & XPL) clay fragments, rounded and clean, 100 μm; 3) occasional chalky fabric zones; 4) occasional zones with strong organic staining (bioturbation); 5) micrite coatings and infills, rarely root replacement infills; 6) iron stained zones (50–300 μm); mottles.

Iron Age lynchets (site 192)

Lithics

Julie Gardiner

A small quantity of struck flint was recovered, with only 31 flakes from seven contexts recorded, all unretouched (Table A4.5). The raw material is good-quality chalk flint, in a fairly fresh state with a dense white patination. The pieces are hard-hammer struck, generally fairly large and thick there are some blade-like examples present. They suggest the *ad hoc* use and discard of good flint, fortuitously recovered from the ditch.

Table A4.5. *The lithic types present at Gussage Cow Down site 192.*

Contexts	252	253	254	255	256	257	276	Total
Primary	0	0	0	1	3	0	0	4
Secondary	1	0	1	3	3	2	1	11
Tertiary	2	1	1	9	2	1	1	17
Total	**3**	**1**	**2**	**13**	**8**	**3**	**2**	**31**
Blade-like	2		1		1	1		16%
Weight (g)	41	4	19	242	176	112	51	645

Soil micromorphology

Helen Lewis

The micromorphological descriptions of the samples taken from Gussage Cow Down lynchet system 192 are as follows (Table A4.6):

Sample 273: Buried soil under Iron Age bank F.122

Both samples have the same fabric; one has a potsherd in it. Both have poorly sorted angular to sub-angular chalk rock fragments making up 20–25% of the area.

Structure & porosity: sub-angular and angular blocky (3000–9000 μm), accommodated, well developed. 10–15% porosity: cracks & channels (50–500 × 1200–5000 μm, partially accommodated and accommodated, straight), occasional chambers (10,000 μm).
Mineral components: quartz grains, chalk fragments and micrite dominate; also occasional shells, flint, calcite grains & earthworm granules; clays are masked by light sesquioxide staining.
C:f ratio: 15:85: very coarse, coarse & medium sand 5%, fine sand 10%, very fine sand 15%, silt 40%, clay 30%. Clay loam.
Organic components: <5–10%, mainly 'punctuations' and small black fragments (<100 μm), as well as charcoal (<250 μm); occasional root remains, rare spores & fungal sclerotia.
Groundmass: light brown (PPL & XPL), orange-yellow (RL), crystallitic; open porphyric-related distribution.
Pedofeatures: 1) micritic clay coatings (50–100 μm thick) on peds and pore edges; 2) pot fragment: red to very dark reddish-brown (PPL & XPL), strong red (RL), with quartz sand and sand-sized flint rock fragments, as well as apparent 'grog' as temper. Overall texture is coarse & medium sand 5%, fine sand 20%, very fine sand 15%, silt 10%, clay 50%. There are frequent zones of red clay (up to 1000 μm); possible organic components are <5%, 'punctuations'. Possible 'grog' fragments are sub-angular, very dark red to black (PPL & XPL), very dark red (RL), with 200–500 μm quartz grains in a clay matrix; 3) In the thin section without pot, there is one ped (700 μm in size) with 30% organic components, comprising fine angular charcoal fragments and 'punctuations', all <200 μm.; 4) micrite crystals lining pores; 5) iron 'mottles' <100 μm in size; 6) mite excrements (<100 μm, ovoid) in some pores.

Sample 275/1 (bank to buried soil transition)

This sample has three layers: 1.5–2 cm of 'melted' chalk and soil, over 3 cm of soil fill with frequent chalk rock fragments (30–40%, <0.5 cm), over 2 cm of 'melted' chalk and soil, with occasional flint (1–1.5 cm).

Structure: chalky fills have apedal crack structure; soil fill has crumb and sub-angular blocky structure (<4000 μm).
Porosity: 5–15% in chalky fills (higher in lower chalky fill), 20% in soil fill: cracks (100–250 × 1000–7000 μm, straight, unaccommodated and accommodated), channels in soil and lower chalky fill (150–500 × 2000–5000 μm, partially accommodated, straight) and occasional chambers in soil fill (5000 μm).
Mineral components: chalk, quartz sand, occasional flint, rare glauconite, occasional calcitic earthworm granule.
C:f ratio: where soil aggregates in all layers: 5:95: very coarse, coarse & medium sand 0%, fine sand 5%, very fine sand 15%, silt 70%, clay 10%.
Organic components: <5–5% in chalky fills, black angular fragments (<100 μm); <5% in soil fill, as in others but with occasional obvious charcoal (<200 μm).
Groundmass: mixed grey and light reddish-brown (PPL & XPL), creamy yellow and orange (RL); crystallitic, particularly grey and micritic in chalky fills.

Table A4.6. *Gussage Cow Down Iron Age bank and ditch: micromorphological descriptions (H. Lewis).*

Context	Structure	Porosity	Mineral components	C:f ratio	Organic comp	Groundmass	Pedofeatures
[253] bank	Apedal crack & crumb	5–15% channels	Chalk, quartz, flint, glauconite, calcite	5:95 Silt	<5–5% 'punctuations', some charcoal	Grey & light reddish-brown (PPL & XPL), creamy yellow & orange (RL); crystallitic	'Dusty' clay coatings; calcareous coatings & infills; micrite coatings; amorphous iron staining; subrounded orange clay fragments
[260] buried soil	Crumb & sub-angular blocky	10–20% cracks & channels	Quartz, shell, calcite, chalk, flint, micrite	15–85% Loam	As above	Light reddish-brown (PPL & XPL), orange-brown (RL); stipple speckled & crystallitic	Coatings & hypocoatings; red clay fragments; calcareous zones; micrite coatings; iron staining; mite excrements
[263] buried soil	As above	As above	As above	As above	5%, as above, also roots & sclerotia	As above	As above
[275]	Sub-angular & angular blocky	10–15% cracks & channels	Quartz, chalk, micrite, shell, flint, calcite	15:85 Clay loam	<5–10% 'punctuations', charcoal, roots, spores, sclerotia	Light brown (PPL & XPL), orange-yellow (RL); crystallitic	Micritic clay coatings; pot fragment; charcoal & organic inclusion rich soil fragment; micrite coatings; mite excrements

Pedofeatures: In upper chalk layer: 1) clay coatings, 'dusty' brown (PPL), orange-brown (XPL), 100–200 μm; 2) chalky coatings and infills up to 200 μm thick; 3) occasional infills of organic soil similar to soil layer fabric; 4) micrite in pores; 5) occasional amorphous iron staining on chalk rock fragments. In soil layer: 1) coatings of material similar to fabric of upper chalk layer, up to 250 μm thick; 2) zones of chalky fabric from upper chalk layer; 3) rare orange (PPL & XPL) low birefringent clay fragments, subrounded, 100–150 μm; 4) micrite coatings. In lower chalk layer: as in upper chalk layer, but frequent micrite-rich clay coatings and frequent micrite in pores.

Sample 275/3: Upper and lower buried soil

One main horizon, but with variations over depth, mainly with respect to chalk rock fragment inclusions: upper thin section has 20–30% chalk fragments which show size sorting over depth (increasing from 0.3–0.6 cm in the upper 3 cm to 0.5–1.5 cm in the lower 3 cm of the thin section). Also, the soil becomes increasingly brown in colour with depth.

Structure and porosity: Upper: crumb and sub-angular blocky (4000 μm), with 10–20% porosity (interpedal cracks & channels 100–300 × 700-6000 μm, curved and zigzag, unaccommodated and partially accommodated; some straight channels with depth).

Mineral components: quartz grains, earthworm granules, shell fragments, chalk and flint rock fragments, micrite in groundmass and as pedofeatures, rare linear fragments of red (PPL & XPL) clay (see pedofeatures).

C:f ratio: 15:85%: very coarse & coarse sand <5%, medium sand 5%, fine sand 5%, very fine sand 15%, silt 40%, clay 35%.

Organic components: <5–5%, rare charcoal with structure (250–1000 μm), 'punctuations' and small angular black fragments (20–250 μm), rare orange and brown (PPL) very fine root fragments (<100 μm). With depth organics are 5%, and include occasional larger root fragments (200 × 1000 μm) and rare fungal sclerotia.

Groundmass: mixed zones of stipple speckled and crystallitic, very light reddish-brown (PPL & XPL), becoming light reddish-brown with depth (PPL), orange-brown (RL), open porphyric-related distribution.

Pedofeatures: 1) subtle coatings and hypocoatings on peds of material that is the same as the groundmass, 20–100 μm thick, but with finer texture (5% very fine sand/coarse silt, 25% fine silt; 70% clay). 2) red clay fragments (PPL & XPL), 20 × 200 μm, clean but non-laminated, iron stained with low birefringence; 3) occasional zones with chalky greyish-brown (PPL & XPL), yellow (RL), crystallitic fabric (only seen in upper thin section); 4) micrite on peds and in pores; 5) light iron-oxide staining on groundmass; 6) occasional mite excrements in channels (100–150 μm, ovoid and round).

4.2. Fir Tree Field shaft and associated features (1992–96)

Bone artefacts from the associated pits

Martin Green

A small number of bone artefacts were found in the pits located outside the shaft, as follows (Fig. A4.1):

1. F.2, L5: Pin or awl (100 × 13 mm) made from a split rib from a medium to large mammal. End broken and partly weathered but some polish still extant at tip and sides.

2. F.3, L5: Pin (78 × 8 mm) made from part of a shaft of a large mammal bone. Worked to a fine point. Large areas of polish still present over much of the surface.

3. F.4, L5: Tip of weathered bone pin (37 × 5 mm). Few traces of polish remaining.

4. F.8, L4: Small bone pin or awl (54 × 8 mm) made from a metapodial of a medium sized mammal. Some remains of polish remain.

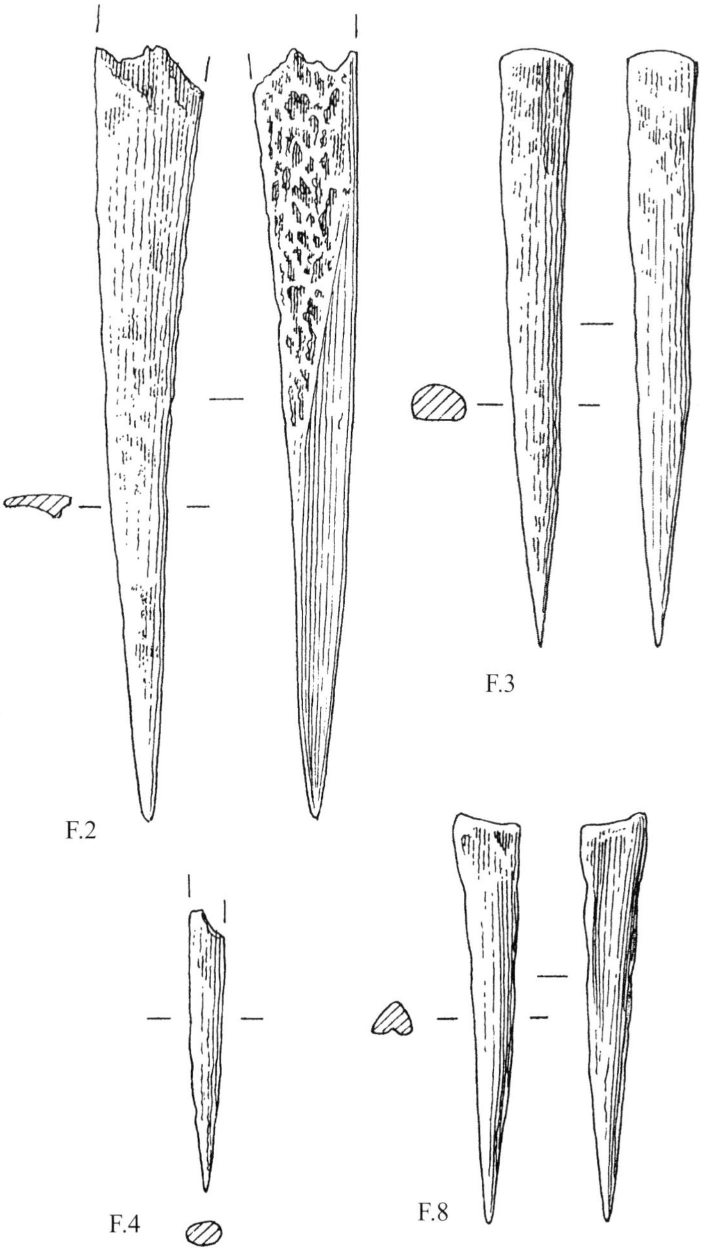

Figure A4.1. *Fir Tree Field Shaft: bone pins and awl. Actual size (N. Griffiths).*

Not illustrated:

5. F.8, L4: Cow rib bone 'knife' (184 × 28 mm). The tip of the rib is sharpened on both sides to form a rounded end for about 75 mm. The remaining portion is very weathered and is probably incomplete.

Lithics from the shaft

Martin Green

The total assemblage from the shaft consists of over 6800 artefacts from a series of layers (Table A4.7; Figs. A4.2–A4.5) extensively radiocarbon dated from the later fifth to later third millennia BC. With the exception of one obvious outlier date, OxA-8008, a coherent set of results relating to their stratigraphic relationships was obtained (Allen & Green 1998) (Fig. 3.14). The 2.8 m deep weathering cone produced 10 dates spanning this sequence with a further seven dates in the late fifth millennium BC obtained from the fill of the main shaft. The seven dates, covering about 4 m of rubble from 7 to 3.05 m in depth, span fewer than 500 radiocarbon years and the probability distributions indicate a span of 200–300 years is more likely. They are very late, or transitional, Mesolithic in date.

Layer 2 (2a & b)

This final silting horizon produced 141 sherds of Beaker pottery, four sherds of Early Bronze age urn and 20 sherds of Peterborough Ware (see Cleal below). A radiocarbon date of 2460–2040 cal. BC (OxA-7985) was obtained from a pig femur found near the base of the layer.

Over 900 lithic artefacts were retrieved from the layer which included a retouched, utilized and tool (RUT) component of 3.8%. Some residual material is likely to be present as suggested by the finds of a chisel arrow-

Table A4.7. *Fir Tree Field (FTF 92): flints from the shaft and roundhouse.*

Feature	Layer	Cores	Core frags.	Flakes	Chips	Retouched	Utilized	Scrapers	Other tools
Shaft									
F.1	2	22	14	458	389	3	1	23	b. & t. arrow; 2 ch. arrow; 3 borers; 2 knives
F.1	4	34	2	611	387	6	1	9	ch. arrow; disciodal knife denticulate; 3 hammers, m-intermediate; 2 borers; 2 axe fragments (1 part-polished); chopper (tabular flint slab); 2 obl. arrows
F.1	5	80	37	852	495	10	4	9	3 saws; obl. arrow; 2 borers; hammer; comb. tool
F.1	6	41	19	277	288	-	2	2	axe roughout; polished axe frag.
F.1	6a	21	12	204	276	1	-	1	polished axe frag.; fabricator; 3 microliths
F.1	6b	6	5	305	281	4	-	3	-
F.1	7	20	6	202	196	-	-	-	8 microliths, m- intermediate
F.1	7a	16	4	167	144	-	-	-	m-intermediate
F.1	6/7a	19	4	190	141	-	-	2	-
F.1	7/7a	9	5	77	44	1	-	1	microlith, m/burin
F.1	8	-	1	20	9	-	-	-	-
F.1	primary rubble	33	4	202	35	-	-	-	borer, hammer
Roundhouse									
PH 1	-	-	1	1	-	-	-	-	-
PH 2	-	-	1	-	-	-	-	-	-
PH 4	-	-	1	-	-	-	-	-	-
PH 6	-	-	2	-	-	-	-	-	-
PH 7	-	-	2	1	-	-	-	-	-

head and Peterborough sherds. However, a large part of the assemblage conforms to Beaker characteristics including a barbed and tanged arrowhead, a fragment of an unfinished example broken during preparation of a notch for a barb, two knives and a neatly bifacially worked borer. Amongst the 23 scrapers were nine of well-made 'thumb' type and a further eight roughly made on small, thick and irregular flakes of the form described from the nearby Beaker pit F.4 (Figs. 3.16 & A4.5).

Layer 4 (separated from layer 2 by a sterile rubble layer 3)
This layer, possibly representing plough washed material, produced 17 sherds of Beaker, ten of Peterborough Ware and a single Plain bowl sherd. A radiocarbon date of 3310–2920 cal. BC (OxA-8006) was obtained from a sheep tibia found towards the base of the layer. Statistically indistinguishable from the two dates obtained from the underlying layer 5, the bone may have been disturbed from the underlying layer.

Almost 1100 lithic artefacts were forthcoming from this layer and as suggested by the pottery, may be a partly mixed assemblage. There was an RUT percentage of 2.8 which included three scrapers of the 'thumb' type already described. Characteristic later Neolithic forms include two oblique arrowheads, a chisel arrowhead and a Y-shaped axe (Fig. A4.2, nos. 7, 9 & 10). A fine edge polished knife (Fig. A4.2, no. 8) is also a form known from third-millennium BC contexts. Two axe fragments included a heavily reworked polished portion and the cutting edge of a roughly chipped axe with abrasion and shattering caused by use. An unusual find recovered at the base of the layer was a large tabular slab (160 × 115 × 22 mm), bifacially flaked along one side, end and part way along the other to a pear-shaped outline. Much bruising is apparent along its long flaked edge and end suggesting it was used as a chopping implement. Three hammerstones, with the type of bruising most likely produced from pounding rather than flaking flint, were also recovered.

Layer 5
This dark organic-rich layer represents a stabilization horizon of some duration, with radiocarbon dates of 3350–3030 cal. BC (OxA-7986), 3310–2910 cal. BC (OxA-8007) and 2570–2300 cal. BC (OxA-8008) which can probably be disregarded as it is the only one of six dates to be out of sequence in the whole shaft. Corroborative support for the remaining two dates was provided by an accompanying ceramic assemblage of 50 Peterborough Ware sherds and three sherds of Plain bowl. Almost 1500 flints were retrieved from this layer

including a 2% RUT component. Significant amongst these were an oblique arrowhead (Fig. A4.2, no. 6), in light of the dates an early association for this form, two serrated blades and a serrated flake. The nine scrapers recovered consisted of two broken, one 'thumb', a steeply retouched core-scraper with the remainder of standard convex type. An unusual implement was the combination flake tool (Fig. A4.5, no. 5).

Layer 6

This silt-rich layer produced seven sherds of Peterborough Ware and an assemblage of 631 flints with a very small RUT element of under 1%. However, amongst these were two axes (Fig. A4.5, nos. 2 & 4). The first was a small unfinished chipped example unusually made from a piece of tabular flint and the other was a part polished fragment which had been reworked with a tapering butt only to break a second time. A small core with clear deliberate scoring on an area of cortex was an unusual find (Fig. A4.4).

Layer 6a

This silt-rich layer produced a single sherd of Plain bowl associated with an *in situ* hearth which gave a radiocarbon date of 4240–3980 cal. BC (OxA-7981). An assemblage of 520 worked flints were retrieved from this layer which included a 1.3% RUT element. However, one of these was a butt end of a polished ?flint axe (Fig. A4.5, no. 3) which currently is the earliest dated polished axe from the UK mainland (D. Field pers. comm.). Amongst the remainder was a bifacially flaked fabricator of lentoid section with some remaining frost pitting surviving from the original nodule on the dorsal surface, a scraper made from a cortical flake struck from a nodule of gravel flint and three microliths.

The deeply patinated microliths were probably residual finds relating to the Mesolithic activity recorded in layer 7 and below. The three microliths consisted of:

1. anciently broken 24 mm rod blunted down both sides;
2. anciently broken 16 mm rod retaining both bulb and striking platform and blunted down both sides;
3. 24 mm ?rod blunted down one side and trimmed along one-third of the opposite side. Trace of bulb remaining and tip broken anciently.

Layer 6b

Although producing no pottery, this layer did give radiocarbon determinations of the very early fourth millennium BC (3980–3710 cal. BC (OxA-8009); 4050–3790 cal. BC (OxA-8010)). An assemblage of 604 worked flints included an RUT element of a little over 1%. Four of these were retouched flakes with the remaining three scrapers. These consisted of a large crudely retouched end scraper of 90 × 45 mm, an equally large convex scraper of 87 × 50 mm, and a smaller convex scraper made on a cortical flake from a nodule of gravel flint very similar to the example found in layer 6a above.

Layer 7

This layer produced no pottery but there were two radiocarbon dates of 4340–3990 cal. BC (OxA-8000) and 4240–3990 cal. BC (OxA-7987) obtained from two bones from a wild animal assemblage (see Maltby below). An assemblage of 433 lithic artefacts came from this layer with an RUT component of ten artefacts or just over 2%. This component consisted almost entirely of microlithic pieces — eight microliths and a micro-intermediate (Fig. A4.3). The remaining artefact, classified as a tool in this instance was a tranchet axe-sharpening flake (Fig. A4.2, no. 1) found at the layer 7/7a interface. Seven of the microliths were found in a tight cluster, probably originally representing a hafted tool or weapon, and are described in detail below.

The remaining microlithic pieces are described as follows:

8. tip portion of a narrow 1 mm rod blunted down one side and broken anciently;
9. bladelet/micro-intermediate 29 mm notched near tip on right side probably for micro-burin removal.

In light of the wild faunal assemblage from this layer, the microlithic component of the RUT element, including the likely hafted group, this layer is clearly Mesolithic in date. The assemblage currently represents the latest dated Mesolithic activity recorded in England or Wales (Barton & Roberts 2004).

Layer 7a

Of the 332 lithic artefacts recovered from this layer only a single retouched piece was found. It is an unfinished microlith, 29 mm in length, partly blunted down both sides at the bulbar end.

Layer 6–7a

Owing to a partial collapse of the section face some material from these two layers became mixed. 356 lithic artefacts were recovered which included only two retouched pieces, both convex scrapers.

Layers 7–7a

Owing to a second partial section collapse, some material from these two Mesolithic layers also became mixed. 139 lithic artefacts were recovered which included three microlithic pieces and an unusual scraper, as follows:

1. tip of bladelet 12 mm obliquely trimmed across end;
2. bulbar micro-burin 14 mm with notch on right edge;
3. portion of narrow rod 18 mm blunted down one side and broken in past at both ends.
4. 'Scraper' made on a short thick flake, 36 × 40 × 17 mm, with exaggerated bulb. Heavily battered right hand side and end. Some battering of striking platform and marginal battering of other side.

Layer 8

Layer 8 was a minor rubble lens located just below the weathering cone and sandwiched between two organic-rich lenses. A date of 4340–4040 BC (OxA-8011) was obtained from charcoal. Only 30 lithic artefacts were found in this layer and did not include an RUT element.

Primary rubble fills

276 lithic artefacts were recovered from the excavated rubble fills of the shaft from depths of between 3.4 and 9 m. Six late fifth-millennium BC dates were obtained from these levels including two obtained from roe deer skeletons that were presumably pit fall victims. However the skeleton found at a depth of 5 m had a large cortical flake lying on some of the bones indicating that there was some associated human activity, perhaps scavenging taking place. Analysis of the flakes reveals that about two-thirds of the worked flint from these levels was produced from fresh nodules which had either fallen out of the walls of the shaft or had perhaps been dug out. Much evidence for nodule sampling was present and included three minimally flaked nodules with single conjoining flakes for each. One of these was retrieved at a depth of 9 m, the lowest point from which any trace of human activity was recovered. Only two tools were found, a hammerstone and a crude point or borer.

Catalogue of illustrated flints (Fig. A4.2)

1. L7/7a interface: Tranchet axe-sharpening flake. The pointed 'pick-like' end has been skilfully removed by a transverse blow which would have produced a successful cutting edge for the axe. Associated radiocarbon date of 4340–3990 BC (OxA-8000).

2. L6: Small unfinished chipped axe made from a piece of tabular flint. Both ends retain portions of the natural surfaces of the original tabular broken slab. The natural surface at the broader end has been used as a platform to reduce this part to a cutting edge but remains unfinished probably due to a flaw in the flint that was revealed at this point.

3. L6a: Broken pointed butt end of a polished axe of slightly asymmetrical outline. The material is an off-white coarse-grained cherty ?flint with traces of fossil inclusions which is likely to have been imported. Remains of two deep flake scars or flaws in the original material remain on the right side. Associated with a radiocarbon date of 4240–3980 cal. BC (OxA-7981) this is at present the earliest date polished axe from the UK mainland (D. Field pers. comm.).

4. L6: Portion of a polished flint axe re-chipped to form a tapering butt. The re-flaking has removed all traces of polish on one side with the other retaining traces of the original flake scars within remaining polished areas.

5. L5: Combination flake tool consisting of a partly serrated left-hand edge, two deeply bifacially worked notches and a point/borer end.

6. L5: Oblique arrowhead. Clearly associated with Peterborough Ware and two radiocarbon dates within the late fourth millennium BC. This is an early dated example of its type.

7. L4: Oblique arrowhead. The right side retaining part of the ventral surface of the original flake including a major part of the large bulbar scar. Although edge worked around the whole perimeter of the dorsal surface this arrowhead clearly belongs to the oblique class.

8. L4: Large discoidal edge polished knife. Bifacially worked from a large flat flake which retains its striking platform but with its bulb removed by invasive flaking. This form of knife is often associated with third-millennium BC contexts (Smith 1965, fig. 48, F.152).

9. L4: Oblique arrowhead. Steeply blunted along one side with marginal retouch on the opposite edge. The ventral surface is unretouched and retains both bulb and striking platform. This does not appear to be unfinished and perhaps represents an example of a 'symbolic arrowhead' as discussed by Brown (in Brown 1991, 130 & fig. 6.5) regarding similar forms found at Wyke Down 1 henge.

10. L4: 'Y'-shaped axe made from a large flake with parts of the ventral surface surviving on the right side. A form common in the later Neolithic surface industries of the area (Green 2000, fig. 41) and elsewhere in flint bearing regions (cf. Smith 1965, fig. 48, no. 155; Edmonds 1995, fig. 65).

Description of microlith group in layer 7 (Fig. A4.3)

Seven microliths were found in a group during excavation near the base of layer 7 and close to the interface with layer 7a. From the very tight position and similar dappled grey patina, it seems likely that this group entered the shaft as a hafted tool or weapon. A radiocarbon date was obtained from a nearby red deer bone of 4240–3990 cal. BC (OxA-7987). This group at present represents the latest dated microliths from England and Wales (Barton & Roberts 2004, 347).

1. Narrow rod, 32 × 4 × 3 mm. Blunted completely down the left side and about 75% of the right side. Anciently broken at distal end.

2. Narrow rod, 29 × 4 × 3 mm. Steeply blunted down both sides and anciently broken at distal end.

3. Narrow rod, 32 × 4 × 3 mm. Steeply blunted down both sides.

4. Narrow rod, 28 × 5 × 3 mm. Steeply blunted down both sides.

5. ?Rod, 23 × 5 × 2 mm. Blunted down left side and anciently broken at bulbar end.

6. ?Rod, 27 × 6 × 2 mm. Blunted down left side.

7. Tip of narrow rod, 15 × 3 × 2 mm. Steeply blunted down both surviving sides. Anciently broken at distal end.

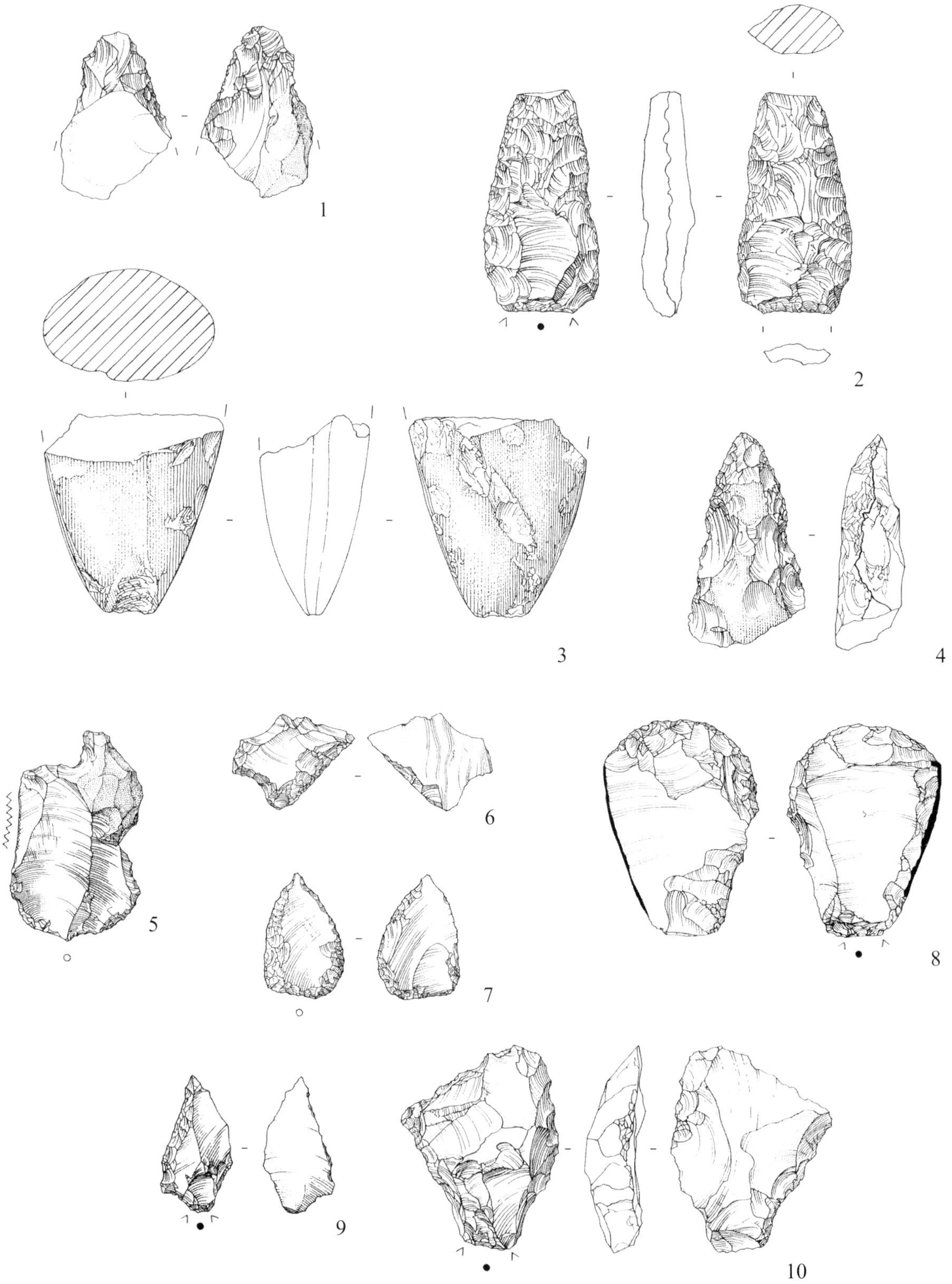

Figure A4.2. *Fir Tree Field shaft: flint tools 1–10. Scale 1:2 (J. Richards).*

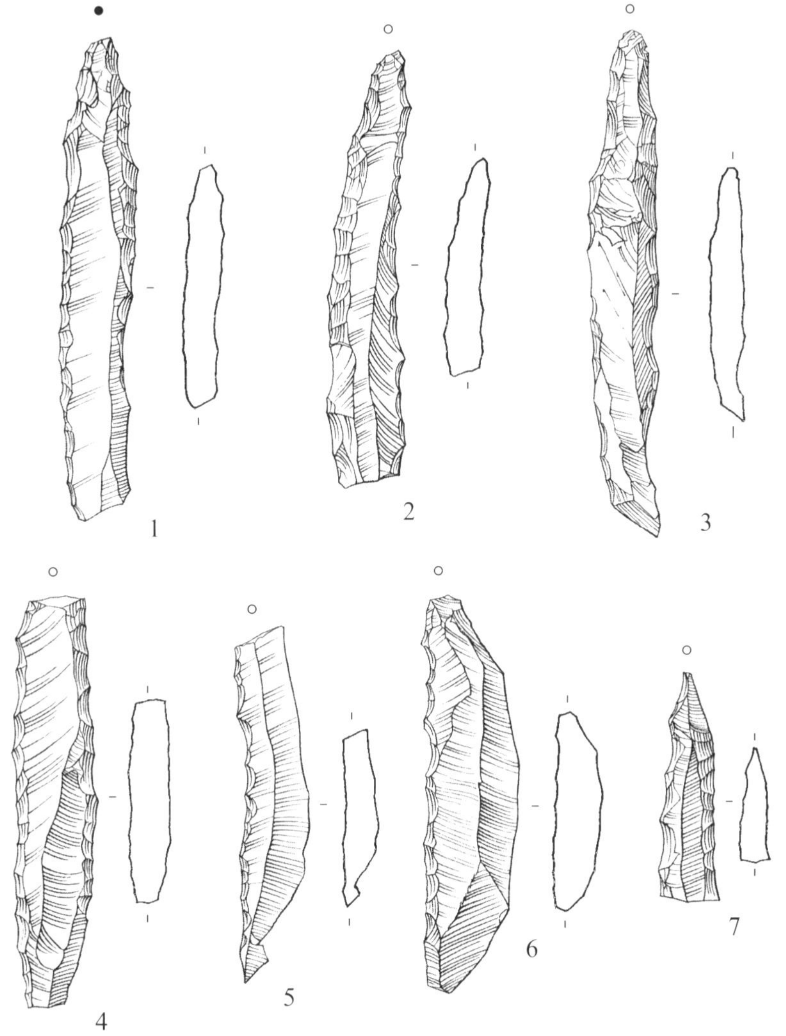

Figure A4.3. *Fir Tree Field shaft: flint microlith group from layer 7, nos. 1–7. Scale 2:1 (outlines 1:1) (J. Richards).*

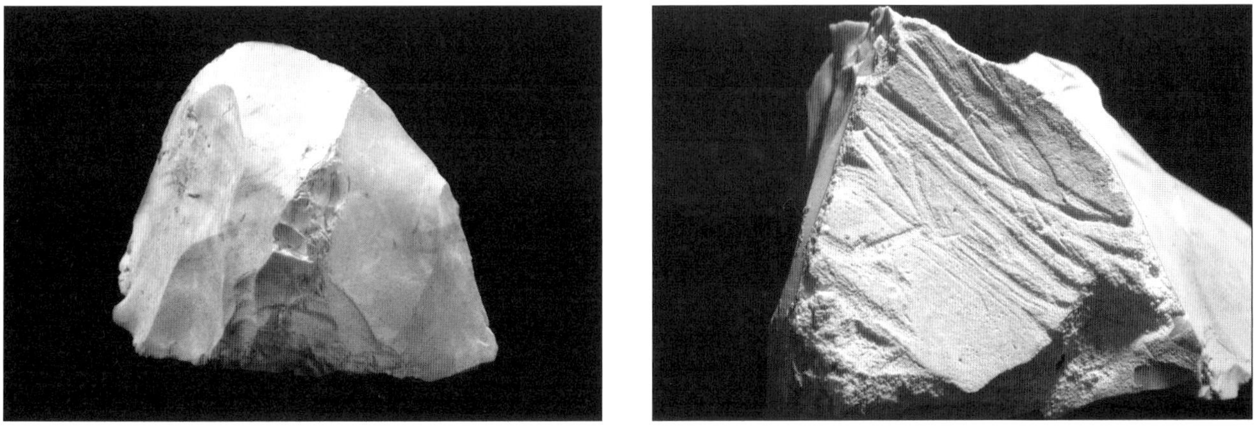

Figure A4.4. *Two views of small flint core from layer 6 with scored lines in cortex (D. Cousins).*

Stone objects from the shaft
Martin Green

Heavier elements of formerly more widespread Tertiary deposits, which today lie principally on the southern and eastern fringes of our study area, are commonly found within the ploughsoil. The most common consist of flint pebbles and two forms of sarsen. Coarse ferruginous sandstone, sometimes with flint pebble inclusions, is known locally as heathstone and *in situ* deposits are known to occur near Cranborne at Creech Hill. At least three pieces of this material are known from the study area, which bear Neolithic or Bronze Age art, two of which are published (see Bradley this volume; Lewis *et al.* 2000; Longworth & Haith 1992). The other piece was found near Badbury Rings (D. Smith pers. comm.). A finer-grained silica rich sarsen also occurs, particularly in the valley gravel deposits. Rarer materials noted within the surviving Tertiary deposits on Pentridge Hill include very occasional lumps of Upper Greensand chert and pebbles of southwestern origin. This source of locally obtainable 'exotic' rocks was clearly exploited by prehistoric people.

The stone objects recovered from the Fir Tree Field shaft are as follows:

S1. L2/2a: Small chip of ferruginous sandstone. Wt 1 g.

S2. L2/2a: Cubular small coarse-grained quartz-rich biscuit coloured sandstone fragment. One surface smoothed and polished from use as a quern or rubber. Wt 36 g.

S3. L2/2a: Fine-grained quartz-rich reddish-brown sandstone fragment. Probably a siliceous sarsen. Wt 9 g.

S4. L2/2a: Roughly triangular split fragment of a coarse-grained pinky-grey sandstone with a single large 8 mm quartz inclusion. The upper surface with a gloss from use as a quern or rubber. Wt 181 g.

S5. L5: Roughly flaked sub-circular, cricket-ball-sized lump of ferruginous sandstone with two small areas of the smooth natural surface surviving. A zone has been subsequently pecked to form a semi-circular edge (30 mm in diameter), possibly for use as a pounder. See a similar artefact found at the base of the shaft at Monkton-up-Wimborne (Fig. A4.23, no. S1). Wt 795 g.

S6. L6a: Ferruginous sandstone lump edge flaked to form a rough disc (max. diameter 85 mm) of plano-convex section. Some abrasion to the edge suggests it may have been used for pounding. Wt 247 g.

S7. L7/7a: Large tabular pebble of ferruginous sandstone. No signs of use or damage. Wt 622 g.

Worked chalk objects
Martin Green

The only piece recovered from the shaft (not illustrated) was retrieved from a deeply stratified chalk rubble horizon which had a radiocarbon date from a roe deer skeleton at the same level of 4330–4000 cal. BC (OxA-8012; Table 1.1).

C1. Shaft rubble, at a depth of 7 m: Sub-circular 'bun'-shaped lump of plano-convex section (80 × 75 × 35 mm). The worn convex surface displays a series of shallow cut/chop marks resulting from the shaping. Wt 201 g.

In addition, two pieces (not illustrated) were found in the Beaker pit F.2:

C2. F.2, L5: Sub-triangular shaped lump (75 × 52 × 32 mm). One face and longer edge smoothed by careful shaping. Wt 100 g.

C3. F.2, L5: Small sub-rectangular shaped lump (42 × 33 × 17 mm), Scrape marks resulting from shaping visible on all surfaces. Both faces bear central scoped depressions about 8 mm deep on one face and 4 mm on the other, possibly resulting from an unfinished attempt at perforation. Wt 21 g.

Lithics from the associated pits
Martin Green

Three large Beaker pits in a 'cloverleaf' arrangement (F.2–F.4) were found some 14 m to the west of the shaft with a further Beaker pit (F.5) located about 7 m northeast of this group (Figs. 3.12, 3.16–3.18 & A4.5). Two Grooved Ware pits (F.6 & F.8) lay 22 m to the northwest and 27 m to the southwest of the shaft respectively (Figs. 3.12, 3.15 & 3.16).

F.2

The retouched, utilized and tool (RUT) component from this pit comprised 5.6% of the assemblage, 2% (10) of which were scrapers (Table A4.8). Five of these were of standard convex form, four thumb (two inversely retouched) and a crude convex scraper made from a thick, squat, irregular flake. Seven hammerstones/pounders were recovered (1.4% of the assemblage), six of which were from layer 5. These six were all made from frost damaged surface flint and had degrees of battering along parts of angular edges, which was clearly the result of a pounding action and not from flaking flint. They are identical in form to those more commonly found on sites of Middle Bronze Age (see Green 2000, fig. 76) to Early Iron Age date.

F.3

The RUT component of this pit was greater and formed 7.5% of the assemblage. Over 5.9% (15) comprised scrapers (Table A4.8). One-third of these were of standard convex form, three were crude convex scrapers made from thick, squat irregular flakes, two side scrapers (one inversely retouched), one crude thumb scraper and four were broken and unclassifiable.

F.4

The RUT component from this pit comprised 11% of the assemblage (Table A4.8). This consisted primarily of a high number of scrapers (51; 9.8%). All except one of these were found in the basal fill of the pit, layer 5, and were subjected to more detailed analysis. A surprising number, 18 (36%) were broken to some extent. Of these, eight were unclassifiable fragments, leaving 42 either intact or sufficiently intact for analysis. Length, breadth and thickness measurements produced an average size of 36 × 34 × 10 mm, a small size for a flint bearing region but not uncharacteristic for Beaker scrapers.

However, this group does reveal some unusual characteristics. Both the percentage of side scrapers (31%) and the amount of inverse retouch (e.g. Fig. A4.5, no. 4) on scrapers of all types (36%) is surprising. Equally surprising is the number that are crudely made, often with minimal retouch on small thick and irregular flakes (see Fig. A4.5, nos. 6 & 8) and the number that are broken (25%). Clearly these flakes have been carefully selected, as amongst the waste flakes from the layer are many that would have provided greater potential for larger and better made scrapers.

The remaining tools from this feature include two hammerstones, a knife and a Type C chisel arrowhead (after Clark 1934) from layer 2. As a few sherds of Peterborough Ware were found in all four of the Beaker pits, it is likely that this arrowhead is also residual.

F.5

The RUT component of this pit comprised only eight artefacts or 5.3% of the assemblage (Table A4.9). Amongst the tools was an end scraper made from a crested blade, a denticulate also made from a blade and a flake with one edge polished by heavy use. Amongst the waste was a blade of Upper Greensand chert.

F.6

The RUT element of this Grooved Ware pit consisted solely of two remarkably different chipped flint axes, one of unique form (Fig. A4.5; Table A4.9). The latter was one component of what appears to be a structured deposit placed on the floor of the pit.

1. The axe is well made and has a pronounced splayed cutting edge. The deeply concave sides expand outwards to form a second splayed cutting edge where it had broken. The axe was undoubtedly designed to be symmetrical. Although rather different Seamer and Duggleby axes (Manby 1988) do have expanded cutting edges and occupy a similar timespan, they are not symmetrical and are part polished. Although the axe of Seamer type found with the burial at Whitegrounds (Brewster 1984) does partly expand towards the butt it is not sufficient to make it symmetrical. However, this example is the best dated of its form and produced a date of 2750 bc which when calibrated gives a date range of *c.* 3500–2900 cal. BC. The two dates obtained from the antler from two of the Grooved ware pits previously uncovered in Fir Tree Field (Barrett *et al.* 1991a) were somewhat later (i.e. pit 11a: 2910–2500 cal. BC; pit 32: 2880–2490 cal. BC).

2. This axe was made from a frost-shattered nodule originally of oval section. Economically made, the flaking was restricted to forming a cutting edge and to waisting one side from two angles, presumably to facilitate hafting. This form of axe, economically flaked on a frost-fractured piece of flint, is a common type found in the later Neolithic surface industries of the area. Frost-split nodules resembling large flakes are common on clay-with-flints deposits and were deliberately selected for axe manufacture, as shown by the many found which still retain part of the original frost-split surface. These expedient tools were simply produced with a minimum of skill and effort and could easily be replaced if broken in a region where workable surface flint is abundant. One particular form the 'Y'-shaped axe (Green 2000) is often produced from split nodules which have a natural cutting edge formed by intersecting frost fractures. By utilizing these naturally steep edges, a serviceable axe can be made by chipping a narrow butt on the piece for hafting.

F.8

The RUT element of the lithic assemblage from this Grooved Ware pit was small (1.7%), comprising only ten pieces (Table A4.9). Of the six finished tools, four were serrated blades, two burnt (one a broken segment), and one double-sided. The remaining two tools were both convex scrapers, one well made and the other marginally retouched.

Catalogue of illustrated flint (Fig. A4.5)

1. F.6, L4: Splayed flint axe described above.

2. F.6, L4: Roughly flaked axe described above.

3. F.4, L5: Small invasively retouched scraper of 'thumb' type.

4. F.4, L5: Small inversely retouched scraper of 'thumb' type.

5. F.4, L5: Small convex scraper retouched practically around the whole perimeter with the convex end produced at the bulbar end by removal of bulb and striking platform.

6. F.4, L5: Small crudely retouched 'thumb' scraper produced from a thick flake.

7. F.4, L5: Similar crudely retouched 'thumb' scraper similar to 6 above.

8. F.4, L5: Small side scraper made from an angular, partly frost weakened, thick flake.

9. F.4, L5: Small marginally retouched 'thumb' scraper produced on a short squat flake.

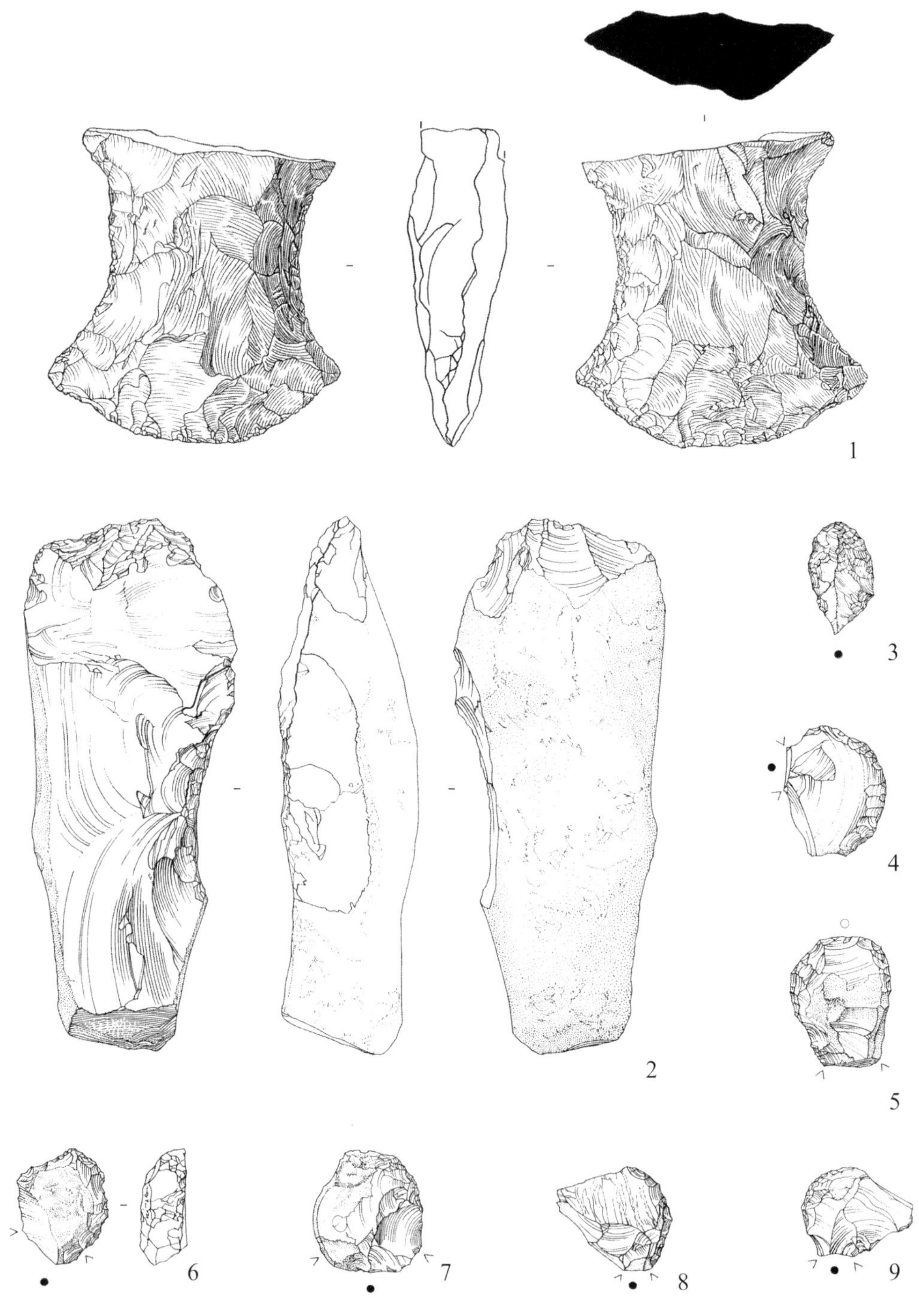

Figure A4.5. *Fir Tree Field shaft: flint axes and scrapers 1–9. Scale 1:2 (J. Richards).*

Table 4.8. *Fir Tree Field (FTF 95): flints from Beaker pits F.2–F.4.*

Feature	Layer	Cores	Core frags.	Flakes	Chips	Retouched	Utilized	Scrapers	Other tools
F.2	2	5	1	29	8	-	1	2	denticulate
F.2	3	-	-	4	13	1	1	3	-
F.2	3/3a	4	-	17	8	-	1	-	-
F.2	3/3a/3b	4	-	18	4	-	-	-	hammerstone
F.2	3b	1	-	13	1	-	-	1	
F.2	4	3	-	-	-	-	-	-	-
F.2	5	13	12	167	132	2	2	4	6 hammerstones
Totals		**30**	**13**	**248**	**166**	**4**	**5**	**10**	**8**
F.3	2	3	-	14	5	1	-	1	-
F.3	3	10	-	47	8	2	-	2	-
F.3	4	-	3	5	2	-	-	-	-
F.3	4	-	3	5	2	-	-	-	-
F.3	4a	1	-	13	8	-	-	3	-
F.3	4b	-	-	6	-	-	-	1	-
F.3	5	10	4	65	30	-	1	8	-
Totals		**24**	**7**	**150**	**53**	**3**	**1**	**15**	**-**
F.4	2	-	1	18	3	-	-	1	ch. arrow
F.4	3	3	-	22	1	-	-	-	-
F.4	3/4	-	-	6	-	-	-	-	-
F.4	4	1	-	1	-	-	-	-	-
F.4	5	10	9	225	163	1	2	50	knife, 2 hammers
Totals		**14**	**10**	**272**	**167**	**1**	**2**	**51**	**3**

Table 4.9. *Fir Tree Field (FTF 95): flints from pits F.5, F.6 and F.8.*

Feature	Layer	Cores	Core frags.	Flakes	Chips	Retouched	Utilized	Scrapers	Other tools
F.5	2	1	2	35	18	-	-	-	edge polished flake, denticulate
F.5	3	3	-	41	7	4	-	1	-
F.5	4	1	2	9	11	-	-	-	-
F.5	6	1	1	10 (1 ugc)	1	-	1	-	-
Totals		**6**	**5**	**95**	**37**	**4**	**1**	**1**	**2**
F.6	2	7	1	84	80	-	-	-	-
F.6	3	-	22	5	-	-	-	-	-
F.6	4	9	2	70	60	-	-	-	rough axe, splayed axe
Totals		**18**	**3**	**176**	**145**	**8**	**2**	**2**	**2**
F.8	2	1	4	11	12	2	-	-	-
F.8	4	11	1	151	389	1	1	2	4 saws
Totals		**12**	**16**	**162**	**401**	**3**	**1**	**2**	**4**

Stone objects from the associated pits

MARTIN GREEN

Only seven pieces of stone were incorporated into the huge shaft and at least four of these bore evidence of human modification. It seems probable that most, if not all, of the pieces described below (not illustrated) were intentionally placed within the pit infillings.

Catalogue (not illustrated)

S8/9. F.2, L3/3a: Two flint pebbles (48 & 45 mm in diameter) derived from tertiary deposits.

S10. F.2, L3b: Angular fragment of coarse light grey sandstone. Wt 77 g.

S11. F.2, L5: Angular fragment of coarse-grained pinkish-grey sandstone. Two opposed flat faces with traces of wear from use as a rubber or quern. Wt 150 g.

S12. F.2, L5: Two broken joining fragments of a soft fine-grained sandstone, possibly sarsen. Partly blackened by heat. The upper convex surface has traces of wear from use as a rubber or quern. Wt 188 g.

S13. F.2, L5: Fragment of a coarse-grained pinkish-grey hard quartz-rich sandstone. Wt 65 g.

S14. F.2, L5: Small fragment of an iron-rich partly oxidized nodule of ?pyrite. Wt 4 g.

S15. F.2, L5: Segment of a marcasite nodule with smooth outer surface. Wt 25 g.

S16. F.2, L5: Small lump of coarse-grained ferruginous sandstone. Wt 14 g.

S17. F.3, L2: Fragment of coarse-grained pinkish-grey hard quartz-rich sandstone. Traces of polish on one face suggest it was once part of a quern or rubber. Wt 84 g.

S18. F.3, L3: Small fragment of coarse-grained ferruginous sandstone. Wt 4 g.

S19. F.3, L3: Flint cast, with much of the mineralized shell surviving, of a fossil sea urchin of the genus *Echinocorys*.

S20. F.3, L4a: Flint cast, with a large natural frost pit at one end, of a fossil sea urchin of the genus *Echinocorys*.

S21. F.3, L5: Flint pebble (30 mm) derived from Tertiary deposits.

S22. F.3, L5: Fragment of coarse-grained pinkish-grey hard quartz-rich sandstone. One face has traces of wear from use as a quern or rubber. Wt 127 g.

S23. F.4, L5: Lump of coarse-grained pinkish-grey hard quartz-rich sandstone. Wt 346 g.

S24. F.4, L5: A small chip of sandstone similar to S23 above. Wt 1 g.

S25. F.4, L5: Smooth rounded kidney shaped nodule of maracasite. Wt 41 g.

S26. F.4, L5: Small segment from a similar nodule to S25 above. Wt 5 g.

S27. F.4, L5: Small chalk cast of a fossil sea urchin of the genus *Micraster*. Part of the mineralized shell survives on the posterior surface.

S28. F.4, L5: Part of the mineralized shell (50 mm max. diameter) of a fossil oyster of the genus *Platyceramus*.

S29. F.5, L2: Recently broken fragment of fine-grained ferruginous sandstone. No signs of use. Wt 76 g.

S30 & 31. F.6, L3: Two fragments of heavily weathered coarse-grained ferruginous sandstone. Wt 49 g.

S32. F.6, L3: fragment of a decayed fine-grained dark brown ferruginous sandstone. One face slightly convex and smooth, possibly from use. Wt 15 g.

S33. F.6, L3: Fragment of an iron-rich nodule, possibly marcasite. Wt 6 g.

S34. F.6, L4: Fragment similar to S32 above but with no signs of use. Wt 67 g.

S35. F.6, L4: Fragment similar to S32 above. One small area with possible trace of wear. Wt 19 g.

S36. F.8, L4: Broken lump of fine-grained ferruginous sandstone. Wt 193 g.

S37. F.8, L4: Small fragment of coarse-grained ferruginous sandstone. Wt 16 g.

S38. F.8, L4: Flint pebble (50 mm) derived from Tertiary deposits.

S39. F.8, L4: Small fragment of a coarse decayed quartz-rich sandstone with part of one flat surface with traces of use wear. Wt 8 g.

Pottery from the shaft

Rosamund Cleal

The ceramics from the shaft come from the uppermost layers and span the fourth and third millennia cal. BC in date (Figs. A4.6 & A4.7). The majority belong to the Beaker tradition, but there is a sizeable Peterborough Ware assemblage and one earlier Neolithic plain bowl. Grooved Ware is absent from the shaft, in contrast to the shaft's immediate vicinity in which there were pits containing Grooved Ware.

Earlier Neolithic

Three rim sherds, and two body sherds, from layers 6a, 5, and 4, probably belong to a single earlier Neolithic Bowl (Fig. A4.7, P35). The form is simple, a vertical upper body springing, presumably, from a rounded base. The rim is barely modified, only one of the sherds showing what appears to be a deliberate rolling over the rim, but it is not pronounced and the other rims (one with just a slight out-turn) seem, on the grounds of fabric and colouring, to belong to the same vessel. Although layer 6 is the lowest layer in the shaft to produce Neolithic pottery, there are no features of P35 which suggest that it is particularly early in the fourth millennium BC. It is clearly not carinated, which is a feature of much early fourth-millennium BC pottery, and, strengthening the case for a slightly later date, it could quite clearly fit within the broadly mid-fourth-millennium assemblage at Windmill Hill (e.g. Smith 1965, P57 & P68, which are similar in size as well as form). The form does, however, also occur in the pre-enclosure pottery from Windmill Hill (Smith 1965, P24), allowing that an earlier than mid-millennium date is possible. A cereal-grain impression (or possibly a seed impression) occurs on the exterior of the rim sherd from layer 5.

Peterborough Ware

Sherds of Peterborough Ware occurred in layers 4, 5 and 6 and at the interface between 5 and 6a. Only five sherds, all probably belonging to a single vessel, were found in layer 6 (represented by P1 & P36). The lack of any extension of the rim profile on the exterior as well as the extensive use of twisted cord on the interior of the neck and the relatively fine fabric (compared to many Mortlake fabrics) suggest that this is, in the conventional classification, an Ebbsfleet Ware vessel. The sooting on the interior shows that it has been used for cooking, carbon deposition in the upper interior body being a result of the charring of food deposited from the surface of the cooking liquid.

A Fengate Ware vessel seems to be represented by the rim P2 from layer 5 (Fig. A4.6). Seven fairly substantial sherds (weighing a total of 104 g; one illustrated as P3 in Fig. A4.6) from the same layer almost certainly belong to the same vessel and show that the body was covered with oblique lines of twisted cord impressions.

Beaker

The deepest occurrence of Beaker sherds in the shaft is in layer 4, where there is also a considerable amount of Peterborough Ware. P5 (Fig. A4.6) could be either Peterborough Ware or Beaker, although the fact that it is well made suggests the latter as more likely. Other, unillustrated, sherds are small and mostly plain; at least one base-angle sherd of a Beaker is present.

The majority of the Beaker sherds are from layer 2 and 2a and are illustrated as P6–P34 (Fig. A4.6). Both fine and 'coarse' or rusticated Beaker vessels are represented. In traditional terms the comb-decorated Beakers such as P6–P9, P15 and P24 (Fig. A4.6) would be classified as long-necked and remain so in the more recent Needham classification (Needham 2005, 195). P24, for instance, is a close parallel to a vessel illustrated by Needham from West Heslerton in Yorkshire (Needham 2005, fig. 9.3) and dated to 2200–1980 cal. BC on associated human bone (at 2 sigma) (Needham 2005, table 5), a date not inconsistent with that from layer 2 in the shaft. Needham demonstrates that Long Necked (LN) Beakers like these are used in graves from fairly early in the Beaker using period, at least from the twenty-second century and perhaps earlier (Needham 2005, 195). In the case of the shaft, however, the interest lies not so much in the presence of LN Beakers, but in the fact that the group appears to be a settlement assemblage, and these are relatively rare in central southern England (relative to, for instance, East Anglia).

P10 is unlikely to be long-necked, as the upper body is neither upright nor convex, and the same may be said of P27 (Fig. A4.7). Not enough survives of the non-LN Beakers from the shaft to be certain of their form, but they certainly include at least two finger-rusticated vessels (P11/14/32 & P13), one thick-walled Beaker coarseware vessel (P33), and one vessel which, although its superficial appearance suggests Grooved Ware (P29) is very unlike the Grooved Ware from Fir Tree Field and is much more likely to be a form of coarse Beaker.

One unusual feature of the Beaker assemblage, apart from its rarity as a non-funerary assemblage, is that it includes at least one Beaker tempered with common flint inclusions (P16–P19) (Fig. A4.6). This is an extremely rare fabric to find in Beakers and is interesting given the occurrence of a sherd in a similar fabric from the vicinity of the shaft, in association with a round house.

Fir Tree Field shaft ceramic sequence

In the years since the shaft was first excavated the understanding of fourth-millennium cal. BC ceramics in southern England has made some progress (summarized in Cleal 2004), and it now seems unlikely that the vessel P35 (Fig. A4.7) is from very early in the fourth millennium BC, despite the earliness of the radiocarbon date from a layer containing one of the vessel's sherds. The fact that two sherds of this vessel are also found in higher layers with Peterborough Ware also indicates that it was redeposited rather than found *in situ* in those layers (i.e. in 5 and the base of 4). The association of the single sherd which was found in layer 6a (the right-hand sherd in the illustration) must remain unclear while there is only one radiocarbon date, and that from a single bone which could have been redeposited; there is no pottery from the hearth in 6b, for which the dating — early in the fourth millennium cal. BC — seems much more secure.

Sherd from the house in Fir Tree Field

A single plain body sherd was found on the surface of posthole 1 from this structure. There are shallow depressions on the surface but the sherd is not obviously decorated. The fabric is hard but brittle with a hackly fracture and the sherd is about 7 mm thick. There is common, well-calcined flint (<2 mm) and no other inclusions are visible. The exterior is orange, the interior pale brown and the sherd worn. Among the shaft pottery the closest parallel for this is P19 (Fig. A4.6), a comb-decorated Beaker sherd. If the sherd were dark-coloured rather than orange on the exterior it would be easily lost within a Middle Bronze Age assemblage of the local area, but if the house were Middle Bronze Age it would be usual to find it associated with a considerable ceramic assemblage. The possibility remains, therefore, that the structure is associated with the Beaker period use of the immediate area.

Illustrated sherds

Figure A4.6

Layer 6

P1. Layer 6 (–80 cm A35 S12 <40 cm): Peterborough Ware. Body sherd decorated with horizontal lines of twisted-cord impressions (Z-twisted impressions, therefore S-twisted cord) in a slightly laminated fabric with sparse flint (<3 mm), occasional chalk and sparse fine sand. Exterior pale orange-brown, core and interior black; some traces of black residue adhering on interior. Fair condition. Four other body sherds of the same vessel were associated with it and

Figure A4.6. *Fir Tree Field shaft: pottery. Scale 1:2 (M. Brudenell after R. Cleal).*

one badly preserved sherd with a concave neck may also belong to the same vessel.

Layer 5

P2. Layer 5 (E 1/2): Peterborough Ware (Fengate sub-style). Rim sherd (and one in worse condition not illustrated) of a vessel with a bevelled rim and applied clay used to form a collar, decorated with shallow irregular impressions and almost certainly with pits in the neck which have pushed up bosses on the interior; there are twisted-cord impressions on the rim top. The fabric is hard, with sparse flint (*c.* 5%) some of which is sharply angled and not noticeably heat-crackled while other flint inclusions are clearly calcined. There is also fine sparse to moderate sand. The surfaces are fair to worn and there are patches of sooting/residue on the interior below the rim, and sooting on the rim top.

P3. Layer 5 (E 1/2): Peterborough Ware. Seven body sherds of one vessel (one illustrated) decorated with short lengths of twisted-cord impressions. Fabric as for [P1] and could be part of the same vessel. Exterior of illustrated sherd pale orange, core black, interior orange-brown. Edges slightly abraded, surfaces good condition.

P4. Layer 5 (A36 S12): Peterborough Ware. Body sherd decorated with horizontal twisted cord in a fabric with sparse flint (<3 mm, most <1 mm) and moderate fine sand. Exterior pale brown, core and interior black. Condition fair. This appears to be a different vessel from P1 and P3.

Layer 4

P5. Layer 4 (E 1/2): Probably coarse Beaker. Body sherd decorated with plastic single fingernail impressions in a soft fabric with a hackly fracture containing sparse well-calcined flint (<2 mm, most <1 mm). Exterior pale brown, core bi-coloured by surface colours, interior grey-brown. Condition fair.

Layer 2

P6–P9. Layer 2 (E 1/2): One rim (rim diameter not determinable) and four body sherds (three illustrated) of a Beaker in a hard fabric with a smooth fracture containing sparse well-calcined flint (<3 mm), sparse fine sand and possibly some sparse grog (<2 mm; difficult to distinguish from the matrix). Surfaces pale orange, core obscured.

P10. Layer 2 (E 1/2): Two rim sherds (diameter not determinable) of a Beaker in a hard, well-fired fabric with a compact fracture containing moderate fine sand, including some dark grains. Exterior dark orange-brown and orange-brown, interior as exterior with some buff, core dark grey/black. Condition fair.

P11. Layer 2 (E 1/2): Beaker. Body sherd with horizontal cordon with finger marks on one side probably from pinching it into place. The fabric is hard with a smooth fracture and contains moderate fine sand; there may be some grog present but it is difficult to distinguish it from the matrix. Surfaces orange-brown, core black. Condition fair.

P12. Layer 2 (E 1/2): Plain rim sherd, probably of a coarse Beaker, or possibly of an Early Bronze Age tradition although it is not diagnostic of a recognizable style (Food Vessel, Collared Urn). The fabric is coarse with a hackly to blocky fracture and a smooth, almost soapy feel. There is some fine sand present but the main inclusion type is clearly grog which is very difficult to distinguish from the matrix; a few blocky voids may represent dissolved chalk. Surfaces pale orange-brown. Condition fair.

P13. Layer 2 (E 1/2): Beaker body sherd with paired non-plastic fingernail decoration in a hard fabric with a smooth fracture containing moderate fine sand and dark grains likely to be iron oxides or glauconite, some of which are <2 mm; there is also rare fine flint (<1 mm). Exterior orange-brown, core dark grey, interior pale brown. Edges worn.

P14. (S4 on cleaned surface): Beaker body sherd of same vessel as (P11); it appears that the cordon in this case is an applied rib; a likely decorative scheme for this type of Beaker is that the lower body had horizontal cordons while the neck was decorated with ribs, often finger-pinched. Fabric, colour and condition as (P11). There are at least two more sherds of this vessel from layer 2A.

P15. Layer 2 or 2A: Two body sherds (one not illustrated) of a Beaker in a hard fabric with a hackly fracture containing sparse flint (<4 mm); a slight soapy feel to the fabric may indicate that there is grog present, but it is not distinguishable from the matrix. Surfaces pale brown. Condition slightly worn.

P16–P18. Layer 2 (E 1/2): One rim and two body sherds of a Beaker in a hard, coarse fabric with a hackly fracture containing moderate to common well-calcined fine flint (<4 mm, most <2 mm) and sparse find sand. Exterior orange-brown, interior grey-brown. Condition fair. Five other decorated body sherds from the same context appear to belong to this vessel, as also, probably, does P19.

P19. Layer 2A (S1 Cleaned surface): Two Beaker body sherds, joining along an ancient break, probably belonging to the same vessel as P16–P18. The fabric is hard but brittle, with common (*c.* 25%) flint, well-rounded and crushed (<4 mm, most <2 mm). Exterior orange, interior grey, core obscured. Condition: slightly worn on the exterior.

P20. Layer 2A (S1): Beaker. Rim sherd decorated with rectangular-toothed-comb impressions in a hard fabric with a hackly fracture and containing sparse to moderate flint (<3 mm) unevenly distributed.

P21. Layer 2A (S1): Beaker. Body sherd almost certainly belonging to the same vessel as P20, although the frequency of flint in the fabric seems to vary throughout the vessel, with less towards the rim. Exterior orange-brown, core and interior pale brown.

P22. Layer 2A (S1): Beaker. Rim sherd decorated with comb impressions which have been impressed when the vessel was still reasonably wet as they are clearly impressed and some are distorted. The fabric is hard and compact with sparse small flint (<2 mm) and rare chalk fragments; there is also sparse fine sand. Orange throughout. Condition fair.

P23. Not illustrated. Layer 2A (S5 on cleaned surface): Beaker body sherd, possibly part of the same vessel as P16–P19. The fabric is hard with a hackly fracture and contains common well-calcined flint (<4 mm, most <2 mm) and moderate sand. Exterior pale orange-brown, interior grey-brown, core orange exterior half, interior half grey. Condition fair.

P24–P25. Layer 2A (A25 S1): Beaker rim sherd with a diameter of approximately 220 mm and one body sherd probably of the same vessel, in a hard well-fired fabric containing moderate fine sand and possibly some grog. Exterior orange-brown, interior pale brown, core bi-coloured orange and black. Surfaces fresh, edges fair condition.

P26. Layer 2A (NW quadrant. A37 S1): Single Beaker body sherd in a hard fabric with a hackly fracture containing moderate small flint (<3 mm, most <1 mm). The orientation of the sherd is not clear; the motif may be a floating lozenge, or might be the apex of a triangular motif. Exterior pale orange, interior grey, core orange. Condition fair.

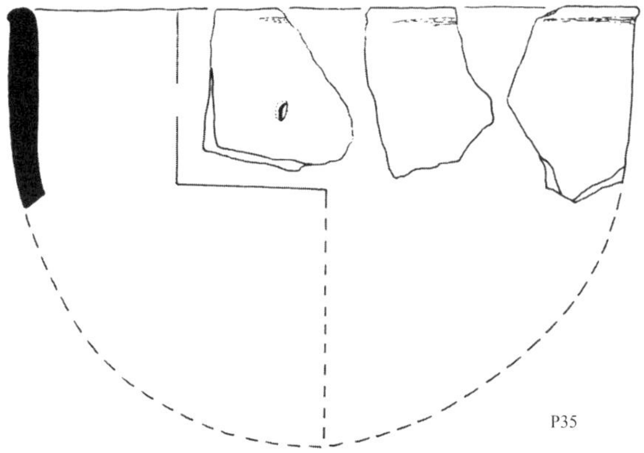

P35

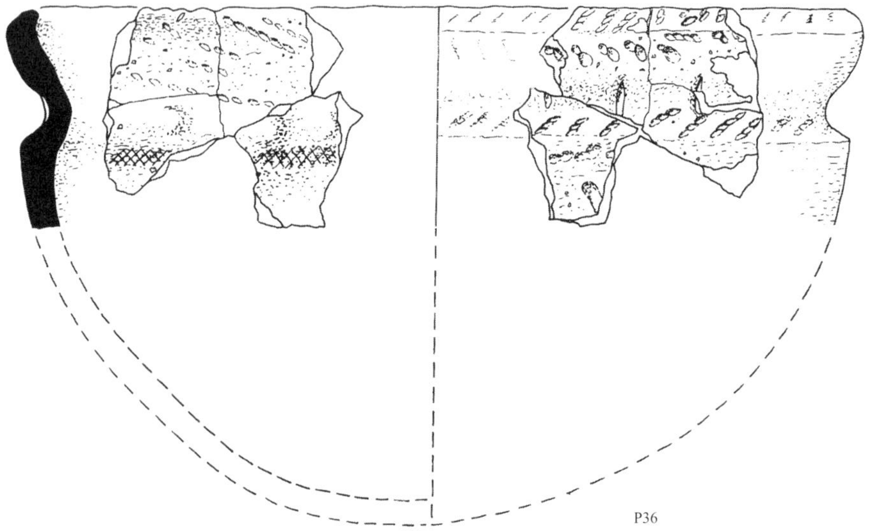

P36

Figure A4.7. *Fir Tree Field shaft: pottery. Scale 1:2 (M. Brudenell after R. Cleal).*

P27–P28. Layer 2A (NW quadrant): Rim and body sherd probably belonging to the same Beaker, in a hard, compact fabric with a smooth fracture containing moderate fine sand including a lot of dark grains. Red throughout. Fair condition.

P29. Layer 2A (A25 S1): Two rim sherds, probably coarse Beaker, decorated with grooves. The fabric is hard, compact and contains moderate fine grog which is difficult to distinguish from the matrix, with some find sand. Exterior brown, core black, interior dark grey. The sherds are slightly worn. The sherds may be coarse Beaker, the other possibility being Grooved Ware, but they do not fit obviously into the Grooved Ware assemblages from the same field.

P30. Layer 2A (NW quadrant): Body sherd of a Beaker in a hard compact fabric containing moderate fine sand, moderate small grog (<2 mm) and rare flint. Exterior orange, interior pale brown, core bi-coloured orange and black. Worn condition.

P31. Layer 2A (S1): Beaker. Body sherd with grooved decoration in a hard compact fabric with some sand. Exterior surface almost fresh, interior missing.

P32. Layer 2A (S1): Body sherd of the same vessel as P11 and P14 from layer 2, and others of the same vessel are not illustrated. Fabric as for P11 but with slightly more dark grains. Surfaces orange-brown, core black. Condition fair.

P33. Layer 2A (W1/2 A25 S1): Rim sherd (diameter not determinable) probably of a coarse Beaker, in a soft fabric with a hackly fracture; grog is present but difficult to see (<3 mm, probably moderate in frequency). Surfaces pale brown. Condition worn.

P34. Layer 2B (E 1/2): Beaker body sherd (orientation not certain) in a soft fracture with a hackly fracture, containing sparse grog and flint, and moderate fine sand. Exterior buff, interior grey-brown, core bi-coloured orange and dark grey. Condition fresh to fair.

Figure A4.7

P35. Layer 6A (one sherd; right in the illustration); L5 (one sherd; left in the illustration); base of L4 (one sherd; central in the illustration): Neolithic Bowl with a diameter at the rim of about 15 cm. Three rims sherds, not conjoining but clearly belonging to a single vessel with a rim which varies in form from slightly rolled over to simple

Table A4.10. *Fir Tree Field (FTF 92) shaft: total number of sherds of each ceramic type from each layer.*

Layer	Plain bowl	Peterborough	Beaker	EBA urn
2/2a	-	20	141	4
4	1	10	17	-
5	3	50	-	-
6	-	7	-	-
6a	1	-	-	-
Total	**5**	**87**	**158**	**4**

upright. The fabric is hard and compact with approximately 10% (by surface area) of well-sorted shell (although the maximum dimension is 8 mm, most inclusions are fine, <1 mm); no other inclusions are visible. The sherds are dark-coloured throughout, with dark grey-brown to dark grey surfaces and dark grey cores. The condition of the sherds varies slightly, with that from Layer 6A freshest, with surfaces almost fresh and breaks slightly worn; that from L5 is the most worn, but even so is still in fair condition, and the sherd from L4 has a good exterior surface but worn interior.

P36. Layer 6 (E 1/2): Two rim sherds and two body sherds, all conjoining, of a Peterborough Ware bowl with a diameter at the rim of approximately 22 cm. The fabric is fairly hard, coarse and slightly laminated, containing sparse to moderate (approximately 5–10% by area) poorly sorted flint inclusions (<6 mm, most <2 mm); most of the flint is angular and poorly calcined or not noticeably calcined. Some sparse fine dark grains are also present. Exterior pale brown to pale red-brown, interior pale brown, core black. The exterior is worn, with the core colour showing through in some places, but the edges and interior are in fair condition, sooting surviving in the interior angle and, on one sherd, just below it. The decoration is difficult to determine accurately: on the exterior it is worn and on the interior appears to have only been very lightly impressed. The impressions on the interior appear to be of 'fat' short twisted-cord impression; in the neck there are finger- or thumb- impressions, with the nail impressions surrounded by slighter impressions from the finger or thumb tip. On the interior there are longer twisted-cord impressions, but so lightly impressed that the segments appear separated by large gaps.

Animal bones from the Fir Tree Field shaft and associated pits

Mark Maltby

Recording methods

All animal bones from the excavations were recorded onto a data base (Microsoft Access), which forms part of the site archive. Where appropriate, the following information was recorded on each fragment: species; area of body; anatomical element; part and zone of element present; proportion of element present; gnawing damage; surface condition; fusion data; tooth-ageing data; butchery marks; metrical data; other comments. Counts of fragments (Number of Individual Specimens - NISP) included all identified limb bone shaft fragments, dorsal ends of ribs and vertebral bodies. Unidentified fragments were not recorded in detail but their numbers were noted and assigned where possible into large mammal (ULM - cattle/red deer-sized) and medium-sized mammal (USM - sheep/pig/roe deer-sized) categories (Tables A4.11–A4.17). Sheep/goat includes six elements diagnostic of sheep but no element of goat was positively identified. The bovid category consists of domestic cattle and possibly aurochs (wild cattle). Tooth eruption and tooth wear descriptions followed the method of Grant (1982). Measurements taken are mainly those recommended by von den Driesch (1976). Trochlea height measurements of humeri are as described by Legge & Rowley-Conwy (1988, 124).

Animal bones from the shaft (F.1)

Later Mesolithic contexts

Primary shaft filling at 7 metres: The first of two roe deer skeletons was found at a depth of about 7 m. A total of 68 elements were recorded (Tables A4.11 & A4.16; Fig. A4.8). These counts are slightly inflated by the recording of different parts of the same bones (radius, pelvis, femur, tibia, calcaneus, metatarsal) that were retrieved in slightly different locations. All of the major limb bones were recovered along with about half of the smaller bones of the feet. However, the vertebral column is represented only by some of the lumbar and sacral vertebrae. None of the thoracic or cervical vertebrae nor any of the ribs were found. A small portion of the left mandible survived but no other cranial elements were retrieved (Table A4.16). Although parts of the body are missing, there is no evidence for butchery or gnawing and it is likely that this roe deer entered the shaft as a complete animal, and possibly was a victim of a fall during the Late Mesolithic period. Subsequent weathering has affected the coherence of the articulated skeleton, although much of it did survive as a loosely articulated group amongst the chalk rubble.

Ageing data were obtained from the mandible and the state of epiphyseal fusion of the bones. The mandible still possessed the deciduous fourth premolar, which indicates that it belonged to an immature animal. Both the first and second molars were fully erupted. The second molar is only just in wear, however, indicating it probably belonged to a deer of about a year old. This age estimate is supported by the fusion data, which shows that only the earliest-fusing points have fused (acetabulum; scapula; distal humerus; proximal radius). Some of the first and second phalanges possess fused proximal epiphyses but others are unfused. All the later-fusing limb bone epiphyses are also unfused. Metrical data were recorded on some of the fused epiphyses and the results are stored in the bone archive. The distal articular breadth of a humerus (von den Driesch 1976, BT) measures 25.9 mm; the proximal breadth of a metacarpal (Bp)

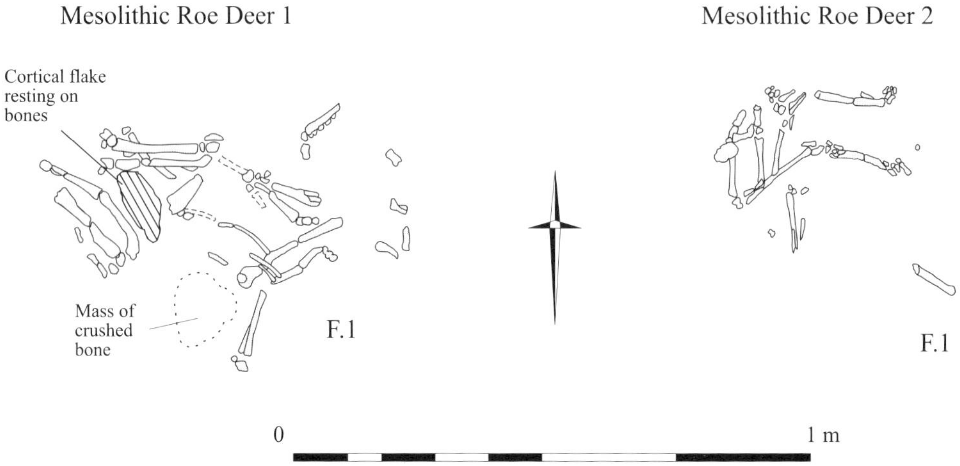

Figure A4.8. *Fir Tree Field shaft: the Mesolithic roe deer skeletons 1 & 2 within the shaft fill (N. Griffiths after M. Green).*

measures 21.4 mm and an astragalus has a greatest lateral length (GLl) of 30.7 mm.

A left femur found slightly above the skeleton belonged to a second, younger, roe deer. This femur is not from the second skeleton found at 5.2 m discussed below. No evidence for butchery or gnawing was found on this bone, which belonged to a juvenile animal.

Two bone fragments of large mammal were also recovered from this depth. One has been identified as the proximal end of a red deer femur. The epiphysis survives but is unfused indicating that it belonged to an immature animal.

Primary shaft filling at 5.2 m: A second roe deer skeleton was discovered at this depth and represents another animal that was probably a victim of a fall during the late fifth millennium BC. Again element counts (82) are slightly inflated by the recording of different parts of the same bones (radius, femur, tibia, calcaneus, metapodial) in slightly different locations. Again the integrity of the articulated skeleton has been disturbed by the collapse of the sides of the shaft. Slightly more of this skeleton has been recovered than the one found at 7 m. Apart from the right pelvis, all the major limb and girdle bones are represented, albeit often in an incomplete and fragmented state. In addition, a sufficiently large number of the small bones of the limb extremities have been recovered to indicate that all the feet are represented. Several vertebrae and ribs have also survived, as have parts of both maxillae and the right mandible (Table A4.16). There is no evidence for butchery or gnawing and the skeleton represents the remains of a complete animal.

This skeleton belonged to a younger roe deer than the one found at 7 m. The jaws possess deciduous premolars in a fairly early stage of wear and the mandibular first molar has only very slight enamel wear on its anterior cusp. This suggests the skeleton belonged to a calf of about six months old. Of the fusion points, only the scapula has partly fused, whereas all the other early-fusing points (acetabulum; distal humerus; proximal radius) are unfused.

Primary shaft filling at 4.5 m: Two bones were recovered from this depth. A section of large mammal rib shaft was found. Although unidentified, its large size suggests it probably belonged to an aurochs. A left scapula of a roe deer was also recovered. This fused specimen did not belong to either of the skeletons found lower down the shaft. There is slight evidence of gnawing damage, indicating that this bone was accessible to a carnivore before final deposition. The glenoid has a breadth (von den Driesch 1976, BG) of 19.9 mm.

Primary shaft filling at 3.05 m: The distal half of a left red deer metatarsal has survived in a fragmented state. It belonged to an immature animal as its distal epiphysis is unfused. Possible axial cut marks on the shaft have been noted and illustrated (Green 2000, 26).

Top of primary shaft filling (Layer 8): All but one of the 19 bones recorded belong to a skeleton of a toad, presumably another victim of a fall. A large mammal limb bone shaft fragment is more likely to have belonged to red deer than an aurochs based on its size but it does not retain any morphologically diagnostic features to confirm this identification.

Late Mesolithic contexts
Layer 7b–8: Two elements were recorded, one of which consists of an unidentified mammal fragment. How-

ever, a substantial portion of a bovid calcaneus was recovered. This was heavily eroded and could not be measured but its large size suggests it belonged to an aurochs rather than a domesticated animal.

Layer 7a: Only two unidentified large mammal fragments were recovered, one of which was burnt.

Layer 7: Sixty animal bone fragments were recorded, of which 19 were identified (Table A4.11). Thirteen of these belong to red deer. The elements represented (Table A4.17) include three eroded fragments of antler. One of the three tibia fragments has been slightly damaged by gnawing. No evidence of butchery was noted but all the bones are fragmented. Five bovid bones were identified. Three of these were recorded as aurochs on the basis of their large size. A slightly weathered right scapula has a greatest glenoid process length (GLP) of 89.3 mm and a glenoid cavity length (LG) of 73.7 mm, comfortably in the size range of aurochs. The distal half of a left tibia possesses a maximum breadth (Bd) of 79.5 mm and a depth of 63.6 mm, which is again larger than domestic cattle. The shaft of this specimen has been broken transversely and there is a possible impact scar just below the break that was perhaps made when the bone was broken open to extract marrow. The third bone consists of a shaft fragment of a right tibia, which although not measurable, is also sufficiently robust to be identified as aurochs. Shaft fragments of a humerus and metatarsal are from much smaller bovids although they have been tentatively identified as aurochs. However, as both bones are porous, it is not possible to determine with total confidence whether they belonged to a wild or domestic calves. Similarly, it is not certain whether a humerus from a neonatal mortality belonged to a domestic pig or a wild boar.

Early Neolithic contexts

Layer 6b: Nineteen animal bone fragments were recovered, of which only four could be identified (Table A4.11). A central portion of a bovid left calcaneus is not measurable but appears from its small size to have belonged to a domestic animal. Similarly, a right femur shaft is more likely to have belonged to a domestic pig than a wild boar based on its small size. Wild species are, however, represented by a roe deer calcaneus and a left pubis of a red deer.

Layer 6a: The 54 animal bone fragments consist mostly of unidentified fragments, mainly of large mammal (Table A4.11). Red deer provide five of the seven identified elements. These include three small portions of antler, two of which survive in a very fragmentary state whilst the third consists of a tine. A maxilla fragment belongs to a juvenile animal, which has only its deciduous premolars in wear and the first molar not fully erupted. A small portion of a red deer left ischium was also found, conceivably from the same pelvis found in layer 6b. A shaft fragment of a bovid metatarsal and distal epiphysis of a pig left femur are both likely to have belonged to domestic animals, judging by their relatively small size.

Early Neolithic/Late Mesolithic contexts

Layers 6–7b: A total of 75 animal bone fragments were recovered, mainly from grey silts. Most of the fragments are eroded and unidentifiable, although the majority are from large mammals (Table A4.11). At least 13 of the 15 identified fragments are from wild mammals. Red deer elements were again the most commonly identified. The seven elements include three pelvis fragments, all of which include part of the acetabulum. They could, however, all have belonged to one individual. Shaft fragments of a red deer tibia and metatarsal were also found along with a small piece of antler and a largely complete second phalanx, which has been slightly damaged by gnawing. Roe deer are represented by shaft fragments of a right humerus and a left radius. The humerus is slightly charred.

Four large bovid elements were identified as aurochs. These include a slightly charred distal end of a humerus with a fused epiphysis. An unfused proximal end of a left tibia is a very robust specimen, which may have belonged to a sub-adult male aurochs. A fused distal end of a right tibia was measured. The breadth (Bd) measures 76.6 mm and the depth 57.7 mm, again comfortably beyond the range of domestic cattle recorded on British Neolithic sites. A large lumbar vertebra possesses fused cranial and caudal epiphses and belonged to an adult animal. In addition, an unidentified large mammal rib shaft is from a very large bovid, probably an aurochs. However, a bovid metatarsal fragment belongs to a much smaller animal, almost certainly a domesticated animal. A small fragment of pig mandible was also recovered, although it is not possible to determine whether this was from a domestic or wild animal.

Middle–Late Neolithic contexts

Layer 5: This layer produced the largest faunal assemblage (354 fragments) from the shaft. However, many of these are slightly eroded and very fragmented, including 246 unidentified elements (Table A4.11). Cattle provide 48 of the 108 identified elements (Table A4.13). These include 12 loose teeth, their abundance also being indicative of the fairly poor preservation of

Table A4.11. *Animal bones recorded in Fir Tree Field shaft F.1.*

Species	7 m	5.2 m	4.5 m	3.05 m	8	7b/8	7a	7	6b	6a	6–7b	5	4	3	2a	2	Total
Cattle									1	1	1	48	14			13	**78**
Sheep/Goat													1			1	**2**
Pig								1	1	1	1	34	8		1	4	**51**
Aurochs						1		5			4	2					**12**
Red deer	1			1				13	1	5	7	8					**36**
Roe deer	68	82	1						1		2	2	2				**158**
Weasel/Stoat												14					**14**
Toad					18												**18**
Total identified	*69*	*82*	*1*	*1*	*18*	*1*	*0*	*19*	*4*	*7*	*15*	*108*	*25*	*0*	*1*	*18*	**369**
ULM	1		1		1		2	33	5	23	45	129	28		2	11	**281**
USM									4	3	6	72	23			13	**121**
UM						1		8	6	21	9	44	22	1	1	38	**151**
UB												1					**1**
Total undentified	*1*	*0*	*1*	*0*	*1*	*1*	*2*	*41*	*15*	*47*	*60*	*246*	*73*	*1*	*3*	*62*	**554**
Total	**70**	**82**	**2**	**1**	**19**	**2**	**2**	**60**	**19**	**54**	**75**	**354**	**98**	**1**	**4**	**80**	**923**

Counts are of number of individual specimens (NISP); ULM = unidentified large mammal;
USM = unidentified medium-sized mammal; UM = unidentified mammal; UB = unidentified bird

cattle bones. However, bones from most areas of the body are represented, including several vertebrae, which do not usually survive well (Table A4.13). A distal metatarsal has a maximum breadth (Bd) measurement of 57.8 mm, and two astragali have lateral lengths (GLl) of 68.5 mm and 70.3 mm. A first phalanx has a peripheral length (Glpe) of 62.8 mm and a proximal breadth (Bp) of 27.9 mm. Although these are quite large specimens, they all fall within the range of domestic cattle recorded on other British Neolithic sites such as Windmill Hill (Grigson 1999). The metatarsal has evidence of slight charring on the shaft just below the break, perhaps as the result of heating the bone to release the marrow. A mandible bears a number of cut marks on the lingual aspect of the ramus. Although the majority of the bovid bones from this layer are considered to be from domestic cattle, at least two elements, the lateral half of a centro-quartal and a diastema of a mandible are large enough to have belonged to aurochs.

Pig elements were the second most commonly identified. A total of 34 elements were recorded (Table A4.15). Eighteen of these are loose teeth, again indicating poor survival of the bones. Six fragments of scapula were recovered. Two mandibles possess a first molar in an early stage of wear and an unerupted second molar, indicating that they belonged to pigs of around 6–12 months of age. A maxilla belonged to an older animal that possessed at least the first two molars in wear. A third mandible has all three molars in wear, although the third molar has only slight wear (Grant 1982, stage b). This specimen would have belonged to an animal probably killed in its third year. The greatest length of the third molar measures 38.6 mm, which indicates it belonged to a large animal. It is slightly smaller than the largest measured specimen (39.3 mm) from Durrington Walls (Albarella & Payne 2005). Although unmeasurable, a lower canine of a male and a scapula were also from quite large animals and it is possible that some of the pig remains belonged to wild boar, although it is probably impossible to distinguish these from large domestic pigs. Several other elements are from smaller animals and are likely to have been from domesticated pigs.

Eight elements of red deer were recorded from layer 5 (Table A4.17). These include three antler fragments, two of which are tines. A scapula has a greatest glenoid process length (GLP) of 63.8 mm and a glenoid length (LG) of 51.7 mm. Roe deer are represented by portions of humerus and femur. Fourteen bones belong to the skeleton of a small mustelid. Because it is from an immature animal, it is not absolutely clear whether it belonged to a stoat or a weasel, although comparison with modern reference material suggests that it is more likely to have been a young weasel. No evidence of carcass processing was found on the bones and it may have been a natural mortality, which may have burrowed or fallen into the pit.

Late Neolithic contexts

Layer 4: A total of 98 animal bone fragments were recovered, of which 73 are unidentified. Most of the bones are eroded. Cattle bones provide 14 of the identified elements (Table A4.11). All but three of these are loose teeth, which indicates the poor preservation of the bones. The presence of both deciduous and worn permanent teeth indicates both young and adult cattle are represented. None of the elements appear large enough to have belonged to aurochs. All but one of the pig elements are also loose teeth (Table A4.15). Sheep/goat is for the first time definitely represented by a single tooth. Roe deer are represented by a tooth and an astragalus.

Beaker contexts
Layer 3: Only one unidentified mammal fragment was recorded.

Layer 2a: Only four fragments were recovered, of which the only identified fragment was the pig femur submitted for radiocarbon dating.

Layer 2: Although 80 animal bone fragments were recovered, these were poorly preserved and only 18 were identified (Table A4.11). Thirteen of the elements belong to cattle, all but one of which are loose teeth (Table A4.13). Again both immature and adult domestic cattle are represented. Three of the four pig elements are also loose teeth. A worn lower third molar attests to the presence of an adult sheep/goat. No elements of wild animals were identified.

Animal bones from other features (Table A4.12)
Beaker pits
F.2: This produced the second largest animal bone assemblage from the pits consisting of 348 fragments. However most of the assemblage was poorly preserved with high levels of erosion amongst the bones in the upper layers. Fifty-six fragments are burnt. These were mainly found in layer 5, particularly amongst some ashy material. Only 94 elements were identified (Tables A4.12; Figs. 3.15, 3.16 & A4.9). These include several bones of small mammals and amphibians, mainly found near the base of the feature indicating that the pit may have remained open for a short period. Species represented include types of both frog and toad, short-tailed vole and mole. All could be pitfall victims, although the mole could have burrowed into the pit fills and the toad could have been hibernating.

Seventy-four elements of larger mammals were identified, all of which belonged to domestic species. Cattle are the most commonly represented with 36 fragments. Loose teeth account for nine of these elements indicating only moderate survival of the bones (Table A4.13). A cattle tibia has a distal breadth (Bd) of 57.8 mm and a distal depth (Dd) of 44.0 mm. Two first phalanges have proximal breadths (Bp) of 27.3 mm and 31.8 mm. These all fall well within the range of domestic rather than wild cattle. A mandible has heavy wear on the third molar (Grant 1982a, stage l) and belonged to a mature individual. In contrast, a fairly complete metatarsal has an unfused distal epiphysis and is porous and belonged to a calf. A hyoid has knife cuts probably made during the removal of the tongue. An accessory carpal also bears knife cuts possibly made during skinning and disarticulation of the front foot.

Sheep/goat is represented by a total of 30 elements (Table A4.14). These include articulated remains recovered from layer 4a. These consist of 17 fragments of vertebrae of a young lamb or kid. Most of the lumbar and thoracic vertebrae are represented, although they survive in a fragmented state. The bodies of most of these vertebrae have not fused indicating they belong to an animal that was probably under six months old. There is no evidence of butchery marks nor any gnawing or weathering damage. This suggests that at least a portion of this carcass was discarded without processing for meat and it may have been deposited soon after death. A calcaneus and metatarsal found at the base of layer 3 could have belonged to one immature sheep. Both bones are from the right side of the body and the calcaneal tuber and the distal metatarsal are both unfused, indicating an age of less than two years. A mandibular fourth deciduous premolar from layer 4a/b and a distal portion of humerus from layer 5 both have morphological characteristics indicative of goat. The humerus that a distal articular breadth (BT) of 28.8 mm and a maximum trochlea height of 17.5 mm. Eight elements of pig were identified including three loose teeth (Table A4.15).

F.3: This produced the largest faunal assemblage from the pits of 413 fragments, of which 280 ele-

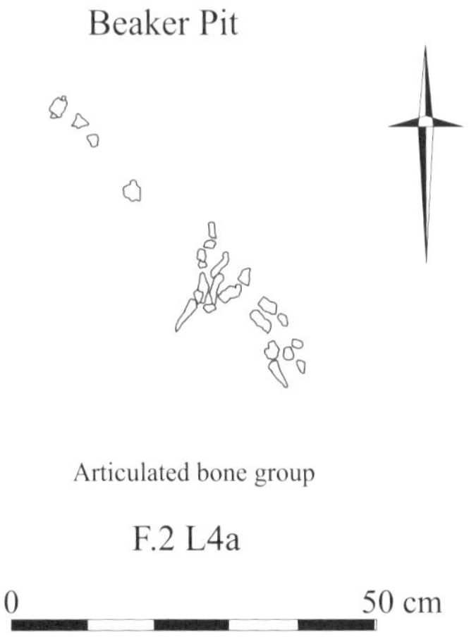

Figure A4.9. *Fir Tree Field shaft: the articulated bone group at the layer 4a/b interface in Beaker pit F.2 (N. Griffiths after M. Green).*

ments were identified. However, 222 of these belong to small mammals and amphibians found mainly in the lower part of the fills. These probably include pitfall victims. Species represented include short-tailed vole, woodmouse, common shrew, and types of frog and toad (Table A4.12). Of the 58 elements of larger mammals, 43 belonged to pig, of which 27 belong to an associated bone group found in layer 4a. These represent the remains of a young piglet. Parts of the left mandible and maxilla have survived. Both possess deciduous premolars that have erupted but are not in wear indicating that the pig was less than two months old. Bones of all four legs are represented, although they are all incomplete. The left forelimb is represented by the radius and ulna; the right forelimb by the scapula, humerus, radius and ulna. The left hindlimb is represented by the tibia and calcaneus; the right by the femur and tibia. Part of the left pelvis was also retrieved. Three incomplete metapodials could have come from any of the limbs. No phalanges were recovered. The atlas and parts of at least seven other vertebrae were found. At least three ribs are represented. There is no evidence for butchery, weathering or gnawing. Although only a partial skeleton was recovered, it is probable that this animal entered the pit whole. Whether it fell into the pit alive or was deposited as a dead body cannot be determined. A scapula and humerus, also from layer 4a, belonged to an even younger pig. The bones are very porous and small. They could be from a well-developed foetus or perhaps a perinatal mortality. A pig radius and two unworn deciduous premolars in Layer 4b are also from one or more neonatal animals. A pair of tibiae from layer 5 could also be from a foetus or perinatal mortality. An older pig is represented by a distal end of a humerus in layer 5. The epiphysis has just fused, indicating it belonged to an animal culled probably early in its second year. It has been butchered, bearing oblique incisions on the medial aspect of the distal end made during separation from the radius and ulna. Similar fine incisions were observed on the medial aspect of a rib head made during separation from the spine. A pig of over two years of age is represented by the fused distal half of a metapodial.

Thirteen cattle elements were found in F.3 representing all areas of the body (Table A4.13). Neonatal calves are represented by the distal half of a tibia from layer 5 and an unworn deciduous premolar from layer 4. Adult cattle are represented by two fused vertebrae. A humerus has a distal articular breadth (BT) of 79.8 mm and a maximum trochlea height of 46.5 mm. This falls comfortably within the size range of domestic cattle.

No sheep/goat elements were found in this pit. A roe deer tooth, a complete shed antler from a two-year-old buck and a small burnt fragment from the base of a red deer antler were the only other identified elements recorded (Tables A4.16 & A4.17).

F.4: This feature produced 169 animal bone fragments, of which 126 are unidentified. Most of these and many of the identified elements are eroded and preservation is generally quite poor. Small numbers of amphibian and rodent bones were found in the lower fills (Table A4.12). Cattle are the most commonly identified, represented mainly by hind leg and foot elements (Table A4.13). Some of the foot bones may have been associated. There are a pair of third phalanges in layer 5 and a first and second phalanx from layer 4/5 articulate. Calves are represented by the porous shafts of a femur and tibia. Adult cattle are represented by a caudal vertebra, a distal femur and a rib, all of which have fused epiphyses. A cattle metacarpal has a proximal breadth (Bp) of 52.4 mm and a proximal depth (Dp) of 32.8 mm. Two first phalanges have proximal breadths (Bp) of 27.2 mm and 28.4 mm, the former having a greatest peripheral length (Glpe) of 61.3 mm. All three bones fall within the size range of domestic cattle.

Fourteen pig elements also include several foot bones (Table A4.15). A left maxilla possesses deciduous teeth in an early stage of wear but also an unerupted first molar. This jaw belonged to a piglet of probably about three–four months of age. The other pig bones belonged to older animals, although most of these were probably from immature rather than adult animals.

Sheep/goat were represented by two fragments of a radius and the distal part of a right humerus. The humerus and one of the radii have morphological characteristics diagnostic of sheep. The radius has a fused distal epiphysis and belonged to an adult animal. The distal humerus is also fused and has evidence for exostosis extending from the lateral surface of the articulation. This pathological condition is more likely to develop in older animals. It has a distal articular breadth (BT) of 27.3 mm (not directly affected by the pathology) and a maximum trochlea height (HT) of 17.9 mm. No elements of deer were recovered from this pit.

F.5: Only 12 animal bone fragments were recovered, nine of which remain unidentified (Table A4.12). All the bones are eroded. A pig tooth and the shafts of a sheep/goat left humerus and left tibia were the only elements identified (Tables A4.14 & A4.15).

Grooved Ware pits

F.6: Although this feature produced 211 animal bone

Table A4.12. *Animal bones recorded in the pits at Fir Tree Field.*

	Pit						
Species	**F.2**	**F.3**	**F.4**	**F.5**	**F.6**	**F.8**	**Total**
Cattle	36	13	17		5	8	**79**
Sheep/Goat	30		3	2			**35**
Pig	8	43	14	1	20	8	**94**
Red deer		1			1	1	**3**
Roe deer		1					**1**
Mole	1						**1**
Common shrew		2					**2**
Short-tailed vole	2	39					**41**
Woodmouse		12					**12**
Rodent	1	96	2				**99**
Frog	1	6					**7**
Toad	4	19	4				**27**
Frog/Toad	11	48	3				**62**
Total identified	*94*	*280*	*43*	*3*	*26*	*17*	**463**
ULM	98	33	42	1	62	48	**284**
USM	66	48	33	7	34	16	**204**
UM	90	51	51	1	89	55	**337**
UB		1					**1**
Total undentified	*254*	*132*	*126*	*9*	*185*	*119*	**825**
Total	**348**	**412**	**169**	**12**	**211**	**136**	**1288**

Counts are of number of individual specimens (NISP); ULM = unidentified large mammal; USM = unidentified medium-sized mammal; UM = unidentified mammal; UB - unidentified bird.

fragments, many were eroded, burnt and fragmented, which meant that only 26 could be identified (Table A4.12; Figs. 3.15 & 3.16). Most of the 105 charred fragments, found mainly in clusters in layer 4, could not be identified. No rodent or amphibian bones were recovered, which may suggest that this pit did not remain open for any significant period. Sheep/goat is absent from the identifiable material, which is comprised mainly of pig. Its 20 elements are from all areas of the body (Table A4.15). They include a right radius and ulna from the same animal. The proximal radius is unfused, indicating they belonged to an animal under a year old. The only measurable pig bone is a scapula, which has a greatest glenoid process length (GLP) of 36.8 mm. This is larger than the average size of such bones from Durrington Walls (36.7 mm) but within the upper end of the range recorded there (42.7 mm). A second large scapula, unfortunately unmeasurable, was also found in this pit.

Five elements of cattle were identified (Table A4.13). They include the posterior half of a left mandible located near a splayed flint axe (Green 2000, 72). The mandible bears slight signs of charring on the underside of the body of the jaw. Cut marks were observed on the lateral aspect of the body running parallel to and below the cheek teeth. These were possibly made during the removal of cheek meat. The mandible belonged to a sub-adult animal with the fourth deciduous premolar

Table A4.13. *Cattle element counts, Fir Tree Field.*

	F.1	**F.1**	**F.1**	**F.1**	**F.1**	**F.1**	**F.2**	**F.3**	**F.4**	**F.6**	**F.8**	
Element	**6b**	**6a**	**6–7b**	**5**	**4**	**2**						**Total**
Horncore				3								3
Skull frag.				1			3		1		3	8
Maxilla							1	1		1		3
Mandible				1		1	2			1		5
Hyoid							1					1
Loose teeth				12	11	12	9	2			2	48
Scapula				2			1					3
Humerus				5	1			2				8
Radius				2			1				1	4
Ulna								1				1
Pelvis				1				1	1	1		4
Femur				1				1	3			5
Tibia				2			2	1	2			7
Carpals							1					1
Astragalus				3								3
Calcaneus	1			1			1					3
Other tarsals							1					1
Metacarpal				3	1				1			5
Metatarsal		1	1	5			1					8
Phalanx 1				1			4	1	2		1	9
Phalanx 2							1		1			2
Phalanx 3				1			1		2			4
Sesamoids								1	1		1	3
Atlas							1					1
Cervical V				1				1				2
Thoracic V				1			2			2		5
Lumbar V					1							1
Sacrum				1			1	1				3
Caudal V									1			1
Ribs				1			2		2			5
Total	**1**	**1**	**1**	**48**	**14**	**13**	**36**	**13**	**17**	**5**	**8**	**157**

Counts exclude elements identified as aurochs.

Table A4.14. *Sheep/goat element counts, Fir Tree Field.*

	F.1	**F.1**	**F.2**	**F.4**	**F.5**	
Element	**4**	**2**				**Total**
Loose teeth	1	1	2			4
Humerus			2	1	1	4
Radius			3	2		5
Ulna			2			2
Tibia					1	1
Calcaneus			1			1
Metatarsal			1			1
Phalanx 1			2			2
Thoracic V			10			10
Lumbar V			5			5
Unid V			2			2
Total	**1**	**1**	**30**	**3**	**2**	**37**

Table A4.15. *Pig element counts, Fir Tree Field.*

	F.1	F.1	F.1	F.1	F.1	F.1	F.1	F.1	F.2	F.3	F.4	F.5	F.6	F.8	
Element	**7**	**6b**	**6a**	**6–7b**	**5**	**4**	**2a**	**2**							**Total**
Skull frag.										2			2		**4**
Maxilla					1					1	1				**3**
Mandible				1	3			1		1			1	1	**8**
Loose teeth					18	7		3	3	2		1	2	4	**40**
Scapula					6					1			2	1	**10**
Humerus	1				2					2	1				**6**
Radius										3			1		**4**
Ulna									1	2	1		2		**6**
Pelvis										1					**1**
Femur		1	1		2		1			1	1		1		**8**
Patella											1				**1**
Tibia					1	1			1	4					**7**
Fibula									1				2		**3**
Carpals										1			1		**2**
Astragalus											1				**1**
Calcaneus										1					**1**
Metacarpal											1				**1**
Metatarsal											1				**1**
Metapodial										5			3		**8**
Peripheral Mp											1		1	1	**3**
Phalanx 1									1	3			1	1	**6**
Phalanx 2					1				1	1	1				**4**
Phalanx 3											1		1		**2**
Atlas										1					**1**
Cervical V											1				**1**
Unid V										7					**7**
Ribs										4	2				**6**
Total	**1**	**1**	**1**	**1**	**34**	**8**	**1**	**4**	**8**	**43**	**14**	**1**	**20**	**8**	**145**

still present, although quite heavily worn. All three molars were in wear, although there was only slight wear on the anterior cusp of the third molar (Grant 1982a, mandible wear stage MWS = 29). The jaw belonged to an animal probably culled in its third year. A maxilla belonged to a much younger calf. This possessed deciduous premolars in an early stage of wear whereas the first molar was only just erupting through the bone. This jaw possibly belonged to an animal of about six months old. A small, slightly charred, portion of pelvis probably belonged to a male animal. A fragment of thoracic vertebra was also charred.

A red deer left tibia was the only other element identified. It consists of the distal end, which has only just fused, indicating it belonged to a sub-adult animal.

F.8: This pit produced 136 animal bones, of which only 17 were identified (Table A4.12). The bones are quite poorly preserved and include 55 that have been burnt. Eroded fragments are also common. Much of the material is very fragmented.

No rodents or amphibians were recovered from this pit, again suggesting that it did not lie open for a long period. Eight cattle fragments (Table A4.13) include an unworn deciduous tooth of a young calf and an unworn permanent fourth premolar of a sub-adult animal. A temporal fragment is slightly charred. A first phalanx from the base of the pit has a greatest peripheral length (GLpe) of 70.3 mm and a proximal breadth (Bp) of 28.2 mm and is probably from a large domestic animal. The proximal half of a right radius was also found on the base of the pit, next to the phalanx and partly resting on a naturally perforated large nodule of flint (Green 2000, 72). The radius is heavily weathered indicating it lay exposed to the elements for a considerable time. A broken rib had been worked at one end to a rounded point, perhaps for use as a knife (see p. 266). Given the absence of any small mammal and amphibian bones in this feature, it is more likely that it lay elsewhere prior to final deposition at the bottom of this pit. The presence of carnivore gnawing damage also supports this hypothesis.

Eight pig elements were also identified (Table A4.15). They include a mandible with a first molar in an early stage of wear from a pig of about 6–9 months old. A burnt fragment of scapula was also identified. The only other identified element is a red deer antler tine found at the base of the pit alongside the flint nodule (Green 2000, 72).

Discussion

The faunal assemblage from the shaft and adjacent pits totals 2211 fragments. However, 1379 (62%) of these are unidentified (Tables A4.11 & A4.12), which significantly decreases the amount of information that can be derived from it. In addition, apart from the roe deer skeletons from the lower part of the shaft, much of the identified material is fragmentary and eroded, which restricts the availability of ageing, metrical and butchery data. Nevertheless, it is possible to provide a partial account of the history of the species that were inhabiting the area around the shaft and their exploitation.

The later Mesolithic

It would appear that at least two young roe deer fell into the natural fissure sometime in the late fifth millennium BC. Although neither skeleton is complete, enough bones survive in articulation to indicate that the animals entered the shaft complete and probably alive. Their presence in a landscape of closed but light forest is unsurprising and young inexperienced animals would presumably have been at greater risk of being victims of such accidents.

The discovery of over 300 worked flints in the shaft fills below the weathering cone, to a maximum depth of 9 m helps explain how individual bones entered the shaft. Indeed, a large flake struck from the outside of a flint nodule was found lying on top of part of a roe deer found at 5 m depth (Fig. 3.15). The butchered red deer metatarsal and a largely complete femur of a third roe deer, found near the lower of the two skeletons, suggests some scavenging was taking place, both for flint and food, during these phases of rubble accumulation towards the end of the fifth millennium BC. The fragmentary nature of other red deer, roe deer and aurochs bones indicates not only their presence but possibly their exploitation for meat as well. Perhaps surprisingly, no evidence for wild boar was found amongst the small number of bones in these fills.

Whereas the evidence for the hunting of wild mammals in the lower fills of the shaft is scanty, there is much clearer evidence for the exploitation of these animals in layer 7 of the shaft. In addition to the presence of Mesolithic weaponry (microliths), there is a significant increase in the density of animal bones. Elements mainly of red deer but also of aurochs were found. At least one aurochs bone displays evidence of marrow extraction. A neonatal pig and two calf bones are also represented. It cannot be determined whether these are from wild or domestic animals. The likelihood is that they were from wild animals. However, a few domestic cattle bones have been identified in Late Mesolithic faunal assemblages in Ireland (Schulting *et al.* 2004) and it is not beyond the realms of possibility that these bones are from domestic cattle.

Early Neolithic

Very few identifiable bones were recovered from layers 6a and 6b of the shaft. Wild species such as roe and red deer are still represented,

Table A4.16. *Roe deer element counts, Fir Tree Field.*

	F.1	F.1	F.1	F.1	F.1	F.1	F.1	F.3	
Element	**7 m**	**5.2 m**	**4.5 m**	**6B**	**6–7B**	**5**	**4**		**Total**
Antler								1	1
Maxilla		2							2
Mandible	1	1							2
Loose teeth							1	1	2
Scapula	2	2	1						5
Humerus	2	2			1	1			6
Radius	3	6			1				10
Ulna	2	2							4
Pelvis	3	1							4
Femur	4	3				1			8
Patella	1	1							2
Tibia	5	4							9
Carpals	7	5							12
Astragalus	2	2					1		5
Calcaneus	3	3		1					7
Centroquartal	2	2							4
Other tarsals	3	1							4
Metacarpal	2	2							4
Metatarsal	3	2							5
Metapodial	1	1							2
Phalanx 1	7	6							13
Phalanx 2	4	3							7
Phalanx 3	2	4							6
Sesamoids	4	2							6
Cervical V		3							3
Thoracic V		6							6
Lumbar V	4	4							8
Sacrum	1	1							2
Unid V		7							7
Ribs		4							4
Total	68	82	1	1	2	2	2	1	160

Table A4.17. *Red deer element counts, Fir Tree Field.*

	F.1	F.1	F.1	F.1	F.1	F.1	F.1	F.3	F.6	F.8	
Element	**7 m**	**3.05 m**	**7**	**6b**	**6a**	**6–7b**	**5**				**Total**
Antler			3		3	1	3	1		1	**12**
Skull frag.			2								**2**
Maxilla					1						**1**
Loose teeth			2				2				**4**
Scapula							2				**2**
Humerus			1								**1**
Pelvis				1	1	3	1				**6**
Femur	1		1								**2**
Tibia			3			1			1		**5**
Metatarsal		1	1			1					**3**
Phalanx 2						1					**1**
Total	**1**	**1**	**13**	**1**	**5**	**7**	**8**	**1**	**1**	**1**	**39**

although antler as well as bone fragments of red deer are present. The bovid and porcine elements are of a size small enough to be considered to be from domesticated cattle and pigs. It is unfortunate that most of the bones from this part of the shaft could only be assigned to layers 6–7b, which could contain mixed Late Mesolithic and Early Neolithic material. The identified elements from that assemblage belong mainly to wild species. As in layer 7, red deer is the most commonly represented with aurochs and roe deer also present. The most likely explanation would be that most of the bones in this assemblage are from the Late Mesolithic. However, bones small enough to be from domestic cattle and pigs were also found. Perhaps the remains do come from a period of transition when foragers began to supplement game with meat obtained from newly introduced domesticated animals.

Middle–Late Neolithic

The faunal remains from layer 5 of the shaft are associated with Peterborough Ware. It is an assemblage dominated by cattle and pig, with evidence for carcass processing on some of the cattle bones. Metrical data and more subjective observations suggest that most of the cattle and pig bones are from domestic animals, although small numbers of elements of aurochs and perhaps wild boar are represented, as are red and roe deer. Although sheep/goat elements are notable by their absence, most of the bones deposited into the shaft in this period were those of domestic mammals. This perhaps indicates that there was a much greater reliance on these animals for meat than in the earlier Neolithic, although it is difficult to make broad generalizations from a small assemblage from such an unusual feature.

The Late Neolithic material from layer 4 of the shaft is very poorly preserved. However, domestic cattle and pig elements continued to be the most frequently deposited. Roe deer is the only wild species definitely present and the absence of red deer and aurochs perhaps indicates a further move away from the exploitation of wild species. Sheep/goat remains appear for the first time, albeit only represented by a single tooth.

The faunal assemblages from the Grooved Ware pits F.6 and F.8 have a lot in common. They both contain a large number of unidentified fragments, many of which have been burnt. They both contain what appear to be carefully placed elements near their base. The absence of bones of amphibians and rodents suggests that the pits were filled in fairly rapidly. Pig and cattle are the main species represented in both pits, although pigs are more abundant in F.6. The very limited ageing data reveal that animals of various ages are represented. One element of red deer was identified in each pit.

The assemblages from these pits are also similar to those recovered from earlier excavations of Grooved Ware pits in Fir Tree Field, 140 m to the southwest (Legge 1991). Apart from the exceptional assemblage in Pit 11a, which was dominated by cattle and also contained elements of red deer, roe deer and bear as well as pig and sheep/goat, most pits produced more pig elements than cattle. Several of these pits also contained placed bone deposits sometimes associated with artefacts. The association of relatively high percentage of pigs in sites with Grooved Ware has been noted on a number of Late Neolithic sites, particularly at Durrington Walls (Albarella & Serjeantson 2002).

No elements of sheep/goat were identified in either F.6 or F.8. As noted in the discussion of the assemblage from Monkton-up-Wimborne Late Neolithic pit circle and shaft complex, sheep/goat are absent or poorly represented in most of the Neolithic assemblages from this area. This may appear surprising since it has been demonstrated that sheep/goat were present in the area from the Early Neolithic. Preliminary results from the causewayed enclosure at Hambledon Hill indicated that they contributed 25 per cent of the identified bones (Legge 1981). However, this appears to be an unusually high percentage when compared with other Neolithic sites. In an assemblage obtained from another local causewayed enclosure at Whitesheet Hill, sheep/goat provided only 5 per cent of the hand-collected bone excluding associated bone groups (Maltby 2004) in an identified assemblage of 290 elements that was dominated by pig (68 per cent). In the assemblages discussed in this volume, sheep/goat are not found in the Monkton-up-Wimborne shaft until the latest Neolithic levels (layer 4) and there only in very small numbers. No sheep/goat elements were found in any of the Neolithic pits in the circle around the central pit and shaft. They are also absent from the Fir Tree Field shaft below the latest Neolithic layer. Elsewhere in the vicinity, sheep/goat bones were not found in the small assemblage of 30 identified elements from the Cursus and provided only three of the 64 bones from the Wyke Down henge and eight of the 142 elements in the Grooved Ware pits elsewhere in Fir Tree Field (Legge 1991). If the bone assemblages in any way represent the relative numbers of animals available in the locality, it would appear that sheep were not being exploited in anything like the numbers of cattle and pig during the Middle and later Neolithic. Sheep/goat are also poorly represented in several Neolithic assemblages near Stonehenge (Maltby 1990a).

The Beaker period

Faunal assemblages associated with Beaker pottery were found in upper layers of the shaft and in four pits. Apart from the absence of any identified elements of wild species, the bone assemblage from layers 2 and 2b of the shaft is similar to that from the Late Neolithic deposits (which also contained a few Beaker sherds at the top of layer 4). Most of the small number of identified elements belonging to cattle (13 fragments) with pig (4) and a single sheep/goat element also present.

Apart from F.5, all the pits produced small mammal and amphibian remains. If the numbers of finds to any extent reflect the length of exposure prior to infilling, F.3 lay open for a longer period than F.2 and F.4. The presence of both woodmouse and short-tailed vole indicates that both woodland and more open scrub or grassland were located in the vicinity. F.3 also produced a species profile that is similar to the Grooved Ware pits. No sheep/goat elements were identified and pig bones were the most frequently recovered. However, although these include at least two bones from butchered animals, many of the pig bones belonged to very young animals that were probably not processed. These include substantial parts of the skeleton of a piglet of under two months of age. Domestic cattle elements were also found in some numbers as well as parts of a red deer antler and a roe deer bone.

F.4 did produce sheep/goat elements but only in small numbers compared with cattle and pig. Sheep/goat are better represented in the other two pits. F.5 produced only three identified elements but two of these are of sheep/goat. The more substantial assemblage from F.2 includes a partial skeleton of a young sheep/goat and both sheep and goat are represented. Although sheep/goat elements outnumber those of pig, cattle elements were the most frequently identified in this pit.

Therefore, although there were some changes in the relative abundance of species deposited in features of Beaker date, there is a lot of variation between features. Sheep/goat appear more consistently than in Neolithic deposits, but are not present in every feature. Pig continues to be prevalent in some features but not in others. Several reasons could account for these variations. Environmental evidence indicates that the area around the shaft began to be cleared of woodland around this period as part of a more general opening up of the landscape for agricultural expansion. By the Middle Bronze Age in this area faunal assemblages associated both with settlement and burials have evidence for more sheep/goat and cattle and a marked decrease in the abundance of pig (Legge 1991). Perhaps, at the beginning of this period of this transition, we should expect faunal assemblages to become more variable. Variations in species abundance in the shaft pits may indirectly reflect vegetation change within the Beaker period but dating evidence is not sufficiently refined to test this and we should be cautious about the dangers of over-interpreting variations from such small samples.

Unfortunately, it is not feasible to carry out detailed ageing analysis on this assemblage. It is clear from the detailed summaries above that both young and mature cattle, sheep, roe deer, red deer and aurochs are represented. Very few of the pig bones are from adult animals and there are the remains of several very young animals in one of the Beaker pits.

Several articulated groups of bones were recovered from the excavations. However, there is no clear evidence that any of them represent deliberate deposition of sacrificed animals. Indeed, the roe deer appear to be victims of falls. The weasel/stoat and toad also probably entered the deposits unaided by humans and it is even possible that the piglet in F.3 was a victim of a fall, although it is perhaps surprising that it could not escape from a depth of 70 cm. This leaves the remains of the lamb or kid in F.2 as the only case where deliberate deposition is perhaps a more likely explanation for the presence of the partial skeleton.

Butchery data are very limited but they have provided glimpses of evidence for the filleting of meat and marrow procurement in the Mesolithic, Neolithic and Beaker periods on both wild and domestic species. None of the antler fragments have evidence for working or wear, although most are so poorly preserved that such evidence was difficult to observe.

Summary of the molluscan data

Michael J. Allen

The molluscs from the shaft

Snails were analysed from a series of nine spot samples from 'soil' lenses within the chalk rubble fill of the main shaft (pre-4470–4240 cal. BC to 4320–4000 cal. BC), and a contiguous column of 23 samples was (taken by Roy Entwistle) through the weathering cone and covering the Late Mesolithic (4340–4040 cal. BC; OxA-8011, 5355±45 BP) to Beaker period (2460–2040 cal. BC; OxA-7985, 3775±45 BP) (Allen & Green 1998). Shells were well preserved in the thin soil lenses averaging over 300 per sample. Numbers increased significantly in the soil fills of the weathering cone reaching 2338 per sample with highest numbers just below and just above the Peterborough Ware phase old land surface. The earliest molluscan assemblages from soil lenses in the shaft rubble (pre-4470–4240 cal. BC; OxA-7991, 5500±55 BP) and those from the top of shaft rubble (pre-4340–4040 cal. BC; OxA-8011, 5355±45 BP) all indicate remarkably open conditions probably of open dry long grassland with some shrubs (including hazel charcoal), possibly typical park woodland (cf. Vera

2000). There is no evidence of ancient woodland in either the snail or the charcoal evidence.

At the base of the weathering cone (base of layer 7a) rate of infill starts to slow, and colonization of shade-loving species probably representing local long grassland but also possibly some scrub can be seen at about 4340–4040 cal. BC. However, these locally shady conditions are removed and replaced by bare earth and short grassland established in the lower fills of the weathering cone; upper portion of layer 7a (4240–3990 cal. BC). This broadly relates to the find of Mesolithic rod microliths (hafted).

Shade-loving conditions re-establish in layer 7/6b (4240–3990 cal. BC) and 6b (3980–3710 cal. BC) and an open woodland with ash and hazel seems likely, with long and dense grassland in the Mid Neolithic Peterborough Ware activity. Cleared open grassland and possibly arable conditions are rapidly established in conjunction with the onset of post Peterborough Ware layers (layers 4 & 3), which presumably relate to the Grooved Ware phase (3310–2920 cal. BC). Although no Grooved Ware was recovered from the shaft, a series of Grooved ware pits surrounded it. These open conditions continued through the Beaker 'occupation' phase deposits (layer 2/2a) at 2460–2040 cal. BC.

Different phases of shade-loving conditions and vegetation regeneration have different characteristic and will be presented in more detail elsewhere (Allen in prep.)

The molluscs from the associated pits

Surrounding the shaft was a series of Grooved Ware and Beaker pits (Green 2000; see above). A series of 23 samples was taken as short sequences from three Grooved Ware pits (pits 6, 8 & 9) and three Beaker pits (pits 2, 4 & 5). The Grooved Ware pits represent open downland environments, but probably ones with long ungrazed grassy swards, and possibly shrubs. In contrast the Beaker pits show significantly increased shell numbers and more open dry conditions typical of grazed but dense grassland swards, again there is hint of shadier (?shrubby) environments in the vicinity. These environments, not surprisingly, reflect those depicted from the long sequence in the shaft. Subtle differences between pits might represent different microenvironments created by different and varying pit histories within both the Grooved and Beaker groups. Other differences between the two sets of pits will be defined by further analysis.

4.3. Fir Tree Field Middle Bronze Age enclosure *by Martin Green*

Site summary

The results of this excavation have been reported on in Barrett *et al.* (1991a, 183–211). The enclosure was ditched on three sides and defined by a fence-line on its fourth. Inside there were the remains of six roundhouses, a large rectilinear building, a possible four-post structure and several short fence lines. Although broadly Middle Bronze Age in date, not all the structures were necessarily contemporary. A new section was cut through close to the southern terminal of the rectilinear enclosure in 2004 to obtain a profile for snail analysis (Green 2000, fig. 73).

Soil micromorphology

HELEN LEWIS

Two thin sections were taken by Richard Macphail in 1996 through the present day topsoil in Fir Tree Field. This was an organic A horizon, with two mixed fabrics, with fabric 2 as the main component of the lower thin section, developed on the chalk substrate (Table A4.18).

Structure & porosity: Fabric 1: crumb (<3000 μm), occasional sub-angular blocky (<3000 μm), with 20–30% porosity, mainly packing pores and channels (100 × 1000 μm and 250–400 μm diameter). Fabric 2: apedal channel, <5% porosity (channels <150 μm diameter).

Table A4.18. *Fir Tree Field section: micromorphological descriptions (H. Lewis).*

Context	Structure	Porosity	Mineral components	C:f ratio	Organic components	Groundmass	Pedofeatures
A horizon	Crumb to sub-angular blocky	20–30%, packing pores & channels	Quartz; chalk; shell; earthworm granules; micrite	10:90 Clayey silt loam	5–10%, very fine charcoal & black round fragments	Brown (PPL), very dark brown to undifferentiated (XPL); stipple speckled to undifferentiated	Zones of crystallitic fabric; micrite coatings
Lower A horizon	Apedal channel	<5% channels	As above	As above	10% as above	Orange (PPL), very dark reddish-brown (XPL), stipple speckled, some granostriation	Micrite coatings

Mineral components: Both fabrics: quartz, chalk, shell, earthworm calcitic granules, calcite grains, micrite.
C:f ratio: Both fabrics: 10–90: very coarse & coarse sand 0%, medium sand 5%, fine sand 5%, very fine sand 5%, silt 45%, clay 40%.
Organic components: Fabric 1: 5–10%, mainly <250 μm sized, often linear, often charcoal, and <150 μm black round fabrics; Fabric 2: as fabric 1, but only 10%.
Groundmass: Fabric 1: brown (PPL), very dark brown to undifferentiated (XPL), brown (RL), with zones of crystallitic fabric (micritic); Fabric 2: orange (PPL), very dark reddish-brown (XPL), orange (RL), stipple speckled with occasional granostriated zones.
Pedofeatures: Both fabrics: micrite on pore edges.

Molluscan data
MICHAEL J. ALLEN

See Chapter 3, pp. 171–3 and Table 3.17, p. 173.

4.4. Wyke Down Henge 2 and Grooved Ware settlement (WD2 96)

Lithics
MARTIN GREEN

A total flint assemblage of 5917 pieces of worked flint was recovered from the excavations from three main contexts: the henge ditch (2817), twenty-two pits (2595) and postholes (505) (Table A4.19; Figs. 3.25–3.32, A4.10 & A4.11). The retouched, utilized and tool (RUT) component was slightly higher in the pits (5.1%) and postholes (4.4%) to the henge ditch (3.6%).

The henge ditch
The western side of the henge was completely excavated, but with only one of three sections completed on the eastern side. The two terminals on the eastern side were only partly excavated, that on the northeast to the top of the primary fills and in the southeastern terminal only layer 2 was removed. The ditch fill consisted of four main contexts — the upper secondary layer 2, the lower secondary layer 3, the upper primary fills, layers 3a to 4c, and the lower primary fills of layers 5 and 6 (Fig. 3.25 & 3.26).

Layer 2 consisted of a mid-brown decalcified silt and produced 1007 flint artefacts. A particular concentration of cultural material was noted in trench H which included 43% of the total flintwork recovered from layer 2. A retouched, utilized tool (RUT) component of 4.7% was present which included 20 scrapers or 42% of the RUT component. This group included 13 of convex form, six side and a single end. A chisel and four oblique arrowheads (Fig. A4.11, nos. 10 & 11) included a type H from the southwest terminal (Fig. A4.11, no. 8), a type recovered in large numbers from Durrington Walls (Wainwright & Longworth 1971), and a possible 'symbolic arrowhead' as described by Brown (in Barrett *et al.* 1991b, 130 & fig. 6.5). Four fabricators included a small example with much use wear at either end, a long (122 mm) narrow example of lozenge section with no visible signs of wear, another of 'D' section (77 mm) with a plunging fracture at one end and a fragment. Two flake borers included one with an attenuated point (45 mm). A single serrated flake was also recovered.

Layer 3 consisted of grey chalky silts and produced 994 artefacts. The concentration of cultural material noted in Trench H in layer 2 also continued into layer 3 and included 36% of the flintwork recovered from the layer. An RUT component of 3.9% included 11 scrapers (28%), all of convex type. Arrowheads included two of chisel form (Fig. A4.11, nos. 6 & 7) and two obliques including a crude 'symbolic' example of the type described elsewhere (see above). Six serrated blades were recovered and included a particularly long example (86 mm) with a serrated edge of 60 mm).

In the upper primary fill layers 3a to 4c, the rate of soil accumulation was much faster and the number of incorporated artefacts much less than in the secondary fills above. 437 flint artefacts were recovered from these levels and included an RUT element of 2.7%, one-third of which comprised scrapers, two of convex form and two end scrapers, including one with a serrated edge (Fig. A4.11, no. 4). Other significant finds included three chisel arrowheads, one well made (Fig. A4.11, no. 3), and the other two of crude workmanship, an elaborate 'tanged' fabricator (Fig. A4.11, no. 5) and a serrated blade.

In the primary chalk rubble fills of layers 5 and 6, 379 artefacts were recovered of which only three were retouched or utilized (0.8%). These included two scrapers (Fig. A4.11, no. 2) and another of the crudely made 'symbolic' arrowheads (Fig. A4.11, no. 1) of the type already described.

Pits of the Grooved Ware settlement
Twenty-two pits were found scattered outside the henge principally to the west (Figs. 3.21–3.24), most of which produced significant numbers of sherds of Grooved Ware (see Cleal below). The excavated area was arbitrary and the settlement area undoubtedly continues beyond the areas stripped. 2595 flint artefacts were recovered at an average per pit of 118 (Table A4.20). However, the actual numbers varied from 649 (in F.22) to 6 (in F.12 & F.13) (Table A4.20). F.22 was by far the richest pit discovered and not only produced

Table A4.19. *Lithic types present in the Wyke Down 2 henge ditch.*

Ditch/trench no.	Layer	Cores	Core frags.	Flakes	Chips	Retouched	Utilized	Scrapers	Other tools
Trench A	L2	2		12					
Trench B	L2			46	30				
Trench C	L2	7		94	24	2	1		fab., borer, saw
Trench D	L2	3		42	6			1	fab., ch. arrow
Trench E (E TERM)	L2	2		22	10				borer
Trench E (W TERM)	L2	7		68	27	1		1	
Trench F	L2	1		28	12		1		knife frag
Trench G	L2	10		50	18			1	obl. arrow
Trench H	L2	12		283	115	1		17	3 obl. arrows, hammer, fab. 2, notched flk
Trench I	L2	1		25	7	1	1		
Trench A	L3			24	3			2	Saw
Trench B	L3	1		32	33				
Trench C	L3	7		68	2	2	2	1	
Trench D	L3	4		105	20	2	1	1	2 ch. arrows
Trench E (E TERM)	L3	3		11	1				
Trench E (W TERM)	L3	1		39	5				
Trench F	L3	8		99	27	3		3	
Trench G	L3	4		74	30	1		1	
Trench H	L3	11		261	65	2		4	2 obl. arrows, 2 borers, bifacial pointed tool, small axe-like tool, 5 saws
Trench I	L3	1		16					notched flake
Trench A	L3/4 interface			13					
Trench A	L4a, 4b, 4c			43	2			1	
Trench B	L3a	2		43	21				
Trench C	L4	2		18	4				fab.
Trench C	L4a	1		8	7				
Trench D	L4a			10					ch. arrow
Trench D	L4b	3		29	4			1	ch. arrow
Trench E (E TERM)	L4	2		14	3				ch. arrow (rough-out)
Trench E (W TERM)	L3b			4	1				
Trench E (W TERM)	L4			4					
Trench F	L4			28	12				
Trench F	L4a	2		18					
Trench G	L4–4c	5		38	11			2	
Trench H	L4a	3		64	17	1	1		scraper/saw, saw
Trench A	L5			9					
Trench B	L5	3	3	50	53				
Trench C	L5			31	2				
Trench D	L5	3		24	3				
Trench E (W TERM)	L5			8					
Trench F	L5		1	36	19			1	
Trench G	L5	3		10	3				
Trench G	L6	7		40	42			1	obl. arrow (unfin)
Trench H	L5	1		20	5				

Table A4.20. *Lithic types present in the pits of the Wyke Down 2 Grooved Ware settlement. PC = Portland Chert.*

Feature no.	Layer no.	Cores	Core frags.	Flakes	Chips	Retouched	Utilized	Scrapers	Others
F.1	L2	8		65	22	1	1		3 obl .arrows (1 unf), notched flake
F.2	L2	1	1	37	21	2		1	knife
F.2	L3			10	3				
F.3	L2			7	4				
F.3	L3		1	17	12				borer
F.3	L5	2		11	7				
F.4	L2	2		73	18	1		1	
F.4	L3	2		71	20			2	saw
F.4	L4	6		24	19				
F.5	L2			1	1			1	
F.5	L3	1		5	5				
F.5	L5			10	8				
F.6	L2	4	2	96	59			6	
F.6	L3	1		20	17				obl. arrow (unf)
F.6	L4			6	3				
F.6	L5			21	14	9			hammer/ball
F.7	L2	3		58	22			1	saw, denticulate
F.7	L3	3		34	32			1 (frag.)	obl. arrow (frag.)
F.7	L4	1		27	36			1	
F.7	L4a	2		22	10				
F.7	L4c			11	4		1	1	
F.8	L2	11		73	66		3	2	2 obl. arrows, 2 saws
F.9	L2			6	1				
F.9	L3			2	1				
F.10	L2			5	4				
F.10	L3a	1		9 (1 in PC)				2	
F.10	L5A	1		5					hammer
F.11	L2			17 (1 in PC)	3				2 notched flakes
F.11	L3			1 (PC)	1				
F.11	L4	6		24	19				
F.11	L6			1 (PC)					
F.12	L2	1		5					
F.13	L2			1	1	1			
F.13	L3			3					
F.14	L2	3		16	6	2			knife
F.15	L2	2		53	28			10	knife
F.16	L2	1		15	12	2	1	2	saw
F.16	L4			6	4				
F.16	L5			10	10				
F.16	L6	1			5				
F.17	L2			14	7			1	
F.17	L3/4								saw
F.17	L4			2					
F.17	L6			15	7	1	1	1	
F.18	L2	3	1	61	41	1		1	borer, saw, trans. arrow (frag.)
F.18	L3			40	15				
F.18	L4	1		11	6				
F.18	L5	6	1	19	16	1		1	
F.19	L2	3		31	42	4	2	2	saw, scraper/saw
F.19	L3			2					
F.20	L2	2		26			1		
F.21	L2	2		18	7				
F.22	L2	8		110	42	1	1	7	obl. arrow, 2 dentic.'s
F.22	L3	16	6	258	138	1	3	6	2 saws, trans. arrow frag.
F.22	L4	1		22	19	1	2		obl. arrow

Table A4.21. *Lithic types present in the postholes of the Wyke Down 2 Grooved Ware settlement.*

Feature no.	Part of	Cores	Core frags.	Flakes	Chips	Retouched	Utilized	Scrapers	Others
PH1	Isolated			22	24				
PH2	Isolated	1		7	5				
PH3	Isolated	1		3					
PH4	Fence			6	4				
PH5	Fence	1		6					
PH6	Fence	1		8	4	1			
PH26	Fence			3	2				
PH7	Isolated				1				
PH8	Isolated	1		30	9				pick, obl. arrow (r'out)
PH10	House 1			18	31				hammer
PH11	House 1			14		1			
PH12	House 1			9	7			1	2 saws
PH13	House 1			9	4				
PH14	House 1			19	6				saw
PH15	House 1 L2			3	3				
PH15	House 1 L3			29	28			1	
PH16	House 1			1					
PH19	House 1	2		5	1				
PH39	House 1 (stake)			1					
PH28	Isolated			50	33		2		obl. arrow, waisted tool
PH29	Isolated			5	11				
PH30	House 2			2	4				
PH31	House 2			1					
PH32a	House 2			7				3	
PH32b	House 2			25	9			5	
PH32c	House 2			2	1				
PH35	House 2			1					
PH49	Isolated			1	2				

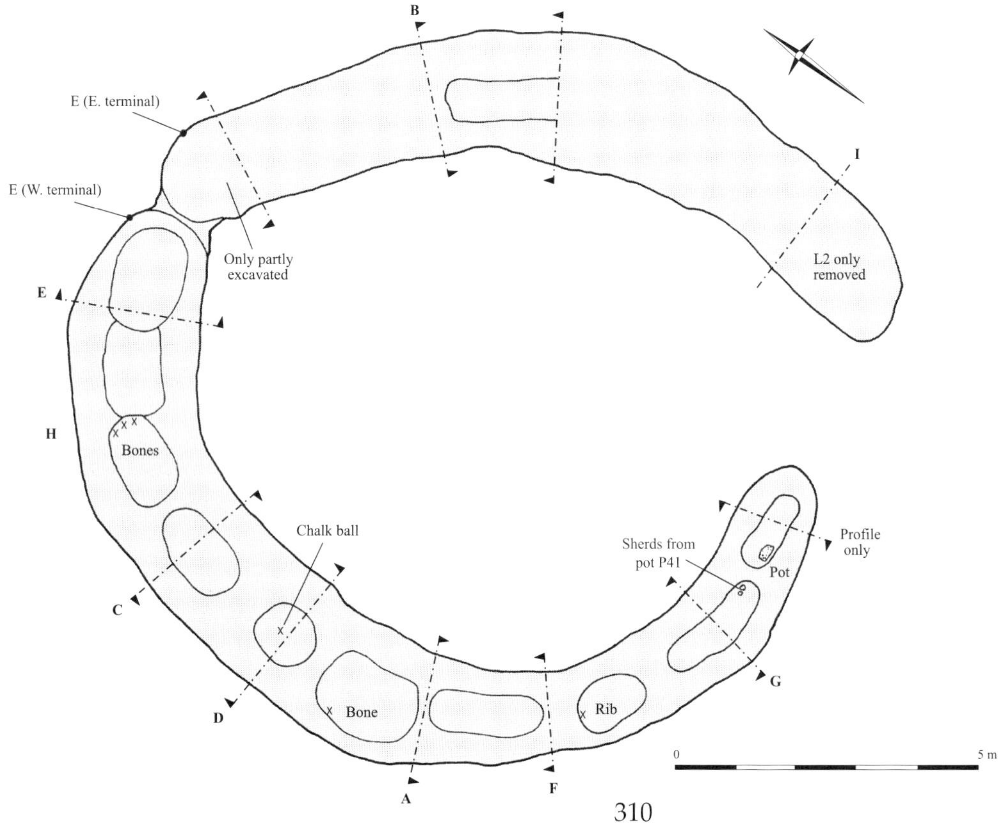

Figure A4.10. *Wyke Down henge 2: plan and trench layout (N. Griffiths after M. Green).*

quantities of animal bone and pottery, but also a quarter of the flints recovered from the pits. The varying quantities of artefacts between the pits partly reflects relative pit size. In general, the flint recovered from the pits appear to represent fairly mundane assemblages of domestic character with little in the way of 'special deposits' as recorded in Fir Tree Field (see Appendix 4.2). However, there was the occasional exception such as the dog skull and part-polished flint 'ball' found on the floor of F.6 (Green 2000, fig. 44d). This latter artefact (53 mm in diameter) had originally been heavily battered by pounding and subsequently part polished to a fine smooth surface. Its juxtapositioning with the dog skull suggests it was an artefact of symbolic significance reminiscent of the chalk balls sometimes recovered from henge monuments. Bradley & Thomas (1984) record six 'chalk' balls from the excavations at Maumbury Rings, although it is clear from the photograph illustrating their article that at least one of the balls, if not more, is polished flint. Another exception was the structured deposit recorded in F.8 (Fig. 3.31). A number of natural flints including one of phallic form had been placed in this pit together with sherds of Grooved Ware. Amongst the worked flint were two oblique arrowheads, one finely made and the other crude, possibly representing a symbolic arrowhead, four convex scrapers and two serrated blades. Other unusual groupings of flint artefacts include ten scrapers and a knife in the small pit F.15 just to the north of building 2, and three oblique arrowheads (Fig. A4.11, nos. 15 & 16; and one of crude type) from the large shallow pit F.2, one of a pair to the south of the henge. With these exceptions, the flintwork from the pits appears to be of largely domestic character.

Catalogue of illustrated flint (Fig. A4.11)

1. Trench G, L6 on base of ditch: Unfinished oblique arrowhead with crude battered bifacial edge retouch on right side and trimmed at the base of the dorsal surface. This may be another example of a 'symbolic' arrowhead as discussed by Brown (1991, 130 & fig. 6.5) and similar to other crude forms found in the Wyke Down henge 1 ditch (see also nos. 9, 19 & 23–4 below).

2. Trench F, L5 on base of ditch: Large convex scraper. The neatly retouched convex end with clear signs of use polish.

3. Trench D, L4b: Narrow chisel arrowhead with slightly splayed cutting edge.

4. Trench H, L4a: End scraper with one side serrated and with signs of use damage.

5. Trench C, L4: Bifacially flaked fabricator, narrowed towards butt end to produce an almost tanged form, perhaps to facilitate hafting. The anciently damaged tip with no remaining signs of use polish.

6. Trench D, L3: Chisel arrowhead with splayed cutting edge, one corner of which is anciently broken.

7. Trench D, L3: Chisel arrowhead with splayed cutting edge.

8. Trench G, L2: Oblique arrowhead with pronounced elongated barb (Clarke's Type H) and with the point anciently broken. Large numbers of this type were recorded at Durrington Walls (Wainwright & Longworth 1971, fig. 74).

9. Trench H, L3: Very poorly made oblique arrowhead. See comments for 1 above.

10. Trench H, L2: Fine, completely bifacially flaked oblique arrowhead.

11. Trench H, L2/3 interface: Oblique arrowhead of similar form to 10 above and found nearby. So similar are these that it is tempting to see them as being made by the same person.

12. F.4, L3: Serrated blade with traces of gloss on the edge.

13. PH 8: Small roughly flaked pick made from a broken part-polished axe fragment. Only a very small area of part polish survives close to the right hand battered edge. This edge was probably deliberately battered to facilitate hafting of the original axe, a feature that has been observed on a number of polished axes found in the area.

14. F.2, L2: Flake knife with extensive inverse retouch and utilization damage on the cutting edge.

15. F.1, L1: Oblique arrowhead with a pronounced curving profile, found with 16 (below).

16. F.1, L1: Oblique arrowhead with a pronounced lump along one side caused by the knappers failure to successfully reduce the piece. Tip of barb broken anciently. Found with 15 (above).

17. F.4, L3: Unfinished discoidal knife broken in manufacture.

18. F.6, L2: Combination tool, scraper/point. Convex scraper at distal end and bifacially edge retouched to a point at the bulbar end.

19. F.8, L2: Oblique arrowhead with crude blunting retouch. Possibly another example of a non-functional or 'symbolic arrowhead' (see 1 above).

20. F.8, L2: Well-made oblique arrowhead with extreme tip broken in antiquity.

21. F.19, L2: Combination tool, scraper/saw. Convex scraper with heavily utilized saw edge on right side. Part of striking platform perimeter on dorsal side exhibits use polish.

22. F.22, L2: Convex scraper.

23. F.22, L4: Poorly retouched oblique arrowhead with a curving profile. Possibly another example of a 'token' arrowhead (see nos. 1, 9 & 19 above).

24. F.22, L3: Poorly retouched oblique arrowhead made from a flake with a pronounced hinge fracture, part of which survives. Also possibly a 'token' arrowhead (see nos. 1, 9, 19 & 23 above).

25. F.22, L3: Extensively retouched convex scraper of square outline with traces of use polish. The multiple flake scars on the dorsal surface and partially facetted butt suggests the original flake was struck from a disc/tortoise core.

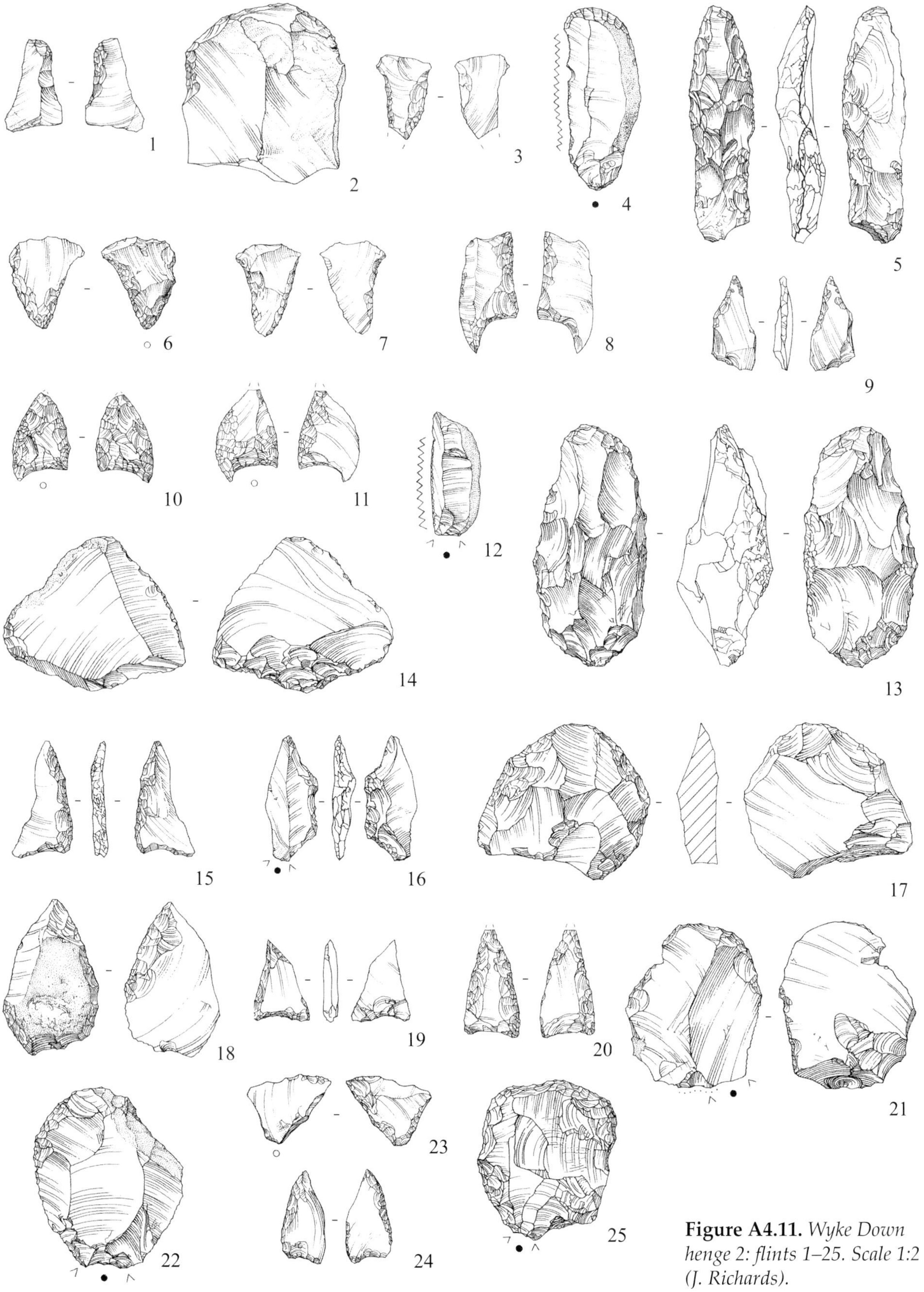

Figure A4.11. *Wyke Down henge 2: flints 1–25. Scale 1:2 (J. Richards).*

Stone objects

Martin Green

See the general comments made in Appendix 4.2 for the Fir Tree Field shaft site.

Catalogue (not illustrated)

S1. Trench A, L3/4 interface: Chip of fine-grained quartz-rich mid-brown sandstone. Wt 2 g.

S2. F.1, L2: Small lump of medium-grained iron-rich heathstone. Wt 2 g.

S3. F.1, L3: Large lump of coarse-grained heathstone. Wt 420 g.

S4. F.7, L4a: Small lump of fine-grained iron-rich heathstone. Wt 41 g.

S5. F.7, L4c: Small fragment of medium-grained iron-rich heathstone. Wt 2 g.

S6–8. F.8, L2: A structured deposit in this pit (Fig. 3.30) included three selected natural flints: S6: a 'phallic'-shaped gravel nodule (160 × 35 mm); S7: an oval flint pebble derived from tertiary deposits (58 × 49 mm); S8: a pointed oval gravel nodule (95 × 65 × 57 mm).

S9. F.12, L3: Small lump as described for S5. Wt 6 g.

S10. F.15, L2: Small flint pebble (20 mm) derived from Tertiary deposits.

S11. F.18, L2: Small lump of hard fine-grained iron-rich heathstone. Wt 37 g.

S12. F.19, L2: Small lump of coarse-grained heathstone. Wt 7 g.

S13. F.22, L3: Small lump of medium-grained iron-rich heathstone. Wt 8 g.

S14. F.22, L3: Lump of medium-grained iron-rich heathstone. Wt 42 g.

S15. F.22, L3: Tabular fragment of siliceous sarsen (19 mm thickness) broken by heat. One face very smooth perhaps from use wear. Wt 69 g.

S16. PH 2, L2: Small lump of soft fine-grained iron-rich heathstone. Wt 7 g.

S17. PH 6: Fragment of a chalk ammonite revealing weathered suture impressions. Wt 48 g.

S18. Building 1, PH 14, L2: Small chip as described for S16. Wt 1 g.

S19. Building 1, PH 19, L2: Oval flint pebble (69 × 57 mm) derived from Tertiary deposits.

S20. Building 2, PH 31: Large heat spall from a lump of siliceous sarsen with a rounded outer natural surface. Wt 200 g.

S21–8. Building 2, PH 32b, L2: Seven small fragments of medium-grained iron-rich heathstone. Gross wt 51 g.

S29. Building 2, PH 32b, L2: Chip of Upper Green sandstone. Wt 6 g.

S30. Building 2, PH 32b, L2: Fragment broken from a large lump of siliceous sarsen with a dark undulating weathered crust. Wt 57 g.

S31. Building 2, PH 32b, L2: Soft heat shattered lump of siliceous sarsen in a number of disintegrating fragments. Gross wt 92 g.

S32. Building 2, PH 32b, L2: Large fragment of medium-grained iron-rich heathstone. Wt 134 g.

S33. PH 49, L2: Large 'phallic'-shaped chalk flint nodule (173 × 48 mm).

Pottery of Wyke Down 2 henge

Rosamund Cleal

The following is a summary report, giving an overview of the ceramic assemblages, with a catalogue of the illustrated sherds. The pottery was selected for illustration by the excavator and was illustrated before the compilation of the catalogue, so that there are in one or two cases discrepancies between the illustration and the description, but in these cases I have tried to make the catalogue entry as accurate as possible. The aim of this section is to characterize the assemblages, to discuss in brief the site context in which the material was found, and to relate the ceramics to the local, regional and national context at the time.

The ditch

The majority of the pottery from the henge ditch segments (Figs. A4.12 & A4.13) is Grooved Ware, with a small minority of sherds belonging to the Peterborough tradition, and with two Beakers represented.

Peterborough Ware

Five vessels are represented by sherds, one of them by several rim sherds (at least P1–P3), one by a single rim, P11, and the others by body sherds P5, P6, P7 (Fig. A4.12). Both vessels with surviving rims (P1/2/3 and P11) (Fig. A4.12) have unthickened or barely thickened rims with oblique slashes across the tops. P11 has a fingernail impression with a slight depression around it, probably accidental, in an unpronounced neck. Both vessels are in fabrics which are unlike those of the Grooved Ware from the site, and they almost certainly belong to either the Ebbsfleet or Mortlake substyles of the Peterborough tradition (P11 has the superficial appearance of Fengate Ware but is probably not assignable to that sub-style). The Ebbsfleet Ware vessels from Windmill Hill illustrated in Smith (1965, fig. 32), are comparable in rim form, if not in decoration to P1/2/3, while a geographically closer comparison may be to P140 from Handley Hill barrow 27 (Cleal 1991, P140). Fabrics with quartz inclusions are a minor feature of the Peterborough tradition (Cleal 1995). The body sherds are probably from three separate vessels: P6 may be from a Fengate Ware vessel as linear fingernail impressions are particular characteristic of that sub-style. The sherds of most of the vessels were found in all levels of the henge ditch (layers 2–5), in sections C, F and H, where they must either have been redeposited or represent a fairly late use of Peterborough Ware and one contemporary with the deposition of Beakers (see below).

Grooved Ware

The amount of Grooved Ware from the henge ditch is small, only between about six and ten vessels in most cases represented by no more than a handful of sherds. The single exception is a substantial part of a vessel from the ditch terminal in section G.

The fabrics are limited in range and typical of Grooved Ware fabrics in Wessex. In particular they show the characteristic features of a tendency to use recipes with multiple inclusions, and a preference for calcareous inclusions, particularly chalk and shell (Cleal 1995, 190). As demonstrated in Cleal (1995), this is not solely dictated (or even mainly dictated) by availability of local materials as there are strikingly different preferences between traditions which are clearly unrelated to local geology (particularly, for example, the use of grog in some traditions and not in others). The nature of the shell has not been established, but in both cases — marine or fossil — the pottery, clay or inclusions or all of them must have been brought from outside the immediate area of Wyke Down.

In terms of Longworth's (1971) definition of sub-styles the pottery shows most features of the Durrington Walls sub-style (Wainwright & Longworth 1971), with its typical subdivision of vessels into panels, such as the vessel with applied ribs, P20/21/24 (Fig. A4.12). Longworth (1971) also includes twisted cord as a diagnostic feature of the Durrington Walls sub-style, as occurs on the rim P9. Three other vessels appear to be plain or with minimal decoration — P4, P10 and P31 — and others survive too partially to establish the overall decorative scheme. None of the vessels seem likely to have been very large, although the wall thickness of P20/P21/P24 does suggest a substantial vessel.

The vessel P41 (Fig. A4.13) from the terminal pit is exceptional in terms of its decorative scheme as it includes features diagnostic of the Durrington Walls, Woodlands and Clacton sub-styles. The overall decorative scheme of the vessel is within the Durrington Walls sub-style, with panels demarcated by vertical dividing lines in incision filled with fingernail impressions. Within the scheme, however, are triangular areas (and one lozenge) filled with impressed dots which are typical of the Clacton sub-style. The paired fingernail impression-filled bar chevron within one of the panels is unusual and seems to be without parallel in the Durrington Walls sub-style. Around the rim top are applied pellets, which are diagnostic of the Woodlands sub-style, while the decorated base is absolutely atypical of southern English Grooved Ware (and rare everywhere). *Internal* decoration of the base is a feature of some Scottish Grooved Ware, as for example at Links of Noltland, Westray, Orkney (Sheridan 1999, ill. 12.7) and a much simpler form of decoration, consisting of lines of twisted cord or incision ringing the base on the exterior occur on two Grooved Ware vessels from post pits of a timber circle at Knowth, Ireland. It may be that external basal decoration was a local preference to Wyke Down (i.e. because one of the vessels from the houses also has it, P139/P140 from Building 2).

Beaker

Nine of the illustrated sherds (P17/18/19/26/37) represent a single Beaker, two of them rim sherds (Figs. A4.12 & A4.13). This is a very well made and fired vessel, with a hard fabric and thin walls. The vessel clearly carries zones of oblique comb impressions, mainly fine square-toothed comb, the zones being bounded with fine twisted cord impressions. One sherd (P37) shows opposed comb impressions separated by twisted cord. This decorative scheme is typical of Cord Zoned Maritime Beakers, of which only a handful occur in the British Isles, mainly in funerary deposits. These include Avebury (West Kennet Avenue grave by stone 29a; Smith 1965, P350 — although Smith queries whether twisted cord is present in this case, and the sherds are very worn and unclear), Wilsford G54 (grave) (Smith 1991, fig. 12, P6), Brean Down, Somerset (Clarke 1970, fig. 112) and Christchurch (Clarke 1970, fig. 313). It also seems to occur, although it is not specifically referred to in the pottery report, in the upper levels of the henge ditch at Mount Pleasant, in a context not dissimilar to that at Wyke Down 2. (P177 from Site IV would appear likely to be CZM, although the cord bounding the zones is of 'chain-plaited lines', so either must be crochet or plaited cord: Longworth 1979, 109, fig. 50.) At least three other Beakers seem to be represented in the Wyke Down ditch; P14–P16 and P22 from section I, P38 from section E and P39 from section D. All of these are from the same layer (2) as the CZM vessel, from which there was also Peterborough Ware.

Catalogue of illustrated pottery from the henge ditch

General notes

The pottery was examined after the selection of sherds had been made and illustrated and no count was made of the entire collection. Only a minority of featured sherds (i.e. with decoration or physical features such as ribs or bases) are not illustrated and the selection gives a good impression of the complete collection;

it is likely that very few vessels are not represented among the illustrated sherds and several vessels are certainly illustrated by more than one.

The sherds were examined with a ×10 hand lens and frequency of non-plastic inclusions was estimated by eye by surface area with the help of comparative charts. Terms used are: rare (about 1–2%), sparse (*c.* 3–5%), moderate 10–15%, common 20–30%. Dark grains are in most cases almost certainly iron oxides, but they may be, or include, other minerals, particularly glauconite. All sherds are Grooved Ware unless otherwise specified.

Pottery is numbered consecutively within the site; 'P' numbers do not indicate that sherds belong to separate vessels (i.e. separate non-joining sherds of the same vessel may have separate 'P' numbers).

Figure A4.12

P1. Section H L2: Peterborough Ware. Almost certainly from the same vessel as P2 and P3 and probably of an Ebbsfleet sub-style vessel with a slightly variable profile. Slightly laminated fabric with sparse to moderate unevenly distributed blocky flint (<6 mm) with sparse fine to coarse sand and some dark grains. Surfaces pale orange-brown, core black. Condition fair; some black residue adhering on interior surface.

P2.: Section H L2/3: Peterborough Ware. Rim sherd of same vessel as P1 and P3.

P3. Section H L3: Peterborough Ware. Rim sherd in a fabric with some large blocky flint (<7 mm) and some smaller angular flint (<2 mm); some sand is also present. There are broad shallow grooves across the top of the rim. Surfaces are pale brown, top of rim black. Condition fair.

P4. Section H L2: Rim sherd, possibly from the same vessel as P31, containing moderate coarse sand and rare, ill-sorted chalk inclusions which may be natural (one is a pebble 9 mm × 6 mm); dark grains are also present. The sherd is plain but there is a large tripartite impression on the exterior which may be accidental; the interior has a patch of dark matter adhering. Exterior pale brown, core black, interior mid brown with black area. Condition fair.

P5. Section H L2/3: Peterborough Ware. Sherd from just below a shoulder angle (the beginning of the angle is just visible) with oblique whipped-cord impressions in a sandy fabric with sparse to moderate small angular flint (<2 mm) and coarse sand. This is not the same vessel as the rim sherds P1–P3. Exterior brown, core black, interior orange-brown. Condition weathered.

P6. Section C L4/5: Peterborough Ware, probably of the Fengate sub-style. Body sherd in hard fabric with large (<7 mm) well-calcined flint inclusions and sparse fine sand. Decoration is of two lines of overlapping fingernail impressions; orientation is not certain. Exterior orange; core and interior black. Condition fair.

P7. Section CL4/5: Peterborough Ware. Body sherd, probably from the base of a thick rim; very similar fabric to P3 but on form does not seem to be same vessel. The fabric is slightly laminated and contains moderate large blocky flint (<6 mm) and some smaller angular flint; some fine to coarse sand is present. Surfaces pale brown, core is dark grey. Condition fair.

P8. Section H L2: Although this decorated body sherd could be 'lost' within an assemblage of 'domestic' Beaker the slightly shelly fabric is similar to much of the Grooved Ware from the site and local area. The impressions on the exterior are almost cuneiform. Fabric contains visible sparse dark grains (<1 mm), moderate sand (fine and coarse) and sparse shell (<3 mm); some grog is almost certainly present but is difficult to distinguish from the matrix. Exterior: dark brown; core black; interior: orange. Surfaces and breaks worn.

P9. Section H L2: Rim with a reconstructable diameter in fabric containing moderate coarse sand. The decoration is of coarse Z-twisted cord impressions (i.e. of S-twisted cord). Running from the rim down one broken edge is what appears to be a longitudinal perforation through the vessel wall; in the lower part of the sherd another perforation seems to be running alongside the upper one, as if perforations had been run in from both directions. The diameter of the perforations is about 5 mm at maximum, in a wall thickness of 11 mm maximum. The sherd is dark grey-brown throughout and is in fair condition.

P10. Section H L2: Rim. Possibly Grooved Ware (uncertain identification). Plain rim sherd of a thick-walled vessel in a fabric which probably contains grog, and rare small flint. The finish is very irregular. Surfaces pale brown, core black. Condition fair. This may be plain Grooved Ware although the thick, flat-topped rim is not typical. (The likeliest other possibility is that it may belong to a local Middle Bronze Age tradition.)

P11. Section F L4: Peterborough Ware. (The sherd is illustrated incorrectly in that it is shown inverted and with a continuation line whereas it is actually a rim with a thick almost collared neck and fingernail impressions on the top (not shown); the impression which is illustrated is not from a thumb or fingertip as shown but a fingernail impression with a very slight depression to one side of it, probably caused accidentally when the nail was pushed in.) The form would seem to suggest the Fengate sub-style, but the suggestion of the collar is slight and the fabric would be highly unusual in that sub-style. It is more likely that this belongs to the Ebbsfleet sub-style and is one of the rarer types with a 'bottle'-type neck. The fabric is strikingly unusual within the Wyke Down assemblage, containing common large chunks of angular quartz (<4 mm), some of which are clustered and cemented together into a matrix, suggesting a sandstone (technically a grit as the quartz is angular) which has partly disaggregated naturally or as the result of crushing; there are also some dark grains. Exterior: mid-brown, interior and core black. Fair surfaces and breaks. There is what appears to be a tarry deposit on the exterior.

P12, P13. Ditch C L4: Two body sherds of one vessel in a fabric with common coarse sand and decorated with tear-drop shaped impressions which vary slightly; there is also a shallow grooved line not showing in the illustration. Exterior and interior surfaces dark brown, core black. Condition fair.

P14, P15, P16 and P22. Section I L2: Beaker. Four sherds (and two not illustrated from same context) of a Beaker in a hard but brittle fabric with few visible inclusions but almost certainly including grog and some sand. Brown throughout, with some sherds fired to orange-brown on the exterior surface. The decoration is of short incised lines, arranged in a zone of zigzag or triangular motif apparently around the base and including a ladder zone in crude comb (another separate zone appears to show an ermine motif (P16 not illustrated very clearly). Although the decorative motifs and techniques seem diverse the sherds do seem to belong to one vessel. The sherds are in fair condition, although the interior of the base angle sherd is more worn than the other surfaces.

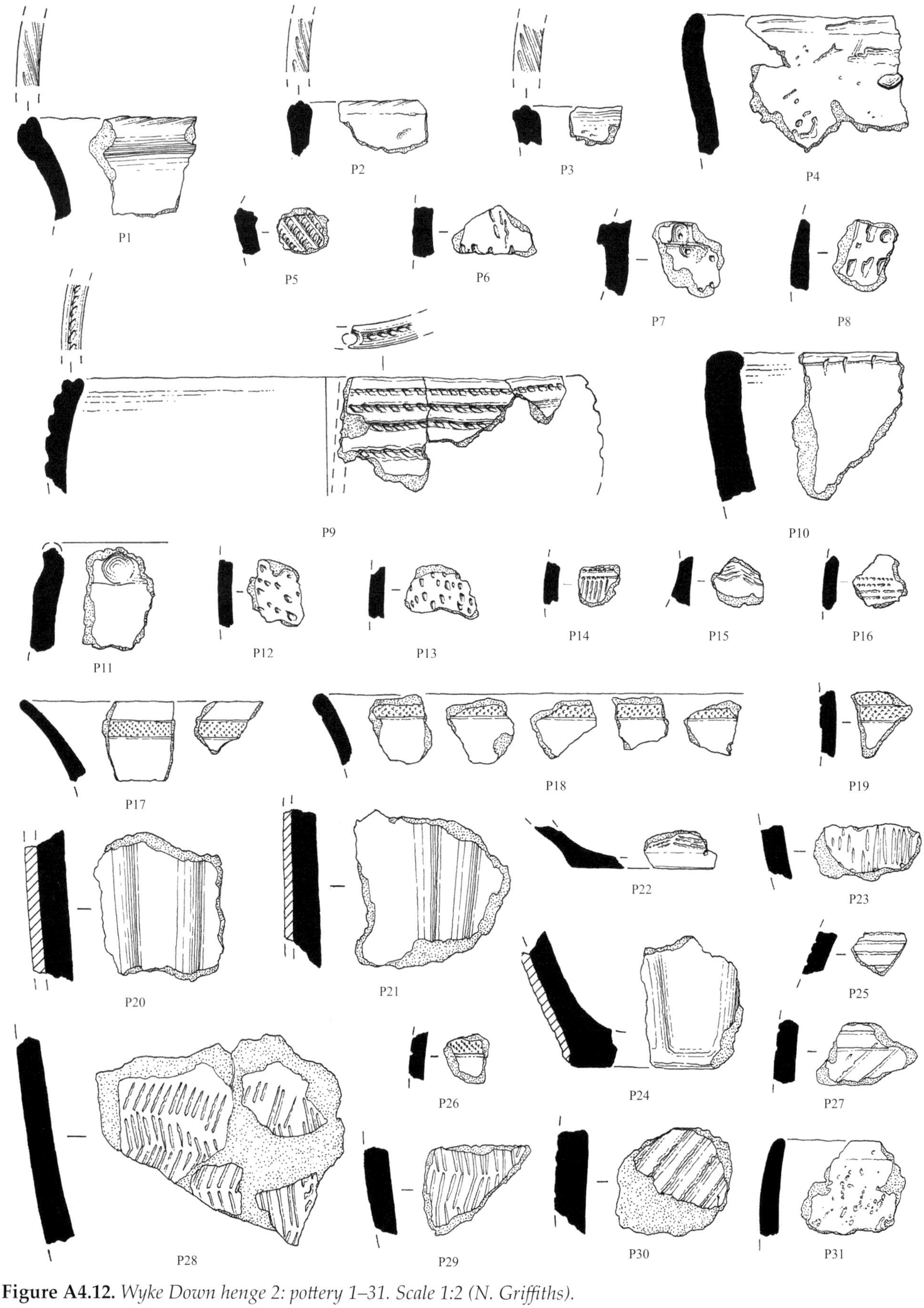

Figure A4.12. *Wyke Down henge 2: pottery 1–31. Scale 1:2 (N. Griffiths).*

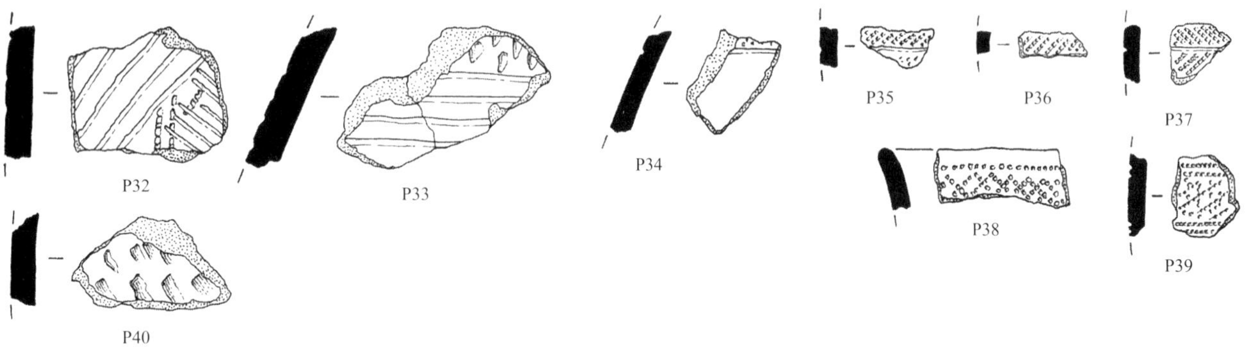

Figure A4.13. *Wyke Down henge 2: pottery 32–40. Scale 1:2 (N. Griffiths).*

P17, P18, P19 and P26. Section E L2 and C L2: Beaker. Two rim sherds, body sherds and several not illustrated of a single Cord-Zoned-Maritime Beaker decorated with square tooth comb and twisted cord impressions. The comb impressions are only oblique one way (i.e. are not cross hatching) from top right to bottom left and are of comb with teeth *c*. 1 mm square. The rim is slightly less out-turned than illustrated. The zones are cord-defined (fine, z-twisted impressions), and in one case have comb impressions (P19) in the opposing direction. The nature of the line separating the two lines of opposing impressions in P19 is not clear but may be twisted cord. The sherds of P18 are not close to the rim, as shown, but from further down the vessel wall. The vessel appears to be well-made and fired but in a fabric which varies slightly: the rim appears finer, with sparse coarse sand and very little visible grog; lower down the body the coarse sand is more frequent and the grog more easily distinguished from matrix; there are also rare flint fragments and dark grains. Exterior and interior surfaces orange-red; core black. The condition of the sherds varies, although all are worn to some degree. The body sherds of P18 show a more weathered interior surface than exterior and all edges range from fair condition to worn. See also P34–P37 from Section H.

P20. Section H L2: Body sherd with applied ribs in a soft fabric probably containing grog, although it is very difficult to distinguish it from the matrix; some sparse sand is also present, some sparse small shell may also be present. Surfaces pale orange, core black. Exterior fair condition, edges and interior weathered.

P21. Section H L2: Body sherd with applied ribs possibly from same vessel as P20, but if so the fabric of the vessel is varied. Soft fabric with some grog, sparse small (<2 mm) shell and sparse sand. Exterior orange, interior dark grey brown and core black. Exterior fair, edges weathered and interior very weathered or degraded (but no pitting).

P22. Section C L2: Described with P14, above.

P23. Section D L4A: Body sherd; part of same vessel as P28.

P24. Section H L2: Base angle sherd, probably from the same vessel as P20. Orange surfaces. Fair condition. Underside of base very sandy — coarse sand grains — as if it had stood on a sandy surface during manufacture or drying.

P25. Section H L4A: Body sherd in fabric with shell and sand.

P26. Section C L2: Beaker. Described with P17 above.

P27. Section G F.20: Body sherd, probably of same vessel as P32; weathered condition.

P28, P29. Section D L4A: Two body sherds of one vessel in a fabric with sparse shell (<3 mm; some voids <6 mm), sparse to moderate sand. Exterior brown, core black, interior brown. Some residue adhering. Worn exterior, fair edges and interior surface.

P30. Section C L4: Body sherd, possibly of same vessel as P28, with shell and chalk. Surfaces pale brown. Condition fair.

P31. Section H L2: Plain rim sherd with few visible inclusions (sparse sand, possibly some grog). Exterior pale brown, core dark grey; interior is badly degraded, with pitting. Some damage on the exterior (i.e. not decoration). Condition: exterior fair apart from damage, edges worn, interior degraded.

Figure A4.13

P32. Section E F.20 L2: Body sherd (orientation not certain) in a soft sandy fabric with common fine to coarse sand and common grog (<3 mm), with rare small (<2 mm) chalk and shell. The decoration is of shallow grooves (the appearance in the illustration of comb is caused by one set of diagonal lines cutting through an existing line when the clay was still damp and is clearly not comb impressed and is Grooved Ware). The sherd is weathered on both edges and surfaces. Exterior pale brown, core black, interior orange-brown.

P33. Section D L4A: Body sherd in a fabric with moderate shell (<9 mm). Exterior mid-brown, black core, dark brown interior. Fair condition.

P34, P35, P36, P37. Section H L2: Beaker. Four body sherds of the same Beaker vessel as P17–P19, P26. P37 almost certainly has a line of twisted cord, very indistinct and worn, between the two rows of angled opposed impressions. P35 is dark grey on the exterior and may have been burnt on that side. All are weathered.

P38. Section E L2: Beaker. Poorly finished rim sherd in a hard but brittle fabric containing rare large sub-rounded fragments of creamy grey stone (?limestone; not tested with acid) (<5 mm maximum dimension), some fine to coarse sand, and rare flint(<2 mm). Surfaces orange, core black. Condition: weathered. Decoration is of irregular (and now worn) comb, probably square-toothed.

P39. Section D L2: Beaker. Body sherd with crosshatched zone with the impressions vertical in one direction and oblique in the other; they are apparently made with different combs, that used for the oblique impressions having smaller teeth. The fabric contains sparse calcareous inclusions (<2 mm), some grog, and sparse coarse sand. Exterior orange, core dark grey and orange core, black interior surface. Surfaces and edges are worn.

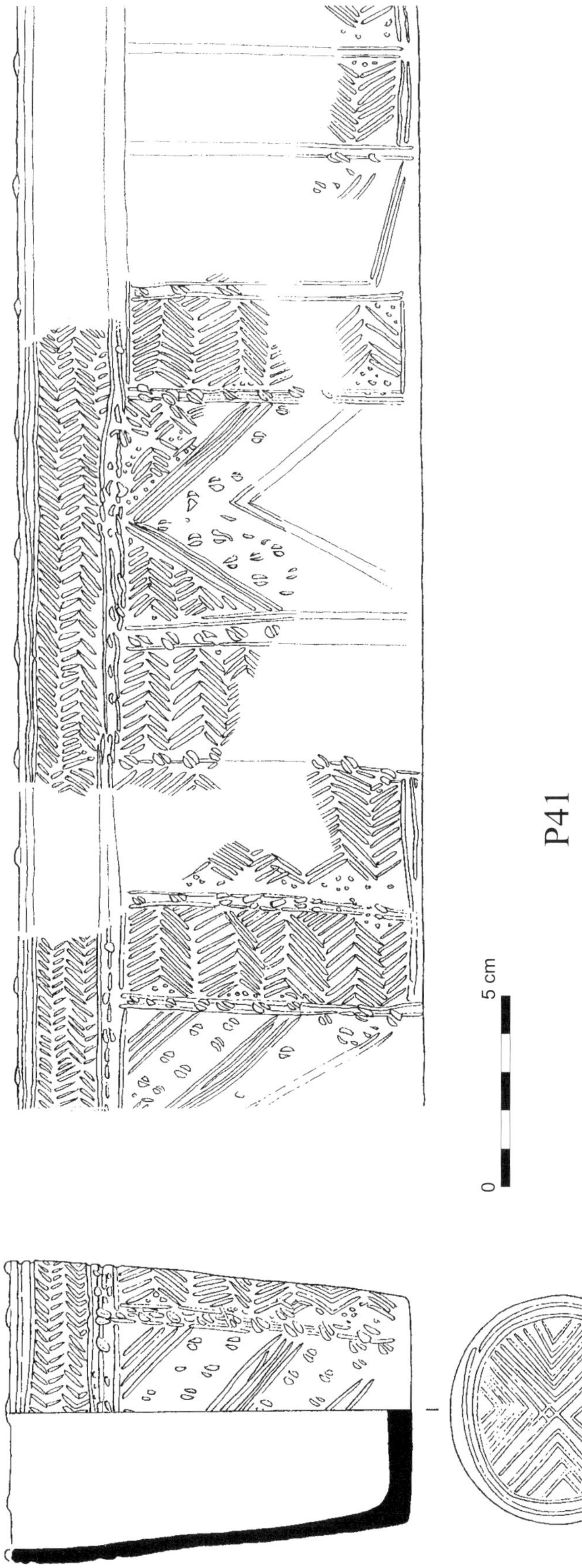

Figure A4.14. *Vessel P41 from southwest terminal of Wyke Down henge 2 (N. Griffiths).*

P40. Section D L3: Body sherd in hard fabric with sparse sand and sparse small (<2 mm) shell. Decoration is of single slightly plastic fingernail impressions. Exterior reddish-brown, core black, interior mid brown with some sooting/dark matter adhering. Condition weathered.

Figure A4.14
P41. Base of SW terminal of henge ditch, Section G: Large part of Grooved Ware vessel (whole base plus approximately one third of the body) in a hard, well-made fabric with sparse to moderate shell (most <3 mm) and with some very fine shell fragments showing in fresh breaks. No other inclusions are certainly visible, except a moderate amount of very fine dark grains, which are only visible in the light-coloured areas. The upper third has sooting and discolouration on the exterior and interior and black residue adhering around the interior of the base. It is a better made and finished vessel than most of the Grooved Ware on the site.

Summary of the molluscan data

Michael J. Allen

The henge ditch

Two sequences through the north and south ditches of the Wyke Down 2 henge were both broadly similar to each other and essentially paralleled that from Wyke Down henge 1 (Entwistle & Bowden 1991): open downland, probably grazed and trampled grassland, with local shadier and denser vegetation (i.e. lush dank grass and tall herbaceous plants) developing in the upper ditch fills when the surrounding area may have been tilled.

The Grooved Ware pits

In contrast, the two Grooved Ware pits sampled (pits 6 & 7) indicated local shady micro-habitats probably largely reflecting the pit microhabitats themselves. But they also indicate the presence of these refugia in the immediate vicinity. This contrasts markedly with the contemporaneous assemblages from Grooved Ware building 1 (postholes 15a & 18) and building 2 (postholes 34 & 35). Although there may be taphonomic issues relating to precisely when the snails arrived in the posthole (which during the life of the building would have been occupied by the post), the assemblages were predominately very open and indicated very short, grazed and trampled, dry calcareous grassland. Again these analyses tend to indicate the variety of contemporaneous micro-habitats within an excavated area.

Summary of the faunal remains analysis

Abigail Rothwell & Mark Maltby

The faunal remains recovered from the Wyke Down henge 2 ditch and the pits and postholes of the associated Grooved Ware settlement were the subject of a dissertation by Rothwell (2002), and so only the salient results are briefly described here.

Results

Only 16% of the total number of bone fragments recovered were identifiable (3548 fragments from 102 contexts) with a fairly restricted assemblage of ten species present (Table A4.22). In general the assemblage is in a poor state of preservation with much erosion of the surface of the bones, and as a result very little information was retrievable about butchery, sex and age.

The pits of the settlement adjacent to the henge produced the most fragments, the most that were identifiable and the most diverse population (Table A4.22). In contrast, the henge ditch exhibited a less diverse assemblage than the surrounding features with only four taxa identified: pig, cattle, sheep/goat and roe deer. The fragments of bone from the post holes of the two structures were even more restricted with only cattle and roe deer identified.

Pig was the most abundant species in the assemblage with 36 per cent of the identified fragments and an estimated minimum number of eight individuals (Table A4.22). Most of these (83.4 per cent) were from the pits associated with the Grooved Ware structures rather than the henge ditch. The most abundant elements of the assemblage were loose teeth, with skull and limb bones. Although the fusion information was limited, it was suggested that the majority of pigs killed were young animals with over half of them being less than a year old.

Cattle formed 32 per cent of the identified species with a minimum number of individuals of seven (Table A4.22), of which 44.8 per cent came from the henge ditch and were dominant in the henge ditch segments. Again, teeth and tooth fragments were the most abundant body elements present throughout, with no large dominance of meat-bearing bones. Although fusion information was again very limited, it was suggested that the cattle were sub-adult to adult animals. Only two bones (a radius and a scapula) exhibited evidence of butchery in the form of incisions.

Sheep/goat are only present in small numbers and mostly in the pits associated with the two structures rather than the henge ditch (Table A4.22). There was a minimum number of two individuals present (Table A4.22) with the humerus being the most common body element.

A small number of other species were identified that occurred in very limited numbers. These included a near complete dog skull found with a polished flint ball in the base of the henge ditch, six roe deer fragments in both the henge ditch and associated pits, and one red deer antler and three fragments of red deer, a single mandible of fox, a few (<1 per cent) mice and field vole fragments, a few frog/toad remains and two unidentified fragments of fish bone.

Table A4.22. *The numbers and percentages of identified/unidentified fragments of animal bone in the Wyke Down 2 henge ditch and associated features.*

	Henge ditch		Features		Postholes		Totals	
Element	**NISP**	**NISP %**	**NISP**	**NISP %**	**NISP**	**NISP %**	**NISP**	**NISP %**
Pig	34	28	171	40	-	-	**205**	**36**
Cattle	82	67	100	23	4	67	**186**	**32**
Sheep/goat	3	2.5	16	4	-	-	**19**	**3**
Roe deer	3	2.5	2	<1	1	17	**6**	**1**
Red deer	-	-	3	<1	-	-	**3**	**<1**
Field vole	-	-	46	11	-	-	**46**	**8**
Frog/toad	-	-	102	24	-	-	**102**	**18**
Mouse	-	-	4	<1	-	-	**4**	**<1**
Dog	-	-	1	<1	-	-	**1**	**<1**
Fox	-	-	1	<1	-	-	**1**	**<1**
Fish			2	<1			**2**	**<1**
Total	**122**		**450**		**6**		**578**	
% identified	17%		21%		2%		**16%**	
No. of unidentified fragments	595		2140		237		**2972**	

Interpretative notes

In comparison with the excavations of other Wessex henge monuments, the number of identified fragments is high with the only the large henges of Durrington Walls and Mount Pleasant having more. Nonetheless, this may reflect the relatively high proportion of the henge ditch being dug at Wyke Down and preservation conditions as much as depositionary practices. In both the Wyke Down 1 (Legge 1991, 56–7) and 2 henges, cattle predominate in the ditches. But the distinction between the henge ditches and the surrounding features are not always distinguished in other faunal reports, except for Coneybury, and so comparisons are somewhat restricted. Coneybury, unlike Wyke Down 2, had far fewer bones present in the adjacent features and they were predominantly cattle (Maltby 1990a, 150).

In contrast, pig was the most abundant animal represented in the pit fills associated with the Grooved Ware structures adjacent to the henge. Pigs were also common, and sheep/goat less common, from all of the Late Neolithic sites/samples investigated in the study area (see Maltby this volume). In addition, pig is acknowledged to be a common occurrence at other Grooved Ware sites (Thomas 1991, 22), and in particular at the great henge of Durrington Walls (Grigson 1982b; Albarella & Serjeantson 2002) and at Mount Pleasant (Harcourt 1979, 215). It has been suggested that pigs were better ecologically adapted than cattle for exploiting wooded environments (Grigson 1982b, 310), but that they may have been associated with feasting (Albarella & Serjeantson 2002) at these sites. However, the Wyke Down pig assemblage does not show a dominance of meat-bearing bones, although this could have been skewed by selective deposition and/or poor preservation resulting in an observable absence of butchery evidence. This was similarly the case for cattle where there was no dominance of meat bearing bones. For both the pig and cattle bones, the majority appear to have come from the ditch segments located opposite from the terminals of the henge ditch, which could indicate that there was some depositionary factor at work.

Human bone

Jacqueline I. McKinley

Only one deposit with the fragmentary remains of disarticulated human bone was found in the ditch of Wyke Down henge 2 (trench A, layer 3/4) interface.

Methods

The degree of erosion to the bone was recorded following McKinley (2004b, fig. 6). Age was assessed from the stage of skeletal development (Bass 1987; Scheuer & Black 2000) and the patterns and degree of age-related changes to the bone (Buikstra & Ubelaker 1994). Standard measurements were taken where possible (Brothwell & Zakrzewski 2004) and non-metric traits recorded (Finnegan 1978).

Results

The redeposited fragments of an adult right humerus shaft were recovered from the ditch of Wyke Down henge 2. The fairly gracile shaft was heavily eroded by root action (Grade 4–5; McKinley 2004b, fig. 6).

The pottery

ROSAMUND CLEAL

Structures and associated features

As with the henge ditch, the pottery from the area of the structures is almost entirely Grooved Ware, with the only exceptions being a single Peterborough Ware sherd and a small number of Beaker sherds.

Peterborough Ware

A single rim, P91 (Fig. A4.17), probably of Ebbsfleet Ware, was found in layer 3 of the pit F.3. The rim form is not thickened or expanded and it probably belonged to a vessel with a 'bottle-shaped' neck. The fabric, being flint-gritted, is relatively hard and it is possible that it is a redeposited piece.

Beaker

Three sherds (two conjoining), P119 and P128 (Fig. A4.18), are the only Beaker sherds identified, and unlike the CZM Beaker material from the henge these are not identifiable as belonging to a particular Beaker tradition. They were found in pit 22, a pit containing fairly large pieces of Grooved Ware, including some with traits diagnostic of the Durrington Walls sub-style and a vessel with unusual decoration but which seems to belong with the other Grooved Ware in terms of fabric (P124/126/P131) (Figs. A4.18 & A4.19).

Grooved Ware

The Grooved Ware is in terms of fabric fairly consistent. Very few vessels contain no shell, and most vessels are in a fabric with sparse to moderate shell mixed with other inclusions, some of which may be added and some naturally occurring in the clay. The commonest non-shell inclusions are sand and chalk; a few sherds contain dark grains which are probably iron oxides, but could include glauconite. A chalky rock which may be chalk, or possibly limestone, occurs rarely (P122) (Fig. A4.18); one possibly sandstone fragment was also identified (also in P122). Flint is also a rarely occurring inclusion. The nature and source of the shell has not been identified, but there are no obvious fragments of limestone matrix or non-shell fossils and although this does not preclude the inclusions being naturally occurring fossil shell it renders it slightly less likely. Marine shell (i.e. from 'live' sources in the Late Neolithic) have been identified elsewhere in Wessex (Cleal *et al.* 1994). If the shell were to be fresh marine (see Ladle below), this would imply that virtually all the pottery could have been made locally if the shells were transported (e.g. as food), although if the dark grains were glauconite this would imply a source in the Greensand, for instance around the Vale of Wardour, 10 km to the northwest. Fossil shell could also have been found in some of the clays found in that direction and at not much greater distance, such as the Kimmeridge or Oxford Clays (Melville & Freshney 1982; British Geological Survey 1979).

In terms of the conventional sub-styles the Grooved Ware from the Wyke Down structures, as from the henge, is assignable to the Durrington Walls sub-style, although there are some vessels which show diagnostic traits of the Clacton sub-style. Two vessels have a decorative scheme arranged around incised panels (P44, which also has at least one applied rib, and P46) (Figs. A4.15 & A4.16) and a number of sherds have applied ribs fulfilling the same function (P45, P66, P123, P125, P130, P132, P136, P141) (Figs. A4.16, A4.18 & A4.19). In three cases — P43, P78 and P98 (Figs. A4.15 & A4.17) — the surviving sherds do not show any applied ribs, but the presence of applied horizontal cordons hints strongly at the vessels having had the classic Durrington Walls style division into upper third bounded below by a horizontal cordon and the lower two thirds divided into rib-bounded panels. Two vessels, represented by P121 (Fig. A4.18) and P135 (Fig. A4.19), have closely spaced ribs, which allow no space for decorative infill between them; although this is relatively rare it does occur at Durrington Walls (Longworth 1971, in Wainwright & Longworth 1971, P45, P46). Two vessels certainly have circular motifs, in the form of concentric grooved circles (P66 and P74) and two more almost certainly also have (P42 and P112/P113) (Figs. A4.13–A4.18 & 169). P79 (Fig. A4.17) has a deep circular impression, probably based on a thumb or finger impression but which has been made deliberately round, and this also occurs at Durrington Walls (Longworth 1971, in Wainwright & Longworth 1971, P168); and cord occurs on five vessels (with a sixth, P95, only possibly decorated with cord, around the rim top) from the area of the structures and pits (P85, P90 (probably Grooved Ware) (Fig. A4.17), P127 (Fig. A4.18), P129 (Fig. 169) and P124/P126/P131) (Figs. A4.18 & A4.19); in P124/P126/P131 it is an unusual arrangement which is so unlike the normal use of cord on Durrington Walls Grooved Ware that it seems hardly appropriate to take it as representing the sub-style and there are no close parallels for it known to the writer.

In addition to the Durrington Walls traits it is also noteworthy that there is a use of rectilinear motifs on the non-henge Grooved Ware; these are rare in Wessex Grooved Ware (and generally). Two vessels show this: P48 from F.18 (Fig. A4.16) and P106 from F.4 (Fig. A4.18). The decoration on the exterior (i.e. underside) of the base on P139/P140 (Fig. A4.19) is also highly unusual (and seems less complex than the saltire on the vessel from the henge ditch, P41: Fig. A4.14).

The number of circular motifs is an unusual feature of the assemblage. These are not a common feature of Grooved Ware assemblages, with only approximately six or seven vessels occurring, for instance, in the much larger assemblage from Durrington Walls (although more have emerged in the recent excavations) (Longworth 1971, in Wainwright & Longworth 1971, figs. 58–9), and it is difficult to establish how many vessels are represented by the sherds). They are known also from Woodhenge, although the motifs vary, but are absent from the henge at Mount Pleasant. If it is at all a preference at sub-regional level then this could be taken to suggest that the makers and/or users of the Wyke Down Grooved Ware had more contact with the users of the Avon valley henges than with those using the monuments around the Stour valley, unless the differences are due to chronology. Here, however, the lack of a well-defined chronology creates considerable difficulties. Both Wyke Down henges 1 and 2 appear to be relatively early in the use of Grooved Ware, and their primary use appears to have been over before the middle of the third millennium BC. On present dating evidence this would make them earlier in use than the Durrington Walls Grooved Ware, but the recent and ongoing campaign of research at the latter site is indicating that the history of that site is more complex than previously thought, which in turn makes it more difficult at present to assess the relative chronologies of these two areas.

If at least some comparisons can be found for the circular motifs in other monumental assemblages the same cannot be said for the rectilinear motifs as there are none at Durrington Walls, Mount Pleasant or Woodhenge. Only at Marden, in the Vale of Pewsey, is there anything rectilinear from a henge, and that appears to be a Greek key pattern rather than a closed form. Both Hamilton (1997, 115) and Smith (1985, 175) discuss rectilinear patterns, the former suggesting that rectangular or key patterns may be a popular feature in north Wiltshire, although also acknowledging that Smith (1985) had identified the widespread though rare occurrence of such motifs (e.g. at Lawford, Essex, Carnaby Top, Yorkshire, and at Fengate, Cambridgeshire).

Relationships between Wyke Down 1 and 2, and with Fir Tree Field pits

The assemblages from the two monuments (and from the Wyke Down 2 structures and pits) share some features and on the grounds of the ceramics alone would seem unlikely to be widely separated in time. Neither of the monuments produced large amounts of Grooved ware, but in both cases this did include large portions of some vessels. Circular motifs occur at both sites (Wyke Down 1, P187), as does decoration involving large areas of herringbone (Wyke Down 1, P186, P192 and possibly other vessels) and there are minor occurrences of plastic fingernail impression at both sites (Cleal 1991). The excavator has pointed out that one vessel may be represented by sherds from both Wyke Down 1 and 2 (Cleal 1991, P186 and from F18 at Wyke Down 2 (P52) (Fig. A4.16)); it is indeed highly likely that both are from the same vessel, but the only conclusive evidence would be of conjoining sherds and this has not yet been searched for. There are also cases at each site of vessels with 'basket-weave' type infill, as on P45 and P46 (Fig. A4.16) from features at Wyke Down 2 and P199 at Wyke Down 1 but as this is a relatively common decorative scheme it should not perhaps be taken as a strong indication of the sites having potters or sources of pottery in common. There is also a single instance of a fairly unusual decorative motif — of single round impressions between grooved lines — which occurs at both sites: in feature F6 at Wyke Down 2 (P108) (Fig. A4.18) in a shelly fabric and in a fabric with no visible inclusions at Wyke Down 1 (P201) (Cleal 1991).

In contrast, there are also differences between the sites, as for instance in the occurrence of a shallow bowl with interior geometric decoration at Wyke Down 1 (Cleal 1991, P188) but none at Wyke Down 2. The range of fabrics displayed at the two sites is also different. Wyke Down 1 vessels are in a variety of fabrics, few including shell and several having no visible inclusions (which may indicate that they are grog-tempered as it is unlikely that there are no inclusions at all as the pottery would be unlikely to survive firing); at Wyke Down 2, in contrast, the majority of vessels include some shell, and grog seems a minor inclusion.

The Grooved Ware from Fir Tree Field pits, approximately 1 km to the southwest, although like the Wyke Down ceramics dating to the first half of the third millennium rather than to the second half, does not show any particularly striking similarity to that from the pits from Wyke Down. Only in the case of P25 from Fir Tree Field (Cleal 1991) with panels filled with oblique multiple lines and single impressions between applied ribs is there any clear similarity, in this case with P41 (Fig. A4.14), the large portion of a single vessel from the excavated terminal of the Wyke 2 henge ditch. The fact that this is quite an unusual method of filling panels, not seen, for instance at Durrington Walls, could be taken as an indication that all three sites had some commonality of potting tradition or source of pottery.

Minor features

Two features are worth comment: pre-firing pits/perforations and post-firing perforations. Two vessels have pre-firing perforations — P9 (Fig. A4.12) and P60 (Fig. A4.16). In both cases the perforations are into the vessel wall longitudinally, rather than through it directly from exterior to interior. In both cases the deep perforations run down into the wall from the rim top and are quite clearly deliberate. In the case of P9 it seems that another perforation had been made, presumably from the exterior on the lower body (as it would be difficult to do it from the interior) also into the wall, so that for some of their length the perforations were running parallel. A very similar practice seems to have occurred in P60.

One other occurrence of small pits may also be deliberate: in P79 (Fig. A4.17) there are two small 3–4 mm deep holes 2 mm in diameter behind a thumb- or finger-circular motif, there is no obvious fortuitous reason for them and it seems clear that they were made deliberately. In all these cases the intention is unclear, as the holes would be fairly inconspicuous on the finished vessel. The only similar occurrence known to the writer is from Woodhenge, where there is a perforation on the interior, behind the hollowed out circle in figure 26.4 (Cunnington 1929). It is perhaps fanciful, but not impossible, that the holes were made so that some of the contents of the pot were absorbed by the vessel itself (the makers not having the benefit of the work of organic chemists on absorbed residues), for reasons perhaps not entirely functional (in our terms). It also seems likely to be more than co-incidence that in two cases the perforations on the interior, although different, both occur on the interior of hollowed out finger-or thumb circular motifs. One sherd, P46 (Fig. A4.16), has a pre-firing pit which was clearly made after the decoration as it cuts through it, and P66 (Fig. A4.16) has traces of the edge of a mending hole on one edge. Mending holes are common on Grooved Ware, relative to other Neolithic and Early Bronze Age ceramics, and would seem to indicate that vessels retained some importance, functional or otherwise, even after damage (Cleal 1988).

It should also be noted that Mukherjee *et al.* (2004) conducted a stable isotope analysis of organic residues in 23 sherds of Grooved Ware pottery from the Wyke Down 2 henge ditch and associated structural features. Although the lipid residues were quite poorly preserved, all of the lipid extracts were identified as degraded animal fats. The Grooved Ware vessels appear to have been used to process a variety of commodities such as ruminant dairy and porcine fats.

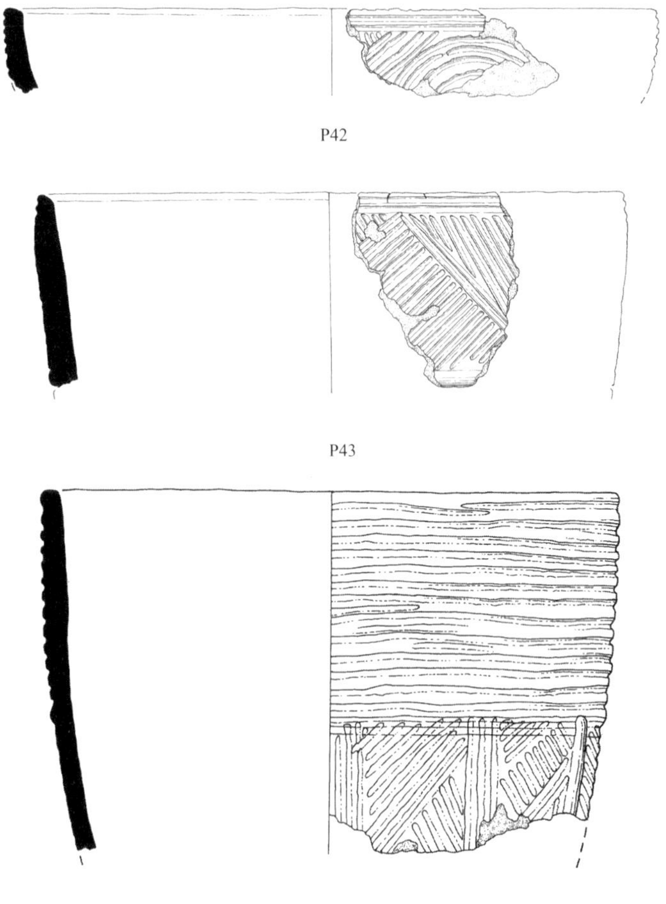

Figure A4.15. *Wyke Down henge 2: pottery 42–4. Scale 1:4 (N. Griffiths).*

Catalogue of illustrated pottery for Wyke Down 2 structures and other features

Figure A4.15

P42. On surface of posthole PH 31. Building 2: Large rim sherd in a shelly fabric with sparse to moderate shell (<7 mm, most <4 mm); no other inclusions are visible. The sherd is thick, measuring 14 mm thick at 40 mm below the rim. Exterior probably dark brown (though the exterior has been consolidated so the colour is distorted), core black, interior brown. Condition fair but the sherd is fragile.

P43. PH 6 of fence line: Large rim and body sherd. Shelly fabric — mostly <5 mm, though some larger; no other inclusions are visible. The sherd is very thick: at 50 mm below rim it is 15 mm thick All the decoration is of grooving. Exterior pale to mid brown, core black, interior treated so colour not identifiable. Condition fair.

P44. PH 15 L3. Building 1: Very large single body sherd in a fabric with moderate to common shell (most <5 mm); there are some sparse and fine dark grains. The decoration is largely of well-defined grooving but with one applied rib. Surfaces orange-brown except for a large burnt patch on interior which may have formed *in situ* (i.e. because the vessel would have to have been lying on its side and it seems more likely that this was when it was already broken). Good condition, the edges are only very slightly less than fresh.

Figure A4.16

P45. F18 L5 — 48 cm at edge of pit in dark layer beneath rubble: Large body sherd containing sparse shell <3 mm and some sand. Exterior: orange-brown, core: dark grey, interior pale brown; exterior slightly worn, edges fair, interior fair.

P46. F18 L5: Body sherd in a shelly fabric with some sand. The decoration is of well-formed grooves but there is also a deep impression — a pit 5 × 4 mm apparently made before firing but after the decoration had been completed as it impinges on the grooved infill. The motif on that side of the upright division may be a lozenge, but there is insufficient to be certain and it may be chevrons. The surfaces are orange; the general condition is fair with the exterior surface appearing fresh.

P47. F.17 L2: Body sherd in a shelly fabric with some sand and possibly grog. The appearance of multiple lozenges is probably deceptive; in this case it seems probable it is herringbone or chevrons. The sherd is orange-brown throughout and worn.

P48. F.18 L3: Rim in a fabric with some shell. The motif may be multiple squares; in the ladder motif the impressions are more rounded than shown. Pale brown surfaces, dark grey core. Exterior fair, interior worn.

P49. F.18 L6: Body sherd with horizontal groove; the fabric contains some sand. Exterior orange; fair condition.

P50. F.18 L5: Body sherd in fabric with moderate small shell and some sand. The decoration is incision. Exterior: orange, core: black, the interior is obscured by black residue. The sherd is in fair condition.

P51. F.17 L3/4 interface: Body sherd with paired plastic fingernail impressions. The vessel is well-made in a shelly fabric, with sparse shell (<5 mm maximum dimension (most <3 mm)); some coarse sand and grog is also present. Exterior orange, core black, interior brown. The exterior surface is fresh, the edges fair and the interior abraded/degraded.

P52. F.18 L6: Body sherd in a well made fabric with sparse very small inclusions which may be shell or possibly chalk (they are impossible to identify with certainty at ×10 magnification). Surfaces pale orange. Condition fair. The decoration was made when the clay was plastic as the grooves have ridges. It is possible that this is the same vessel as P186 from Wyke Down 1 (M. Green's initial observation); this appears quite likely on the grounds of fabric, decoration and general appearance but no search has yet been made for conjoins.

P53. F.18 L5: Body sherd containing some sand and shell. There is some sooting on the exterior. The sherd is dark grey and in fair condition.

P54. F.18 L5: Rim with groove running around the top in a moderately shelly fabric. There is some sooting trapped in the deep grooves on the exterior. Exterior: dark grey, interior: orange and dark grey. The diameter cannot be certain as there is very little of the circumference surviving. Fair condition.

P55. F.16 L4: Body sherd in a sandy fabric with some grog (<2 mm). Exterior brown, core black, interior obscured by chalky deposit. Condition fair.

P56. F.16 L4: Body sherd in a soft fabric with moderate sand. Brown exterior surface, core and interior black. Surfaces and edges worn.

P57. F.16 L4: Body sherd in a well-made fabric with sparse small shell (<2 mm) with some sand, mainly fine dark grains. Could possibly be same vessel as P52, although the herringbone seems finer in this sherd.

P58. F.16 L5: Body sherd in a slightly sandy fabric with some fine shell (<1 mm). Exterior brown, core and interior black. Condition fair.

P59. F.16 L4: Body sherd in a slightly sandy fabric with some small shell (<2 mm). Exterior brown, core black, interior obscured by chalk. The lower part of the sherd shows the scar of a missing rib/cordon, or possibly of a lug. Condition fair.

P60. F.16 L4: Small rim sherd in a slightly sandy fabric with small shell (<1 mm). The angle of the rim is uncertain but could be upright. The decoration is of broad shallow grooves, but there is also one clear perforation 4 mm in diameter showing in one edge running from just beneath the rim on the interior, obliquely down to the exterior, exiting 10 mm below the rim and being 11 mm in total; another perforation or deep impression, 2 mm in diameter, is visible in the other broken edge of the rim but this is extending down into the vessel wall to a depth of at least 5 mm. This seems a similar practice, although on a smaller scale, to that on vessel P9 from the henge. Exterior and rim top dark brown, core and interior black. Condition worn.

P61. F.16 L5: Body sherd in a soft slightly sandy fabric with moderate sand and moderate small shell (<2 mm); some grog may be present but it is difficult to distinguish it from the matrix. The orientation of the sherd is not certain: it appears from the exterior to be shouldered, but the curvature of the sherd is not consistent with this — it may be a worked up rib or cordon set at an angle with the adjoining zone on one side filled with oblique grooving. Exterior and core orange; core has black residue adhering. Surfaces fair, edges worn.

P62. F.16 L5: Soft body sherd with no visible inclusions. Exterior and core orange-brown, interior dark grey possibly with a very thin layer of black residue adhering. Condition fair.

P63. F.16 L5: Body sherd in a slightly sandy fabric with sparse small shell (<1 mm); no other inclusions visible but it could be the same vessel as P65. Exterior mid-brown, core black; interior obscured by chalky residue. Condition fresh.

P64. F.16 L4: Body sherd which may belong to the same vessel as P66. Sparse shell, one inclusion 15 × 10 mm). Exterior red-brown, core and interior dark grey. Fair condition.

P65. F.16 L5: Body sherd in a slightly sandy fabric with sparse fine shell (<3 mm) and some fine grog. The motif may be circular, executed in wide shallow grooving. Surfaces dark brown, core black. Condition fresh.

P66. F.16 L6: Part of a vessel with grooves and impressed decoration; the circular motif is of concentric circles rather than a spiral. The fabric is shelly, with sparse shell (most inclusions <2 mm (max. dimension 10 mm)); the vessel is quite fine and well-made. Exterior: dark greyish-brown, core black, interior brown. The exterior is fresh, the edges fair, but the interior slightly degraded. There is some sooting running vertically down the middle of the circular

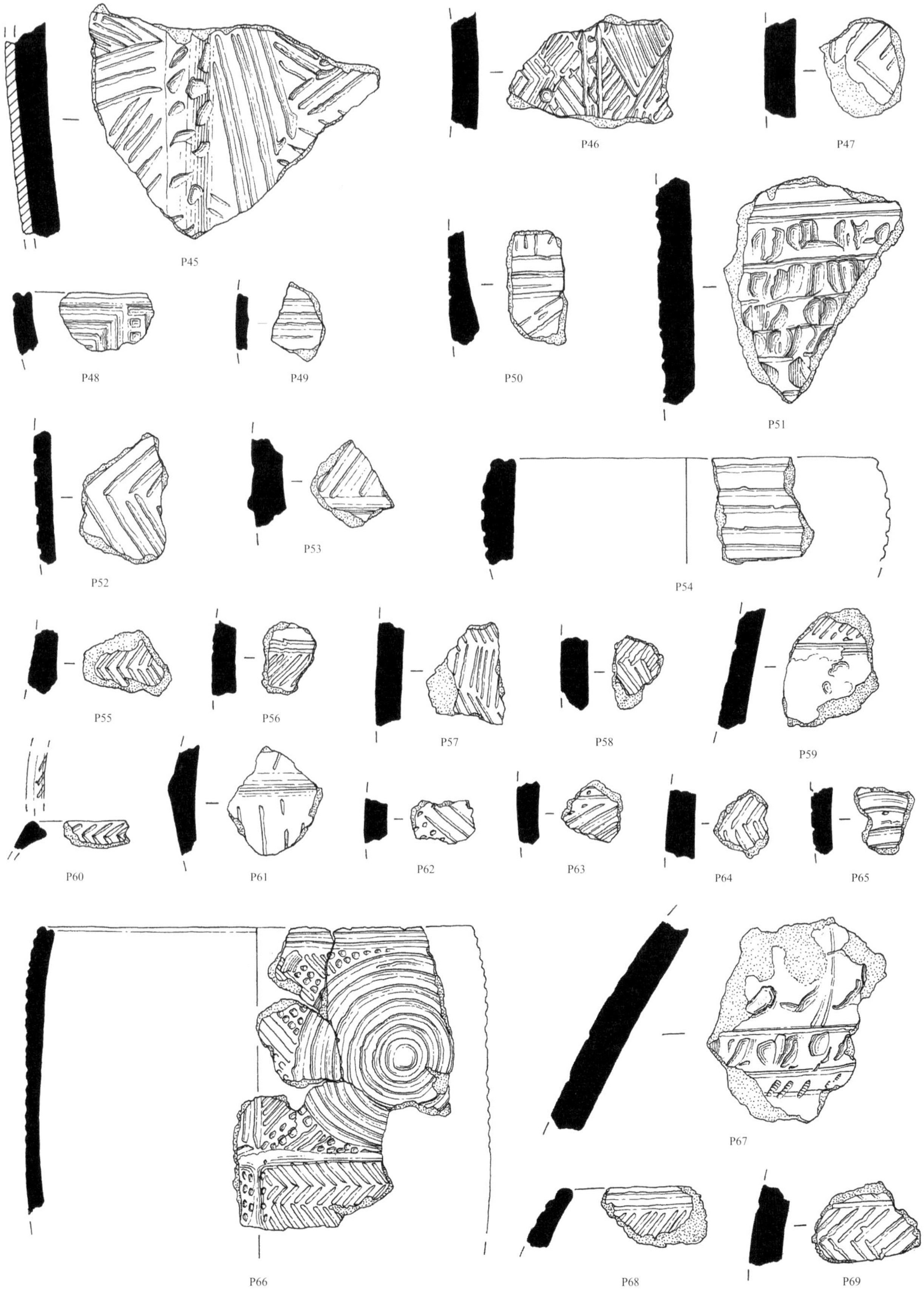

Figure A4.16. *Wyke Down henge 2: pottery 45–69. Scale 1:2 (N. Griffiths).*

motif. There are traces, on the left-hand edges (as illustrated), of the beginnings of a mending hole.

P67. F.16 L2: Large body sherd, part of which (the upper as illustrated) has decayed and been consolidated after excavation. The fabric is coarse and shelly (<3 mm). The sherd is probably illustrated the wrong way up (as paired plastic fingernail impressions usually 'point' to the top of a vessel); in the better-preserved part of the sherd there are shallow grooved lines between rows of pairs of impressions. Exterior orange-brown, interior dark brown, core bi-coloured orange (exterior third, black). Condition fresh but degraded.

P68. F.16 L2: Probably a rim sherd, although the rim top is very worn; the rim angle as illustrated is not reliable and the vessel wall could be more upright. The fabric is slightly sandy, with some dark grains, some very small shell and small grog (<1 mm). Surfaces brown, core black. The sherd is in poor condition, with surfaces and edges worn, and the interior in particular is degraded.

P69. F.16 L2: Body sherd in a fabric with sparse fine shell (most <2 mm), sparse fine grog (<2 mm) and some sand. Exterior grey-brown, interior pale brown, core black. Condition fair, but exterior degraded in patches.

Figure A4.17

P70. F.16 L2: Body sherd in a fabric with sparse small shell (<2 mm) but voids may be missing larger pieces or missing chalk inclusions; there is also some sand present. The decoration is of broad horizontal grooves and finer grooved herringbone. Surfaces grey-brown, core black. Condition fair.

P71. F.16 L2: Body sherd in a fabric with some sand and sparse voids (<4 mm) which may represent chalk with some shell. The decoration is grooved; the rounded impressions shown at the top of the sherd in the illustration are voids.

P72. PH 12 L2. Building 1: Rim sherd in a fabric with moderate small shell and some coarse sand. Below a zone of horizontal grooves are oblique impressions. Orange surfaces. Condition worn.

P73. PH 12 L2. Building 1: Body sherd in a sandy fabric. The decoration is of lightly incised or grooved lines. Worn. Exterior orange, interior dark grey.

P74. PH 12 L2. Building 1: Body sherd with decoration similar to P66 although this is not the same vessel. The fabric contains some chalk, and a small number of creamy grey inclusions which may be another inclusion type (both <2 mm); there is also sparse small shell and common small angular grog (2 mm); some small round voids, which could be missing grog or another inclusion type. The fabric is very unusual in the Wyke Down 2 assemblage. The decoration is of grooved circles (i.e. rather than spirals, as shown by the two inner circles which are complete). Exterior reddish-brown; core and interior black. Condition fair.

P75. PH 12 L2. Building 1: Body sherd in a hard fabric with sparse shell and some sand. The decoration is of irregular impressions and one grooved line. Brown surfaces. Fair condition.

P76. PH 12 L2. Building 1: Rim in a fabric with some shell and decorated with a grooved line below the rim. Black throughout; fair condition.

P77. PH 12 L2. Building 1: Body sherd, possibly from same vessel as sherds P83 and P86, although if so the sand content of the vessel varies as this sherd is slightly sandier than the other two. Exterior surface orange. Condition fair.

P78. PH 15 L3. Building 1: Body sherd in fabric with moderate to common shell (<5 mm). The decorative scheme and orientation of the sherd is not clear. Pale brown exterior, black core and interior. Fair condition.

P79. PH 12 L2 –15 cm. Building 1: Body sherd in a fabric with sparse shell (<3 mm) and sparse dark grains. The decoration is of a deep circular motif which is clearly impressed by a thumb but is twisted round to make the motif more neatly circular. Above the circle is a zone of horizontal grooved lines and to the left a vertical panel of horizontal grooved lines bounded by a vertical grooved line. Below the circle is a zone in which the decoration is reminiscent of a technique at Wyke Down 1. Some clay seems to have been applied (and this probably includes the area pressed by the thumb impression) as this seems raised above the grooved area; within this there is a motif created by small circular impressions. On the interior, just below where the circle is on the exterior, are two small deep holes (*c.* 2 mm diameter each, 3 mm apart and 3–4 mm deep) which appear to be deliberate. Mid brown surfaces, black core. Condition fair.

P80. PH 15 L3. Building 1: Body sherd in a fabric with moderate to common shell (<7 mm).The decoration is of an applied rib (missing from the illustration) and oblique grooved lines. Exterior brown, interior black. Fresh condition.

P81. PH 15 L3. Building 1: Body sherd in a fabric with moderate shell (<4 mm, most <2 mm). The decoration is of an applied rib and very deep grooving. Exterior orange, interior orange with a patch of dark grey. Condition fair. (M. Green has noted that it seems to be from the same pot as the very large sherd P44; this seems likely (as P44 is from the same posthole), but the sherds have not yet been compared.)

P82. PH 15 L3. Building 1: Body sherd in a fine compact fabric with some chalk (<2 mm, most <1 mm) and sand. Surfaces pale grey with dark core showing through. Condition fair.

P83. PH 15 L3. Building 1: Body sherd, probably from same vessel as P77 and P86 as fabric and decoration are very similar. Pale brown surfaces. Condition fair.

P84. PH 15 L3. Building 1: Body sherd in a fabric with very common shell (*c.* 25%); the shell is relatively well-sorted (<5 mm, but most <3 mm); this is a different fabric to the majority of the sherds in shelly fabrics from this site. The decoration is of broad well-defined grooves, cut when the vessel was still quite plastic as there are ridges to each side. Exterior orange, core bi-coloured by surfaces, interior dark grey. Condition fair.

P85. PH 15 L3. Building 1: Body sherd in a sandy fabric with sparse shell. The decoration appears to be short lengths of twisted cord. Brown throughout. Worn.

P86. F.14 L2: Body sherd in a soft fabric with some chalk, shell and sand. The decoration is of very shallow grooves. Surfaces pale orange. Condition fair.

P87. F.14 L2: Body sherd (illustration is 90° to the correct orientation) in a fabric with very common (*c.* 40%) shell. Decoration is of closely spaced applied vertical ribs. Pale brown surfaces, dark grey core. Exterior fair, edges and interior worn.

P88. F.14 L2: Small rim fragment in a compact fabric with some small chalk and probably some grog. Surfaces pale orange, core black. It is possible this is not Grooved Ware, but it could be within the range. There are no deliberate marks on the exterior as the fingernail impressions appear accidental. If it is not Grooved Ware it may be

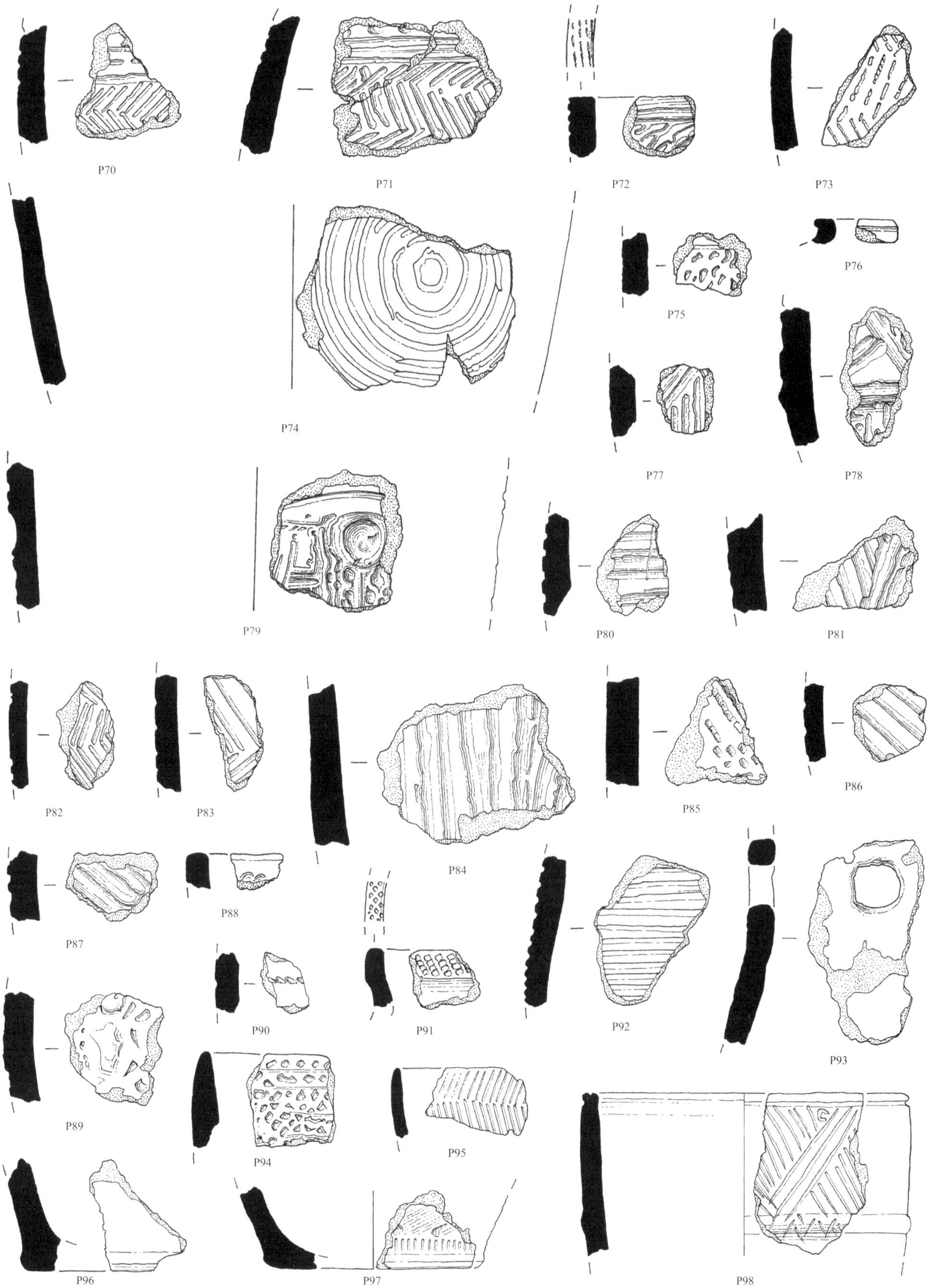

Figure A4.17. *Wyke Down henge 2: pottery 70–98. Scale 1:2 (N. Griffiths).*

coarse Beaker (although it is thick for Beaker pottery, at 8 mm) or Collared Urn. Condition fair.

P89. PH 28 L2: Body sherd in a sandy fabric. There are traces of a rib which has partly come off — the rest seems rusticated. Exterior and core dark grey; interior orange-brown. Condition fair.

P90. F.14 L2: Body sherd in a similar fabric to P88. There is some grog visible and also possibly shell. Pale orange throughout. The decoration is of coarse twisted cord. Although it is possible that this belongs to another tradition (possibly Collared Urn, as suggested by the excavator) both fabric and decoration, although not typical, are within the range of Grooved Ware.

P91. F.3 L3: Peterborough Ware. Rim sherd with whipped cord impressions obliquely across the rim top as well as on the exterior and interior. Fabric contains moderate flint (<2 mm) and common coarse sand.

P92. PH 4 –15 cm: Body sherd in a fabric containing moderate small shell and common coarse sand. Is possible this is just below the rim and is the upper third of a vessel. Surfaces pale brown; core black. Worn condition.

P93. F.14 L.2: Body sherd with sparse chalk (<3 mm) and moderate coarse sand. The sherd is featureless apart from a large perforation 16 mm in diameter. Although the hole looks superficially like a pickaxe or fork tine hole, it is not modern damage as the edges are not fresh; however , neither do they look like the original surface (i.e. it does not look like a pre-firing feature), nor is the hole counter sunk like mending holes. It is slightly more angled along one edge. Orange surfaces. Condition is worn/degraded.

P94. PH 8: Rim sherd in a hard sandy fabric. The decoration is of jabs/impressions creating a rusticated effect. Surfaces reddish-brown. Fair to worn.

P95. F.19 L2: Rim sherd in a soft fabric with sparse shell and sand. Orange throughout; condition worn. The rim top is almost completely worn away, but it is just possible that there is a line of twisted cord along it (not shown in the illustration).

P96. F.10 L3: Base angle sherd (and one plain sherd of the same vessel) in a sandy fabric. Exterior orange, core dark grey, interior brown. Condition fair.

P97. F.19 L2: Base angle in coarse fabric with sand, some shell and some grog. The decoration is of upright vertical incised lines above the base, with oblique lines above. Dark brown throughout. Condition fair.

P98. F.19 L2: Rim sherd in a sandy fabric with some small shell (<2 mm); some grog may also be present. Exterior reddish-brown; surfaces treated. Condition worn.

Figure A4.18
P99. F.19 L2: Not seen.

P100. F.19 L2: Small body sherd in a fabric with shell and sand. The impressions are deep and slightly irregular. Surfaces grey-brown, core black. Fair condition.

P101. F.19 L2: Body sherd in a fabric with some shell. Orange throughout. Worn condition.

P102. F.19 L2: Body sherd in a fabric with some shell. The impressions are single (i.e. they are not comb impressions and the sherd is not Beaker). Black throughout. Fair condition.

P103. F.19 L2: Body sherd in a fabric with some shell and sand. The decoration is of plastic fingernail rustication. Brown exterior, core and interior dark grey. Condition fair.

P104. F.4 L2: Body sherd in a fabric with some shell and sand. The decoration is rusticated (plastic fingernail). Surfaces orange-brown, core dark grey. Worn condition.

P105. F.4 L4: Body sherd in a fabric with sparse to moderate shell and some sand. The decoration includes a zone with geometric grooving above a zone with some fingernail rustication bounded by grooves. Exterior mid brown, core and interior black. Fair condition.

P106. F.4 L3: Body sherd in a fabric with sparse shell (<4 mm, most <1 mm) and some fine dark grains; there is also some fine sand. Exterior: buff, core black, interior dark brown. Fresh exterior surface, fair edges; the interior surface seems to be degraded. The motif appears to be multiple squares or rectangles in grooving, alongside cuneiform impressions.

P107. F.6 L2: Body sherd in a fabric with few visible inclusions, although there are rare small shell and flint inclusions (<2 mm); there may be grog present but it is difficult to distinguish from the matrix. The orientation of the sherd as drawn is unlikely to be correct — it is more likely to have a vertical applied rib, which has a single non-plastic fingernail impression. Exterior red-brown, core and interior black. Surfaces and edges worn.

P108. F.6 L4: Body sherd in a shelly fabric; it could be part of the same vessel as P110 and both could be part of same vessel as P66 from Feature 16. Decoration is of shallow grooves and round impressions. Fair condition.

P109. F.6 L5: Body sherd in a shelly fabric with some sand. The illustrated angle has been based by the illustrator on a curved line bounding the herringbone: it may be a horizontal line at this angle, but it is perhaps more likely that the motif is curvilinear, as a vessel wall at that angle would be unusual in Grooved Ware. Surfaces pale brown, core dark grey. The surfaces and edges are worn.

P110. F.6 L5: Body sherd in a fabric with some shell. Exterior brown, interior dark brown with some black residue adhering. The decoration is of broad shallow grooves and shallow round impressions. Fair condition.

P111. F.6 L2: Body sherd of same vessel as P112 and P113. Decoration is of deep broad grooves with the edge of an applied cordon (or possibly a lug or boss) showing at one edge. Fair condition.

P112. F.6 L2: Body sherd, probably belonging to the same vessel as P111 and the rim sherd P113 from layer 5, in a fabric with moderate chalk (<6 mm, most <3 mm) and some coarse sand; on the interior surface sand is coarse and common. Exterior pale brown; core black; interior black. Surfaces and breaks are in fair condition. The decoration is shallow grooving, in a pattern including a curvilinear motif, probably concentric circles.

P113. F.6 L5: Rim sherd probably of the same vessel as P112, although if so the quality of the vessel's fabric is variable as this sherd contains sparse chalk (<2 mm) and some fine sand; some dark grains are also present. Exterior and interior pale greyish brown. Surfaces and edges fair condition.

P114. F.6 L2: Body sherd in a fabric with soft pale grey inclusions, probably calcareous, and moderate coarse sand. Exterior reddish-

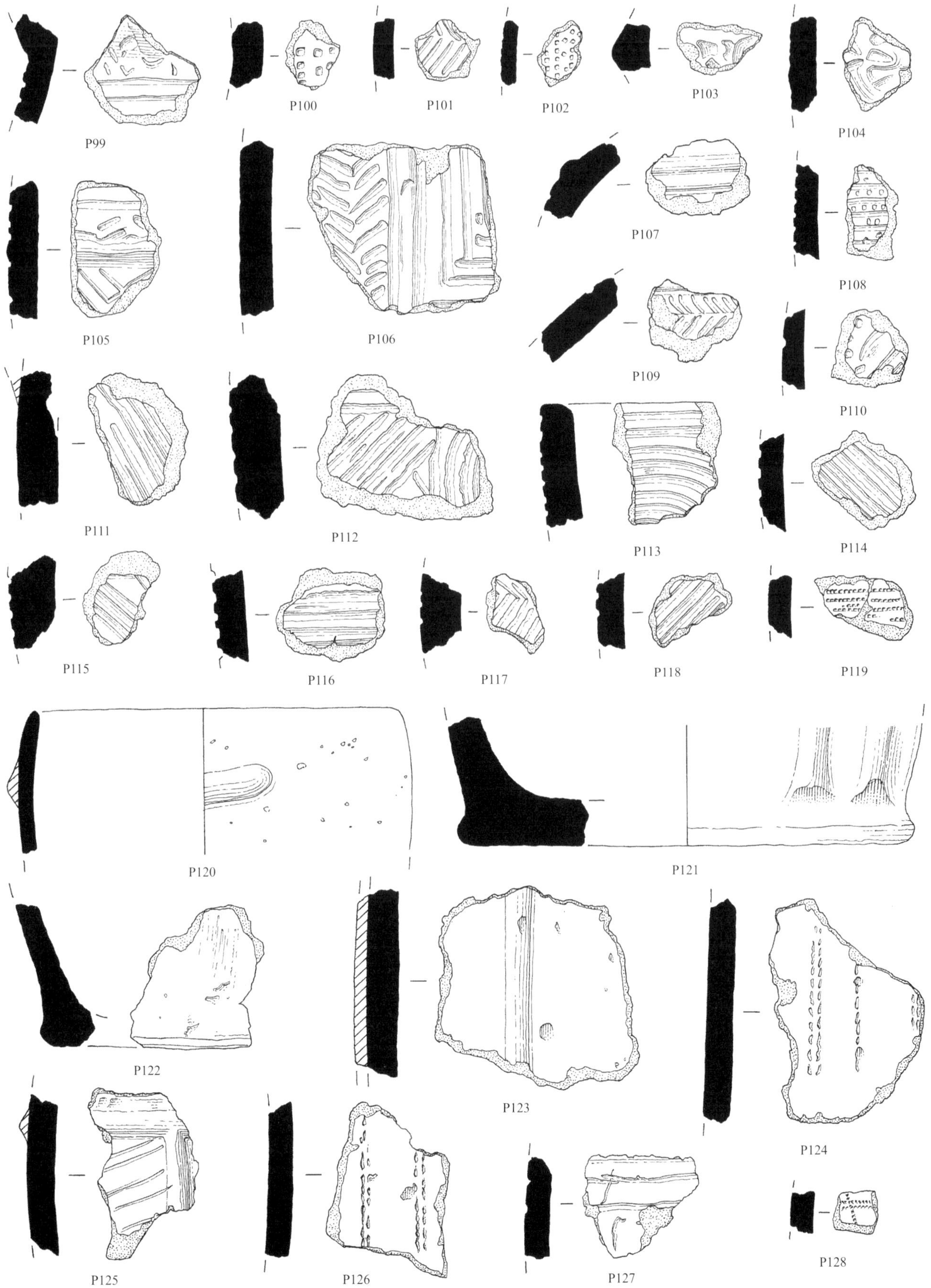

Figure A4.18. *Wyke Down henge 2: pottery 99–128. Scale 1:2 (N. Griffiths).*

brown, core and interior black, with some black residue adhering to the interior surface. Condition fair.

P115. F.6 L5: Body sherd, probably of the same vessel as P112 etc. Condition fair.

P116. F.7 L3: Body sherd in a fabric with some small sparse shell and some sand. Orange surfaces. Worn condition.

P117. F.7 L4C: Body sherd in a fabric with some shell and sand. Pale brown surfaces, dark grey core. Fair condition.

P118. F.7 L2: Body sherd, probably of the same vessel as P116. Worn condition.

P119. F.22 L2: Beaker body sherd decorated with square-toothed-comb impressions in a fabric with flint, grog, calcareous inclusions and sand. Exterior orange-brown, interior grey-brown. Surfaces fair, edges worn.

P120. F.22 L3: Two rim sherds of a plain vessel with an applied lug. The fabric contains rare to sparse calcareous inclusions, probably chalk (<2 mm), rare to sparse flint (<2 mm) and some sand. Condition: fresh surfaces, breaks and interior weathered; the interior of one of the sherds has been degraded, while the other has not.

P121. F.22 L3: Part of the base of a vessel with ribs; it is possibly the base to the same vessel as P132 as the fabric is similar to P132 with the addition of one large black rounded pebble (10 mm). There is black residue over the interior around the base angle.

P122. F.22 L3: Part of the base of a vessel in a fabric with calcareous inclusions (<3 mm, most <1 mm), rare grey inclusions of unidentified rock, and one red sandy inclusion which appears to be an iron-rich sandstone. Exterior: orange, core bi-coloured by surface colours, interior largely obscured by black residue. Condition: fresh surfaces, edges fair.

P123. F.22 L3: Body sherd with applied rib. The fabric contains calcareous inclusions, probably chalk (some voids), some grog, and sparse sand. Exterior: pale orange-brown, core: dark grey, interior: dark grey. Condition: surfaces fresh, edges fair.

P124. F.22 L3: Part of the same vessel as P126 and P131 in a fabric with rare chalk and sparse sand inclusions. The decoration is of single lines of lightly impressed cord accompanied by faint impressions alongside which may be the result of a fingernail being lightly impressed into the surface while the potter was pressing the cord onto the surface or another cord impression. Exterior: pale orange, interior black, core bi-coloured by surface colours. Condition: surfaces and edges fair.

P125. F.22 L3: Body sherd with applied cordons; few inclusions visible but some shell and grog is present. Exterior and interior dark grey. Condition fair.

P126. F.22 L3: Part of the same vessel as P124 and P131; these are more clearly paired impressed lines; one of the pairs may be a form of more complex cordage than simple twisted cord but the impressions are too indistinct to determine its form. Condition: fair exterior surface and edges, fresh interior.

P127. F.22 L3: Body sherd with three horizontal rows of Z-twisted cord impressions (i.e. of S-twisted cord] and a curvilinear impression which is similar to the 'horseshoe' impressions found on Peterborough Ware; this sherd, however, is not Peterborough Ware and in fabric and colour it is consistent with the Grooved Ware from the site. The fabric contains sparse chalk fragments (<3 mm) and some grog. There is a black residue on the interior surface. The exterior is dark grey-brown. Condition: surfaces fresh, edges fair.

P128. F.22 L2: Beaker body sherd decorated with a fine square-toothed comb in a hard fabric containing some grog and sand. Surfaces orange, core grey-brown. Both edges and surfaces are worn.

Figure A4.19

P129. F.22 L3: Rim in a fabric with sparse chalk fragments (<2 mm), sparse to moderate sub-angular grog (<2 mm) and rare shell (9 mm max. dimension). The exterior surface is slightly ridged and the horizontal lines below the rim on the exterior are coarse twisted cord (Z-twisted impressions, therefore S-twisted cord). There is sooting on the exterior below the rim and in another area lower down. The exterior surface and edges are fresh and the interior degraded.

P130. F.22 L3: Body sherd in a fabric with sparse coarse sand, sparse small chalk (<2 mm) and some grog (difficult to distinguish from matrix). Exterior yellow-brown, interior black, core bi-coloured as surface colours. Fresh exterior, fair edges; interior surface has some residue adhering. The decoration is of applied ribs and broad grooves; the grooves to the left of the left-hand rib have internal striations.

P131. F.22 L3: Body sherd belonging to same vessel as P124 and P126. The fabric has small blocky chalk inclusions (<10 mm, most <2 mm), sparse sub-angular grog (<6 mm, most <3 mm) and sparse to moderate coarse sand; there are moderate small voids where either small chalk inclusions or sand grains, or both are missing. The fabric is very degraded, the excavator suggesting that this may have been caused by the ashy nature of the feature fill. The sherd may be the other way up from that illustrated. The decoration is of vertical lines of Z-twisted cord impressions (i.e. from S-twisted cord); the adjacent pair of cord impressed lines are not plaited. Exterior mid brown, core dark brown, interior obscured by post-excavation treatment.

P132. F.22 L3: Body sherd containing rare to sparse large chalk inclusions (<5 mm, most <3 mm) and sparse coarse sand; there may also be grog but if so it is difficult to distinguish from the matrix. Exterior pale orange-brown, core dark grey, interior black; there is a black residue adhering to the interior surface. Fair surfaces and edges.

P133. F.8 L2: Body sherd in shelly fabric (moderate small inclusions). Exterior orange, interior and core black. Condition fair.

P134. F.8 L2: Rim sherd of a shallow bowl in a fabric with sparse shell. The vessel could be a small earlier Neolithic bowl but as the fabric is within the range for Grooved Ware it could be a plain shallow Grooved Ware vessel. Black throughout. Fair condition.

P135. F.8 L2: Body sherd coarse fabric with moderate shell with applied ribs. Exterior orange, core and interior black. Fair condition.

P136. F.8 L2: Body sherd in very poor condition and in a very friable fabric containing sparse flint and sand. The decoration is of two very shallow ribs with shallow incision filling zones between. Exterior and core orange, interior black.

P137. PH 30. Building 2: Body sherd in a fabric with some shell and sand. The decoration is of wide deep grooves which may be part of a simple herringbone pattern in which the strokes have run into each other, but alternatively, because the angled groove does seem to have been made in one action, it is possible that it is part of a lozenge motif (the angle could not be from a square or rectangle). Orange throughout. Condition fair.

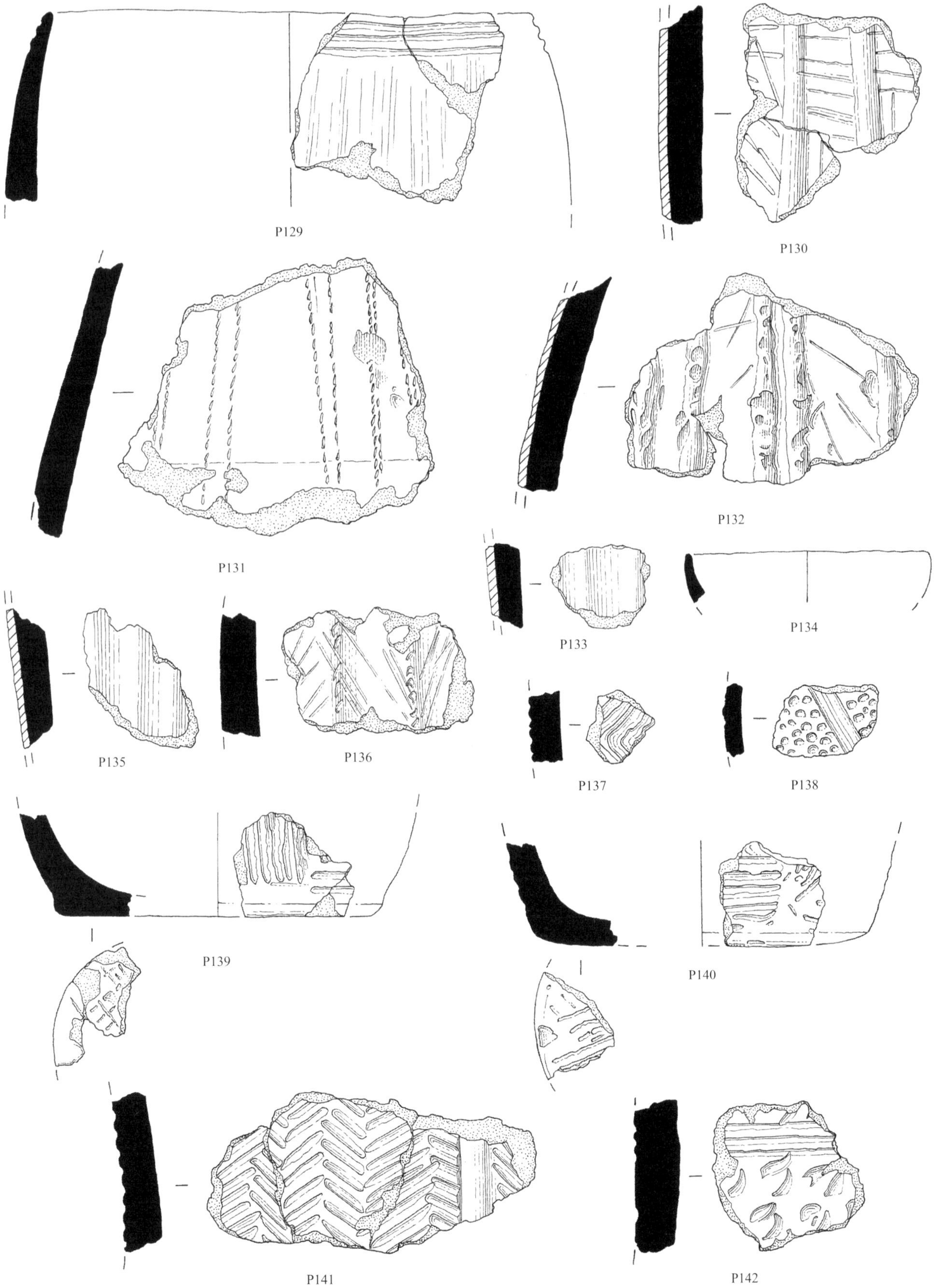

Figure A4.19. *Wyke Down henge 2: pottery 129–42. Scale 1:2 (N. Griffiths).*

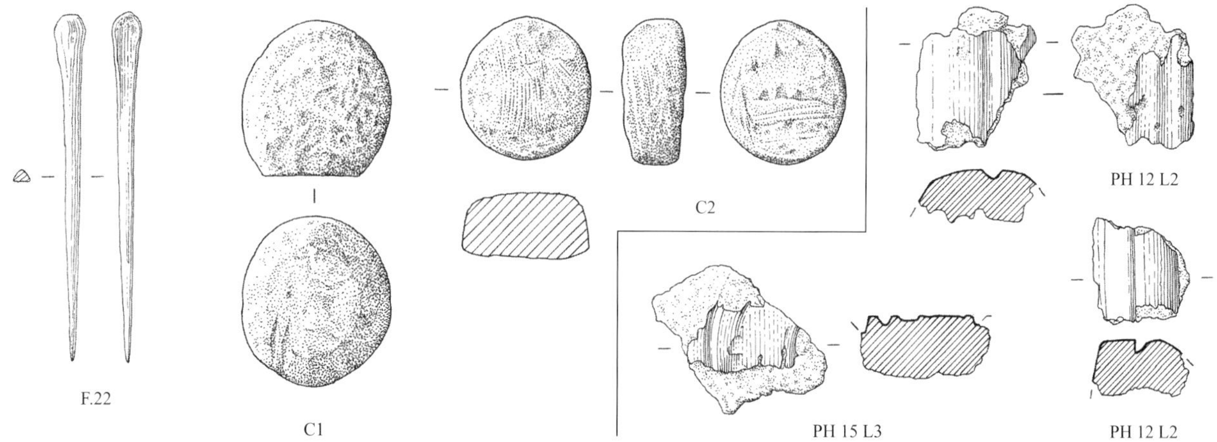

Figure A4.20. *Wyke Down henge 2: bone pin in pit F.22 (left); and the chalk objects C1 from the henge ditch section C (left middle) and C2 from the pit F.18 near building 2 (right middle), and the decorated wall-plaster fragments from building 1 (right). Note the right side of fragment PH 12 L2 shows internal wattle impressions. Scale 2:1 (N. Griffiths).*

P138. PH 32B L2. Building 2: Body sherd in a shelly fabric with sparse to moderate small shell (<4 mm, most <2 mm). The sherd may be illustrated at an incorrect angle as there is what appears to be a vertical applied rib with a groove alongside it, separating panels filled with round impressions, possibly made by reed as each is more pronounced around the circumference than in the middle. The sherd is in poor condition and has been consolidated after excavation.

P139. PH 30. Building 2: Base angle sherd, part of same vessel as P140 with same fabric, colouring and condition. The interior surface is degraded or worn, as in P140.

P140. PH 30. Building 2: Base and lower body sherd in a fabric with some shell (<5 mm, most <3 mm) and some sand; from the same vessel as P139. The decoration is of panels filled with grooving, separated by a slight rib, probably worked up rather than applied. There is grooving on the underside of the base but it is not possible to determine the decorative motif. Surfaces grey brown, core black. Condition fair.

P141. PH 31. Building 2: Large body sherd in a fabric with moderate small shell (<3 mm) and possibly sparse sand. The sherd is thick (14 mm; excluding the applied rib). Exterior pale brown; core black; interior damaged (degraded); exterior surface fresh, edges fair to worn.

P142. PH 32B L2. Building 2: Thick body sherd in fabric with moderate shell (<3 mm, but some very fine) and some grog (difficult to distinguish from the matrix). The sherd is very thick (15 mm). The decoration is of a zone of paired plastic fingernail impressions below a zone of oblique grooved decoration and separated from it by horizontal grooving. Surfaces pale brown, core dark grey. Exterior fair, edges and interior worn.

Worked bone

Martin Green

Just one worked bone artefact was recovered from pit F.22, as follows (Fig. A4.20):

F.22, L4: Slender bone pin (91 × 6 mm) made from a split long bone of a medium-sized mammal. The slightly swollen head from part of the surviving articular end.

Catalogue of worked chalk objects

Martin Green

Illustrated objects (Fig. A4.20)

C1. Base of henge ditch, trench C: Well-made golf-ball-sized (43 mm) ball. Weathered surface. Portion sheared off, probably during excavation, produces a flat base. Wt 69 g.

C2. F.18, L2/3 interface: Well-made disc (34 × 15 mm). Clear scrape marks resulting from shaping on all surfaces. One face bears three closely spaced engraved lines which terminate in a point. Wt 22 g.

Objects not illustrated

C3. F.22, L4: Small sub-rectangular lump (45 × 29 × 22 mm) the narrow end cut to a smooth edge. One face with a long (30 mm) scoop bearing clear scrape marks and a smooth groove towards one edge. The opposite edge bears a series of closely spaced coarser grooves. Wt 31 g.

The remaining pieces described are not indisputably modified by human agency. However, they are included for consistency of record.

C4. F.6, L5: Small natural angular lump of triangular section (61 × 50 × 30 mm). One face bears a series of scratched parallel lines partly enclosed by a curving line on one side. Wt 61 g. Found in association with a dog skull and a partly polished flint ball on the base of the pit. The scratches may have been made by an animal but the relationship with the undoubted special deposit makes this less likely.

C5. F.6, L2: Small weathered discoidal lump (45 × 17 mm). Two-thirds of the edge has apparently been smoothed/worn to a rounded profile. One face also bears a small area partly criss-crossed by engraved lines. Wt 30 g.

C6. PH 3: Pea-sized ball (6 mm). Careful examination suggests this is not a fossil making it likely to have resulted from human modification. Wt 2 g.

Table A4.23. *Wyke Down 2: contexts for clay and chalk-based building materials.*

Contexts	Weight of chalk (g)	Weight of clay (g)
Postholes of Building 1	732	512
Internal feature PH 16/19	879	80
Pits with Building 1	2	60
Postholes of Building 2	-	799
Pits with Building 2	-	103
Henge ditch secondary fills	-	141
Isolated pit F.22	4	-
Totals	**1617**	**1695**

Table A4.24. *Wyke Down 2: contexts, description and sizes of the analysed pieces of chalk building material.*

Context	Description	Dimensions (mm)
1: PH 13, L3	Type 2 chalk-based fragment with internal impressions	26 × 15 × 7
2: PH 15, L2	Type 2 chalk-based fragment with smoothed outer surface	28 × 19 × 15
3: PH 15, L3	Type 1 clay-based fragment with no finished surfaces	54 × 35 × 11
4: PH 15, L3	Type 1 clay-based fragment with eroded smooth surface	44 × 29 × 19
5: PH 16	Type 1 chalk-based fragment with no finished surfaces	21 × 19
6: PH 16	Type 2a chalk-based fragment with fine smoothed surface	19 × 18 × 10
7: F.16	Type 1a clay-based fragment with finer consistency	23 × 23 × 8

C7. PH 32A, L2: Small sub-pyramidal weathered lump (42 × 46 × 30 mm). The base may have been deliberately smoothed and the top has a depression, 20 mm in diameter, with traces of scratch marks resulting from either human or animal agency. Wt 49 g.

Clay and chalk-based building material

Martin Green *with* Vanessa Muros

During the excavations significant amounts of both clay- and chalk-based building material was uncovered particularly from the postholes forming Buildings 1 and 2 (Figs. 3.21–3.24; Tables A4.23 & A4.24). Much of this had once clearly formed part of a thin render applied to the walls of the buildings. This material was of two main types: 1) a clay-based daub of fairly coarse consistency with occasional organic impressions and 2) a chalk-based thin (*c.* 10 mm) plaster or render with clear parallel organic impressions of this wattle likely to be either reed or willow withies (Fig. A4.21). Within these two main types were fragments of a much finer clay-based render (1a) and an equally fine (2a) chalk-based render of much denser consistency than type 2 and with no organic impressions. Both of these types

Figure A4.21. *Close-up view of the plaster fragment from posthole F.15 in Building 1 showing the wattle impressions (Wessex Archaeology).*

often reveal portions of smooth and strikingly flat outer finished surfaces. A further type 2b consist of a fairly coarse thick (up to 45 mm) chalk daub with evidence of finished surfaces including one fragment with two planes evident.

Although fragments of clay-based daubs of type 1 have been occasionally uncovered in Neolithic contexts the remarkable chalk-based renders just described appear to be unique in a British Neolithic context. Close examination of some of the better preserved fragments reveal extraordinary detail. This includes evidence of at least two phases of rendering, deep incisions on outer surfaces (Fig. A4.20), probably representing decoration and traces of a final coating possibly similar to a lime wash. Quantifying by context clearly shows the close relationship of this material to the two buildings. Interestingly chalk based plaster of types 2 predominates in the postholes of building 1 made up substantially by numbers of small fragments. Much less clay-based daub occurred within this building and consisted in the main of one substantial lump (493 g) found in PH 15 immediately below a very large sherd of Grooved Ware. The shallow conjoined feature PH 16/19 within building 1 (Fig. 3.23) contained the entire amount of the chalk-based plaster type 2a. Although the shape and shallowness of this feature suggests an oven base, the fine type 2a plaster with very flat finished surfaces found within, may suggest a different function. The postholes of building 2 on the other hand, revealed only clay-based daub of type 1. The majority of this (707 g) came from the three intercutting postholes 32a/b/c forming the southwest corner of the internal square structure and included a right-angle corner fragment.

It is tempting from the evidence provided by the building materials and structure size to see building 1 as a structure with finely finished and decorated walls with building 2 as an ancillary structure of coarser construction. Owing to the remarkable nature of this material, six samples from Building 1 and one from pit F.16 were submitted for analysis at the Institute of Archaeology by Vanessa Muros and a summary of her results follows.

Examination of all samples at magnifications of ×10, ×20 and ×40 through a binocular microscope allowed the selection of three (nos. 1, 5 & 6) for examination using scanning electron microscopy and the grain sizes were compared to modern lime plaster. All three samples proved to be very granular in structure with grains much longer and more rounded than those in modern lime plaster. However, this result was inconclusive and there followed a series of seven analyses to further understand the composition of the samples.

These tests involved:

1. Testing for carbonate compounds using an acid spot test which revealed that all samples contained some type of carbonate compound, most probably calcium carbonate.
2. Testing for sulphate compounds using another acid spot test which produced negative results in all samples.
3. Examination with energy dispersive x-ray analysis (EDX). Fragments from samples 1 and 6 produced calcium, iron, magnesium, aluminium, potassium, silica, sulphur, chlorine and manganese. The sulphur found in the analysis was determined as an impurity due to the negative results produced in the sulphate tests.
4. Examination using transmitted and polarized light microscopy. A thin section was taken from sample 6 and revealed features such as calcium carbonate crystals, various inclusions, impurities and the remains of fossils. The identification of these micro-fossils would be one determining factor in determining whether this was re-carbonated lime as they are absent in burnt limestone. No micro-fossils were found in the three samples examined using SEM, but a larger sample size would be desirable.
5. Ultra-violet examination was used to detect any signs of pigment on all samples, but this analysis was inconclusive.
6. All samples were tested for pH, and the results were similar to the pH of calcium carbonate.
7. X-ray diffraction (XRD) was used to identify the compounds present in sample 6 and help identify whether the material was made from re-carbonated lime. Several compounds were looked for, with no calcite, aragonite, calcium hydroxide, calcium sulphate and magnesite present, but calcium oxide was present. These results were unexpected. The sample was not identified as a calcium carbonate compound although it did show that some calcium oxide was present. This may mean that the sample was heated at some point in its past. However with no calcium hydroxide present, this indicates that the sample is not a true lime plaster because it was never slaked.

Marine shells

Mike Ladle

A small number of mainly fragmentary marine shells were recovered from features forming part of the Grooved Ware settlement including postholes of both buildings and in some cases, on the bases of pits, suggesting selective deposition. These are as follows (not illustrated):

S1. Building 1, PH 15, L3: Great scallop *Pecten* maximum. Part of lower (right) valve with remnants of five ribs and growth rings. Part of a structured deposit with large Grooved Ware rim sherd and a large lump of clay daub.

S2. Building 2, PH 13, L2: Shell possibly scallop *Pecten* sp. Two small pieces very thin and worn with undulations or ribs apparent.

S3. Building 2, PH 32b, L2: Fragment of common oyster *Ostrea edulis*. Part of upper (right) valve with adductor muscle scar and umbone (beak).

S4. F.4, base of pit: Common oyster *Ostrea edulis*. Almost complete lower (left) valve, very thick with adductor muscle scar and umbone.

S5. F.6, L5, near base of pit: Scallop *Pecten* sp. Fragment of shell (ear and part of one rib) from right (lower valve).

S6. F.7, L3: Oyster. Small fragment.

S7. F.19, L2: Common oyster *Ostrea edulis*. Part of lower (left) valve with adductor muscle scar and remnant of hinge.

S8. F.9, L2: Oyster. Four small fragments of shell.

S9. F.19, L5: Oyster. Small fragment.

S10. F.22, L3: Shell fragment possibly cockle *Cardium* sp. Small piece of shell edge, ribbed and much abraded.

These shells could all have come from the beaches of south Dorset, within a day's walk of the site. Cockles would have been picked up from a sheltered, shady, intertidal sea shore; the scallops live below the tide marks and were probably washed up alive after a storm or picked up as dead shells from the beach and the oysters could have been collected on very low tides or by diving in the shallow water of lagoons or inlets such as Poole Harbour or the Fleet near Weymouth.

4.5. The Wyke Down barrow group (1998–2000)

Lithics

JULIE GARDINER

Barrow 36

Three hundred and ninety struck flints were recovered from this barrow: 21 (5.4%) from the barrow mound, 320 (82%) from the ditch and 49 (12.6%) from other contexts (Table A4.25). The raw material appears to be small rounded nodules of surface chalk flint with a thick, fairly smooth, buff-brown cortex, and most pieces have dense white cortication.

Apart from one scraper and two borers, all the material is waste, comprising 90.3% flakes and blades, with 7 cores and one core fragment (2.1%) and 30 irregular pieces (7.6%). The largest collection is from layer [68], a lower tertiary fill in the ditch, which produced 199 flints, including 4 cores and all three retouched tools. The cores are all crudely worked, with one single-, one double- and two multi-platform types weighing 43–106 g. The single scraper is on a very thin flake with steep retouch and the two borers are very roughly made.

Layer [67], above [68] produced 84 struck flints, including three crude single platform cores and a trimming. Many of the flints from this layer are burnt and 1044 g of unworked burnt flint was also recovered from this layer.

Overall the flintwork consists of small, squat, thick flakes displaying many hinge and step fractures and the use of hard-hammer technique throughout. The flint itself may well have been quarried from the barrow bank or gathered from the weathering ditch and seems to represent the casual discard of material used expediently at a time considerably later than the construction of the barrow but before the ditch had completely filled in, presumably in the mature Bronze Age.

Barrow 44

Excavation of barrow 44 produced 169 struck flints, of which 105 (62.1%) (Table A4.25) are from the mound of the barrow and the remainder from the ditch and topsoil over it. With the exception of one scraper all the material is waste.

The flintwork from the barrow mound is predominantly from the turf core and must be considered to be either earlier than, or contemporary with, the construction of the monument. The raw material is locally available chalk flint which occurs as irregularly shaped, elongated, rounded nodules with a fairly thick buff-brown cortex. The flints are virtually all corticated dense white but there is some red staining on the cortex as well as some adhering clay which suggests that the raw material may have come from a nearby source on clay-with-flints.

The standard of workmanship is poor with thick, squat, chunky flakes and irregular pieces of waste exhibiting a high incidence of hinge and other fractures. One of the cores is little more than a poor-quality nodule with a few removals. The other has a single platform producing predominantly blades though it was clearly abandoned because of the large number of step fractures that were produced. The single scraper, from the junction between the turf core and underlying soil [98/9] is neatly worked but on an irregular flake. Overall, the material appears to be of mature Bronze Age date.

Flintwork from the barrow ditch is all from the upper fills or topsoil. It seems to be a very *ad hoc* collection of roughly worked pieces using fairly poor quality surface flint. The size and quality of the material suggest that it was picked up from the surface rather than deriving from the barrow ditch or mound where rather better material is likely to have been found. It is all undiagnostic and clearly post-dates the use of the barrow.

Barrows 34 and 41

A total of 849 struck flints were recovered from the mound surfaces and the fills of the ring-ditches (Table A4.26). Of these 216 (25%) came from contexts associated with the denuded mound 41, and the remaining 633 from contexts associated with barrow 34. Less than 1% of the total assemblage consists of retouched implements and only 1.7% of cores or tested nodules. Only six retouched implements were recovered: three scrapers, a denticulated piece and a bifacially retouched piece, and an irregular flake with some invasive retouch. Almost the entire assemblage consists of discarded material in the form of flakes, irregular waste, chips and broken pieces. About 78% of the flints were struck from nodules freshly quarried from the chalk, presumably during the digging of the barrow ditches. With very few exceptions, all the quarried flints are in very fresh, sharp condition with no edge damage or post-depositional breaks. Essentially this material is very similar to the material that was recorded from the Dorset Cursus excavations in 1984 (Gardiner 1985; 1991). As the likely raw material would have been encountered during

Table A4.25. *The lithic types present at Wyke Down barrows 36 and 44.*

Barrow 36	Mound & OLS				Ditch			Treehole	Trample	Valley trench	Other contexts			
Context	**85**	**86**	**87**	**72**	**66**	**67**	**68**	**88**	**?**	**103**	**73**	**80**	**83**	
Obj. no.	**92**	**80**	**83**	**89**	**103, 104**	**114**	**95, 115, 130**	**98, 99**	**101**	**119**	**90**	**100**	**88**	**Total**
Flake	12	2	6	0	37	65	167	12	1	10	3	16	2	**333**
Blade	0	0	0	1	0	9	6	0	0	0	0	0	0	**16**
sub-total	*12*	*2*	*6*	*1*	*37*	*74*	*173*	*12*	*1*	*10*	*3*	*16*	*2*	***349***
Core	0	0	0	0	0	3	4	0	0	0	0	0	0	**7**
Core trim/frag.	0	0	0	0	0	1	0	0	0	0	0	0	0	**1**
Irreg. waste	0	0	0	0	0	6	19	4	1	0	0	0	0	**30**
Scraper	0	0	0	0	0	0	1	0	0	0	0	0	0	**1**
Borer	0	0	0	0	0	0	2	0	0	0	0	0	0	**2**
Total	**12**	**2**	**6**	**1**	**37**	**84**	**199**	**16**	**2**	**10**	**3**	**16**	**2**	**390**
Weight (g)	95	48	5	1	447	1255	2650	67	89	19	8	116	2	**4802**
Burnt (g)	0	0	0	0	11	1044	21	0	163	88	6	0	0	**1333**

Barrow 44	Mound							Ditch				Field edge	
Context	**107**	**96**	**97**	**98**	**98/9**	**99**	**100**	**104**	**105**	**108**	**109**		
Obj. no.		**93**	**87**	**81**	**96**	**91, 94**	**82**	**108**	**110**	**109, 117**	**116**	**106**	**Total**
Flake	23	11	15	4	20	32	9	9	8	7	7	2	**147**
Blade	0	0	0	0	0	3	0	0	0	1	1	0	**5**
sub-total	*23*	*11*	*15*	*4*	*20*	*35*	*9*	*9*	*8*	*8*	*8*	*2*	***152***
Core	0	0	0	0	0	0	2	0	0	1	0	0	**3**
Irreg. waste	1	0	3	1	0	4	0	3	0	1	0	0	**13**
Scraper	0	0	0	0	1	0	0	0	0	0	0	0	**1**
Total	**24**	**11**	**18**	**5**	**21**	**39**	**11**	**12**	**8**	**10**	**8**	**2**	**169**
Weight (g)	209	96	276	160	177	1549	232	167	11	210	164	20	**3271**
Burnt (g)	0	17	47	0	0	0	0	0	1	0	0	27	**92**

Table A4.26. *The lithic types present by context at Wyke Down barrows 34 and 41.*

Barrows 34 and 41											
Context	**Site sub-div.**	**Arch. component**	**Cores/tested nodules**	**Irreg. waste**	**Flakes**	**Chips**	**Broken**	**Tools**	**Total**	**Weight**	**Burnt no./wt**
1	mound 34	1	0	0	1	0	0	0	1	1	0
		61	0	0	1	0	0	0	1	10	0
2	mound 41, tr. 1	54	1 frag.	0	0	0	1	0	2	39	1/15
		55	0	0	0	0	3	0	3	9	1/7
		56	0	0	3	0	0	0	3	11	0
		57	0	0	2	0	0	0	2	3	0
3	mound 41, tr. 1	46	0	0	0	0	1	0	1	5	0
		49	0	0	0	0	0	0	0	0	1/6
4	mound 41, tr. 2	60	1	1	3	0	0	0	5	74	1/2
5	mound 41, tr. 2	44	0	0	0	0	0	0	0	0	1/3
		62	1 frag.	1	15	0	5	0	22	112	5/25
6	mound 41, tr. 1	50	0	0	1	0	0	0	0	5	0
		58	0	1	3	0	0	1	5	90	1/4
7	mound 41, tr. 1	52	0	1	16	0	0	0	17	144	0
8	mound 41, tr. 2	47	0	0	4	0	0	0	4	10	0
		48	0	1	0	0	0	0	1	2	0
		51	0	1	2	0	1	0	4	12	0
9	mound 41, tr. 2	40	0	0	61	0	5	0	66	777	1/25
10	mound 41, tr. 2	63	1	3	15	0	0	0	19	411	0
11	mound 41, tr. 2	45	0	1	0	0	0	0	1	28	0
		53	1 on flake	1	2	0	0	0	4	170	0

Table A4.26. *(cont.)*

Barrows 34 and 41											
Context	**Site sub-div.**	**Arch. component**	**Cores/tested nodules**	**Irreg. waste**	**Flakes**	**Chips**	**Broken**	**Tools**	**Total**	**Weight**	**Burnt no./wt**
19	mound 34, tr. 3	30	0	0	3	0	0	1	4	38	
		32	0	0	2	0	3	1	6	43	
20	mound 34, tr. 3	13	0	0	5	0	0	0	5	20	2/16
		78	0	0	9	0	1	0	10	134	
21	mound 34, tr. 3	6	1	0	4	0	0	0	5	70	2/44
22	mound 34, tr. 3	7	0	0	6	0	0	0	6	7	
23	mound 34, tr. 4	41	0	0	8	0	0	0	8	24	
		75	0	1	41	0	1	0	43	340	1/2
		76	0	0	0	0	0	0	0	0	3/61
24	mound 34, tr. 4	5	0	0	3	0	0	0	3	9	
		14	0	0	0	0	0	0	0	0	8/168
		35	2	12	101	9	1	0	125	2019	
25	mound 34, tr. 3	34	0	0	4	0	1	0	5	62	
		38	0	4	77	18	5	0	104	364	
		77	0	6	134	18	3	0	161	828	
26	mound 34, tr. 3	8	0	0	2	0	0	0	2	44	
27	mound 34, tr. 5	4	0	1	2	1	0	0	4	12	1/3
		69	0	0	7	0	1	0	8	36	1/3
28	mound 34, tr. 5	18	2	2	13	0	1	0	18	270	2/18
		72	0	0	3	0	1	0	4	14	
29	mound 34, tr. 5	16	0	0	6	0	0	0	6	32	
		17	0	0	0	0	0	0	-	-	2/23
		28	0	0	1	0	0	0	1	1	
30	mound 34, tr. 3	31	1	0	0	0	0	0	1	314	
31	mound 34, tr. 3	19	0	0	3	0	0	0	3	22	
		20	0	0	2	1	0	0	3	23	
32	mound 34, tr. 5	22	0	1	1	0	0	0	2	21	
		37	0	0	20	0	0	0	20	238	1/4
		67	0	0	1	0	0	0	1	5	
33	mound 34, tr. 5	15	0	0	2	0	0	0	2	19	
		21	0	1	0	0	1	0	2	9	
34	mound 34, tr. 5	26	0	0	9	0	1	0	10	146	
		29	0	0	2	1	0	0	3	5	
35	mound 34, tr. 4	3	0	0	3	1	0	0	4	18	1/2
		39	0	1	5	0	0	0	6	470	2/194
41	mound 34, tr. 6	71	1	0	16	0	0	1	18	300	
42	mound 34, tr. 6	23	1	1	5	0	1	0	8	346	
		24	0	0	0	0	0	0	0	0	6/35
		70	0	0	3	0	0	0	3	34	1/218
43	mound 34, tr. 6	12	0	0	0	0	0	0	0	0	4/19
		25	1	0	4	0	0	0	5	35	
50	mound 41, tr. 7	64	0	0	3	0	0	0	3	8	
51	mound 41, tr. 7	73	0	0	2	0	1	0	3	23	
52	mound 41. tr. 7	63	0	1	3	0	0	0	4	57	
53	mound 41, tr. 7	74	0	1	36	6	1	2	46	480	1/148
Total			**14**	**43**	**691**	**55**	**40**	**6**	**849**	**8234**	**50/1045**
% total			1.7	5.1	81.4	6.5	4.7	0.7			

the construction of the monuments (flint in the chalk and in any drift deposits — see below), and the waste material discarded in a completely fresh condition, it can be proposed that the knapping of flint was exactly contemporary with the construction of the barrows. It seems most likely that the flint knapping represents no more than fortuitous working of suitable nodules encountered during ditch digging. The paucity of 'proper' cores is explained by the expedient, brief and unsystematic working of the discarded nodules, and the lack of tools strongly suggests that any useful items were taken away for use elsewhere.

Soil micromorphology

Helen Lewis

The micromorphological descriptions for the ring-ditch fills, barrow mounds and buried soils are summarized in Tables A4.27–A4.29.

Barrow 34

Context 21, Samples 67 & 68 (Mound turves)

An entire single turf in the turf stack, with mottling on either side of it marking the turf edges.

Structure: apedal channel with some crumb and sub-angular blocky (1000 µm).
Porosity: 10%, channels (150–600 × 1200–30,000 µm, straight and zigzag, partially accommodated).
Mineral components: quartz, flint, very rare chalk, rare angular plagioclase and possibly potassium feldspars, amphibole, epidote, and possibly garnet and zircon (Chris Doherty pers. comm. 2006), very rare bone fragments (250 µm); clays masked by staining but appear 'dirty' reddish-brown (XPL), dark brown (PPL).
C:f ratio: 15:85: very coarse and coarse sand 5%, medium sand 5%, fine sand 5%, very fine sand 25%, silt 25–30%, clay 30–35%.
Organic components: 5%, black angular fragments (<50–300 µm), occasionally obviously charcoal, and 'punctuations'; occasional root fragments.

Table A4.27. *Wyke Down ring ditch fills: micromorphological descriptions (H. Lewis).*

Context	Structure	Porosity	Mineral components	C:f ratio	Organic components	Groundmass	Pedofeatures
B44 [102]	Apedal channel & angular blocky	10–15% cracks & channels	Quartz, chalk, shell, calcite, micrite	20:80 Silt loam	5–10%, roots, charcoal, 'punctuations', sclerotia	Greyish-brown (PPL), brown (XPL), light reddish-brown (RL); crystallitic	Micritic clay coatings; rounded red clay fragments, iron mottles
B44 [103]	Crumb & blocky	10–25% packing pores	As above, rare bone	10:90 Clayey silt	5%, as above but rare charcoal	Brown to reddish- or golden brown (PPL & XPL), orange (RL), stipple speckled & crystallitic	As above; amorphous organic-stained zones

Table A4.28. *Wyke Down round barrow mounds: micromorphological descriptions. (H. Lewis).*

Context	Structure	Porosity	Mineral components	C:f ratio	Organic components	Groundmass	Pedofeatures
B34 [31]	Crumb, some columnar & blocky	15–20%, packing pores & channels	Quartz, chalk, shell, feldspars, calcite grains	15:85 Silty clay	5% 'punctuations' & small fragments; rare charcoal	Reddish-brown (PPL), very dark reddish-brown (XPL), orange (RL), stipple speckled & granostriated	'Dusty' clay coatings, micritic zones, amorphous organic staining, fine iron & manganese mottles
B34 [21]	Apedal channel, some crumb & sub-angular blocky	10% channels	Quartz, flint; rare chalk, feldspars, amphibole, epidote; rare bone	As above	5% charcoal, 'punctuations', roots	Dark brown (PPL), dark reddish-brown (XPL), reddish-brown (RL), stipple speckled to undifferentiated	'Dusty' clay coatings; fragments of iron rich clay; amorphous organic staining, occasional iron mottles
B41 [6]	Apedal channel & angular blocky	15% channels	Quartz, flint, rare calcite, glauconite, feldspars, amphibole, epidote, rare bone	10:90 Clayey silt	<5–5% roots & 'punctuations'	Light brown to brown (PPL), very dark reddish-brown (XPL), orange (RL), stipple speckled	Frequent clean clay coatings; occasional 'dusty' clay coatings; sand & silt infills (vermiform); rounded clay fragments; iron staining & mottles
B36 [72]	Angular-sub-angular blocky; flattened angular peds at base	10–25% channels	Quartz, flint, chalk, micrite, rare calcite, glauconite, feldspars, amphibole, epidote	10:90 Silty clay	5% 'punctuations', rare charcoal	Brown to reddish-brown (PPL), very dark reddish-brown & orange (XPL), orange (RL); granostriated & stipple speckled	'Dusty' clay coatings; occasional rounded clay fragments; calcareous fabric zones; micrite coatings, frequent iron mottles; light amorphous sesquioxide staining; mite excrements
B44 [107]	Angular & sub-angular blocky	15–30% channels	Quartz, flint, chalk, micrite, shell	15:85 Clay loam	<5%, 'punctuations', charcoal, roots	Reddish-brown (PPL), very dark reddish-brown (XPL), orange red (RL); granostriated & stipple speckled	'Dusty' clay coatings & infills; calcareous zones; micrite coatings; frequent iron & manganese mottles

Table A4.29. *Wyke Down round barrows: micromorphological descriptions of the buried soils (H. Lewis)*

Context	Structure	Porosity	Mineral components	C:f ratio	Organic components	Groundmass	Pedofeatures
B34 [36]	Angular blocky & crumb	25% packing pores & channels	Micrite, quartz, chalk, flint, calcitic earthworm granules	10:90 Clayey silt	<5% 'punctuations'	Reddish brown (PPL), light reddish-yellow (XPL), orange-red (RL), crystallitic	Micritic clay coatings, depleted zones (re. calcium carbonate) & calcareous zones, occasional iron mottles
B41 [13]	Apedal channel	10–20% packing pores & channels	Quartz, flint, rare calcite, glauconite, feldspars, amphibole, epidote, rare bone	15:85 Clay	5% 'punctuations' & charcoal	Light reddish-brown to orange (PPL), orange-yellow (XPL), orange (RL), reticulate & granostriated	Frequent clean clay coatings; 'dusty' clay coatings & hypocoatings; occasional calcareous zones; amorphous iron staining; iron mottles
B36 [83]	Crumb & sub-angular blocky	15–25% channels	As above	10:90 Silty clay	5% 'punctuations', rare charcoal	As in context [72] above; becoming grano- & linear striated with depth	'Dusty' clay coatings; calcareous zones; iron & manganese mottles
B36 [91] & [87]	Angular blocky & columnar	10–15% channels & cracks	As above	As above	As above, rare roots	Yellow-light reddish-brown (PPL), very dark reddish-brown (XPL), light red & orange (RL); granostriated	As above
B44 [100]	Angular-sub-angular blocky	15–20% channels	Quartz, flint, chalk, micrite, shell, rare glauconite	20:80 Clayey silt	5–10% roots, charcoal, 'punctuations'	Greyish-brown (PPL), brown (XPL), light reddish-brown (RL); crystallitic	Micritic clay coatings; red clay fragments; iron mottles

Pedofeatures: 1) 10 μm thick clay coatings on many voids, yellow with low birefringence, 'dusty' to 'dirty'. Also 'dirty' and silty coatings on rock fragments and voids of similar clay, but 50–100 μm thick, and some are hypocoatings; 2) 500–1500 μm sized angular blocky fragments of light reddish-brown (PPL), very dark reddish-brown (XPL), iron-stained clay (c:f 0:100: very fine sand 5%, silt 30%, clay 65%), with <2% organic components ('punctuations'). Groundmass has cross-parallel and granostriated clays. This appears to be a fragment of material similar to the 'drift' clay soils seen under Barrow 41; 3) zones of earthworm sorting; 4) amorphous organic staining on much of the groundmass; 5) occasional iron mottles up to 3000 μm, frequent mottles <2000 μm; 6) zone (<1000 μm) in groundmass with pellets (20 μm, ovoid), apparently infilling old pores.

Context 31, Samples 18, 19 & 66 (Turves over chalk core/upper buried soil)

Field description: very dark brown (10 YR 2/2) silty clay loam with rare chalk fragments. The more organic-stained nature of this profile with depth represents the presence of the buried soil, which appears to be more organic-stained than the turves or than context 36 (lower buried soil). In samples 18 and 19 there is no clear horizon distinction between contexts 31 and 36, except that the lower few cm of sample 18 have this more organic nature. A boundary is seen between the two contexts in sample 66. Sample 66 has two stone lines (flint and chalk), showing earthworm sorting. These are also size sorted: one line comprises mainly 0.5–1 cm sized stones; the other comprises 0.3 cm.

Structure & porosity: crumb (3000 μm) and some fine columnar and blocky structure. 15–20% porosity, mainly packing pores and channels (200–2000 × 1000–10,000 μm, curved and crooked, partially accommodated and accommodated).
Mineral components: quartz, chalk, shell, feldspar, calcite grains.
C:f ratio: 15:85: very coarse and coarse sand 5%, medium sand 5%, fine sand 5%, very fine sand 20%, silt 30%, clay 35% (in calcareous zones 35% silt, 30% clay).
Organic components: 5%, black angular and rounded fragments (<100 up to 400 μm), rare charcoal at base.
Groundmass: reddish-brown (PPL), very dark reddish-brown (XPL), light orange (RL), stipple speckled with some granostriation.
Pedofeatures: 1) clay coatings 100 μm thick, light brown (PPL), orange-brown (XPL), 'dusty'; 2) some micritic zones (<200 μm), especially as channel infills, reddish-brown (PPL), light reddish-brown (XPL), light orange (RL), stipple speckled with frequent micrite; 3) iron and manganese mottles <100 μm); 4) some very organic-stained zones.

Context 36, Sample 20 (Lower buried soil)

A–A/C of a rendzina soil, described in the field as very dark brown (10 YR 2/2) silty (clay) loam with occasional chalk fragments. The sample is very mixed in fabric, with calcareous and non-calcareous fabrics. Some of this mixing is definitely related to soil faunal activity. However, such mixing was also typical of tilled rendzinas from Butser Ancient Farm in Hampshire, where it was not a major feature of untilled comparatives (Lewis 1998b). This is a physically disrupted soil, but whether that relates to monument construction, cultivation or other influences is uncertain.

Structure: angular blocky and crumb (<5000 μm).
Porosity: 25%: packing pores and channels (200–2000 × 3000–10,000 μm, curved, partially accommodated and unaccommodated; occasional chambers (*c.* 5000 μm).
Mineral components: micrite and quartz grains, chalk, flint, occasional calcitic earthworm granules.
C:f ratio: 10:90: very coarse sand 0%, coarse and medium sand 5%, fine sand 5%, very fine sand 15%, silt 45%, clay 30%.
Organic components: <5%, fine angular and rounded black fragments (<150 μm).
Groundmass: reddish-brown (PPL), light reddish-yellow (XPL), orange-red (RL), crystallitic (micritic) fabric.
Pedofeatures: 1) micrite rich clay coatings, greyish-red (PPL & XPL), up to 150 μm thick; 2) zones relatively depleted in calcium carbonate, although with micrite occasionally, with very dark reddish-brown (XPL) stipple speckled groundmass but same as main groundmass otherwise (including PPL colour); 3) zones that are 'chalky', greyish-red (PPL), light greyish-red (XPL); 4) occasional iron mottles.

Barrow 36

Context 72, Samples 120, 122A–B & 104 (Barrow mound)

Turf mound; angular flint gravel (0.5–2.5 cm) seen at base in sample 104.

Structure & porosity: mixed angular to sub-angular blocky (<10,000 μm), with zones of loose crumb and apedal channel; structure becomes more apedal with depth. In thin section 120, at the base of context 72 there is a line of flattened angular peds (3000 × 7000 μm), suggesting compaction; similar features have been found associated with compaction zones at the edges of ard ploughmarks (Lewis 1998b), but there are no clear indicators of cultivation here; perhaps they relate to compaction of the mound.
Porosity: 10–25%, mainly channels (100–3000 × 500–20,000 μm, straight, partially accommodated and accommodated); occasional chambers (3000 μm) with mite excrements.
Mineral components: quartz grains dominate, occasional flint, occasional to rare chalk, micrite (with depth only in pores as pedofeature), rare calcite grains, occasional glauconite; clays are dusty yellow-orange with low birefringence, and are often masked with sesquioxide staining. Occasional angular plagioclase and possibly potassium feldspars, amphibole, epidote, and possibly garnet and zircon (Chris Doherty pers. comm. 2006).
C:f ratio: 10–90: very coarse sand 0%, coarse and medium sand 5%, fine sand 5%, very fine sand 15%, silt 35–40%, clay 35–40% (more clay with depth). Clay loam.
Organic components: 5%, mainly 'punctuations' and <500 μm black rounded and angular fragments, and root fragments; rare charcoal (<300 μm), becoming extremely rare with depth.
Groundmass: granostriated with stipple speckled zones; brown to reddish-brown (PPL), very dark reddish-brown with orange clays (XPL), orange (RL); close and open porphyric-related distributions.
Pedofeatures: 1) clay coatings and infills (including old infills & hypocoatings) of 'dusty' reddish-brown (PPL), orange (XPL) low birefringent clay, 100 μm thick; 2) occasional very dark reddish-brown (PPL & XPL) rounded clay fragments (<200 μm); 3) occasionally chalky (micritic) fabric, greyish-brown (PPL), light gold (XPL), crystallitic, with c:f 15:85 (coarse sand 5%, medium sand 5%, fine sand 5%, very fine sand 10%, silt 50%, clay 35%), <5% organic matter, mineral components as main groundmass; 4) occasional 250 μm sized, rounded zones of very dark brown (PPL), undifferentiated (XPL) strongly organic-stained fabric (bioturbation indicator); 5) micrite pore coatings; 6) frequent iron oxidation mottles 50–250 μm, moderate manganese mottles (<400 μm). The boundary between contexts 72 & 83 is marked by a line of manganese and iron mottles in thin section 120; 7) light amorphous sesquioxide staining on groundmass; 8) mite excrements in chambers.

Context 83, Sample 120, 121

Apparent boundary between contexts 72 & 83 is marked by a horizontal line of manganese mottles (black XPL, PPL, RL), 5000 μm thick, becoming 1000 μm in places. This appears to be the rooting line of a turf topsoil.

Structure & porosity: coalesced crumb and sub-angular blocky, with some apedal channel structure; 15–25% porosity, channels as in context 72.
Mineral components: as in context 72.
C:f ratio: 10:90: very coarse sand, coarse sand and medium sand 5%, fine sand 5%, very fine sand 10%, silt 40%, clay 40%.
Organic components: 5%, as in context 72, charcoal up to 500 μm.
Groundmass: as in Context 72, but becoming grano- and linear striated with depth.
Pedofeatures: 1) 10–100 μm thick clay coatings, occasionally 'dusty' yellow (XPL), low birefringence, light reddish-brown (PPL); 2) patches of calcareous soil (up to 2500 μm) very fine sand 5%, very fine sand 5%, silt 55%, clay 35%); 3) iron and manganese mottles as in context 72.

Contexts 91 & 87, Sample 124 (Trench 2, base of turf mound/top of buried soil)

Although no visible difference was seen between the two sampled contexts, but under the upper 1.5 cm there is a 'line' of manganese and iron mottles. This must represent the top of the old land surface beneath the barrow mound. In the upper part of this is also rounded flint gravel (0.5 cm) — presumably the upper <2 cm is context 91, and the rest of the profile is context 84. The soil material is otherwise the same, and they will be described together here. The soil appears to be turf, but with disturbance indicators ('dusty' to 'dirty' clay coatings).

Structure: angular blocky (5000–10,000 μm) with occasional columnar (5000 × 25,000 μm), occasional apedal channel zones.
Porosity: 10–15%, channels and cracks (150–600 × 500–3000 μm, straight, partially accommodated and accommodated).
Mineral components: quartz, angular plagioclase and possibly potassium feldspars, amphibole, epidote, and possibly garnet and zircon (Chris Doherty pers. comm. 2006), micrite and calcite, flint and chalk rock fragments; clays are yellow (PPL), reddish-brown (XPL).
C:f ratio: 10:90: very coarse, coarse and medium sand 5%, fine sand 5%, very fine sand 15%, silt 45%, clay 30%.
Organic components: 5%, black angular fragments <150 μm, and root remains.
Groundmass: granostriated, yellow to light reddish-brown (PPL), very dark reddish-brown (XPL), light red and orange (RL).
Pedofeatures: 1) occasional iron-rich 'dusty' and 'dirty' clay coatings and infills, up to 250 μm thick, reddish-brown (PPL), very dark reddish-brown (XPL), with sand grains, occasional hypocoatings of the same; 2) zones of reddish-micritic fabric; 3) mottles (iron and manganese (<150) throughout, but particularly in boundary zone (as described above); occasional iron-replaced root remains.

Barrow 41

Context 13, Samples 54, 55 & 56

Buried lower A(–Bt), strongly influenced by and growing on a clay-rich drift or fan deposit. In the field this was reddish-brown (5 YR 4/4) to dark brown (10 YR 3/3) silty clay to silty clay loam with flint cobbles and chalk rubble. Some areas look like a 'palaeosol' with rubified clay in pores and in the groundmass, including some rounded fragments of the same, mixed in with what appears to be topsoil material. The context includes mineralogical indicators suggesting the fan is a glacial outwash fan (Chris Doherty pers. comm. 2006).

Structure: apedal channel to coalesced crumb, some vermiform, occasional sub-angular blocky (5000–15,000 μm).
Porosity: 10–20%, interpedal packing pores and channels (100–400 × 1000–20,000 μm), straight and curved, partially accommodated).
Mineral components: quartz, angular plagioclase and possibly potassium feldspars, amphibole, epidote, and possibly garnet and zircon (Chris Doherty pers. comm. 2006); chalk (15–40% of the samples), flint, rare bone fragments (<200 μm) and possible degraded bone (apatite); rare calcitic earthworm granules; clays are orange (PPL & XPL) with low to moderate birefringence.
C:f ratio: 15:85: very coarse and coarse sand 5%, medium sand 5%, fine sand 5%, very fine sand 15%, silt 20%, clay 40%.
Organic components: 5%, 'punctuations' and black fragments 50–150, up to *c.* 500 μm; much appears to be charcoal.
Groundmass: reticulate and granostriated, light reddish-brown to orange (PPL), orange-yellow (XPL), orange (RL); close porphyric-related distribution.
Pedofeatures: 1) very frequent clean clay coatings, up to 250–500 μm thick, microlaminated, with some 'dusty' laminations; clay colour as in groundmass; 2) 'dusty' yellow-orange clay coatings and hypocoatings, <150 μm thick.; 3) occasional zones (3000 μm) as main groundmass, but 'chalky' with frequent micrite; 4) staining with amorphous iron; 5) iron mottles (<200 μm).

Context 6, Samples 69 & 70

One horizon; there is a line of 0.5 cm sized flint gravel at 6–8 cm depth in sample 69, occurring just above a series of vermiforms (earthworm sorting).

Structure: apedal channel, with occasional angular blocky peds.
Porosity: channels (100–4000 × 40,000–100,000 μm, partially accommodated and accommodated, straight and zigzag).
Mineral components: As context 13, flint rock fragments; clays are stained with amorphous iron oxide, very 'dusty', orange and red (PPL & XPL).
C:f ratio: 10:90: very coarse and coarse sand 5%, medium and fine sand 5%, very fine sand 25–30%, silt 35–40%, clay 25%.
Organic components: <5–5%, occasional roots (up to 2500 × 10,000 μm), and angular fine fragments (<150 μm).
Groundmass: light brown to brown (PPL), very dark reddish-brown (XPL), orange (RL), stipple speckled, some poro- and granostriated zones.
Pedofeatures: 1) frequent clean clay coatings in pores, microlaminated, orange (PPL & XPL), up to 200–300 μm thick; 2) occasional dusty to 'dirty' clay coatings, <100 μm thick, yellow & orange (PPL & XPL); 3) infills of silt and sand, with <5% clay (possibly vermiform); 4) occasional clay fragments, rounded, 100–300 μm, light brown (PPL), orange-red, stipple speckled (XPL); 5) zones of iron staining and mottling, up to 1000 μm in size, plus frequent iron mottles 200–1000 μm.

Barrow 44

Context 107, Sample 45 (mound turf)

This context has a stone line in its lower 1 cm of chalk and flint rock fragments (0.3–2 cm), revealing earthworm sorting in the base of the mound.

Structure & porosity: angular and sub-angular blocky (3000–5000 μm), 15–30% porosity, channels (250–2500 × 1000–10,000 μm, curved and straight, unaccommodated and partially accommodated.
Mineral components: quartz grains, flint and chalk fragments, occasional micrite as pedofeature, occasional shell; clay is orange with low birefringence, often masked by staining.
C:f ratio: 25:85: very coarse sand 0%, coarse and medium sand 5%, fine sand 10%, very fine sand 25%, silt 35%, clay 25%. Clay loam.
Organic components: <5%, fine black fragments and occasional charcoal <500 μm, 'punctuations', occasional root remains.
Groundmass: reddish-brown (PPL), very dark reddish-brown (XPL), orange-red (RL), granostriated with some stipple speckled; coarse porphyric-related distribution.
Pedofeatures: 1) clay coatings as in Barrow 36, 120–122 and old infills of the same; 2) calcareous (chalky) zones as in 36, 120–122, occasional, up to 3000 μm in size; 3) micrite in some pores, sometimes as coatings 100–200 μm thick; 4) frequent iron mottles <200 μm, some manganese mottles of the same size, and several peds have frequent iron and manganese mottles (<50 μm) on one edge only, in linear zones 250–300 μm wide (possibly marking turf rooting depths)

Context 100, Sample 45 (turves/buried soil)

Structure & porosity: angular to sub-angular blocky, 15–20% porosity, as in 107, but mainly straight channels.
Mineral components: as in 107, rare glauconite grains.
C:f ratio: 5:95: very coarse, coarse and medium sand 0%, fine sand 5%, very fine sand 15%, silt 40%, clay 40%. Clay.
Organic components: as in 107.
Groundmass: light brown (PPL), very dark brown (XPL), orange (RL), stipple speckled with granostriated zones; close porphyric-related distribution.
Pedofeatures: as in 107, but lines of mottles on edges of peds are thinner (<200 μm). Also, occasional small rounded clay fragments (<200 μm) reddish-brown (PPL & XPL), iron stained; rare calcareous fabric inclusions.

Context 102, Samples 136/1–2 (ring-ditch fill)

Structure: apedal channel and angular blocky (<10,000 μm), some crumb with depth (<5000 μm).
Porosity: 10–15%, cracks and channels (50–250 × 100–1000 μm, straight, accommodated and partially accommodated).
Mineral components: quartz grains, chalk rock fragments (<10,000 μm), shell, earthworm calcitic granules, calcite grains, frequent micrite.
C:f ratio: 20:80: very coarse sand 5%, coarse sand 5%, medium sand 5%, fine sand 5%, very fine sand 10%, silt 40%, clay 30%.
Organic components: 5–10%, occasional roots, moderate charcoal (<200–1000 μm), black and brown angular and subrounded fragments (<150 μm), occasional fungal sclerotia.
Groundmass: greyish-brown (PPL), brown (XPL), light reddish-brown (RL); crystallitic; open and closed porphyric-related distributions.
Pedofeatures: 1) micrite-rich clay coatings on aggregates (<150 μm); 2) zones of very 'chalky' soil, similar to the main groundmass, but very grey in colour and micrite rich; 3) <50 μm sized red clay fragments, rounded, low birefringence; 4) iron mottles (<100 μm).

Context 103, Samples 136/3–4 (ring-ditch fill)

Structure: loose crumb with zones of angular and sub-angular blocky (<5000 μm).
Porosity: 20–25% in upper part, decreasing to 10–15% in lower part; mostly packing pores and vughs between peds and chalk gravel, with occasional channels as in context 102.

Mineral components: as in context 102; near the base a tiny bone fragment was found (<150 μm), and one larger fragment of possible altered bone was seen (*c.* 2000 μm).
C:f ratio: 10:90: very coarse sand <2%, coarse and medium sand 5%, fine sand 5%, very fine sand 10%, silt 45%, clay 35%.
Organic components: 5%, as in context 102, but rare obvious charcoal, occasionally up to 2000 μm in size.
Groundmass: brown to reddish- or golden brown (PPL & XPL), orange (RL), stipple speckled and crystallitic, becoming crystallitic with depth.
Pedofeatures: 1) zones with frequent micritic clay coatings and infills (<150 μm), increasing in frequency with depth; 2) coatings on rock fragments of groundmass material (<250 μm) thick; 3) occasional earthworm size sorted zones (*c.* 700 μm zones); 4) micrite coatings and infills; 5) organic-stained zones (earthworm sorting); 6) iron mottles (<150 μm); 7) mite excrements in pores.

Effects of past and present ploughing on prehistoric monuments: case studies from Wyke, Bottlebush and Gussage Downs

Charles French

As a result of the test interventions in the both the Dorset Cursus and Bronze Age barrows of Wyke and Bottlebush Downs (Figs. 3.33–3.57) new data were observed that are relevant to the history of erosion of prehistoric monuments by past and present arable agriculture. This has been reported on as part of a larger study (French 2002), but a summary of the main findings is reiterated here (Table A4.30).

Results

The test excavation of four barrows of the Wyke Down group (SAM 299 c, f, g & i), a length of the Dorset Cursus on Bottlebush Down (SAM 310), and long barrow 294 on Gussage Cow Down (Figs. 3.1 & 3.33), has permitted the quantification of erosion caused by modern ploughing on the preservation of these partially upstanding monuments, and in some cases has shown the probable extent of much earlier, probably prehistoric, mound erosion.

Barrow 41

Barrow 41 is much destroyed and denuded, but to a large extent this had already occurred in the past. The monument survives as a small, low (*c.* 80 cm high), turf mound in the centre of the area enclosed by a ring-ditch (Fig. 3.47). The ring-ditch is about 3.5 m wide, 1.6 m deep, with an internal diameter of about 19 m and an external diameter of about 26 m. The inner third of ring-ditch backfill was composed of densely packed, clean chalk rubble which would appear to be a deliberate backfill deposit. One could speculate that this chalk substrate material had acted as either a cap to the turf core, or as a circular bank on the inner edge of the ring-ditch, which had been deliberately put back into the ditch after a short period. Thus this barrow

Table A4.30. *Observed amounts of soil truncation at four barrows of the Wyke Down group.*

	Barrow site			
Truncation stage	**41**	**34**	**36**	**44**
Surviving mound height	0.8 m	1.58 m	2.45 m	1.33 m
Truncation step across ring-ditch	-	-	10 cm	7 cm
Truncation step on berm	-	on north: *c.* 30 cm; on south: gentle slope over 6 m	15 cm	18 cm
Truncation on outer edge of mound	*c.* 25 cm	(as above)	27 cm	53 cm
Total truncation	*c.* 25 cm	*c.* 30 cm	52 cm	78 cm

has probably previously been two slightly modified versions of a disc barrow.

There is a difference of about +25 cm between the lower level of the chalk rock on the outer edge of the ring-ditch and a plinth of chalk subsoil beneath the surviving mound (Table A4.30). The mound itself has become elongated along the east–west direction of the property boundary between the Shaftesbury Estate to the north and Martin Green's land to the south (Fig. A2.10).

Barrow 34

Barrow 34 exhibited a more complicated and diverse sequence than barrow 41. Despite the presence of a large antiquarian trench on its northern side (which may be early twentieth century in date, and is unfortunately unrecorded), the mound of this barrow appears to be substantially intact and standing to a height of about 1.6 m above the present-day surface (Figs. 3.39–3.42). On its northern side, the mound rises abruptly from the flat level of the present-day ploughsoil, which has been ploughed for a 45-year period (P. Rymer & M. Green pers. comm.). In contrast, the southern side, which has not been ploughed for the past 15 years and is under established grassland (M. Green pers. comm.), exhibits a more gentle sloping aspect between the edge of the surviving mound and the inner edge of the outer ring-ditch. This mound talus slope exhibits a 56 cm thickness over a distance of 6 m (Table A4.30).

Barrow 36

Significantly, the berm area of barrow 36 between the inner edge of the ring-ditch and the outer edge of the surviving mound exhibited three steps in the upper level of the chalk subsoil, rising as one moves in towards the centre of the barrow. The first step was of 10 cm increase between the outer and inner edges of the ring-ditch, then about 1.5 m in from the inner edge of the ring-ditch there was a second step of 15 cm, and then another metre in there was a step of 27 cm at the outer edge of the surviving mound. This gives a total figure of 52 cm of truncation over a width of about 5 m in the level of the chalk subsoil on the east side of the mound (Figs. 3.44 & 3.45; Table A4.52). As is the case for barrow 34, the field in which this barrow sits is known to have been ploughed for 45 years, which has apparently caused the significant denudation and re-shaping of this mound.

Barrow 44

Sample excavation of barrow 44 revealed very much the same picture as for barrow 36, but the mound and buried soil were less well preserved (Fig. 3.50). This barrow mound survives to a maximum height of 1.33 m and has suffered severe rabbit disturbance and possibly an antiquarian trench being placed in its northwestern quadrant. As was the case for barrow 36, it too exhibits three steps in the level of the chalk between the inner edge of the ring-ditch and the outer edge of the surviving edge of the barrow mound. In this case, there are steps of 7, 18 and 53 cm giving a total figure of 78 cm of apparent loss of chalk rock surface through ploughing (Table A4.30).

The Dorset Cursus in Trench 4

Four sample sections were excavated through the northern and southern Cursus ditch and bank on the *c.* 1 km stretch leading to the Bottlebush Down terminal (Figs. 3.33 & 3.51–3.57). Unlike those sections previously excavated by Richard Bradley (Barrett *et al.* 1991a, figs. 2.12 & 2.13), these new sections revealed fills that were mainly dominated by clean chalk rubble. Moreover, the surviving bank was much reduced in height (i.e. less than 30 cm) and appears to have been ploughed in pre-Roman times. From the 2001 field investigation it looks as if the Bottlebush Down length of both cursus banks was deliberately slighted back into the ditch shortly after the monument was constructed. This is in contrast to the other fill sequences observed further to the south in the same monument on Wyke Down and in Firtree Field by Bradley (Barrett *et al.* 1991a) where the homogeneous organic silt loam and fine chalk rubble fills suggested a natural infilling over a lengthy time period. These observations in the northern sector of what is believed to be the first major length of Dorset Cursus would suggest a complete change in its importance and function in the landscape within the later Neolithic. In addition, it may imply a complete re-aligning of the fields and boundaries in this part of the downland landscape.

As an important addendum to this work, the sample excavations enabled the assessment of the preservation of this important linear field monument. Although only part of the southern bank and ditch of the cursus is scheduled in this stretch from the terminal northwards, the whole length is quite poorly preserved. Modern ploughing, post-seventeenth-century AD estate boundaries and recent scrub woodland development have all played a hand in the destruction of the northern and southern banks. In addition, there were several ard marks preserved beneath the Roman Ackling Dyke and the Cursus northern bank in Trench 1 (Fig. 3.54), which may also indicate some later prehistoric slighting and denudation of what was left of the cursus bank in places. A major observation of this field season was that both Cursus banks in the *c.* 1 km length investigated were severely slighted, probably in the Late Neolithic period and again in preparation for the construction of the Roman Ackling Dyke in the first century AD. Indeed, where the southern Cursus bank appears to be very prominent and well preserved (and scheduled) on the northern part of Bottlebush Down, there is the same slighted bank present, accentuated by the later creation of a property boundary, and plough down-cutting creating a false contour on its eastern side (Fig. 3.57).

Long barrow 294

When Grinsell (1959) recorded this barrow in the earlier 1950s, it was observed to be at least three feet high (or 90 cm). When we excavated a sample trench from the north–south ditch across what we thought was the surviving mound, it turned out to be little more than a very slight contour in the underlying chalk giving the perception of a *c.* 25–35 cm rise of surviving mound (Figs. 3.2 & 3.4). Crudely, this amounts to *c.* 60 cm of mound denudation over 60 years, or an average of about 1 cm each year.

Conclusions

Ploughing in both prehistoric and recent times has resulted in varying degrees of destruction of upstanding scheduled monuments in the Wyke, Bottlebush and Gussage Down areas of Cranborne Chase. These conclusions have been fully attested by the topographical and geophysical surveys and test excavations undertaken during this project.

Both the northern and southern banks of the Dorset Cursus have been affected by the deliberate back-filling of the ditch with bank material as well as pre-Roman ploughing. The cursus banks occasionally survive to a thickness of about 25–30 cm where they are protected either by other monuments or later field boundaries. Where they are unprotected and have been ploughed for at least the past 45 years, no earthwork survives, there is only a dissipated linear scar visible in the field today. There is almost no survival of the southern cursus bank at the Bottlebush Down terminal despite the presence of an apparent earthwork, which is in fact an abrupt contour change in the underlying chalk rock, created through ploughing and boundary creation/maintenance.

The Wyke Down barrow group consists of only seven upstanding barrow mounds out of a total of at least 29 ring-ditch sites (Fig. 3.34). The ploughed-out barrow mounds are just visible in the field under the right light conditions as very slight rises in the field, but are not otherwise detectable.

Ploughing since the latter 1950s of the scheduled and upstanding barrows has resulted in the sculpting of the round barrow mounds on Wyke Down resulting in ovoid to sub-rectangular to sub-square shapes to the once circular mounds, and the near total destruction of one long barrow on Gussage Down.

In addition, this same ploughing has created truncation steps around at least four barrows, with up to *c.* 25–78 cm of chalk subsoil having been scoured away over about 45 years. This has particularly affected the ring-ditch, berm and outer edges of the mounds, and is particularly marked in barrows 36 and 44. The resulting impression is of very substantial earthen mounds, which are very prominent in the landscape, but excavations demonstrate that tens of centimetres of this height is actually preserved plinths of chalk rock which contrast with the relatively denuded landscape.

In one case at barrow 36, the conversion of former arable land to pasture for the last 16 years has dramatically slowed this truncation process on its southern side. It has resulted in a less abrupt and more gently sloping talus edge to the surviving barrow mound, which better preserves the constructional aspects of the outer edge of the mound and reflects the natural collapse and erosion of the mound over time.

Although much destruction of upstanding mounds can be proven to have occurred in antiquity, severe sculpting of the shape and the reduction in size of monuments has occurred as a result of modern ploughing over very short time periods of less than 50 years, and especially in the last couple of decades. This has demonstrated that the scheduled status of these prehistoric monuments is insufficient to really protect the monuments if they reside in fields used for arable agriculture.

4.6. The Dorset Cursus on Wyke and Bottlebush Downs (2001)

Lithics

Julie Gardiner

Sections across the north and south Cursus ditches in 2001 produced a quantity of struck flint. Trenches 1 and 2, across the north ditch, produced a combined total of 138 struck flints, all of which are waste material (Tables A4.31 & A4.32). Of this total, 126 (91%) are flakes, with 8 (6.2%) irregular pieces of waste and 4 (2.8%) cores/ fragments and tested nodules. In Trench 1, worked flint was recovered from primary, secondary and tertiary fills, though only in small quantities, with the largest concentration (41 = 54% of total from trench) in context [138], a tertiary fill representing weathering of the ditch edge. Flint was recovered from a primary fill [188] and a tertiary fill [141] in trench 2 with 1 flake from a pollen sample (sample number 187) taken from context 178, a stabilization horizon in the lower third of the ditch.

At least two main periods of flintworking are represented here. Material from the uppermost fills — [137] and [141] — is crude and roughly worked in rolled and stained chalk flint, with much edge damage. There are a number of hinge fractures and the flakes are generally quite heavy, squat and have broad platforms with pronounced bulbs of percussion. This material clearly post-dates the use of the cursus and is likely to be of mature Bronze Age date.

The remainder of the flintwork, in contrast, consists of small groups of flakes in very fresh condition in the good-quality chalk flint that was frequently encountered during the original construction of the monument. However, very little of the flint comes from the primary silt, most is from fairly high in the ditch and seems, therefore, to represent activities secondary to the original use of the monument. It appears that a few nodules were occasionally gathered from either the ditch fill or, perhaps more likely, as they weathered out of the upper ditch sides, were worked nearby and then a small amount of material was thrown back into the ditch. Knapping did not, apparently, take place within or very close to the ditch as only small quantities are represented with little irregular waste and virtually no small chips. Although the quantities are small, these groups of flints are most likely Late Neolithic or Early Bronze Age in date on technological grounds. Tertiary silts [138] and the buried soil under the adjacent Ackling Dyke [140] both produced irregular lumps of non-local stone, possibly constituents of 'hard core' from Roman road construction and indicate some disturbance of the upper ditch fill.

Trenches 3 and 4, across the south cursus ditch, produced three times as many flints as the north ditch sections. This material is entirely comparable with that from the north ditch with almost 95% waste flakes, 2.7% irregular waste and 2% cores/tested nodules. In addition there are two retouched items: a scraper and a flake with some probable invasive retouch (though the edges are also plough damaged). Both of these, however, are from the overlying topsoil [200].

Once again, the ploughsoil [192], [200] and uppermost ditch fills [193], [201] produced crudely worked flints that are well rolled and plough damaged. Pieces are heavy, irregular, hard hammer struck with hinge fractures, broad platforms and prominent bulbs of percussion. As in the north ditch, the remainder of the flintwork consists of small groups of fresh flakes. The largest group, from [202], is very fresh with clean sharp edges. It includes four rather roughly worked cores weighing 86–204 g, each with only a few removals, and appears to be a deliberate dump of knapping debris.

Table A4.31. *The lithic types present in Trenches 1 and 2, north Cursus ditch.*

	Trench 1							Trench 2			
Context	**137**	**138**	**140**	**162**	**163**	**164**	**Total**	**141**	**178/<187>**	**188**	**Total**
Primary	1	11	0	2	3	2	**19**	3	0	0	**3**
Secondary	2	10	1	3	0	2	**18**	9	1	3	**13**
Tertiary	3	17	4	3	4	1	**32**	39	0	2	**41**
Sub-total	*6*	*38*	*5*	*8*	*7*	*5*	**69**	*51*	*1*	*5*	**57**
Irreg.	0	2	0	0	1	3	**6**	2	0	0	**2**
Core	0	0	0	0	0	0	**0**	0	0	1	**1**
Core frag./trim	0	1	0	0	0	0	**1**	0	0	0	**0**
Tested nodule	0	0	0	0	0	0	**0**	2	0	0	**2**
Sub-total	*0*	*3*	*0*	*0*	*1*	*3*	7	*4*	*1*	*1*	5
Ret.	0	0	0	0	0	0	**0**	0	0	0	**0**
Total	**6**	**41**	**5**	**8**	**8**	**8**	76	**55**	**1**	**6**	62
Chips	0	5	0	0	0	0	**5**	0	0	0	**0**
Weight (g)	80	409	33	42	123	70	**757**	701	10	322	**1033**
Burnt (g)					5		**5**	82			**82**

Table A4.32. *The lithic types present in Trenches 3 and 4, south Cursus ditch.*

	Trench 3						Trench 4					
Context	**192**	**193**	**194**	**195**	**196**	**Total**	**200**	**201**	**202**	**205**	**206**	**Total**
Primary	4	3	0	2	0	**9**	0	17	21	3	2	**43**
Secondary	8	6	1	7	0	**22**	4	38	34	11	0	**87**
Tertiary	1	4	1	7	2	**15**	1	54	143	7	0	**205**
Sub-total	*13*	*13*	*2*	*16*	*2*	**46**	*5*	*109*	*198*	*21*	*21*	***335***
Irreg.	2	1	0	1	0	**4**	0	0	6	0	1	**7**
Core	0	1	0	0	0	**1**	0	0	4	0	0	**4**
Core frag./trim	0	0	0	0	0	**0**	0	1	0	0	0	**1**
Tested nodule	0	0	0	0	0	**0**	0	2	0	0	0	**2**
Sub-total	*2*	*2*	*0*	*1*	*0*	*5*	*0*	*3*	*10*	*0*	*1*	***14***
Ret.	0	0	0	0	0	**0**	1+?1	0	0	0	0	**2**
Total	**15**	**15**	**2**	**17**	**2**	**51**	**7**	**112**	**208**	**21**	**3**	**351**
Chips	0	0	0	0	0	**0**	0	0	0	0	0	**0**
Weight (g)	229	338	8	284	13	**872**	164	1887	2904	456	60	**5471**
Burnt (g)								61				

Pottery

KASIA GDANIEC

Three sherds of Beaker/Early Bronze Age pottery were recovered from a single context, all three occurring in the upper tertiary fills of the south Cursus ditch in Trench 4. Two small sherds provide a useful relative Beaker date whilst the third can be considered broadly contemporary on form and fabric.

The three sherds are described as follows (not illustrated):

153 & 154. Trench 4, F.108 [201]: Two sherds: 1 × 2.2 g, 1 × 2.0 g. Partially oxidized fabric, reddish-brown external surfaces and black internal surface and core. White fine and medium quartzite/flint grits temper the fabric. Thin-walled: 5 mm thick. External surface decoration consists of evenly spaced rows of square and rectangular toothed comb impressions. Found in one wheelbarrow load sample for dry sieving. These were comb-impressed Beaker sherds.

155. Trench 4, F.108 [201]: One sherd: 27.5 g. Partially oxidized fabric of coarse vessel, 15 mm thick. Reddish-brown surfaces and dark grey/black core. Temper: occasional very fine-fine flint grits. Well-abraded edges suggest significant weathering of relatively durable sherd. Undecorated. This is either a sherd of domestic beaker or an Earlier Bronze Age coarse vessel.

Molluscan data

MICHAEL J. ALLEN

See Chapter 3, pp. 158–64 and Tables 3.10–3.12, pp. 162–4.

Soil micromorphology descriptions

CHARLES FRENCH & HELEN LEWIS

The micromorphological descriptions for the buried soils and ard standstill horizons in the ditch fills for the Dorset Cursus investigations are summarized in Tables A4.33 and A4.34.

Trench 1

F.101, context 162, sample 177

Structure: two main fabric units; Fabric 1: upper 4–6.5 cm.

Table A4.33. *Cursus ditch fills: micromorphological descriptions (C. French).*

Context	Structure	Porosity	Mineral Components	C:f ratio	Organic components	Groundmass	Pedofeatures
F.101 [162]	Sub-angular blocky	10%, vughs	Frequent chalk rock fragments; micrite; quartz sand	7:93 Silt loam	<10%, very fine amorphous & charcoal	Pale yellowish-brown (XPL), pale greyish-brown (PPL) crystallitic	Rare intrusive fabric, reddish-brown-black with fine sand
F.101 [137]	Sub-angular blocky to pellety	10% channels & vughs	Rare chalk rock fragments; quartz sand; micrite; golden brown clay	10:90 Silt loam	5%, as above	Pale golden brown (XPL), brown (PPL) crystallitic	Strong amorphous sesquioxides on 50% of fabric
F.101 [182]	Pellety to irregular & sub-angular blocky	10–20% irregular vughs, rare channels	As above, occasional chalk rock fragments	7:93 Silt loam	As above	Yellowish-brown (XPL), greyish-/reddish-brown (PPL) crystallitic	20% irregular zones of amorphous sesquioxides
F.108 [196]	Irregular blocky	10–40% vughs & channels	As above, 50% chalk rock fragments	5:95 Silt loam	As above	Greyish-brown (PPL), brown (XPL), crystallitic	Fabric zone of large chalk fragments & micrite, with pale greyish-brown (PPL), brown (XPL) matrix
F.108 [197]	Irregular blocky	10–50%, vughs, complex packing pores	As above	As above	<5%, as above	As above	Irregular to discontinuous infills of fine chalk fragments in zones of dense chalk fragments with grey (PPL) & pale greyish-white (XPL) matrix
F.108 [194]	Irregular blocky	20–25% channels & vughs	Rare fine chalk rock fragments; quartz sand; micrite; clay	5:95 Silt loam	5–10% fine charcoal, organic staining, 'punctuations'	Brown (PPL), reddish-brown (XPL), crystallitic	Channel infills with reworked fabric
F.108 [207/8]	Apedal	As above	Frequent chalk rock fragments; quartz sand; micrite	As above	As above	Pale greyish-white (PPL & XPL), and brownish-black (PPL), reddish-brown (XPL), crystallitic to stipple speckled	Strong amorphous sesquioxide & organic matter impregnation
F.108 [210]	Apedal	20–25%, vughs & complex packing voids	Frequent chalk fragments; micrite; quartz sand	5:95 Silt loam	10%, 'punctuations'	Reddish-brown (PPL), yellowish-brown (XPL); stipple speckled to crystallitic	Amorphous sesquioxide impregnation
F.108 [206]	Apedal	20% as above	As above	As above	10%, amorphous	Reddish-brown (PPL), yellowish-brown (XPL); stipple speckled to crystallitic	As above; also a second fabric: as main fabric but fewer chalk fragments & frequent zones of micrite, pale greyish-brown (PPL), pale brown (XPL)
F.108 [202]	Apedal to weak channel	As above	Frequent micrite; 20% chalk fragments; quartz sand	As above	As above	Pale reddish-brown (PPL & XPL); stipple speckled	Strong amorphous sesquioxide impregnation

Table A4.34. *Cursus bank: micromorphological descriptions (C. French).*

Context	Structure	Porosity	Mineral components	C:f ratio	Organic components	Groundmass	Pedofeatures
F.100 [155]	Sub-angular to irregular blocky, some pellety	<30%, irregular vughs	50% chalk fragments <3 cm, subangular to subrounded; 50% micrite	2:98 Silt (loam)	<10%, 'punctuations', very fine charcoal	Greyish-brown (PPL), pale yellowish-brown (XPL)	Occasional zones of amorphous sesquioxide impregnation

Components: 70% chalk fragments, sub-rounded to sub-angular, <1.5 cm; 25% micrite and 5% very fine quartz sand in amorphous to irregular mass in between; pale yellowish-brown (XPL), pale greyish-brown (PPL).
Porosity: few (10%) fine (<1 mm) vughs within micritic very fine sand; Fabric 2: lower 4.5–7.5 cm; Structure: small (<7 mm) sub-angular blocky.
Components: 50–60% chalk fragments, sub-rounded; < 8 mm; coarse/fine ratio: 7/93; coarse fraction: 2% medium and 5% fine quartz sand, 100–750 um, sub-angular to sub-rounded; fine fraction: 10–20% very fine quartz sand, 50–100 um; <73% micrite; yellowish-brown (XPL), greyish-brown/pale brown (PPL).
Organic component: <10% very fine to fine, amorphous/carbonized organic matter, <100 um; contains minor amounts (<2%) of third intrusive fabric, a strong reddish-brown (XPL) to black (PPL), small irregular to sub-angular blocky micritic fine sand.

F.101, context 137, sample 161

Structure: small, sub-angular blocky to pellety; *Porosity*: 10% intrapedal channels, <1 mm wide, <3 cm long, all orientations, weakly serrated; 10–20% interpedal vughs, <1 mm.
Organic components: 5% very fine amorphous/carbonized organics, <50 um; amorphous organic staining of 50% of fabric.
Components: <2% chalk fragments, sub-rounded, <2 mm; coarse/fine ratio: <10/90; coarse fraction: 3% medium and 7% fine quartz sand, sub-rounded to sub-angular, 100–750 um; fine fraction: 10–20% very fine quartz sand, 50–100 um, sub-rounded; <10% silt; 55% micrite; <2% pure clay; pale golden brown (XPL), brown (PPL).
Pedofeatures: *Amorphous*: 50% of fabric strongly impregnated with amorphous sesquioxides.

Trench 2

F.101, Context 182, samples 179 and 190

Structure: mixture of chalk fragments and one fabric in small, irregular pellety to irregular sub-angular blocky, <5 mm.
Porosity: <2% intrapedal channels; 10–20% intrapedal irregular vughs.
Organic components: 5% very fine amorphous/carbonized organics, <50 um.
Components: *c.* 60–70% chalk fragments, <1.5 cm, sub-rounded to sub-angular; coarse/fine ratio: 7/93; 2% medium and 5% medium quartz sand, 100–750 um, sub-rounded to sub-angular; 13% very fine quartz sand, 50–100 um; <5% silt; 70% micrite; occasional flint fragments; shells; yellowish-brown (XPL), greyish-/reddish-brown (PPL); crystallitic groundmass.
Pedofeatures: *Amorphous*: *c.* 20% irregular zones of amorphous sesquioxides.

Trench 3

F.108, context 196, samples 195 and 193

Structure: two fabric units; Fabric 1: upper 5.5 cm. Small (<5 mm), irregular blocky aggregates with fine channels.
Porosity: 10% fine intrapedal channels; up to 40% vughs.
Components: 50% chalk fragments, <1.5 cm, sub-rounded to sub-angular; coarse/fine ratio: 5/95; coarse fraction: 5% fine quartz, 100–250 um, sub-angular to sub-rounded; fine fraction: 10% very fine quartz, 50–100 um, sub-rounded; <5% silt; 75% micrite, <2% pure clay; brown (XPL), greyish-brown (PPL); Fabric 2: lower 5.5 cm in large sub-rectangular zone; large <2 cm, chalk fragments and micrite; yellowish-brown (XPL), pale greyish-brown (PPL).

F.108, context 197, sample 194

Structure: two fabric units; Fabric 1: upper 6.5–9.5 cm.
Porosity: 10–50% small (<1 mm) vughs to irregular complex packing.
Components: 50% chalk fragments, irregular to sub-angular, <1 cm; with 50% micrite interspersed; pale greyish-white (XPL), pale reddish-brown (PPL); <5% very fine amorphous/carbonized organics; Fabric 2.
Porosity: <10% vughs, with irregular to discontinuous infills of fine (<1 mm) chalk fragments
Components: dense chalk fragments, <1 cm; pale greyish-white (XPL), grey (PPL).

F.108, context 194 upper, sample 192/1

Structure: one fabric, large irregular blocky.
Porosity: 10% intrapedal channels, <1 mm wide, <1 cm long, irregular, weakly serrated; often infilled with reworked fabric with irregular vughs; 10% vughs, sub-rounded to irregular, <1 mm.
Organic component: general amorphous organic impregnation of fine fabric; 5% fine charcoal, <100 um.
Components: same fabric as in lower sample; reddish-brown (XPL), brown (PPL).

F.108, context 194 lower, sample 192/2

Structure: one fabric in irregular aggregates, 5–0.5 mm.
Porosity: 5% channels, <250 um wide, <1 cm long, irregular, all orientations, weakly serrated; 20% vughs, from small sub-rounded to large irregular, with discontinuous infill of small aggregates of main fabric.
Organic components: 10% very fine amorphous organic punctuations, <50 um; <5% fine carbonized organics, 50–100 um.
Components: <5% fine chalk fragments, <8 mm, sub-angular; coarse/fine ratio: 5/95; coarse fraction: 5% fine quartz, 100–250 um, sub-rounded to sub-angular; fine fraction: 10% very fine quartz, sub-rounded, 50–100 um; 75% micrite; <5% silt; <2% pure clay; pale brown (XPL/PPL).

Trench 4

F.108, contexts 207/8, sample 198

Structure: one apedal fabric.
Components: abundant (50–80%) chalk fragments, <1.5 cm, irregular to sub-rounded to sub-angular; interspersed with small (<2 mm), irregular aggregates of very fine quartz sand and micrite; pale greyish-white (XPL/PPL) with reddish-brown (XPL) to brownish-black (PPL) aggregates; two irregular lenses of micrite at base and near top of slide, <5 mm thick.
Pedofeatures: *Amorphous*: aggregates are strongly impregnate with amorphous sesquioxides and amorphous organic matter.

F.108, context 210 upper, sample 225/1

Structure: two fabric units; Fabric 1: upper 3.5 cm; weathered chalk fragments, <1 cm, sub-angular to sub-rounded; Fabric 2: lower 5 cm
Porosity: 20–25% irregular vughs and complex packing voids.
Components: 50–80% weathered chalk fragments, <1 m, sub-rounded; and 20–50% irregular aggregates of micritic sand/silt; 5% fine quartz sand; 2% very fine quartz sand; 10% silt; 83% micrite; with 10% very fine organic punctuations; greater to lesser impregnation with amorphous sesquioxides.

F.108, context 210 lower, sample 225/1

Similar to fabric 2 of sample 225/1 above, but lesser chalk fragments (25–50%) and greater (50–75%) aggregates of micritic sand/silt.

F.108, context 206, sample 199

Structure: two fabric units; Fabric 1: upper 1.5–3.5 cm.
Porosity: 20% vughs, irregular to complex packing, <2 mm.
Components: 60–80% dense, irregular chalk; 20–40% micrite with <5% pure clay, yellowish-brown (XPL), reddish-brown (PPL); with some irregular zones of amorphous sesquioxide impregnation; 10% very fine amorphous organic matter, all replaced by amorphous sesquioxides; Fabric 2: similar to fabric 1, except less chalk fragments (<50%), and greater (>50%) aggregates of micrite with <15%

fine to very fine quartz sand and minor (<5%) silt components; <30% irregular vughs to complex packing, <4 mm; pale brown (XPL), pale greyish-brown (PPL).

F.108, context 202, sample 197
Structure: apedal with fine, irregular, weakly developed, channel structure.
Components: 20% chalk fragments, <1 cm, sub-rounded, evenly distributed within micritic fabric, in irregular, dense aggregates of very fine quartz sand (5%) and micrite, pale reddish-brown (XPL); micrite strongly impregnated with amorphous sesquioxides.

Cursus bank
Trench 1, F.100, context 155, sample 162
Structure: sub-angular to irregular blocky to pellety, <4 mm.
Porosity: <30% irregular vughs.
Components: 50% chalk fragments, <3 cm, sub-angular to sub-rounded; 50% micritic sand/silt; coarse /fine ratio: 2/98; coarse fraction: 2% fine quartz & rare calcite sand, 100–250 um, sub-rounded to sub-angular; fine fraction: 5% very fine quartz sand; <5% silt; 88% micrite; minor (10%) zones of amorphous sesquioxide impregnation; occasional shell & flint fragments; occasional calcitic earthworm granules; <2% very fine charcoal; pale yellowish-brown (XPL), greyish-brown (PPL); crystallitic groundmass.

4.7. Ackling Dyke on Wyke/Bottlebush Downs

Soil micromorphology descriptions
CHARLES FRENCH & HELEN LEWIS

The micromorphological descriptions for the buried soils and ard marks associated with the Dorset Cursus and Ackling Dyke investigations are summarized in Tables A4.35 and A4.36.

Pre-Ackling Dyke activity (in Trench 1, Wyke Down Dorset Cursus excavations, 2001)
Trench 1, context 158, sample 180: fill of ard mark under bank
Structure: fill of open V-shaped ard mark contains micritic soil fabric that is similar to context 140, sample 181 (see below); ard mark, *c.* 3 cm deep and *c.* 6 cm wide at upper surface of the cut, and base of ard mark *c.* 1 cm wide; cut into weathered chalk and micrite with <10% irregular vughs. The ard-mark cut is distinguished by a channel running along the edge of the cut (250–500 μm wide), with frequent calcareous (chalky) pellets (150 μm, rounded) and occasional chalky clay coatings and infills (up to 500 μm in size, on some peds at the edge of the cut). The channel also contains some root remains. Other characteristics typical of ard marks include a higher porosity in the fill (10–20%, channels & cracks) than the soil under/outside the cut (porosity 10%, channels & vughs), and one infilled planar void/root channel running perpendicular to the cut, 1000–2000 μm thick × 10,000 μm long.

Trench 2, context 177, sample 163
Structure: one fabric with weak to moderately well-developed columnar to irregular/sub-angular blocky peds.
Porosity: 10% fine, discontinuous channels, <5 mm wide and <5 cm long; 10% irregular vughs, <1 mm.
Components: coarse/fine ratio: 10/90; coarse fraction: 5% medium and 5% fine quartz sand, sub-rounded to sub-angular, 100–750 μm; fine fraction: 15% very fine quartz sand, 50–100 μm, sub-rounded; <5% silt; 80% micrite; reddish-brown (XPL), brown (PPL).
Pedofeatures: *Amorphous*: minor amorphous sesquioxide impregnation and amorphous organic staining.

Trench 2, context 141, sample 178
Structure: moderately well-developed sub-angular blocky with columnar/vertical aspect.
Porosity: 10% fine channels; 20% vughs with discontinuous infills of pellety aggregates within them.
Components: coarse/fine ratio: 5/95; coarse fraction: 5% fine quartz sand, 100–250 μm, sub-rounded to sub-angular; 20% very fine quartz sand, 50–100 μm, sub-rounded; <5% silt; 65% micrite; <2% pure clay; one fragment of burnt limestone was identified, typical of *c.* 600°C heating, beginning to change to lime (identified by Chris Doherty pers. comm. 2005).
Pedofeatures: *Amorphous*: strong amorphous sequioxide impregnation of whole fabric.

Trench 2, context 177/9, sample 188
Structure: one fabric; small irregular blocky, <5 mm; with zones of pellety microstructure in voids, up to 40% of fabric.
Porosity: fine channels, <1 mm wide, weakly serrated; few (<5%) vughs, elongate to irregular, <500 μm.
Components: coarse/fine ratio: 20/80; coarse fraction: 5% coarse sand size chalk, 1–2 mm, sub-angular; 5% medium quartz sand and chalk fragments, 500–1000 μm, sub-rounded to sub-angular; 10% fine quartz sand, 100–500 μm, sub-rounded; fine fraction: 20% very fine quartz sand, 50–100 μm, sub-rounded; <5% silt; 45% micrite; reddish-brown (CPL), dark brown (PPL).
Organic component: 10% fine organic punctuations, <50 μm.
Pedofeatures: *Amorphous*: all impregnated with amorphous sesquioxides.

Buried soil beneath Ackling Dyke
Trench 2, context 170, samples 185 and 186: buried soil under Ackling Dyke
Structure: two fabric units; Fabric 1: upper 2 cm; heterogeneous mixture of 50–75% fine chalk fragments, <2 mm, and 25–50% sub-rounded to irregular with micritic calcium carbonate, <250 μm, pale yellowish-brown (PPL); Fabric 2: lower 11 cm.
Components: main fabric (80% of total groundmass) composed of 30–40% fine chalk fragments, <4 mm, sub-angular to sub-rounded; 30–40% soil fabric; coarse fraction: 10% fine and 10% very fine quartz sand, 50–250 um; 70% micrite, 5% silt, 5–10% organic punctuations, <1% limpid clay coating micrite; reddish-brown (XPL/PPL).
Porosity: 20–30% interpedal channels; small blocky structure.
Pedofeatures: *Amorphous*: general amorphous sesquioxide impregnation of whole fabric; *Fabric*: with intrusive, minor (<20% of total groundmass) fabric composed of irregular blocky aggregates, <1 cm, of micrite and very fine chalk fragments interspersed within main fabric; rare stone, <2 cm, sub-rounded.

Trench 1, context 140, sample 181
Structure: two fabric units: Fabric 1: upper 9 cm.
Components: 25–40% chalk fragments, <1.5 cm, angular to sub-rounded, in all orientations, intermixed with small, sub-angular blocky peds composed of coarse fraction: 2% medium, 10% fine and 10% very fine quartz sand, 50–500 μm, sub-rounded to sub-angular; fine fraction: 70% micrite, <5% silt; <5% organic punctuations; yellowish-brown (XPL), brown (PPL); crystallitic groundmass;

Table A4.35. *Buried soils/leveling over North Cursus ditch under Ackling Dyke: micromorphological descriptions (C. French).*

Context	Structure	Porosity	Mineral components	C:f ratio	Organic components	Groundmass	Pedofeatures
TR2 [177]	Columnar to irregular or sub-angular blocky	20% fine discontinuous channels & vughs	Micrite; quartz sand	10:90 Silt loam	<10% 'punctuations'	Brown (PPL), yellowish- & reddish-brown (XPL); crystallitic to stipple speckled	Occasional amorphous sesquioxide impregnation & staining
TR2 [141]	Well-developed sub-angular blocky with some columnar	10–30% fine channels & vughs	As above, 2% pure clay	5:95 Silt loam	As above	As above	Occasional discontinuous infills of pellety aggregates; strong amorphous sesquioxide impregnation of whole fabric
TR2 [177/9]	Irregular blocky, some pellety	5–10%, fine channels & vughs	Occasional chalk fragments; quartz sand; micrite	20:80 Sandy silt (loam)	10% 'punctuations'	Dark brown (PPL), reddish-brown (XPL); stipple speckled to crystallitic	Amorphous sesquioxide impregnation of whole fabric

Table A4.36. *Buried soils and ard-mark fills under Ackling Dyke: micromorphological descriptions (C. French).*

Context	Structure	Porosity	Mineral components	C:f ratio	Organic components	Groundmass	Pedofeatures
TR2 [170]	Fine blocky	20–30%, channels	Frequent chalk fragments; micrite; quartz sand	10:90 Silt (loam)	5–10% 'punctuations'	Pale yellowish-brown, becoming reddish-brown with depth (PPL & XPL), crystallitic	Rare limpid clay coatings; general amorphous sesquioxide impregnation; occasional intrusive fabric of irregular blocky micrite & chalk zones seen with depth
TR1 [140]	Sub-angular blocky	As above	Moderate chalk fragments (becoming larger with depth); frequent micrite; quartz sand	12:88 Silt (loam)	<5% 'punctuations'	Brown (PPL), yellowish-brown (XPL); greyish-white (PPL & XPL) in lower 2 cm; crystallitic	None noted
TR1 [158] Ard-mark fill	As above	As above; at cut <10% irregular vughs	As above	As above	As above	As above	As above

occasional shells and calcitic earthworm granules. Fabric 2: lower 2 cm; large, sub-angular chalk fragments, 1–1.5 cm; greyish-white (PPL/XPL); crystallitic groundmass.

Tree hollow fill, sample 126

Structure: crumb, some coalesced (<5000 μm).
Porosity: 10–30%, channels (150–300 × 1000–2000 μm, straight, un-accommodated) and packing pores.
Mineral components: micrite, calcite & quartz grains; frequent chalk rock fragments (*c.* 60% of slide area), flint, rare shell, rare calcitic earthworm granule.
C:f ratio: very coarse sand 0%, coarse, medium and fine sand 5%, very fine sand 5%, silt 85% (mostly micrite, some quartz), clay 10%.
Organic components: 5%, black angular fragments (50–100 μm), occasional charcoal (<500%); organic components decrease with depth.
Groundmass: grey (PPL), light grey-gold (XPL), grey-yellow (RL); open porphyric-related distribution.
Pedofeatures: 1) rare orange clay (PPL & XPL) fragments <150 μm, rounded; 2) micrite infills and coatings; 3) rare iron mottles (<200 μm) and iron staining.

4.8. Monkton-up-Wimborne Neolithic enclosure (MUW 00)

Lithics
JULIE GARDINER

One hundred and thirty-four struck flints were recovered from the two ditch sections. Ditch section 3–4 produced only 16 pieces (409 g), virtually all of them from topsoil/upper tertiary layers 1 and 7. The section through the east ditch produced 118 struck flints (1973 g) from most levels within the ditch.

Most of the flint from the ditches is in a very fresh, almost mint, condition with a thick chalky white to smooth buff cortex and very little sign of any edge damage. It was probably encountered during the original digging of the ditches. Material from the topsoil/upper tertiary layers in section 3–4 (contexts 1 & 7) and contexts 2 and 3 in the east ditch contrasts is generally smaller, more rolled with stained cortex and with more evidence of accidental edge damage and breakage.

Only three recognizably retouched pieces were recovered: a scraper and a barbed-and-tanged arrowhead from layer 2 and another possible (broken) scraper from layer 5a, all in the east ditch. Five cores, including one heavy flake from layer 8 from which several further removals have been made, and one core-trimming flake were also present and the remainder of the assemblage (94%) consists of debitage — flakes, blades and irregular waste.

Three of the four cores and the core-trimming flake come from layer 5a, a secondary silt in the east ditch. One of these is a neat prismatic core, with mostly bladelike flakes removed from a single platform. It weighs 148 g. A second core (86 g) resembles a 'tortoise' core in appearance though it has only been used for the removal of small flakes from a single platform. The third is a small, elongated nodule with a few removals of blades struck from a single platform (79 g). The core trimming appears to have come from a core producing blades. The core from layer 2 is a broken nodule from which a few removals have been made.

Simple metrical analysis of 110 complete flakes indicated that 34 blades and bladelike flakes (length: breadth ratio of 3:5 or less) are present. This constitutes 30.9% of the measured sample and is a clear indication of the Early Neolithic date of the bulk of the assemblage.

Molluscan data
MICHAEL J. ALLEN

See Chapter 3, pp. 164–5 and Table 3.13, p. 166.

4.9. Monkton-up-Wimborne pit/shaft complex (MUW 97)

Lithics
MARTIN GREEN

The total assemblage from the site consists of over 6200 artefacts, principally found in three main contexts: the F.1 flat-bottomed pit, chalk platform (Fig. 3.63) and shaft (Tables A4.37–A4.39; Figs. 3.64–3.67). The remainder were obtained from the eight excavated pits of the outer circle, sections through the ring-ditch and a number of smaller features (Table A4.39; Fig. 3.65).

F.1 flat-bottomed pit
Flint was recovered from nine layers within this feature: layers 2, 3, 4, 6, 6a, 7, 9, 9a and 11. These consisted of: layer 2 upper secondary, layer 3 lower secondary, layer 4 upper primary and the remainder lower primary fills.

Layer 2 contained a mixed assemblage and it clearly contains residual material as well as lithics contemporary with the lower levels of this layer. These lower levels have been dated by the associated pottery to the Middle Bronze Age. The retouched, utilized, tool element (RUT) was 3.5% and included arrowheads of various dates including barbed and tanged, oblique and chisel, the latter made from Portland chert.

Layer 3 also contained a mixture of residual material indicative of gradual accumulation over time in the shaft. This assemblage is dated by associated Early Bronze Age pottery and a radiocarbon determination of 1530–1400 cal. BC from a burial within this layer. The RUT element was 5.3% and amongst the likely residual material were two oblique arrowheads (Fig. A4.23, nos. 3 & 4) and part of a serrated blade. Contemporary Early Bronze Age lithics include three barbed and tanged arrowheads (Fig. A4.23, nos. 5–7), and probably at least half of the scrapers which were small, well made and of convex form (Fig. A4.22, nos. 12 & 14).

Layer 4 had an RUT component of 3.2% which contained a number of implement types particularly characteristic of later Neolithic industries in the area

(Green 2000, 65–8). This included a small pseudo-tranchet axe (Fig. A4.22, no. 10), a tranchet axe-sharpening flake (Fig. A4.25, no. 11), two serrated blades and a heavy borer. An unusual form was the small bifacially flaked discoidal tool (Fig. A4.22, no. 9) similar to another found in layer 3 (Fig. A4.22, no. 13).

In the lower primary layers (6, 6a, 7, 9, 9a & 11), the RUT element was low (2.6%) but included six small squat flakes (counting as 1% of the RUT element) struck from a polished flint axe. These are almost certainly struck from the same axe but do not conjoin. Two small chisel arrowheads of petit tranchet type are a form often associated with Peterborough Ware, as they are in this case.

The shaft

In the secondary layers 7a–e, amongst the small RUT element were two exceptional arrowheads (Fig. A4.23, nos. 1 & 2), perhaps indicative of structured deposition within the shaft. The large fine leaf-shaped arrowhead was found in layer 7a in close association to a deposit containing two pieces of worked chalk (see catalogue C14 & C15), human bone fragments including part of a skull and a large flint pebble derived from Tertiary deposits. The unusual elongated chisel arrowhead of petit tranchet form in layer 7c was in close association with two pieces of worked chalk (see catalogue nos. C12 & C13), a red deer antler beam, a cow skull (Fig. A4.30d), and human skull fragments. A small pseudo-tranchet axe of the type already discussed (Fig. A4.22, no. 4) was found in layer 7b near another piece of human bone.

In the lower primary layers 10 and 10a–d, the cortex analysis of this assemblage clearly shows a preference for knapping fresh flint dug from the chalk. Much of the material consists of primary knapping waste from the reduction of nodules with a thick fresh cortex. A thin layer of nodular flint was removed from the base of the shaft and sporadic smashed nodules were present in the shaft walls, suggesting much of the freshly procured flint was derived from the digging of the shaft and knapped *in situ*. Two conjoining flakes were noted in layer 10b, lending weight to this suggestion. The RUT element was very small, only 1.3%. Amongst the finished tools were two knives (Fig. A4.22, nos. 3 & 5), a scraper and a crude pick made from a fragment of a fresh tabular nodule (Fig. A4.22, no. 2). Found on the floor of the shaft the battered condition of this implement suggests it may have been used to extract some of the exposed flint.

The chalk platform layers 12–16

Although constructed in an episodic way, the unweathered condition of the chalk rubble within the layers forming the platform indicates that little time elapsed between each addition. Therefore the flint assemblage has been examined as a single assemblage. The RUT element was very small, only 1.6%, but included a thick, crudely blunted, chisel arrowhead of petit tranchet type, a heavy borer made from a fresh piece of tabular flint (Fig. A4.22, no. 1) and a burin. Although burins are a characteristic Mesolithic form, they are known within the predominantly later Neolithic surface industries of the area and a further example was excavated from another Neolithic context within the Down Farm ring-ditch (Barrett *et al.* 1991a).

The outer pit circle

Eight of the fourteen pits were completely excavated and produced little in the way of incorporated material of any kind. However, small amounts of struck flint were recovered from all of the pits with some striking spatial patterning. Of the 421 artefacts recovered from the secondary levels of the pits, 27% was recovered from F.2 and 30% from F.15, pits which flanked the eastern entranceway. Conversely, only 5% was recovered from F.9 and 1% from F.8, pits which flanked the western entranceway. Of the 244 flint artefacts recovered from the primary levels, 37% was recovered from F.15, 24% from F.2, 5% from F.9 and less than 1% from F.8. It has to be noted that F.8 was the shallowest pit in the circle (Fig. 3.75) that was excavated, but the other three were of similar size to each other. Clearly the eastern approach to the site was treated differently from the west. This was also suggested by the distribution of the RUT element which was only 2% or 12 artefacts from all excavated pits. Four of these were recovered from the western pits including a flake from the cutting edge of a polished flint axe from F.15. The pits flanking the western entrance only produced one artefact, a scraper from F.9. F.12, the largest pit, excavated, produced only three artefacts but included a kite-shaped arrowhead (Fig. A4.23, no. 8) in the secondary levels and a portion of a fabricator or rod in the primary fills.

Other features

Very small amounts of worked flint were found in other features recorded both within F.1 and externally. The two postholes F.16 and F.17 beyond the eastern entranceway produced 16 and 33 pieces of worked flint respectively. These included a chisel arrowhead (Fig. A4.23, no. 9) and a knife in F.17. Two sections excavated through the ring-ditch only produced 20 flakes, 10 chips and a core, although a chisel arrowhead in Portland chert was recovered from the ploughsoil above.

Table A4.37. *Monkton-up-Wimborne (MUW 97): flints from F.1 and the shaft.*

Feature	Layer	Cores	Core frags.	Flakes	Chips	Retouched	Utilized	Scrapers	Other tools
F.1 NEQ	2	2	1	160	68	-	-	2	knife, pounder/rubber, b. & t. arrow
F.1 NWQ	2	16	2	414	215	6	7	8	oblique arrow, borer, 2 hammers
F.1 SEQ	2	8	1	199	88	5	-	6	ch. arrow (pc), borer, pick/axe butt, burin
F.1 SWQ	2	7	1	224	126	1	-	7	2 borers, pick/axe butt
Totals	**33**	**5**	**997**	**497**	**12**	**7**	**23**	**14**	
F.1 NEQ	3	14	1	202	87	1	1	10	3 b. & t. arrows, ch. arrow, borer
F.1 NWQ	3	2	2	151	69	3	6	5	-
F.1 SEQ	3	8	4	234	99	3	1	7	2 hammers/pounders
F.1 SWQ	3	15	1	128	51	-	3	5	hammer, discoid, saw, 2 obl. arrows
Totals	**3**	**39**	**8**	**715**	**306**	**7**	**11**	**27**	**12**
F.1 NEQ	4	8	-	107	45	1	-	3	-
F.1 NWQ	4	9	3	175	108	2	2	-	tranchet axe-sharpening flake
F.1 SEQ	4	5	1	72	37	1	2	4	4 pseudo-tranchet axe, discoid, saw, knife
F.1 SWQ	4	-	-	223	115	4	1	2	borer, saw
Totals	**4**	**22**	**4**	**577**	**305**	**8**	**5**	**9**	**7**
F.1 NEQ	6a	7	1	94 (2 cnjn)	19	1	-	1	6 small flakes from polished axe, 2 hammers, fabricator, borer
F.1 NWQ	6a	8	4	272	203	6	1	-	hammer, borer, 2 ch. arrows
Totals	**6a**	**15**	**5**	**366**	**222**	**7**	**1**	**1**	**14**
F.1 NEQ	6	3	-	53	15	-	-	-	blunted flake
F.1 NWQ	6	-	-	5	7	-	-	-	-
Totals	**6**	**3**	**-**	**58**	**22**	**-**	**-**	**-**	**1**
F.1 NEQ	9a	-	-	38	8	-	-	-	-
F.1 NWQ	9a	-	-	28	1	1	-	-	-
Totals	**9a**	**-**	**-**	**66**	**9**	**1**	**-**	**-**	**-**
F.1 NEQ	9	1 (cnjns. flake)	15	-	1	-	-	-	-
F.1 NWQ	9	3	-	2	-	-	-	-	-
Totals	**9**	**4**	**-**	**17**	**-**	**1**	**-**	**-**	**-**
F.1 NWQ	11	-	-	8	-	-	-	-	-
Shaft	7	8	1	89	36	1	-	1	2 hammers
Shaft	7a	11	-	138	43	-	1	1	leaf arrow
Shaft	7b	-	-	27	5	-	1	-	pseudo-tranchet axe
Shaft	7c	15	4	92	24	-	1	-	hammer, denticulate
Shaft	7d/e	1	1	19	5	-	-	-	ch. arrow
Totals	**sec. fills**	**35**	**6**	**365**	**113**	**1**	**3**	**2**	**7**
Shaft	10, S1	7	1	30	5	1	-	1	-
Shaft	10, S2	10	2	94	12	1	-	-	-
Shaft	10a, S1	20	6	160	60	1	-	-	-
Shaft	10a, S2	2	-	26	6	-	-	-	-
Shaft	10b	17	4	84 (cnjn)	16	-	-	-	-
Shaft	10c	5	-	16	2	-	-	-	2 knives
Shaft	10d	10	2	29	4	-	1	-	pick
Totals		**71**	**15**	**429**	**105**	**3**	**1**	**1**	**3**

Table A4.38. *Monkton-up-Wimborne (MUW97): flints from the platform.*

Feature	Layer	Cores	Core frags.	Flakes	Chips	Retouched	Utilized	Scrapers	Other tools
F.1	12	1	-	18	1	-	-	-	-
F.1	13	3	-	70	17	1	-	2	hammer, burin
F.1	14	1	-	16	1	-	-	-	-
F.1	14/16 interface	-	-	10	-	-	-	-	-
F.1	15	-	-	25	10	-	-	-	-
F.1	16	16 (1 cnjn. with flake)	5	425	115	4	1	-	ch. arrow, borer
Totals		**21**	**5**	**564**	**144**	**5**	**1**	**2**	**4**

Table A4.39. *Monkton-up-Wimborne (MUW 97): flints from other features and the outer circle of pits (F.2, F.5, F.6, F.8, F.9, F.12, F.13, F.15).*

Feature	Layer	Cores	Core frags.	Flakes	Chips	Retouched	Utilized	Scrapers	Other tools
F.18	2	-	-	1	-	-	-	-	-
F.19	2	-	-	3	8	-	-	-	-
F.22	2	-	-	4	1	-	-	-	-
F.23	rubble	-	-	1	-	-	-	-	-
F.2	2	2	-	26 (1 pc)	2	1	-	-	quartzite hammer
F.2	3	3	-	56	24	-	-	-	saw
F.2	4	-	-	4	-	-	-	-	-
F.2	6	4	2	2	48	1	-	-	-
Total		**9**	**2**	**134**	**27**	**1**	**-**	**-**	**2**
F.5	2	3	-	13	-	-	-	1	-
F.5	3	-	-	3	-	-	-	-	-
Total		**3**	**-**	**16**	**-**	**-**	**-**	**1**	**-**
F.6	2	3	1	27	4	-	-	2	-
F.6	3	-	-	10	7	-	-	-	-
F.6	6	-	-	11	-	-	-	-	-
Total		**3**	**1**	**48**	**11**	**-**	**-**	**2**	**-**
F.8	2	-	-	4	-	-	-	-	-
F.8	5	-	-	1	-	-	-	-	-
Total		**-**	**-**	**5**	**-**	**-**	**-**	**--**	
F.9	2	3	-	5	1	-	-	-	-
F.9	3	1	-	9	-	-	-	-	-
F.9	6	2	1	7	-	-	-	1	-
Total		**6**	**1**	**21**	**1**	**-**	**-**	**1**	**-**
F.12	2	5	1	22	7	1	-	-	-
F.12	3	4 (1 ugc)	-	10	1	-	-	-	kite-shaped arrow
F.12	4	-	-	1	2	-	-	-	-
F.12	5	-	-	8	1	-	-	-	-
F.12	6	6	-	12	1	-	-	-	fabricator frag.
Total		**15**	**1**	**53**	**12**	**1**	**-**	**-**	**2**
F.13	2	-	-	13	9	1	-	-	-
F.13	3	-	-	7	-	-	-	-	-
F.13	4	-	-	3	-	-	-	-	-
F.13	6	1	-	10	1	-	-	-	-
Total		**1**	**1**	**-**	**33**	**10**	**1**	**-**	**-**
F.15	2	2	2	46	17	-	-	-	-
F.15	3	-	-	42	17	-	-	-	polished axe flake
F.15	4	-	-	1	-	-	-	-	-
F.15	6	1	1	46	3	-	-	-	-
Total		**4**	**3**	**168**	**47**	**1**	**-**	**-**	**-**
F.16	2	3	-	6	6	-	1	-	-
F.17	2	-	-	18	7	-	-	-	-
F.17	3	1	-	5	-	-	-	-	knife, ch. arrow
PH7	2	-	-	1	-	-	-	-	-
Ring-ditch	2, S1	1	-	15	7	-	-	-	-
Ring-ditch	2, S2	-	-	5	3	-	-	-	-
Ring-ditch	Ap								ch. arrow (chert)

Illustrated flint (Fig. A4.22)

1. Platform section K–L, L16 (from knapping cluster): Pointed tool/borer. Bifacially flaked from a fresh tabular piece of flint, probably exposed during the digging of the shaft.

2. Shaft, L10d (basal fill): Crudely worked pick economically flaked from a fresh tabular piece of flint, probably exposed during the digging of the shaft. Pronounced crushing along one edge with marginal crushing on the opposite edge.

3. Shaft, L10c (in 'earthy' lens at 5.7 m): Knife, bifacially flaked along the right edge with bulbar end snapped off. Sharp left edge partly formed by a frost weakness within the original nodule and with traces of utilization.

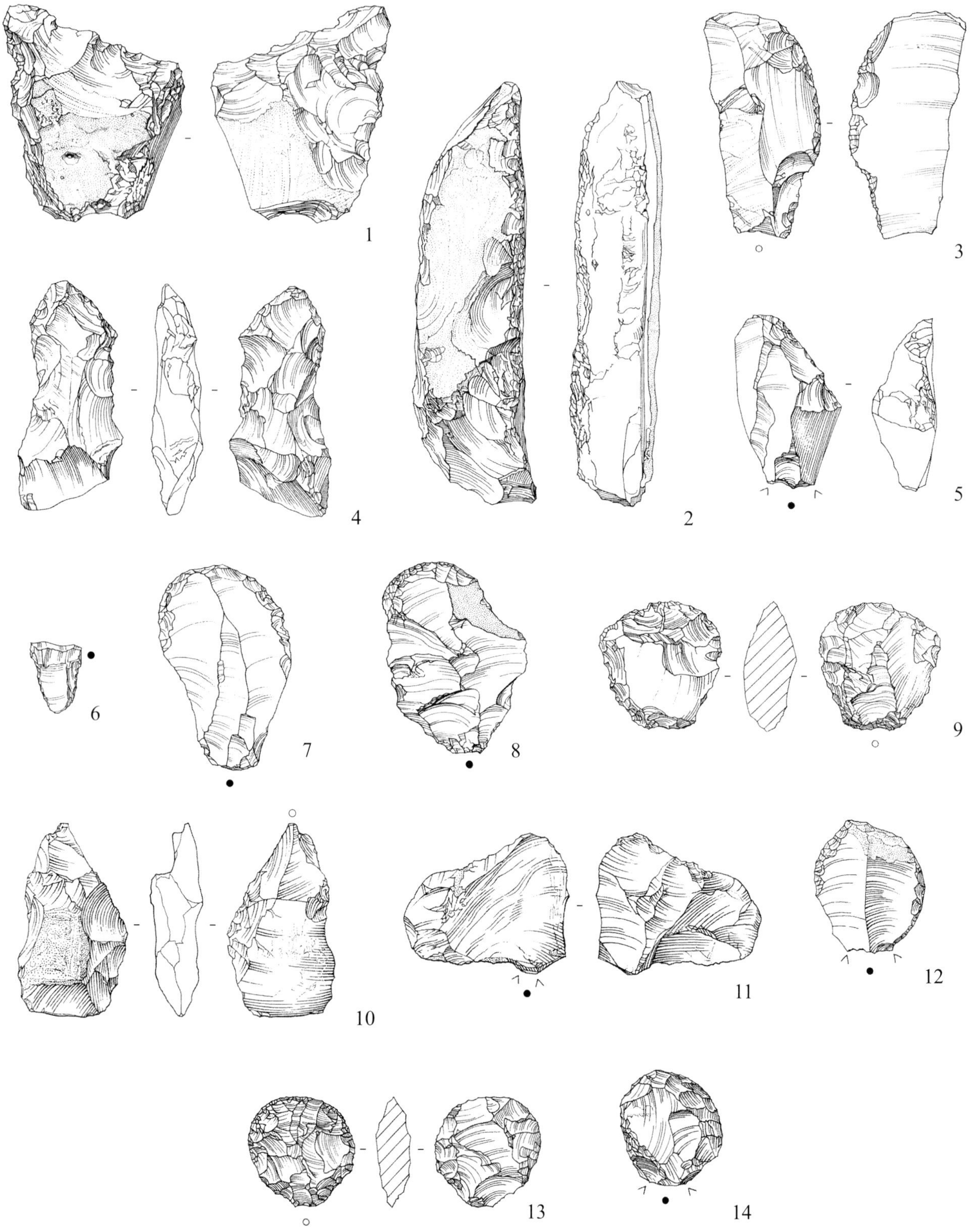

Figure A4.22. *MUW 97: flints 1–14. Scale 1:2 (J. Richards).*

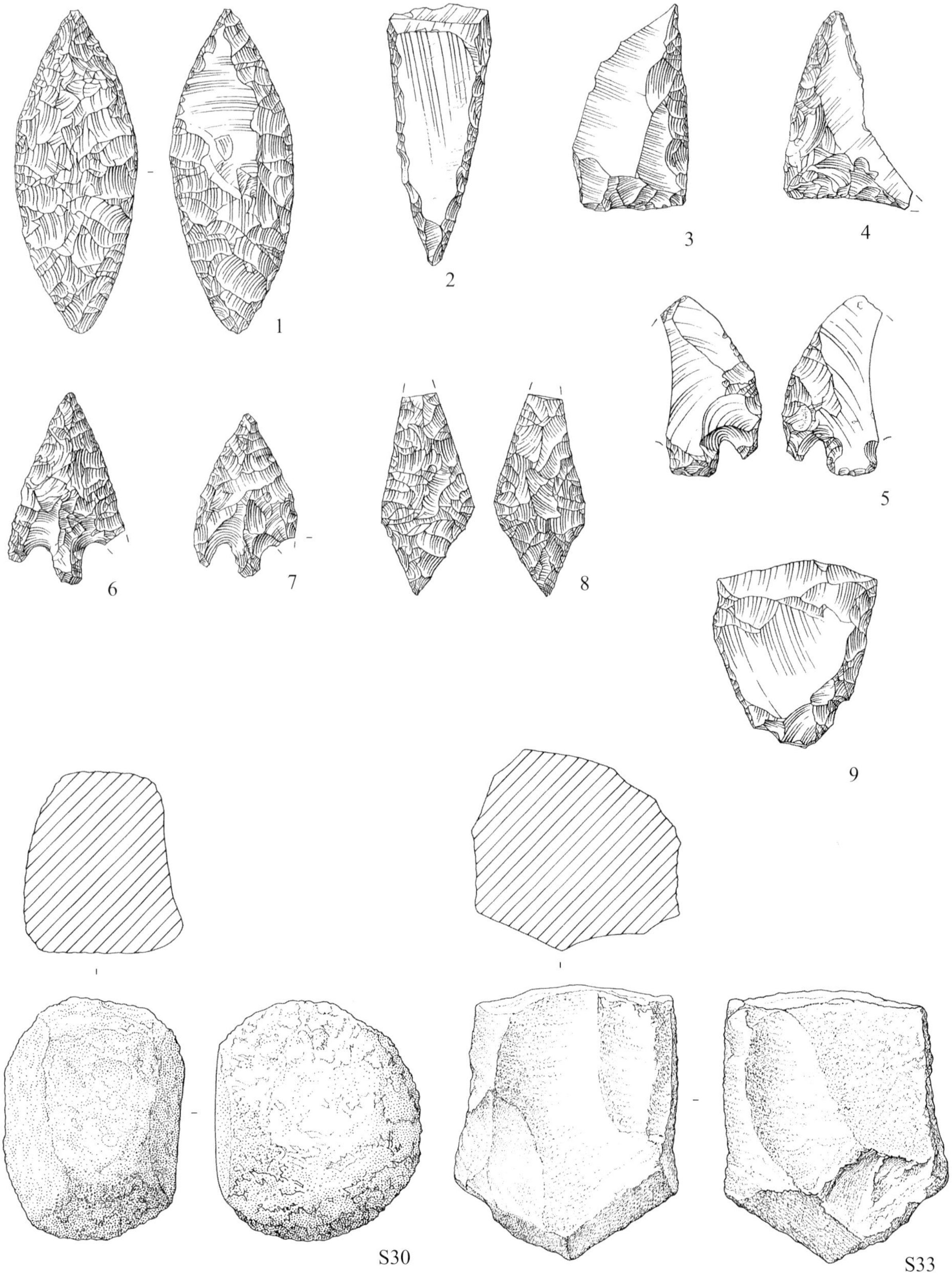

Figure A4.23. *MUW 97: flint arrowheads 1–9 above, and a pecked sandstone 'ball' (S1) and flaked lump of Upper Greensand chert (S2) below. Scales 1–9 1:1, Scale 30 & 33 1:2 (J. Richards).*

4. Shaft, L7b: Small pseudo-tranchet axe probably made from a large flake, no traces of which survive.

5. Shaft, L10c (in earthy lens at 5.7 m): Backed knife. Part of right side obliquely blunted to tip. Sharp left edge with faint traces of utilization. Found close to 3 above.

6. F.1, NWQ, L6a (on base of pit): Small type A chisel arrowhead, steeply blunted down both sides.

7. F.1, NEQ, L4: Convex scraper made from a prepared 'blade-like' flake with a facetted butt. The convex end neatly retouched.

8. F.1, NEQ, L4: Convex scraper made from a thin flake. The dorsal surface exhibiting previous removals terminating in hinge fractures. The convex end neatly retouched.

9. F.1, SEQ, L4: Bifacially flaked discoidal tool. Part of the pronounced convex ventral surface of the original flake survives on the left side. Similar to 13 below.

10. F.1, SEQ, L4: Small pseudo-tranchet axe made from a large flake struck from a frost-weathered surface nodule. Part of the ventral surface of the flake remains and the 'tranchet' cutting edge is also formed from part of the original flake edge. A frost weakness has caused an awkward step fracture at the butt.

11. F.1, NWQ, L4: Tranchet axe-sharpening flake.

12. F.1, SEQ, L3: Convex scraper. Neatly retouched convex end partly into cortex.

13. F.1, SWQ, L3: Bifacially flaked discoidal tool similar to 9 above.

14. F.1, NEQ, L3: Convex scraper of 'thumb' type. Neatly retouched practically all around.

Illustrated arrowheads (Fig. A4.23)

1. Shaft, L7a: Fine bifacially flaked leaf arrowhead with part of the original flake surface surviving on the right side.

2. Shaft, L7c: Unusually elongated Type A chisel arrowhead, neatly blunted down both sides to a pointed base. The chisel edge-damaged on the ventral surface probably from use.

3. F.1, SWQ, L3 (base): Oblique arrowhead bifacially flaked along the right edge and base.

4. F.1, SWQ, L3: Oblique arrowhead bifacially flaked along the left edge. The base partly formed by a step fracture on the underside which has been reduced by the retouching on the upper surface.

5. F.1, NEQ, L3: Broken unfinished barbed and tanged arrowhead. The basic outline had been completed by marginal bifacial retouch including the notch for the right barb. The piece appears to have split when pressure was applied to form the left notch.

6. F.1, NEQ, L3: Barbed and tanged arrowhead completely bifacially flaked. The right hand barb broken anciently.

7. F.1, NEQ, L3 (at 40 cm): Barbed and tanged arrowhead completely bifacially flaked. The right-hand barb broken anciently.

8. F.12, L3: Kite-shaped arrowhead completely bifacially flaked. The base noticeably asymmetric and the tip broken anciently.

9. F.17, L2: Type B chisel arrowhead retouched down both sides of the dorsal surface. The ventral surface unretouched.

Catalogue of worked chalk objects

Martin Green

Illustrated objects (Fig. A4.24)

C1. Shaft, L10C/D interface: Large elaborately decorated block with carefully worked 'cup'-shaped depression. Wt 10,659 g. Discussed in detail by Bradley (this volume pp. 378–9, Fig. A4.29).

C2. Shaft, L10B: Cylindrically perforated tabular fragment. No evidence of modification other than perforation. Wt 41 g.

C3. Shaft, L10A: Small carved lump of oval section with a slightly protruding oval boss at one end. One face bears an incomplete incised circle. Wt 72 g.

C4. Platform, L12, section K–L: Fragment of carved oval object broken during an attempt at perforation. Remains of hour-glass drilling survives on both faces. Wt 27 g.

C5. Shaft, L10A: Carefully worked and centrally perforated 'doughnut'-shaped object, broken anciently. One element of a structured deposit consisting of part of a pig skull, flint pebble and a second chalk object (see C9 below). Wt 74 g.

C6. Shaft, L10: Fragment of a carved and perforated circular object, broken anciently. Possibly part of a spindle whorl.

C7. F.1, NEQ L4: Small carved ball. Weathered surface. Wt 9 g.

C8. Shaft, L10: Carefully worked and smoothed 'marsh-mallow' shaped object. Wt 20 g.

Objects not illustrated

C9. Shaft. L10A: Roughly carved squared lump with one face possibly the base, left unworked. Tool marks clearly visible with three deeper horizontal incisions (max. length 45 mm) on one face near base. Part of a structured deposit with C5 above. Wt 430 g.

C10. On the base of the shaft, L10D: Large rectangular block (240 × 140 × 105 mm) dressed on two faces. A semi-circular recess (max. diameter 110 × 45 mm depth) carved towards one end bears clear tool marks. On the opposite side, opposite the recess is a shallow vertical groove. Wt 3628 g.

C11. On the base of the shaft, L10D: Natural sub-rectangular lump (115 × 75 × 55 mm) with a deep (10 mm) curving groove (max. length 95 mm) incised across one face. Wt 258 g.

C12. Shaft, L7C: Natural angular lump of lozenge-shaped section with a conical depression (max. depth 25 mm, width 17 mm). Possibly an antler pick mark. Wt 94 g.

C13. Shaft, L7C:Natural trapezoidal tabular fragment (78 × 64 × 14 mm) with incised lines on both faces. One face has two lines (max. length 45 mm) crossing each other at a shallow angle. The other has a number of incisions which partly outline a sub-circular form partially within which is a more deeply incised sub-triangle. Wt 66 g.

C14. Shaft, L7A: Weathered natural sub-rectangular fragment (75 × 53 × 20 mm). One face engraved with three diagonal lines, the shortest of which is stopped by a short horizontal line. Wt 66 g.

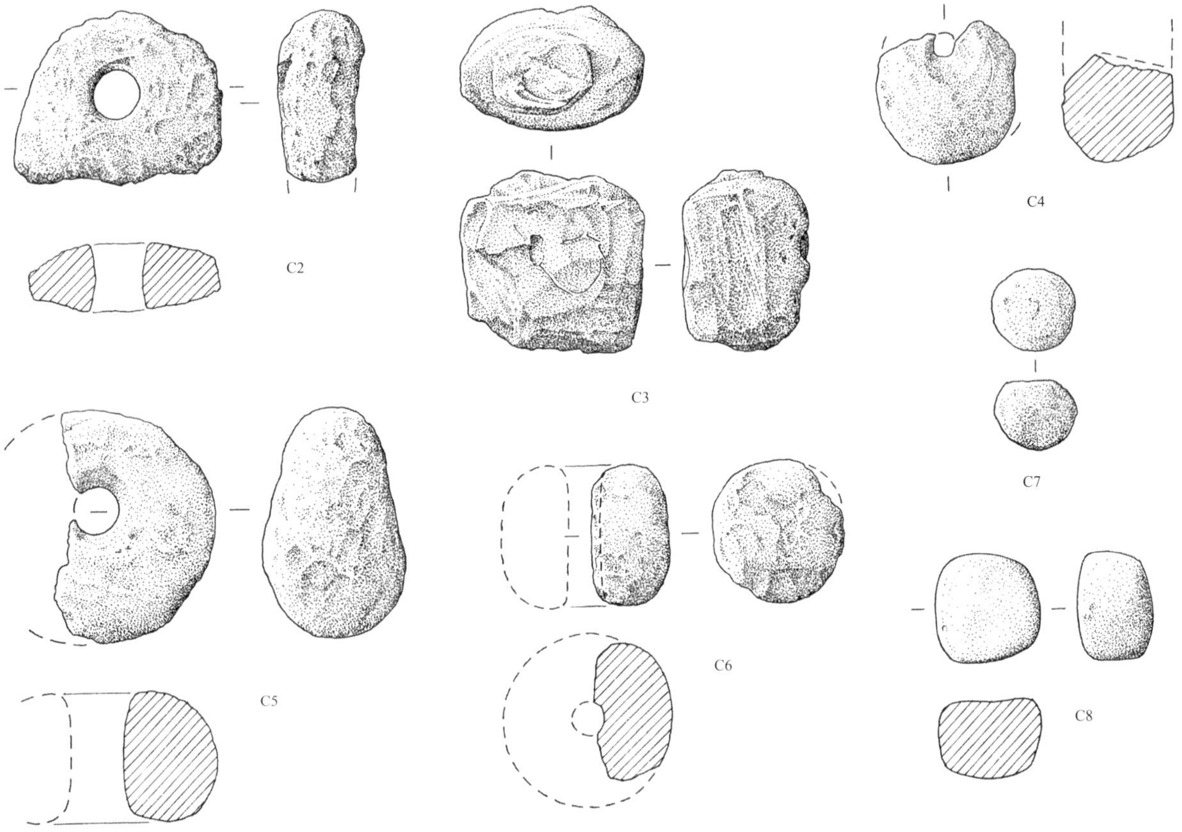

Figure A4.24. *MUW 97: chalk objects C3, C5 & C6 from layers 10a, C2 in 10b, C4 in layer 12 and C7 in layer 4. Scale 2:1 (N. Griffiths).*

C15. Shaft, L7A. Natural rectangular burnt fragment (82 × 70 × 18 mm). One face engraved with three horizontal lines and a short diagonal line. Wt 10 g.

C16. Platform, L12. Large weathered natural trapezoidal block (165 × 160 × 55 mm). One edge scored by a series of seven short (max. 50 mm), deep vertical lines. Wt 1296 g.

C17. Platform, L16. Large natural sub-rectangular block (237 × 165 × 95 mm). One face exhibiting an oval depression (48 × 34 × 22 mm). A second smaller shallow lentoid-shaped depression 32 mm long lies close to the first and may have been produced by an antler pick. Wt 256 g.

C18. F.8 outer pit circle, L6. Natural weathered sub-rectangular lump (100 × 90 × 57 mm) with two deeply grooved vertical lines extending from one edge across the lump for a distance of 70 mm. Two further short diagonal lines lie between the two vertical lines with a maximum length of 35 mm. This piece is very similar to C16 and is also of the same hard blocky chalk characteristic of the deeper levels reached in the shaft. Wt 399 g.

C19. F.9 outer pit circle, L5. Very large rounded 'football'-sized lump. Traces of numerous pock-marks including a clear group of four in an area 20 × 30 mm do not appear to be of natural origin. The roughly rounded form suggests it has been deliberately shaped but the weathered surface makes it difficult to be sure by how much. The lump formed part of a structured deposit with a touching incomplete cattle tibia. Wt 9865 g.

As mentioned above, large blocks/lumps of chalk were also found close to the bases of other pits forming the outer circle, including F.5 and F.6. This blocky chalk only occurred below the level of the F.1 pit at depths cut through by the digging of the shaft. Chronologically the deliberate incorporation of these pieces on the bases of a number of the pits forming the outer circle suggests contemporaneity between the shaft and the pit circle.

The remaining pieces in the catalogue (C20–C31) are likely to have been modified incidentally during the original extraction of chalk rather than deliberate modification.

C20. Platform, L16. Natural sub-rectangular block (200 × 115 × 30 mm). Two deep grooves 75 and 40 mm respectively, cut into one edge, were probably made by the use of an antler pick. A natural perforation pierces one end. Wt 1474 g.

C21. Platform, L16. Natural sub-rectangular block (248 × 135 × 60 mm). One face has three pock-marks up to 15 mm in diameter. One pock-mark has an eroded hollow, perhaps artificial, running from it to the edge of the block. Wt 2130 g.

C22. Platform, L16. Natural sub-rectangular block (140 × 125 × 110 mm). One face scored with three deep diagonal grooves (max.

length 125 mm) and an area of pock-marks, probably resulting from quarrying with an antler pick. Wt 1378 g.

C23. Platform, L16: Small rectangular natural lump (diameter 57 mm) with a single tapering hollow (10 × 10 mm). Almost certainly an antler pick mark. Wt 108 g.

C24. Platform, L16: Natural sub-rectangular block (157 × 112 × 90 mm). One eroded face scored by a deep groove (max. length 120 mm) and two adjacent tapering hollows. Probably the result of removal by antler pick. Wt 1092 g.

C25. Platform, L16: Large natural lozenge-shaped block (330 × 200 × 100 mm). One eroded face bears several deep (45 mm max.) hollows, probably the result of removal by antler pick. Wt 3515 g.

C26. Platform, L16: Natural sub-rectangular block 9133 × 92 × 50 mm) of triangular section. The smooth face bears a deep (12 mm max.) tapering groove 72 mm long. Probably the result of quarrying by antler pick. Wt 481 g.

C27. Platform, L16: Natural sub-rectangular block (98 × 85 × 79 mm). One eroded face bears a deep groove terminating in two hollows probably resulting from removal by antler pick. Wt 437 g.

C28. Platform, L16: Large natural sub-rectangular block (290 × 190 × 110 mm). One face has a single shallow pock-mark. Probably the result of quarrying by antler pick. Wt 5668 g.

C29. Platform, L16: Roughly triangular natural block (160 × 150 × 60 mm). One face with two depressions, one clearly a chop mark from a stone axehead. The other elongated oval depression is of less clear origin. Wt 1311 g.

C30. Platform, L16: Rough lozenge-shaped natural lump (120 × 90 × 60 mm). One face has an elongated tapering depression (max. 50 mm) which has been formed by a sideways blow. Possibly produced by a blow from a narrow bladed (15 mm) stone axe. Wt 351 g.

C31. F.23 grave: Natural sub-rectangular block (130 × 128 × 75 mm). One face with a long (100 mm) groove horizontal with the edge of the block. Traces of two further parallel grooves lie close to the edges of the block. A small area has been affected by rodent gnawing. Wt 952 g.

The Neolithic and Beaker pottery

Rosamund Cleal

It did not prove possible to examine the Monkton-up-Wimborne pottery in as much detail as the Wyke Down 2 pottery. The majority of the material is Peterborough Ware, including all the pottery from the shaft.

Peterborough Ware

The three illustrated sherds (P6, P8 & P10) (Fig. A4.25) from the shaft clearly belong to the Peterborough tradition, but are atypical. Upright rims with little or no expansion are a feature usually only of the Ebbsfleet sub-style, but the rim P6 is not typical of these Ebbsfleet rims and it seems more likely that it should be seen simply as idiosyncratic. Irrespective of sub-style, it clearly does belong loosely within the tradition because of its coarse fabric and twisted cord decoration, as do the two other sherds, although they too are displaying a type of cord impression for which there are no clear parallels within the local area.

The Peterborough Ware vessels from the surrounding feature are much more typical and clearly belong within the Mortlake sub-style, as indicated by their t-shaped rims. P1 and P2 are similar enough to each other to suggest that they might be the product of the same potter or family. Peterborough Ware, including several Mortlake Ware vessels, has been found within a few kilometres, particularly in the Handley Hill area, 2–3 km to the north, but there are no vessels which are particularly close parallels (Cleal 1991, figs. 7.11, 7.12 & 7.15).

Beaker

There is relatively little Beaker from the feature, but one rim at least is of interest in that it is a cordoned vessel. Cordons are particularly found on Beakers of Needham's Tall Mid-Carinated type, a type current mainly in the two centuries leading up to the turn of the third and second millennia BC (Needham 2005, 188 and figs. 6 & 13). A single sherd appears to be from a cord-impressed Beaker (P15). Although these can be as early as twenty-third or even twenty-fourth century BC, it is also quite possible that the vessel was in contemporary use with the cordoned Beaker P14.

Illustrated and unillustrated pottery catalogue

(Fig. A4.25)

P1. F.1 NEQ L9: Mortlake Ware. Single rim to shoulder sherd and one separate sherd from the neck in a soft laminated fabric with sparse small angular flint (<3 mm); the fabric has a slightly 'soapy' feel that may indicate the presence of some calcareous material, possibly chalk, but identification is difficult because there is a calcareous deposit adhering in some places. The decoration is oblique opposing approximately straight twisted-cord impressions on the rim and the same or similar impressions above and below the shoulder. On the interior lip of the rim are small rounded impressions, which may be knot impressions, and two opposing rows of twisted-cord impressions, coarser than those on the exterior. All the twisted cord on the vessel is fairly coarse, and individual strands are visible within the impressions, particularly in the lower row on the rim exterior. Exterior pale grey-brown, core dark grey, interior pale brown except around the upper row of cord impressions below the lip of the rim, and around the interior shoulder angle, where there is sooting. Exterior fair (although the zone just above the shoulder seems slightly more worn than the rest), edges and interior fair.

P2. F.1 NEQ L9: Mortlake Ware. Three conjoining rim sherds (one sherd broken into three probably along post-depositional breaks) in a coarse laminated fabric with large inclusions of quartz or quartzite (creamy grey to yellowy-grey opaque glassy sub-rounded to sub-angular inclusions <8 mm max. dimension; there is no structure visible within the inclusions); there is also one blocky fragment at least 8 mm × 8 mm of an unidentifiable rock, partly obscured by post-depositional calcareous concretion and some fine sand is also present. The decoration is of two rows of well-defined whipped cord 'maggots' on the rim top (each with five segments) and small

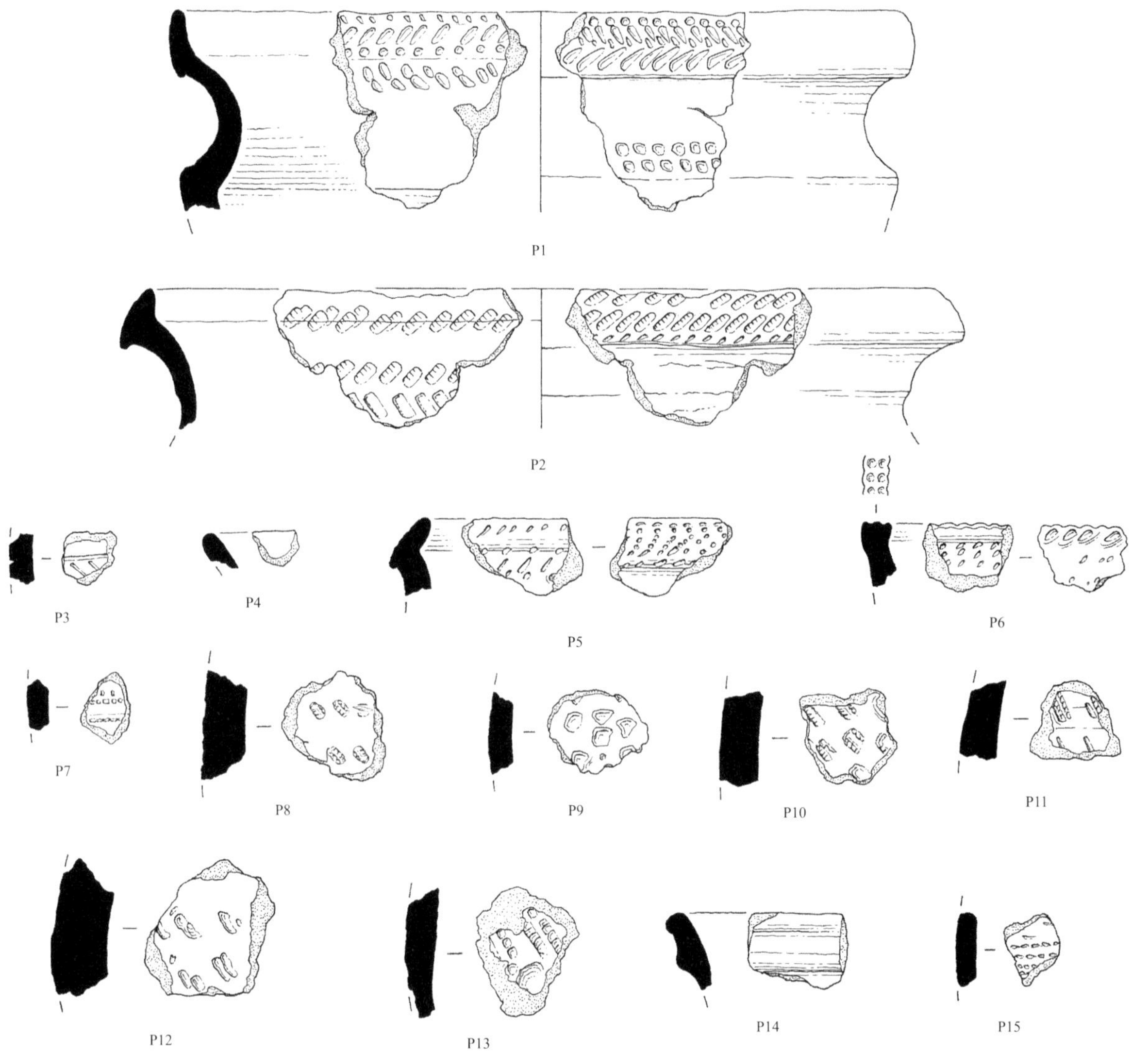

Figure A4.25. *MUW 97: Neolithic and Early Bronze Age sherds. Scale 1:2 (N. Griffiths).*

rounded impressions along the outer edge of the rim top which could be knots. On the interior there is a row of similar 'maggots' below the lip of the rim, and two opposing oblique rows further down. Exterior and interior dark brown, core black. Condition fair.

Other sherds from the same context, not illustrated

Shoulder angle sherd from vessel with well-defined shoulder angle, and with multiple rows of whipped cord 'maggots' both above and below the shoulder. The fabric is similar to P2 but the decorative scheme does not seem compatible as to accommodate a plain zone as in P2 and two rows of 'maggots' the neck would have to be much deeper than is usual in Mortlake Ware; it is likely, therefore, that it represents another vessel.

One small body sherd with well-defined cord 'maggots', probably belonging to vessel P1 or P2 or to the vessel represented by the shoulder sherd.

There are approximately ten other very small body sherds, fragments and crumbs from this context, all belonging to the vessels described.

P3. F.2 L2 Outer pit circle: Small body sherd with incised decoration, probably in a ladder pattern. The sherd is too small to define fabric but certainly includes some shell. Surfaces pale brown, core dark grey. The sherd is in worn condition. The sherd could be identified as Grooved Ware on the grounds of the shelly fabric and colouring; however, incision (rather than broader grooving) is relatively rare, and the pattern appears much more like the patterns found on Beakers. While shelly fabrics are rare in Beakers they are not completely unknown and clear oxidized colours are not always present. Although the sherd could belong to either tradition, it seems slightly more likely that it is from a Beaker than from a Grooved Ware vessel.

Not illustrated. F.2 L3: Two small body sherds Peterborough Ware — flint . These seem to have plastic decoration, possibly fingernail, so may belong to a vessel not otherwise represented.

P4. F.1 SEQ L3 with secondary burial 'Adam' : Very small plain rim, Beaker. Fabric with sand, flint and probably some grog. Orange-red throughout.

P5. F.1 SEQ Base of L3: Peterborough Ware (Mortlake) Rim sherd in a fabric with moderate flint (<4 mm). The decoration is of twisted-cord impressions; two opposing rows on the rim exterior, one row below the rim on the interior, and a row of possible knot impressions on the inner lip of the rim. Surface grey-brown. Core black. Condition worn.

P6. F.1 shaft L7C: Peterborough Ware. Rim sherd in a laminated fabric with moderate blocky to angular inclusions, cream to pale grey; they may be flint (or possibly chert) but are not typical. The decoration is of twisted cord; the impressions on the rim top are finer than those on the interior and exterior surfaces. Surfaces pale orange; core orange-brown. Condition fair.

P7. F.1 SEQ L3: Beaker body sherd in a fabric with flint, sand and grog. The decoration is of horizontal lines of rectangular-toothed comb impressions bordering a zone with vertical impressions; it is not clear what has made the vertical impressions. Exterior pale brown. Condition worn.

P8. F.1 Shaft L7C: Peterborough Ware. Body sherd from the same vessel as P10; the inclusions are the same blocky reddish pieces, plus at least one creamy grey inclusion with a conchoidal fracture which must be flint or chert. The pattern of very fine horseshoe impressions and 'stray' impressions (which do not show in the illustration) is so similar from impression to impression that it is possible that it was made by some complex cordage or textile. Exterior brown, core dark grey, interior orange-brown. Condition fair.

P9. F.5 L2 Outer Pit Circle: Body sherd, probably Beaker, in a slightly sandy fabric with some fine flint (<1 mm). The impressions vary but the majority are rounded, possibly executed with a stick end. Exterior and core orange, interior brown. Worn.

P10. F.1 Shaft. Earthy lens near base. –5.95 m L10D: Peterborough Ware, from the same vessel as P8. Thick body sherd in a fabric with large blocky inclusions (<8 mm × 8 mm); some may be a reddish-flint, as at least one conchoidal fracture is visible, but there seem also to be quartz or quartzite inclusions present. The decoration is of three rows of fine, well-impressed horseshoe impressions in fine twisted cord (Z-twist impressions), but there are also 'stray' linear twisted impressions which are so lightly impressed that they may be a by-product of the potter's manipulation of the cord to produce the horseshoes or of some complex cordage, as suggested above. Exterior mid-brown, core not visible (calcareous deposit) interior orange-brown. Fair condition.

P11. F.1 SEQ L3: Peterborough Ware. Sherd from the same vessel as P8 and P10. Colouring as for P8 and P10. Condition worn to very worn.

P12. F.1 L3 NWQ: Peterborough Ware. Thick body sherd (16 mm) with very abraded twisted-cord impressions (coarse impressions with only two or three twists). The fabric contains moderate to common quartz or quartzite fragments, some very large (<8 mm × 5 mm). Exterior orange; core black, interior dark brown. Condition worn.

P13. F.2, –50 cm near base of L3. Outer Pit Circle: Peterborough Ware. Body sherd with whipped-cord impressions; the deep rounded impression appears to be from a lost inclusion. The fabric contains small angular flint (most <1 mm) and some fine sand. Exterior and interior pale grey-brown.

P14. F.1 NEQ L3: Beaker. Rim sherd with a cordon below the rim, in a sandy fabric with common fine to coarse sand, including dark grains and sparse rare reddish inclusions which are iron oxides. Orange surfaces. Surfaces and edges worn.

P15. F.1 NEQ L3: Beaker. Body sherd in a fabric with fine sand, much of which is of dark grains. The decoration is of horizontal lines of twisted cord impression (Z-twisted impressions). Exterior orange-red, interior orange-brown. Condition worn.

And not illustrated
P21. F.1 NEQ L3: Body sherd with the superficial appearance of Beaker, but it is almost certainly not Beaker; it is in an uncharacteristically very hard fabric, and may be as late as first millennium BC in date, or even later

P22. F.1 NEQ L3: Body sherd in a fabric with some grog; this may be Early Bronze Age in date.

P23. F.3: Beaker.

P24. F.1 NEQ L2: Body sherd, not Neolithic or Beaker; it is possibly first millennium BC in date.

P25. F.1 NEQ L2: Beaker body sherd with some flint and grog, decorated with rectangular comb.

P26. SEQ L2: Beaker body sherd, containing common sand and some ?flint; the sherd is very worn.

Post-Beaker pottery
Martin Green

Pottery of post-Beaker to Romano-British date was recovered from the secondary fills of the central pit F.1 and is described below (Fig. A4.26).

There were 52 sherds (219 g) of Romano-British coarse wares recovered from the top of layer 2 to a maximum depth of 18 cm. This was not unexpected as an area of Romano-British settlement is known within the field some 300 m to the northeast at SU 01851505.

Also, 15 Iron Age coarse ware sherds (41 g) were recovered from the top of layer 2 to a depth of 30 cm. An apparently unenclosed pit group visible on aerial photographs (NMR 4509/62 & 4509/65-6) lies just 150 m to the east at SU 01831460. This is probably part of an Iron Age settlement complex. Also, just 500 m to the south in the adjoining field is the large oval enclosure MUW 96 that was sampled in this survey and is reported on below (this volume).

35 Middle Bronze Age sherds (180 g) were also recovered from the same layer 2 to a maximum depth of 30 cm, with single sherds in layer 3 (5 g) and layer 4 (1 g) probably representing displacements due to animal burrowing. Three ring-ditches are known to be situated within a few metres to the north of this site (NMR 4509/65). The closest and smallest of these was excavated during our survey (see above), and produced pottery of Mid–Late Bronze Age date. It is possible that the other two ring-ditches may be of comparable date.

A further 32 sherds (118 g) of indeterminate Bronze Age date were found in the same layers

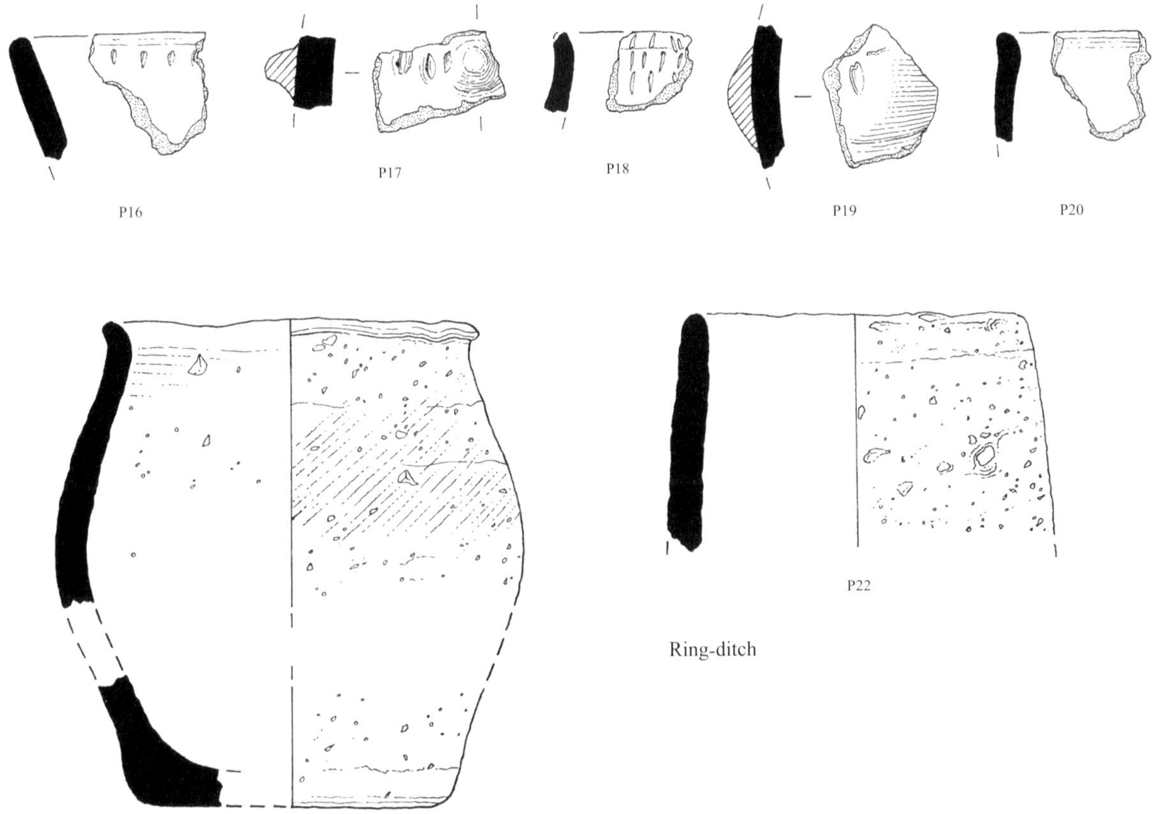

Figure A4.26. *MUW 97: Middle Bronze Age sherds from the ring-ditch. Scale 1:2 (N. Griffiths).*

2/3/4. In addition, there was some Early Bronze Age pottery present, with 6 sherds (56 g) from near the base of L2 (at 42–50 cm), 16 sherds from layer 3 (22 g) and 1 sherd (6 g) from layer 4.

Catalogue of illustrated Middle Bronze Age pottery (Fig. A4.26)

P16. F.1, SEQ, L2, S3: MBA jar. Rim sherd of convex profile with three fingernail impressions. Black/dark brown fabric with sparse grits.

P17. F.1, SWQ, L2, S5: Globular urn. Body sherd with lug and three fingernail impressions. Mid-brown exterior, dark brown interior with small flint grits.

P18. F.1, SWQ, L2, S3: MBA small jar/cup. Rim sherd of externally angled profile with fingernail impressions on rim and body. Dark brown fabric with coarse flint grits.

P19. F.1, NEQ, L2, S2: Globular urn. Body sherd with lug and fingernail impression. Mid-brown exterior, black interior with small flint grits.

P20. F.1, NEQ, L2, S2: MBA jar. Rim sherd with flat top and slightly everted profile. Black/dark brown fabric with small flint grits.

P21. Ring-ditch, T1: M/LBA jar. Thirteen sherds (381 g) of a globular jar with everted rim. Light brown very coarse fabric with large flint grits and some areas blackened internally.

P22. Ring-ditch, T1: M/LBA jar. Weathered rim sherd of a straight sided plain rimmed jar. Light brown very coarse fabric with large flint grits identical to P21 above.

(A further six body sherds (62 g) (not illustrated) were recovered from this same ditch section, and belong to either P21 or P22, as was a single plain body sherd of a M/LBA vessel with a sandy fabric.)

Faunal remains

Mark Maltby

Recording methods

The methods of identification, recording and quantification of animal bones are the same as those employed in the analysis of the assemblages from the Fir Tree Field shaft (see Maltby in 4.2 above). Also, note that the bovid category consists of domestic cattle and possibly aurochs (wild cattle). As will be discussed below, although most of the elements are from domestic cattle, some measurable elements are large enough to have belonged to the larger wild cattle.

Animal bones from F.1

This remarkable feature produced a total of 3414 animal bone elements, of which 1134 were identified to species (Tables A4.40–A4.46). There are significant

Table A4.40. *Animal bone recorded in F.1.*

	Layer																		
Species	**10d**	**10c**	**10b**	**10a**	**10**	**7d/e**	**7d**	**7c/d**	**7c**	**7b/c**	**7b**	**7a**	**7**	**6a**	**6**	**4**	**3**	**2**	**Total**
Cattle	3	1			7	3	1	1	8	5	2	9	6	18	39	40	86	23	**252**
Sheep/Goat														3		4	28	22	**57**
Pig	19			11	4	1	1		4	5		4	10	2	2	16	23	2	**104**
Dog	1								1				3	3	2		1		**11**
Dog/Fox												1							**1**
Red deer	1			3	1	2	1		8	2	1	3	1		2	5	1		**31**
Roe deer									3		2	4				1	1		**11**
Short-tailed vole	1		1	1	2														**5**
Woodmouse				1															**1**
Rodent			9		8				1	2		1							**21**
Frog	14	9	24	19	1		1												**68**
Toad	9	7	46	70	4	2			25			4							**167**
Frog/Toad	42	19	101	158	4	3	1		58	1		5					2		**394**
Owl											1	1							**2**
Thrush family									1			1	1			2			**5**
Small bird	1			1	1	1													**4**
Total identified	**91**	**36**	**181**	**264**	**32**	**12**	**5**	**1**	**109**	**15**	**6**	**33**	**21**	**26**	**45**	**68**	**142**	**47**	**1134**
ULM	1	2		12	9	5			8	10		31	20	114	117	106	223	43	**701**
USM		5		9	7				9	15		36	20	15	11	46	156	78	**407**
UM	39	4	20	40	28				20	11		80	41	145	102	187	372	83	**1172**
Total undentified	**40**	**11**	**20**	**61**	**44**	**5**	**0**	**0**	**37**	**36**	**0**	**147**	**81**	**274**	**230**	**339**	**751**	**204**	**2280**
Total	**131**	**47**	**201**	**325**	**76**	**17**	**5**	**1**	**146**	**51**	**6**	**180**	**102**	**300**	**275**	**407**	**893**	**251**	**3414**

Counts are of number of individual specimens (NISP)
ULM = unidentified large mammal; USM = unidentified medium-sized mammal; UM = unidentified mammal
Cattle = mainly domestic cattle but perhaps includes aurochs. See text for discussion.

Table A4.43. *Sheep/goat elements from F.1, Monkton-up-Wimborne pit/shaft complex (MUW 97).*

	Layer				
Element	**6a**	**4**	**3**	**2**	**Total**
Skull frag.			1		1
Mandible			1		1
Loose teeth	2	3	18	20	43
Humerus		1			1
Radius	1		1		2
Pelvis			1		1
Femur			1		1
Tibia			2	1	3
Metacarpal			1		1
Metatarsal			2	1	3
Total	**3**	**4**	**28**	**22**	**57**

Table A4.44. *Dog elements from F.1, Monkton-up-Wimborne pit/shaft complex (MUW 97).*

	Layer							
Element	**10d**	**7c**	**7a**	**7**	**6a**	**6**	**3**	**Total**
Mandible*		1	1					2
Loose teeth				3			1	4
Scapula					1			1
Humerus						1		1
Radius					1			1
Ulna					1			1
Tibia	1					1		2
Total	**1**	**1**	**1**	**3**	**3**	**2**	**1**	**12**

* includes dog/fox in layer 7a.

Table A4.45. *Red deer elements from F.1, Monkton-up-Wimborne pit/shaft complex (MUW 97).*

	Layer													
Element	**10d**	**10a**	**10**	**7d/e**	**7d**	**7c**	**7b/c**	**7b**	**7a**	**7**	**6**	**4**	**3**	**Total**
Antler	1	3	1	2	1	8	1	1	2	1	2	4	1	28
Maxilla							1							1
Humerus												1		1
Pelvis									1					1
Total	**1**	**3**	**1**	**2**	**1**	**8**	**2**	**1**	**3**	**1**	**2**	**5**	**1**	**31**

Table A4.46. *Roe deer elements from F.1, Monkton-up-Wimborne pit/shaft complex (MUW 97).*

	Layer					
Element	**7c**	**7b**	**7a**	**4**	**3**	**Total**
Antler				1	1	2
Mandible		1				1
Humerus			1			1
Carpals	1					1
Metacarpal		1	1			2
Metatarsal			1			1
Metapodial	1					1
Phalanx 1			1			1
Phalanx 2	1					1
Total	**3**	**2**	**4**	**1**	**1**	**11**

Table A4.41. *Pig element counts from F.1, Monkton-up-Wimborne pit/shaft complex (MUW 97).*

Element	10d	10a	10
Skull frag	1	1	
Maxilla			
Mandible	2		
Loose teeth		7	
Scapula			
Humerus		1	
Radius			
Ulna			
Pelvis		1	
Femur			2
Tibia	1		
Fibula			
Astragalus			
Calcaneus	1		
Phalanx 1			
Phalanx 2			
Phalanx 3			
Cervical V	1		
Thoracic V	9	1	
Lumbar V	4		2
Sacrum			
Vertebra frag.			
Total	**19**	**11**	**4**

Table A4.42. *Cattle element counts from F.1.*

	Layer																	
Element	**10d**	**10c**	**10a**	**10**	**7d/e**	**7d**	**7c/d**	**7c**	**7b/c**	**7b**	**7a**	**7**	**6a**	**6**	**4**	**3**	**2**	**Total**
Horn core					1			2		1	1							5
Skull frag.								1				1		4	1	1		8
Mandible				1	1	1	1	1				1	1	8	3	6	1	25
Loose teeth								1	2		1		5	7	15	50	17	98
Scapula											3	1		1	1		1	7
Humerus								1	1	1			1		1	10	1	18
Radius				2									1		2	1		5
Ulna				1												1	1	2
Pelvis															1		1	3
Femur				1								2	1	6	1	1		11
Tibia					1									1	1	2		5
Astragalus									1						1	1		3
Calcaneus															1	1	1	3
Centroquartal																1		1
Other tarsals															1	1		2
Metacarpal				1								1		1		2		5
Peripheral Mc				1														1
Metatarsal											2				3	3		8
Phalanx 1								1			1		2	1	2	1		8
Phalanx 2														2		1		3
Phalanx 3														2				2
Sesamoids													2	2				4
Atlas (VC1)														1	1			2
Axis (VC2)											1			1		1		3
Cervical V								1	1						1			3
Thoracic V	3	1											2		2			8
Ribs													3	2	2			7
Total	**3**	**1**	**0**	**7**	**3**	**1**	**1**	**8**	**5**	**2**	**9**	**6**	**18**	**39**	**40**	**84**	**23**	**250**

Humerus in Layer 7B = cattle/aurochs.

variations in species represented in different layers and the animal bone assemblage within each of these will be described separately. Species totals (NISP) are given in Table A4.40 listed from the lowest (10d) to the highest layer (2). Table A4.47 includes the species totals for other features within the pit F.1.

Layer 10d

A total of 131 elements were retrieved from the lowest fills of the shaft (Table A4.40). The most significant bones recovered consist of 19 pig bones scattered on the floor at the bottom of the shaft. The bones represented consist mainly of thoracic and lumbar vertebrae (Table A4.41) but also include a substantial portion of a skull, the anterior portions of a pair of mandibles, a cervical vertebra and the complete right tibia and calcaneus. Although strictly speaking these form three associated groups of bones consisting of the head, lower spine and hock, it is likely that all these bones belonged to one animal. All three groups are clearly from an immature skeleton. All the vertebral and limb bone epiphyses are not fused. Both mandibles and maxillae have provided tooth eruption evidence. The deciduous premolars are in early wear, with the mandibular fourth deciduous premolars at Grant (1982) wear stage d. The first molars have fully erupted but only have slight wear on their enamel (Grant stage a). The second molar is visible in the crypt of the left mandible but has not erupted (Fig. A4.30a). This indicates that the jaws belonged to a piglet of about six months of age (Bull & Payne 1982). If it is assumed that the pig was born in the spring and that deposition took place shortly after its death, it can be argued that the bones were deposited in the autumn. The bones survive in good condition with little evidence of weathering and gnawing. Both epiphyses of the tibia were found along with the unfused diaphysis (shaft), which suggests that they were still attached by cartilage. The same applies to many of the vertebral epiphyses. This suggests that the bones were deposited in a fairly fresh condition and had not been left lying around to be scavenged. There is evidence, however, that the carcass has been processed. The left mandible bears incisions located on the lateral surface at the junction of the ramus and the body of the jaw (Fig. A4.30a). These were probably made during the process of removing the cheek meat. One of the thoracic vertebrae has knife cuts on the lateral articular surface made when the rib was detached (Fig. A4.30b). In addition there is evidence of slight charring on the frontal region of the skull, showing that it had been contact with fire. Therefore the pig bones were deposited subsequent to dismemberment, filleting and possibly cooking.

A further associated bone group in this layer consists of three cattle thoracic vertebrae found at the edge of the base and the wall of the shaft (Table A4.42). The vertebral epiphyses were fused indicating that the bones were from an adult animal. Their size suggests they were from a domestic animal. No butchery marks were observed.

Another unusual find at the base of the shaft was the complete right tibia of an adult dog (Table A4.44). No evidence of processing or other post-mortem damage was found. It had a greatest length of 207 mm, which converts into an estimated shoulder height of 61 cm (Harcourt 1974), indicating the bone probably belonged to a large domestic dog rather than a wolf.

Nearly all of the bones found slightly above these deposits at depths between 5.90 m and the base of the shaft belonged to frogs and toads (Table A4.40), indicating that the shaft lay open for a period, during which time these amphibians fell into the shaft. Many of the unidentified fragments probably also belonged to amphibians. The short-tailed vole found at the base of the shaft is also likely to have been a victim of a fall. A tibiotarsus of a small bird species, possibly a skylark, was also recovered. The only identified element likely to have been deposited by humans at this level consisted of a red deer antler tine (Table A4.45).

Layer 10c

Bones found within this layer provided an assemblage of 47 bones, of which all but one of the 36 identified specimens belonged to frogs and toads found within an earthy layer at 5.60 m. Just below this (at 5.70 m) lay part of another cattle vertebra along with a few unidentified fragments of large and medium-sized mammals (Table A4.40). The epiphyses of the vertebra were not fused and therefore it belonged to a younger animal than the one associated with the vertebrae found at the base of the shaft. The bone bears evidence of incisions on one of its lateral processes made when separated from the ribs.

Layer 10b

All the 181 identified bones from this layer found at depths between 4.30 m and 5.00 m consisted entirely of amphibian or rodent bones (Table A4.40). Relatively large numbers of toad and frog bones were recovered along with a few bones of rodent, which definitely included short-tailed vole. All the unidentified fragments could also have belonged to amphibians or small mammals. All these bones are from pitfall victims. There is no evidence for deliberate deposition of bones in this layer, during a time when the shaft was left open.

Layer 10a

Most of the bones found in this layer again belonged to toads and frogs (Table A4.40). There were also a few bones of small mammal (short-tailed vole; woodmouse) and a tibiotarsus of a small immature bird. However, in addition to these pitfall victims, which are testament to the fact that the shaft remained open for some time, there are a few bones of larger mammals, which were deposited by human agency. At a depth of 3.6 m was a red deer antler pick with clear evidence of wear. This was an unusual example in that the pick was manufactured from a section from the top of the antler instead of the base. Significant wear was observed on the trez tine and the section of beam above this was worn smooth where it had been held repetitively during use (Fig. A4.30c). Another smaller fragment of red deer antler, also displaying signs of wear, was found at 3.5 m, and a third piece of antler beam was also recovered, although this was eroded and may have been redeposited. Eleven elements of pig were recorded (Table A4.41). Five of these consist of loose mandibular deciduous teeth (incisors, canine and premolars), conceivably from the same juvenile animal. Two unworn lower incisors could have belonged to an older but still immature pig. More substantial remains of pig included the back quarter of a skull (found at 3.70 m), the shaft of a right humerus and most of a right pelvis (found at 3.50 m), which showed evidence of canid gnawing. Finally, the dorsal part of a thoracic vertebra was recovered at a depth of 3.30 m. Incisions very similar to those found on the thoracic vertebra in layer 10d were noted on the lateral process.

Layer 10

A total of 76 bones were recorded from this layer (Table A4.40). Although small numbers of bones of rodents, amphibians and small birds were recovered, they did not form the major part of the assemblage, in contrast with layers 10a–d. Species identified were short-tailed vole, toad, frog and a passerine species similar in size to a greenfinch (Table A4.40). Twelve bones of larger mammals were identified. Cattle are represented by seven elements (Table A4.42). A small portion of the distal end of a left radius was found at the base of the layer. The fused distal epiphysis indicates it belonged to an adult. It has a maximum distal breadth of 76.7 mm, which falls within the range of overlap between domestic and wild cattle, although it is more likely to have belonged to a domestic male. On the other hand, an unfused and porous left ischium indicates the presence of a young calf. An animal of similar age is represented by a porous shaft of a right humerus. A shaft of a right metacarpal, a peripheral metacarpal, and small fragments of a humerus and mandible were also found. All these elements are also likely to have belonged to domestic cattle.

At least two pigs are represented by right femora. One (from 3.1 m) belonged to a fully-grown animal with a distal epiphysis that had just fused, indicating it was from an animal of about three to four years old. The proximal end of the bone was destroyed by gnawing, indicating the bone was accessible to dogs at some point prior to final deposition. The second pig femur is from a much younger animal. The bone is porous and the distal epiphysis is not fused. Two associated lumbar vertebrae were at a similar stage of development and it is feasible that all three bones belonged to the same piglet. The presence of both young calves and pigs indicates the animals involved were killed in the summer, and deposition may also have taken place in that season.

A small portion of the eroded beam of a red deer antler was recovered from a depth of 2.30–2.50 m. The unidentified fragments (Table A4.40) are also probably mainly from cattle and pig and included three burnt fragments.

Layer 7d/e

Only 17 bones, including five of unidentified large mammal, were recorded (Table A4.40). Five amphibian bones and a chaffinch-sized passerine tibiotarsus belong to species unlikely to have been deposited by human agency. The remaining six bones are from cattle, pig and red deer. The cattle elements consist of parts of a fragmented horn core, the diastema of a right mandible of a calf and the unfused proximal end of a right tibia of an older but still immature animal. Another fairly complete thoracic vertebra of an immature pig was also found. Two red deer antler fragments were recovered. One was from a tine; the second consisted of a charred, eroded and fragmented section of the beam.

Layer 7d

Only five bones all from a depth of about 2.20 m were recorded (Table A4.40). They include two amphibian bones. Another piece of eroded and burnt red deer antler could be from the same specimen recorded in Layer 7d/e. Two jaws with some surviving cheek teeth were also recovered. A cattle mandible possesses deciduous premolars in fairly early stages of wear. It may have belonged to the same jaw recorded in layer 7d/e. The fourth deciduous premolar is at Grant (1982) wear stage h. The first molar has erupted through the bone but has not reached full height (Grant eruption stage 4), suggesting it belonged to a calf of around six months old (Halstead 1985). A pig maxilla has a fully erupted fourth deciduous premolar, which was in wear. However, the first molar has not erupted and this indicates it was from a piglet of around 3–6 months old. If deposition occurred shortly after death, it suggests that the jaws were deposited in the summer and/or autumn.

Layer 7c/d

A complete, probably domestic, cattle right mandible was recovered at the junction of these two layers. All three molars are erupted and in wear but the permanent fourth premolar has not erupted. Tooth wear stages are h, d and b for the first, second and third molars respectively, giving a mandible wear stage (MWS) of 29 (Grant 1982). This suggests that the mandible belonged to a sub-adult animal of perhaps 30–36 months of age (Hambleton 1999, 65).

Layer 7c

The assemblage of 109 bones from this layer again contains a large number of amphibian bones, although only toad was specifically identified. One bone each of a rodent and a member of the thrush family were also retrieved. Elements of at least three species of domestic mammal and two species of deer were also deposited (Tables A4.40, A4.45 & A4.46). Although eight fragments of red deer were recorded, they all consist of eroded and sometimes burnt pieces of antler, many of which could have belonged to the same animal. In contrast, roe deer is represented by three foot bones (Table A4.46). They probably belonged to one very young animal, as all the bones are very porous. The layer also produced the complete mandible of an adult dog. Eight cattle bones were identified. The most impressive of these is a substantially complete cattle skull. The premaxillae and the anterior of the maxillae are missing but the rest survived largely intact. Enough of the maxillae survived to indicate that all three molars were worn and that the skull belonged to an adult. Both horn cores are still attached and enough of the left one survives to provide an estimated length of the outer curvature of *c.* 270 mm to be made. The circumference of the base measures *c.* 202 mm. These dimensions confirm that the skull belongs to a domestic animal. The specimen falls in the lower end of the range suggested by Grigson (1982a, 28) to be from domestic males. The horn core section is fairly conical as indicated by the small difference between the maximum basal breadth (66.4 mm) and the minimum basal breadth (62.6 mm).

A complete cattle horn core was also recovered still attached to a portion of the frontal bone. It has an outer curvature length of 228 mm, a maximum basal breadth of 78.8 mm, a minimum basal breadth of 62 mm and the circumference of the base measures 242 mm. The shape and size of the horn core again indicates that it was probably from a male domestic animal. Another fragmentary cattle horn core was also recovered. A first phalanx of a very young calf was identified. An eroded fragment of mandible, a gnawed cervical vertebra, a small portion of the proximal end of a left humerus and a worn mandibular third molar complete the cattle assemblage in this layer. The four pig bones include the proximal end and shaft of a left humerus of a juvenile. This bone probably belonged to the distal epiphysis that was found in the same layer. A shaft of a right femur and part of a vertebra were also identified as pig. A few fragments of unidentified large and medium-sized mammal probably belonged to cattle and pig respectively. If the roe deer and calf bones were deposited soon after death, this would have probably occurred in the late spring.

Layer 7b/c

Fifty-one fragments were recorded at this junction, of which 15 were identified. Only one amphibian and two rodent bones were present (Table A4.40). The cattle assemblage consists of two worn upper premolars, probably from the same maxilla, a left astragalus and a complete fused cervical vertebra. All these elements are eroded. The astragalus has a greatest medial length (GLm) of 69.9 mm. Although GLm measurements were not recorded, extrapolations from the greatest lateral length measurements (GLl) of specimens at Windmill Hill (Grigson 1965, 157; 1999, 240–41) and Star Carr (Legge & Rowley-Conwy 1988, 131) show that this specimen is larger than most astragali from Windmill Hill but substantially smaller than the aurochs astragali from Star Carr. It possibly belonged to a male domestic animal. An unfused proximal epiphysis of a left humerus displays gnawing damage. Therefore elements of both immature and adult cattle are represented. Three pig loose teeth may have belonged to the same immature animal. A left calcaneus and part of a sacrum are not fused and also belonged to immature pigs. Another burnt and eroded fragment of red deer antler was found. A maxilla of a red deer was also recovered. All three premolars and the first and second molars survived. Their wear indicates that the jaw was from an adult.

Layer 7b

Only six bones were recovered (Table A4.40). A fragmentary portion of cattle horn core was found at a depth of 1.70 m. The distal half of a bovid left humerus, with slight gnawing damage was also recovered. The distal articular (trochlea) breadth measures 93.1 mm. This is therefore a large specimen, which either belonged to a large domestic male or perhaps an aurochs. The specimen is slightly smaller than the aurochs bones from Star Carr (96.8 mm & 97.5 mm) (Legge & Rowley-Conwy 1988, 130). It is slightly larger, however, than the female aurochs skeleton from Ullerslev, Denmark (89 mm) (Degerbøl & Fredskild 1970, 107). It is substantially larger than most of the domestic cattle bones found on British sites such as Hambledon Hill, Windmill Hill and Durrington Walls (Legge 1981, 177). A few specimens of comparable size were recorded at these sites. Legge (1981) favours the interpretation that they were from large male domestic cattle but recognizes that they overlap in size with female aurochs. Harcourt (1971) interprets specimens of 90 mm and 102 mm at Durrington Walls as aurochs but also recognizes the overlap in size between wild and domestic cattle.

The bones of other wild species are definitely present in this layer. A substantial portion of the beam of a red deer antler was located at a depth of 1.80 m. Most of a roe deer left mandible was also found. All six cheek teeth are still embedded in the jaw. They are all fully erupted and in wear indicating the mandible belonged to an adult. In contrast a right metacarpal is both unfused and very porous and belonged to a very juvenile roe deer. The proximal portion of a tibiotarsus of an owl, probably a tawny owl, completed a small but distinctive assemblage in this layer.

Layer 7a

A total of 180 animal bone fragments were recovered (Table A4.40). However, 147 of these are unidentified and there is a notable increase in the proportion of eroded fragments in this layer. At least nine species of mammal and bird were identified, among which cattle elements are the most common (Tables A4.40 & A4.42). These include three scapulae from at least two animals. A right scapula has a glenoid breadth (GB) of 50.7 mm, comparable with some of the largest specimens from Windmill Hill (Grigson 1965, 154). Two left scapulae include one where the glenoid has been destroyed by gnawing. There is no clear evidence that any of these scapulae were used as shovels. A left horn core base has a maximum breadth of 73.8 mm, a minimum breadth of 58 mm, and a basal circumference of 222 mm, slightly smaller than the specimen in layer 7c. It therefore is also probably from a domestic steer or bull, although it is impossible to differentiate male domestic cattle from female aurochs from basal measurements of horn cores alone (Grigson 1982). A mandibular deciduous fourth premolar and an axis (second cervical vertebra) with an unfused caudal epiphysis are from immature animals. Two right metatarsal fragments were identified. These conceivably could have been from the same animal. Both fragments are eroded, and the distal portion is fused indicating the animal was probably adult. The proximal half of the bone has a maximum breadth (Bp) of 48.1 mm, falling comfortably within the range of domestic cattle recorded on other British Neolithic sites such as Windmill Hill (Grigson 1965). A complete first phalanx has a greatest peripheral length (GLpe) of 68 mm and a proximal breadth (Bp) of 33.6 mm. This is a large specimen, slightly smaller than the largest recorded at Windmill Hill (Grigson 1965, 158) and Durrington Walls (Harcourt 1971, 341) but larger than 23 and 39 other specimens measured in those assemblages respectively. Although this is probably from a large domestic male, the possibility that it belonged to an aurochs cannot be ruled out.

Four bones of pig were also recorded in layer 7a. A small fragment from the shaft of a left femur with an unfused distal epiphysis indicates the presence of an immature animal. An unfused proximal epiphysis could have belonged to the same pig. A small fragment of a left pelvis with a fused acetabulum and the distal portion of a left humerus were also found. The distal epiphysis of the humerus was just fusing indicating the pig was over a year but less than two years old (Bull & Payne 1982).

Two fragmentary portions of red deer antler were found. Both are eroded and one is charred. A red deer pelvis was also recovered, consisting of the area around the fused left acetabulum, indicating the animal was probably over a year old. Four eroded fragments of roe deer were also identified. These consisted of part of the shaft of a right metacarpal, the proximal half of a first phalanx, a small fragment of a metatarsal shaft and the distal quarter of a left humerus. The fusion evidence could indicate that elements of at least two roe deer are present with a kid of under a year old represented by the unfused proximal phalanx and an older animal represented by the fully fused distal humerus. This specimen has a distal articular breadth (BT) of 25.4 mm and a trochlea height (HT) of 21.2 mm.

The left mandible of a canid was recovered. The front three premolars were present and in wear. Although the specimen is quite gracile and similar in size to a fox, insufficient diagnostic areas survived to determine whether the jaw belonged to a fox or a dog. Nine amphibian bones and a rodent bone indicate the presence of a few pitfall victims (Table A4.40). Two species of bird were represented. The distal half of a tibiotarsus of an owl (cf. tawny owl) could have belonged to the same bird as the tarsometatarsus found in the adjacent layer 7b. Part of an ulna of a thrush species was also recorded.

Layer 7

The assemblage from this layer is similar in nature to that from layer 7a. Nearly all of the 102 elements are eroded and consist predominantly of small unidentified fragments (Table A4.40). Only 21 elements were identified. Pigs provide ten of these. A maxilla included a worn deciduous premolar of an immature animal of under 18 months old (Bull & Payne 1982). Three mandibular loose teeth were also identified, consisting of a permanent canine of a sow, and a deciduous and permanent incisor. Postcranial elements of pig (Table A4.41) include shafts of left and right humeri, the acetabulum and part of the ischium of a left pelvis, most of a right femur shaft, whose distal end was damaged by gnawing, a small portion of fibula and a first phalanx. The proximal end of the phalanx is not fused, indicating an age of under two years, whereas the fused acetabulum belongs to a pig probably over a year old (Bull & Payne 1982).

The six elements identified as cattle consist of a small fragment of skull, a mandibular ramus, the fused glenoid and part of the blade of a right scapula, shafts of left and right femora and a small portion of a metacarpal shaft. The scapula bears butchery marks consisting of longitudinal incisions running along the lateral aspect of the blade indicative of filleting. None of these elements appear large enough to have belonged to wild cattle.

Three worn loose teeth of dog probably belonged to the same jaw of an adult animal. A tine of a red deer antler was also found. Another ulna of a species of the thrush family was also recovered.

Layer 6a

It was noted during excavation that the layer to the northern edge of the platform contained quite large quantities of animal bone (Table A4.40), many of which are burnt. Post-excavation analysis of faunal material from this layer and layer 6 confirmed this impression. Three hundred fragments include 37 that are charred. The assemblage is generally poorly preserved with all fragments showing evidence of erosion and it is very fragmented, with 274 of the 300 fragments being unidentified. Where it was possible to estimate size, most of the unidentified fragments were assigned to the large mammal category and probably belonged to cattle. This mirrors the small assemblage of 26 identified bones, which includes 18 cattle elements (Table A4.42).

The cattle assemblage from layer 6a includes a diastema of a mandible. Although no teeth survive, the porous nature of the bone suggests it belonged to a calf. Five cattle molars, most of them in wear, were also recovered. The largest surviving element recovered from this layer is a substantial portion of a left humerus with a fused distal epiphysis. The shaft of a left radius belonged to a very young calf. A fragment of the distal epiphysis of a femur was also found. Two complete first phalanges were also recovered. Two proximal sesamoids were found in close association with one of these and belonged to the same foot. The proximal epiphyses of both phalanges are fused and have greatest peripheral lengths of 72.2 mm and 71.3 mm, the latter having a proximal breadth (Bp) of 38 mm. These are larger than any of those published from the Keiller excavations at Avebury (Grigson 1965) or the Wainwright excavations at Durrington Walls (Harcourt 1971). A specimen of slightly larger size was found in the later excavations at Windmill Hill and tentatively identified as aurochs (Grigson 1999, 231–2 & 242). These either belonged to large domestic males or female aurochs. Three dorsal ends of ribs and two unfused thoracic vertebrae complete the cattle assemblage.

A third phalanx and the distal half of a right tibia were the only elements of pig identified. The tibia is from a large animal, having a distal depth of 29.1 mm. This layer also produced the first definite identification of sheep/goat, consisting of two loose teeth and the shaft of a right radius. Parts of a right scapula, radius and ulna could all have belonged to the same adult dog. No bones of amphibians or rodents were found in this layer.

Layer 6

The bone assemblage was derived mainly from the northeast and northwest quadrants of F.1. All the 275 fragments are eroded and 26 are charred. Unidentified fragments, mainly from large mammals, form the bulk of the assemblage (Table A4.40). Thirty-nine of the 45 identified elements belong to cattle. Head elements are well represented (Table A4.42). Eight mandible fragments include one that is slightly charred. All but one are from the left side and derive from at least three individuals. One fragment from the body of the jaw was from a large (possibly wild) animal but no measurements could be taken. Three mandibles provide tooth-ageing evidence. The most complete specimen possesses all its permanent cheek teeth, all of which are worn. All three molars are at tooth wear stage k, providing a mandible wear stage of 45. This indicates the mandible was from a mature adult. The charred right mandible has surviving first and second molars, both at wear stage k, and belonged to an animal of similar maturity. A third mandible belonged to a younger animal. None of the premolars are fully erupted and the first molar is at wear stage h. This stage of dental development suggests the animal died at around three years of age. All seven loose teeth are mandibular premolars and molars, all of which could have belonged to the mandibles also found in this layer. All four skull fragments are from different parts of the cranium and there is no evidence that they belonged to more than one animal.

Six small fragments of cattle femur were recorded. At least five of these are from the right side of the body and at least three individuals are represented. Only one provides ageing evidence. The distal epiphysis was just fused, indicating it belonged to an animal of about four years old or a little older. A shaft of a left tibia and part of the blade of a right scapula were also found. Most of a left metacarpal was recovered. This was a large specimen with a proximal breadth (Bp) of 70.4 mm and a proximal depth of 43.1 mm. Again the breadth dimension is comparable with the smaller aurochs metacarpals from Star Carr (Legge & Rowley-Conwy 1988, 130). However, although it was not possible to measure its greatest length, the metacarpal was relatively short and its stockiness suggests that it belonged to a bull. This would indicate it belonged to a domestic rather than a wild bovid. Seven cattle phalanges and sesamoids were recovered (Table A4.42), some of which probably belonged to the same foot. These include a first phalanx with a greatest peripheral length of 72 mm and a proximal breadth of 38.1 mm. This again falls within the range of overlap between large domestic male cattle specimens and female aurochs. A pair of large second phalanges probably belonged to the same foot. An atlas and axis were also recorded.

The anterior half of a left pig mandible provided tooth ageing evidence. The first two molars are in wear (wear stages f & c, respectively) and the permanent fourth premolar is just in wear. The third molar was not found but it is unlikely that it would have been fully erupted. The mandible possibly belonged to a pig of about 18 months old. The only other pig element identified was the porous neck and blade of a right scapula of a juvenile animal.

Dog was also represented by two bones. A left humerus and left tibia have both survived largely complete. The humerus has a fused distal epiphysis with a maximum distal breadth (Bd) of 29.1 mm, belonging to a medium-sized dog. Both fused ends of the tibia survived but fragmentation meant that no length measurement could be taken. The distal breadth (Bd) measured 20.1 mm, representing an animal of similar size and it is feasible that both bones belonged to the same adult dog. Two fragments of red deer antler were recovered, one of which is charred.

Layer 4
A large number of small and very eroded fragments characterize this and subsequent upper fills of F.1. Although 407 fragments were recorded, only 68 were identified (Table A4.40). Cattle elements (40) dominate the identified component of the assemblage. The sample is biased towards loose teeth and other sturdy elements of the skeleton (Table A4.42). The more complete bones include the proximal half of two left metatarsals with proximal breadths (Bp) of 55.3 mm and 54.5 mm. These are large specimens, most likely to have belonged to male domestic animals. A complete left astragalus has a greatest lateral length (GLl) of 62.6 mm, a greatest medial length (GLm) of 55.9 mm and a distal breadth (Bd) of 41.4 mm, measurements that indicate it belonged to a smaller domestic animal, most possibly a cow. A first phalanx has a greatest peripheral length of 74.5 mm and a proximal breadth of 38.2 mm. This is a large specimen and again could be assigned to a very large domestic bovid or a female aurochs. Another large bovid is represented by a small portion of fused distal tibia, with a distal depth (Dd) measurement of 62.2 mm. This is larger than all the specimens from Keiller's excavations at Windmill Hill, where the largest specimen measured only 55 mm (Grigson 1965, 157) and comparable in size with aurochs specimens from Star Carr (Legge & Rowley-Conwy 1988, 131). Two ribs, two thoracic vertebrae and a cervical vertebra found at the base of layer 4 in the northwest quadrant could all have belonged to the same immature animal.

Other species represented in layer 4 include pig, sheep/goat, red deer, roe deer and a member of the thrush family (Tables A4.40–A4.46). Pig is the most common of these with 16 elements. Loose teeth are again prominent (Table A4.41) indicating poor preservation. A pair of mandibles was found in the northwest quadrant. The deciduous premolars and the first two molars were in wear but the third molars have not erupted. Tooth-wear stages of e and a were recorded for the first and second molars respectively. The specimens both have a mandible wear stage of 18. They are slightly less developed than the specimen from layer 6 and were probably from a slightly younger pig, perhaps between 14 and 18 months old. Three of the four sheep/goat elements are loose teeth (Table A4.43). Four of the five red deer elements consist of small antler fragments. There is also a distal portion of a right humerus. Its greatest distal breadth (Bd), distal articular breadth (BT) and greatest height of the trochlea (HT) measured 61.9 mm, 56.1 mm and 43.7 mm, respectively. This specimen is comparable in size to the largest from Star Carr (Legge & Rowley-Conwy 1988, 127). The likelihood is that the humerus was from an adult stag. A cast roe deer antler was found in the northwest quadrant.

Layer 3
Although this layer produced the largest assemblage in terms of fragments counted (893), it is very poorly preserved and consists mainly of small fragments of unidentified mammals. Only 142 elements were identified (Table A4.40). These are comprised mainly of cattle (86), whose bones have a better chance of survival than those of smaller mammals in heavily eroded assemblages. Even the cattle assemblage is poorly preserved as indicated by the high percentage of loose teeth in the assemblage (Table A4.42). However, a few bones are worthy of comment. A first phalanx has a greatest length of 85 mm, which is probably from an aurochs. A centroquartal (navicular cuboid) has a maximum breadth (GB) of 85.0 mm and is also almost certainly from a male aurochs. In contrast, a metatarsal with a maximum proximal breadth (Bp) of 44.3 mm is definitely from a domestic animal. Ageing evidence is extremely limited. Two limb bones are porous and belonged to calves but the chances of survival of such bones is likely to have been decreased by the heavy erosion. Most of the loose teeth are permanent teeth but do include some worn deciduous incisors and premolars from immature cattle.

Sheep/goat elements (28) were the second most common species identified (Table A4.40) but most of these are loose teeth (Table A4.43). Most of the remaining elements consist of heavily eroded shafts of limb bones, which have provided very little fusion and metrical data. A radius and tibia have evidence of charring. A sheep mandible has deciduous premolars and the first molar in wear (stage f) but the second molar has not fully erupted. It has a mandible wear stage of 16 (Grant 1982) and probably belonged to an animal that was a little over a year old. A fragment of frontal bone has evidence that the sheep was horned.

Pig elements (23) are outnumbered by sheep/goat in contrast with nearly all the lower layers in F.1. Again, loose teeth are the most common element recorded (Table A4.41). Two maxillae still possessed deciduous premolars and therefore belonged to pigs of less than 18 months old. An incisor is the only element of dog recognized in this layer. Two antlers were recovered. The first consists of a tine of a red deer. The second is a more complete cast antler of a roe deer. Two amphibian bones were also found.

Layer 2

The bone assemblage is very similar in its poor preservation to that from layer 3. All the bones were eroded and only 47 out of 251 fragments were identified to species. Sheep/goat and medium-sized mammal fragments are better represented than in layers 3–4 perhaps indicating their increasing importance. However, the problems of preservation severely restrict the quality of the evidence. All but two of the 22 sheep/goat elements and 17 of the 23 cattle elements are loose teeth (Tables A4.42 & A4.43). The only two element of pig identified are also loose teeth (Table A4.41). No other species were identified.

Animal bones from other features

Small quantities of animal bones were recovered from 12 other features providing a total of 140 fragments, of which 59 were identified (Table A4.47). Generally preservation was poor with all bones apart from those in the grave F.23 showing evidence of erosion.

Features within the central pit F.1

Although best considered as part of the sequence of infilling of F.1 discussed above, small numbers of animal bones were found in four features within it. The scoop F.18, cut into the floor of F.1, produced only four unidentified mammal fragments. Pits F.19 and F.22 were cut into the surface of the platform and sealed by layer 3. The former provided 17 fragments but only two teeth were identified. These were deciduous incisors of cattle and pig. F.22 produced no identified elements (Table A4.47). Grave F.23 produced bones of only amphibians and rodents. Frog, toad and short-tailed vole were all identified. Their presence indicates that the grave cut was left open for some time.

Outer pit circle, postholes and ring-ditch

Six of the eight excavated pits produced small quantities of animal bones. F.2 contained 14 fragments, 13 of which were unidentified and mainly belonged to large mammals. The proximal portion of a cattle left metatarsal was identified. The only element from F.5 is a worn cattle lower third molar. F.6 produced an unidentified mammal fragment and a fragmentary portion of red deer antler from its base. F.9 contained ten unidentified fragments and the proximal half of a cattle right tibia. The proximal epiphysis has fused indicating the bone belonged to an adult animal. Although quite large, it is probably from a domestic animal. F.12 contained two unidentified large mammal fragments, seven unidentified mammal fragments and the distal end of a cattle right humerus. The maximum trochlea height (HT) measures 46.6 mm, which indicates it was from a domestic animal. F.13 produced six unidentified fragments and a cattle tooth fragment. The large posthole F.16 produced a cattle upper molar.

Only one unidentified large mammal fragment was recovered from the ring-ditch.

Discussion

Given the nature of the shaft, it is highly likely that many of the animal bones were selected for disposal. Although casual disposal or accidental loss of animal bones cannot be ruled out in some cases, structured deposition needs to be considered for the majority of the larger mammal remains found in the lower layers of the shaft. The first significant deposition seems to have been of parts of a young pig (layer 10d). This, however, was not an animal that was sacrificed and buried as a complete carcass. This carcass has been at least partially processed, with butchery marks providing evidence for segmentation and filleting. The scorch marks on the skull may indicate that the meat from this animal was roasted, possibly on a spit. Subsequently, parts of the pig were deposited on the floor of the shaft. Although the remains could be viewed simply as food-preparation waste with these bones discarded after meat had been removed from them, the decision to place them at the bottom of the deep shaft is likely to have been dictated by motives beyond casual waste disposal. Its flesh had been cooked and presumably consumed before selected portions of the defleshed carcass were deposited. It is tempting to suggest that the pig was slaughtered, butchered and consumed as part of some significant event or ceremony and that the deposition in some way symbolically marked this event or its closure. Ageing evidence suggests that this event took place in the autumn. Although no butchery marks were found on the cattle thoracic vertebrae found on the edge of the bottom of the shaft, these may also mark the remains of another animal processed and deposited in a similar manner. In addition, the butchered thoracic vertebra of cattle and pig in layers 10c and 10a respectively, indicate similar processing prior to deposition. It is also not easy to dismiss the complete tibia of an adult dog in layer 10d as a bone that was casually discarded. It was well preserved and the bone is likely to have been selected for deposition.

The same conclusion would apply to other dog bones deposited at higher levels.

The large number of amphibian and small mammal bones above the bones on the floor of the shaft suggests that these remains were not covered by rapid infilling. Indeed, the high percentages of such bones in most of the lower layers of the shaft indicate that the shaft must have remained open for a substantial amount of time, during which other depositions of small amounts of animal bone were made. Unfortunately, the rate of deposition of such pitfall victims is not known, so the period the shaft remained open cannot be determined by the numbers of such bones. Both frogs and toads move around on dry land particularly during the summer months. Most of the identified small mammal bones belong to short-tailed vole, indicating fairly open grassland conditions. Although it is assumed that most of the remains of small mammals and amphibians represent pitfall victims, the identification of owl bones is a reminder of another possible source for the deposition of such bones.

Table A4.47. *Animal bones recorded in other features, Monkton-up-Wimborne pit/shaft complex (MUW 97).*

	(Within F.1)				**Other features**								
Species	**F.18**	**F.19**	**F.22**	**F.23**	**F.2**	**F.5**	**F.6**	**F.9**	**F.12**	**F.13**	**F.16**	**RD**	**Total**
Cattle		1			1	1		1	1	1	1		7
Pig		1											1
Red deer							1						1
Short-tailed vole				4									4
Woodmouse													0
Rodent				2									2
Frog				19									19
Toad				7									7
Frog/Toad				18									18
Total identified	**0**	**2**	**0**	**50**	**1**	**1**	**1**	**1**	**1**	**1**	**1**	**0**	**59**
ULM		5	2		8				2			1	18
USM		6	2		5								13
UM	4	4	3	15			1	10	7	6			50
Total undentified	**4**	**15**	**7**	**15**	**13**	**0**	**1**	**10**	**9**	**6**	**0**	**1**	**81**
Total	**4**	**17**	**7**	**65**	**14**	**1**	**2**	**11**	**10**	**7**	**1**	**1**	**140**

Counts are of number of individual specimens (NISP).
ULM = unidentified large mammal; USM = unidentified medium-sized mammal;
UM = unidentified mammal;
RD = ring-ditch.

Ageing evidence from young mammal bones can indicate the approximate time of year of their death. If their remains were deposited in the shaft shortly thereafter, their remains can indicate the season of deposition. As noted above, this could indicate that the pig at the bottom of layer 10d was deposited in the autumn. The porous cattle and pig bones in layer 10 are likely to have belonged to animals that died sometime during the summer. The cattle and pig mandibles in layer 7d belonged to animals probably killed in the late summer or autumn. The very porous cattle and roe deer bones in layer 7c are from animals that probably died soon after their birth in the spring. Unfortunately, it is very difficult to pinpoint the season of death of older immature and adult animals.

The red deer antler remains include a largely complete pick but also some isolated tines and a number of small fragmented sections of beam. Many of the small fragments could have belonged to the same antlers. Several of the antler fragments were charred, perhaps indicating heat was applied during working. Structured depositions of red deer antler picks and other sections of antler have been recorded on many British Neolithic sites. These include causewayed camps such as Windmill Hill (Grigson 1999) and Etton (Armour-Chelu 1998), flint mines at Grimes Graves (Clutton-Brock 1984) and later Neolithic sites including Durrington Walls (Harcourt 1981) and Stonehenge (Serjeantson & Gardiner 1995). It is likely that the presence of the antlers in the shaft is of more significance than simply the discard of processing waste or broken picks.

In addition to antler, a few bones of roe deer and red deer were identified mainly between layers 7a and 7e (Tables A4.45 & A4.46). Most of the bones of wild mammal were confined to these layers and this may indicate that bones of such species were preferentially selected for deposition. This may also include fox and aurochs. Although metrical analysis of bovid bones was very limited, the assemblage produced several bones of large animals that fall within the size range of aurochs.

However, the most positive identifications of aurochs are from specimens found in upper layers of F.1. The largest measurable bovid specimens from layers 6 and below all fall within ranges that have been described elsewhere as both male domestic and/or female aurochs by various authors. Unfortunately it has not been possible to positively distinguish all of these bones using traditional metrical analysis. However, the complete skull and horn core from layer 7c and the metacarpal from layer 4 all have morphological characteristics that suggest they belonged to large domestic males rather than wild cattle. The other large bones could therefore also belong to large domestic cattle. It could be that there was a preference for the deposition of male cattle body parts. If this interpretation is correct, the bias towards male cattle (at least amongst the measurable bones) sets this assemblage apart from most of the larger Neolithic cattle assemblages from southern Britain, in which bones of female domestic cattle tend to be found much more frequently than those of males (Legge 1981; Grigson 1999; Armour-Chelu 1998). Deposition of clearly recognizable male skeletal parts of bulls, particularly their skull and horns may have been considered desirable. Deer antlers too come from males and a humerus of a red deer stag was also found. Unfortunately it is not possible to sex other deposited bones, particularly those of immature animals, and so any universal interpretation of selected deposition along the gender lines must remain extremely tentative.

Given the unusual nature and context of the assemblage, it is almost impossible to be certain whether the animals represented reflect general dietary, husbandry and hunting practices. In terms of species representation, cattle and pig are the main two domestic species identified in the lower layers. Both species are generally well represented in Neolithic samples from southern England. Assemblages with higher percentages of pig tend to be found in later Neolithic samples with Durrington Walls (Harcourt 1971; Albarella & Serjeantson 2002; Albarella & Payne 2005) being the best-known example. Pig and cattle were the most common mammals found in the Late Neolithic pits from Fir Tree Field (Legge 1991). Therefore the preponderance of cattle and pig remains in the shaft is not unusual for the period and locality. The absence of sheep/goat remains in the lower layers is also not unprecedented. They are poorly represented at Fir Tree Field (Legge 1991), Durrington Walls (Harcourt 1971) and other Neolithic sites in the area around Stonehenge (Maltby 1990a). Generally, on English sites, it is not until the Bronze Age that sheep/goat became one of the most common species, as demonstrated on the Down Farm Middle Bronze Age settlement (Legge 1991) and in the upper layers of F.1. The rare appearance of aurochs in these later deposits is also not unprecedented, as they probably did not become extinct from Britain before the end of the second millennium BC.

Similarly beyond noting that juvenile, immature and adult cattle were represented and that most of the pig remains were from immature animals, there are insufficient data to discuss whether the mortality data are typical of contemporary assemblages.

Summary of the molluscan data

Michael J. Allen

A very large programme of 70 samples of 2000 g or 1500 g were taken and processed from the Monkton-up-Wimborne 'ceremonial' hilltop complex which consisted of the 10 m diameter and 1.5 m deep 'pit' (F.1) surrounded by a ring of 14 unevenly spaced oval pits (Green 2000). Within the central pit was a 7 m deep shaft. Unlike that in Fir Tree Field, this shaft pit was not natural, and had been humanly excavated, the upcast of which had been dumped on the floor of the large main pit. On this and under a flint cairn was a Bronze Age flexed burial. These features provided a series of sequences covering the Middle Neolithic (Peterborough Ware) phases. Samples included a buried soil material beneath the shaft upcast, a sequence through the shaft, sequences through the main pit fill and samples through one of the surrounding oval pits (pit 12).

Snails were abundant with up to 2298 per sample, with many samples (16%) having in excess of 1000 snails and 39% having in excess of 500. It is clear that this monument was constructed in very open downland, which is not a surprise in view of the proximity of the Monkton-up-Wimbourne Neolithic mortuary enclosure (MUW 00) in the same field. The soil material beneath the shaft upcast and within the pit indicates open short-grazed dry downland. The upcast itself provided soily micro-habitats which were exploited my shade-loving species, indicating the proximity of shadier and more mesic habitats within the vicinity. This is confirmed by the assemblages from the shaft infill. Although indicating open downland at the base, the fills show rapid migration of shade-loving species, the majority probably reflecting the shady microhabitats of the feature itself, but confirming the presence and proximity of suitable shady ?woody habitats in the vicinity.

The infill of the main pit (F.1) post-dates 3500–3100 cal. BC (OxA-8035; 4585±50 BP), and shows mixed assemblages reflecting a locally overgrown hollow, which became a trap for windblown silts. The deposition of silty stone-free deposits in the top of the pit promoted growth of grassy and herbaceous vegetation locally, but the presence of the silts themselves indicates an open disturbed and probably tilled landscape in the vicinity. So this monument, in its latter infilling stages, may have been an abandoned and overgrown hollow on the top of Monkton-up-Wimborne. Assemblages from pit 12 (one of the oval, irregularly spaced, features surrounding the central pit F.1), in contrast indicate very open dry grassland and arable landscapes, confirming the highly local overgrown microhabitats within the central pit itself.

Human remains

Jacqueline I. McKinley

The skeletal remains of four Neolithic individuals, buried together in a single grave (F.23) (Figs. A4.27 & A4.28), were received for analysis, together with those of a Bronze Age individual buried in the same locality. The Neolithic grave formed part of a currently unique Neolithic monument, comprising a large (*c.* 10 m diameter), flat-based, circular pit feature (F.1), cut over 1 m into the chalk natural and surrounded by a concentric circle of pits (Figs. 3.61–3.67). The grave was cut into the north-northeast side and base of F.1, the southeast quadrant of the latter also being cut by a deep shaft. Seven fragments of disarticulated human bone were recovered from the upper shaft fills. The Bronze Age burial was excavated from below a flint nodule cairn within the upper levels of the layers sealing F.1 and its associated features; disarticulated human bone was recovered from the same level.

Methods

Age was assessed from the stage of tooth development (Beek 1983) and ossification/epiphyseal bone fusion (Gray 1977; McMinn & Hutchings 1985; Webb & Suchey 1985), the length of long bones (Bass 1987), pattern of degenerative changes in the pubic symphyses (Brooks 1955), assessment of the sternal-end ossification in the ribs (Iscan *et al.* 1984; 1985), tooth wear patterns (Brothwell 1972) and the general degree of other age-related changes to the bone (Bass 1987). Sex was ascertained from the sexually dimorphic traits of the skeleton (Bass 1987; Schutkowski 1993).

Platymeric, platycnemic and robusticity indices were calculated according to Bass (1987), and cranial index according to Brothwell (1972). Stature was estimated using Trotter & Gleser's regression equations (1952; 1958). Pathological lesions and morphological variations were recorded and diagnoses suggested where appropriate. Anatomical terminology according with Gray (1977) and McMinn & Hutchings (1985).

Results

A summary of the results is presented here (full details are held in the Down Farm Museum archive).

Disturbance and condition

The Neolithic skeletons are in very good condition with only slight root marking to a few bones from skeleton D. The grave fill comprised chalk rubble and very little soil had infiltrated the cranial cavities or accumulated around the skeletons. The weight of the chalk rubble pressing on the skulls has resulted in some warping, particularly in the juvenile crania which are also quite heavily fragmented. The condition of the bone fragments from the shaft was variable, most being in good condition, though one fragment was slightly root marked. The Bronze Age skeletal remains (skeleton E) were heavily root marked, as was the redeposited bone from the same level, indicative of the proximity of these deposits to the ground surface.

Skeletal indices and morphology

Various indices were calculated for the two adults, the Neolithic female (skeleton C) and the Bronze Age male (skeleton E). Both crania are in the dolichocrany range, at 71.9 and 71.7 respectively. The estimated stature of the female is 1.59 m (*c.* 5' 2¼") and that of the male 1.63 m (*c.* 5' 4¼"). The platymeric indices (degree of anterior-posterior flattening of the proximal femur) are very close, at 81.3 and 81.1 respectively, both in the platymeric range. The platycnemic indices (mesolateral flattening of the tibia) are 64.1 (mesocnemic) and 71.5 (eurycnemic).

The Bronze Age male was of a robust build with strong muscle attachments, particularly in the humeri. The sexual dimorphism between the two adults was marked, the male having strongly masculine skull and pelvic traits. The femoral robusticity indices were calculate at 11.7 for the female and 13.2 for the male.

Demographic data

The Neolithic grave (F.23) held the *in situ* remains of four individuals: three juveniles, ranging in age from *c.* 5–10 years, and one adult female (*c.* 30–45 years). Further Neolithic remains were recovered from the shaft, representing a minimum of one adult. The origin of the latter is unknown, but the relatively 'fresh' condition of most of the fragments implies that they were not exposed to the elements for long prior to redeposition. Since the shaft does not represent an original component of feature F.1, it is possible that its excavation may have disturbed a mortuary deposit.

The Bronze Age remains represent those of an adult male *c.* 30–45 years. At least three Bronze Age ring-ditches, possibly burial monuments, have been found in the immediate vicinity (*c.* 20 m to the north), and this skeleton may not represent the lone burial it currently appears. The redeposited bone at the same level (sub-adult/adult *c.* 15–25 years) may also be of Bronze Age date, though the remains may have been disturbed from an earlier deposit.

Pathological lesions and morphological variations

Dental disease: Mild-heavy dental calculus was observed in one juvenile and both adult dentitions, the Neolithic female being the greatest affected with

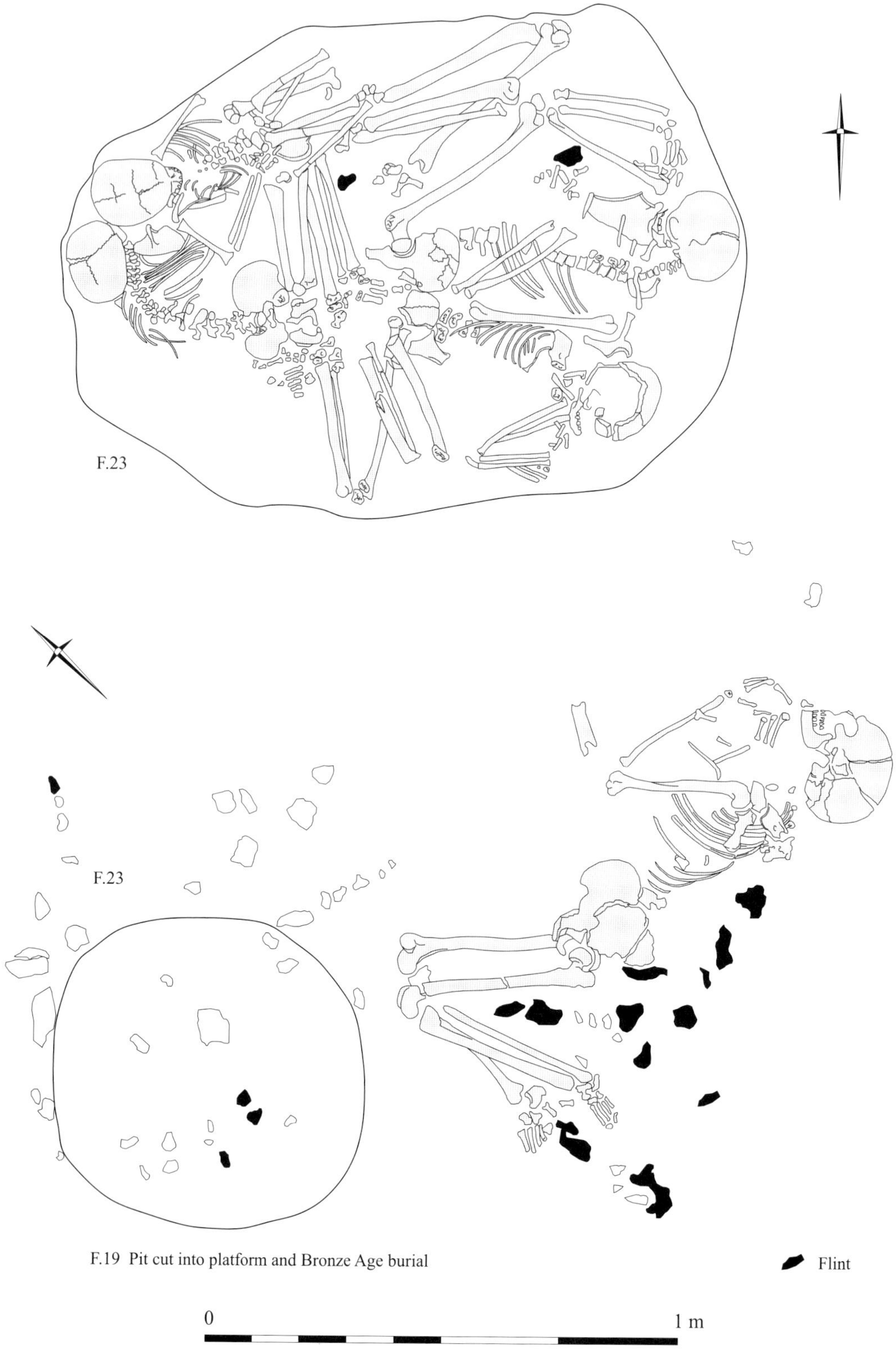

Figure A4.27. *MUW 97: multiple burial F.23 (above) and the semi-crouched burial of 'Adam' (below) (K. Faxon).*

Figure A4.28. *MUW 97: near vertical view of the multiple burial F.23, looking northwest (D. Webb).*

deposits 2 mm thick extending almost to the occlusal surface in the mandibular M3. Both adult dentitions showed evidence of alveolar reabsorption, which was particularly marked in the distal mandible and maxilla of the Bronze Age male (up to 6.3 mm reabsorption). No teeth had been lost *ante mortem*.

No carious lesions were present in the Neolithic dentitions, but 47% of the maxillary and 6% of the mandibular teeth from the Bronze Age adult male were carious. Half the lesions were in the molar teeth, where they are generally most common (Hillson 1990, 294), with maxillary lesions also occurring in an incisor and several premolars. Most of the lesions appeared to have been cervical in origin, though in one case destruction of the tooth was too severe to be able to judge the origin of the lesion. Occlusal tooth wear was particular heavy in this individual, especially in the maxillary teeth where almost all of the anterior tooth crowns had been worn away. The exposure of the pulp cavity and subsequent infection in the maxillary left I1 was as a result of excess occlusal wear. Infection from the carious lesions had tracked down into many of the tooth sockets resulting in the development of a dental abscess. Half of the six maxillary abscesses (37.5%) were associated with carious lesions, as were the two (12%) mandibular lesions. Infection at the apex of the mandibular right M1 had tracked into the adjacent M2 socket and exited buccally via a 3.7 mm sinus. The lesion at the apex of the maxillary incisor exited labially via a sinus and also appeared to extend into the floor of the nasal cavity. Lesions in the maxillary left M1 socket and right P1–M2 sockets, all exited via sinuses (up to 10 mm diameter) through the palate, and reactive new bone in at least the anterior two-thirds of the palate indicates the spread of infection into the soft tissues. A small fistula, 2.2 mm diameter, from the abscesses in the right of the maxilla into the antrum, does not appear to have any associated new bone. An abscess, with no associated carious lesions, was noted in the mandibular right deciduous molar sockets in Skeleton A. The infection had spread into the adjacent soft tissues over the buccal body of the mandible.

The rates of dental diseases seen in the Bronze Age male are considerably higher than the averages recorded for the period of 4.8% caries and 1% abscesses, though an average of 13.2% *ante mortem* tooth loss has also been recorded whereas none had occurred in this instance (Roberts & Cox 2003, 82–4). The excessive wear seen in the dentition of the Bronze Age male suggests a coarse diet. The high rate of caries indicates a poor level of nutrition, probably based on a predominantly vegetable or cereal diet. The level of dental hygiene also appears to have been low, and with such a level of infection in the teeth and soft tissues of the palate, eating can hardly have been a pleasant experience. It is also probable that the appalling condition of the dentition would have had an adverse effect on the rest of the digestive tract. The condition of the Bronze Age male's dentition appears in stark contrast to that of the Neolithic female, particularly since they appear to fall into the same — albeit broad — age band. The rates of dental disease for the Neolithic are generally fairly low at 2.5% caries, 4% abscesses and 10% *ante mortem* tooth loss (Schulting pers. comm. from a sample of 21 British sites) but Skeleton C had no caries or abscesses and does not exhibit the excess wear seen in Skeleton E. Whilst there may be other predisposing factors (Hillson 1990, 287), the implication is for a different diet with a higher level of nutrition, probably comprising predominantly meat or fish, possibly supported by a relatively good level of dental hygiene.

Deficiency disease: Cribra orbitalia is manifest as pitting in the roof or one or both orbits and is believed to result from a metabolic disorder connected with childhood iron deficiency anaemia (Robledo *et al* 1995). The condition was observed in the three Neolithic juveniles (60%); light grade porotic in skeleton D, medium grade cribotic in skeleton A and heavier grade trabecular in skeleton B (Robledo *et al.* 1995, fig. 1). Although the condition is not totally absent in populations with a protein-rich diet (Robledo *et al.* 1995), the higher level of nutrition provided generally has a positive effect.

The tibiae were all x-radiographed to check for the presence of Harris Lines. A few, very faint lines were evident at the distal end of all the Neolithic tibiae, ranging from *c.* 9 in skeleton B, to 2–4 in skeleton C. None were noted in the Bronze Age skeleton. These lines, indicators of periods of arrested growth, may form in response to various factors e.g. illness or nutrition stress; those observed here do not indicate any significant detrimental incidents.

Trauma: Changes on the spine and supraspineous fossa of the left scapula from the Bronze Age skeleton E appear to have resulted from direct trauma to the bone. The lateral end of the spine is absent (50 mm), the central — 19 mm —remnant being remodelled with smooth, well-organized bone, except on very medial border where it is slightly uneven (*rhomboid minor* attachment). The superior portion of the medial border is angled slightly more acutely to normal and x-radiograph shows an area of translucency across the angle between the medial and superior borders, crossing the remnant base of the spine, which may be indicative of a healed fracture. A series of smooth-margined, sub-circular lesions are located in the supraspineous fossa; two medial lesions extend from the superior angle to the spine (max. 30.9 × 20 mm); two, smaller lateral lesions at the base of the coronoid process; a further small lesion, evident from the x-radiograph, being adjacent to the medial angle. The *rhomboid minor* attachment is located at the medial end of the spine, and the point of fracture corresponds with the medial end of the *trapezius* attachment.

Fractures to the scapula are not common and generally result from direct violence (Adams 1987). The type of fracture indicated here does not fall in to any of the categories shown by Adams (1986, fig. 111). A direct blow to the spine of the scapula would have to be both forceful, to overcome the cushioning effect of the mass of protective soft tissue in this area, and well-directed not to affect any other bones. The supraspineous area may have been affected via transmitted stress from a blow, the force of which appears to have been taken by the superior-medial part of the scapula, fracturing the bone from the inferior-medial end of the spine to the mid-point on the superior border. The efficient functioning of the shoulder muscles must have been affected as a result of this trauma.

Infection: The posterior shaft at the dorsal end of the left 2nd distal phalanx from skeleton C was slightly thickened with raised, uneven contours. The lesions are indicative of an infection, possibly osteomylitis.

Tumours: A slightly raised area of bone was observed in the slightly damaged exocranial surface of the right dorsal parietal from skeleton A, forming a slight 'boss' *c.* 11 mm in diameter. X-radiographs of the skull show a *c.* 25 mm diameter area of translucency comprised of several small destructive lesions (maximum 4.4 mm) within the medullar, predominantly towards the dorsal side of the affected area. A similar, but less pronounced area of translucency, *c.* 26.3 mm diameter, was noted crossing the coronal suture between the anterior right parietal and the frontal bone. There is no associated new bone formation. The rest of the skeleton was also x-radiographed, and no associated or similar lesions were apparent.

The lesions are suggestive of a malignant bone tumour of some form. Although secondary tumours are more common than primary tumours of the bone, the former generally occur in older adults (Adams 1986, 76). In some respects — location, size, destructive nature — the lesions are similar to those seen in (primary) myeloma (Adams 1986, 75; Roberts & Manchester 1997, 191). Although solitary lesions have been recorded, multiple lesions are more usual (Roberts & Manchester 1997, 191) and more than one area of the skeleton is often affected (Adams 1987). Similar lesions may result due to a metastatic (secondary) tumour, Ortner & Putscher (1985, 394) stating that 'In the cranial vault, the most marked destruction is in the dilpoë, while the outer and inner tables are long spared'. With both types of tumour, the lack of associated lesions elsewhere in skeleton A, but particularly, the age of the individual, make such diagnoses debatable. If the suggested diagnosis of malignant tumour is correct, this would prove to be one of the earliest manifestations of the condition as yet recorded (Roberts & Manchester 1997, 192; Ortner & Putscher 1985, 395–8).

Joint disease: Degenerative disc disease (Rogers & Waldron 1995) is largely related to age and reflects 'wear-and-tear', being concentrated in the area of the vertebral body most affected by stress. Lesions were limited to the lower thoracic in the female (9%), being more dispersed in the male (22%). Schmorl's nodes (Rogers & Waldron 1995) were noted only in the male skeleton (9%).

The aetiology of osteoarthritis is complicated, including the effects of age, mechanical alteration through activity or injury and genetic predisposition (Rogers & Waldron 1995). Slight-mild lesions were noted at five sites in the female and three in the male skeleton. Osteophytes (irregular growths of new bone along joint margins) may occur alone, the prevalence increasing with age, or in association with other lesions where they may be indicative of disease such as osteoarthritis (Rogers & Waldron 1995). Lesions were noted at 15 sites in the Neolithic female and 22 sites in the Bronze Age male skeleton, in both cases lesions were predominantly in the vertebral body surface margins and costo-vertebral facets, but involving more extra-spinal sites in the male. Pitting and other destructive lesions may develop in response to a number of conditions and it is not always possible to ascertain the specific cause of individual lesions (Rogers & Waldron 1995). Most of these lesions were seen in joint surfaces, with others peri- or juxta-articular and are most likely to represent the early stages of joint disease. Similarly it is not always possible to be conclusive with respect to the causes of exostoses (bony growths at tendon and ligament insertions on the bone) which include advancing age, traumatic stress, or various diseases. Lesions were noted at only one site in the female and nine sites in the male.

The rates of joint lesions were not high for either adult, however, the significance of this observation, and the differences between the two individuals in sites and rates of occurrence, are debatable. Are the latter indicative of the difference between Neolithic and Bronze Age, male and female, or both? With single instances of this nature there is a limit to what may be implied, since any observation may be significant simply to one or other individual rather than the population from which they were drawn. It must suffice to say that neither individual appears to have suffered undue physical stress —other than that implied by the fracture to the left scapula of the male.

Morphological variations represent variations in the skeletal morphology and may, with other predisposing factors, indicate genetic relationships within a 'population' (Berry & Berry 1967). There are, however, problems both with the applicability of recording methods and the uncertain heritability of traits (Tyrrell pers. comm.). Some traits have been attributed to developmental abnormalities, for instance, extra sutural ossicles (Brothwell 1972, 95–8) and 'squatting' facets', the latter most often attributed to being a response to prolonged periods spent in a squatting posture (Brothwell 1972, 92; Molleson 1993, 156).

Discussion

Much of the Neolithic human bone from Britain has been recovered as disarticulated bones and fragments, commonly originating from long barrows or causewayed enclosures and discrete graves containing *in situ* burials are relatively rare (Kinnes 1979, 126; Pryor 1984; Barclay & Halpin 1998a, 31–3). The location of this particular grave, and the composition of the burial within it — all four individuals apparently deposited as part of a single event — cannot but suggest an act of more ritual significance than that attendant simply on the burial of the dead. A specific link between the individuals is implied —given the composition of ages, one would expect at least one of the juveniles to be the child of the adult female, and indeed one of was confirmed as the child of the adult (J. Montgomery pers. comm.). It should not be assumed, however, that these individuals necessarily represent 'sacrificial victims', as it is possible that all succumbed to an infectious disease at the same time.

Disarticulated bones fragments from F.1 and the shaft

Miscellaneous fragments from the shaft. All shown on same Skeleton Sheet.

NWQL3 (S2)
From layer 3: Shown on same record sheet as those from shaft F.1 due to lack of context data in recording leading to assumption that all was from there.
SKULL: Small molar.
AGE: older subadult/young adult.
CONDITION: Heavily root marked and worn.

SEQL7A
UPPER LIMB: Phalange.
1.50 slice, 1.50 from the centre rail.
COMMENT: adult bone. No pathology. Fresh appearance but root marked.

L7C
SKULL: Vault fragments including thick frontal. Endocranial sutures fused, exocranial fusing.
AGE: adult bone.
CONDITION: Fresh appearance. Fresh breaks with no joins.

SEQ L7A
SKULL: Vault; upper sutures open.
AGE: adult bone.
CONDITION: Good.

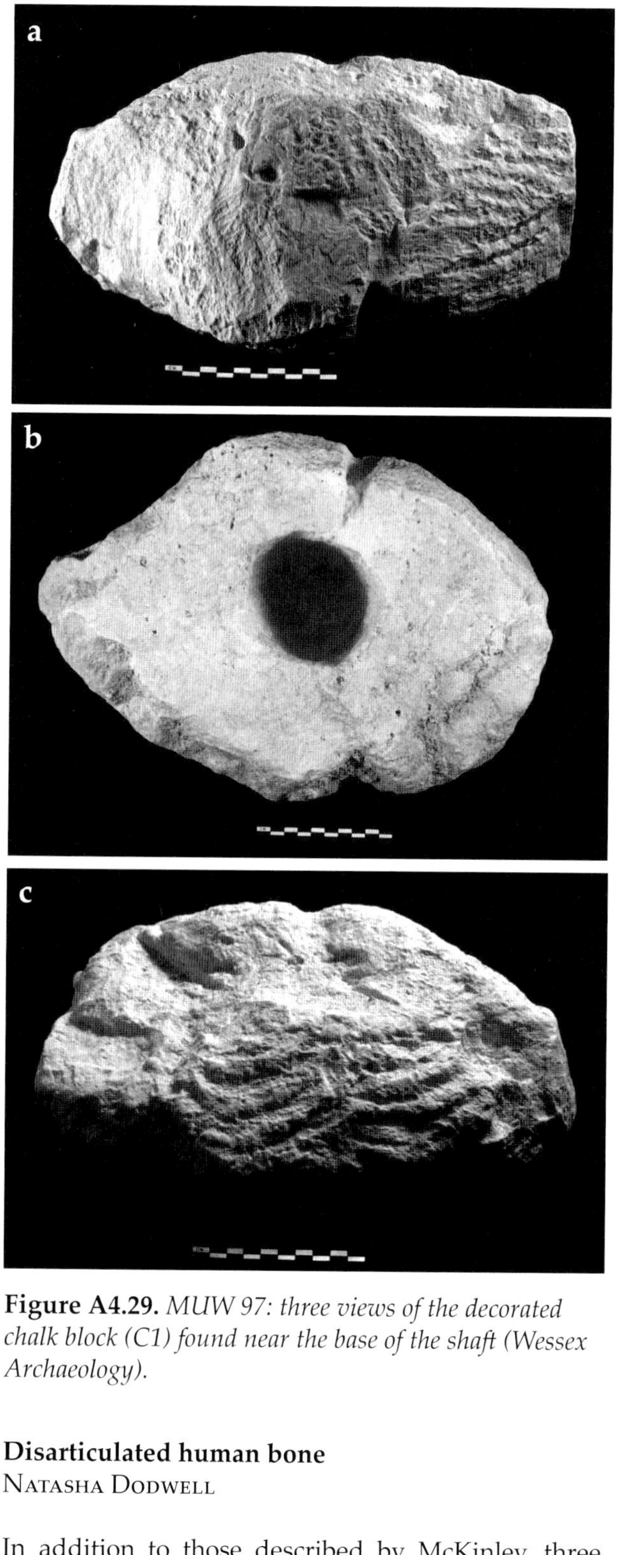

Figure A4.29. *MUW 97: three views of the decorated chalk block (C1) found near the base of the shaft (Wessex Archaeology).*

Disarticulated human bone

Natasha Dodwell

In addition to those described by McKinley, three further fragments of disarticulated human bone were recovered from layers 7a and b of the fill of the large central shaft F.1. They were examined macroscopically and recorded using methods outlined by Knüsel & Outram (2004) and McKinley (2004b). The assemblage consists of the mid shaft of a right radius and the lateral third of a left clavicle whose articular surface is missing (from F.1, shaft layer 7a) and the distal half of a proximal phalange (hand/foot) from F.1 layer 7b. The bones are white, presumably the result of their chalky burial environment, the cortical bone is well-preserved and some slight brown veining, the result of root action was noted on the radius shaft. The breaks at the ends of the bones are ancient but it is not possible to determine whether they are intentional or not. They are all adult sized and represent the remains of a minimum of one individual.

The decorated chalk object

Richard Bradley

The object from the base of the shaft at Monkton-up-Wimborne is most unusual. It consists of a large, irregular block of quarried chalk measuring 34 cm by 25 cm in maximum dimensions and is 20 cm thick (Fig. A4.29). It has a roughly flat underside with a central hole between 8 and 9 cm across (A4.29b). This comes to a rounded base at a depth of 10 cm. The interior of the hole is smooth and it seems possible that it has provided the socket for a post or 'handle'.

The other side of the stone is markedly irregular and its upper surface may have flaked away. Around the edges and on the sides, however, there are clear signs of decoration and other treatment. This takes five distinct forms:

1. In some cases the surface of the chalk seems to have smoothed and/or scraped. This happens in two main areas. These are adjacent to one another, but the orientation of the scratches is different.
2. Beside one of these smoothed areas is another zone where the surface of the stone seems to have been prepared by pecking. The individual peck marks are quite widely spaced, creating a dimpled effect. Under strong lighting it appears that this technique may have been employed more widely.
3. One edge of the stone is emphasized by two sets of parallel lines or grooves (Fig. A4.29a). These are on either side of a distinct ridge and are laid out at an acute angles to one another. On one side there are three wide grooves, and, on the other, are seven more which are significantly narrower. Under strong lighting it seems possible that this scheme was continued by at least four fainter lines, little trace of which remains.
4. A similar scheme is found elsewhere on the surface of the stone. Here six parallel grooves occur in a compact area, supplemented by two or three others extending from them at an acute angle. All these grooves are quite wide and are emphasized by prominent peck marks.

5. Finally on one edge of the stone there are two sets of nested arcs (Fig. A4.29c), again characterized by wide grooves formed by pecking. One consists of three such grooves and the other of three or, possibly, four. Both were conceived as a single design and the two groups are joined together.

These observations suggest that originally the stone had a rather domed profile and a flat base which was perhaps supported on a post. The top part of the stone has probably been lost, but enough remains to suggest that the upper surface had been deliberately shaped by pecking, scraping and/or deliberate smoothing. The decorative scheme may originally have extended over a larger area than it does today. Had the object been supported on an upright post that decoration might have covered the entire area that was visible.

This object does not have any parallel in southern England, but that is perhaps not surprising since chalk carvings would have little chance of surviving unless they were protected soon after they were made. In that respect they can be compared with the scratched designs found on the face of a ditch at the Neolithic enclosure of Flagstones (Woodward 1988).

The designs at Flagstones have been compared with those in Breton megalithic art, but the object at Monkton-up-Wimborne has more features in common with decorated monuments in Britain and Ireland. The clearest comparisons are with Irish megalithic art, which was made by a similar technique (Shee Twohig 1981, chap. 4; O'Sullivan 2006). Indeed, the wide pecked grooves are very like those found on well preserved monuments. The suites of parallel lines certainly feature on Irish passage tombs, but it is the nested arcs that are more diagnostic. That comparison is particularly relevant since the radiocarbon dates from the Dorset shaft are similar to those from these Irish monuments which O'Sullivan (2006) assigns to the period between about 3400 and 3000 BC.

This remarkable find is a reminder that the distribution of decorated stonework in the British and Irish Neolithic may be the result of differential preservation and that certain motifs and decorative schemes may have been much more prevalent than appears today. It also suggests long-distance contacts between Ireland and Wessex, two of the regions with the most complex and varied monuments in Neolithic Europe.

Stone objects

Martin Green

Catalogue of the stone objects found in the shaft and associated pits, with comments from the Fir Tree Field shaft site pertaining (see Green in App. 4.2 above) (Fig. A4.23, nos. S30 & S33, remainder unillustrated).

S1. F.1, NEQ, L2, S1: Small lump of coarse-grained quartz-rich sandstone with one smooth face with evidence of wear from use as a rubber or quern. Wt 37 g.

S2. F.1, SWQ, L2, S1: As above. Wt 6 g.

S3. F.1, SEQ, L2, S1: Fragment of a whetstone of plano-convex section, made from a fine-grained dark brown sandstone. Wt 18 g.

S4 & S5. F1, SEQ, L2, S1: Two small lumps of a hard coarse-grained quartz-rich medium brown sandstone. No visible signs of use. Wt 29 g and 25 g.

S6. F.1, NWQ, L2, S2: Small natural lump of soft coarse-grained heathstone. Wt 37 g.

S7. F.1, NWQ, L2, S1: Fragment of pinkish-grey coarse-grained quartz-rich sandstone with no visible signs of use. Wt 30 g.

S8. F.1, SEQ, L2, S3: Small natural lump of soft coarse-grained heathstone. Wt 6 g.

S9. F.1, SEQ, L2, S5: As for S8 above. Wt 4 g.

S10. F.1, SEQ, L3. As for S8 above. Wt 56 g.

S11 & S12. F.1, NEQ, L3: Two lumps similar to S8 above. Wt 92 and 19 g.

S13 & S14. F.1, SEQ, L3: Lump of fine-grained dark brown heathstone (126 g) and a smaller lump of coarse-grained heathstone. Wt 38 g.

S15. SEQ, L3: Broken natural tabular lump of hard medium-grained heathstone. Wt 310 g.

S16. SEQ, L4: Partly rounded lump of coarse-grained pinkish micaceous sandstone. Wt 94 g.

S17 & S18. NWQ, L4: Two lumps of soft coarse-grained hathstone. Wt 97 and 4 g.

S19. NWQ, L4: Fragment of an Upper Green sandstone quern or rubber. Wt 155 g.

S20. NWQ, L6: Large lump of dark brown iron-rich sandstone. Angular fresh breaks suggest the piece has been knapped. Wt 499 g.

S21. NWQ, L6: Tabular fragment (17 mm thick) of coarse biscuit-coloured sandstone with one face smoothed from use as a quern or rubber. Wt 144 g.

S22. NWQ, L7: Small hard natural lump of iron-rich sandstone. Wt 43 g.

S23. Shaft L7a: Large flint pebble with a number of natural depressions derived from Tertiary deposits. Part of a structured deposit within layer 7a. Wt 1226 g.

S24. Shaft L7c: Angular lump of off-white siliceous sarsen. A broken fresh surface at on end suggests it may have been knapped. Part of a structured deposit within layer 7c. Wt 354 g.

S25. Shaft L10: Fragment of medium-grained iron-rich sandstone broken from a larger lump. Wt 86 g.

S26. Shaft L10a, at 3.8 m: Large broken flint pebble derived from Tertiary deposits. Wt 283 g.

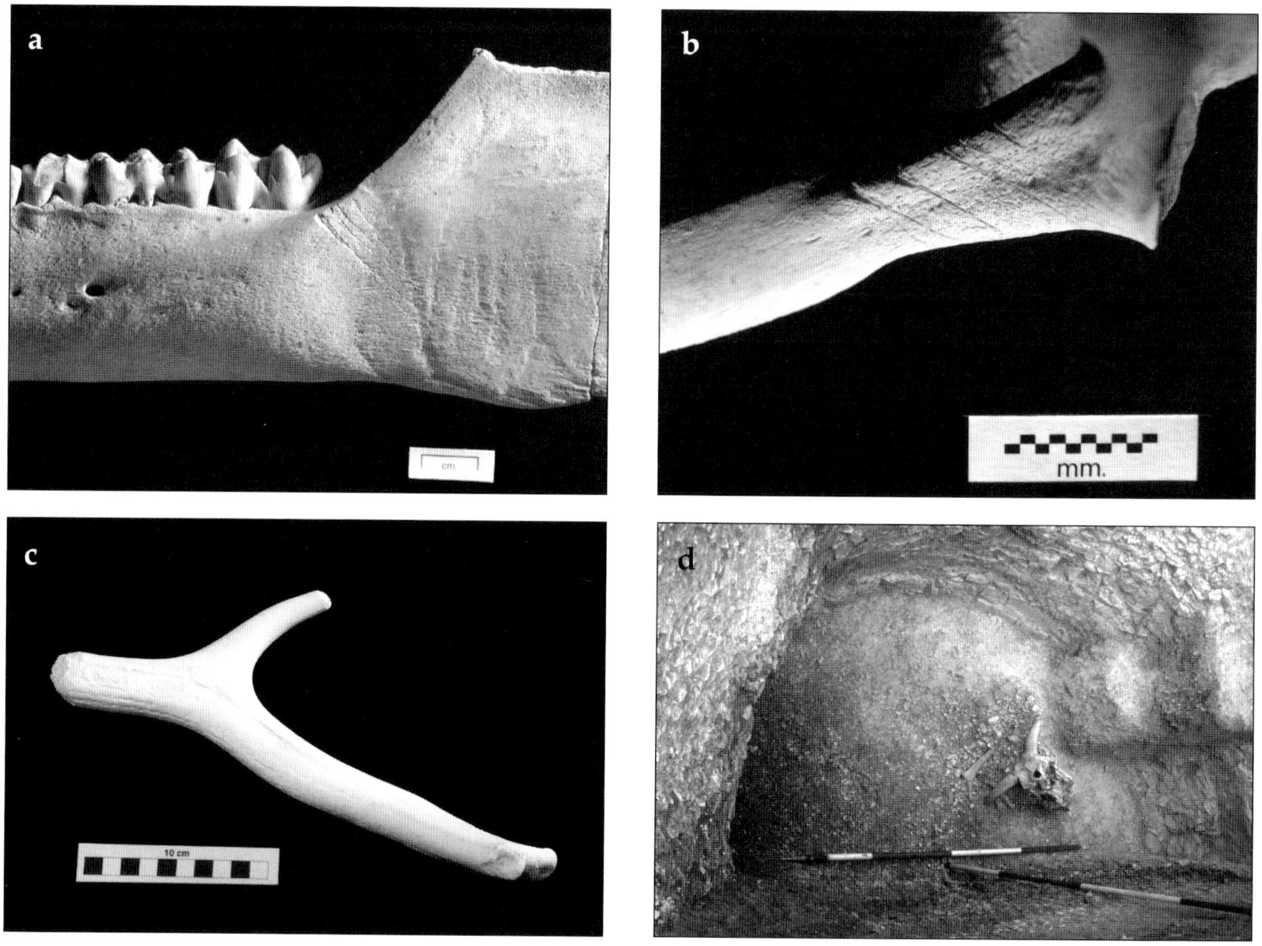

Figure A4.30. *MUW 97: a) & b) Pig bones from the base of the shaft bearing clear butchery marks from a flint knife; c) red deer antler from the shaft; d) the cow skull and associated red deer antler* in situ *in the shaft (D. Cousins & M. Green).*

S27. Shaft L10b: Natural flint pebble derived from Tertiary deposits. Wt 236 g.

S28 & S29. Base of the shaft, L10d: Iron-stained lump of gravel flint (482 g) and flint pebble derived from tertiary deposits. Wt 253 g.

S30. Lump of coarse-grained iron-rich heathstone pecked into a rough ball (Fig. 176, no. 15). Found in base of the shaft with animal vertebra. Note a similar specimen was found in the Fir Tree Field shaft (see S5, above this volume). Wt 418 g.

S31 & S32. F.12, L2: Two small lumps of fine-grained iron-rich heathstone. Wt 25 and 3 g.

S33. F.12, L3: Knapped lump or core of coarse-grained Upper Greensand chert. One surviving small smoothed area suggests it had originally been part of a quern or rubber (Fig. 175, no. S2).

S34. F.15, L2: Small lump of fine-grained iron-rich heathstone. Wt 6 g.

S35. F.2, L2: Small lump of Upper Green sandstone. Wt 19 g.

4.10. Monkton-up-Wimborne pond barrow (MUW 02)

Lithics

Martin Green

The excavation of the pond barrow and associated features produced over 6000 worked flints (Table A4.48; Figs. 3.68–3.73 & A4.31), the vast majority of which were found within the layers infilling the excavated hollow of the barrow. The remainder were found in a few features external to the barrow and from the two cremation pits found within (Fig. 3.75). These will be discussed in turn.

F.3 and F.4

These two shallow pits to the east pre-dated the barrow and proved to be Beaker in date with F.3 producing a decorated sherd (Fig. A4.32, no. 1). The

Table A4.48. *Monkton-up-Wimborne pond barrow (MUW 02): flints.*

Feature	Layer	Cores	Core frags.	Flakes	Chips	Retouched	Utilized	Scrapers	Other tools
F.1 NEQ	2	8	3	331	214	4	4	4	borer, hammer, saw
F.1 NWQ	2	29	6	367	225	8	3	7	knife
F.1 SEQ	2	13	6	443 (1 pc)	241	10	5	10	fabricator, denticulate
F.1 SWQ	2	7	1	219	112	5	2	3	borer
Total		**57**	**16**	**1360**	**792**	**27**	**14**	**24**	**7**
F.1 NEQ	3	92	18	630	237	11	-	17	ch. arrow, 2 hammers, Y shape
F.1 NWQ	3	27	8	254	140	2	1	3	2 borers, 2 fabricators
F.1 SEQ	3	51	12	388	167	13	9	10	denticulate, hammer, Y shape
F.1 SWQ	3	33	11	209	69	1	3	5	part polished axe frag.
Total		**203**	**49**	**1481**	**613**	**27**	**13**	**35**	**12**
F.1 NEQ	4	16	6	287	167	-	-	2	-
F.1 NWQ	4	7	2	200	123	3	-	2	-
F.1 SEQ	4	6	1	160	87 (1 pc)	1	2	5	hammerstone
F.1 SWQ	4	2	2	81	29	2	1	1	-
Total		**31**	**11**	**728**	**406**	**6**	**3**	**10**	**1**
F.3	2	-	-	17	-	-	-	-	-
F.4	2	-	-	9	2	-	-	1	-
F.5	2	5	1	8	3	-	-	-	1
F.7		-	-	1	3	-	-	-	burnt fabricator
F.8	3	2	2	9	39	-	-	-	-
F.8	4	8	2	12	27	1	-	1	-

former only produced 18 worked flints and the latter 12. Amongst the group from F.3 there was a finely retouched knife/scraper (Fig. A4.31, no. 1) and from F.4 a thumb scraper produced on an irregular flake struck from a gravel flint nodule.

F.5

F.5 is one of a pair of postholes forming a possible entrance on the eastern edge of the barrow. This posthole produced 19 worked flints including a small burnt convex scraper and a hammerstone.

F.7

The primary cremation pit located near the centre of the barrow produced only five pieces of worked flint. However, amongst the cremated mass of bone in layer 5 was a heavily calcined fabricator of 'D'-shaped section (75 × 16 × 14 mm). Parts of both sides exhibit areas of crushing from use although no characteristic use polish is visible at either end. The bulbar surface, however, does bear traces of use polish. This tool is clearly a 'pyre good' and was associated with a radiocarbon date of 1890–1610 cal. BC.

F.8

This funnel-shaped cremation pit produced 103 worked flints including unusually a disc core in layer 2, which is most likely residual, and a scraper in layer 4.

F.1 pond hollow

The final accumulation of the pond hollow (layer 2) comprised a decalcified dark brown silt and was associated with mainly Middle Bronze Age ceramics (see Green below). A total assemblage of 2299 worked flints were recovered from this layer which included a retouched, utilized, tool (RUT) component of 3.2% (Table A4.48). Amongst this percentage, a third (24) were scrapers. As suggested by the pottery, this layer contains a mix of material from the later Neolithic to Middle Bronze Age in date. Twenty-one of the scrapers were of convex form, two end and a thumb scraper. Only three of the convex scrapers were of the roughly retouched type more characteristic of Middle Bronze Age industries in the area. Amongst the remaining RUT element, a serrated flake and a blade knife (81 × 30 × 10 mm) are likely to be later Neolithic with the two fabricators and three borers probably of the Early Bronze Age.

Within the carefully laid flint platform (layer 3) of the pond hollow, much animal bone, Middle Bronze Age pottery and 2433 worked flints were deposited. The latter amount included an RUT element of 3.6% of which 35 (39%) were scrapers. Fourteen of these, including four made from frost shattered natural 'flakes,' were of roughly retouched type more likely to be of Middle Bronze Age date with the remainder of standard convex form with a single thumb scraper. Amongst the remaining RUT element were several

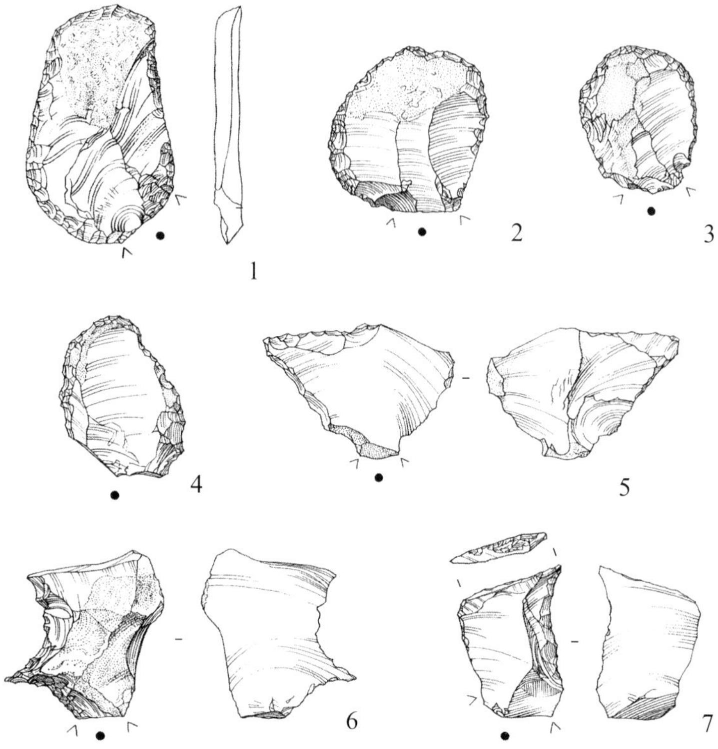

Figure A4.31. *MUW 02 pond barrow: flints. Scale 1:2 (J. Richards).*

likely to be residual including two 'Y'-shaped tools, a heavily reflaked portion of a part polished axe that has been reused as a core with opposed platforms, a chisel arrowhead, two borers and two fabricators.

In layer 4, the chalky wash primary infill of the pond hollow, there were 1199 worked flints and sherds of mainly Early Bronze Age pottery. Amongst the flints there was an RUT element of 1.9% of which ten (43%) were scrapers (Fig. A4.31, nos. 2–4). Three small flake borers (Fig. A4.31, nos. 5–7) were also found and are likely to be of contemporary Early Bronze Age date.

Catalogue of illustrated flint (Fig. A4.31)

1. F.3, Beaker pit: Knife/scraper. Carefully retouched around almost the entire perimeter.

2. F.1, L4: Convex scraper. Neatly retouched around most of perimeter. Bulbar end snapped off.

3. F.1, L4: Convex scraper. Neatly retouched around perimeter.

4. F.1, L4: Convex scraper. Neatly retouched around perimeter.

5. F.1, L4: Borer, inversely retouched from two angles.

6. F.1, L4: Borer, retouched from two angles to form point on left side.

7. F.1, L4: Borer, retouched from two angles to form point at top of right edge.

Prehistoric and Romano-British pottery

Martin Green

Significant numbers of sherds were recovered from the fills of the pond barrow which range in date from Beaker to Romano-British in date. Furthermore, two collared urns containing cremations were discovered in pits F.2 and F.8 (Fig. A4.32, nos. 5 & 6), and a single sherd of Beaker (Fig. A4.32, no. 1) from the outlying pit F.3.

17 sherds (144 g) of Romano-British coarse wares were recovered from the upper fill of F.1, layer 2. A further 4 abraded sherds (20 g) came from the layer 2 interface with the flint platform (layer 3). An area of Romano-British settlement is situated some 200 m to the east at SU 01851505.

137 sherds of Middle Bronze Age pottery (788 g) were also recovered from the upper fill of F.1, layer 2. A further 52 sherds (321 g) came from the flint platform (layer 3) and just 2 sherds (6 g) from the flint platform/layer 4 interface.

5 sherds of Early Bronze Age pottery (85 g) were recovered from layer 2 of F.1, with 7 sherds (180 g) from layer 3, and 10 sherds (237 g) from the layer 3/4 interface, and 4 sherds (24 g) from layer 4. Finally, 4 small abraded sherds of Beaker (4 g) were recovered in a residual context within layer 2.

Catalogue of illustrated pottery (Fig. A4.32)

Early Bronze Age

P1. F.3: Beaker body sherd.

P2. F.1, NEQ, L2: Beaker body sherd.

P3. F.1, SEQ, L2: Beaker body sherd.

P4. F.1, SEQ, L2: Beaker body sherd (with additional small body sherd with traces of two horizontal rows of cord impressions with a short diagonal impression in between (not illustrated)).

P5. F.2 cremation: Finely potted thin-walled collared urn with base remove by ploughing. Found inverted in pit outside the eastern edge of the pond barrow F.1. Twisted cord hurdle pattern on collar and short diagonal slash marks on top of rim. The shoulder bears stab marks in groups of either two or three. This latter decoration is a most unusual trait.

P6. F.8 cremation: Plain collared urn. Found inverted in funnel-shaped pit within the pond barrow F.1. Similar profile to another plain-collared urn found at Christchurch and recorded by Longworth (1984). Of coarser manufacture than P5 above.

P7. F.1, NEQ, L3: ?Food vessel. Body sherd decorated with row of short vertical grooves. Light brown exterior, black interior, groggy fabric.

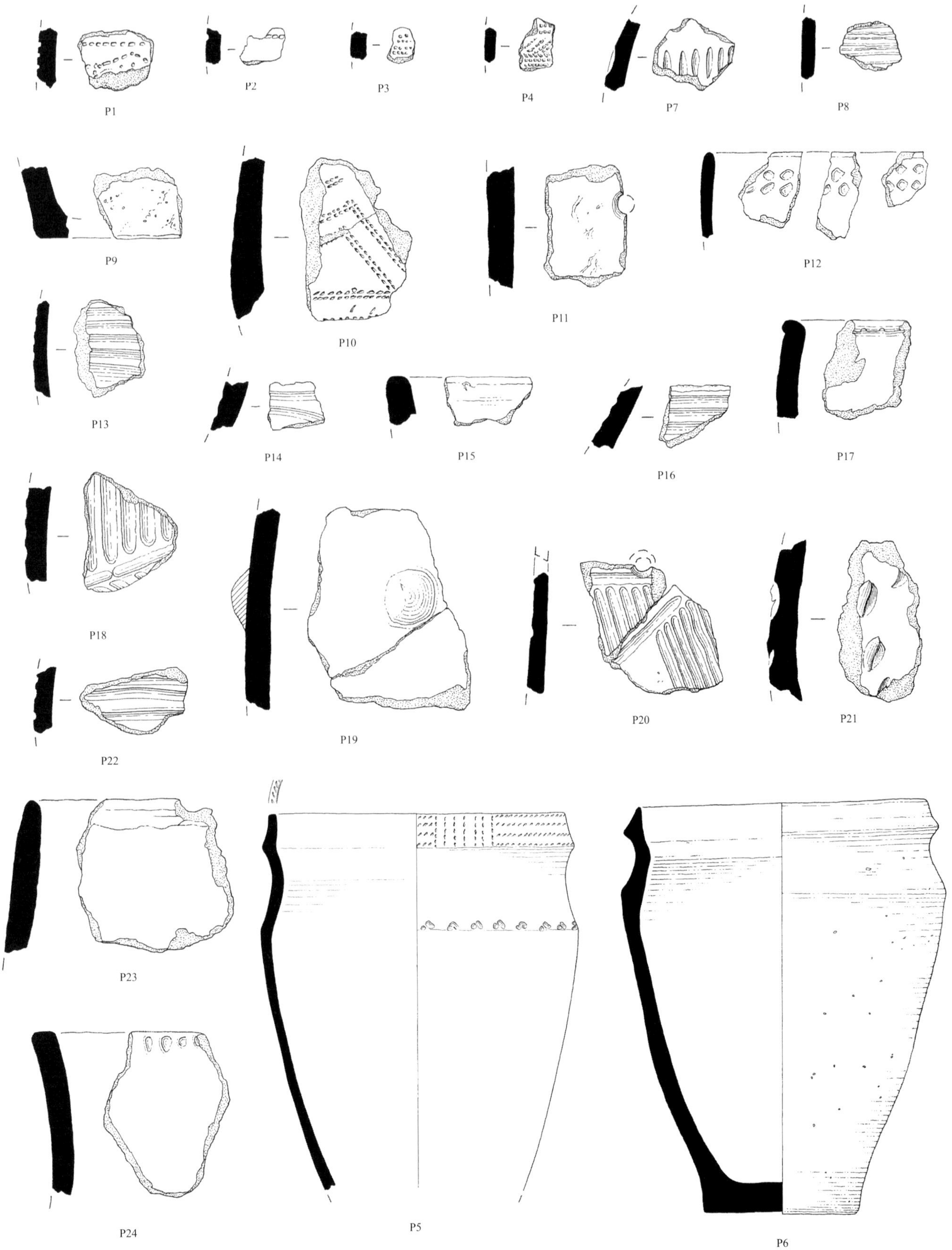

Figure A4.32. *MUW 02 pond barrow: the Beaker and Bronze Age sherds, and the collared urns from F.2 and F.8. Scales P1–P24 1:2, P5–P6 1:4 (N. Griffiths).*

P8. F.1, NEQ, L2: ?Food vessel. Body sherd decorated with a series of narrowly spaced grooves. Dark brown exterior, black interior, groggy fabric.

P9. F.1, SWQ, L4: EBA urn. 2 basal sherds (only one illustrated), dark brown groggy fabric with soapy texture.

P10. F.1, SEQ, L3: Collared urn. Body sherd (probably from collar) decorated with double rows of twisted-cord impressions. Some internal smearing visible. Dark brown groggy fabric.

P11. F.1, NEQ, L4: EBA urn. Body sherd with part of a perforation. Light brown groggy fabric.

Middle Bronze Age

P12. F.1, NEQ, L3: MBA cup. 3 rim sherds decorated with two rows of shallow impressions immediately below the rim and 5 body sherds (not illustrated) from the same vessel. Fine black fabric throughout with occasional flint grits.

P13. F.1, SWQ, L2: Globular urn. Worn body sherd decorated with four horizontal shallow grooves. Mid-brown exterior, black interior with fine flint grit.

P14. F.1, SEQ, L2: Globular urn. Body sherd decorated with two shallow grooves, the lower one gently curving. Light buff fabric with fine flint grit.

P15. F.1, SEQ, L2: ?Bucket urn. Rim sherd with slightly convex profile. Dark brown exterior, black interior with a few large coarse flint grits.

P16. F.1, NWQ, L2: Globular urn. Body sherd with three shallow horizontal grooves. Black fabric throughout with fine flint grit.

P17. MBA jar. Rim sherd of everted profile. Mid-brown exterior, light brown interior with shell temper and soapy texture. This vessel is similar in form to that recovered from the ring-ditch at MUW 97 (see Fig. 178, P21).

P18. F.1, NWQ, L2: Globular urn. Body sherd zonally decorated with a series of shallow grooves. Black hard fabric with much fine flint grit.

P19. F.1, NWQ, L3: ?Globular urn. Body sherd with vestigial lug. Dark brown/black exterior, black interior with much fine flint grit.

P20. F.1, SEQ, L2: Globular urn. Body sherd zonally decorated with a series of shallow grooves and with remains of a perforation. Dark brown/black exterior, black interior. Similar to P18 (above) but a finer-quality vessel.

P21. F.1, NEQ, L2: Bucket urn. Body sherd with fingernail decoration. Dark brown exterior, black interior with coarse flint grits.

P22. F.1, SEQ, L2: Globular urn. Body sherd decorated with shallow horizontal grooving. Buff fabric throughout. Similar to P14 (above).

P23. F.1, NEQ, L2: ?Bucket urn. Rim sherd bearing scar of former applied everted rim. Dark brown/black fabric with fine flint grit.

P24. F.1, NEQ, L2: ?Bucket urn. Flat-topped rim sherd with shallow impressions immediately below rim. Black fabric with fine shell and a little flint temper.

Animal bones

MARK MALTBY *with* VICTORIA FORD & KATE MASON

The small sample of animal bones that was retrieved from the excavation was recorded onto a computer database, and the reference collection housed in the School of Conservation Sciences, Bournemouth University, was used to identify bones to species using the following criteria (where appropriate):

- context;
- species;
- element;
- quantity;
- zones present;
- percentage of surviving bone;
- fusion data;
- tooth-aging data;
- measurements;
- porosity;
- condition;
- side of body;
- and any other relevant comments were recorded.

Preservation

The preservation of the bones was very poor and nearly all the bones and teeth were heavily eroded and weathered. The severity of the taphonomic decay was such that any butchery marks will have been obliterated. There were two burnt bone fragments. No records of animal gnawing were recorded but again the severe erosion is likely to have destroyed any evidence.

The assemblage was extremely fragmentary. This is reflected by the high proportion of unidentified fragments, comprising 72% of the total sample, and also in the generally small size of the surviving bone in the identifiable material. Very few (6%) of the identifiable bones survived complete in comparison to 45% of the bones consisting of 10% or less of the complete bone. There were a large number of bones that displayed evidence of modern breaks, some of which had been glued back together.

Species identified

The following species were identified:

- cattle;
- sheep/goat;
- pig;
- dog;
- red deer;
- mole;
- small mammal;
- frog/toad.

Only cattle and sheep/goat were found in appreciable numbers (Tables A4.49 & A4.50).

Faunal remains from layers 2–4 were recorded by quadrant and the total from each of these is given in Table A4.49. The majority of the bones (67%) were found in layer 3, 29% in layer 2, whereas only a small number were found in layer 4. With regard to quadrants, half the bones were found in the northeast quadrant (NEQ), 21% in the southeast quadrant (SEQ), 19% in the northwest quadrant (NWQ) and 10% in the southwest quadrant (SWQ). It appears therefore that more bone survives in the eastern half than the western. Although there were some differences in the relative abundance of cattle and sheep/goat between quadrants and layers there was no consistent pattern in this variation as such variations are as likely to reflect differential preservation of bones as they are any variation in the disposal practices.

The cattle sample

Overall cattle were the most common species represented in the assemblage, comprising in total 137 specimens (51% of the identified sample), the total included 57 (42%) loose teeth (Tables A4.49 & A4.50). The high incidence of teeth is indicative of poor preservation. The most common bone elements (radius, tibia and humerus) are all relatively dense elements and would be expected to survive in such conditions better than more fragile bones such as vertebrae and ribs. It is possibly significant that other dense elements such as the metapodials and mandible are less commonly represented, perhaps indicating a preference for the disposal of upper limb bones. However, the presence of large numbers of loose teeth indicates

Table A4.49. *Animal bone elements from the MUW 02 pond barrow.*

Element	Cattle	Sheep/ Goat	Pig	Dog	Red deer	Mole	Small mammal	Frog/ Toad	Total
Horn/Antler	1				1				**2**
Maxilla inc. MXG		1							**1**
Skull frag.	2	1							**3**
Mandible inc. MNG	9	2	2						**13**
Loose teeth	57	95	5	1	1		1		**160**
Humerus	11	1							**12**
Radius	13	5				1			**19**
Ulna	2								**2**
Scapula	2								**2**
Pelvis	4							1	**5**
Femur	5	1							**6**
Patella									
Tibia	12	13							**25**
Fibula									
Carpals	1								**1**
Astragalus	4								**4**
Calcaneus	1								**1**
Centroquartal									
Metacarpal	3	1							**4**
Metatarsal	5	4							**9**
Metapodial	1								**1**
Lat. metapodial									
1st phalanx	2								**2**
2nd phalanx									
3rd phalanx	1								
Atlas									
Axis									**0**
Cervical vert.									
Thoracic vert.	1								
Lumbar vert.									
Sacrum									
Ribs									
Total	137	124	7	1	2	1	1	1	**274**

Table A4.50. *Animal species numbers present by context from the MUW 02 pond barrow.*

Context	Cattle	Sheep/ Goat	Pig	Dog	Red deer	Mole	Small mammal	Frog/ Toad	Unid.	Total
F1NEQL2	17	20	1		1				53	92
F1NEQL3	64	47	6			1	1	1	264	384
F1NEQL4	3				1				15	19
F1NWQL2	13	5							125	143
F1NWQL3	3	3							27	33
F1NWQL4		1							5	6
F1SEQL2	4	5							31	40
F1SEQL3	16	32		1					101	150
F1SEQL4	3	2							9	14
F1SWQL2	1								10	11
F1SWQL3	10	9							63	82
F1SWQL4	2								2	4
F8L3									3	3
F8L4	1								1	2
Total	**137**	**124**	**7**	**1**	**2**	**1**	**1**	**1**	**709**	**983**

that a much larger number of jaws were originally deposited than indicated by the number of mandibles and maxilla that have survived.

There was very limited ageing evidence. Three mandibles belonged to adult animals, two of which had heavy wear on the molars suggesting they belonged to quite elderly animals. Younger cattle were represented by teeth from two jaws. The first consisted of a mandible in which the second molar was in an early stage of wear. This animal probably died in its second year. Maxillary teeth of a calf under six months of age were also found. It is inevitable that the preservation conditions will have militated against the survival of bones of younger animals. It is notable that there are no unfused bones represented, as these are much more likely to have been destroyed than older fused specimens.

Measurements were possible on six cattle bones. This limited information indicated that the cattle represented belonged to relatively small domestic animals.

Sheep/goat sample
A total of 124 element of sheep/goat were identified representing 46% of the identified sample. Two teeth could be identified specifically as sheep. No elements of goat were positively identified. Contemporary ovicaprid assemblages usually contain a majority of sheep bones. The elements represented are dominated by loose teeth which comprise 77% of the sheep/goat assemblage. This indicates that sheep/goat bones were even more severely affected by the poor preservation conditions than cattle. Other elements of sheep/goat are restricted almost entirely to the densest parts of the major limb bones (Tables A4.49 & A4.50). Fusion data are restricted to the presence of two fused proximal radii. No jaws were found with surviving teeth. Some information about the ages of death can be obtained from loose teeth but this can be biased by problems of differential recovery and survival. Both adults and immature sheep were represented.

Other species
None of the other species were present in large numbers. Only one mandible and six teeth of pig were recovered and the teeth may all have belonged to one mandible in layer 3 (NEQ). Overall pig provided only 3% of the identified material, a very low figure more typical of Bronze Age rather than Later Neolithic samples in the region, though it should also be noted that pig bones do not survive well in these conditions.

Dog was represented by one maxillary tooth in layer 3 (SEQ), the size of the tooth indicating that it belonged to a dog rather than a wolf.

Red deer was represented by a fragment of antler in layer 4 (NEQ) and an upper premolar in layer 2 (NEQ).

The mole, frog/toad and small mammal bones may represent later intrusions into the deposits. These were all found in layer 3 (NEQ).

Absent from the assemblage were identified elements of horse, bird and fish. It is unlikely that bird and fish bones would have survived given the preservation conditions. Although domestic horse bones have been found in some Late Neolithic and Early and Middle Bronze Age assemblages, they rarely contribute a large part of the samples attributed to these periods.

Cremated human bone
JACQUELINE I. MCKINLEY

Cremated bone from two Early Bronze Age cremation graves (7 & 8) and the *in situ* remains of an urned burial of the same date from a third grave (2) was received for analysis. All three graves were associated with a pond barrow (feature 1), from the Middle Bronze Age fills of which redeposited cremated bone was recovered. Bone from a total of 15 contexts and sub-contexts was subject to analysis, together with the *in situ* fill of the urned burial.

Methods
The urned burial from grave 2 was excavated by the writer. The fill was removed in a series of nine 20 mm spits from the rim down (note that this vessel had been buried inverted), each spit being removed in quadrants with the exception of spit 9 which, being at the level of the broken base, contained very little bone and was subject to collapse. Annotated scale drawings were made at each level and digital photographs were taken. These additional sub-contexts were maintained throughout analysis to allow details of the burial formation process to be ascertained.

Osteological analysis followed the writer's standard procedure for the examination of cremated bone (McKinley 1994a, 5–21; 2000a). Age was assessed from the stage of skeletal and tooth development (Beek 1983; Scheuer & Black 2000), and the general degree of age-related changes to the bone (Buikstra & Ubelaker 1994). Sex was ascertained from the sexually dimorphic traits of the skeleton (Gejvall 1981; Buikstra & Ubelaker 1994).

Results
A summary of the results is presented in Table A4.51 and below (the detailed recording notes are held in the Down Farm Museum archive).

Disturbance and condition
Two of the three burials (from graves 7 & 8) were

undisturbed. The base of the inverted urned burial from grave 2 had been removed as a result of plough damage, but the vessel is largely intact and no bone loss or additional bone fragmentation is likely to have resulted from this limited damage.

The bone from the burials is generally in very good condition and trabecular bone — the first to be destroyed under conditions adverse to good bone survival (McKinley 1997a, 245; Nielsen-Marsh *et al.* 2000) — is well represented within the assemblage. A few bone fragments from spits 3 and 5/6 within the burial from grave 2 are either slightly root worn or slightly chalky in appearance; the root marking being confined to less well oxidized bone fragments (see below). Only one fragment of — probably animal — bone, from spit 1 could be described as eroded and it may be that this fragment was intrusive having originally derived from a different burial environment. The redeposited bone from feature 1 was all slightly eroded and root marked with a chalky appearance.

Demographic data

A minimum of three adults, covering a broad age range, were identified (Table A4.51). Two of the adults were confidently sexed as females, the third also possibly being female. The bone recovered from the final Middle Bronze Age silting within feature 1 was clearly redeposited. Fragments of redeposited Early and Middle Bronze Age pottery were also recovered from this layer and it is possible that the bone derived from a disturbed and redeposited burial. Alternatively, it may have originated from one of the same cremations as represented within one of the three identified burials, e.g. from a disturbed pyre site or deposit of pyre debris.

Pathology

Pathological lesions were observed in remains from all three burials, being most diverse, extensive and heavy in those of the elderly female from grave 8 (Table A4.51). Lesions indicative of osteoarthritis (Rogers *et al.* 1987; Rogers & Waldron 1995, 32–46) were observed in 12 spinal and non-spinal joints from grave 8. Degenerative disc disease — a condition resulting from the breakdown of the intervertebral disc largely related to age and reflecting 'wear-and-tear' (Rogers & Waldron 1995, 26–7) — was seen in six vertebral bodies from all areas of the spine but most commonly in the weight-bearing lumbar region where there was also anterior vertebral body collapse probably related to osteoporosis. Other lesions — marginal osteophytes (new bone) and pitting in articular surfaces — reflective of age related wear-and-tear or the early stages of joint disease were also seen in several joints. The impression gained of this elderly female is of a tough physical existence rendering the individual frail and subject to the various aches, pains and stiffness of old age, but living within a supportive community conscious of her increasing frailty.

Schmorl's nodes (destructive lesions resulting from a rupture in the intervertebral disc) were observed in four lumbar vertebrae of the young adult female from grave 2. Such lesions commonly reflect strain in the lower back from heavy lifting in the young individual.

Non-specific infection in the form of periosteal new bone was observed in a fragment of fibula shaft from grave 7. The form of the new bone indicated it was healing or healed.

Table A4.51. *Summary of analysis of cremated bone. (KEY: rpd = redeposited pyre debris; C = cervical; T = thoracic; L = lumbar; prox. = proximal. NB: Includes only those pyre goods observed by the osteologist.)*

Feature	Deposit type	Bone weight	Age/sex	Pathology	Pyre goods
1	redeposited	32.3 g	subadult/adult >13 yr		
2	urned burial	1033.3 g	adult *c.* 20–23 yr female	Schmorl's nodes — 4T; destructive lesion — 1T; morphological variation — 5 wormian bones	3 g animal bone
7	unurned burial + rpd	729 g	adult *c.* 35–50 yr ??female	*ante mortem* tooth loss?; dental caries; periosteal new bone — fibula; exostoses — iliac crest, patellae	fabricator
8	urned burial	1132.2 g	adult >50 yr female	periodontal disease; osteoarthritis — bi-lateral temporo-mandibular, C2, 3C, 1T, left wrist (scaphoid & lunate), distal femur & prox. tibia; degenerative disc disease — 2C, 1T, 3L; vertebral body collapse — 3L; pitting — medial clavicle, right patella, left prox. tibia; osteophytes — atlas, axis, L articular process, left scapula, right distal radius, distal finger phalanx, left patella, right prox. tibia, left talus; exostoses — right patella, left prox. tibia (lateral)	1.9 g ?animal bone

Pyre technology and cremation ritual

The majority of the bone from each burial is white in colour, indicative of full oxidation of the bone (Holden *et al.* 1995a,b). Variation in colour, illustrating incomplete oxidation, was, however, observed in bone fragments from each burial. The bone from grave 7 showed least variation with only one blue fragment of femur shaft. Most variation was seen in the bone from grave 2 where several fragments of radius and femur (though still a minority) were observed as blue throughout or in the interior surfaces; odd fragments of tibia, fibula, carpals and innominate were also blue. Where fragments could be sided (some radius, innominate and carpals), the right side was affected. A few bone fragments from grave 8 were also blue (right ulna and lunate) with more fragments showing the pre-white grey colouring including various facial bone (mostly the right side), clavicle and innominate fragments (both left), and femur. Numerous factors may be instrumental in affecting oxidation of the bone and some skeletal elements may be more prone to poorer oxidation due to their dense soft tissue coverage (femur) or potential peripheral position on the pyre (e.g. head and hands: McKinley 2004a, 293–5). The preferential siding of poorly oxidized fragments in two cases suggests something may have been cutting off/curtailing the oxygen supply to these parts of the corpse during cremation, e.g. fuel directly overlying bone or leather/skins covering parts of the body. Alternatively, a strongly veering wind may have shifted the heat of the pyre away from one or other side and/or end. Conversely, several fragments of skull vault and fragments of lower limb shaft from both graves 7 and 8 exhibited extensive warping, several vault fragments having almost curled round on themselves. This observation suggest a high temperature was maintained in at least parts of these pyres after exposure of the bone to the effects of dehydration.

The weight of bone from the burials ranges from 729 g to 1132.2 g, the lowest weight being recovered from the unurned burial. These weights are between *c.* 46% and 71% of the average expected weight of bone from an adult cremation (McKinley 1993). Both the range and the average of 964.8 g places the weights within the upper range recovered from Bronze Age burials and at the lower end of the consistently high range of weights recovered from primary barrow burials (902–2747 g, average 1525.7 g: McKinley 1997b). It has been suggested that the time and effort taken in collecting and burying such a substantial proportion of the bone in these primary deposits is reflective of the 'high regard' in which the individuals were held by their community, though the ritual significance of the mortuary deposit — irrespective of the individuals — was probably also a factor (McKinley 1997b).

The maximum size of bone fragment recorded is 149 mm from the burial in grave 8, which is commensurate with the size of fragments from modern crematoria prior to cremulation (McKinley 1993). The maximum fragments from the other two burials were 143 mm from grave 7 and 107 mm from grave 2. The maximum fragment from the redeposited material in feature 1 was substantially smaller at 45 mm. The largest fragments within the burial from grave 2 were generally found in the central area of the fill between spits 4–7. The majority of fragments from all deposits (*c.* 75–91%) were recovered from the 10 mm sieve fraction. A higher proportion of the bone from the lower spits (4–9) of the burial from grave 2 were recovered from the 10 mm sieve fraction than within the upper spits. There are a number of factors which may affect the size of cremated bone fragments the majority of which are exclusive of any deliberate human action other than that of cremation itself (McKinley 1994b). In this instance the lack of disturbance to the burials will have resulted in less fragmentation of this highly brittle material along the dehydration fissures formed in cremation whilst the bone was in the ground, maintaining a closer link to the size of fragments at the time of deposition.

Identifiable elements from all areas of the skeleton were included in each of the burials and there is no evidence to suggest the preferential recovery of certain elements for burial. The small bones of the hands and feet and fragments of tooth root were common in each of the burials (9–11 tooth roots, 25–49 carpal/tarsals, metacarpal/tarsals and phalanges). These elements are relatively common in Bronze Age burials and it has been suggested that this may reflect a collection procedure which included raking-off the upper levels of the burnt-out pyre to recover bones, rather than hand collection of individual fragments from the surface, thereby enhancing the chance of recovering such small bones.

Pyre goods, in the form of a small quantities of cremated animal bone (*c.* 0.2–0.3% of the total bone weight in each case), were recovered from the two urned burials. A survey of Bronze Age burials has shown that an average of *c.* 16% contain small quantities of animal bone, with sheep/goat/pig being the most common species (McKinley 1997b).

Burial-formation process

The unurned burial in grave 7 was made in the lower 0.15 m of the grave cut. The deposit was lifted as a single sample, consequently the distribution of the archaeological components within the fill is unknown. Although the deposit appeared to comprise a homogenous mix of fuel ash and cremated bone, excavation of such deposits as two or more sub-samples (halves, quadrants and spits) has demonstrated an uneven distribution of bone within the fills illustrating the 'burial' was made as a concentration (possibly within an organic container) with pyre debris deposited separately, generally over the burial into which it subsequently infiltrated (McKinley 1998; 2000b).

Eighty per cent of the bone from grave 7 was contained within the lower 0.15 m of the cut (i.e. the burial plus redeposited pyre debris), the remainder being equal divided between the two grave fills — layer 4 (0.26 m deep) and layer 3 (0.2 m deep) — deposited above it; a bone fragment from layer 3 joined one from the burial. The distribution of bone within these two grave fills is again unknown as they were collected as single deposit. Was the inclusion of bone within these fills deliberate or incidental? The bone may have been deliberately scattered within the grave as it was being backfilled. Alternatively, more bone may have been recovered from the pyre site for burial than was in the event considered necessary (fuel ash was not recorded in either of these layers) and the excess bone may have been dumped on one side to subsequently be incidentally incorporated in the grave fill.

Once the osseous-free soil had been cleared from the rim of the vessel from grave 2 (excavated from rim down though inverted at the time of burial), a slightly concave fill protruded below the level of the rim by *c.* 15 mm in the centre. This indicated that a soft organic cover (?textile) had been attached over the rim, the weight of the bone causing it to bulge down slightly in the centre on inversion. The lower-most 40 mm of fill (i.e. rim level) contained only 5.5% of the bone from the burial, the majority (*c.* 87%) of the bone being in the central 100 mm of vessel, 60–160 mm from rim. The horizontal distribution of the bone showed it to be concentrated towards centre in lower-most 80 mm, suggesting that prior to inversion it had formed a cone, the shape being maintained by the larger bone fragments as the vessel was rapidly inverted prior to burial, with no shaking to rearrange the fragments. The upper-most levels (spits 7–9) within the damaged (inverted) base showed the bone concentrated towards one side, suggesting the vessel may have been tipped-over during insertion of the first few handfuls of bone. There was no apparent ordered deposition of the remains within the vessel either horizontally or vertically, fragments from different skeletal areas (skull, axial skeleton, upper and lower limb) being distributed throughout. Several fragments of bone from different spits and quadrants were found to join (e.g. humerus from spits 2d and 5/6b and fragments of cervical vertebra from spits 5/6c and 7a).

Summary of the molluscan evidence

Michael J. Allen

It is very rare to have suitable contexts to sample for molluscan analysis from pond barrows, and the four samples taken from the pond barrow fill and cremation pit, provide evidence rarely seen from this type of barrow. Although only a shallow sequence was sampled the assemblages were rich and dominated by open country species (*Vallonia* spp., *P. muscorum*, *H. itala*) suggesting the existence for long-established open-country conditions. The presence of a number of shade-loving and catholic species (*P. pygmaeum*, *D. rotundatus*, *A. nitudula*), may suggest that the longer more unkempt grass grew within the pond barrow than in the local grazed or tilled environment surrounding it.

4.11. Monkton-up-Wimborne Early Iron Age enclosure (MUW 96)

Lithics

Martin Green

The primary fills produced 75 pieces of struck flint which included a retouched, utilized tool (RUT) component of 5% which comprised four utilized flakes. The secondary fills produced 185 pieces which included an RUT element of 3.8%. This included four finished tools — a scraper, crude point/borer, denticulate and a pounder with a heavily bruised angular edge. This latter implement is characteristic of later prehistoric flint industries in the area but has now been convincingly recorded in contexts as early as Beaker (see this volume p. 286). The primary fill of the small ditch produced a further 12 pieces including a notched flake.

Although there is likely to be a residual element amongst this assemblage it is clear that the majority, which is fresh and unabraded, relates to Middle Bronze Age to Early Iron Age phases. Much of it is crudely struck with hinge fractures common amongst the flakes. All types within the RUT component are of the few tool forms currently known to continue into the later Bronze and earlier Iron age in this area. Five flint pebbles found in the ditch were probably derived from Tertiary residues remaining in the area and may have been selected for use as sling stones.

Pottery catalogue (not illustrated see p. 130 for summary)

Martin Green

Layer 2, spit 1 (0–15 cm): Sherd from pedestal based vessel. Black burnished ware (BBW). LIA–RB. Wt 40 g.

Layer 2, spit 1: Five body sherds of BBW. LIA–RB. Wt 18 g.

Layer 2, S1: Body sherd of hard flint gritted fabric, dark brown. ?EIA. Wt 7 g.

Layer 2, spit 2 (15 cm above base of layer): Abraded rim sherd of BBW. LIA–RB. Wt 5 g.

Layer 2, S2: Nine small body sherds of BBW. LIA–RB. Wt 30 g.

Layer 2, S2: Body sherd of light brown groggy fabric known to be of early RB in the area. Wt 8 g.

Layer 2, S2: Body sherd of hard sandy fabric. Light brown exterior, darker interior. LIA–RB. Wt 9 g.

Layer 2, S2: Body sherd of soft sandy fabric. Mid-brown with shell temper. EIA–MIA. Wt 5 g.

Layer 2, S2: Body sherd of rusty brown soft sandy fabric. EIA–MIA. Wt 7 g.

Layer 4/5 interface: Wheel-thrown body sherd of fine hard sandy fabric. Dark brown exterior, lighter interior. Possibly oxidized BBW. LIA–RB. Wt 12 g.

Layer 4/5 interface: Body sherd of hard heavily shell-tempered fabric. Dark brown/black. EIA. Wt 6 g.

Layer 5: Body sherd of soft sandy thick (16 mm) fabric, occasional shell temper. MIA–LIA. Wt 13 g.

Layer 5: Body sherd of soft sandy fabric wit grits. Light brown throughout. EIA–MIA. Wt 6 g.

Layer 8: Flat-topped rim sherd with groove along edge. Fine sandy fabric, dark brown/black exterior, mid-brown interior. ?BBW. LIA–RB. Wt 2 g.

Layer 8: Body sherd of fine dark brown/black sandy fabric. ?BBW/ LIA–RB. Wt 11 g.

Layer 8: Two small body sherds of a hard dark brown/black corky fabric. Organic impression in one sherd. EIA–MIA. Wt 5 g.

Layer 8: Body sherd of fine mid-brown sandy fabric with much smaller temper. EIA–MIA. Wt 2 g.

Layer 9: Body sherd with applied cordon (8 mm). Hard flint-gritted fabric mid-brown throughout. From MBA barrel urn (see Barrett *et al.* 1991a, fig. 8.7, no. 13). Wt 14 g.

Layer 9: Three small body sherds of a soft orangey-brown sandy fabric with occasional grits. MIA–LIA. Wt 11 g.

Layer 10: Three body sherds of a hard dark brown/black hard corky fabric. EIA–MIA. Wt 9 g.

Layer 10: Two body sherds of a hard orangey-brown fabric with a little fine flint grit. Wt 5 g.

Layer 10: Five body sherds of a soft sandy fabric with much shell temper. Orangey-brown exterior, dark brown interior. EIA. Wt 11 g.

Layer 10: Two body sherds of a soft sandy dark brown fabric with a little flint grit. ?MIA–LIA. Wt 6 g.

Layer 10: Body sherd of a hard fine fabric with ?haematite coating. EIA. Wt 2 g.

Layer 10: Body sherd of a hard coarse flint-gritted fabric. Mid-brown exterior, black interior. MBA. Wt 3 g.

Layer 10/11 interface: Body sherd of hard coarse flint-gritted fabric. Orangey-brown exterior, dark brown interior. MBA. Wt 2 g.

Layer 6 (external shallow ditch): Body sherd of soft sandy fabric with much shell temper. Orangey-brown exterior, mid-brown interior. EIA. Wt 10 g.

Animal bones

Mark Maltby

Twenty-three animal bone fragments were found in the enclosure ditch, of which fourteen of the elements were identified. Nearly all of these were eroded. Apart from two bones of a toad found in the lowest fill, all the identified elements belonged to domestic mammals. No butchery marks were observed, such marks may in some cases have been obliterated by the erosion.

Four cattle elements were recorded. A metatarsal and scapula belonged to significantly smaller cattle than those found in the area in the Neolithic. The scapula has a greatest length of the glenoid process (von den Driesch 1976, GLP) of only 56.1 mm. A fragment of humerus and a lower molar were also identified as cattle. Two loose teeth and a fragment of skull belonged to sheep/goat. An unfused proximal epiphysis of a femur and a fragment of fibula belonged to pig. Horse is represented by a third phalanx and dog by the distal half of a radius, which has a maximum distal breadth (Bd) of 21.6 mm, and a pelvis. The presence of dogs is also attested by gnawing marks on the cattle metatarsal.

The mammals represented are those which have been the most commonly recorded on contemporary later prehistoric settlements in southern England. Domestic horse was introduced into Britain towards the end of the Neolithic period and the species is often found in small numbers on Bronze Age sites but with increasing frequency on Iron Age settlements. Sheep and cattle are usually the most common species encountered on Iron Age sites in Wessex (Hambleton 1999).

A note on the molluscs

Michael J. Allen

In close proximity to the Monkton-up-Wimborne ceremonial complex (MUW 97) lies a large Iron Age sub-oval enclosure with a 2 m deep V-shaped ditch cutting an earlier ?later Bronze Age ditch. Analysis here indicated that both the later Bronze ditch and Iron Age enclosure were constructed, and existed, in very open dry downland conditions of grazed pasture or arable. This may provide some context for the local nature of the Middle Neolithic and later micro-environments of the ceremonial complex on the hilltop.

4.12. Knowlton South Circle and Environs

Stratigraphic descriptions of the henge ditch fills

Michael J. Allen

The following pedological descriptions follow Hodgson (1976) (Table A4.52).

Context and description

3 Ploughwash — Yellowish-brown (10YR 5/6) calcareous very silty clay with weak coarse blocky structure, few small and medium chalk stones, rare medium flints, smooth gradual boundary.

14 Ploughwash/colluvium — Brown (10YR 5/3) to greyish brown (10YR 5/2) silty clay with moderate medium blocky structure and common to many small and medium rounded chalk pieces, clear smooth boundary.

17/15 Tertiary fill — Yellowish-brown (10YR 5/4) silty clay with common small rounded chalk pieces (15) becoming slightly darker with depth dark yellowish-brown (10YR 4/6) in context 17. Typical colluvial tertiary fills.

20 Stabilization — Dark yellowish-brown (10YR 4/4) almost stone-free silty clay with moderate medium sub-angular structure, 0.5% fine and medium macropores. Stabilization at top of secondary fill with some possible evidence of *in situ* pedogenesis.

25 Secondary fill — Yellowish-brown (10YR 5/4) to light yellowish-brown (10YR 6/4) silty clay with very weak medium sub-angular structure, many small chalk pieces; some fine calcareous coatings (hand lens) on weak interpedal faces, clear boundary; probably wash, possibly derived from the bank.

27 Secondary fill — Yellowish-brown (10YR 5/4) calcareous silty clay with medium prismatic structure, 0.1% medium macropores, common to many small and medium chalk fragments. The lower portions of this include finer fills from north including brown (10YR 5/3) silty clay with common medium and large flints and many fine chalk pieces (39) and a highly calcareous very pale brown (10YR 7/4) to light yellowish-brown (10YR 6/4) silty clay with common to abundant small well-rounded chalk pieces and rare medium subrounded chalk pieces (37).

43 Upper Primary fill — Pale brown (10YR 6/3) silty clay with no structure discernible due to abundant subrounded and sub-angular medium chalk rubble, common large chalk fragments and rare medium flints, clear smooth boundary. Weathered (smaller chalk fragments) upper primary fill, with lenses of finer light yellowish-brown (10YR 6/4) calcareous silty with common to many small and medium chalk pieces (context 42) washing in from the north.

Table A4.52. *Additional descriptive criteria, Knowlton South Circle henge ditch.*

Descriptive parameter	Decision
Texture	Texture is described after Bullock *et al.* (1985). There is an additional division in some cases between coarse (20–50 μm) and fine (2–20 μm) silt
Coarse/fine limit	A 100 μm limit was used to describe coarse:fine ratios (coarse/very coarse/medium/fine sand = coarse; very fine sand/silt/clay = fine)
Frequency (percentages)	All percentages are visual estimates of proportion of area, often in reference to an abundance chart (Bullock *et al.* 1985, 24–5)
Light sources	Fabrics were described in plane-polarized light (PPL), cross-polarized light (XPL) and reflected light (RL) only
Horizon names	All horizon names follow Bridges (1978)
Calcium carbonate precipitation pedofeatures	Following Courty *et al.* (1989, 175; after Folk 1959), micrite is <10 μm, microsparite is 10–50 μm, and sparite is >50 μm in size

45 Humic soil — Dark brown (10YR 3/3) humic silty clay loam with fine weak crumb structure, 0.1% fine macropores, and rare small and very small chalk pieces, clear to sharp smooth boundary. There is no conclusive indications (field inspection) for *in situ* pedogenesis indicating that this is a stabilization horizon with the infill sequence. Probably largely A horizon soil material derived from the original soil through which the henge ditch was cut. Collapsed due to weathering of chalk under the soil horizon.

46: Primary fill — Medium and large sub-angular chalk fragments with little matrix, but some silty calcareous marl and very pale brown (10YR 7/3) matrix. Clear lenses of up to 25 mm thick but mainly 8 mm to 13 mm of fine rounded chalk pieces and silty calcareous matrix exist; these may represent annual winter silting episodes (cf. Bell *et al.* 1996).

Soil micromorphology descriptions

Helen Lewis

Four thin sections from two profiles from the Knowlton South Circle were described following Bullock *et al.* (1985), and also using Fitzpatrick (1993), one from the bank-buried soil sequence (KWC 50A1–A2), and one from the early ditch fills (KWC40A1–A2). Further description of criteria used is given in Table A4.52.

The micromorphology results are summarized as follows (Table A4.53):

Context 7/8/9: Base of chalk bank with topsoil inclusions through bioturbation (earthworms and rooting). This layer is compact and somewhat cemented; it is calcareous, with frequent micrite in the groundmass and micrite coatings in most modern pore spaces. In some areas a sort of chalk 'slurry' appears to have formed, seen surrounding aggregates of both defined types. This is especially frequent at the base of the layer, where the bank context meets the underlying soil horizon.

Context 6: The groundmass suggests a topsoil layer, and structure and fabric pedofeatures suggest mixing by earthworms and other soil fauna both during growth and post-burial. The rooting structure seen is sometimes marked by infills (mostly calcium carbonate precipitation), and shows a network of primarily fine (250–500 μm) channels evenly distributed, with general vertical orientation. This suggests the soil was once a grass/turf-covered soil, as suggested by the mollusc interpretation (Wessex Archaeology 1994, 5–6). 'Dusty' to 'dirty' clay pedofeatures are seen throughout. These characteristics might relate to the plough intrusion mentioned by Burrow & Gale (1994b, 17, 20–21), although could indicate other types of disturbance, such as clearance or disruption related to bank construction.

Context 26: Lower A horizon material, showing slight iron enrichment. The base of this horizon sees a small accumulation of possible cultural material (charcoal, bone) in a 1 cm sized zone. Such materials are seen at no other location in the profile. The location of these inclusions at this depth may not be significant culturally (possibly more a reflection of bioturbation), but their location together in such a small area is interesting; perhaps this small patch has been introduced from or abuts a deposit representing domestic or funerary activity. These types of materials are often commonly found in ancient cultivated soils, perhaps supporting an interpretation of this buried soil being tilled. However, these materials are not generally characteristic of the profile as represented in these thin sections, and there are no clear microscopic indicators of ancient cultivation.

Context 34/bA–B: While said to be A–B material, this layer might be better named as A–C material. It appears to become more calcareous with depth, and presumably overlies the chalk geology, representing the top of the weathering zone at the surface of the rock, where it interacts with the overlying soil profile. It has been influenced by earthworm and rooting activity, and sees some light iron accumulation, as seen in Context 26.

Context 43: Three soil fabrics are mixed through bioturbation, and show degrees of influence of weathering chalk. Fabrics 2 and 3 occur as peds, and as mammilated excrements and old infills within pore spaces. Fabric 1 appears to be the main soil fabric, both on the basis of frequency of occurrence as well as relationship with chalk fragments.

Context 45: This horizon is very mixed, with the same fabrics as in context 43, Fabric 3 being dominant until the very base of the thin section, at which point Fabric 1 becomes dominant. Fabric 2 is only expressed as infills, showing the continued influence of overlying chalk weathering on this horizon. Although earthworms and rooting clearly account for most of the mixing, some disturbance may also be indicated, probably related to its location within a ditch, and from deposition of overlying materials.

KWC 94 <50> Samples A1–A2

Sample A1

Two clear layers are visible: 3.5–4 cm of chalk bank material (contexts 7/8/9), consisting of angular and

Table A4.53. *Knowlton Circle bank [7], [8], [9], buried soil [6], [26], [34], primary ditch turf slump [45] and fills [46] & [47]: micromorphological descriptions (H. Lewis).*

Context	Structure	Porosity	Mineral components	C:f ratio	Organic components	Groundmass	Pedofeatures
[7/8/9]	Sub-angular blocky	10–15% channels & cracks	Frequent chalk rock fragments; quartz; micrite; occasional feldspar; flint & sandstone fragments	30–35:65–70 Sandy silt loam	<5–5% 'punctuations' & very fine charcoal	Mixed: greyish-brown/reddish-brown (PPL), orange-yellow (XPL) zones of crystallitic & grano- or porostriated	Calcareous aggregates & infills; organic aggregates; iron nodules; moderate iron & amorphous organic staining; micrite coatings; decalicified zones; excrements
[6]	Channel, crumb & blocky	15% channels	As above; micrite mainly as pedofeature	30:70 Sandy clay loam-clay loam	5% as above	Very dark reddish-brown (PPL & XPL), stipple speckled	Calcareous zones & infills; 'dusty' clay coatings; micrite coatings; iron & organic staining; iron nodules; excrements
[26]	Crumb	20% channels	As above; 1 bone fragment	35–40:60–65 Sandy silt loam	As above, charcoal mainly in one zone at boundary with 34	Greyish-brown (PPL), light yellowish-brown (XPL), crystallitic	Zone with more frequent charcoal & bone fragment; micrite coatings; sesquioxide staining; excrements
[34]	As in 6	25–30% channels	As in 7/8/9	As in 7/8/9	As in 6	As above	As in 7/8/9
[43]	Mixed channel/vugh, blocky & crumb	15–25%, channels	Chalk, sandstone, flint rock fragments; quartz; micrite; some feldspar	40:60 Sandy silt loam	<5% 'punctuations' & very fine charcoal	Mixed: reddish-brown/greyish-brown (PPL), orange-brown/yellow/dark brown (XPL), stipple speckled/crystallitic	'Dusty' clay coatings & infills, some laminated; micrite coatings; amorphous sesquioxide staining; iron nodules; decalcified zones; excrements
[45]	Mixed channel, crumb & sub-angular blocky	15–20% channels	As above; micrite mainly as pedofeature	40–45:55–60 Sandy silt loam	5–10% as above	Mixed, as above	As above

subrounded chalk fragments (0.2–1.5 cm in size), light yellow (calcareous) crumb aggregates up to 0.8 cm in size, and very fine organic crumb aggregates (0.1–2 cm). This layer overlies 5 cm of a relatively stone-free organic soil horizon (context 6), with occasional very fine chalk fragments and red-brown flints/cherts, both up to 0.2 cm in size. The boundary between the layers is sharp and straight. At the base of the thin section there are slightly larger chalk fragments (0.3–0.4 cm) and flint fragments (grey & red-brown, 0.5–1.5 cm in size), possibly indicating the boundary with Context 26. Macroscopically, the thin section shows the lower part of a bank as described in the field (loosely packed chalk fragments in calcareous soil), and buried A horizon (topsoil) material with a fine crumb structure, organized on a larger scale into channel structure, which fits the field description of a worm-worked topsoil horizon of a brown earth. Midway down context 6 there are slight indications of a possible feature, with diagonal orientation of a rock fragment in a line with a 1 mm thick zone of similarly oriented micro-aggregates. This, however, could not be further resolved microscopically.

Sample A2

Two clear layers are distinguished, based on organic matter and calcareous components, as well as abundance and type of inclusions. The upper 2.5–3 cm is mixed organic and calcareous soil aggregates with moderate angular and subrounded chalk fragments (up to 1.5 cm in size) and rare fine flint and charcoal inclusions (<0.5 cm) (apparently context 26). This overlies a horizon of mainly calcareous material with frequent subrounded chalk fragments of 0.5–1 cm size (context 34 and/or buried A–B material). The boundary between contexts 26 & 34/bA–B is somewhat diffuse and appears to be slightly diagonally oriented. This orientation is reflected in the angles of two larger chalk fragments in context 26. Chalk fragments in context 34/bA–B are diagonally oriented in the opposite direction. Context 26 appears to mark the lower part of the buried A horizon mentioned above (context 6).

Context 7/8/9

Structure: mixed poorly to moderately developed sub-angular blocky structure (<0.5 cm) with rare crumb, and crack/channel structure.
Porosity: 10–15%, mostly interpedal and intrapedal channels (100–500 μm wide, up to 20,000 μm long, straight and curved, partially accommodated), and cracks (10–100 μm wide, 200–5000 μm long, straight and zigzag, accommodated to partially accommodated).
Mineral components: Most grains are quartz and micrite, with occasional feldspar and undifferentiated fragments. Fine fraction material is mainly micritic. Clays, while present throughout, are mainly visible in organic aggregates; these are medium reddish-brown (PPL) orange-yellow (XPL), and honey-yellow (RL). In horizon 34/bA–B there are frequent very fine (50 μm) rounded iron nodules. Weak and moderate iron staining (stronger in the lowermost horizon) is seen, especially on more organic aggregates. Rock fragments — frequent chalk fragments, 0.2–2 cm in size, sub-angular, angular and subrounded. Other calcareous coarse inclusions are occasional calcitic earthworm granules, and fossil shell fragments (eroded out from the chalk). Other rock fragments include rare very fine (up to 500 μm) red (iron-rich) flint/chert fragments, grey angular flint fragments (up to 1500 μm), sandstone fragments (up to 500 μm).
C:f ratio 30–35:65–70 — very coarse sand 10%, coarse sand 5–10%, medium sand 5%, fine sand 10–15%, very fine sand 15%, silt 40–45% (of which 25–30% micrite), clay 10%. Sandy silt loam.
Organic components: <5–5%, mostly 'punctuations' and very fine angular isotropic particles, including some with visible pore spaces that are probably carbonized remains (all are up to 50 μm in size). Organic horizon aggregates have strong organic staining on most of the groundmass.
Groundmass: calcareous aggregates and infills are crystallitic (micritic) with low to moderate birefringence, grey to greyish-brown (PPL), golden (XPL) and gold-yellow to white-yellow (RL); open porphyric distribution. Organic aggregates are stipple speckled to slightly crystallitic (micritic), slightly reddish-brown (PPL), dark brown (XPL) and light yellowish-brown (RL); close porphyric distribution. Some places show poro- and granostriation of medium reddish-brown (PPL) orange-yellow (XPL), and honey-yellow (RL) clay zones. These are in the groundmass as well as in very fine pores.
Pedofeatures: 1) clay associated with or masked by micrite coatings is present in all horizons, but the lower part of the profile appears to have been very well reworked, and textural separation of primary components is not obvious; 2) infills & coatings of unsorted very fine sand, silt (including micrite) and clay, frequently overlain by fine micrite coatings. Clay characteristics are masked by micrite; 3) zones depleted with regard to calcium carbonate in the groundmass. These appear to represent root holes, although some may represent mixing caused by faunal activity, and/or depletion related to localized moisture regimes; 4) coatings and infills of micrite; 5) mixing of fabrics has produced pedofeatures of calcareous organic soil infilling channels; also of decalcified soil material within zones of calcareous material, and *vice versa*. These fabric inclusions appear to mainly be related to earthworm mixing and rooting action; 6) extremely fine iron nodules; 7) earthworm excremental features and zones of mite excrement.

Context 6

Structure: channel structure with frequent fine (1–3 mm) peds, predominantly crumb and angular blocky, composed of rounded, coalesced faecal aggregates. Typically faunal excremental and bioturbated fabric, with a larger structure related to rooting.
Porosity: 15%, mostly channels (50–3000 μm wide, up to 40,000 μm long, straight and curved, partially accommodated), and vughs (50–250 μm).
Mineral components: as above, but very rare chalk, and micrite seen mostly as infills in pore spaces; although some micrite is seen in the groundmass, the latter is largely non-calcareous. Also rare red (iron-rich) flint/chert up to 0.5 cm in size, rounded to subrounded. There is weak to moderate iron staining on the main groundmass, and rare very fine iron nodules (up to 200 μm). Clays are as described above for organic aggregates. Near the base of the layer is an ironstone fragment.
C:f ratio: 30:70 — very coarse sand <5%, coarse sand 5%, medium sand 5%, fine sand 10–15%, very fine sand 20%, silt 30–35% (of which 10–15% micrite), clay 15–20%. Sandy clay loam to clay loam.
Organic components: 5%, as above, but some larger charred remains (up to 200 μm). The groundmass is strongly organic stained. Rare spores.
Groundmass: The main groundmass is very dark reddish-brown (PPL & XPL), reddish-yellow (RL), and stipple speckled, with a close porphyric distribution. Frequent zones of calcium carbonate-rich, crystallitic material are found, up to *c.* 0.4 mm in size. These are yellow-brown or grey-brown (PPL), light golden brown (XPL) and light reddish-yellow (RL). Some of these zones appear to originate from earthworm and rooting action. Others are infills in old pore spaces and coatings on peds. Apart from these zones, the general groundmass is decalcified/non-calcareous, with strong organic staining, suggesting a topsoil.
Pedofeatures: 1) & 2) as in Context 7/8/9; 3) zones with very fine very dark brown (PPL), orange (XPL) and yellow (RL) dusty clay coatings (20–50 μm thick, laminated). These are found coating grains in the groundmass, and in the modern pore spaces, and are often associated with fine micrite zones or coatings; 4) coatings and infills of micrite; 5) as 5) in Context 7/8/9; 6) extremely fine iron nodules, organic staining; 7) earthworm excremental features & zones of mite excrement.

Context 26

Structure: crumb (0.2–0.5 cm).
Porosity: 20%, mostly interpedal and intrapedal channels (200–2000 μm in diameter, up to 10,000 μm long, curved, partially accommodated), with some cracks (30–60 μm wide, 1000–2000 μm long, straight, partially accommodated) and chambers (1000–1500 μm in size).
Mineral components: as for Contexts 7/8/9 above, but mostly fine chalk fragments (0.3 mm); also rare grey flint, up to 1.5 cm in size, sub-angular. Clays are mainly masked by micrite and moderate iron staining on the main groundmass. One fragment of bone is present near the base of the layer. This is 1000 μm long, 250–400 μm wide, very light brownish-white (PPL), greyish-white (XPL) and translucent to white (RL).
C:f ratio: 35–40:60–65 — very coarse sand 5%, coarse sand 10%, medium sand 10%, fine sand 10–15%, very fine sand 10–15%, silt 45% (of which 30–35% micrite), clay 5%. Sandy silt loam.
Organic components: As in Context 6, but charred remains mainly seen at the boundary with Context 34/bA–B, where there is a small area with several larger fragments (200–2500 μm) of carbonized plant remains, with a small bone fragment (see 'Mineral Components'). At the top of the horizon a live root fragment is seen, suggesting that despite its buried location some deeper relatively recent rooting has affected the horizon. Moderate organic staining has affected much of the groundmass.
Groundmass: the main groundmass is crystallitic (micritic) and lightly iron and organic-stained (becoming less strongly stained with depth). It is greyish-brown (PPL), light reddish-brown (XPL), light yellow-orange (RL), and has an open porphyric distribution.
Pedofeatures: 1) coatings and infills of micrite; 2) as 5) in Context 7/8/9; 3) extremely fine iron nodules, one larger oblong nodule with embedded sand and silt grains (see 'Mineral Components');

light iron staining ; 4) earthworm excremental features & zones of mite excrement.

Context 34/bA–B
Structure; mixed channel, angular blocky structure (0.3–1 cm), crumb and vermicular (0.2–0.8 cm).
Porosity: overall 25–30%, interpedal porosity 15–30%, mainly packing pores, channels (200–4000 μm wide, up to 20,000 μm long, partially accommodated, curved and straight), often partially infilled with fine aggregates. Intrapedal porosity is <5–10%, mainly channels (100–500 μm in diameter), and rare cracks (20–50 μm wide, 1000 μm long, straight and zigzag, partially accommodated).
Mineral components as in Context 7/8/9.
C:f ratio: 30–35:65–70 — very coarse sand 10%, coarse sand 5–10%, medium sand 10%, fine sand 5%, very fine sand 10%, silt 50–55% (of which 40–45% micrite), clay 5%. Sandy silt loam.
Organic components: 5%, as in Context 7/8/9.
Groundmass: as in Context 26, but more influenced by chalk (more grey (PPL) and greyish-yellow (RL), less iron stained). Some aggregates have the same groundmass as Context 26.
Pedofeatures: 1) as 3) in Contexts 7/8/9; 2) coatings and infills of micrite; 3) as 5) in Context 7/8/9; 4) extremely fine iron nodules; light iron staining; 5) earthworm excremental features & zones of mite excrement.

KWC 94 <40> Samples A1–A2

General thin-section description

<40> A1 — Two horizons: 0–2 cm of chalk fragments mixed with soil; 2–9.5 cm of organic soil with a crumb and channel structure, and a bedded, broken chalk fragment at the base. The upper layer appears to be Context 43 material, with the rest of the section being *Context 45* (slumped turf/stabilization layer).

<40> A2 — One visible horizon: 0–7.5 cm — a relatively stone-free organic soil horizon, with strong rooting and vermicular indicators, with fine crumb peds and a fine channel structure. This appears to be a grassland A horizon. There are a few stones <0.5 cm in size. This entire thin section appears to be *Context 45*.

Context 43
Structure: mixed channel and vugh where micrite cementing chalk fragments; where soil aggregates are present these are sub-angular blocky and crumb (2 mm), with some vermiform.
Porosity: 15–25%, mostly channels (200–2000 μm wide, 1000 μm long, unaccommodated and partially accommodated), vughs (300–1000 μm) and packing pores.
Mineral components: grains are mostly quartz, micrite and some feldspar. Rock fragments are mostly chalk with occasional sandstones, chert/flint (grey and red), and possible ironstone.
C:f ratio: 40:60 — very coarse sand 20%, coarse sand 5–10%, medium sand 5–10%, fine sand 10%, very fine sand 10%, silt 35% (of which 20–25% micrite), clay 10–15%. Sandy silt loam.
Organic components: <5%, mostly 'punctuations' and very fine carbonized remains (<50 μm).
Groundmass: three fabrics mixed together, and described here in order of visual percentage of area: 1) moderately to strongly calcareous soil fabric, crystallitic (micritic), medium reddish-brown (PPL), light orange-brown (XPL), light orange-yellow (RL), iron stained as well as containing a great deal of micrite in its groundmass, and with a close porphyric-related distribution; 2) extremely calcareous soil fabric, crystallitic (micritic), light greyish-brown (PPL), light yellow-golden (XPL), creamy yellowish-orange (RL), apparently iron stained, and with an open porphyric-related distribution; 3) non-calcareous soil fabric, stipple speckled, medium reddish-brown (PPL), dark brown (XPL) and orange-red (RL), organic stained, and with a close porphyric-related distribution.
Pedofeatures: 1) dusty to slightly dusty clay coatings are associated with or masked by micrite coatings. Clay characteristics are generally masked by micrite and/or organic and iron staining. The characteristics of these coatings, as well as their thickness etc., are generally the same as seen in Profile <50>, except that they are much more frequent, being present in most modern pore spaces, and as infills in old pores. Some show thick laminations, from 50 to 300 μm thick, with reddish-grey (PPL), light yellow-brown (XPL) and light orange-yellow (RL) clay layers with amorphous iron/organic staining and micrite inclusions alternating with grey (PPL), light greyish-brown (XPL), whitish-yellow (RL) layers that appear to be primarily composed of micrite coated with amorphous clay/iron/organic matter; 2) in the more calcareous aggregates there are zones depleted with regard to calcium carbonate in their groundmass. These appear to represent root holes, although some may represent mixing caused by faunal activity, and/or depletion related to localized moisture regimes; 3) coatings and infills of micrite, sometimes forming bridging structures across peds and grains, and cementing soil aggregates in some places; 4) extremely fine iron nodules, light iron staining and strong organic staining; 5) earthworm excremental features.

Context 45
Structure: mixed channel and crumb structure, with very fine sub-angular blocky (2000–5000 μm).
Porosity: 15–20%, mostly channels (200–500 μm wide, 5000 μm long, accommodated to partially accommodated, straight and curved; occasionally in upper part of the context 1000–3000 μm wide, up to 40,000 μm long, partially accommodated, straight and zigzag); rare chambers (10,000 μm).
Mineral components: grains are as above, but micrite mainly in calcareous zones and as pedofeatures. Rock fragments: sandstones, occasional chalk, chert/flint (grey and red), possible ironstone.
C:f ratio: 40–45:55–60 very coarse sand <5%, coarse sand 5%, medium sand 10%, fine sand 15%, very fine sand 10%, silt 40% (of which 10% micrite), clay 10–15%. Sandy silt loam. Lower — 20–25:75–80 — very coarse sand <5%, coarse sand 5%, medium sand 5%, fine sand 10%, very fine sand 10–15%, silt 40–45%, clay 15–20%. Sandy silt loam to clay loam.
Organic components: 5–10%, mostly 'punctuations', with occasional fine carbonized remains (50–500 μm), becoming smaller with depth.
Groundmass: mixed Fabrics 1 and 3 from above. For the most part, more organic Fabric 3 is dominant, but Fabric 1 becomes more frequent with depth; at the base of the thin section, most of the soil material is this fabric; perhaps this is Context 46/47 material?
Pedofeatures: 1–4) as above, but less organic staining; 5) earthworm excremental features and zones of mite excrement, especially in chambers.

The decorated heathstone from Knowlton

Richard Bradley

The two slabs found at Knowlton are both of local heathstone. These stones had been ploughed out of the field immediately to the south of the Wimborne Minster road and taken to the Knowle Hill farmyard where the farmer, Robert Antell, showed them to C. French and H. Lewis in the spring of 2000 (Lewis *et al.* 2000).

The larger stone is decorated with four concentric rings (Fig. A4.33). Its companion is undecorated but

Figure A4.33. *Close-up of the concentric pecked ring marks on one of the heathstones from Knowlton South Circle (C. French).*

the edges of the slab seem to have been trimmed to a curving edge. Again it is difficult to suggest a context for these finds as rock carvings are mainly a feature of upland Britain. In this case it seems possible that both stones formed part of a burial cist. Two decorated stones of similar character have been found in what was probably an Early Bronze Age round barrow at Winterborne Came on the Dorset Ridgeway, although they are antiquarian finds and are not particularly well dated (Warne 1866, 37). Also, another heathstone slab with weathered concentric rings is known to have been found close to Badbury Rings (M. Green pers. comm.). It is quite possible that the decorated stone from Knowlton is older than these examples, as there is circumstantial evidence that this tradition of rock art began during the Neolithic period. Some of the decorated cist slabs associated with Bronze Age burials in northern England and Scotland seem to have been reused, and the practice of carving concentric circles without a central cup mark is first documented in Irish megalithic art of the later fourth millennium BC (Bradley 1997, chaps. 4 & 9). Another feature which recalls the decoration associated with passage tombs is the absence of a radial line leading from the centre of the design. This is not sufficient evidence to suggest the original context of these stones. The best that can be said is that they could date from any time between about 3400 and 2000 BC.

Soil micromorphology descriptions from the Knowlton environs sites

HELEN LEWIS

The micromorphological descriptions for the soil profiles are summarized in Table A4.54.

Farringdon Clump TPG-3

Structure: One horizon of apedal channel structure, with 10–15% porosity (channels, 50–1000 μm diameter × 40,000 μm maximum length, partially accommodated, straight & crooked).
Mineral components: quartz grains and quartz sandstone fragments, flint fragments.
C:f ratio: 20:80: very coarse sand 5%, coarse sand 5%, medium sand 5%, fine sand 5%, very fine sand 30%, silt 30%, clay 20%. sandy (clay) loam.
Organic components: 5%, <50 μm angular & rounded black fragments.
Groundmass: light reddish-brown (PPL), very dark brown to black (XPL), stipple speckled with granostriated zones.
Pedofeatures: 1) frequent slightly 'dusty' clay coatings in modern pore space, some microlaminated, orange (PPL), orange-red (XPL), 50–100 μm thick. Also some very clean coatings, mainly latest in laminated coatings. Rare coatings are very dusty to dirty; 2) old channels filled with brown (PPL & XPL) 'dusty' clay coatings, 50–200 μm thick; 3) occasional iron mottles 500 μm in size, rarely showing root pseudomorphs; 4) rare ovoid brown (PPL), very dark brown (XPL) organic 'pellets' 500 × 300 μm, with 5% very fine sand and silt, rest clay.

Farringdon Clump: Pit Profile A

Structure: one horizon with mixed apedal channel, crumb and sub-angular blocky structure (5000 μm), 5–10% porosity, mainly channels and cracks (150–1000 × 1500–70,000 μm, partially accommodated to accommodated, straight and crooked). Some of these are obviously root channels.
Mineral components: frequent chalk rock fragments, with some chert; quartz grains; shells; one metal fragment (haematite with alteration to goethite-like material (Chris Doherty pers. comm.); frequent earthworm calcitic granules; micrite as pedofeatures.
C:f ratio: 30:70: very coarse sand 5%, coarse sand 5%, medium sand 15%, fine sand 5%, very fine sand 5%, silt 40%, clay 25%. Loam.
Organic components: 5%, frequent 'punctuations' and 100–200 μm angular black (PPL) fragments. Occasional possible charcoal; rare definite charcoal. Rare ovoid seeds/spores, 10–20 μm, brown (PPL). Occasional 250–500 μm brown (PPL) organic components with cellular structure in root holes (modern roots).
Groundmass: crystallitic, light greyish-brown (PPL & XPL).
Pedofeatures: 1) occasional coatings 100–250 μm thick on rock fragments of main groundmass material; 2) rare red (PPL & XPL) clay fragments, 100–150 μm, subrounded and rounded, clean; 3) thin micrite coatings in modern pores; 4) excremental pedofeature of same fabric as main fabric, but in crumb microaggregates 150 μm in size, coalescing in larger pores.

Charlton Dairy: Profile 1/samples 1–2

Two horizons, an A horizon in 1/2 and B horizon in 1/1 (NB thin sections are mislabeled).
Structure: A — sub-angular blocky and crumb, 3000–5000 μm, 5–20% porosity, channels and cracks (100–1000 × 1000–10,000 μm, straight, partially accommodated), rare linear voids (200 × 1000 μm) near rock fragments; B — sub-angular to angular blocky, 3000–5000 μm, some very poorly developed. Porosity as in A.
Mineral components: Flint gravel makes up almost 40–50% of visible area in both horizons. This is angular to sub-angular, 5000–25,000 μm in size. Grains are quartz, with rare glauconite. Clays are orange

Table A4.54. *Faringdon Clump Pit and Charton House sections: micromorphological descriptions. (H. Lewis.)*

Context	Structure	Porosity	Mineral components	C:f ratio	Organic components	Groundmass	Pedofeatures
Faringdon Clump							
FCL TPG-3	Apedal channel	10–15%, channels	Quartz sand & sandstone; flint	20:80 Sandy (clay) loam	5% 'punctuations' & very fine angular fragments	Light reddish-brown (PPL), very dark brown to black (XPL), stipple speckled.	Frequent 'dusty' clay coatings, occasional clean clay coatings & 'dusty' clay channel infills; occasional iron mottles; rare ovoid pellets
FCL 03 Pit Profile A	Mixed apedal channel, crumb & sub-angular blocky	5–10%, channels & cracks	Frequent chalk fragments, occasional chert; quartz sand; shells; earthworm granules	30:70 Loam	As above, occasional charcoal	Light greyish-brown (PPL & XPL), crystallitic	Metal fragment (iron); micrite; occasional groundmass coatings on rock fragments; rare clean clay fragments
Charlton House							
A horizon	Sub-angular blocky & crumb	5–20%, channels & cracks	Frequent flint gravel; quartz sand, rare glauconite; clays are orange (PPL & XPL)	25:75 Loam	5% 'punctuations & very fine rounded black fragments	Brown (PPL), dark brown to dark reddish-brown (XPL), stipple speckled	Occasional 'dusty' clay coatings & infills; occasional earthworm sorting & depletion zones
B horizon	Sub-angular to angular blocky	As above	As above	30:70 Sandy clay loam	As above, occasional brown cellular fragments (roots)	Strong reddish-brown (PPL), reddish-brown (XPL), stipple speckled & granostriated	Frequent 'dusty' clay channel infills; faunal depletion zones

Table A4.55. *Lithic types by field from the Knowlton environs field-walking programme.*

Flint type	AH	KH	FCL	BRO	BD	BRO/BD
Waste	797/97%	296/97%	522/92%	707/93%	986/96%	1693/95%
Core tools	1/ 4.8%	0	3/ 6.5%	9/16.7%	2/ 5%	11/11.6%
Flake tools	20/ 95.2%	8/100%	43/93.5%	45/83.3%	36/ 90%	2/2.1%
Arrowheads	0	0	0	0	2/ 5%	82/77.3%
Total	**818**	**304**	**568**	**761**	**1026**	**1788**
Scrapers	17/81%	5/62.5%	39/84.8%	39/72.2%	27/67.5%	66/69%
Cores/frags. of total	8/1%	6/1.97%	14/2.46%	14/1.8%	23/2.2%	39/2.2%
Bl/Fl ratio	1:46	1:259	1:30.5	1:71	1:290	1:126
No. categories (22)	11	10	11	17	12	19
No. tool categories (13)	5	3	4	10	7	11

Scrapers	AH	KH	FCL	BRO	BD	BRO/BD
'Normal'	14/82.3%	5/100%	29/74.4%	26/66.7%	27/100%	53/80.3%
End	1/5.9%	0	3/7.7%	4/10.3%	0	4/6.1%
Side	0	0	1/2.6%	1/ 2.6%	0	1/ 1.5%
Hollow	1/5.9%	0	3/7.7%	8/20.5%	0	8/12.1%
Core/scraper	0	0	2/5.1%	0	0	0
End/hollow	0	0	1/ 2.6%	0	0	0
Horseshoe	1/5.9%	0	0	0	0	0
Total	**17**	**5**	**39**	**39**	**27**	**66**

Flakes	AH	KH	FCL	BRO	BD	BRO/BD
Flakes	696/94%	240/92.7%	369/80.7%	435/67.7%	779/89.3%	1214/80.1%
Bladelike	13/1.8%	0	33/7.2%	34/5.3%	0	34/2.2%
'Heavy'	31/4.2%	19/7.3%	55/12.0%	174/27.1%	93/10.7%	267/17.6%
Total	**740**	**259**	**457**	**643**	**872**	**1515**

(PPL), reddish-orange (XPL).
C:f ratio: A — 25:75: very coarse sand 15%, coarse & medium sand 5%, fine sand 5%, very fine sand 10%, silt 40%, clay 25%. Loam. B — 30:70: very coarse sand 20%, coarse & medium sand 5%, fine sand 5%, very fine sand 15%, silt 20%, clay 35%. Sandy clay loam.
Organic components: A — 5%, <50 μm rounded black fragments & 'punctuations'; B — 5%, 50–150 μm black angular and rounded fragments, 'punctuations', and occasional brown cellular fragments (roots).
Groundmass: A — medium brown (PPL), dark brown to very dark reddish-brown (XPL), stipple speckled. B — strong reddish-brown (PPL), reddish-brown (XPL), stipple speckled with granostriated zones.
Pedofeatures: 1) A — occasional clay linings in voids and on grains, 'dusty' brownish-orange (PPL & XPL), 100 μm thick; 2) A — rare zones (*c.* 1000 μm sized) with clay infills in a rounded/circular orientation. These appear to be infilled channels. Clays are orange (PPL & XPL), 'dusty', 20–250 μm thick; 3) B — frequent old channel clay infills 200–300 μm thick zones in the groundmass, red (PPL & XPL), slightly 'dusty'; 4) B — rounded zones (*c.* 500 μm sized) of non-iron stained, organic-stained main fabric. These appear to be faunal in origin. A — occasional earthworm sorted zones, with depletion pedofeatures re. organic matter, clay & very fine silt.

Lithics from the Knowlton environs field walking and test pitting
Julie Gardiner

The discussion of the flint data is detailed by Gardiner above (pp. 138–45), so only the summary table (Table A4.55) of the flint recovered is presented here.

4.13. Deer Park Farm

Micromorphological analysis of the soil profile
Charles French

A *c.* 22 cm long soil block was taken by Martin Green from the deep sounding in the northeast corner of trial hole 1 in January, 1998. The field profile beneath the present-day ploughsoil revealed a *c.* 20 cm thick orange clay overlying a grey clay layer which became progressively sandier down-profile until pure grey sand was reached at a depth of 70 cm below the ploughsoil. The whole profile is developed on a geological transition zone between the chalklands of Cranborne Chase and clays mixed with gravels of the Reading Bed series (Fig. 2.1). The orange clay horizon was sampled for soil micromorphological analysis. Two large thin sections were made using the method of Murphy (1986), and the slides described using the terminology of Bullock *et al.* (1985), with the results summarized below and in Table A4.56.

The surviving profile at Deer Park Farm appears to be a thin, modern ploughsoil and a clay- and iron-rich B horizon. Effectively the middle to upper part of this argillic brown earth profile is missing, that is there is no evident Eb horizon, which could be explained through past agricultural practices leading to thorough horizon mixing, and/or erosion and truncation through water and wind agencies. The base of the ploughsoil (Ap) was composed of grey, dense, very fine sandy clay with evidence of residual organic matter in the form of iron-replaced plant tissue, stem and root fragments. Up to about one-third of the soil fabric was affected by strong impregnation with amorphous sesquioxides.

This overlay *c.* 6 cm of yellowish-brown, apedal, relatively dense, very fine sandy clay. The abundant clay component (35%) was composed of non-laminated 'dusty' or impure clay which exhibited irregular, strong zones of amorphous sesquioxide impregnation. These characteristics are suggestive of a weathered horizon subject to alternate wetting and drying, or a Bws horizon. Occasionally the voids are completely infilled with heavily iron-impregnated fine groundmass material, which is suggestive of soil disturbance in the overlying profile, most probably associated with recent agricultural activities.

This horizon overlay a mottled to striated orange/reddish-brown, very dense, fine to very fine sandy clay with very strong amorphous sesquioxide impregnation in irregular, vertical zones. The predominant, non-laminated, dusty clay component (>35%) exhibited birefringent striations and cross striations. In addition, there were occasional, non-laminated, pure clay coatings on grains and in the groundmass (<10%). In general, the 'dusty' clay component is very heavily impregnated with amorphous sesquioxides. These textural pedofeatures are suggestive of an iron-rich, rather weakly developed, palaeo-argillic horizon or Bts horizon (Avery 1980; Limbrey 1975, 134–7). This orange clay soil was associated with minute charcoal flecks and a flint assemblage suggestive of the late glacial/early Holocene transition or Allerød to earlier Younger Dryas periods (*c.* 12,000–10,700 BP). A thermoluminescence date for this soil of 10,700 BP provided corroboration for this.

The detailed thin-section descriptions
Charles French

Profile 1/1–1/2
Three units or horizons were present in this upper sample through the palaeosol, as follows:

Unit 1 (0–2 cm)
Structure: apedal; dense.
Porosity: *c.* 15% vughs, small, <500 μm; few (10%) channels resulting

Table A4.56. *Deer Park Farm: micromorphological descriptions (C. French).*

Context	Structure	Porosity	Mineral components	C:f ratio	Organic components	Groundmass	Pedofeatures
Ap	Apedal	15% vughs & channels	Quartz sand	10:90 loam	5%, roots	Reddish-brown (PPL), golden to dark brown (XPL), weakly striated	Frequent non-laminated 'dusty' clay in groundmass and as coatings; amorphous sesquioxide impregnation
Bws	As above	10–15%, mostly vughs, rare channels	As above	As above	As above	As above, with strong sesquioxide impregnation	Strong amorphous sesquioxide impregnation and nodules
Bts	Apedal, massive	<2% vughs	As above	20:80 Sandy clay loam	<1%, rare replaced plant tissue	Reddish-yellow (PPL), reddish-orange (XPL), random striated to cross striated	Strong sesquioxide impregnation; occasional non-laminated clean clay; frequent non-laminated 'dusty' clay

from roots/stems, especially at base of unit, horizontal and vertical, <4 cm long, <4 mm wide.
Organic component: occasional (5%) roots/stems with <50% cell structure survival, partially replaced by fine groundmass material, irregular, all sizes from 1 mm+.
Mineral components: limit 100 μm; coarse/fine ratio: <10/>90; coarse fraction: 10% fine quartz, 100–250 μm, sub-rounded to sub-angular; fine fraction: 35–40% very fine quartz, 50–100 μm, sub-rounded to sub-angular; 15% silt; 35% clay; golden to dark brown (CPL), reddish-brown (PPL), strong yellowish-brown (RL).
Groundmass: fine and related: closed porphyric; very weakly striated.
Pedofeatures: *Textural*: very abundant (35%) non-laminated, dusty clay, mainly in groundmass and as coatings of grains, and occasionally in voids, gold to golden brown (CPL), weak to moderate birefringence; *Amorphous*: whole fabric generally impregnated with amorphous sesquioxides; *c.* 30% of groundmass exhibits irregular zones of greater impregnation with sesquioxides.

Unit 2 (2–8 cm)
Structure and mineral components: as for Unit 1, except for
Porosity: few (<5%) channels, <1.5 cm long, <1 mm wide, vertical, straight to irregular; 10% vughs, <250 um, sub-rounded.
Pedofeatures: *Amorphous*: few zones of completely infilled large vughs/channels with heavily amorphous sesquioxide impregnated fine fabric, *c.* 2–5 mm; whole groundmass exhibits strong amorphous sesquioxide impregnation; 10% sesquioxide nodules, sub-rounded, <250 μm.

Unit 3 (8–21.6 m)
Structure: massive, apedal.
Porosity: few (<2%) vugh, sub-angular to sub-rounded, <250 μm.
Organic components: very rare (<1%) plant tissue replaced by amorphous iron and/or dusty clay.
Mineral components: limit 100 μm; coarse/fine ratio: 20/80; coarse fraction: 20% fine quartz sand, 100–250 μm, sub-rounded to sub-angular; fine fraction: 20% very fine quartz sand, 50–100 μm, sub-rounded; <15% silt; >45% clay; reddish-orange (CPL), reddish-yellow (PPL & RL). In zones with strong sesquioxide impregnation (especially in 1/2) the coarse fraction is reduced to 5% fine quartz sand, with 25% very fine quartz sand, 20% silt and 55% clay.
Groundmass: fine: random striated to occasionally cross-striated; related: closed porphyric.
Pedofeatures: *Textural*: occasional to many (<10%) non-laminated, pure clay, gold to reddish-orange (CPL), moderate birefringence; very abundant (>35%) non-laminated dusty clay; weak to moderate birefringence, reddish-orange (CPL); *Amorphous*: whole groundmass impregnated with amorphous sesquioxides, with up to 75% very heavily impregnated in irregular, vertical zones.

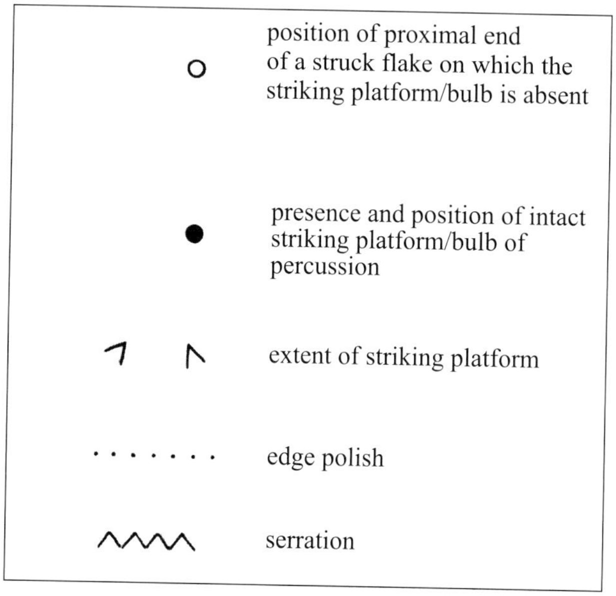

Figure A4.34. *Key to the flint drawings (M. Green).*

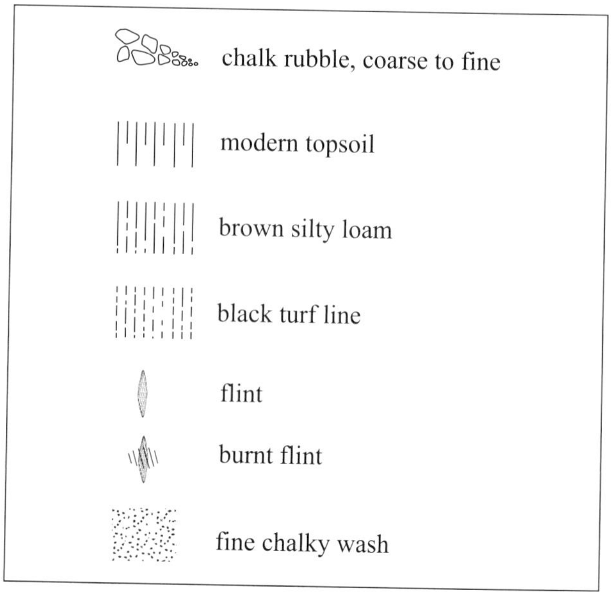

Figure A4.35. *Key to section drawings (M. Green).*